Materials Horizons: From Nature to Nanomaterials

Series Editor

Vijay Kumar Thakur, School of Aerospace, Transport and Manufacturing, Cranfield University, Cranfield, UK

Materials are an indispensable part of human civilization since the inception of life on earth. With the passage of time, innumerable new materials have been explored as well as developed and the search for new innovative materials continues briskly. Keeping in mind the immense perspectives of various classes of materials, this series aims at providing a comprehensive collection of works across the breadth of materials research at cutting-edge interface of materials science with physics, chemistry, biology and engineering.

This series covers a galaxy of materials ranging from natural materials to nanomaterials. Some of the topics include but not limited to: biological materials, biomimetic materials, ceramics, composites, coatings, functional materials, glasses, inorganic materials, inorganic-organic hybrids, metals, membranes, magnetic materials, manufacturing of materials, nanomaterials, organic materials and pigments to name a few. The series provides most timely and comprehensive information on advanced synthesis, processing, characterization, manufacturing and applications in a broad range of interdisciplinary fields in science, engineering and technology.

This series accepts both authored and edited works, including textbooks, monographs, reference works, and professional books. The books in this series will provide a deep insight into the state-of-art of Materials Horizons and serve students, academic, government and industrial scientists involved in all aspects of materials research.

Review Process

The proposal for each volume is reviewed by the following:

1. Responsible (in-house) editor
2. One external subject expert
3. One of the editorial board members.

The chapters in each volume are individually reviewed single blind by expert reviewers and the volume editor.

Tamanna Bhuyan · Yugal Kishore Mohanta ·
Manika Khanuja · Ramesh Namdeo Pudake
Editors

Multifunctional Quantum Dots

Biological, Environmental and Energy Applications

Springer

Editors
Tamanna Bhuyan
Department of Applied Biology
University of Science and Technology
Meghalaya
Baridua, Meghalaya, India

Yugal Kishore Mohanta
Department of Applied Biology
University of Science and Technology
Meghalaya
Baridua, Meghalaya, India

Manika Khanuja
Centre for Nanoscience
and Nanotechnology
Jamia Millia Islamia (A Central University)
Jamia Nagar, Delhi, India

Ramesh Namdeo Pudake
Amity Institute of Nanotechnology
Amity University Uttar Pradesh
Noida, Uttar Pradesh, India

ISSN 2524-5384 ISSN 2524-5392 (electronic)
Materials Horizons: From Nature to Nanomaterials
ISBN 978-981-92-0171-6 ISBN 978-981-92-0172-3 (eBook)
https://doi.org/10.1007/978-981-92-0172-3

This Springer imprint is published by the registered company Springer Nature Singapore Pte Ltd.
The registered company address is: 152 Beach Road, #21-01/04 Gateway East, Singapore 189721, Singapore

Preface

This book, entitled *Multifunctional Quantum Dots—Biological, Environmental, and Energy Applications*, aims to provide a comprehensive and interdisciplinary insight into the fabrication, functionalization, and various applications of quantum dots (QDs) across three significant arenas—biomedicine, environmental science, and energy. The transformative potential of QDs in biosensing, the food and agriculture industry, environmental monitoring, and next-generation energy devices has been emphasized. Considering the expanding research landscape, this book envisions the multifunctionality of quantum dots as a versatile platform capable of addressing global challenges in health, sustainability, and clean energy. The chapters in this book cover both fundamental concepts and real-time applications, reflecting the insights and expertise of leading scientists in the field.

Important Features of the Book are Enumerated

- A concise introduction showcasing the unique material structures, as well as the physical and chemical properties of quantum dots, is presented at the beginning of a few chapters, followed by a detailed discussion on various synthesis and characterization techniques in subsequent chapters. The later sections of the chapters introduce the integration of miniaturized quantum dots into a wide range of biosensing platforms, biodiagnostic applications, and nano-based drug delivery systems.
- Eco-friendly quantum dot formulations have also been prominently emphasized for their potential applications in agriculture and the food industry, particularly in improving productivity, ensuring food safety, and enhancing overall efficiency across the agri-food value chain. The fate, toxicity, and bioaccumulation potential of quantum dots are also covered in later chapters, along with their comprehensive risk assessment on the environment and human health.
- The subsequent chapters encompass the multifunctional role of quantum dots in advanced energy technologies for engineering next-generation energy storage

and conversion devices. Additionally, their potential to revolutionize photovoltaic technology, along with the role of optimizing quantum dot materials in addressing existing challenges within renewable energy systems, has been thoroughly reviewed in the selected chapters.
- The closing chapters underscore the expanding horizons of QD technology, including their integration in microfluidic systems, where their exceptional optical properties improve lab-on-a-chip technology applications. The latter sections outline the life cycle assessment (LCA) of quantum dots, stressing the significance of sustainable practices in QD development for the cutting-edge applications and next-generation quantum dot-based technologies.

After reading the book, readers will gain an understanding of how the operating mechanisms of various types of quantum dots enable their diverse applications across different fields of interest. Our goal is to bridge the cross-disciplinary knowledge gap and offer a comprehensive resource for scientists, students, and industry professionals striving to fully leverage the potential of QDs. As the future becomes more reliant on nanoscience, the adaptability of quantum dots demonstrates the transformative potential of materials science in shaping a better world.

Lastly, the editorial team expresses heartfelt gratitude to all authors whose expertise and dedication have made this book a reality, and thank you for your invaluable insights. Special thanks to the reviewers for their critical feedback, which greatly enhanced the quality of this book. We are also grateful to the institutions and organizations that supported this work, fostering research and innovation in quantum dots. We thank the team from Springer for your continuous support, and our readers, whose curiosity and commitment inspire progress in the field of quantum dots.

Baridua, Meghalaya, India — Tamanna Bhuyan
Baridua, Meghalaya, India — Yugal Kishore Mohanta
New Delhi, India — Manika Khanuja
Noida, Uttar Pradesh, India — Ramesh Namdeo Pudake

About This Book

Quantum dots (QDs) are luminescent semiconductor nanocrystals (NCs) that have been extensively studied owing to their distinct electronic, optical, and structural properties. Drawing on the knowledge of its expert authors, this book compiles a comprehensive review of the latest research in quantum dot fabrication, properties, and applications. Beginning with an introduction to the fundamentals of QDs, the subsequent sections provide a detailed overview of some imminent new directions in biomedical, environmental, and energy applications. Organized into four parts, the first part comprises an introduction to QDs, their fabrication, and characterization techniques for successful luminescence. The second part presents the main aspects of biologically oriented applications of QDs, including bioimaging, bioassay, biodiagnostics, drug research, and agriculture. The third part will cover important features of the quantum dot-based detection of environmental pollutants, their fate, and their accumulation in the environment. The fourth part focuses on the recent achievements of quantum dots in energy storage, photovoltaic (PV) technology, and building energy devices. The concluding part of the section will offer a productive discussion on the challenges of emerging quantum dot technology and the regulatory policies associated with quantum technology.

The past and recent research papers are assembled in a way that depicts the chronological developments and, respectively, summarizes potential future directions. This book will aid scientists in continuing to study QDs by providing information about methods and protocols, helping to expand their research. This book will help academicians and researchers continue to study QDs while also assisting them in advancing their research.

Contents

Editors and Contributors

About the Editors

Dr. Tamanna Bhuyan is currently working as an Assistant Professor and a researcher at the Department of Applied Biology, University of Science and Technology Meghalaya. She received a Ph.D. (Nanotechnology) from the Indian Institute of Technology Guwahati in 2020 in the area of Micro/Nanorobots for Therapeutics and was also a recipient of the Best Ph.D. Thesis Award. She has been a DST INSPIRE Fellow and has published more than 16 research papers in reputed journals. These papers have garnered a total of 1263 citations with an h-index and i10 index of 10. She has authored three book chapters, holds two patents, and her research work has been featured in **Nature India** and **India Biosciences**. She, along with her team, bagged SITARE-GYTI (Gandhian Young Technological Innovation) Appreciation 2021 for the work "Microfluidic Immunosensor for Point-of-Care-Testing of Beta-2-Microglobulin in Tear", and National Level Women Scientist Award 2024 conferred by Microbiologists Society, India. She has been awarded projects sponsored by DST SERB, the Ministry of Electronics and Information Technology (Government of India) and New Generation Innovation and Entrepreneurship Development Centre project funded by NSTEDB-DST (Government of India) executed at USTM. She is also an expert reviewer for reputed journals. Dr. Bhuyan is also working as the Coordinator of the Microbiologist Society in India for Meghalaya, India.

Dr. Yugal Kishore Mohanta has more than six years of post-Ph.D. teaching and research experience and currently holds the position of Assistant Professor at the Department of Applied Biology within the School of Biological Sciences, University of Science and Technology, Meghalaya (USTM), India. Dr. Mohanta's current area of research pertains to nanobiotechnology and sustainable development goals. Dr. Mohanta has international research experience as both a Ph.D. visiting scholar and a postdoctoral fellow at KU Leuven, Belgium, and the University of Nizwa, Sultanate of Oman, respectively. Currently, he is running an international research network project scheme grant sponsored by **Xjenza Malta**, the Government of Malta, in

2024–2025. Dr. Mohanta has authored 110 research publications in peer-reviewed journals, which are indexed in the Science Citation Index Expanded (SCIE). These papers have high impact factors and have garnered a total of 4400 citations, having an h-index of 36 and an i10 index of 81. Additionally, Dr. Mohanta has made contributions to 27 book chapters, five edited books for Springer Nature, CRC Press, and Elsevier-Academic Press, and authored numerous articles on scientific topics for national science publications. He is an expert reviewer and editorial board member for journals of international repute. Dr. Mohanta currently also holds the position of team leader at the Nano-biotechnology and Translational Knowledge Laboratory and served as the principal investigator for the SERB-DST project, a start-up research grant funded by the Government of India in New Delhi. Dr. Mohanta is also working as the Chief Coordinator of the Microbiologist Society, India, for the northeast states of India.

Dr. Manika Khanuja is currently working as an Assistant Professor at the Centre for Nanoscience and Nanotechnology, Jamia Millia Islamia University, New Delhi. She completed her Ph.D. from IIT, Delhi in 2009 on Palladium Nanoparticles and Bimetal Layers for Enhanced Hydrogenation Properties. She has global experience in Nanoscience and Nanotechnology for SDGs, particularly clean water, affordable energy, and good health. She received the David Livingstone Fellowship from the UN for innovative research in water purification, disease diagnosis, and energy storage. Recently, she held various research grants, including Nano mission (DST), SERB, UGC, and BRICS, and published over 100 research publications, 10 books, and 12 book chapters. She has delivered 80 invited talks on national and international stages and is currently supervising five Ph.D. scholars, with 10 Ph.D. degrees already awarded. Her research interest is in harnessing advanced materials for different applications and specializes in creating nanomaterials for water disinfection, sensing, healthcare, and energy storage.

Ramesh Namdeo Pudake is an Assistant Professor at Amity University Uttar Pradesh, Noida one of the top-ranked private universities in India. Dr. Pudake received his Ph.D. from China Agricultural University, Beijing, PRC. After his Ph.D., he is engaged in research in a wide range of topics in biotechnology and nanotechnology. He has also worked in the Department of Agronomy at Iowa State University, IA, USA, on functional genomics. Currently, he is focusing on research on different applications of nanotechnology in agriculture. Dr. Pudake has published more than 50 research papers/reviews in reputed journals, edited ten books with Springer, Elsevier, and CRC Press, and authored more than 30 book chapters. He also has a scholarship award from the Chinese Government for pursuing his doctoral studies. He is an expert reviewer and editorial board member for journals of repute, including Nature Scientific Reports.

Contributors

R. Atchudan School of Chemical Engineering, Yeungnam University, Gyeongsan-si, Republic of Korea

Dipankar Bandyopadhyay Centre for Nanotechnology, Indian Institute of Technology Guwahati, Assam, India;
Jyoti and Bhupat Mehta School of Health Sciences and Technology, Indian Institute of Technology Guwahati, Assam, India;
Department of Chemical Engineering, Indian Institute of Technology Guwahati, Assam, India

Nilambar Bariha Department of Fire and Safety Engineering, Rajiv Gandhi Institute of Petroleum Technology, Sivasagar Campus, Sivasagar, Assam, India

M. Boopalan Pachaiyappa's College for Men, Kancheepuram, Tamil Nadu, India

Subhash Chand Department of Chemistry, Career Point University, Hamirpur, Himachal Pradesh, India;
Department of Chemistry, ABV Government Degree College Sunni, Shimla, Himachal Pradesh, India

Dimpal Chauhan Department of Chemistry, Government College Jukhala, Bilaspur, Himachal Pradesh, India

Murthy Chavali Department of Research and Development, MIT World Peace University (MIT-WPU), Pune, Maharashtra, India

Manjusha Chourasia Ministry of Electronics and Information Technology, New Delhi, India

Sanbed Das Department of Chemical Engineering, Indian Institute of Technology Bombay, Mumbai, Maharashtra, India

Santanu Das Department of Ceramic Engineering, Indian Institute of Technology (Banaras Hindu University), Varanasi, India

Srijita De Centre for Nanotechnology, Indian Institute of Technology Guwahati, Assam, India

Rahul Deka Jyoti and Bhupat Mehta School of Health Sciences and Technology, Indian Institute of Technology Guwahati, Assam, India

Omkar Suresh Deshmukh Department of Chemical Engineering, Indian Institute of Technology Guwahati, Assam, India

V. P. Devarajan Department of Physics and Centre for Research and Development, K.S. Rangasamy College of Technology, Tiruchengode, Namakkal, Tamil Nadu, India

Saurabh Dubey Centre for Nanotechnology, Indian Institute of Technology Guwahati, Assam, India

Tarab Fatima Center for Nanoscience and Nanotechnology, Jamia Millia Islamia, New Delhi, India

C. Revathi Ganesh Chellammal Women's College of the Pachaiyappa's Trust, Guindy, Chennai, Tamil Nadu, India

Tanusree Ghoshal Department of Chemical Engineering, Indian Institute of Technology Guwahati, Assam, India

Bina Gidwani Department of Pharmaceutical Quality Assurance, Columbia Institute of Pharmacy, Raipur, Chhattisgarh, India

Rahul Kumar Goswami Department of Pharmaceutical Chemistry, Columbia Institute of Pharmacy, Raipur, Chhattisgarh, India

Akansha Gupta Department of Biotechnology, IILM University, Greater Noida, Uttar Pradesh, India

Simran Gupta Department of Medical Devices, National Institute of Pharmaceutical Education and Research - Kolkata, Kankurgachi, Kolkata, West Bengal, India

Radhika Babasaheb Jadhav Department of Medical Biotechnology and Stem Cells and Regenerative Medicine, Centre for Interdisciplinary Research, D.Y. Patil Education Society (Deemed to be University), Kolhapur, Maharashtra, India

Veenu Joshi Center of Basic Science, Pt. Ravishankar Shukla University, Raipur, CG, India

Nayanjyoti Kakati Department of Chemical Engineering, Indian Institute of Technology Guwahati, Assam, India

Bharat Kale Department of Research and Development, MIT World Peace University (MIT-WPU), Pune, Maharashtra, India

Govinda Kapusetti Department of Medical Devices, National Institute of Pharmaceutical Education and Research - Kolkata, Kankurgachi, Kolkata, West Bengal, India

P. Siva Karthik Department of Chemistry, University College of Engineering Panruti, Panruti, Tamilnadu, India

Ranjit Kate Department of Research and Development, MIT World Peace University (MIT-WPU), Pune, Maharashtra, India

Shweta Kaushal Department of Chemistry, Career Point University, Hamirpur, Himachal Pradesh, India;
Centre for Nano-Science and Technology, Career Point University, Hamirpur, Himachal Pradesh, India

Manika Khanuja Center for Nanoscience and Nanotechnology, Jamia Millia Islamia, New Delhi, India

Kuldeep Kumar Department of Chemistry, Career Point University, Hamirpur, Himachal Pradesh, India;
Centre for Nano-Science and Technology, Career Point University, Hamirpur, Himachal Pradesh, India

Ravi Kumar Department of Chemistry, School of Sciences, Baddi University of Emerging Sciences and Technology, Baddi, HP, India

Sunny Kumar ChBE, Georgia Institute of Technology, Atlanta, GA, USA

Ganesh Lonkar Department of Research and Development, MIT World Peace University (MIT-WPU), Pune, Maharashtra, India

Ankita Mahadik Department of Medical Devices, National Institute of Pharmaceutical Education and Research, Kolkata, India

Jiwajyoti Mahanta Department of Chemical Engineering, Indian Institute of Technology Guwahati, Assam, India

Surjendu Maity Duke University School of Medicine, Duke University, Durham, NC, USA

Aishwarya Joji Mathew Department of Chemistry, CHRIST University, Bangalore, India

Honey Mittal Department of Physics and Environmental Sciences, Sharda School of Engineering and Science, Sharda University, Greater Noida, India

Marcos Jefferson Moreno Baque Departamento de Materiales, Escuela Politécnica Nacional, Quito, Ecuador

Priya Namdeo University Institute of Pharmacy, Pt. Ravishankar Shukla University, Raipur, CG, India

Arpita Pandey Tiwari Department of Medical Biotechnology and Stem Cells and Regenerative Medicine, Centre for Interdisciplinary Research, D.Y. Patil Education Society (Deemed to be University), Kolhapur, Maharashtra, India;
Global Innovative Center of Advanced Nanomaterials, College of Engineering, Science and Environment, University of Newcastle, Callaghan, NSW, Australia

Ravindra Kumar Pandey Department of Pharmacognosy, Columbia Institute of Pharmacy, Raipur, Chhattisgarh, India

Prathu Raja Parmar Department of Chemical Engineering, Indian Institute of Technology Guwahati, Assam, India

Avantika Patil National Institute of Pharmaceutical Education and Research - Kolkata, Kankurgachi, Kolkata, West Bengal, India

Reetika Pechetty School of Sciences, Woxsen University, Hyderabad, India

P. Periasamy Department of Physics, Nehru Institute of Engineering and Technology, Coimbatore, Tamil Nadu, India

Harsh Pipil Manav Rachna Centre for Advanced Water Technology and Management (MRCAWTM), Manav Rachna International Institute of Research and Studies, Faridabad, Haryana, India

Sudhagar Pitchaimuthu Research Centre for Carbon Solutions, School of Engineering and Physical Sciences, Institute of Mechanical, Processing and Energy Engineering, Heriot-Watt University, Edinburgh, UK

Alina Pruna Instituto Universitario de Investigación de Tecnología de los Materiales, Universitat Politècnica de València, Valencia, Spain

Siddharth Ramesh Talla Department of Medical Devices, National Institute of Pharmaceutical Education and Research, Kolkata, India

Nelly Ma. Rosas Laverde Departamento de Materiales, Escuela Politécnica Nacional, Quito, Ecuador

Ankita Sahu University Institute of Pharmacy, Pt. Ravishankar Shukla University, Raipur, CG, India

Richa Sharma Department of Chemistry, S.D.W.G. Government College Beetan, Una, Himachal Pradesh, India

Shiv Shankar Shukla Department of Pharmaceutical Quality Assurance, Columbia Institute of Pharmacy, Raipur, Chhattisgarh, India

Amit Kumar Singh School of Science (Biotechnology), Woxsen University, Hyderabad, Telangana, India;
Department of Mechanical Engineering, George Mason University, Manassas, USA

Mohammad Talha Nagori Department of Medical Devices, National Institute of Pharmaceutical Education and Research, Kolkata, India

Sakshi Tiwari University Institute of Pharmacy, Pt. Ravishankar Shukla University, Raipur, CG, India

M. J. Umapathy College of Engineering Guindy, Anna University, Chennai, Tamil Nadu, India

Tania C. Valdiviezo-Abarca Departamento de Materiales, Escuela Politécnica Nacional, Quito, Ecuador

Yeshaswini Veeramachaneni School of Sciences, Woxsen University, Hyderabad, India

Manisha Verma Department of Pharmaceutical Quality Assurance, Columbia Institute of Pharmacy, Raipur, Chhattisgarh, India

Anuja Vibhute Department of Medical Biotechnology and Stem Cell and Regenerative Medicine, Centre for Interdisciplinary Research, D. Y. Patil Education Society (Deemed to be University), Kolhapur, Maharashtra, India

T. P. Vinod Department of Chemistry, CHRIST University, Bangalore, India; Centre for Renewable Energy and Environmental Sustainability, CHRIST University, Bangalore, India

Rajangam Vinodh Green Hydrogen Lab (GH2Lab), Institute for Hydrogen Research (IHR), Université du Québec à Trois-Rivières (UQTR), Trois-Rivières, Québec, Canada

Amber Vyas University Institute of Pharmacy, Pt. Ravishankar Shukla University, Raipur, CG, India

Shivani Yadav Department of Environmental Engineering, Delhi Technological University, New Delhi, India

Part I
Introduction

Chapter 1
Understanding Quantum Dots

Bina Gidwani, Ankita Sahu, Priya Namdeo, Sakshi Tiwari, Veenu Joshi, Shiv Shankar Shukla, Ravindra Kumar Pandey, and Amber Vyas

Abstract Quantum dots (QDs) are tiny particles, measuring between 2 and 10 nm. They are made of crystalline substances, either inorganic or organic semiconductors. They have excellent chemical properties, photostability, and a high quantum yield, and they emit light that can be changed based on their size. This chapter provides a comprehensive synopsis of recent developments in the production, characteristics, and uses of quantum dots. This chapter looks at the most recent quantum dots (QDs) developments, including their advantages, disadvantages, characterization, and also explores various synthesis methods of QDs, like bottom-up, top-down, colloidal, etc. Furthermore, the chapter also updates on patents and marketed products of quantum dots. The chapter subsequently examines the various applications of quantum dots in biomedical imaging, sensors, and primarily in drug delivery.

Keywords Drug delivery · Quantum dots · Imaging · Sensors · Synthesis · Characterization

1 Introduction to Quantum Dot

Quantum dots are synthetic entities that contain charge, ranging in size from a single electron to thousands. Their diameters typically vary from nanometres to several microns, with modern nanofabrication methods enabling meticulous regulation of their dimensions, morphology, and interactions. Mark Reed originally introduced the term "quantum dot" in 1986, designating a completely confined zero-dimensional entity within a top-down classification system for quantum dots in solid materials.

B. Gidwani · S. S. Shukla · R. K. Pandey
Columbia Institute of Pharmacy, Raipur, CG, India

A. Sahu · P. Namdeo · S. Tiwari · A. Vyas (✉)
University Institute of Pharmacy, Pt. Ravishankar Shukla University, Raipur, CG, India
e-mail: ambervyas@gmail.com

V. Joshi
Center of Basic Science, Pt. Ravishankar Shukla University, Raipur, CG, India

T. Bhuyan et al. (eds.), *Multifunctional Quantum Dots*, Materials Horizons: From Nature to Nanomaterials, https://doi.org/10.1007/978-981-92-0172-3_1

Quantum dots (QDs) are colloidal fluorescent semiconductor nanoscale crystals that were first developed in the early 1980s by Louis Brus at Bell Laboratories and Alexander Efros and Alexei I. Ekimov at the A.F. Ioffe Physical-Technical Institute in St. Petersburg [1, 2]. Quantum dots (QDs) are semiconductors and nanocrystals formed from materials of the periodic groups of II-VI (like CdTe and CdSe) and III-V (like InP). Fluorescent biological labeling, imaging, solar cells, composites, and detection are some of the many potential uses for these man-made semiconductor nanoparticles due to their exceptional optical, electrical, and photophysical characteristics. In the context of fluorescence resonance energy transfer, they are also useful as donors. At sufficiently small sizes, these semiconductor particles exhibit quantum confinement phenomena, which limit the available energy levels for the electron–hole pairs. The relationship between the power and wavelength of light determines the particle's optical characteristics, which can vary significantly depending on its size. As a result, by adjusting the size of the quantum dots, multiple particles of different colors can be created to emit or absorb specific wavelengths of light [3]. The remarkable abilities of quantum dots (QDs) and the significant study that has been conducted on their potential for research advancement and growth in the area have influenced this section to offer fundamental insights into the fundamental characteristics of QDs. In order to get an understanding of the basic challenges that are associated with the manufacturing process of quantum dots (QDs), it emphasizes the numerous synthesis techniques for the creation and characterization of QDs. QDs' potential as cancer imaging and therapeutic tools is discussed in this chapter to fill the knowledge gap & lead to new discoveries. This chapter also summarizes the application in quantum dots.

The discovery of QDs stemmed from a hope to comprehend and control the behavior of semiconductor materials at the nanoscale. From the time when Alexey I. Ekimov made the initial finding of quantum dots in 1981, there has been a growing interest among researchers in this specific form of nanomaterial, which possesses distinctive optical and electrical characteristics.

QDs are remarkable because their small size results in a physically enclosed electron cloud, a phenomenon known as quantum confinement. QDs have evolved over the last two decades, allowing for precise control of QD nano-sized particles as well as the use of QDs in sophisticated applications. They have transformed the quality of display screens and health care devices, which are particularly used in treating serious diseases, including leukemia, cancer. So, their distinct qualities addressed standard materials' limitations, such as efficiency and tunability. Quantum dots are a prerequisite for creating cutting-edge technologies such as quantum computers, biosensors, and flexible electronics [4]. The journey of quantum dots, spanning from 1981 to 2023, is shown out in Fig. 1.1, which lists the different subsequent innovations [5–7].

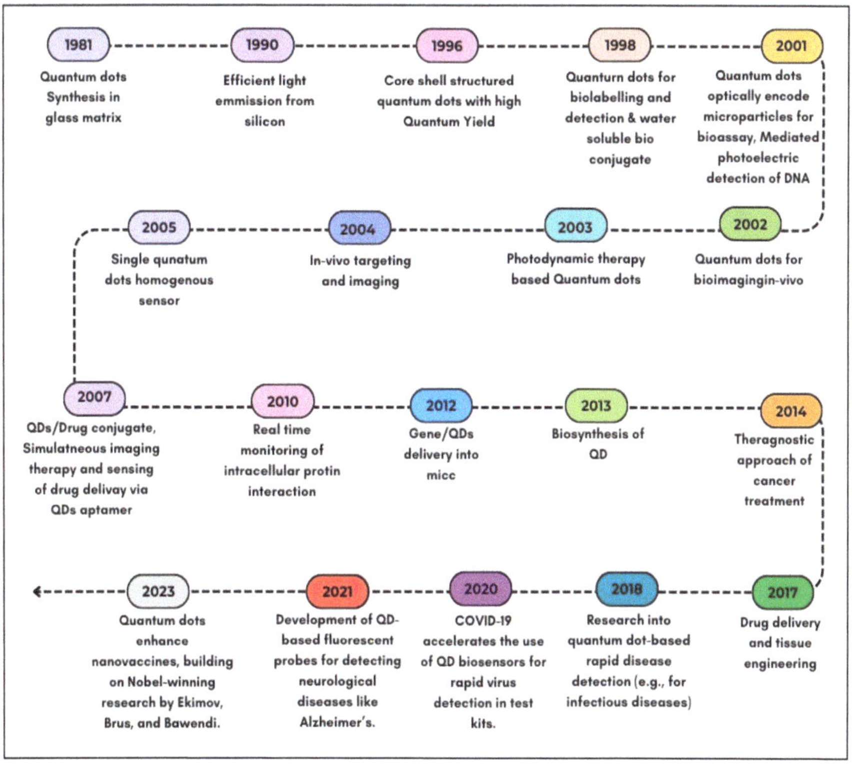

Fig. 1.1 The various successive inventions of quantum dots from 1981 to 2023

2 The Basic Structure of QDs

The core–shell-surface architecture of quantum dots (QDs) allows for very stable light and chemically driven activation of surface and luminescence quantum yield, as well as other desirable properties. QD size ranges from 2 to 10 nm and usually consists of 200–3,000,000 atoms, but only has 100 free electrons or fewer [2]. QDs are made up of a semiconductor core that is overcoated with a shell and capped to increase solubility in aqueous buffers. The inorganic core is responsible for the essential optical characteristics and semiconducting features [8]. The core is mostly made up of CdSe (cadmium-selenium), CdTe (cadmium-tellurium), InP (indium phosphate), and InAs (indium arsenide) in a crystal structure that determines the excitation wavelengths and fluorescence emission. CdSe is an excellent choice for optoelectronics and bioimaging, whereas CdTe and InAs are commonly employed in infrared detectors and solar cells. The major drawback associated with these materials is that no functional modification or binding is possible on the core material. To overcome this problem, QDs are coated with a shell material, which helps in reducing its toxicity, making it water-soluble, and provides a surface for attachment of various

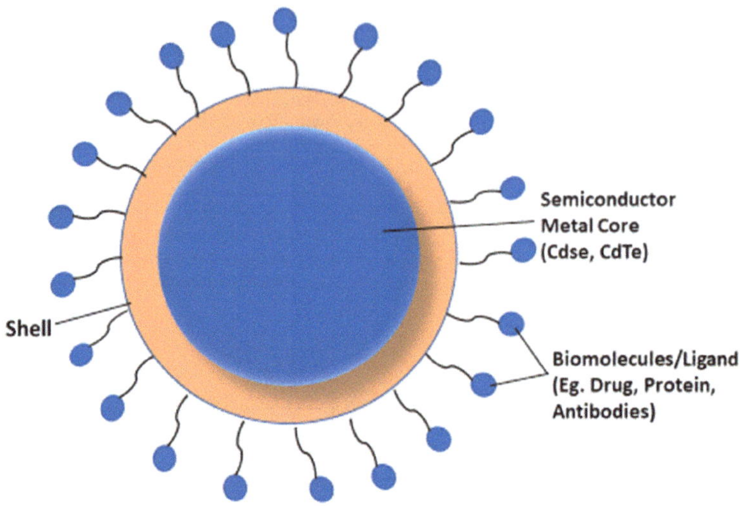

Fig. 1.2 The illustration represents the configuration of quantum dots. Cadmium Selenide constitutes the inner core, while the outside shell is composed of Cd1-xZnxS, all encased in silica to prevent oxidation. The outer shell functions as an absorber

ligands. Before ligands were attached to the active QD-core surface, it was passivated with another inorganic shell. The surrounding shell structure also stabilizes the core. Zinc sulfide (ZnS) makes up most of the shell and affects decay rates, photostability, and quantum yield of fluorescence [9, 10]. To make sure that QDs remain stable, soluble, and functional, they are often covered with an outer layer or protective shell. The exterior layer is constructed of biocompatible material, such as POSS-PCU and PEG (polyhedral oligomeric silsesquioxane-polycarbonate urethane, polyethylene glycol) [11–13]. When created, QDs are hydrophobic as they lack surface functionalities which produce hydrogen bonds; however, molecules that are hydrophilic or polymers can be included to improve water dispersibility. For instance, ionisable functional groups attached to amphiphilic polymers might enhance the water stability of quantum dots (QDs). Figure 1.2 displays a simplified depiction of a quantum dot (Hamidu and [14].

3 Classification of Quantum Dots

Semiconductor groups II–IV, IV–VI, or III–V form the core of most quantum dots (QDs) (e.g., CdSe, CdTe, PbSe, GaN, GaAs, InAs, and InP), which are encased in a semiconductor shell with a wide bandgap, such as ZnS. Quantum dots can be composed of heavy metals or silicon, each exhibiting distinct properties. Figure 1.3 illustrates the many classes of quantum dots. Quantum dots may be classified into

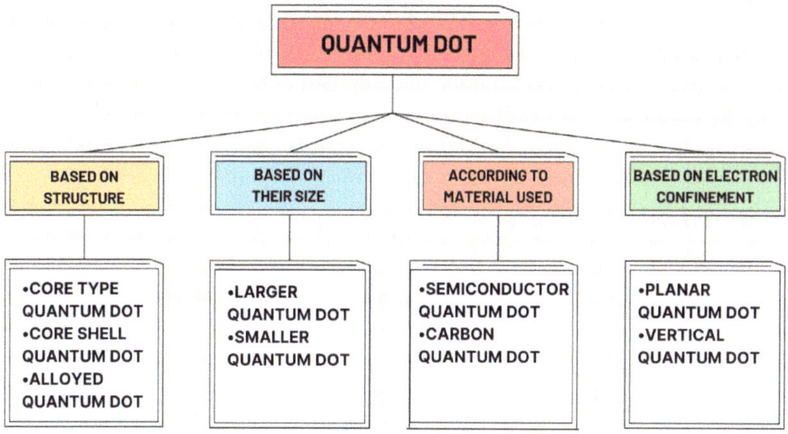

Fig. 1.3 Classification of quantum dots

four primary types based on composition, arrangement, size, processing material, and manner of confinement.

(a) Composition and structure
(b) Size
(c) The material used for preparation
(d) The type of electron confinement

3.1 Considering Their Structure/Composition

(i) **Core–shell Quantum Dots**: Core–shell semiconductor nanocrystals typically comprise a core material surrounded by a second semiconductor material with a higher band gap. Quantum yield is improved by coating QDs with shells. It makes non-radiative recombination sites more flexible to different processing conditions for different applications by making them passive. This approach has been extensively studied as a means for tuning the optical-physical properties of quantum dots [15].

(ii) **Core-type Quantum Dots**: This category of nanocrystal typically consists of a single component material characterized by a consistent internal composition, particularly chalcogenides—sulfides, selenides, or tellurides of cadmium, zinc, or lead. The luminescence exhibited by these particles is a result of the recombination of electron–hole pairs. The photo and electroluminescence properties could be improved by adjusting the size of the core-type nanocrystals [16].

(iii) **Alloyed Quantum Dots**: It includes a homogeneous combination of several materials instead of discrete, independent areas. Two semiconductors with different band gaps create particles with novel properties that set them separate from the original materials. This enhancement significantly boosts the material's effectiveness, particularly in terms of its optical and electronic properties. Consequently, they have unique and supplementary compositions in addition to their standard ones [17].

3.2 Based on Their Size

They are separated into two categories, which are referred to as larger QDs and smaller QDs.

(i) **Larger Quantum dots—Their** diameter ranges from 5 to 6 nm, and they emit longer wavelengths characterized by orange or red emissions. For instance, LED and Photovoltaic are applicable in this QD.
(ii) **Smaller Quantum Dots**—Their diameter measures around 2–3 nm, and they emit shorter wavelengths, resulting in blue or green-colored emissions. With the introduction of a ligand, a carrier molecule, and a polymer coating, there is potential for an increase in size. These quantum dots find applications in imaging, biological studies, and the development of biosensors [18].

3.3 According to the Material Used

Two types of quantum dots exist: semiconductor and carbon-based.

(i) **Semiconductor Quantum Dots**: QD semiconductors were discovered in 1981. Cadmium, selenide, zinc sulfide, and other inorganic elements make up semiconductor quantum dots and three-dimensional nanoparticles. These are round, crystalline, and < 6 nm in size. They are long-lived, size-dependent, and excitation-independent photoluminescent. These substances exhibit amphiphilic characteristics, possess an extensive surface area, and demonstrate excellent photostability, semiconductor functionality, and colloidal attributes. They demonstrate a more significant benefit in terms of functionalization. Semiconductor quantum dots, often referred to simply as quantum dots, are synthesized using a bottom-up approach. These applications are prevalent in analytical sensing, biomedical imaging, and catalysis, with a lesser focus on drug delivery therapy [16].
(ii) **Carbon Quantum Dots**: Carbon quantum dots, or carbon dots, were initially identified in 2004. These materials are composed of a carbon-based structure with sp2 hybridized carbon and possess dimensions below 10 nm. They possess hydrophilic properties and have little toxicity. Carbon quantum dots absorb ultraviolet light in the range of 260 to 350 nm and can be synthesized using both top-down and bottom-up approaches. Their varied functionality and passivity properties, together with chemical inertness and non-photo blinking, render

them appropriate for biological imaging, drug administration, and analytical imaging applications [5].

3.4 *Based on the Electron Confinement Type*

There are three main types of quantum dots: planar, vertical, and self-assembled.

i. **Planar Quantum Dots**: QDs that are planar are also referred to as lateral QDs. As a result of the heterostructures of GaAs/AlGaAs, a two-dimensional electron gas, also known as a two-dimensional electron cloud, is generated. The 2DEG-heterostructure has metal gates patterned on it. Negative gate bias raises the electrostatic potential. This depletes the 2DEG, which splits into a planar dot structure in the center under certain biasing conditions. Gate voltage affects the form and height of planar QDs, which are sensitive to external perturbations [17].
ii. **Vertical Quantum Dots**: Layers of heterostructure are vertically sandwiched around a thin, zero-dimensional pillar. The GaAs disc is positioned in the middle of two AlGaAs barriers that are very thin. A negative bias is applied to the gate electrode that surrounds the pillars. This causes electrons in the outer portion of the pillar to be depleted, which in turn causes the quantum disc to shrink significantly. Consequently, the electrons are restricted to a very narrow zone in the middle of the pillar as a consequence of this arrangement. Because of the huge band gap material, the tunneling barriers are high; as a result, Vertical QDs are less sensitive to disturbances that come from the outside. In a direction that is perpendicular to the [17] 2DEG, the flow of current is vertical. Quantum dots are used in place of the inversion channel that is present in ordinary MOSFETs. As a result, the single-electron transistor is the most widely used application of planar and vertical quantum dots. The electrostatic confinement of planar and vertical quantum dots typically results in dimensions of approximately 100 nm, while the structural confinement of these QDs is on the order of 10 nm [17].
iii. **Self-Assembled Quantum Dots**: The mismatch in the lattice of two materials, such as a few monolayers of InAs on a GaAs substrate, causes a strain to be produced, which ultimately leads to the breakdown of the two-dimensional layered structure. This results in the random distribution of self-crystalline islands, which are referred to as self-assembled quantum dots. Generally speaking, self-assembled quantum dots (QDs) have a shape that is either lens-like or pyramidal, and their sizes are roughly 10 nm. There is a great deal of potential for laser applications with these pyramidal QDs [19].

4 Quantum Dot Synthesis

Quantum dot (QD) synthesis comprises techniques for QD fabrication with an emphasis on reproducible nature, scalability, and uniformity. This section provides an overview of many basic methods for preparing QDs.

4.1 *Colloidal Synthesis*

The selection of reactant components, chiefly cations of metals and chalcogens, and their amalgamation in a solution initiates the synthesis process. Warming this solution to temperatures ranging from 100 to 300 degrees Celsius induces nucleation, during which tiny nanocrystals develop by atomic aggregation. The process progresses to the growth phase, significantly impacted by temperature and the amount of reactants, which both control the development rate and, consequently, the size of the nanocrystals. Surface passivation methods are employed to regulate shape and size, utilizing ligands or capping agents such as oleylamine to limit nanoparticle development and "cap" them at a specified size. The solution is ultimately purified to remove any byproducts and unreacted precursors, enabling scientists to extract aliquots during the process to produce quantum dots of diverse sizes [13].

4.2 *Hot-Injection Synthesis Method*

Nobel Prize in Chemistry winner Moungi Bawendi invented hot-injection synthesis in 2023. A precursor solution with the right ingredients is immediately injected into a heated, high-temperature reaction mixture. The precursors react swiftly to form QDs, which cool and solidify quickly. Controlling colloidal QD size and composition during synthesis involves changing precursor type, concentration, reaction time, and temperature. This method efficiently and quickly synthesizes high-quality QD with homogeneous size distribution in huge numbers [13, 20].

4.3 *Bottom-Up Synthesis*

Sol–gel (injection/colloidal techniques) and chemical reduction are examples of bottom-up synthesis, which involves constructing nanoparticles (NPs) from smaller components by aggregating atoms or molecules into NPs. Several distinct self-assembly strategies have been described for the synthesis of quantum dots; these methods can be further classified as either wet-chemical or vapor-phase approaches. Sputtering, liquid metal ion sources, vapor-phase methods like molecular beam epitaxy (MBE), and gaseous monomer aggregation are all examples of what this class encompasses [21, 22]. Bottom-up synthesis has many advantages, such as being suitable for mass production, being eco-friendly, and being of low cost. Microemulsion, sol–gel, electrochemistry, competitive reaction chemistry, hot-solution decomposition, and other similar techniques fall under the umbrella of wet-chemical procedures and are given below:

4.3.1 Wet Chemical Method

By precisely controlling the parameters for either a single solution or a combination of solutions, this technique largely mimics conventional precipitation methods. Nanoparticle nucleation and restricted growth are inherent features of this precipitation mechanism. Additional classifications for nucleation include secondary nucleation, heterogeneous nucleation, and homogeneous nucleation. In this process, homogeneous nucleation is possible when the solute atoms or molecules are uniformly mixed to achieve a critical size due to lack of a pre-existing solid contact. The most frequent wet-chemical processes include microemulsion, hydrothermal, sol–gel, hot-solution decomposition, competitive reaction chemistry, microwaves, or sonic waves, in addition to electrochemistry [23]. Various factors, such as the reaction temperature, electrostatic double layer thickness, stabilizers or micelle formation, precursor concentrations, anionic to cationic species ratios, and solvent, contribute to achieving QDs of the desired size, shape, and composition.

4.3.2 Sol–Gel Techniques

Sol–gel methods have been utilized to synthesize nanoparticles like QDs for years. This technique involves the preparation of a sol, which consists of nanoparticles dispersed in a solvent through Brownian motion, utilizing a metal precursor such as alkoxides, acetates, or nitrates in either an acidic or basic medium. Hydrolysis, condensation (the creation of a sol), and growth are the three primary phases of this process. Metal precursor hydrolyzes in the medium and condenses into a sol, then polymerises into a network. This approach synthesizes II-VI and IV-VI QDs such as ZnO, CdS, and PbS. To make ZnO QDs, for instance, one would combine Zn-acetate with alcohol and sodium hydroxide to form a solution, and then regulate the air ageing process. The fact that it is easy to scale up, cheap, and requires little effort is the process's main selling point. The sol–gel process's wide size dispersion and high defect concentration are its main drawbacks. So this synthesis method is employed rarely [24].

4.3.3 Micro-Emulsion Process

Normal and reverse micro-emulsions synthesize quantum dots (QDs) at ambient temperature. Polar solvents like alcohol can substitute water. The reverse micelle approach, which creates an emulsion from polar water and nonpolar long-chain alkanes, is preferred for QD production. This method has enabled the synthesis of II–VI core and core/shell QDs, including CdS, CdS: Mn/ZnS, ZnS/CdSe, CdSe/ZnSe, ZnSe, and IV-VI QDs Its benefits include accurate QD size control by altering the water-to-surfactant molar ratio, narrower size distribution than sol–gel, and easy QD dispersion. However, low yield and contaminants and faults are negative aspects[25].

4.3.4 Vapor Phase Method

The vapor phase technique for generating quantum dots typically initiates with processes involving the growth of layers in an atom-by-atom manner. During the first stages, self-assembling nano-structured compounds were produced via the hetero-epitaxial development of severely strained substances. Under ultra-high vacuum circumstances, typically around 10–10 Torr or 7.5 10–16 Pa, molecular beam epitaxy (MBE) is used to produce semiconductor nano-structured materials on a heated substrate by depositing layers of elemental, compound, or alloy components.

4.4 Top-Down Synthesis

The most prominent example of top-down synthesis in the modern era is the production of semiconductor chips and wafers by lithography or other techniques. Nanoparticles (NPs) are created in this method by hydrolyzing bigger materials into smaller ones. Using this method, a bulk semiconductor is thinned, which ultimately results in the creation of quantum dots (QDs). Electrochemistry, laser ablation, arc discharge, and hydrothermal/solvothermal/special (chemical) oxidation are some of the paths that may be used for top-down cutting (cleaving) [26, 27]. On the other hand, this method has a number of limitations, some of which include the use of costly materials, the presence of severe reaction conditions, and the duration of reaction durations. To achieve quantum dots (QDs) with a diameter of approximately 30 nm, it is usual practice to employ several processes, including electron beam lithography, reactive-ion etching, and/or wet-chemical etching. Through the use of the quantum confinement effect, it is possible to achieve the required packing geometries in addition to the controlled forms and sizes. Ion beams or laser beams that are focused in parallel have also been utilized in the process of fabricating arrays of zero-dimensional dots.

The primary disadvantage of these processes is the introduction of impurities within the quantum dots, resulting in structural imperfections in the quantum dots. Synthesis on a large scale in industry presents significant challenges. Particle size control is challenging, resulting in low yield, extended reaction times, and a complex process. The reaction time is prolonged; certain raw materials require treatment with a strong oxidant before the reaction. Additionally, the process involves elevated pressure and temperature, which may lead to explosion or combustion. On the other hand, top-down techniques typically need severe reaction conditions, costly materials and equipment, and extended processing periods [21, 28].

4.5 Electrochemical Synthesis

Electrochemical synthesis is an alternate method for the manufacture of quantum dots (QDs). To drive the assembly of precursor molecules into quantum dots (QDs), this technique draws on electrochemical forces. As a result of the applied voltage, the redox duration, and the concentration of the supporting electrolyte, the size of the QDs that are formed may be effectively controlled. By employing $LiClO_4$ in propylene carbonate as an electrolyte, Kalita and his coauthors were able to synthesize graphene quantum dots (QDs) from graphene oxide at ambient temperature. These quantum dots, which had an average diameter of three to five nanometres, were utilized as sensors for measuring the moisture content of the soil [29, 30].

4.6 Biogenic Synthesis

One novel biotechnological approach to the mass manufacturing of quantum dots is biogenic synthesis (QDs). This method synthesizes quantum dots, including CdS, by using live microbes like E. coli as bioreactors. The initial phase involves the detoxification of cadmium ions, which are harmful to living organisms, through their conjugation with cysteine-terminated peptides. To make CdS quantum dots, bacteria are injected with detoxified ions and then allowed to interact with endogenous sulfide ions. The resulting product is then exported and collected. Benefits of this approach include biosafety, cost-effective manufacturing, scalability, and environmental sustainability. The detoxification process's peptide composition also influences the quantum dots' physicochemical characteristics [31].

For the synthesis of eco-friendly nano-scaled materials, especially ZnS nano photocatalysts, plant extracts show great promise. Reasons for this include their ability to minimize the toxic effect of specific undesirable nanoparticles, strong antioxidant potential, and plenty of phenols & flavonoids. A number of secondary metabolites in plants serve as efficient capping agents, stabilizers, and reducing agents, which has led to a rise in the use of plant extracts in the fabrication of nano-sized quantum dots (QDs). The use of aqueous extracts from Ficus johannis, in conjunction with techniques such as microwave and ultrasonic extraction, has been established as a green synthesis method for ZnTe and CdTe QDs (Mahajan et al. 2017) [32].

The Echinops persicus plant extract has been used in a new, sustainable way to produce carbon quantum dots (CQDs), according to a recent study. Based on their surface and structural features that enhance interactions with different pollutants, these CQDs show efficiency and adaptability in pollution management, according to the research [33]. Camellia sinensis extract was employed for the green production of CdS QDs. The CdS QDs underwent two stages of preparation. Initially, a specified amount of the extract was combined with a specific amount of CdSO4 (0.025 M) and incubated for three days in the dark. The second step in making CdS QDs involved adding 0.025 M Na2S and incubating the mixture for another four days. The completed solution was centrifuged at 13,000 rpm for 10 min, resulting in a vibrant yellow tint. The sample underwent three rinses with deionized water to eliminate any residual contamination from the recovered CdS quantum dots. The pellet was then lyophilized to enable further characterization studies [34].

5 Unique Features of Quantum Dots

The unique optical and electrical properties of quantum dots (QDs) make them very appealing for use in a wide range of scientific investigations and medical diagnostics. Bioimaging and administration of drugs are two examples of these many uses.

- Quantum dots resist degradation, enabling longer-term tracking of biological activities compared to conventional imaging probes [6].
- When compared to organic dyes, quantum dots (QDs) have a brightness that is 10–20 times higher and a longer color stability [6].
- QDs' stable fluorophores result from their nano-sized inorganic substance [18].
- Large surface area enhances drug loading and tags nano-carriers in biological systems.

  (M.-X. [35]).

- Quantum dots are excellent for biological systems because they may be surface modified. They can be readily molded into numerous forms, sizes, and bioactive molecule coatings.
- QDs are chemically neutral and work well for imaging both in vitro and in vivo, which means they can be used for theranostic drug transport.
- QDs can encapsulate both hydrophilic and lipophilic drugs and exhibit a long duration of blood circulation.
- Because of the modification in the shell of quantum dots, various changes occur, such as charge, size, and solubility.
- QDs have low cost, good stability, good flowability, and narrow size distribution. They exhibit narrow and sharp peaks with broad excitation spectra [36].
- Optimal for electrochemical analysis due to electrical features such as high conductivity, biological acceptability, photocatalytic efficiency, and high surface area-to-volume ratios.

- Because of the quantum confinement phenomenon, quantum dots have useful optoelectronic capabilities. These characteristics encompass a wide range of absorption spectra, fluorescence levels, photostability, and the capacity to adjust emission based on particle size [6].
- The dimensions and chemically adjustable characteristics are beneficial for selecting an emission wavelength appropriate for a particular experiment. The emission wavelength of cadmium sulfide (CdS) and zinc selenide (ZnSe) dots may be adjusted from blue to near-ultraviolet light [14].
- Among their many desirable and essential properties are their resistance to photobleaching, high signal brightness, large absorption coefficients, light emission, and the ability to stimulate a variety of fluorescence colors [14].
- QDs have a longer fluorescence lifespan (10–40 ns) than organic dyes due to their inorganic composition, stronger emission, and greater signal-to-noise ratio. QDs are 10–20 times brighter than organic fluorophore molecules [36].

6 Characterization of Quantum Dot

Examining the optical characteristics, size, shape, and composition of quantum dots (QDs) is an important part of characterization. Fourier-transform infrared spectroscopy (FTIR), Atomic resonance spectroscopy (AES), nuclear magnetic resonance (NMR), ultraviolet (UV) spectroscopy, chromatography, and transmission electron microscopy (TEM) are some of the spectroscopic and chromatographic methods used by researchers. Furthermore, magneto-tunneling studies, atomic force microscopy (AFM) and scanning tunneling microscopy (STM), are used to measure the size of epitaxial quantum dots (QDs), allowing for optimized attributes and optimization for various applications [37, 38].

6.1 Morphological Characterization

Researchers have employed a variety of methods to analyze quantum dots (QDs), but the two most used are scanning electron microscopy (SEM) and transmission electron microscopy (TEM).

TEM, providing a resolution of 0.1–0.2 nm, is essential for elucidating the ultrastructure of samples and is extensively utilized in scientific research, medicines, and materials science. It facilitates the examination of nanoparticle morphology, yielding insights into form, size, and dispersion. Moreover, it is helpful in characterizing quantum dots, with high-resolution TEM accurately evaluating their fine structure and verifying that these quantum dots are uniformly dispersed in aqueous solutions at a resolution of 50 nm [37].

Atomic Force Microscopy (AFM) can provide high-resolution two- and three-dimensional images of quantum dots (QDs) without causing any damage and without using optical imaging. In contrast to transmission electron microscopy (TEM), atomic force microscopy (AFM) makes use of a probe to determine the sample's topography [37, 39].

6.2 Structural Characterization

Understanding the composition and bonding material features of nanoparticles (NPs) relies heavily on their structural properties. Understanding the bulk characteristics of nanoparticles may be obtained from these qualities. The structural characteristics of NPs may be investigated using important techniques such as energy dispersive X-ray spectroscopy (EDX), Fourier-transform infrared spectroscopy (FTIR), and X-ray diffraction (XRD).

X-Ray Diffraction is a flexible and effective way to characterize quantum dots and determine composition, particle size, crystal structure and phase purity. Based on their diffraction pattern, XRD determines quantum dots' crystalline phases.

Fourier Transform Infrared Spectroscopy—The presence of functional groups on quantum dots' surfaces may be identified via Fourier-transform infrared spectroscopy. Typical preparation methods and materials used to make quantum dots cause a wide range of surface functional group compositions. Surface functional groups are therefore identified and characterized by use of FTIR technology.

The **NMR** strategy is often used to obtain structural information on quantum dots.

EDX analyses QDs' elements or chemicals. A QD's electrons are excited by an X-ray beam, causing them to glow. Measure the X-ray energy to ascertain the QD sample's elements. Quantitative and qualitative assessments can use EDX. A sample's elements may be identified using EDX because various elements emit different X-rays [40].

6.3 Optical Characterization

The optical characterization of quantum dots (QDs) reveals details on their photoluminescent, absorbance, and reflectance characteristics. Because of their unique hues, quantum dots (QDs) may be studied using UV–visible (UV–Vis) and photoluminescent spectroscopy to uncover their optical characteristics.

UV–Visible Spectroscopy measures the absorption and scattering of light by a material, also known as extinction, and is based on Beer-Lambert's law. Carbon quantum dots and other quantum dots benefit greatly from this method since their optical characteristics depend on parameters including size, shape, concentration, aggregation state, and the medium's refractive index, among others. Fundamentally, it relies on chemical compounds interacting with ultraviolet or visible light, with absorption leading to excitation and deexcitation processes and distinctive spectra that help identify and characterize quantum dots [37].

One way to learn about a material's characteristics is through **photoluminescent spectroscopy** (PL), which involves measuring the amount of light the material gives off when exposed to a certain kind of energy. The composition, structure, and defects of the material may be discovered by analyzing the emitted light, which is usually visible or ultraviolet. Nanomaterials' optical and electrical properties may be effectively ascertained using this technology [37].

7 Applications of Quantum Dots

The size-tunable optical features of quantum dots (QDs) have attracted a lot of interest across many fields. They discover use in many different things, including computers, optics, photovoltaics, light-emitting diodes, solar cells, and photodetectors. Drug delivery, targeted therapy, cell labeling, cell tracking, bioimaging, and most importantly cancer therapy are just a few of the many domains where quantum dots (QDs) are indispensable in the biological sciences. This chapter focuses further on QDs, on their applications in bioimaging, cancer therapy, and drug delivery.

7.1 Quantum Computing

Superposition and entanglement are two examples of quantum mechanical phenomena that are utilized by a quantum computer to carry out data operations. Silicon photonics has already demonstrated potential as a platform for quantum computation (Norman et al. 2018). Quantum computers differ from regular computers that use transistors. The development of quantum computers, which are very efficient supercomputers, has been made possible by quantum dots. Quantum computers rely on quantum bits, or "qubits," to operate and store data. Qubits can be in two states at the same time: on and off. This remarkable phenomenon enables considerably swifter information processing and enhanced memory capacity relative to traditional computers. Information is processed on a qubit array via a succession of two-qubit and single-qubit gate combinations. This is analogous to how logic gates act on Boolean bits in classical computing. Quantum computers are considerably faster at addressing certain difficult security problems, and quantum chemistry aids in vaccine development (e.g., COVID-19) and optimization. Lockheed Martin purchased the first commercialized quantum computer device via D-Wave Systems Inc. on May 25, 2011. Traditional computers store information which is called "bits," and each bit has the potential to hold either a 0 or a 1. A quantum computer stores a set of qubits. Quantum computers manipulate qubits using a preset order of quantum logic gates. A quantum algorithm is the sequence of gates to be used [41–43].

7.2 Quantum Dot LASER

The quantum dot laser represents another application of QD technology, offering advantages that surpass those of conventional lasers. QD lasers exhibit reduced temperature sensitivity and a lower likelihood of failure at elevated temperatures, thereby offering enhanced operational flexibility for efficient laser performance. Additional benefits encompass lower threshold currents, enhanced power output, and improved stability in contrast to the restricted performance of traditional lasers. The quantum dot laser is going to be a game-changer for optical networks and data transmission [44].

7.3 Quantum Dot LED (QLED)

Nearly one hundred percent of the energy that is introduced into the system is also the energy that is emitted as light by quantum dots. Because of their great efficiency, quantum dots serve as excellent candidates for applications in lighting and as distinct color pixels in brilliant color flat panel displays. For illumination, quantum dots can be put between two electrically conducting layers. The direct delivery of current to the quantum dots situated between these layers induces luminescence, yielding a highly efficient light source. Among the benefits include emitting wavelengths that may be regulated by having narrow bandwidths, size, and almost perfect emission efficiency. Cutting-edge quantum dot-light-emitting diodes with internal quantum efficiencies close to one [45]. Quantum dot-based QLEDs emit at various wavelengths and can be molded into multiple materials like coatings, paint, plastic, and filters, rendering them applicable in diverse contexts [46, 47]. QLEDs could find use as illuminating light sources in photomedicine shortly. Photomedicine is an emerging medical discipline in which light is used to destroy cancer cells using photosensitizers and singlet oxygen (photodynamic therapy, or PDT) or to enhance cellular function, resulting in favorable therapeutic benefits (photobiomodulation, or PBM). Currently, photomedicine uses bulky and expensive light sources such as lasers or LED arrays. The unavailability of effective, low-cost illumination systems has slowed photomedicine's general clinical acceptance [48].

7.4 White Light Sources

White-light quantum dots provide a more consistent distribution of visible spectrum wavelengths, creating a warmer, yellowish light close to natural sunshine. This contrasts with conventional fluorescent tubes and incandescent bulbs, which can emit considerable quantities of invisible infrared radiation. The infrared emission produces surplus heat, adversely affecting energy efficiency in traditional light sources [44, 49].

7.5 *Quantum Dot Semiconductor Optical Amplifiers*

High-speed signal regenerators, ultrawide band polarization-insensitive high-power amplifiers, and wideband wavelength converters are all being developed using quantum dot semiconductor amplifiers for optical signals. Semiconductor optical amplifiers possess several special characteristics. Broadband gain, high saturation output power, low noise figure (NF), excellent four-wave mixing (FWM) efficiency, and quick gain recovery in a few ps are noteworthy. Quantum dots allowed the fabrication of a semiconductor optical amplifier with over 25 dB gain, less than 5 dB noise, over 20 dBm 3-dB saturation output power across a 90 nm bandwidth, and the greatest penalty-free output power. Among all forms of optical amplifiers, the use of quantum dots enabled the creation of a semiconductor optical amplifier that exhibited the highest gain (> 25 dB), noise figure (< 5 dB), and 3-dB saturation output power (> 20 dBm) over the broadest bandwidth (90 nm). In comparison to all previous semiconductor optical amplifiers, the device displayed an ideal free from penalties output power of 23 dBm [50, 51].

7.6 *Photovoltaics and Solar Cells*

The photovoltaic system is an effective way to capture this plentiful solar energy. The quantum dot-sensitized solar cells, also known as QDSSCs, are among the most effective and affordable alternatives to energy from the sun (Albaladejo-siguan et al. 2021; [52]. Unlike bulk semiconductors, quantum dot photosensitizers have the following advantages: (1) Multiple Exciton Generation (MEG) (2) Good excited-state electron injection into the metal oxide conduction band (3) Allows band gap tuning due to quantum confinement [53].

7.7 *Applications in the Health Sector*

7.7.1 Medical Treatment

Quantum dots can be enclosed in a shell resembling biological body sensors, perhaps corresponding to certain diseases, infections, or other substances. Due to their fluorescent properties, Quantum dots will efficiently identify and bind to the disease in substantial quantities. Compared to the dot's surface size, the number of receptors required is tiny. This gives plenty of room to put additional items on the dot. This may encompass a range of pharmaceuticals that address an illness that the quantum dot was initially designed to identify. S. Parani and coauthor described the alginate biopolymer-stabilized CdTe quantum dot as a medication carrier for cancer treatment [54]. Zhou and coauthors showed that QDs have potential as PSs for PDT in the treatment of pancreatic cancer by decreasing SW1990 proliferation of cells and inducing apoptosis through ROS production [55].

7.7.2 Biosensors

Researchers have employed QDs as labeling materials for biosensors. An in-depth assessment of the enhancement of assays & tiny sensors utilizing quantum dots as components is provided. Due to their special nanomaterial characteristics, QDs are being used in the construction of optical probes for imaging and biological detection, therefore providing great accuracy as well as selectivity in chemical and biochemical sciences. The fusion of quantum dots with electrochemical sensors has resulted in advancements in highly specific as well as effective analytical techniques. QDs can boost electrochemical signal density in the detecting system and provide clear, well-resolved voltage measurement readings. QD-based sensors are ideal for building selective, fast, and accurate detection of analyte instruments [56].

The integration of single-molecule detection with quantum dots facilitates the development of exceptionally sensitive nanosensors. Advantages of single quantum dots (QDs) in nanosensors include low-power consumption, sensitivity, speed, and a high signal-to-noise ratio. Two types of QD-based nanosensors are single QD fluorescence resonance energy transfer (FRET) detection and single QD burst coincidence detection. Nanosensors based on single quantum dots have the potential to detect low-abundance species without the need for nucleic acid amplification. This would make it possible to conduct a real-time study of biological and metabolic processes. In addition to viruses, it can recognize DNA, microRNAs, proteins, enzymes, and other minute molecules [57].

QDs may bind several proteins to their surface and have unique physical and optical properties, making them ideal for biosensing. Some innovative QD-based assays that can improve present DNA and protein detection methods were carried out. For instance, a QD-tagged aptamer (nucleic acids bound to specific chemical targets, including adenosine, thrombin, or cocaine) has been used in an adenine-triphosphate (ATP) detection system. Bagalkot and colleagues developed a simple proof-of-concept QD-aptamer combination that uses fluorescence resonance energy transfer (FRET) to image and deliver anticancer medications to prostate cancer cells and targeted tumor cells [58, 59].

7.7.3 Biolabelling and Cellular Imaging

QDs also have a variety of applications in biolabelling and cellular imaging, both in vitro and in vivo. To process images of living organisms, bioimaging technology utilizes techniques such as X-rays, ultrasound, and magnetic resonance imaging. CDs are more suitable for visualizing biological systems both in vitro and in vivo due to their low toxicity, biocompatibility, and fluorescent properties. Suitable for ex vivo cell-labelling research, scientists have created hybrid functionalized QD-liposome nanoparticles that are uptaken by live cells without cell death. Both in vitro and in vivo, these nanoparticles also demonstrate improved penetration and retention within the tumor interstitium; they may be employed as fluorescent probes [60–63].

Several research studies provided insight into QDs for effective BBB penetration and brain-targeted drug delivery. The successful application of asparagine-glycine-arginine peptide NGR-PEG-CdSe/ZnS QDs for imaging malignant brain tumors, where QDs could cross the BBB and target CD13-overexpressing glioma, and CM-QDs-PEG-PAEA-DDA for cellular brain imaging. CdSe@ZnS/ZnS Core@Multiple Shells 12.7 nm fluorescent imaging probe was made of QDs (CM-QDs-PEG-PAEA-DDA). The probe is highly soluble, stable, and biocompatible, and it enters the brain in 1 h (Tan et al., 2023).

7.7.4 Bioimaging

When it comes to diagnosing serious diseases early on and performing other crucial tasks, bioimaging technology is an essential modern application. It includes methods that do not harm living things in any way while still allowing researchers to see how biological systems work in real time. This method enables the non-invasive collection of data on specimens in three-dimensional structures. Quantum dots (QDs) are state-of-the-art fluorescent probes that have transformed molecular tagging for in vitro and in vivo research. QDs are very suitable for use in diagnostics, biosensing, photodynamic treatment, bioimaging, targeted gene delivery, drug discovery, and targeted gene delivery. Researchers are currently working on novel ways to diagnose cancer early and manage therapy by making use of the special physical and chemical properties of QD probes [40]. Quantum dots like carbon, graphene, and Zinc sulphide are the materials that have many applications in tumor imaging, and the results show that these materials exhibit excellent optical properties, chemical stability, and favorable biocompatibility [64].

7.7.5 QD in Drug Delivery Systems

QD has applications in drug delivery systems. QDs can target the delivery system and easily differentiate between sick and healthy cells through the self-assembly of synthesized polypeptides along with semiconductor core–shell quantum dots using metal affinity as the driving force, and have minimal side effects. Research on drug delivery has recognized QDs for their structural features, such as their extremely tiny size (2–10 nm), which allows for rapid passage of QDs during renal filtration. Several surface-targeting ligands may be included into QD nanoparticles, which allow for prolonged circulation of blood, safety, increased drug loading, and controlled drug release. Also, the passage through cells of oligonucleotide antisense tagged with QDs can be enhanced by carbon nanotubes (CNTs) [65]. Degim and colleagues proposed that ZnO quantum dots can successfully transmit active medicinal molecules through the skin [66]. Curcumin QDs reduce viral infection in multiple ways, indicating their use as carriers for antiviral medications and promoting the targeting of antimicrobial treatments by lowering their toxicity [67]. The following features of QD nano-carriers are required for targeted medication delivery [68].

- Drugs should not interact with the nano-carriers.
- A high level of encapsulation efficiency and medication loading capacity is required.
- The nano-carriers must be prepared and purified in a proper manner.
- The low toxic, biocompatible materials are ideal for nano-carrier application.
- The nano-carriers should have a specified mechanical strength and stability, as well as the right particle size and form, among other things.
- The duration of in vivo stay should be high.

8 Marketed Products of Quantum Dots

Quantum dots have recently become commercially accessible for several pharmaceutical uses. Some of the marketed products of QDs are as follows (Table 1.1):

This segment emphasizes the major market participants and their recent advancements in the biomedical applications of quantum dots (QDs).

With an addressable market size of USD 1 billion in 2022, **NANOCOTM** is a prominent player in the healthcare QD industry. One of NANOCO's marketed products that utilizes QD technology for biosensor medical applications is HEATWAVETM. This instrument covers a large portion of the electromagnetic spectrum (400–1650 nm) and can detect hemoglobin at 575 nm, glucose at 1650 nm, and bilirubin at 455 nm, allowing for the non-invasive measurement of biomolecules in human blood [31].

To further reduce the needless removal of healthy tissues during surgery, QDs are utilized in **VIVODOTS®**, another commercially available product, to map tumorous tissues intraoperatively [71].

NNCrystal US Corporation also owns NN-Labs®, which is a player in the market. Biomedical applications involving in vivo imaging, cellular imaging and molecular labeling probes make use of cadmium-based and cadmium-free quantum dots.

The Japanese company QD LASERTM specializes in developing QD technology for use in healthcare and industry. Innovative VISIRIUM® Technology, developed

Table 1.1 Marketed products of QDs

S. No	Product name	Description	Key features	Wavelength range	Reference
	EviTags™	Water-soluble quantum dots with bioactive surface	Conjugation-ready; carboxyl or amine-functionalized	490–680 nm	[68]
	EviFluors®	Quantum dots conjugated to antibodies and proteins	Ready-to-use; conjugated to Goat anti-mouse, anti-rabbit, anti-rat, Streptavidin, and Biotin	520–680 nm	[69]
	EviDots®	Core & core–shell semiconductor nanocrystal quantum dots	Available as core or core–shell; enhanced stability, improved quantum yield, reduced photo-degradation	490–2100 nm (PbSEviDots: 850–1500 nm)	[70]
	EviComposites™	Quantum dot composites with polymer matrix (e.g., PMMA)	UV-curable form; polymeric solution for integration in plastics	–	[18]

by the company, are 50 g-weight eyeglasses that help those with visual impairments by projecting images onto the retina [72].

9 Regulatory Status of Quantum Dots

Clinical translation of QDs is still in the beginning stages, despite the technology's intriguing properties and many potential uses. Strict standards and restrictions are required in the development of new drug products for them to be safe, effective, and cost-effective under the generally recognized as safe (GRAS) category. Toxicology is a key concern for quantum dot-based biological and pharmacological applications. As the first time the FDA authorized the use of an inorganic substance in a way similar to that of a medicine, the 2011 approval of quantum dot technology enabled the first human clinical trial, marking a major landmark [68]. The EU Restriction of Hazardous Substances limits Cd-based QDs in consumer goods to less than 100 ppm. This Cd restriction is shifting QD sales toward Cd-free alternatives like InP and perovskite QDs [73]. There is still space for more investigation, analysis, preclinical testing, and diagnostic and therapeutic applications. Finally, there are currently more than a dozen clinical trials testing the application of QDs in various nations. Table 1.2 shows a few recent patents on QDs related to the biomedical field.

Table 1.2 Patents on QDs

Title of patent	Assignee company	Patent No
A biosensor for exosome detection based on sulfide colloidal quantum dots and its preparation and application	Huazhong University of Science and Technology Wenzhou Medical University Wenzhou Institute of Advanced Manufacturing Technology of HUST	CN119310287A
Semiconductor nanoparticles, method of producing the semiconductor nanoparticles, and light-emitting device	National University Corporation Tokai National Higher Education And Research System, Osaka University, Nichia Corporation	US-12040433-B2
Preparation of transdermal therapeutic systems containing chlorpheniramine/paracetamol and chlorpheniramine/paracetamol quantum dot	Camlikgamze Onalaybuke	TR2024018865a2
Quantum dot contact lens and preparation method thereof	Eye hospital of wenzhou medical university	CN119335764a
Reagent combination for biosensor, biosensor, and application thereof in cancer marker detection	Nankai University	CN119178795A
Application of carbon nitride quantum dots	Sichuan Xuandu Eye Vision Technology Co, Ltd	CN119033811A
Quantum dot, quantum dot composition, and detection reagent and drug containing quantum dot	Hangzhou Meilian Medical Co., Ltd	CN119101499A

10 Challenges of Quantum Dots

- **Toxicity and Environmental Issues**: Because of their unique physical and chemical characteristics, the toxicity of quantum dots is the primary concern about their use in biological processes. Toxic elements in QDs, such as cadmium or lead, may pose environmental and health dangers. Cadmium telluride quantum dots (CdTe) have been used as fluorescent probes for biological imaging and to monitor targeted drug delivery. By incorporating gelatin in the production process, scientists have successfully reduced the toxicity of these particles, potentially aiding in the creation of other low-toxicity nanoparticle composites. Developing cadmium-free and environmentally friendly QDs is a top priority [69].
- **High Cost**: Expensive production processes for high-quality, uniform QDs hinder their widespread use in cost-sensitive industries like photovoltaics [74].
- **Stability Issues:** QDs can degrade with exposure to light, air, or moisture, impacting their effectiveness and lifetime [75]. It is possible that the degradation of quantum dots (QDs) will reduce their drug delivery efficacy. To enhance their performance, researchers are exploring methods such as modifying the surface

and encapsulation to protect QDs from biological degradation. This will allow them to remain in circulation for longer and increase their bioavailability in the body (Pareek et al. 2025).

- **Scalability**: Moving from lab-scale to commercial-scale manufacturing without sacrificing quality is a challenge
- **Regulatory Barriers:** Toxic substance restrictions, particularly in consumer electronics and medical applications, require alternate formulations and compliance measures. There are a lot more regulations that should be made to meet the growing demand for quantum dots.
- **Preparation Issue**: To improve performance, the semiconductor core material must be protected from degradation and oxidation during the QD preparation process in order to create stable, biocompatible, and water-soluble luminescent nanocrystals. Another problematic issue is the elevated reactivity of QD cores that produce unspecific interactions with macromolecules, which leads to particle aggregation and fluorescence variation [76]. For quantum dots to be widely used in developing fields, these issues must be resolved by breakthroughs such as non-toxic materials, scalable production techniques, and sustainable procedures.
- **Limited Information on Clinical Matters**: Despite preclinical advances, QD research is still in its earliest stages, with few clinical studies [77].

11 Advantages of Quantum Dots

It has been proved by the most recent technical advancements that quantum dots (QDs) have the potential to make significant advances to a wide range of purposes, therefore replacing a number of materials that are complicated, costly, and inefficient.

QDs' wide absorption spectrum and narrow emission profile enable simultaneous fluorescence color stimulation and effective photoluminescence signal spectral deconvolution. This makes them suitable for multiplexing and the detection of many signals. Fluorescence molecular imaging provides real-time visualization, high sensitivity, and a non-invasive diagnostic approach that is distinguished by excellent photostability, a large extinction coefficient, and variable emission wavelengths. QD electrons are restricted in all directions, resulting in higher oscillator strengths for electronic transitions, which can be used to make a variety of electronic devices, like as solar cells, LEDs, and transistors.

The stability and composition of QDs contribute to their increased photostability. Due to QDs' great photostability and brightness, they are ideal for high-sensitivity applications such as fluorescent tagging and live-cell imaging. Their fluorescent features and great resistance to metabolic degradation allow for a broader range of studies to be conducted, ignoring any time constraints. Quantum dots' superior biologic features, such as hydrophilicity, low toxicity, chemical stability, and strong biocompatibility, ensure their potential applications in biosensors, drug delivery, and bioimaging.

Quantum dots are far more resistant to deterioration than other optical imaging probes, allowing them to monitor cell activities over longer periods of time and provide fresh insight into molecular interactions. Because quantum dots are nanocrystals, they give strong contrast when imaging with an electron microscope as scattering rises. Their microscopic size allows them to be injected into a wide range of settings, including liquid mixes, textiles, and polymer matrices [8–10].

12 Disadvantages of Quantum Dots

Quantum dots based on CdSe are extremely toxic, and according to significant research, they cause death of cells by causing damage to DNA. Additionally, they require a polymer shell that is stable. Depending on the application, costly precursors and equipment might raise production costs [13]. A few drawbacks are the long synthesis process, limited yield, and specified size control, as the shell can change optical characteristics [68]. In an experimental setup of a white-light-LED, QDs showed complex degradation behavior, which is very different from that of conventional phosphors [49]. Infrared QDs have a relatively low-quality yield, overall conversion efficiency is lower, which may need to be further improved in the future [68]. Organic QDs are non-toxic but significantly impacted by temperature, pressure, and humidity, while inorganic QDs are highly toxic but very stable. Cd-based QDs are harmful because they aggregate in live cells and destroy them, and their elimination and metabolism are unknown; therefore, their buildup in bodily tissues can cause toxicity [31].

13 Future Prospective

Quantum dots (QDs) have shown a lot of promise in the biomedical area, especially in drug delivery as well as imaging. They can serve as therapeutic platforms due to their huge drug attachment surface area, customizable optical properties, and configurable solubility. As nanotechnology continues to improve, the prospect for the development of QDs seems bright. There is ongoing research to design hydrophilic quantum dots that are luminescent, which can be beneficial in the future. Efforts are currently being carried out to develop a method of labeling cells and biomolecules that is more selective and specific. Research is being done to produce quantum dots with greater biosafety and to examine the interference effect of quantum dots on regular physiology. In a variety of tagging applications, single quantum dots of compound semiconductors have been effectively employed in place of organic dyes. This concept has been advanced by mixing quantum dots of various sizes, which result in varying luminous colors, with polymeric microbeads. The researcher also explored carbon QDs for supercapacitor application, which is still at its early stage; it implies that many avenues have not been explored yet. Cancer is a big issue nowadays, so the early diagnosis and treatment of cancer are at the forefront of cancer QD research. These studies include treatment besides detection for cancer diseases, such as leukemia, ovarian cancer, prostate cancer, breast cancer, and pancreatic cancer. The main factors driving the strong demand for QD technology are growing demands for affordable, better diagnostic sensors, high-definition displays, easier medical imaging, and renewable or low-power energy industry. Consequently, the market forecasts of rising QD production for industrial and consumer goods over the next ten years are supported. Future efforts will focus on improving pharmaceutical responses to minimize quantum dot biological toxicity.

14 Conclusion

Quantum dots are a novel semiconductor technology. In this chapter, classical synthesizing and quantum dot applications in many disciplines are discussed (pharmaceutical, medical/biomedical). Thus, QD applications in various sectors have yielded encouraging results and offered new research and development paths. A brief overview of the basic structure, advantages, disadvantages, and classification is provided. Although there are obstacles to overcome, the prospects for quantum dots (QDs) are promising, and it is anticipated that they will play a significant part in the development of a variety of sectors. Overall, QDs are an exciting area of research and have the potential to bring about significant advancements in various fields. However, its toxicity is one of the key disadvantages that limit its application in biological systems. Now that all of this information has been taken into consideration, it is possible to draw the conclusion that quantum dots (QDs) are particles of nano size that serve as key carriers in the realms of medicine and pharmaceuticals.

References

1. Drummen GPC (2010) Quantum dots—from synthesis to applications in biomedicine and life sciences, 154–63. https://doi.org/10.3390/ijms11010154
2. Park H, Shin DJ, Yu J (2021) Categorization of quantum dots, clusters, nanoclusters, and nanodots. https://doi.org/10.1021/acs.jchemed.0c01403
3. Rosenthal SJ, Chang JC, Kovtun O, Mcbride JR, Tomlinson ID (2011) Review biocompatible quantum dots for biological applications. Chem Biol 18(1):10–24. https://doi.org/10.1016/j.chembiol.2010.11.013
4. Delehanty JB, Boeneman K, Bradburne CE, Robertson K (2009) Quantum dots: a powerful tool for understanding the intricacies of nanoparticle-mediated drug delivery, 1091–1112
5. Gidwani B, Sahu V, Shankar S, Pandey R, Joshi V (2021a) Dots: prospectives, toxicity, advances and applications. J Drug Deliv Sci Technol 61(December 2020):102308. https://doi.org/10.1016/j.jddst.2020.102308
6. Bajwa N, Mehra NK, Jain K, Jain NK, Bajwa N, Mehra NK, Jain K, Jain NK (2016) Pharmaceutical and biomedical applications of quantum dots pharmaceutical and biomedical applications of quantum dots, 1401. https://doi.org/10.3109/21691401.2015.1052468

7. Li M, Huang Y, Shen C, Wang Y, Lin Y, Wang Z, Chen N (2025) Application of quantum dots in cancer diagnosis and treatment: advances and perspectives, 94907163:1–32
8. Singh I (2018) Carbon quantum dots: synthesis, characterization and biomedical applications karbon kuantum noktalari: sentez, karakterizasyon ve biyomedikal 15(2):219–230. https://doi.org/10.4274/tjps.63497
9. Basset FB, Valeri M, Roccia E, Muredda V, Poderini D, Neuwirth J, Spagnolo N et al (2021) Quantum key distribution with entangled photons generated on demand by a quantum dot 1–7
10. Wang Q, Ye F, Fang T, Niu W, Liu P, Min X, Li X (2011) Journal of colloid and interface science bovine serum albumin-directed synthesis of biocompatible CdSe quantum dots and bacteria labeling 355:9–14. https://doi.org/10.1016/j.jcis.2010.11.035
11. Giroux M, Zahra Z, Salawu OA, Burgess RM, Ho KT, Adeleye AS (2023) Assessing the environmental effects related to quantum dot structure, function, synthesis and exposure (949). https://doi.org/10.1039/d1en00712b.Assessing
12. He X, Ma N (2014) Ac Ce P Te T. Colloids Surf, B. https://doi.org/10.1016/j.colsurfb.2014.06.002
13. Kim S, Agarwal K, Rai H, Mondal S (2023) Quantum dots: an overview of synthesis, properties, and applications 3
14. Hamidu A, Pitt WG (2023) Recent breakthroughs in using quantum dots for cancer imaging and drug delivery purposes
15. Mohamed WAA, Abd H, Gawad E, Mekkey S, Galal H, Handal H (2021) Quantum dots synthetization and future prospect applications 1926–1940
16. Gidwani B, Sahu V, Shankar S, Pandey R, Joshi V (2021b) Quantum dots: prospectives, toxicity, advances and applications. J Drug Deliv Sci Technol 61(May 2020):102308. https://doi.org/10.1016/j.jddst.2020.102308
17. Joglekar PV, Mandalkar DJ, Nikam MA, Pande NS, Dubal A (2019) Review article on quantum dots: synthesis, properties and application 7(1):510–515
18. Gidwani B, Vyas A (2015) Inclusion complexes of bendamustine with B -CD, HP- B -CD and Epi- B -CD : in-vitro and in-vivo evaluation. Drug Dev Ind Pharm 00(00):1–11. https://doi.org/10.3109/03639045.2015.1027217
19. Drbohlavova J, Adam V, Kizek R, Hubalek J (2009) Quantum dots—characterization, preparation and usage in biological systems, 656–73. https://doi.org/10.3390/ijms10020656
20. Pan Y, Li YR, Zhao Y, Akins DL (n.d.) Synthesis and characterization of quantum dots: a case study using PbS. https://doi.org/10.1021/ed5009415
21. Crista DMA, Esteves JCG (2020) Evaluation of di ff erent bottom-up routes for the fabrication of carbon dots
22. Kalluri A, Dharmadhikari B, Debnath D, Patra P, Kumar CV (2023) Advances in structural modifications and properties of graphene quantum dots for biomedical applications. https://doi.org/10.1021/acsomega.2c08183
23. Belal F, Mabrouk M, Hammad S, Ahmed H, Barseem A (2024) Recent applications of quantum dots in pharmaceutical analysis. J Fluoresc 34(1):119–138. https://doi.org/10.1007/s10895-023-03276-2
24. Panja A (2023) A review on quantum dots (QDs) and their biomedical applications (i)
25. Karmakar R (2015) Review article quantum dots and it method of preparations—revisited 2:116–142
26. Lu J, Yang J, Wang J, Lim A, Wang S, Loh KP (2009) One-pot synthesis of fluorescent carbon graphene by the exfoliation of graphite in ionic liquids 3(8):2367–2375
27. Nanobiotechnol J, Zhao C, Song X, Liu Y, Fu Y, Ye L, Wang N, Wang F, Li L (2020) Synthesis of graphene quantum dots and their applications in drug delivery. https://doi.org/10.1186/s12951-020-00698-z
28. Wang S, Mccoy CP, Li P, Li Y, Zhao Y, Andrews GP, Wylie MP (2024) Carbon dots in photodynamic/photothermal antimicrobial therapy
29. Kalita H, Palaparthy VS, Shojaei M, Aslam M (2020) Electrochemical synthesis of graphene quantum dots from graphene oxide at room temperature and its soil moisture sensing properties. Carbon 165:9–17. https://doi.org/10.1016/j.carbon.2020.04.021

30. Rocco D, Moldoveanu G, Feroci M, Bortolami M (2023) Electrochemical synthesis of carbon quantum dots (2022). https://doi.org/10.1002/celc.202201104
31. Younis MA (2022) Biomedical applications of quantum dots: overview, challenges, and clinical potential 1951–1970
32. Moradialvand Z, Parseghian L, Reza H (2025) Green synthesis of quantum dots: synthetic methods, applications, and toxicity. J Hazardous Mater Adv 18(November 2024)
33. Ali M, Hamideh N, Milad K, Ali K (2020) Phytochemical properties and antioxidant activity of echinops persicus plant extract: green synthesis of carbon quantum dots from the plant extract. SN Appl Sci. https://doi.org/10.1007/s42452-020-2466-0
34. Shivaji K, Mani S, Ponmurugan P, De Castro CS (2018) Green-synthesis-derived CdS quantum dots using tea leaf extract: antimicrobial, bioimaging, and therapeutic applications in lung cancer cells. https://doi.org/10.1021/acsanm.8b00147
35. Zhao M-X, Zhu B-J (2016) The research and applications of quantum dots as nano-carriers for targeted drug delivery and cancer therapy. Nanoscale Res Lett 11(1):207. https://doi.org/10.1186/s11671-016-1394-9
36. Bad U, Mollarasouli F, Bakirhan NK, Ozkan Y, Ozkan SA (2020) Trends in analytical chemistry role of quantum dots in pharmaceutical and biomedical analysis, and its application in drug delivery 131. https://doi.org/10.1016/j.trac.2020.116013
37. Fawaz W, Hasian J, Alghoraibi I (2023) Synthesis and physicochemical characterization of carbon quantum dots produced from folic acid. Sci Rep 0123456789:1–7. https://doi.org/10.1038/s41598-023-46084-1
38. Fontes A (n.d.) Quantum Dots
39. Oyohwose UA, Omoko VI, Emmanuel E, Bello OP (2023) Synthesis, characterization, and applications of carbon dots: a review 11(2):1–13
40. Magesh V, Sundramoorthy AK, Ganapathy D (2022) Recent advances on synthesis and potential applications of carbon quantum dots 9(July):1–27. https://doi.org/10.3389/fmats.2022.906838
41. Baek JH, Hsin H, Laforge J, Nedelcu D (2004) Quantum computing with quantum dots, 1–22
42. Rahman M (2023) Quantum computing: realization of quantum hardware
43. Wolf Ronald De (n.d.) Quantum computing: lecture notes
44. An Y, Liu C, Li Y, Chen M, Zheng Y, Tian H, Shi R, He X (2022) Preparation of multicolour solid fluorescent carbon dots for light-emitting diodes using phenylethylamine as a co-carbonization agent
45. Lim J, Park Y-S, Kaifeng W, Yun HJ, Klimov VI (2018) Droop-free colloidal quantum dot light-emitting diodes. Nano Lett 18(10):6645–6653. https://doi.org/10.1021/acs.nanolett.8b03457
46. Applications, Display (2020) Quantum dots and their applications: what lies ahead? https://doi.org/10.1021/acsanm.0c01386
47. Zhao J, Bardecker JA, Munro AM, Liu MS, Niu Y, Ding I, Luo J, Chen B, Jen AK, Ginger DS (2006) Efficient CdSe/CdS quantum dot light-emitting diodes using a thermally polymerized hole transport layer
48. Zhu R, Chen H, Wu S, Dong Y, Central Florida, United States (2018) Quantum dot light emitting diodes. https://doi.org/10.1016/B978-0-12-803581-8.10429-1
49. Reineke S, Leo K, Thomschke M (2013) White organic light-emitting diodes: status and perspective 85(September). https://doi.org/10.1103/RevModPhys.85.1245
50. Norman JC, Jung D, Wan Y, Bowers JE, Norman JC, Jung D, Wan Y, Bowers JE (2018) Perspective: the future of quantum dot photonic integrated circuits perspective: the future of quantum dot photonic integrated circuits 030901. https://doi.org/10.1063/1.5021345
51. Yadav M, Chaudhary A (2014) Quantum dots: an introduction 12–13
52. Manton A (2015) Importance 13
53. Shilpa G, Mohan Kumar P, Kishore Kumar D, Deepthi PR, Sadhu V, Sukhdev A, Reddy R (2023) Materials science for energy technologies recent advances in the development of high efficiency quantum dot sensitized solar cells (QDSSCs): a review. Mater Sci Energy Technol 6:533–546. https://doi.org/10.1016/j.mset.2023.05.001

54. Parani S, Lakshmi BS, Pandian K (2012) Biopolymer encapsulation of CdTe quantum dot for in-vitro controlled drug delivery release of 6-mercaptopurine 584:258–262. https://doi.org/10.4028/www.scientific.net/AMR.584.258
55. Zhou W, Coleman JJ (2016) Semiconductor quantum dots. Curr Opin Solid State Mater Sci. https://doi.org/10.1016/j.cossms.2016.06.006
56. Karadurmus L, Ozcelikay G, Vural S, Ozkan SA (2021) An overview on quantum dot-based nanocomposites for electrochemical sensing on pharmaceutical assay 20(June):187–203. https://doi.org/10.22037/ijpr.2021.115279.15291
57. Hu J, Wang Z, Li C, Zhang C, Hu J, Wang Z (2017) Advances in single quantum dot-based nanosensors, 13284–13295. https://doi.org/10.1039/c7cc07752a
58. Bagalkot V, Zhang L, Levy-nissenbaum E (2007) Quantum dot—aptamer conjugates for synchronous cancer imaging, therapy, and sensing of drug delivery based on bi-fluorescence resonance energy transfer
59. Pleskova S, Mikheeva E (2018) Using of quantum dots in biology and medicine
60. Barroso MM (2011) Quantum dots in cell biology. https://doi.org/10.1369/0022155411398487
61. Fluorescent Highly and Semiconducting Polymer (2017) HHS Public Access 52(11):3086–3109. https://doi.org/10.1002/anie.201205133.Highly
62. Ghasemi Y, Peymani P, Afifi S (2009) Quantum dot: magic nanoparticle for imaging, detection and targeting, 156–65
63. Jing HH, Bardakci F, Akgöl S, Kusat K, Adnan M, Sasidharan S (2023) Green carbon dots: synthesis, characterization, properties and biomedical applications
64. Bazmi A, Etefagh RH, Nasab SH, Asefnejad A (2025) Role of quantum dots in the advancement of biomedical engineering: a role of quantum dots in the advancement of biomedical engineering: a general review. https://doi.org/10.48309/AJCA.2025.499976.1769
65. Hanada S, Fujioka K, Futamura Y, Manabe N (2013) Evaluation of anti-inflammatory drug-conjugated silicon quantum dots: their cytotoxicity and biological effect, 1323–1334. https://doi.org/10.3390/ijms14011323
66. Degim IT, Kadioglu D (2013) Cheap, suitable, predictable and manageable nanoparticles for drug delivery: quantum dots 32–38
67. Nair A, Haponiuk JT, Thomas S, Gopi S (2020) Biomedicine & pharmacotherapy natural carbon-based quantum dots and their applications in drug delivery: a review. Biomed Pharmacother 132:110834. https://doi.org/10.1016/j.biopha.2020.110834
68. Kumar A, Misra SK, Prajapati NK (2022) Biomedical applications of quantum dots 29–42
69. Modani S, Kharwade M, Nijhawan M (2013) Quantum dots: a novelty of medical field with multiple applications 5(4)
70. Tope S, Saudagar S, Kale N (2014) The pharma innovation—journal review: therapeutic application of quantum dots (QD) 86–105
71. Abdellatif AAH, Younis MA, Alsharidah M, Al O, Tawfeek HM (2023) Biomedical applications of quantum dots: overview, challenges, and clinical potential. https://doi.org/10.2147/IJN.S357980.
72. Laser QD (2022) With the power of the semiconductor laser,‘ I Can't’ Becomes ‘I Can’
73. Noel KJ, Umashankar MS, Narayanasamy D (2024) Exploring research on the drug loading capacity of quantum dots 16(8). https://doi.org/10.7759/cureus.67869
74. Song JH, Jeong S (2017) Colloidal quantum dot based solar cells: from materials to devices. Nano Conver 1–8. https://doi.org/10.1186/s40580-017-0115-0
75. Moon H, Lee C, Lee W, Kim J, Chae H (2019) Stability of quantum dots, quantum dot films, and quantum dot light-emitting diodes for display applications 1804294:1–14. https://doi.org/10.1002/adma.201804294
76. Nabil M, Megahed F (2024) Quantum dot nanomaterials: preparation, characterization, advanced bio-imaging and therapeutic applications 2467–2484
77. Pareek A, Kumar D, Pareek A, Gupta MM (2025) Advancing cancer therapy with quantum dots and other nanostructures: a review of drug delivery innovations, applications, and challenges, 1–30

Chapter 2
Synthesis and Functionalization of Carbon Quantum Dots

Radhika Babasaheb Jadhav and Arpita Pandey Tiwari

Abstract Carbon Quantum Dots (CQDs) have emerged as a revolutionary class of zero-dimensional carbon-based nanomaterials, known for their exceptional photoluminescence, high aqueous solubility, low toxicity, and versatile surface chemistry. This chapter provides a comprehensive overview of various synthesis strategies for CQDs, covering both top-down approaches, such as laser ablation, arc discharge, and electrochemical oxidation, and bottom-up techniques, including microwave-assisted synthesis, hydrothermal processing, chemical oxidation, and pyrolytic carbonization of organic materials. With the growing emphasis on green chemistry and sustainable nanotechnology, significant attention has been directed toward the development of eco-friendly CQD synthesis methods, particularly those utilizing natural or renewable precursors. Surface functionalization is important in tailoring CQD properties and can be achieved during synthesis or through subsequent surface passivation methods. In parallel, advances in heteroatom doping have further enhanced the functional diversity of CQDs, allowing for improved control over their optical and electronic characteristics. To support these advancements, this chapter also details essential characterization techniques used to determine surface functional groups and morphological features of CQDs, providing insight into their structural and functional attributes. Such engineered modifications facilitate fine-tuning of photoluminescence behavior, contributing significantly to their potential in biomedical fields. Ultimately, this chapter examines current trends, key challenges, and the future landscape of CQD synthesis and functionalization, highlighting their promising applications in bioimaging, biosensing, solar energy harvesting, and optoelectronic device technologies.

R. B. Jadhav · A. Pandey Tiwari
Department of Medical Biotechnology and Stem Cells and Regenerative Medicine, Centre for Interdisciplinary Research, D.Y. Patil Education Society (Deemed to be University), Kolhapur, Maharashtra, India

A. Pandey Tiwari (✉)
Global Innovative Center of Advanced Nanomaterials, College of Engineering, Science and Environment, University of Newcastle, Callaghan, NSW, Australia
e-mail: arpitaptiwari@gmail.com; Arpita.Tiwari@newcastle.edu.au

T. Bhuyan et al. (eds.), *Multifunctional Quantum Dots*, Materials Horizons: From Nature to Nanomaterials, https://doi.org/10.1007/978-981-92-0172-3_2

Keywords Carbon quantum dots · Fluorescence · Luminescent · Biosensing · Bottom up · Nanomaterials · Top down

1 Introduction

The nanotechnology era has witnessed a remarkable rise in popularity in recent years. Central to this field are nanoparticles, which are extremely small particles with dimensions of less than 100 nanometers. These nanoparticles can be made from various materials, like carbon, metals, metal oxides, or organic materials. Their small size endows them with different biological, physical, and chemical properties compared to larger particles. This phenomenon is attributed to several factors, including more chemical reactivity, improved stability, increased mechanical strength, and a large volume-to-surface area. The distinctive attributes of nanoparticles facilitate diverse applications across several sectors. These particles are characterized by significant differences in size, shape, and composition as per material properties. They can be found in a wide range of sizes, from 1 to 100 nanometers, and can take on various shapes, including spherical, cylindrical, tubular, conical, hollow core, spiral, and flat. The surface of nanoparticles may be smooth, uniform, uneven, or wavy. Furthermore, various nanoparticles possess crystalline or amorphous structures, which may consist of single or multiple crystals that are either dispersed or aggregated. With advancements in technology, researchers are now more adept at characterizing nanoparticles and discovering their practical uses. Currently, nanoparticles are present in diverse applications, spanning kitchenware, electronics, renewable energy, and aerospace industries. The potential of nanotechnology suggests a promising and sustainable future [35].

Recently, carbonaceous and carbon-based nanomaterials have attracted considerable interest due to their relevant characteristics. These materials are particularly noted for their biocompatibility, minimal toxicity, outstanding thermal and mechanical properties, and their capacity for easy functionalization. Fluorescent carbons, which are frequently termed carbon dots, are distinguished by their unique characteristics that exhibit strong fluorescence. Additionally, carbon dots are noted for their remarkable stability, minimal toxicity, water solubility, and the ability to undergo derivatization. These unique characteristics facilitate their use in multiple disciplines. As a relatively recent discovery, carbon dots are considered among the most promising nanomaterials identified by humanity, firstly made up of heteroatoms (functional groups) linked to the carbonized core. This group includes various nanoparticles such as Graphene Quantum Dots (GQDs), polymer dots, and carbon nanodots. They are classified depending upon dimensions and characterized by a core–shell configuration where an organic functional group represents as a shell while sp2 hybridized represents as a carbon core [40].

Xu et al. [99] identified Carbon Quantum Dots (CQDs) during separation for single-walled carbon nanotubes from carbon soot through gel electrophoresis. At that time, these entities were not referred to as CQDs but instead labeled fluorescent carbon nanoparticles. Following this initial invention, Sun et al. [85] synthesized carbon nanoparticles with various dimensions (less than 10 nm) by using laser ablation, named as carbon quantum dots. CQDs have emerged as a significant component of nanotechnology and are regarded as prominent figures among the diverse array of carbon-based nanomaterials [40].

The outstanding electronic traits of CQDs allow them to serve as both electron acceptors as well as donors. These unique properties promote the occurrence of various luminescent phenomena, including photoluminescence, electrochemical luminescence, and chemiluminescence. Photoluminescence is particularly recognized as the foremost characteristic of Carbon Dots (CDs). To realize this property, it is essential to adjust the size as well as the chemical groups present on the surface of CDs. Furthermore, doping with other elements is often used as a tool to modify the properties of CDs [24]. CQDs are recognized for their remarkable luminescence, commonly termed as carbon nano-lights. A variety of research has demonstrated that diverse fabrication approaches and carbon sources are used to create carbon dots with varying structural forms. The surface of CQDs is adorned with an array of functional groups, such as carbonyl, ether, epoxy, amino, carboxylic acid, and hydroxyl groups. This assortment of functional groups enhances the hydrophilic nature of carbon dots [82].

Recent advancements in the fabrication of high-quality Quantum Dots (QDs) have created opportunities to explore the phenomena related to the unique characteristics of quantum confinement. QDs are recognized as a pivotal innovation in nanotechnology, specifically concerning semiconductor inorganic crystals having different numbers of electrons that occupy well-defined as well as have discrete quantum states. The atomic structure of Quantum Dots (QDs) resembles that of bulk materials, yet they feature an increased number of surface atoms as a result of their three-dimensional truncation. Furthermore, the diminutive size of these particles enables a broad spectrum of variable element ratios, which may contribute to their fluorescent properties [2].

In this chapter, an in-depth exploration of different synthesis techniques for Carbon Quantum Dots (CQDs) is presented, along with discussions on their functionalization and doping. The chapter emphasizes the distinct characterization methods used to investigate the structure and functional groups of CQDs. Additionally, it outlines the applications of CQDs, identifies the challenges encountered, and discusses prospects and other proposed initiatives [104].

2 Synthesis of Carbon Quantum Dots

Carbon Quantum Dots (CQDs) are produced by using two primary techniques: top-down and bottom-up approaches, as shown in Fig. 1. Top-down techniques include arc discharge, laser ablation, electrochemical processes, and chemical oxidation. By these techniques, CQDs are produced in large quantities. However, they demand strict investigational factors, such as the use of strong acids and arc discharges,

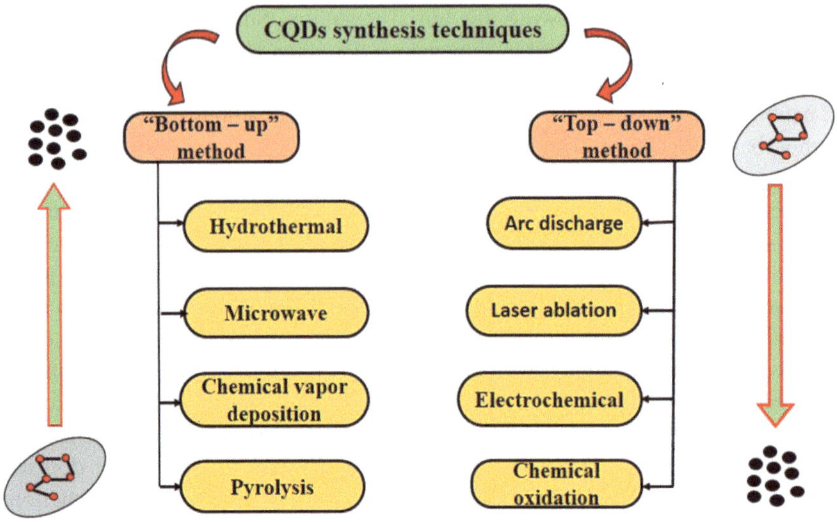

Fig. 1 Schematic representation of the synthesis of CQDs

which involve high costs for research materials, facing barriers in managing size, along with the morphology of synthesized carbon dots. On the other hand, bottom-up strategies involve the application of hydrothermal techniques, microwave-assisted methods, chemical vapor deposition, and pyrolysis. These methods are often utilized for the production of carbon dots, where accurate size control and shape are essential. However, it is crucial to recognize that the process is both complex and time-intensive [38].

2.1 Top-Down Approach

Through a top-down approach, CQDs can be produced with the help of various carbon-based sources, like activated carbon, carbon fibers, graphite, and allotropes, including graphene and Carbon Nanotubes (CNTs), employing multiple techniques. We briefly outline some of the key top-down strategies employed in the synthesis of CQDs.

2.1.1 Arc Discharge

Arc discharge synthesis is a top-down technique utilized to produce Carbon Dots (CDs). It operates by fragmenting larger carbon structures through an electric arc discharge. This method results in the formation of carbon nanoparticles in minimal quantities, and the dust generated from the arc discharge often includes a range of complex elements that are not easily extracted. Many research efforts have focused on the synthesis of Carbon Dots (CDs) through the arc discharge technique. Nonetheless, this method is limited by a low yield, which varies between 2.3 and 8.7%. A high voltage is established between the two electrodes, leading to the occurrence of an electric arc discharge. This discharge promotes the vaporization and fragmentation of the graphite electrode, and some carbon atoms undergo additional rearrangement and bond formation, resulting in the synthesis of carbon dots. Typically, these dots are made up of a core that consists of sp2-hybridized carbon atoms, while their surface is functionalized with an array of chemical groups. As a result, the mixture contains diverse carbon nanostructures, including carbon dots of multiple sizes. The isolation and purification of carbon dots can be achieved through several separation methods, such as centrifugation, filtration, or chromatography. However, the uneven size distribution of carbon dots produced via arc discharge, as depicted in Fig. 2a, poses significant challenges to effective separation and purification. After synthesis of carbon dots through the arc discharge technique, they are characterized with multiple analytical methods to elucidate their properties and structural features. This synthesis method has been commonly applied in the preliminary stages of carbon dot exploration [18].

2.1.2 Laser Ablation Technique

The high-energy laser beam is used in this technique for striking the carbon targets, effectively removing nanosized carbon particles that are subsequently converted into Carbon Quantum Dots (CQDs) through passivation as depicted in Fig. 2b. Notably, the selection of appropriate carbon target materials in this process facilitates the direct formation of CQDs, thereby obviating the requirement for an additional passivation phase. Additionally, the size, shape of Carbon Quantum Dots (CQDs) are adjusted by varying the energy density, pulse width, and wavelength of the laser beam. Laser ablation is characterized by its straightforward operation; hence, it is considered a pioneer technique for CQD synthesis. Nevertheless, it presents several disadvantages, such as the generation of unwanted CQD wavelengths, the necessity for costly equipment, and a lengthy processing time, which hinder its extensive use [98].

Synthesis of CQDs was initially accomplished by Sun et al. [85] via the laser ablation method. This innovative approach involved the laser etching of a combination of graphite powder and cement at 900 °C at 75 kPa in a water vapor environment, leading to the creation of carbon nanoparticles that lacked fluorescence properties. Following this, the solution was mixed in nitric acid and then passivated with Polyethylene Glycol (PEG) 1500 (PEG1500N), lastly treated with acid for the synthesis of different types of CQDs. Li et al. [47] conducted an experiment where they added inorganic solvents to carbon nanomaterials and utilized them as a laser to prepare CQDs in a continuous step method, ensuring the efficiency of the laser etching method. Moreover, altering the organic solvent, CQDs can be synthesized with a range of different agents [98].

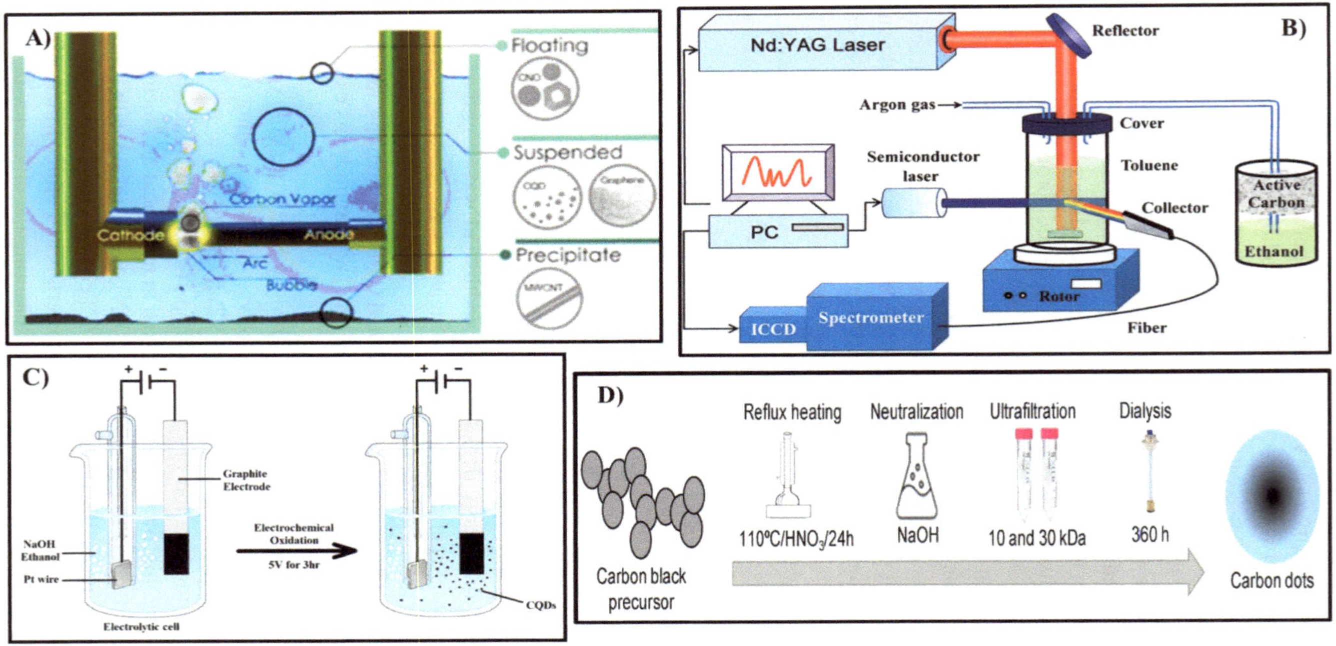

Fig. 2 Synthesis of CQDs by the top-down method. **a** Arc discharge method [18], (Licensed by CC BY 4.0), **b** laser ablation method [98], (Licensed by CC BY 4.0), **c** electrochemical method [57], (Licensed by CC BY 4.0), **d** chemical oxidation method (Gonzalez et al., 2022), (Licensed by CC BY 4.0), Reprinted with permission from the reference

Recently, Cui et al. [12] developed blue-fluorescent Carbon Quantum Dots (CQDs) utilizing inexpensive carbon fiber cloth through a custom-designed double-beam pulsed laser ablation system. During the process, laser beams were split into two rays using a beam splitter to enhance the ablation rate and reduce the overall preparation duration. In contrast, outcomes of single pulse laser ablation, CQDs produced via dual-beam technique exhibit the same size distribution, a high quantum yield up to 35.4%, improved resistance to barrier, thus making them suitable for imaging studies involving HeLa cells. The equipment needed for the laser ablation preparation of Carbon Quantum Dots (CQDs) is costly. Furthermore, the surface of the synthesized CQDs typically requires modification using organic solvents to achieve fluorescence. The incorporation of these organic reagents complicates the purification process of the CQDs [98].

2.1.3 Electrochemical Synthesis

The electrochemical oxidation method is generally associated with the synthesis of CQDs by oxidizing a carbon source electrode through an electrochemical process. Current literature primarily shows attention toward the synthesis of CQDs employing multi-wall carbon nanotube graphite, graphene, and carbon black as the working electrodes, stating advantages including high yield, mild reaction conditions, and the capacity for size control. The initial synthesis of fluorescent CQDs from multi-walled carbon nanotubes was accomplished by Zhou et al. [114] through an electrochemical method. Subsequently, Kang et al. (2010) proposed an alkali-assisted electrochemical technique that yielded CQDs with remarkable optical and electronic properties, thus providing an innovative approach for the high-quality fabrication of these materials. This process entailed the electrochemical oxidation of graphite honeycomb, which was transformed into ultra-fine particles, leading to the generation of CQDs of different sizes. Moreover, it has been noted that highly crystalline CQDs can be synthesized from graphite electrodes through electrochemical oxidation in alkaline conditions. Remarkably, the aqueous solution of CQDs was colorless at first but gradually changed to a bright yellow after being stored at room temperature, potentially due to the oxygenation of surface species over time [91].

Also, CQDs were synthesized utilizing a three-electrode configuration, where the Working Electrode (WE) was a graphite electrode, the Counter Electrode (CE) was either a platinum foil or platinum wire, and the Reference Electrode (RE) was an Ag/AgCl electrode, as shown in Fig. 2c. An alkaline alcohol solution, specifically 0.1 M NaOH/EtOH, served as the electrolyte. The formation of a colorless CQD dispersion was accomplished by applying a consistent potential of 5 V to the graphite working electrode for 3 hours within a nitrogen atmosphere [54]. Throughout the electrochemical oxidation process for CQD generation, the electrolyte maintained its transparency, and there was no significant enlargement of the graphite electrode surface. It is noteworthy that the electrochemical method employed for CQD synthesis utilized alkaline alcohol as the electrolyte, rendering this approach environmentally friendly, cost-effective, and highly reproducible [57].

2.1.4 Chemical Oxidation

The synthesis of Carbon Quantum Dots (CQDs) can be effectively achieved through chemical oxidation, which involves the use of potent oxidants to oxidize and cleave macromolecular carbon materials, as shown in Fig. 2d. This approach is noted for its convenience and efficiency. Recently, extensive efforts have been directed toward synthesizing CQDs from various carbon sources utilizing chemical oxidation techniques. Compounds with amine terminations are applicable as passivation agents for Carbon Quantum Dots (CQDs). The effect of passivation can significantly enhance the fluorescence intensity and Quantum Yield (QY) of the CQDs. The CQDs produced are amorphous, as there was no identifiable lattice structure present. Moreover, the resulting CQDs exhibited a narrow size distribution, falling within the range of 2 to 6 nanometers [42].

2.2 *Bottom-Up Approach*

Bottom-up methodology consists of organic compounds such as Citric Acid (CA) and saccharides, which are commonly utilized as essential building blocks for the synthesis of CQDs. By modifying experimental conditions, researchers can effectively synthesize CQDs of specific sizes in a cost-efficient and scalable manner. The hydrothermal and solvothermal synthesis, microwave-assisted synthesis, chemical vapor deposition synthesis, and pyrolysis, each of which is succinctly described in the following sections [72].

2.3 *Hydrothermal and Solvothermal Synthesis*

Hydrothermal Carbonization (HTC) is an established method to produce new carbon nanomaterials from various initial raw materials. This process is applicable due to its cost efficiency, ecological benefits, low tendency for photobleaching, and nontoxic attributes. In addition, this method generates Carbon Quantum Dots (CQDs) having the same size and displaying significant luminescent attributes. In standard procedure, organic molecules were dissolved in a solvent, and then the solution was transferred into a Teflon-lined stainless-steel reactor. The reaction continues at high temperatures for several hours which results in the formation of carbon cores, namely Carbon Quantum Dots (CQDs) diameter consists within 10 nm, as depicted in Fig. 3a. Zhu and their team [116] indicated that a high quantum yield of 80% attained, which is equal to that of fluorescent dyes. CQDs are derived from citric acid, acting as a carbon source, while ethylene diamine acts as a nitrogen source in the hydrothermal process for the detection of Fe^{+3} in living cells. Rice and glycine act as carbon and nitrogen sources, and then efficient CQDs are produced through this method. These synthesized CQDs act as a probe and detect only Fe^{+3} ions as well as Tetracycline

(TC) antibiotics. Hola and her colleagues synthesized controllable fluorescent CQDs by varying the amounts of citric acid and urea [23]. Furthermore, using biomolecules rich in carbon and nitrogen facilitates the lattice structure of CQDs during the hydrothermal process through changing temperature, leading to a quantum yield of 89.57%. Ease and functionality of nitrogen doping in CQDs lead to its utilization in the fields of bioimaging and biosensing [5].

Besides the hydrothermal method, the solvothermal approach to synthesizing CDs has notable advantages, including its affordability and the requirement for basic equipment. This technique contrasts with the hydrothermal process, which uses water as the solvent, as it employs one or several solvents that are sealed within a Teflon-lined steel autoclave. The interaction of the solvent with the raw carbon source in an environment is characterized by high pressure and temperature. Moreover, low-cost biomass has been efficiently employed for CQDs synthesis by the solvothermal method. Liu et al. [51] have successfully created a sustainable one-step solvothermal method for CQDs synthesis, showing a quantum yield up to 5.7%. This innovative method employed a solution of L-ascorbic acid and glycol, where the mixture was subjected to autoclaving at 160 °C for 4 hours. The generated CQDs demonstrated a significant green-fluorescent emission, thus making them well-suited for cell labeling purposes. Upon completion of dialysis, the purified CQDs indicated a higher quantum yield than a variety of other CQDs. These Carbon Quantum Dots (CQDs), characterized by their multifunctional fluorescence, are suitable for the accurate measurement of pH, Ag^{+}, and Fe^{+3} in solution. This has resulted in the development of a robust pH indicator intended for use in bioimaging [11].

In addition, Mitra et al. [65] harnessed PEG-200 to fabricate uniform CQDs at a temperature of 160 °C for 24 hours. Synthesized CQDs were used for bioimaging applications, attributed to minimal toxicity and remarkable photostability. In another research project, Zheng et al. (2020) synthesized full color-emitting CQDs by the solvothermal method by adjusting concentrations of the reactants in the solvent. They also analysed the possible mechanisms involved in the synthesis, and found that a range of impurities, were functioned as luminescent factors through the ethanothermal process, causing interference with luminescence CQDs. An instance of this can be seen in the molecular fluorophore 8-ethoxy-3H-cyclopenta naphthalen-3-one (ECNO), created during synthesis, which showed a consistent green emission that was not influenced by the excitation wavelength. The CQDs revealed dual emission characteristics, marked by a broad blue emission and a distinct narrow yellow emission, while also exhibiting superior photo and thermal stability relative to ECNO. The clear differences in optical features between ECNO and CQDs imply that the presence of fluorescent molecular impurities might interfere with or change the properties of the synthesized CQDs [11].

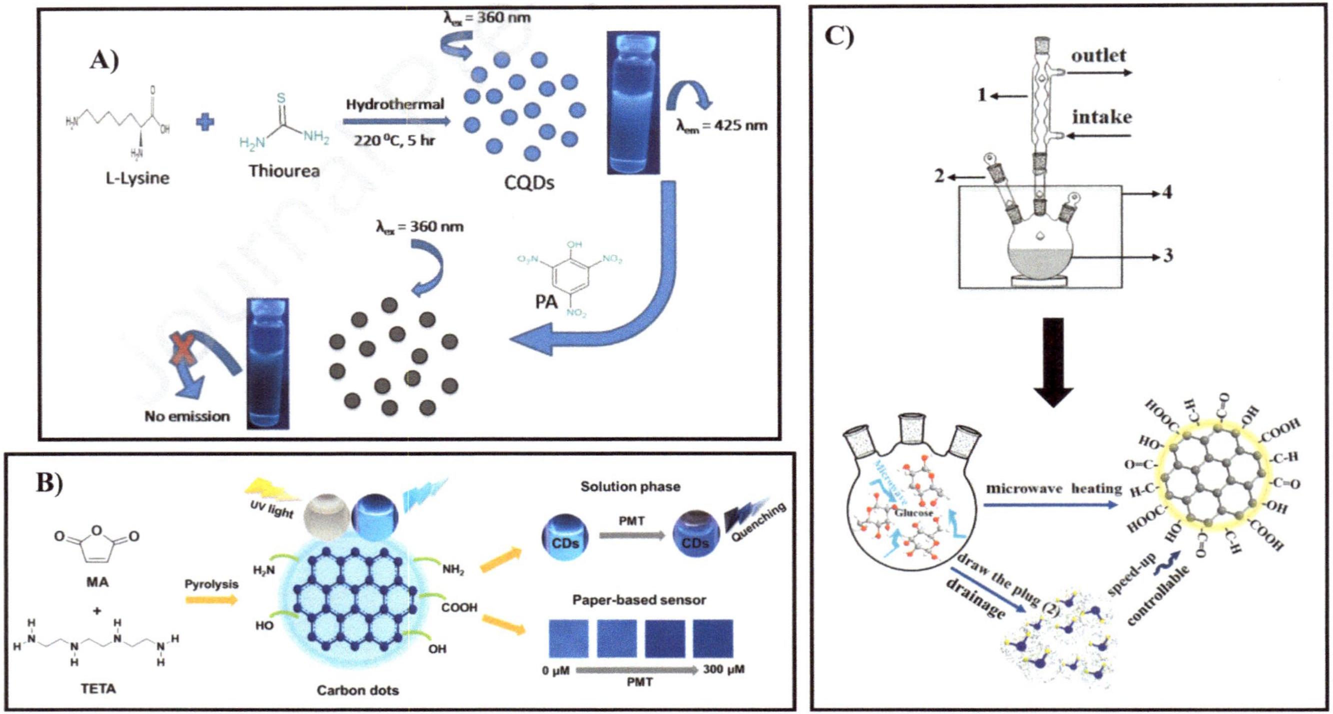

Fig. 3 Synthesis of CQDs by bottom-up method. **a** Hydrothermal method [36], **b** pyrolysis method [98] (Licensed by CC BY 4.0). **c** microwave-assisted method [110], (Licensed by CC BY 4.0). Reprinted with permission from the reference

2.4 *Microwave-Assisted Synthesis*

As a type of carbon material, CQDs possess a remarkable ability to absorb microwaves [9, 97]. When subjected to microwave radiation, CQDs interact vigorously with the microwaves, leading to the formation of multiple "hot spots" on their surfaces. In these regions, temperatures can exceed 1000 °C, and the energy levels are significantly elevated compared to other areas. This effect undoubtedly enhances the occurrence of chemical reactions on the CQDs' surfaces [110].

Transfer of thermal energy to the surfaces of carbon materials occurs through radiation, facilitating energy transmission via electromagnetic radiation during the process of microwave-assisted heating [84]. A predetermined ascorbic acid was added to the polyethylene glycol solution and heated in a microwave reactor at 700 W and 200 °C for one minute [33]. During the exposure to microwave irradiation, the solution exhibited a gradual color change from yellow to brown, which indicates the synthesis of CQDs by dehydration. After cooling, the mixture underwent dialysis for 24 hours. Also, in a recent study, researchers synthesized CQDs from lemon. It was mixed with water and then heated in a microwave for 20 min at 500 W. The product was subjected to dialysis for 24 hours to eliminate any unreacted molecules [89]. The microwave-assisted hydrothermal method is advantageous due to its homogeneous energy distribution, which contributes to the creation of uniformly sized particles. Moreover, the size of these particles can be tailored by varying the heating duration in the presence of a catalyst [2].

In their study, Zhang et al. [110] synthesized Carbon Quantum Dots (CQDs) from glucose using a microwave technique, maintaining a temperature of 150 °C and a power output of 500 W for a period ranging from 7 to 11 minutes. Furthermore, Architha and Shobana [4] successfully produced blue-fluorescent CQDs with a peak quantum yield of 17% by employing a microwave-assisted reflux method, utilizing peppermint as the precursor [98]. Nonetheless, the technique of microwave irradiation faces constraints due to the lack of control over reaction conditions, which demand a short time frame coupled with high temperatures [34].

2.4.1 Chemical Vapor Deposition Synthesis

Chemical Vapor Deposition (CVD) is a prominent method used for CQDs synthesis in recent years. In this process, properties of the resulting product can be controlled by modifying several factors, such as the choice of carbon source, growth duration, hydrogen (H_2) flow rate, and substrate temperature. By chemical vapor deposition method, CQDs were first proposed by Fan et al. [19] utilized methane as a carbon source. Initially, by using a mixture of alcohol and hydrochloric acid, the other chemical groups present on the surface of copper foil were cleaned. Substrate was then heated to a temperature of 1000 °C in a hydrogen and argon atmosphere. Once the hydrogen supply was halted, argon was continuously supplied to remove any remaining hydrogen residues. Methane gas (CH_4) was subsequently supplied in the furnace at a flow rate of 2 mL/min for 3 seconds within an argon environment. The synthesized CQDs displayed a size range of 5 to 15 nm, with a height profile of 1 to 3 nm, indicating successful production of thick-layer CQDs [11].

2.4.2 Pyrolysis Method

The pyrolysis process entails thermal degradation of the precursor at high temperatures, usually 430 °C, under pressure in the absence of oxygen. Moreover, the addition of alkali and a robust acid catalyst enhances the breakdown of carbon precursor into nanoscale colloidal particles. This approach is advantageous due to its practicality, repeatability, and simplicity, along with a high quantum yield. However, a notable challenge is the separation of small precursors from raw material [101].

Liu et al. [52] applied a groundbreaking approach for the synthesis of CQDs, where resol was used as a carbon source along with surfactant-modified silica spheres. Resultant CQDs show blue color fluorescence, characterized as amorphous, with sizes within 1.5 to 2.5 nm, and a quantum yield of 14.7% was recorded. Furthermore, these CQDs demonstrated stability over a wide pH range, specifically between pH 5.0–9.0. Consequently, several investigations were initiated to analyse the synthesis of CQDs using the pyrolysis process. Research was conducted by Pan et al. (2010), who synthesized highly blue-fluorescent Carbon Quantum Dots (CQDs) from Ethylenediaminetetraacetic acid (EDTA) by the pyrolysis approach. The size of the resulting CQDs was 6 nm, with a Quantum Yield (QY) of 40.6% [101].

Following this, in [60], Martindale and associates synthesized fluorescent CQDs by using citric acid at 180 °C, achieving a size up to 6 nm as well. Quantum yields were measured up to 2.3% near 360 nm excitation wavelength. In the year [73], Rong and his team prepared fluorescent Nitrogen-functionalized Carbon Quantum Dots (N-CQDs) via the pyrolysis method using citric acid and guanidinium chloride. The size of the resulting N-CQDs was up to 2.2 nm, quantum yield of 19.2%. These N-CQDs were extensively utilized for metal ions detection, specifically Fe^{+3}, and bioimaging applications. Recently, several carbon quantum dots synthesized by the pyrolysis method have been applied in various fields [101]. In their study, Supchocksoonthorn et al. [87] produced blue light-emitting Carbon Quantum Dots (CQDs) by employing Maleic Anhydride (MA) as the starting material through a one-pot thermal decomposition technique.

2.5 *Green Synthesis*

Green chemistry is an important sector of chemistry that provides innovative solutions to many pressing problems with notable characteristics. The benefits associated with green chemistry include its safety, eco-friendliness, the lack of toxic materials, and its feasibility under normal conditions. Carbon dots have been synthesized using green chemical techniques from an array of natural sources, which encompass chicken eggs, animal materials [78], and various plant species, fruits, and vegetables [61], waste materials like wastepaper and frying oil [95]. The quantity of scholarly articles about carbon quantum dots and green synthesis has experienced exponential growth. Different methods can be employed to achieve the fabrication of carbon dots, include hydrothermal/solvothermal processes, microwave-assisted polymerization, pyrolysis, and carbonization [93]. These methods are prevalent for the synthesis of CQDs as well as offer many advantages [17].

A range of biocompatible and photostable Carbon Quantum Dots (CQDs), measuring between 2.0 and 6.0 nm, was obtained from by-products produced through a one-pot hydrothermal reaction in the bio-refinery process illustrated in Fig. 4 [24]. The main elements of these by-products consisted of degradation products, referred to as auto-hydrolyzate, resulting from the auto-hydrolysis pre-treatment of biomass. The quantum yield was 13%, demonstrating commendable stability in fluorescence, substantial resistance to photobleaching, and adaptability to temperature variations [17].

In the past few years, organic waste such as food and agricultural residues has been harnessed for the production of Carbon Quantum Dots (CQDs). Nitrogen- and oxygen-rich precursors were utilized, with half of the carbon sourced from waste for the valorisation of CQDs [1, 62]. By using walnut shells, CQDs were synthesized successfully. It shows excellent photostability and photoluminescence characteristics when it was compared with metal CDs then resulting in no toxicity in bioimaging applications [10]. Chitin is a biomass waste used to synthesize CQDs by bottom-up approaches. To enhance the longevity of synthesized Carbon Quantum Dots (CQDs)

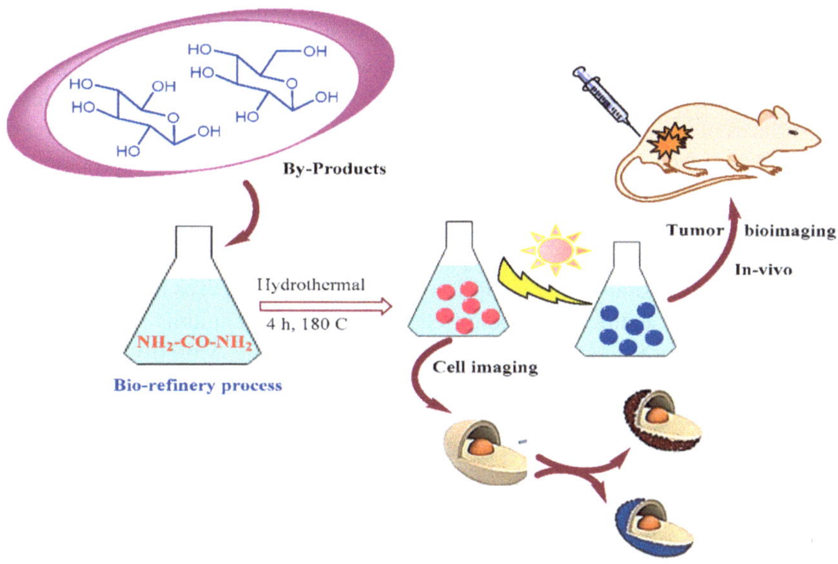

Fig. 4 Green synthesis of CQDs by bio-refinery process. Reprinted with permission from the reference [17]

from chitin, a chitosan gel has been formulated using chitin, ensuring that the material properties remain intact through bottom-up methodologies [15].

Furthermore, this method serves to prevent the reliance on harmful and costly materials, such as metals, to produce this minuscule nanomaterial. In addition, the synthesis of carbon dots from natural resources can significantly contribute to the management of environmental pollution by offering innovative approaches to ecological pollution control. Utilizing agro-waste to produce carbon dots further aids in mitigating environmental challenges associated with agro-waste, particularly by addressing the substantial issues that arise when such waste accumulates in agricultural areas. Numerous nontoxic medicinal plants have demonstrated significant health-related biological activity in both in vitro and in vivo studies against a range of ailments. Nevertheless, it is rare for a plant extract to be utilized as a final product. Depending on the extraction method employed, these extracts may encompass a variety of phytochemicals [29].

In addition, green-synthesized carbon dots have the potential to reduce dependence on conventional medical professionals or medicinal plant sources for therapeutic applications. Consequently, we are confident that the emergence of simpler, cost-effective, and innovative green synthesis methods, along with enhanced properties of CQDs, will lead to the discovery of new and promising applications for these increasingly important nanoparticles [29].

3 Functionalization Techniques for CQDs

The functionalization chemistry is pivotal in shaping the interactions between a nanoprobe and its biological surroundings, as well as its compatibility with various imaging techniques. This functionalization process consists of attaching molecules to the surfaces of nanoparticles. The significant surface-to-volume ratio of these particles facilitates their efficient functionalization to align with applications. A thorough comprehension of the chemistry governing conjugation is vital, as it informs the practicality of diverse functionalization methods [41].

Functionalization is defined as the process of modifying the surfaces of nanoparticles by attaching various biomolecules as shown in Table 1, such as folic acid, biotin, oligonucleotides, peptides, antibodies, etc. This enhancement aims to improve the nanoparticles' characteristics and facilitate precise targeting. Moreover, functionalized nanoparticles are known for their excellent physical properties, including resistance to corrosion, prevention of agglomeration, and non-invasive nature. Considerable research has been undertaken to advance the functionalization of nanoparticles, thereby increasing their overall effectiveness and functionality. The process of functionalizing Nanoparticles (NPs) allows for the introduction of tailored properties that are of particular interest, thereby facilitating advancements in specific imaging modalities. This results in improved quality of image and enhances the overall applicability, functionality of the imaging techniques. Different studies indicate that the functionalization of NPs can be performed in conjunction with other nanoparticles to elevate their performance characteristics [41].

3.1 Surface Passivation

Surface passivation is a critical procedure aimed at enhancing the stability of the surfaces of colloidal quantum dots and increasing their longevity. The bare surfaces of CQDs are at risk of participating in undesirable reactions, which can adversely affect their optoelectronic properties. Implementing surface passivation allows for the formation of a thin protective layer as shown in Fig. 5 that safeguards the CQDs from direct exposure to various impurities. Furthermore, it can lead to an increase in fluorescence intensity. Numerous polymers, as well as organic compounds, play a role as passivation agents. It is crucial that these surface passivation agents are devoid of chromophores in the visible or near UV spectrum, which allows them to remain non-emissive at visible wavelengths as well as confirms the luminescence detected in synthesized CQDs [41].

3.1.1 Non-Covalent Functionalization

Non-covalent interactions can facilitate binding conjugation through the attachment of secondary ligands. The process of non-covalently conjugating biomolecules to nanoparticles is predominantly conducted via affinity-based receptor-ligand systems. Different techniques for non-covalent binding involve electrostatic interactions, π–π stacking, incorporation of molecules, biocompatible films, phospholipids, polymers, and DNA. In contrast to covalent functionalization, non-covalent modification, which is based on electrostatic, complexation, or π interactions between CQDs and modified molecules, not only minimizes the negative impact on the structure of CQDs but also introduces new functional groups, targeting molecules, or metal nanoparticles on the surface. This can add an extra dimension in managing nanoparticle interfacial properties and effectively connects nanoparticles to biological systems [20]. Through π interaction, CQDs are modified to possess extended conjugation domains, which result in the creation of highly consistent fluorescence probes with superior quantum yield, photostability, and biocompatibility [103].

Complexation is defined as the interaction between CQDs and metal ions through the formation of coordination bonds. This interaction is largely facilitated by the presence of several amino, hydroxyl, or carboxyl groups on the surface of CQDs, which can participate in complexation with metal ions. By complexation, when an amphiphilic molecule is affixed to the surface of CQDs, the hydrophobic segment contributes to the enhanced stability of CQDs in organic solvents [103].

Table 1 Represents the different surface modifications of CQDs for various applications

Sr. no	Surface modification	Method of synthesis	Applications	References
1	Coumarin	Hydrothermal method	Cell visualization method	[28]
2	Ethylene-diamine	Hydrothermal method	Lung cancer detection	[16]
3	Nitrogen	Hydrothermal method	Fe^{+3} detection	[74]
4	Titanium dioxide (TiO_2)	One-step microwave-assisted method	Sensitized solar cells	[70]
5	Sulfur and nitrogen	Hydrothermal method	Pharmaceutical applications	[77]
6	Nitrogen	Soft chemical route	Photo degradation	[58]
7	Proline	Hydrothermal method	Salinity stress in grapevine plants	[21]
8	Nicotinamide	One-step microwave-assisted method	Quantification of vitamin	[13]
9	Sulfhydryl	Hydrothermal method	Detection of hg^{+2}	[105]
10	n-butyl amine	Solvothermal method	Nitro explosive detection	[68]
11	Histidine	Hydrothermal method	Manganese detection	[66]
12	Zwitterion dopamine sulfonate	Hydrothermal method	Lubricant additive	[100]
13	Dihydrolipoic acid	Rapid stirring	Highly reactive oxygen species detection	[108]
14	Polylactic acid/ tungsten oxide/amino	Hydrothermal method	Oily wastewater treatment	[69]
15	Ionic liquid	Pyrolysis	Detection of iron and chromium	[94]
16	Chlorophyll	Hydrothermal method	Detection of cervical cancer	[3]
17	Tryptophan and glycine	Hydrothermal method	Detection of Al^{3+} ions	[102]
18	Carbon nitride	Hydrothermal method	Degradation of bisphenol	[64]
19	Hydroxyapatite	Hydrothermal carbonization	Drug delivery	[25]
20	Cysteine	Pyrolysis	Fluorescence quenching	[75]

(continued)

Table 1 (continued)

21	Nitrogen and phosphorous	Micro flow reactor	Bioimaging and environment	[50]
22	Porphyrin	Hydrothermal method	Detection of cu^{+2}	[109]
23	Sodium and potassium	Magnetic stirrer	Enhancing forward osmosis	[32]
24	EDTA	Pyrolysis	Environmental pollutants in soil and water	[83]
25	Photonic crystals	Microfluidic method	Bio coding, sensors, optical displays	[44]

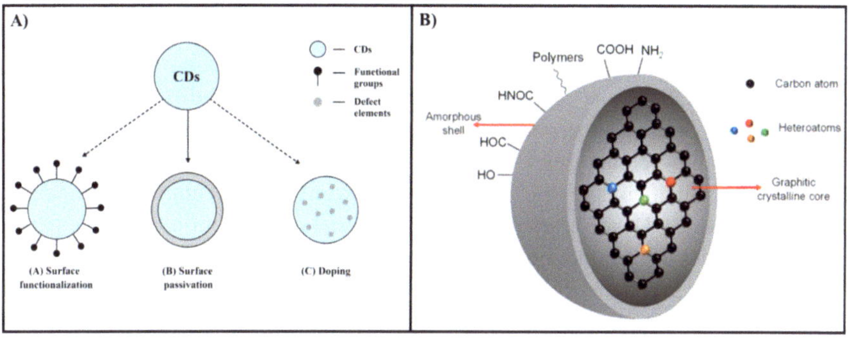

Fig. 5 **a** Represents functionalization, passivation, and doping of CQDs [41]. **b** represents the core shell of CQDs [106]. (Licensed by CC BY 4.0)

3.1.2 Covalent Functionalization

The process of functionalization may also occur via the direct binding of the desired molecule to the reactive ligands located on the surface of nanoparticles, through covalent conjugation. The technique involves a linkage reaction that is assisted a catalyst and is predominantly preferred over nonspecific physisorption, particularly regarding the stability of the functionalization. Covalent modification of CQDs is beneficial to ensure enhanced regulation of their dimensions, form, and physical attributes. This connection allows for the incorporation of additional functional groups or targeting molecules into the pristine carbon dots CQDs through covalent bonds, consequently raising the electron cloud density on the surface of the CQDs [56]. This increase will enhance the conjugate system and transform the fluorescence spectrum of the CQDs. The primary advantage of covalent modification lies in the designability of the reaction, which can be formulated based on the different functional groups present in CQDs and modifying agents. Depending on the reaction involving different functional groups, it can be segmented into amide, sialylation, esterification, sulfonylation, and copolymerization reactions [103].

The amide modification constitutes a reaction between amino compounds and acylating substances. This method is the most prevalent due to the presence of numerous amino or carboxyl groups on the surface of CQDs, which can react with acyl compounds or amino compounds through amidation. One of the most apparent benefits of amide modification is that it can substantially increase the selectivity and quantum yield of CQDs. The process of silylation modification involves a reaction between silane and active hydrogen located on the surface of CQDs. Additionally, it can augment the specific surface area, enabling CQDs to exhibit improved dispersion. This technique can also boost water solubility, selectivity, and markedly lower the cytotoxicity of CQDs, which are widely utilized in the detection of ions and temperature sensing [43].

Esterification methods can significantly extend the application domains of CQDs by introducing various functional components that can interact with these groups. CQDs are generated through the treatment of oxygenates as organic precursors, which generally have a high concentration of carboxyl or hydroxyl groups on their surfaces. The adjusted CQDs show remarkable selectivity and reversibility with respect to targets, which facilitate the potential use of CQDs in detection and cell imaging areas [103]. Amino-containing CQDs can also undergo modification via a sulfonylation reaction, which involves the attachment of various sulfonyl chloride compounds to their surface. CQDs are altered through a sulfonylation reaction, allowing for the introduction of sulfur atoms on their surface. This modification can create energy or emissive trap states that facilitate the capture of photo excited electrons and can also alter the electronic structure of the CQDs. The functional groups present on the surface of carbon dots typically include active hydrogen, which allows for modification through copolymerization reactions with cyclic compounds or compounds containing double bonds. This method allows for the acquisition of large molecular weight CQDs, which is advantageous for use in biological systems [103].

3.2 *Doping of CQDs*

3.2.1 Heteroatom Doping

Investigations aimed at enhancing the performance of CQDs via two key approaches include heteroatom doping and surface functionalization, as these methods improve their optical, electrical, and chemical characteristics. By introducing atomic impurities (e.g., nitrogen, boron, sulfur, phosphorus, etc.) into CQDs, their electronic structures generate n-type or p-type carriers. Thus, it is possible to tailor the optical and electronic properties by utilizing designated types and varying quantities of doping atoms. Heteroatom doping is an effective method for modulating the intrinsic properties of CQDs [63].

3.2.2 Metal Doping and Alloy

In contrast, frequently utilized heteroatoms, metals are more effective as electron donors, featuring a greater number of unoccupied orbitals along with a larger atomic radius. Thus, metal ions doping into Carbon Dots (CDs) significantly enhances opportunities for altering charge transition between the graphene matrix and metal ions. As a result, this produces a more considerable advancement in the physicochemical properties of carbon dots relative to the impact of heteroatom doping. Pioneering the metal ions may transform electronic architecture of Carbon Dots (CDs), leading to variations in energy gap between the Highest Occupied Molecular Orbital (HOMO) and the Lowest Unoccupied Molecular Orbital (LUMO), were underpins the optical behavior of CDs. Compared to pristine CDs, metal-doped CDs demonstrate considerably greater optical absorbance in the visible range. This phenomenon results from charge-transfer absorption involving metal ions and graphite. The process of metal doping enhances the quantum yield of Carbon Dots (CDs), enabling them to produce a spectrum of emissions, including ultraviolet, blue, blue-green, green, and yellow, when they are stimulated by a 360 nm single excitation wavelength. Increased quantum yield of metal-doped CDs brings changes in structural configuration of CDs and enhanced purging of electron-hole pairs, which arise from the introduction of new emission energy traps [88].

The use of metal ions for surface functionalization is regarded as inadequate. As research advances, Feng's research team has put forth a new perspective, suggesting that charge-exchangeable coordinating "functional knots" within the CQDs' backbones could be formed by directly attaching metal dopants to sp2-characterized heteroatoms, rather than through surface functionalization with metal ions, as indicated in 2015 [93]. To date, various Metal ion-doped Carbon Quantum Dots (M-CQDs) have been synthesized, exhibiting not only remarkable optical characteristics but also introducing new intrinsic properties [49]. Until now, several optical sensors have been engineered to detect a range of analytes, attributed to the exceptional photoluminescence of M-CQDs. The fluorescent sensors are utilized based on the principles of fluorescent single-off detection. Their detection mechanism can be divided into two distinct types: (1) Inner Filter Effect (IFE), and (2) the electron transfer process from the excited states of the M-CQDs surface to the analytes [49].

It was found out that the absorption bands associated with Cr(VI) ions, 2,4,6-trinitrophenol (TNP), and tetracycline overlap with the excitation bands of Co-CDs [111] and Tb-CDs [8], as well as Eu-CDs [51], remained unchanged, and the lifetimes of the M-CQDs did not vary with the introduction of the aforementioned three analytes. Consequently, the primary factor contributing to the fluorescence quenching of M-CQDs during the detection of these analytes was found to be highly efficient IFE. The methods proposed demonstrated significant selectivity and sensitivity, achieving a low Limit of Detection (LOD) of 1.17 mM for Cr(VI) ions, 200 nM for TNP, and 0.3 mM for tetracycline [49].

Photoluminescent properties of Carbon Dots (CDs) can be enhanced through the effects of Surface Plasmonic Resonance (SPR) from metal Nanoparticles (NPs). Moreover, metal ion-doped carbon dots exhibit enhanced catalytic activity compared to their unmodified versions. Introduction of metal ions in carbon dots modifies spin density as well as charge distribution, resulting in an increased number of active catalytic sites that effectively bind with redox species during catalytic reactions. Additionally, when metal ions are doped on Carbon Dots (CDs), then it results in increased electrical conductivity and increased their capacity for electron donation and acceptance. This advancement aids in the transfer of electrons process as well as enhances photocatalytic activity. Furthermore, incorporation of magnetic metallic ions such as Gd (III) and Mn (II) allows CDs to play the role of contrast agents for magnetic resonance imaging [88].

3.3 Core–Shell Structure

The phenomenon of in situ growth, marked by heterogeneous nucleation where a nanostructure is synthesized on the surface of another nanomaterial, leads to the development of core–shell architectures. CQDs exhibiting aggregated or cross-linked polymer hybrid nanostructures can serve as a means for modifying core–shell architectures. These materials are characterized by cross-linked polymer structures or shells that are abundant in functional groups or polymer chains, providing them with superior stability, enhanced compatibility, ease of modification, and a wider array of applications. The stability of these materials promotes a robust structural connection between the two components, leading to efficient charge separation at their interfaces. As a result, CQDs and metal nanostructures with contrasting charges are held together through electrostatic interactions, facilitating the Layer-by-Layer (LbL) assembly of core–shell nanohybrids [30].

The structural modification is identified as core–shell architecture, characterized by CQDs that contain a dense carbon core enveloped by a less dense shell. To create a hydrogel with CQDs, certain metallic nanoparticles, including silver nanoparticles, are utilized. The modification of metallic nanoparticles and silver nanoparticles leads to the confinement of electrons as well as holes at the CQDs' surface, were in turn impacts their Photoluminescence (PL) properties. The interaction of CQDs with metallic ions results in the trapping of electrons and holes, thereby diminishing their radiative recombination and resulting in the quenching of PL [30].

4 Characterization Techniques for Synthesized CQDs

Carbon, when found in the elemental state, is a black mineral known for its low fluorescence along with limited water solubility. Nevertheless, CQDs are highly utilized in research due to their high luminescence and good solubility. Due to the structural arrangement, CQDs possess different properties. Functional groups, including carboxyl groups on the CQDs surface, play an important role in providing biocompatibility as well as improving water solubility. For the surface passivation of

CQDs, various materials are used such as biological, polymeric, organic, and inorganic types. The surface passivation mitigates different surface defects, leading to an improvement in the fluorescence properties of CQDs. These materials are characterized by their stable fluorescence, conductivity, and elevated quantum yield [5]. The following are the characterization techniques that reveal the minute details of synthesized CQDs.

4.1 UV-Visible Spectroscopy

Optical characteristics of CQDs can be evaluated through UV-vis and fluorescence measurements. The analysis found that the absorption spectrum, taken from 265 to 450 nm at room temperature (illustrated in Fig. 6a, inset), featured a band centered at 340 nm. The found results were associated with the n-π transition of C=O bonds, along with possible effects from other surface functional groups (Wang et al. 2013; Schneider et al. 2017). Furthermore, it was noted that the fluorescence spectrum, captured across a wide range of excitation wavelengths (325–395 nm), showed the strongest emission when excited at 325 nm, with the emission peak occurring at 454 nm (Fig. 6a). An increase in the excitation wavelength resulted in a red shift emission wavelength, demonstrating excitation-dependent fluorescence characteristics. Previous research (Sharma et al. 2016) suggests that this phenomenon can be attributed to the existence of several distinct electronic states. Additionally, it was found out that such states could stem from a variety of aggregates instead of being linked to a single carbon-based nanoparticle. Other plausible explanations for the excitation-dependent behavior of Carbon Quantum Dots (CQDs) involve the presence of emissive surface traps and variations in the size of the particles [7].

4.2 X-Ray Photoelectron Spectroscopy (XPS)

The total scanning XPS spectrum of the synthesized CQDs, as shown in Fig. 6b, shows the two significant peaks at 284 eV and 532 eV, which are attributed to C1s and O1s, respectively. Following the Gaussian deconvolution of the XPS data, analysis of the high-resolution O1s spectrum (Fig. 6c) revealed the presence of two significant peaks at 529.8 eV and 531.0 eV. These peaks are associated with the C=O and C–O–C/C–OH functional groups, respectively (Borse et al., 2017). Moreover, the analysis of the high-resolution C1s spectrum (Fig. 6d) shows two separate peaks at 284.0 eV and 288.2 eV. The peak at 284.0 eV corresponds to C–C, C=C, and C–H bonds, while the peak at 288.2 eV is related to O=C–O and O=C groups (Ogi et al., 2014). These findings imply that the sample's primary structure is a heterocyclic compound [7].

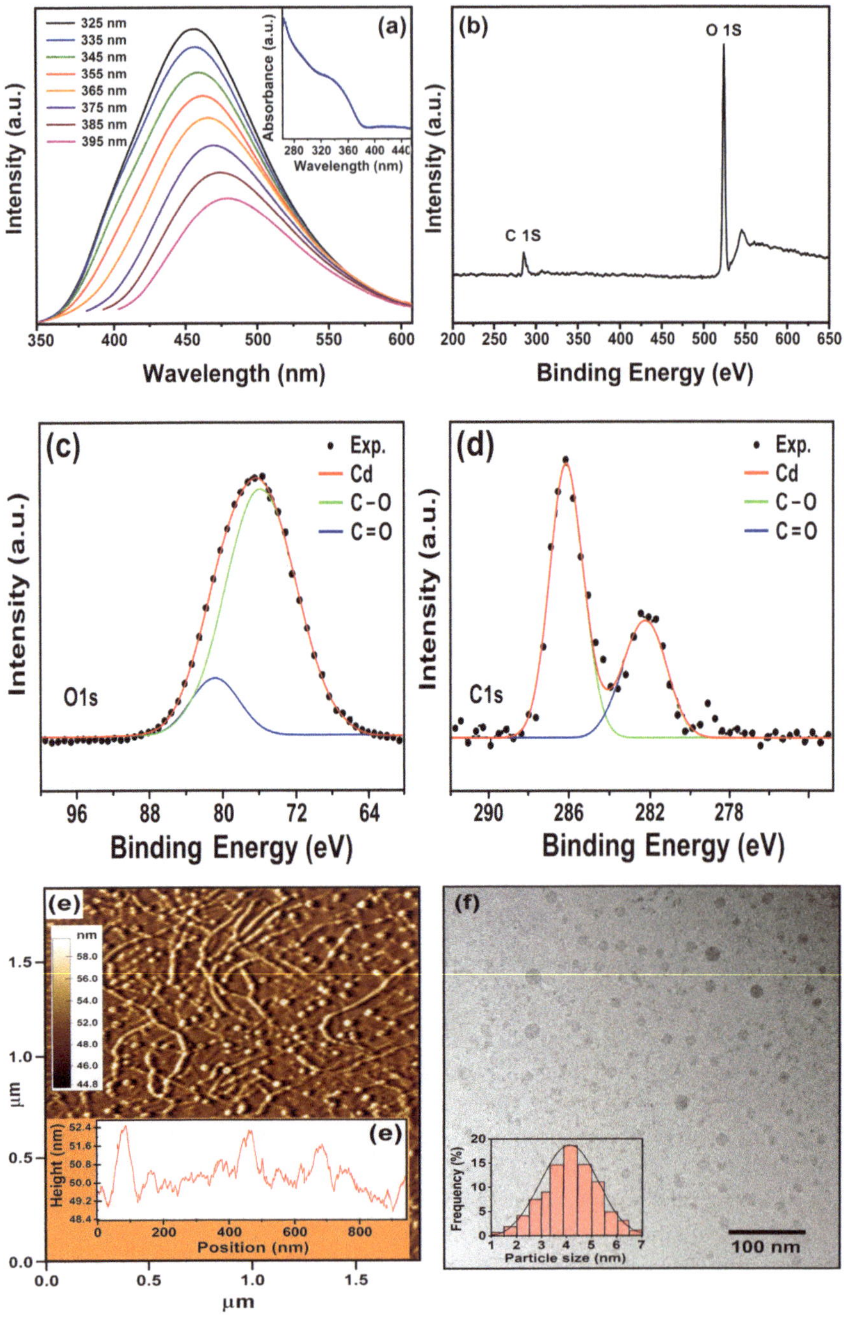

Fig. 6 Represents different characterization techniques for CQDs. **a** U.V. visible and absorption spectrum of CQDs, **b** XPS spectrum of CQDs, **c** deconvolution of XPS data for O1s and **d** deconvolution of XPS data for C1s, **e** and **f** represent AFM and TEM for morphology study. Reprinted with permission from the reference [76]. (Licensed by CC BY-NC-ND 4.0)

4.3 *Atomic Force Microscopy (AFM)*

It was found out that the AFM image shown in Fig. 6e indicates the presence of quasi-spherical Nanoparticles (NPs) measuring between 1.2 and 2.8 nm, as determined by height analysis. Moreover, the micrograph illustrated in Fig. 6f validates the size range and morphology found in the AFM observations. Although the predominant number of particles were within this specified range, there were also some nanoparticles observed with sizes reaching up to 7 nm. It was observed that the CQDs possess a diverse size distribution, ranging from 1.2 to 7 nm. This discrepancy in size can be linked to the synthesis method utilized, which lacked a size-controlling agent. Additionally, it is noteworthy that these findings are in agreement with prior literature about CQDs smaller than 10 nm [7].

4.4 *Fourier-Transform Infrared Spectroscopy (FT-IR) and X-Ray Diffraction (XRD)*

The emission spectra variations found the presence of multiple functional groups on the surface of the Carbon Dots (CDs), which were subsequently characterized and confirmed through FTIR spectroscopy. The surface functional groups identified in the synthesized material are illustrated in Fig. 7a. It was found out that the broad peak at 3300 cm^{-1} relates to the asymmetric stretching vibrations of hydroxyl and secondary amine groups [39]. In addition, the sharp and narrow band found at 1635 cm^{-1} corresponds to the bending vibrations of amide groups and the stretching of carbon–carbon double bonds [14]. A notable peak at 1040 cm^{-1} confirms the presence of epoxy groups [37]. The FTIR findings also found that the CQDs possess hydrophilic properties, which contribute to their improved stability and excellent dispersibility in aqueous environments, attributed to the abundance of surface functional groups. Moreover, the XRD pattern for the CQDs presented in Fig. 7b displays a broad peak at approximately 20°, indicating the presence of small-sized carbon particles and hinting at a possible graphitic structure [76].

5 Applications of CQDs

The introduction of CQDs represents a new category of carbon-based nanomaterials, distinguished by their remarkable optical properties, outstanding biocompatibility, cost-effectiveness in production, and the simplicity of their modification and functionalization processes. These properties highlight their significant potential across a wide array of applications. However, the practical implementation of CQD-based materials necessitates further research. Such developments are likely to greatly expand the potential applications of CQDs. These areas such as

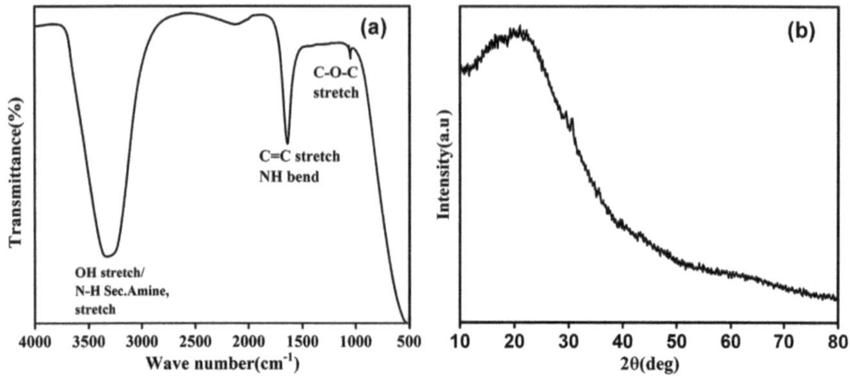

Fig. 7 Represents the characterization technique of synthesized CQDs. **a** represents the FT-IR spectrum; **b** represents the XRD image of CQDs. Reprinted with permission from the reference [76]

bioimaging, biosensing, solar energy conversion, and optoelectronic applications have been thoroughly examined [53].

5.1 Bioimaging

Bioimaging refers to a technique that is employed for non-invasive visualization of biological systems in real time. This approach seeks to minimize interference with living processes as much as possible. It is frequently utilized to extract information about the three-dimensional structure of specimens from an external viewpoint, thus avoiding any medical intervention. In a broader sense, bioimaging includes different techniques for imaging living cells. It covers the study of subcellular structures, entire cells, tissues, and multicellular organisms. CQDs are recognized as viable options for bioimaging due to their enhanced tunable Photoluminescence (PL), lower toxicity, and hydrophilic properties, along with remarkable opto-chemical stability under ultraviolet irradiation. Nevertheless, CQDs primarily produce blue color emissions having limited excitation spectra poses a significant limitation for bioimaging applications [59, 92]. Various studies have demonstrated that CQDs that are doped with heteroatoms can achieve extended emission wavelengths and multi-colored fluorescence, making them ideal for extended duration and real-time cellular imaging, in addition to in vivo bioimaging. Cellular imaging studies indicated that Nitrogen-doped Carbon Quantum Dots (N-CQDs) possess excellent biocompatibility, enabling them to effectively tag Hepg2 cells. In contrast to single hetero-doped carbon quantum dots, those that are co-doped with dual heteroatoms, including nitrogen, phosphorus, or boron, exhibit superior biocompatibility and high quantum yield. The yellow fluorescent nitrogen as well as Boron co-doped Carbon Quantum Dots (N, B-CQDs) were synthesized using a microwave-assisted hydrothermal method with o-phenylenediamine (OPD) and Boric Acid (BA) solution. To produce fluorescence images, HeLa cells were treated with a solution of 50 g/mL of N, B-CQDs [55].

It has been found that the imaging of cancer cells is essential for the prompt detection of cancer and significantly aids in preventing the progression of tumors. Various cancer cell types have employed functionalized Carbon Dots (CDs) for fluorescence imaging, and it has been observed that the mechanism of recognition is primarily based on the interactions between the surface functional groups of CDs and those present on cancer cells. The cancer cell lines that are frequently employed in research consist of HeLa cells for cervical cancer, MCF-7 cells for breast cancer, A549 cells for human lung cancer, HepG2 cells for human liver cancer, and HEK-293 cells, which are tumorigenic human embryonic kidney cells [46]. Nonetheless, the quest to improve the specificity and targeting of contrast agents for imaging cancer cells still shows difficulties, mainly because of the overlapping recognition elements and ligands located on the surface of cancer cells [6].

Research indicates that bioimaging is a robust technique for the real-time depiction of biological processes, making it a highly effective strategy for the diagnosis of cancer cell lines. This study explored the bioimaging potential of functionalized Carbon Quantum Dots (CQDs) that were sustainably synthesized by using human hair. This synthesized CQDs were used to treat on MCF-7 human breast cancer cells for bioimaging study. The treated cells were observed to emit fluorescence when viewed through a blue filter using fluorescence microscopy. The findings are highly encouraging, suggesting that these sustainable functionalized Carbon Quantum Dots (CQDs) may function effectively as biomarkers or bioimaging probes. Moreover, it was found out that the CQDs derived from hair not only displayed robust bioimaging capabilities but also showed anticancer properties, serving as multifunctional biomedical nanoagents with the capacity to both diagnose and suppress cancer through cytotoxic effects. These results indicate that such CQDs possess considerable promise for future clinical applications [31].

5.2 *Biosensing*

CQDs serve as effective carriers for biosensors owing to highly solubility in water, flexibility for surface modification, nontoxic properties, excitation-dependent multicolor emission, outstanding biocompatibility, remarkable photostability good cell permeability. CQD-based biosensors enable the visual detection of various parameters, including glucose [80], cellular copper [115], phosphate [113], iron [116], potassium [96], pH [81], and nucleic acids [47]. It has been found that a novel fluorescence probe derived from yellowish-green emissive CQDs can successfully detect phoxim. This report highlights the initial application of these CQDs, which were synthesized through a basic one-pot hydrothermal treatment of citric acid and p-phenylenediamine. The detection system was found to be based on quenching of CQDs fluorescence by phoxim through the Inner Filter Effect (IFE). The CQDs demonstrated significant sensitivity and selectivity for phoxim, featuring two separate linear detection ranges (1–10 μM and 10–98 μM) and a minimal detection limit of 0.09 M. Additionally, the proposed techniques were shown to be efficient in the analysis of real-world samples, achieving recovery rates of 93% to105.9% [26].

Li et al. [45] revealed that the presence of pyrophosphate and high-energy phosphate bonds in the carbon dots functions as a nuclear growth inhibitor and energy donor. The synthesized CQDs were later applied as fluorescent probes in the design of biosensors targeting hydrogen peroxide (H_2O_2). The combination of Fe^{2+} and H_2O_2 found that the fluorescence of C-dots was progressively quenched as the concentration of H_2O_2 increased. When glucose is recognized by the sensor, it is specifically oxidized by Glucose Oxidase (GOD), leading to the release of H_2O_2 and a progressive decrease in the fluorescence of C-dots. At the same time, they examined the sensor's feasibility by applying the biosensor and the Glucose Oxidase-Peroxidase (GOD-POD) method to determine glucose levels in human serum samples. The results demonstrated consistency between the two methods. Moreover, Carbon Quantum Dots (CQDs) can be combined with aptamers [104].

5.3 *Solar Energy Conversion*

The effective conversion of solar energy into electricity or hydrogen fuel necessitates the use of cost-effective materials that can efficiently absorb photon energy. Currently, two key technologies are found for this purpose: (i) Photoelectrochemical (PEC) cells and (ii) photovoltaic solar cells, both of which are adept at capturing and converting solar energy. High efficiency charge separation in solar cells was found by using heterostructures that exhibit suitable bandgap alignment and precisely engineered interfaces. The combination of semiconductor heterostructures with Quantum Dots (QDs) has shown remarkable effectiveness in promoting efficient charge transfer. These heterostructures, which are based on quantum dots, usually include donor and acceptor components that interface with one another. Upon being illuminated, it was found that the donor materials create electron-hole pairs that are subsequently separated at the junction by the acceptor. The donor should demonstrate a high absorption coefficient across the solar spectrum and have a bandgap that is tailored for optimal sunlight absorption. Moreover, it was identified as crucial that the conduction bands of the donor and acceptor are appropriately aligned to promote efficient charge migration. The acceptor serves as a channel for charge carriers, leading them from the donor to the conductive electrode layer [67].

The exploration of various solar cell architectures that include Quantum Dots (QDs) is promising. This encompasses Photoelectrochemical (PEC) cells that utilize a thin layer of QDs placed in an electrolyte, along with QD-sensitized solar cells. The primary advantage of incorporating QDs into solar cell designs is their tunable bandgap, which facilitates specific light absorption. A significant advancement has been found, revealing that solar cells utilizing Quantum Dots (QDs) could potentially exceed the Shockley–Queisser efficiency threshold. Different types of semiconductors QDs, such as CdSe, CdS, PbS, and InAs, have been engineered to operate in combination with wide bandgap materials. The size-dependent optical characteristics of QDs allow them to capture photons from specific areas of the solar spectrum. Although the integration of Quantum Dots (QDs) with carbon nanostructures in solar cells is still in its preliminary development stages, it was found to be a promising pathway for the progression of next-generation solar technologies. Colloidal Quantum Dots (CQDs) can take on multiple roles within solar cell frameworks, functioning as donors, acceptors, or sensitizers in conjunction with wide bandgap semiconductor materials [67].

These solar cells have gone through three generations till now. Crystalline SiO_2 is the first generation of solar cells [22], while thin film solar cell is the second generation that have been developed [27]. Third generation is the Dye-Sensitized Solar Cells (DSSCs), polymer solar cells, perovskite solar cells, and carbon dots solar cells [79]. Polymer solar cells are also called Organic Solar Cells (OSCs). They can be produced through the roll-to-roll technique, and they are also recognized for their flexibility for devices and their cost effectiveness [90]. The research conducted by Padmanathan and Prakasam [71] led to the development of hybrid carbon dots/titanium dioxide (CQDs/TiO_2) photo electrodes for their use in dye-sensitized photovoltaic solar cells. Carbon Quantum Dots (CQDs) were synthesized using an easy one-step hydrothermal process that incorporated citric acid and urea. Afterward, CQDs/TiO_2 composites were produced by blending CQDs with uncoated TiO_2 nanopowder, followed by ultrasonication. The resulting particle sizes were between 30 and 40 nm. CQDs/TiO_2 samples demonstrated superior surface areas (112 m^2/g) and pore sizes (30–35 nm) relative to bare TiO_2, which directly affects the efficiency of solar cells. The photoanode constructed from CQDs/TiO_2 demonstrated a significant Power Conversion Efficiency (PCE) of 15.5% [112]. have described the production of Luminescent Solar Concentrators (LSCs), utilizing both CQDs and mixed-halide inorganic perovskite QDs simultaneously. The semi-transparent tandem LSC, constructed from CQDs and perovskite quantum dots, showed an external optical efficiency of about 3.05% when subjected to one Sun illumination. This reflects a 27% and 117% improvement in efficiency over the single-layer LSCs that were developed.

5.4 *Optoelectronic Applications*

Carbon quantum dots served as favorite nanoparticles having optoelectronic applications, remarkable optoelectronic properties, and high specific surface areas which encompass electron mobility, long-lived hot-electron lifetimes, rapid electron extraction, tunable bandgap, and stable fluorescence [107].

5.4.1 Light-Emitting Diodes (LEDs)

LEDs play a crucial role in sectors of solid-state lighting and full-color display technologies. The introduction of Colloidal Quantum Dots (CQDs) as a novel fluorescent quantum dot system offers a promising alternative to the expensive phosphors and toxic metal-based semiconductor quantum dots traditionally used in LEDs. This is attributed to tunable and stable fluorescence emission, cost-effectiveness, and environmentally friendly characteristics. Broad emission characteristics of CQDs are highly suitable for the creation of White Light-Emitting Diodes (WLEDs). These lighting devices are viewed as the next generation of illumination solutions, thanks to their advantages, which include energy efficiency, long operational life, and compact design. Therefore, considerable efforts are invested in realizing the potential applications of CQDs within the realm of LEDs [107].

The roles of Colloidal Quantum Dots (CQDs) in Light-Emitting Diodes (LEDs) are broadly divided into two distinct categories: phosphor-converted LEDs (pc-LEDs) and electroluminescent LEDs. For pc-LEDs, blue or ultraviolet LED chips are predominantly utilized as an initial light source to energize CQD-based phosphors optically. On the other hand, electroluminescent LEDs facilitate the injection of electrons into the active emission layer of Colloidal Quantum Dots (CQDs) through external bias voltage, leading to radiative recombination, and are characterized by potentially superior efficiency. Thus, it is crucial to pursue the development of electroluminescent LEDs [107].

6 Conclusion and Future Prospects

The emergence of Carbon Dots (CDs) as promising nanomaterials in various sectors can be attributed to their unique attributes. Although multiple synthesis methods, including physical, chemical, and biological approaches, are available, the biological route is recommended as the preferred option due to its economic viability and environmental sustainability. Moreover, carbon dots produced through biological synthesis show a high degree of biocompatibility, thereby enhancing their applicability in biomedicine, particularly in the realm of drug delivery. In recent years, these carbon dots have attracted significant interest for their potential uses in optoelectronics, bioimaging, sensing, energy storage, and catalysis [18].

Despite the advancements made, numerous challenges and future considerations regarding carbon dots persist. A significant challenge lies in the synthesis techniques currently in use. For carbon dots to be utilized on a larger scale, it is essential to establish efficient and scalable methods for their synthesis. While several approaches, including hydrothermal/solvothermal methods, microwave-assisted synthesis, and pyrolysis, have been explored, achieving precise control over the size, shape, and surface chemistry of carbon dots is still imperative. To achieve reproducibility and reliability, it is imperative to standardize both the synthesis protocols and the characterization methods, which include Transmission Electron Microscopy (TEM), spectroscopy (UV-Vis, fluorescence), and elemental analysis. To convert the potential of carbon dots into tangible applications, it is essential to integrate them into devices and to scale up their production. This undertaking requires tackling challenges associated with device manufacturing, ensuring compatibility with current technologies, and achieving cost-effective large-scale synthesis. For future opportunities, carbon dots present remarkable potential for sophisticated applications such as solar cells, sensors, and catalysts. Sustained research and development will promote the investigation of new applications and the enhancement of carbon dots for particular functions.

References

1. Abbas A, Mariana LT, Phan AN (2018) Biomass-waste-derived graphene quantum dots and their applications. Carbon 140:77–99
2. Abd Rani U, Ng LY, Ng CY, Mahmoudi E (2020) A review of carbon quantum dots and their applications in wastewater treatment. Adv Colloid Interf Sci 278:102124
3. Alam MB, Minocha T, Yadav SK, Parmar AS (2022) Therapeutic potential of chlorophyll functionalized carbon quantum dots against cervical cancer. Chemistry Select 7(48):e202204562
4. Architha NRM, Shobana C (2021) Microwave-assisted green synthesis of fluorescent carbon quantum dots from Mexican Mint extract for Fe3 + detection and bio-imaging applications. Environ Res 199:111263. https://doi.org/10.1016/j.envres.2021.111263
5. Azam N, Najabat Ali M, Javaid Khan T (2021) Carbon quantum dots for biomedical applications: review and analysis. Front Mater 8:700403
6. Bartkowski M, Zhou Y, Nabil Amin Mustafa M, Eustace AJ, Giordani S (2024) CARBON DOTS: bioimaging and anticancer drug delivery. Chemistry–A Eur J 30(19):e202303982
7. Carneiro Cruz AA, Freire RM, Froelich DB, Alves de Lima AC, Muniz AR, Ferreira OP, Fechine PBA (2019) Fluorescence based platform to discriminate protein using carbon quantum dots. ChemistrySelect 4(19):5619–5627
8. Chen BB, Liu ZX, Zou HY, Huang CZ (2016) Highly selective detection of 2, 4, 6-trinitrophenol by using newly developed terbium-doped blue carbon dots. Analyst 141(9):2676–2681
9. Chen J, Liang X, Liu W, Gu W, Zhang B, Ji G (2019) Mesoporous carbon hollow spheres as a light weight microwave absorbing material showing modulating dielectric loss. Dalton Trans 48:10145–10150
10. Cheng C, Shi Y, Li M, Xing M, Wu Q (2017) Carbon quantum dots from carbonized walnut shells: structural evolution, fluorescence characteristics, and intracellular bioimaging. Mater Sci Eng C 79:473–480
11. Cui L, Ren X, Sun M, Liu H, Xia L (2021) Carbon dots: synthesis, properties and applications. Nano 11(12):3419
12. Cui L, Ren X, Wang J, Sun M (2020) Synthesis of homogeneous carbon quantum dots by ultrafast dual-beam pulsed laser ablation for bioimaging. Mater Today Nano 12:100091
13. Dadkhah S, Mehdinia A, Jabbari A, Manbohi A (2022) Nicotinamide-functionalized carbon quantum dot as new sensing platform for portable quantification of vitamin B12 in fluorescence, UV–vis and smartphone triple mode. J Fluoresc 32(2):681–689
14. Das T, Saikia BK, Dekaboruah HP, Bordoloi M, Neog D, Bora JJ, Lahkar J, Narzary B, Roy S, Ramaiah D (2019) BlueFluorescent and biocompatible carbon dots derived from abundant low-quality coals. J Photochem Photobiol B Biol 195:1–11

15. Desmond LJ, Phan AN, Gentile P (2021) Critical overview on the green synthesis of carbon quantum dots and their application for cancer therapy. Environ Sci Nano 8(4):848–862
16. El-Brolsy HMEM, Hanafy NA, El-Kemary MA (2022) Fighting non-small lung cancer cells using optimal functionalization of targeted carbon quantum dots derived from natural sources might provide potential therapeutic and cancer bio image strategies. Int J Mol Sci 23(21):13283
17. El-Shabasy RM, Farouk Elsadek M, Mohamed Ahmed B, Fawzy Farahat M, Mosleh KN, Taher MM (2021) Recent developments in carbon quantum dots: properties, fabrication techniques, and bio-applications. Processes 9(2):388
18. Etefa HF, Tessema AA, Dejene FB (2024) Carbon dots for future prospects: synthesis, characterizations and recent applications: a review (2019–2023). C 10(3):60
19. Fan L, Zhu M, Lee X, Zhang R, Wang K, Wei J, Zhong M, Wu D, Zhu H (2013) Direct synthesis of graphene quantum dots by chemical vapor deposition. Part Syst Charact 30:764–769
20. Georgakilas V, Tiwari JN, Kemp KC, Perman JA, Bourlinos AB, Kim KS, Zboril R (2016) Noncovalent functionalization of graphene and graphene oxide for energy materials, biosensing, catalytic, and biomedical applications. Chem Rev 116(9):5464–5519
21. Gohari G, Panahirad S, Sepehri N, Akbari A, Zahedi SM, Jafari H, Dadpour MR, Fotopoulos V (2021) Enhanced tolerance to salinity stress in grapevine plants through application of carbon quantum dots functionalized by proline. Environ Sci Pollut Res 28:42877–42890
22. Green MA (2002) Third generation photovoltaics: solar cells for 2020 and beyond. Phys E Low-dimensional Syst Nanostructures 14:65–70. https://doi.org/10.1016/S1386-9477(02)00361-2
23. Hola K, Sudolská M, Kalytchuk S, Nachtigallová D, Rogach AL, Otyepka M, Zboril R (2017) Graphitic nitrogen triggers red fluorescence in carbon dots. ACS Nano 11(12):12402–12410
24. Huang S, Li W, Han P, Zhou X, Cheng J, Wen H, Xue W (2019) Carbon quantum dots: synthesis, properties, and sensing applications as a potential clinical analytical method. Anal Methods 11(17):2240–2258
25. Hui KC, Dhanapalan N, Khayrani AC, Imanto T, Sambudi NS (2022) The synthesis of amino-functionalized carbon quantum dots-decorated hydroxyapatite as drug delivery agent. C R Chim 25(G1):295–306
26. Huo X, Liu L, Bai Y, Qin J, Yuan L, Feng F (2022) Facile synthesis of yellowish-green emitting carbon quantum dots and their applications for phoxim sensing and cellular imaging. Anal Chim Acta 1206:338685
27. Imamzai M, Aghaei M, Thayoob YHM, Forouzanfar M (2012) A review on comparison between traditional silicon solar cells and thin-film CdTe solar cells. In: Proceedings of national graduate conference (Nat-grad). Tenaga Nasional Universiti, Putrajaya, Malaysia, pp 1–5
28. Janus Ł, Radwan-Pragłowska J, Piątkowski M, Bogdał D (2020) Coumarin-modified CQDs for biomedical applications—two-step synthesis and characterization. Int J Mol Sci 21(21):8073
29. Jing HH, Bardakci F, Akgöl S, Kusat K, Adnan M, Alam MJ, Gupta R, Sahreen S, Chen Y, Gopinath SC, Sasidharan S (2023) Green carbon dots: synthesis, characterization, properties and biomedical applications. J Funct Biomater 14(1):27
30. John VL, Nair Y, Vinod TP (2021) Doping and surface modification of carbon quantum dots for enhanced functionalities and related applications. Part Part Syst Charact 38(11):2100170
31. Kamble PA, Gambhir RP, Vibhute AA, Parkhe VS, Tiwari AP (2022) A sustainable synthesis of functionalized carbon quantum dots from hair for the applications of cytotoxicity study and in vitro bioimaging
32. Kamel AH, Alsalhy QF, Ibrahim SS, Faneer KA, Hashemifard SA, Jangizehi A, Seiffert S, Maskos M, Shakeri A, Bantz C (2023) Novel sodium and potassium carbon quantum dots as forward osmosis draw solutes: synthesis, characterization and performance testing. Desalination 567:116956

33. Kaur M, Kaur M, Sharma VK (2018) Nitrogen-doped graphene and graphene quantum dots: A review onsynthesis and applications in energy, sensors and environment. Adv Colloid Interface Sci 259:44–64. https://doi.org/10.1016/j.cis.2018.07.001
34. Khan ZM, Rahman RS, Islam S, Zulfequar M (2019) Hydrothermal treatment of red lentils for the synthesis of fluorescent carbon quantum dots and its application for sensing Fe^{3+}. Opt Mater 91:386–395
35. Khan Y, Sadia H, Ali Shah SZ, Khan MN, Shah AA, Ullah N, Ullah MF, Bibi H, Bafakeeh OT, Khedher NB, Eldin SM (2022) Classification, synthetic, and characterization approaches to nanoparticles, and their applications in various fields of nanotechnology: a review. Catalysts 12(11):1386
36. Khan ZM, Saifi S, Aslam Z, Khan SA, Zulfequar M (2020) A facile one step hydrothermal synthesis of carbon quantum dots for label-free fluorescence sensing approach to detect picric acid in aqueous solution. J Photochem Photobiol A Chem 388:112201
37. Khan WU Wang D, Zhang W, Tang Z, Ma X, Ding X, Du S, Wang Y (2017) High quantum yield green-emitting carbon dots for Fe(III) detection, biocompatible fluorescent ink and cellular imaging. Sci Rep 7(1):1–9
38. Kong J, Wei Y, Zhou F, Shi L, Zhao S, Wan M, Zhang X (2024) Carbon quantum dots: properties, preparation, and applications. Molecules 29(9):2002
39. Krishnaiah P, Atchudan R, Perumal S, Salama ES, Lee YR, Jeon BH (2022) Utilization of waste biomass of poa pratensis for green synthesis of N-doped carbon dots and its application in detection of Mn2+ and Fe3+. Chemosphere 286, 131764
40. Kumar P, Dua S, Kaur R, Kumar M, Bhatt G (2022) A review on advancements in carbon quantum dots and their application in photovoltaics. RSC Adv 12(8):4714–4759
41. Kumari A, Bhattacharya J, Moulick RG (2023) Bioimaging applications of carbon quantum dots. In: Carbon quantum dots for sustainable energy and optoelectronics, Woodhead Publishing, pp 239–261
42. Lai C, Lin S, Huang X, Jin Y (2021) Synthesis and properties of carbon quantum dots and their research progress in cancer treatment. Dyes Pigments 196:109766
43. Lei J, Yang L, Lu D, Yan X, Cheng C, Liu Y, Wang L, Zhang J (2015) Carbon dot-incorporated PMO nanoparticles as versatile platforms for the design of ratiometric sensors, multichannel traceable drug delivery vehicles, and efficient photocatalysts. Adv Opt Mater 3(1):57–63
44. Li G, Cheng R, Cheng H, Yu XQ, Ling L, Wang CF, Chen S (2021) Microfluidic synthesis of robust carbon dots-functionalized photonic crystals. Chem Eng J 405:126539
45. Li N, Liu SG, Dong JX, Fan YZ, Ju YJ, Luo HQ, Li NB (2018) Using high-energy phosphate as energy-donor and nucleus growth-inhibitor to prepare carbon dots for hydrogen peroxide related biosensing. Sensors Actuators B Chem 262:780–788
46. Li H, Yan X, Kong D, Jin R, Sun C, Du D, Lin Y, Lu G (2020) Recent advances in carbon dots for bioimaging applications. Nanoscale Horiz 5(2):218–234
47. Li H, Zhang Y, Wang L, Tian J, Sun X (2011) Nucleic acid detection using carbon nanoparticles as a fluorescent sensing platform. Chem Commun 47(3):961–963
48. Li H, He X, Kang Z, Huang H, Liu Y, Liu J, Lian S, Tsang CH, Yang X, Lee ST (2010) Water-soluble fluorescent carbon quantum dots and photocatalyst design. Angew Chem Int Ed 49(26):4430–4434
49. Lin L, Luo Y, Tsai P, Wang J, Chen X (2018) Metal ions doped carbon quantum dots: synthesis, physicochemical properties, and their applications. TrAC Trends Anal Chem 103:87–101
50. Lin L, Yin Y, Li Z, Xu H, Hessel V, Ostrikov KK (2022) Continuous micro-flow synthesis of fluorescent phosphorus and nitrogen co-doped carbon quantum dots. Chem Eng Res Des 178:395–404
51. Liu ML, Chen BB, Yang T, Wang J, Liu XD, Huang CZ (2017) One-pot carbonization synthesis of europium-doped carbon quantum dots for highly selective detection of tetracycline. Methods Appl Fluoresc 5(1):015003
52. Liu Y, Li W, Wu P, Ma C, Wu X, Xu M, Luo S, Xu Z, Liu S (2019) Hydrothermal synthesis of nitrogen and boron co-doped carbon quantum dots for application in acetone and dopamine sensors and multicolor cellular imaging. Sensors Actuators B Chem 281:34–43

53. Liu J, Li R, Yang B (2020) Carbon dots: a new type of carbon-based nanomaterial with wide applications. ACS Cent Sci 6(12):2179–2195
54. Liu M, Xu Y, Niu F, Gooding JJ, Liu J (2016) Carbon quantum dots directly generated from electrochemical oxidation of graphite electrodes in alkaline alcohols and the applications for specific ferric ion detection and cell imaging. Analyst 141(9):2657–2664
55. Liu Y, Li W, Wu P, Ma C, Wu X, Xu M, Luo S, Xu Z, Liu S (2019) Hydrothermal synthesis of nitrogen and boron co-doped carbon quantum dots for application in acetone and dopamine sensors and multicolor cellular imaging. Sens Actuators B Chem 281:34–43
56. Ma H, Liu X, Wang X, Li X, Yang C, Iqbal A, Liu W, Li J, Qin W (2017) Sensitive fluorescent light-up probe for enzymatic determination of glucose using carbon dots modified with MnO_2 nanosheets. Microchim Acta 184:177–185
57. Magesh V, Sundramoorthy AK, Ganapathy D (2022) Recent advances on synthesis and potential applications of carbon quantum dots. Front Mater 9:906838
58. Mandal SK, Paul S, Datta S, Jana D (2021) Nitrogenated CQD decorated ZnO nanorods towards rapid photodegradation of rhodamine B: a combined experimental and theoretical approach. Appl Surf Sci 563:150315
59. Markovic ZM, Labudova M, Danko M, Matijasevic D, Micusik M, Nadazdy V, Kovacova M, Kleinova A, Spitalsky Z, Pavlovic V, Milivojevic DD (2020) Highly efficient antioxidant F-and cl-doped carbon quantum dots for bioimaging. ACS Sustain Chem Eng 8(43):16327–16338
60. Martindale BC, Hutton GA, Caputo CA, Reisner E (2015) Solar hydrogen production using carbon quantum dots and a molecular nickel catalyst. J Am Chem Soc 137(18):6018–6025
61. Mehta VN, Jha S, Basu H, Singhal RK, Kailasa SK (2015) One-step hydrothermal approach to fabricate carbon dots from apple juice for imaging of mycobacterium and fungal cells. Sens Actuators B Chem 2013:434–443
62. Meng W, Bai X, Wang B, Liu Z, Lu S, Yang B (2019) Biomass-derived carbon dots and their applications. Energy Environ Mater 2(3):172–192
63. Miao S, Liang K, Zhu J, Yang B, Zhao D, Kong B (2020) Hetero-atom-doped carbon dots: doping strategies, properties and applications. Nano Today 33:100879
64. Ming H, Wei D, Yang Y, Chen B, Yang C, Zhang J, Hou Y (2021) Photocatalytic activation of peroxymonosulfate by carbon quantum dots functionalized carbon nitride for efficient degradation of bisphenol a under visible-light irradiation. Chem Eng J 424:130296
65. Mitra S, Chandra S, Pathan SH, Sikdar N, Pramanik P, Goswami A (2013) Room temperature and solvothermal green synthesis of self-passivated carbon quantum dots. RSC Adv 3:3189–3193
66. Mohagheghpour E, Farzin L, Ghoorchian A, Sadjadi S, Abdouss M (2022) Selective detection of manganese (II) ions based on the fluorescence turn-on response via histidine functionalized carbon quantum dots. Spectrochim Acta A Mol Biomol Spectrosc 279:121409
67. Molaei MJ (2020) The optical properties and solar energy conversion applications of carbon quantum dots: a review. Sol Energy 196:549–566
68. Mondal TK, Kapuria A, Miah M, Saha SK (2023) Solubility tuning of alkyl amine functionalized carbon quantum dots for selective detection of nitroexplosive. Carbon 209:117972
69. Nugraha MW, Wirzal MDH, Ali F, Roza L, Sambudi NS (2021) Electrospun polylactic acid/tungsten oxide/amino-functionalized carbon quantum dots (PLA/WO3/N-CQDs) fibers for oil/water separation and photocatalytic decolorization. J Environ Chem Eng 9(5):106033
70. Olmos-Moya PM, Velazquez-Martinez S, Pineda-Arellano C, Rangel-Mendez JR, Chazaro-Ruiz LF (2022) High added value functionalized carbon quantum dots synthetized from orange peels by assisted microwave solvothermal method and their performance as photosensitizer of mesoporous TiO_2 photoelectrodes. Carbon 187:216–229
71. Padmanathan S, Prakasam A (2020) Design and fabrication of hybrid carbon dots/titanium dioxide (CDs/TiO_2) photoelectrodes for highly efficient dye-sensitized solar cells. J Mater Sci Mater Electron 31(4):3492–3499

72. Rasal AS, Yadav S, Yadav A, Kashale AA, Manjunatha ST, Altaee A, Chang JY (2021) Carbon quantum dots for energy applications: a review. ACS Appl Nano Mater 4(7):6515–6541
73. Rong M, Feng Y, Wang Y, Chen X (2017) One-pot solid phase pyrolysis synthesis of nitrogen-doped carbon dots for Fe^{3+} sensing and bioimaging. Sensors Actuators B Chem 245:868–874
74. Šafranko S, Stanković A, Hajra S, Kim HJ, Strelec I, Dutour-Sikirić M, Weber I, Bosnar MH, Grbčić P, Pavelić SK, Széchenyi A (2021) Preparation of multifunctional N-doped carbon quantum dots from citrus Clementina peel: investigating targeted pharmacological activities and the potential application for Fe^{3+} sensing. Pharmaceuticals 14(9):857
75. Salahinejad M, Sadjadi S, Abdouss M (2021) Investigating fluorescence quenching of cysteine-functionalized carbon quantum dots by heavy metal ions: experimental and QSPR studies. J Mol Liq 334:116067
76. Selvaraju N, Ganesh PS, Palrasu V, Venugopal G, Mariappan V (2022) Evaluation of antimicrobial and antibiofilm activity of Citrus medica fruit juice based carbon dots against Pseudomonas aeruginosa. ACS Omega 7(41):36227–36234
77. El Sharkasy ME, Tolba MM, Belal F, Walash MI, Aboshabana R (2022) Thiosemicarbazide functionalized carbon quantum dots as a fluorescent probe for the determination of some oxicams: application to dosage forms and biological fluids. RSC Adv 12(22):13826–13836
78. Sharma V, Orejon D, Takata Y, Krishnan V, Harish S (2018) Gladiolus dalenii based bioinspired structured surface via soft lithography and its application in water vapor condensation and fog harvesting. ACS Sustain Chem Eng 6:6981–6993
79. Shi Z, Jayatissa A (2018) Perovskites-based solar cells: a review of recent Progress, materials and processing methods. *Materials* 11:729
80. Shi W, Wang Q, Long Y, Cheng Z, Chen S, Zheng H, Huang Y (2011) Carbon nanodots as peroxidase mimetics and their applications to glucose detection. Chem Commun 47(23):6695–6697
81. Shi W, Li X, Ma H (2012) A tunable ratiometric pH sensor based on carbon nanodots for the quantitative measurement of the intracellular pH of whole cells. Angew Chem Int Ed Eng 51(26):6432–6435
82. Singh I, Arora R, Dhiman H, Pahwa R (2018) Carbon quantum dots: synthesis, characterization and biomedical applications. Turk J Pharm Sci 15(2):219–230
83. Singh L, Sharma T, Singh V (2021) Study of structural and functional properties of fluorescent EDTA@ CQDs synthesized from peanut shells via pyrolysis technique. Mater Today Proc 44:192–198
84. Singh RK, Kumar R, Singh DP, Savu R, Moshkalev SA (2019) Progress in microwaveassisted synthesis of quantum dots (graphene/carbon/semiconducting) for Journal Pre-proof Journal Pre-proof bioapplications: a review. Mater Today Chem 12:282–314. https://doi.org/10.1016/j.mtchem.2019.03.001
85. Sun YP, Zhou B, Lin Y, Wang W, Fernando KS, Pathak P, Meziani MJ, Harruff BA, Wang X, Wang H, Luo PG (2006) Quantum-sized carbon dots for bright and colorful photoluminescence. J Am Chem Soc 128(24):7756–7757
86. Sun M, Han Y, Yuan X, Jing P, Zhang L, Zhao J, Zheng YJ (2020) Efficient full-color emitting carbon-dot-based composite phosphors by chemical dispersion. Nanoscale 12:15823
87. Supchocksoonthorn P, Hanchaina R, Sinoy MCA, de Luna MDG, Kangsamaksin T, Paoprasert P (2021) Novel solution-and paper-based sensors based on label-free fluorescent carbon dots for the selective detection of pyrimethanil. Appl Surf Sci 564:150372
88. Tejwan N, Saini AK, Sharma A, Singh TA, Kumar N, Das J (2021) Metal-doped and hybrid carbon dots: a comprehensive review on their synthesis and biomedical applications. J Control Release 330:132–150
89. Venugopalan P, Vidya N (2023) Microwave-assisted green synthesis of carbon dots derived from wild lemon (citrus pennivesiculata) leaves as a fluorescent probe for tetracycline sensing in water. Spectrochim Acta A Mol Biomol Spectrosc 286:122024
90. Vercelli B (2021) The role of carbon quantum dots in organic photovoltaics: a short overview. Coatings 11(2):232

91. Wang R, Lu KQ, Tang ZR, Xu YJ (2017) Recent progress in carbon quantum dots: synthesis, properties and applications in photocatalysis. J Mater Chem A 5(8):3717–3734
92. Wang X, Yang P, Feng Q, Meng T, Wei J, Xu C, Han J (2019) Green preparation of fluorescent carbon quantum dots from cyanobacteria for biological imaging. Polymers 11(4):616
93. Wang Y, Zhang Y, Jia M, Meng H, Li H, Guan Y, Feng L (2015) Functionalization of carbonaceous nanodots from MnII-coordinating functional knots. Chemistry–A Eur J 21(42):14843–14850
94. Wang W, Zhang J, Tian P, Wang X, Qian W, Huang G, Liu X, Liu B (2021) Ionic liquid-functionalized CQDs as effective "signal-off" fluorescence Nanoprobes for selective and sensitive detection of iron (III) and chromium (VI). Chemistry Select 6(39):10489–10500
95. Wei J, Zhang X, Sheng Y, Shen J, Huang P, Guo S, Pan J, Liu B, Feng B (2014) Simple one-step synthesis of water-soluble fluorescent carbon dots from waste paper. New J Chem 385:906–909
96. Wei W, Xu C, Ren J, Xu B, Qu X (2012) Sensing metal ions with ion selectivity of a crown ether and fluorescence resonance energy transfer between carbon dots and graphene. Chem commun 48(9):1284–1286
97. Xiao J, Fang X, Yang S, He H, Sun C (2015) Microwave-assisted heterogeneous catalytic oxidation of high-concentration reactive yellow 3 with CuFe2O4/PAC. J Chem Technol Biot 90:1861–1868
98. Xie Y, Hu J, Hu Y, Jiang X (2024) A review on strengthening mechanisms of carbon quantum dots-reinforced cu-matrix nanocomposites. Nanotechnol Rev 13(1):20240038
99. Xu X, Ray R, Gu Y, Ploehn HJ, Gearheart L, Raker K, Scrivens WA (2004) Electrophoretic analysis and purification of fluorescent single-walled carbon nanotube fragments. J Am Chem Soc 126:12736–12737
100. Xue S, Cui Y, Chen X, Shi J, Liu S, Ye Q, Zhou F (2023) Zwitterionic dopamine sulfonate functionalized carbon quantum dots as water-based lubricant additive for friction and wear reduction. Appl Surf Sci 615:156411
101. Yadav PK, Chandra S, Kumar V, Kumar D, Hasan SH (2023) Carbon quantum dots: synthesis, structure, properties, and catalytic applications for organic synthesis. Catalysts 13(2):422
102. Yan C, Guo L, Shao X, Shu Q, Guan P, Wang J, Hu X, Wang C (2021) Amino acid–functionalized carbon quantum dots for selective detection of Al^{3+} ions and fluorescence imaging in living cells. Anal Bioanal Chem 413(15):3965–3974
103. Yan F, Jiang Y, Sun X, Bai Z, Zhang Y, Zhou X (2018) Surface modification and chemical functionalization of carbon dots: a review. Microchim Acta 185:1–34
104. Yang HL, Bai LF, Geng ZR, Chen H, Xu LT, Xie YC, Wang DJ, Gu HW, Wang XM (2023) Carbon quantum dots: preparation, optical properties, and biomedical applications. Mater Today Adv 18:100376
105. Yao W, Hua Y, Yan Z, Wu C, Zhou F, Liu Y (2021) Sulfhydryl functionalized carbon quantum dots as a turn-off fluorescent probe for sensitive detection of Hg 2+. RSC Adv 11(57):36310–36318
106. You W, Zou W, Jiang S, Zhang J, Ge Y, Lu G, Bahnemann DW, Pan JH (2024) Fluorescent carbon quantum dots with controllable physicochemical properties fantastic for emerging applications: a review. Carbon Neutralization 3(2):245–284
107. Yuan T, Meng T, He P, Shi Y, Li Y, Li X, Fan L, Yang S (2019) Carbon quantum dots: an emerging material for optoelectronic applications. J Mater Chem C 7(23):6820–6835
108. Zhang R, Fan Z (2021) Redox-controlled fluorescent sensor for highly reactive oxygen species detection and cell imaging based on dihydrolipoic acid functionalized carbon quantum dots. J Lumin 234:117998
109. Zhang X, Hou X, Lu D, Chen Y, Feng L (2023) Porphyrin functionalized carbon quantum dots for enhanced electro-chemiluminescence and sensitive detection of Cu^{2+}. Molecules 28(3):1459
110. Zhang H, Wang H, Wang Y, Xin B (2020) Controlled synthesis and photocatalytic performance of biocompatible uniform carbon quantum dots with microwave absorption capacity. Appl Surf Sci 512:145751

111. Zhang HY, Wang Y, Xiao S, Wang H, Wang JH, Feng L (2017) Rapid detection of Cr (VI) ions based on cobalt (II)-doped carbon dots. Biosens Bioelectron 87:46–52
112. Zhao H, Benetti D, Tong X, Zhang H, Zhou Y, Liu G, Ma D, Sun S, Wang ZM, Wang Y, Rosei F (2018) Efficient and stable tandem luminescent solar concentrators based on carbon dots and perovskite quantum dots. Nano Energy 50:756–765
113. Zhao HX, Liu LQ, De Liu Z, Wang Y, Zhao XJ, Huang CZ (2011) Highly selective detection of phosphate in very complicated matrixes with an off–on fluorescent probe of europium-adjusted carbon dots. Chem commun 47(9):2604–2606
114. Zhou J, Booker C, Li R, Zhou X, Sham TK, Sun X, Ding Z (2007) An electrochemical avenue to blue luminescent nanocrystals from multiwalled carbon nanotubes (MWCNTs). J Am Chem Soc 129(4):744–745
115. Zhu A, Qu Q, Shao X, Kong B, Tian Y (2012) Carbon-dot-based dual-emission nanohybrid produces a ratiometric fluorescent sensor for in vivo imaging of cellular copper ions. Angew Chem Int Ed Eng 51(29):7185–7189
116. Zhu S, Meng Q, Wang L, Zhang J, Song Y, Jin H, Zhang K, Sun H, Wang H, Yang B (2013) Highly photoluminescent carbon dots for multicolor patterning, sensors, and bioimaging. Angew Chem Int Ed 52(14)

Chapter 3
Analytical Methods for Characterization of Quantum Dots

Amit Kumar Singh, Reetika Pechetty, and Yeshaswini Veeramachaneni

Abstract Quantum dots (QDs) are semiconductor nanoparticles with exceptional optical and electronic properties, which have emerged as pivotal players in diverse fields, from medical imaging to quantum computing in the recent past. To fully leverage the potential of QDs in specific applications, in-depth analysis and thorough characterization are indispensable. The characterization of these nanomaterials is paramount for understanding their behavior and optimizing their performance across various applications. Essential information, such as structural, optical, compositional, and size characteristics, forms the foundational data necessary to deepen our understanding of QDs and to identify their optimal applications. This chapter offers a comprehensive overview of the analytical techniques employed in the characterization of QDs, focusing on methods that assess their size, shape, composition, and optical properties. By elaborating on these analytical techniques, this chapter provides valuable insights into the systematic investigation of QDs, fostering a deeper comprehension of their properties and enhancing their application in technology and research. Ultimately, this chapter aims to equip readers with the knowledge necessary to effectively characterize QDs, thereby advancing the development and application of these nanomaterials in diverse scientific and industrial domains.

Keywords Quantum dots · Characterization techniques · Nanomaterials · Optical properties · Analytical methods

A. K. Singh (✉) · R. Pechetty · Y. Veeramachaneni
School of Sciences, Woxsen University, Hyderabad, India
e-mail: amitiitg2011@gmail.com

A. K. Singh
Department of Mechanical Engineering, George Mason University, Manassas, USA

T. Bhuyan et al. (eds.), *Multifunctional Quantum Dots*, Materials Horizons: From Nature to Nanomaterials, https://doi.org/10.1007/978-981-92-0172-3_3

1 Introduction

Quantum dots (QDs) are nanoscale semiconductor particles that exhibit unique electronic, optical, and chemical properties due to quantum confinement effects. These properties, such as size-dependent fluorescence, high quantum yields, and tunable band gaps, make quantum dots highly attractive for diverse applications in fields like optoelectronics, biomedicine, and energy [13]. However, the performance and applicability of quantum dots hinge critically on their structural, chemical, and optical characteristics, necessitating precise and reliable methods of characterization. This chapter explores the array of analytical techniques employed to evaluate quantum dots, encompassing structural analysis, compositional assessment, and functional property measurement. The methodologies discussed provide insights into particle size, shape, crystallinity, surface chemistry, and optical behavior, which are pivotal for optimizing synthesis and tailoring quantum dots for specific applications. By combining traditional techniques such as X-ray diffraction (XRD) and transmission electron microscopy (TEM) with advanced spectroscopic and dynamic approaches like photoluminescence (PL) spectroscopy and time-resolved fluorescence, researchers can achieve a comprehensive understanding of QD properties. Furthermore, the chapter underscores the importance of correlating analytical data with theoretical models to predict and enhance quantum dot behavior. Through this synthesis of methods, the chapter aims to guide researchers in selecting and implementing effective characterization tools to meet the demands of modern quantum dot applications.

2 Particle Morphology Determination

Various advanced techniques are employed to analyze the morphology of QDs, each providing unique insights into the size, shape, and composition of these nanoscale particles. These methods include:

1. Scanning Electron Microscopy (SEM), which, although having a lower resolution than TEM, remains effective for observing the surface morphology and composition of QDs.
2. Transmission Electron Microscopy (TEM), a widely used technique for examining the morphology of QDs, delivers high-resolution images that accurately depict the size and shape of individual quantum dots.
3. X-ray Diffraction (XRD), which is used to determine the crystal structure and phase of materials and can indirectly provide insights into the morphology of QDs, especially concerning their crystallinity.
4. Dynamic Light Scattering (DLS) is a widely used technique that helps ascertain the size distribution of particles in suspension, including quantum dots. This method is beneficial for evaluating the aggregation behavior of quantum dots under varying conditions, providing insights into their morphology.

5. Scanning Probe Microscopy (SPM), which can measure the size, shape, and surface characteristics of QDs at the nanoscale, making it particularly valuable for studying quantum dots on various surfaces.

2.1 *Scanning Electron Microscopy*

Quantum dots (QDs) are characterized using scanning electron microscopy (SEM) to investigate their morphological parameters such as particle size, shape, surface texture, and aggregation behavior. Quantum dots can be viewed with precise detail using SEM, which uses a focused, high-energy electron beam to cross the surface of the sample. Based on the needed resolution and brightness, electron beam generation starts at the electron source at the top of the SEM column, which is usually a tungsten filament or lanthanum hexaboride, or a field emission gun (FEG). Once heat is enough to let electrons escape from the emitter, they are accelerated toward a positively charged anode and reach energies of between 1 and 30 keV. The accelerated electron beam is then focused through a series of electromagnetic condenser and objective lenses to produce a fine probe, which is scanned over the sample surface. As this probe interacts with the quantum dot material, it generates a variety of signals, most notably secondary electrons (SE) and backscattered electrons (BSE). SE imaging is particularly advantageous for examining surface morphology and particle distribution in QD films, whereas BSE imaging provides compositional contrast based on atomic number differences (Fig. 1) [22, 27].

For compositional analysis of quantum dots, SEM is often coupled with energy-dispersive X-ray spectroscopy (EDX) [8]. By dislodging inner-shell electrons from atoms in the sample, high-energy electrons from the primary beam emit characteristic X-rays. By detecting and analyzing these X-rays, the EDX detector can identify the elemental constituents of QDs on a qualitative and quantitative basis. SEM-EDX combined approaches have proven particularly useful in confirming the elemental composition and stoichiometry of core–shell quantum dots (for example, CdSe/ZnS, PbS/CdS) and metallic QDs (Fig. 1) and identifying potential contaminants or by-products associated with the synthesis process [29].

It is often necessary to prepare samples on conductive substrates and coat them with a thin conductive layer (e.g., gold, carbon) to preserve the integrity of QDs under an electron beam and to avoid charging artifacts. A low accelerating voltage and beam current are also commonly employed to minimize beam-induced damage and preserve structural integrity [33].

2.2 *Transmission Electron Microscopy*

The imaging of quantum dots is made possible through the use of transmission electron microscopy (TEM), which passes high-energy electrons through thin specimens.

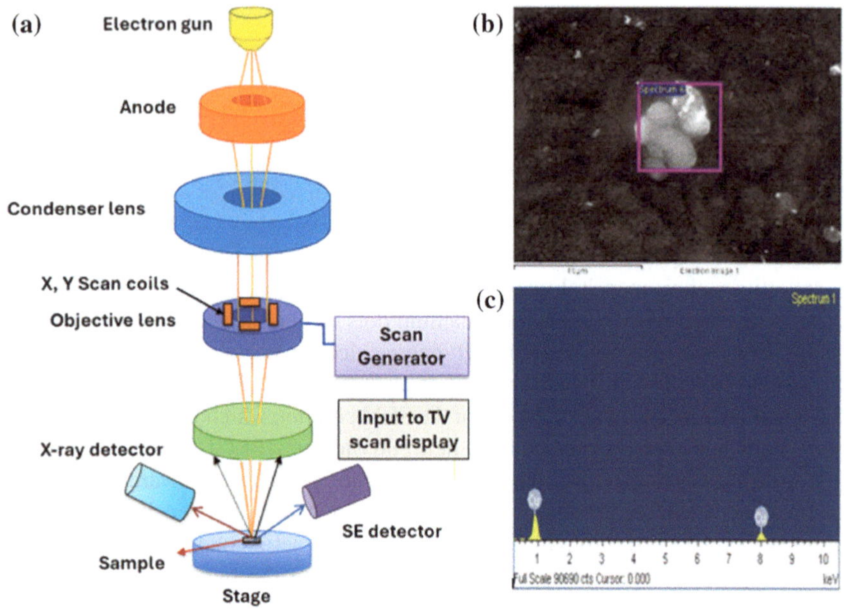

Fig. 1 **a** Basic components of SEM **b** SEM image of the metallic copper QDs (C) EDX spectra of the corresponding metallic copper QDs. Reproduced from Prabhash and Nair [28] with permission under the terms and conditions of the Creative Commons Attribution 4.0 (CC BY) license

A thermionic or field emission gun generates and propels electrons at higher energy levels, which usually range from 100 to 300 kV. Electron beams are focused onto the samples with the use of electromagnetic condenser lenses. Specimens subjected to electron beams thinner than 100 nm in thickness can be penetrated by electrons. During electron beam movement, interaction takes place with the sample's atomic structure. Light scattering arises because of changes in density, thickness, and atomic number. By concentrating transmitted and scattered electrons, objective and projector lenses produce an enlarged picture on a fluorescent screen or digital camera (Fig. 2) [5]. The TEM can produce diffraction patterns, dark-field images, or bright-field images depending on the image mode and aperture setup [21].

Due to the unique spatial resolution of about 0.1–0.2 nm, TEM is an essential tool for the morphological and structural analysis of nanoparticles. Such tools enable direct observation of QDs and their spatial features, which include monitoring their size, shape, and surface characteristics, which helps determine their uniformity and dispersion to a high level of detail. These characteristics assist in employing the QDs in many fields such as optoelectronics, photonics, and biomedical imaging, in which the sizes and shapes of QDs are critically important for functionality due to the effects of quantum confinement.

TEM not only supports high-resolution imaging but also facilitates crystallographic analysis through selected area electron diffraction (SAED) [19]. By adjusting the lens configuration and replacing the objective lens aperture with a diffraction

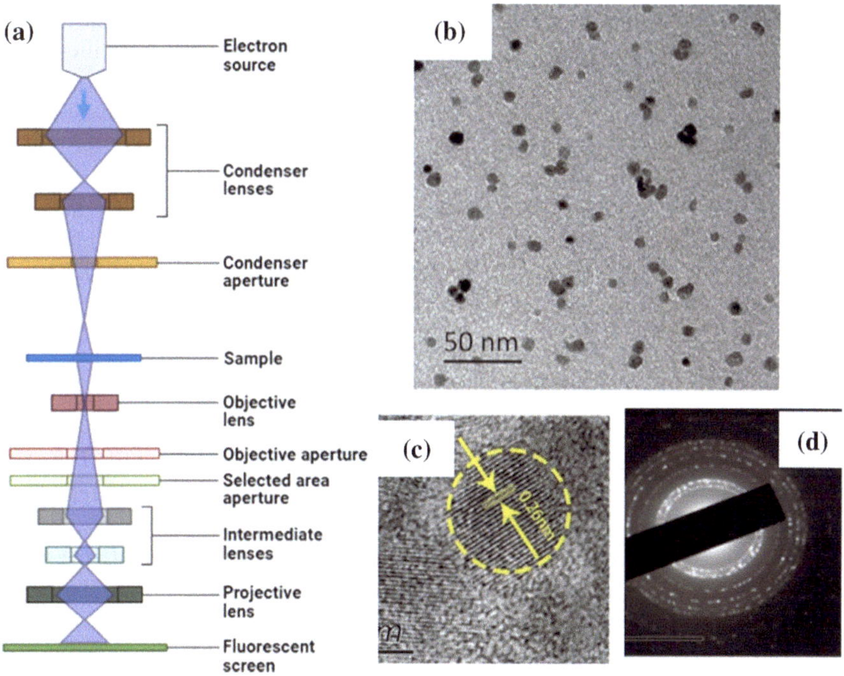

Fig. 2 **a** Basic components of TEM. The illustration was created with licensed Biorender software. **b** TEM image, **c** HRTEM image, and **d** SAED pattern of silane surface modified Zinc Oxide (M-ZnO) QDs. Reproduced from Chen et. al. [5] with permission under the terms and conditions of the Creative Commons Attribution 4.0 (CC BY) license

aperture, TEM can function as an electron diffraction camera. The resulting diffraction patterns provide crucial information about the crystal structure, orientation, and symmetry of the QDs. These patterns can also be used to calculate the crystallite size by analyzing diffraction ring broadening via the Debye–Scherrer Equation [3]. Furthermore, the use of high-resolution TEM (HRTEM) enables direct imaging of atomic planes, allowing for the investigation of interplanar spacings and the detection of crystal defects, stacking faults, and dislocations.

2.3 *Powder X-Ray Diffraction*

Powder X-ray diffraction (PXRD) is one of the most used non-destructive characterization methods for nanoscale materials. Powder X-ray diffraction provides important new perspectives on the structural characteristics of a sample, including its phases, preferred crystal orientations, and other metrics such as average grain size, crystallinity, strain, and defects [13]. Constructive interference of an X-ray beam at specific angles from lattice planes within a sample gives rise to PXRD peaks. The intensities of these peaks are determined by the distribution of atoms throughout the lattice. Consequently, the PXRD pattern serves as a distinctive identifier of the periodic atomic arrangement characteristic of a given material. The information it offers as a bulk approach can be matched with microscopy data to investigate if microscopic observations on a limited number of particles are typical of the bulk of the sample.

The Scherrer Eq. (1) [12, 25] describes the broadening of a peak at a particular diffraction angle (θ), as it relates the crystalline domain size (D_p) to the width of the peak at half of its height (β). The X-ray wavelength (λ) is a constant that depends on the type of X-rays used.

$$D_p = \frac{0.94\lambda}{\beta_{1/2}\cos\theta} \quad (1)$$

The X-ray diffraction pattern of silver sulfide (Ag_2S) QDs at 110, 112, 122, 031, 200, 131, 211 is shown in Fig. 3 [24], which is similar to the reference data (JCPDS-01–075–1061).

Due to the influence of instrumental factors that surpass the actual size of the particles, the Scherrer equation becomes increasingly unreliable toward the region of interest since peak broadening becomes the overriding factor. Crystalline domains exhibiting a size reduction from 50 to 25 nm contribute to peak broadening to a greater degree than what is typically observed. Additional reductions in the size of the crystalline domains significantly enhance peak broadening. When crystalline domain sizes are less than 10 nm, the consequent peak broadening diminishes signal intensity, and peak overlap makes accurate differentiation challenging. Crystalline domain sizes of 5 nm and smaller make the signal-to-noise ratio yield broad signals, resulting in signal loss and increased noise. Beyond this range, the analytical challenge stems from oversized, broad peaks. Dissimilar morphologies among the same material

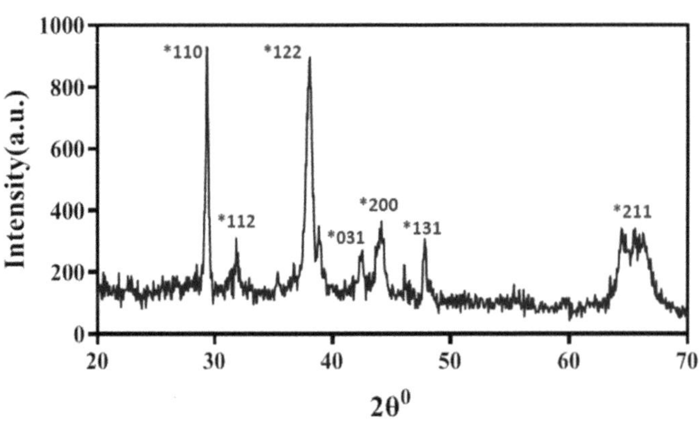

Fig. 3 XRD patterns of silver sulfide quantum dots synthesized via green chemistry using tarragon. Reproduced from Mohammadi et. al. [26] with permission under the terms and conditions of the Creative Commons Attribution 4.0 (CC BY) license

contribute to the widening of XRD peaks. Although the center of the peaks (values of the X and Y axes) remains constant, the intensity of the peak changes (values of the Y axis).

2.4 Dynamic Light Scattering

Dynamic Light Scattering is a technique that measures the scattered light fluctuations in a colloidal suspension to obtain accurate information about nanoparticle size [4]. While electron microscopy can display nanoparticle shapes and sizes, DLS better indicates nanoparticle size and size polydispersity. DLS measures the Brownian motion of particles suspended in a fluid. The technique illuminates these particles with coherent monochromatic laser light, and the fluctuations in the scattered light intensity are analyzed to determine the diffusion of small particles and, thus, the diffusion coefficients of the particles in the fluid. An autocorrelation function for analyzing those fluctuations provides the diffusion coefficient (D). With known temperature (T) and viscosity of the fluid (η), the hydrodynamic particle diameter (d) is estimated using the Stokes-Einstein Eq. (2) [9], where k is the Boltzmann constant.

$$D = \frac{k.T}{3.\,\pi.\eta.\,d} \tag{2}$$

Liu et al. [23] employed TEM and DLS to analyze the size and shape of Carbon Quantum Dots (CQDs). Figure 4a illustrates that the TEM images of the CQDs reveal uniform and monodisperse particles, with sizes ranging from 2 to 4 nm, exhibiting good dispersion. The size distribution of CQDs was assessed via DLS, as illustrated in Fig. 4b, revealing a range of 2–7 nm, with 30% of the CQDs exhibiting a diameter of 3.6 nm. The diameter obtained from DLS was slightly more significant than that from TEM results because the size measured by DLS reflects the average hydrodynamic diameter (d).

This approach significantly aids in characterizing QDs; nonetheless, it presents technical challenges when applied to the surface of quantum dots. Determining the size of quantum dots, particularly the smaller QD variants ranging in 2–10 nm in particle size, is challenging, especially in the presence of larger particles or aggregates in suspension, as signal intensity often diminishes with decreasing particle size. Moreover, the pronounced absorbance band of fluorescent QDs directly influences the size determination using DLS. Fluorescent light is considered noise in the context of DLS since it lacks coherence and adds no valuable information to the measurement. Narrowband filters can selectively filter out fluorescent light to overcome this difficulty. This way, only dispersed light with the same wavelength as the input laser can be analyzed. The quality of correlation functions in DLS measurements, particularly in samples containing QDs, can be significantly enhanced by reducing fluorescent interference.

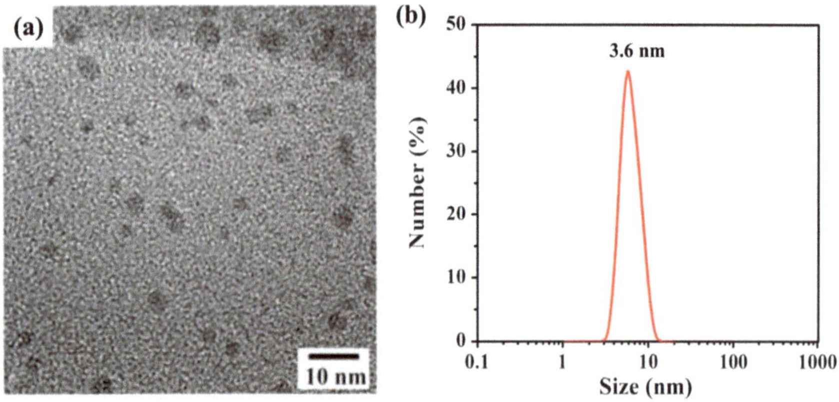

Fig. 4 **a** TEM images and **b** Size distributions of the CQDs samples measured by DLS. Reproduced from Liu et. al. [23] with permission under the terms and conditions of the Creative Commons Attribution 4.0 (CC BY) license

2.5 *Scanning Probe Microscopy*

Atomic Force Microscopy (AFM) is a high-resolution scanning probe microscopy (SPM) technique that allows for the three-dimensional imaging and analysis of surfaces down to the sub-nanometer scale. It operates on the principle of detecting the interaction forces between a sharp probe (tip) and the surface atoms of a sample [15].

The key component of AFM is a microscale cantilever with an ultra-sharp tip, typically made of silicon or silicon nitride, mounted at its free end. This tip is brought extremely close to the sample surface—often within a few angstroms—and raster-scanned across the sample. As the tip interacts with the surface, atomic-scale forces such as van der Waals forces, electrostatic interactions, capillary forces, or contact forces cause the cantilever to deflect. To monitor deflections with great sensitivity, a laser beam is directed onto the rear of the cantilever. A position-sensitive photodetector, such as a four-quadrant photodiode, detects the reflected laser and records the beam's positional changes as the cantilever moves vertically and laterally. This movement is caused by the deflection proportional to the cantilever tilt. The interaction force exerted between the tip and the sample is required to surface map the area of interest. The feedback control loop, which ensures constant interaction force with the AFM tip and sample, is one of the key components of the setup. This is done by first achieving the piezoelectric scanner, whose precision exceeds sub-nanometer measurement units, driving the cantilever upwards and downwards during the scan. A continuous graph outlining the vertical steps necessary to keep the force displays is captured as a topographical image at the sample surface (Fig. 5).

AFM operates under different modes depending on the sample, as well as the information required. They include:

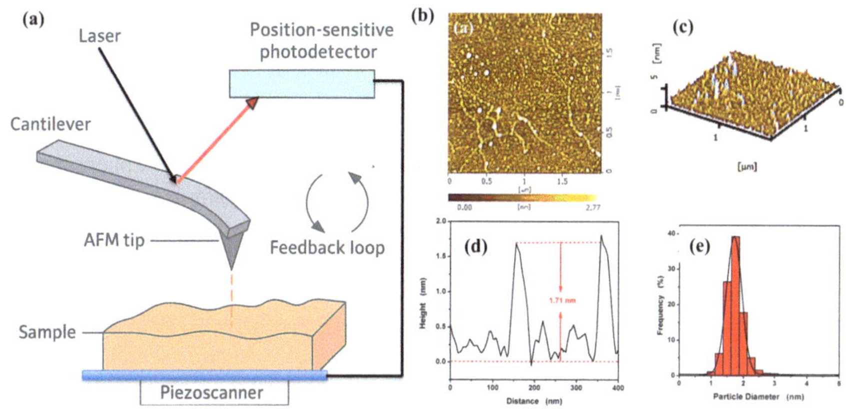

Fig. 5 **a** Schematics of basic components of AFM. The images (**b**) and (**c**) represent 2D and 3D AFM images of Highly luminescent CdS quantum dots capped with thioglycolic acid (TGA@CdS QDs) deposited on silicon substrate, respectively, **d** Height profiles along the line in (**b**). € Statistical analysis of the heights of TGA@CdS QDs measured by AFM. Reproduced from Wang et. al. [31] with permission under the terms and conditions of the Creative Commons Attribution 4.0 (CC BY) license

(a) Contact Mode: The tip is in continuous contact with the surface of the sample. Provides high-resolution imaging but may cause damage to soft samples due to lateral drag forces.
(b) Tapping Mode (Intermittent Contact): The cantilever is oscillated at a frequency near the resonant frequency, which means it will lightly touch the surface at every point. Soft polymers or biological materials can be examined in this mode, which is very delicate and minimizes shear forces.
(c) Non-contact Mode: The tip is positioned above the surface and oscillates with no contact at all, sensing long-range attractive forces. This mode is less damaging but more sensitive to contamination and environmental factors.

AFM is particularly advantageous for the study of quantum dots (QDs)—nanoscale semiconductors whose physical and electronic properties strongly depend on size, shape, and spatial configuration. With AFM, it is possible for researchers to:

- Obtain quantum dot (QD) films or single QD surfaces and their morphologies to the level of precision where their height, diameter, and even roughness can be accurately measured.
- Determine the uniformity and distribution of the sizes, which is important in optical as well as electronic applications such as LEDs, solar cells, and biomedical imaging.
- Understand the pattern of dispersion of QDs on different substrates and identify any agglomeration or clustering that could influence the performance of the device.
- Determine the adhesion and attachment surfaces' mechanical properties by means of force-distance spectroscopy modes, which can define how QDs are bonded to their surroundings or matrices.

- Conduct phase imaging to distinguish different QDs materials or coatings on one surface.

Scanning Tunneling Microscopy (STM) is another SPM method that achieves ultra-high resolution surface characterization, resolving features at the atomic scale on conductors or semiconductors. It is based upon quantum tunneling, a phenomenon where an electron may travel from a sharp conductive tip to a sample surface through space without any physical contact. In STM operation, a sharp metallic tip, frequently constructed from tungsten or platinum-iridium, is positioned at a distance of a few angstroms (0.5–1 nm) from the sample surface. A tunneling current flows as a result of applying a bias voltage to the tip and sample, and this current is known to depend exponentially on the tip-sample separation. The current is said to arise from the tunneling process of wavefunctions belonging to surface atoms, and the tip is known to be extremely sensitive functions of the topographic height of the sample and local density of electronic states (LDOS) of the sample.

Monitored tunneling current can be used to map the surface features as the tip performs a raster scan across the surface, while two principal imaging modes are employed. These modes are:

(a) Constant Height Mode: Changing tunneling current is captured while the tip is held at a fixed height. This mode facilitates quick surface scans but can only be used on perfectly flat surfaces because the tip-surface collision is undesirable.
(b) Constant Current Mode: The tip's vertical position is continuously adjusted by a feedback loop to maintain the tunneling current at a constant value. This mode records the vertical adjustment needed to sustain the current, so precise 3D surface profiles can be calculated.

With a vertical resolution of sub-angstroms (~0.01 nm), lateral resolution of around 0.1 nm, and unparalleled accuracy at surface analysis, STM is among the most powerful tools for atomic-scale surface analysis [23].

Examining quantum dots (QDs) on an atomic scale is made possible by the Scanning Tunneling Microscope (STM). It provides the highest surface and electronic examination of the QDs.

- Topographical Imaging: With the aid of STM, it is possible to get images of up to a nanometer or even a sub-nanometer resolution of QD size, or the shape and area of the QD on material that conducts electricity.
- Electronic Properties via STS: Scanning Tunneling Spectroscopy (STS) is being practiced measuring the tunneling current as a function of the voltage, thus revealing the energy levels, gaps, and quantum confinement of a QD.
- Defects and Surface States: Screening quantum dots for electronic and optoelectronic functionalities with STM allows to identify atomic-level trap states, charge inhomogeneities, and defects that will affect the performance of the QDs.
- QD Arrays and Coupling: STM enables the study of quantum dot (QD) assemblies, revealing the inter-dot spacing and electronic coupling, which is of paramount importance to the design of a device.

Table 1 The table summarizes vital techniques employed to establish the size of quantum dots (QDs)

Technique	Output	Medium	Resolution
TEM	Direct size and shape	Solid	~0.1–0.2 nm
SEM	Surface morphology	Solid	~1–10 nm
AFM	Height, roughness	Solid	~0.1–1 nm (Z)
DLS	Hydrodynamic diameter	Liquid	~1–5 nm accuracy
XRD	Crystallite core size	Powder/film	~1–5 nm

To develop quantum dot technologies, their structural and electronic features need to be probed simultaneously with other measuring techniques. And this is the main advance that STM brings.

Table 1 gives a comparison of major methods for establishing the size of QDs and their resulting outputs, as well as resolution. It aids in method selection according to the requirements of a particular study.

3 Optical Characterization Techniques

The optical characterization methods are important in the study of electronics, photonics, and the structural features of quantum dots due to their size-dependent optical properties stemming from quantum confinement. Their optical behaviors are critical because they permit the control of emission and absorption attributes just by changing the dots' size, especially due to the size-dependent phenomena. Thus, quantum dots might be synthesized for various purposes ranging from bioimaging to optoelectronic devices. This degree of flexibility is fundamental in modern technologies and makes them very useful.

Some of the fundamental optical characterization methods for quantum dots include the following:

- UV-Vis Spectroscopy—Determines the value of the bandgap and estimates the particle size from the absorption spectra.
- Photoluminescence (PL) Spectroscopy—Measures the wavelength and intensity of emission, along with quantum yield.
- Raman Spectroscopy—Provides information on the vibrational modes and the crystal structure.
- Fourier Transform Infrared (FTIR) Spectroscopy—Determines the surface ligands and the bonding surroundings.

3.1 UV-vis Spectroscopy

Researchers investigating semiconductor quantum dots (QDs) benefit from UV-Visible (UV-Vis) spectroscopy due to its simplicity and non-destructive nature, along with the sensitivity to quantum confinement effects. Exemplifying that principle, electrons within a sample can absorb specific energy at wavelengths while being bombarded with UV (200–400 nm) or VIS (400–800 nm) light. In the case of quantum dots, absorption translates to boosting electrons in the valence band to the conduction band, which results in the formation of an exciton (electron-hole pair).

With quantum confinement, size becomes critical to the energy bandgap of QDs, which becomes highly responsive—smaller QDs possess a greater bandgap and capture light within shorter wavelengths, while larger QDs demonstrate red-shifted absorption due to having a smaller bandgap. This property enables inferring the size of the quantum dots, especially in solution-phase systems, through UV-Vis spectroscopy. The average particle size, which can be quantitatively estimated from the bandgap energy, is best represented by the first excitonic peak or absorption edge in the UV-Vis spectrum. This shift can be empirically correlated with particle size using known models or calibration curves for specific materials (e.g., CdSe, PbS, ZnS).

The absorption behaviors of quantum dots follow the **Lambert–Beer law**, expressed as:

$$A = \varepsilon \cdot c \cdot l$$

where:

- A is the absorbance at a given wavelength,
- ε is the molar extinction coefficient ($L{\cdot}mol^{-1}{\cdot}cm^{-1}$),
- c is the concentration of the absorbing species ($mol{\cdot}L^{-1}$), and,
- l is the optical path length (cm).

This law empowers the researchers to not only track the changes in size and bandgap of QD, but also to measure the concentration of quantum dots in a colloidal suspension as long as the molar extinction coefficient is provided.

For instance, in the case of CdSe quantum dots (Fig. 6) [16], empirical relations between the first excitonic absorption peak and the particle diameter make it possible to estimate the size and concentration in one measurement. In addition, the shape and position of the absorption peak reveal information on the surface coverage and the monodispersity of the QD. Well-defined narrow peaks are associated with a uniform size distribution, while broad or shifted peaks are indicative of aggregation, size-polydispersity, or surface defects.

While UV-Vis spectroscopy may lack detailed structural or morphological insight, like shape and crystallinity, it is crucial for monitoring shifts in the optical bandgap, evaluating quantum confinement effects, and assessing the purity of a sample. This

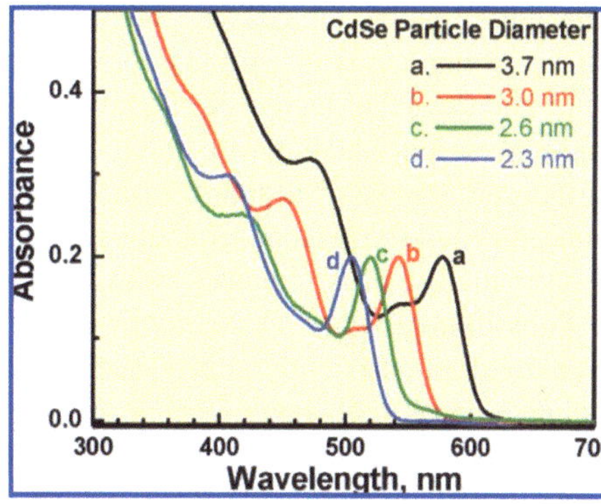

Fig. 6 Absorption spectra of 3.7, 3.0, 2.6, and 2.3 nm diameter CdSe quantum dots in toluene. Reprinted with permission from Kongkanand et. al. [18]. Copyright 2018 American Chemical Society

feature makes it useful for the rapid monitoring of optical transitions, which is important during the synthesis, processing, and use of QDs in devices such as those in optoelectronics, photovoltaics, and bioimaging.

3.2 *Photoluminescence Spectroscopy*

Semiconductor nanomaterials, especially quantum dots (QDs), have their electronic structure and optical quality evaluated through the widely used technique called photoluminescence (PL) spectroscopy. It is an optical characterization technique that explains the underlying processes of light emitted from a material after the substance has been subjected to light [20]. This phenomenon provides a great deal of information on actual quantum dots, including their energy band structure, defect states, and surface characteristics.

To create and plot a PL spectrum, which is a graph or plot of the intensity of light emitted against its wavelength or photon energy, a photographic emission process is induced, which is the reaction of putting a stimulus in the form of light to the material under evaluation. In response to the stimulus, energy is released in the form of light through the process of photoluminescence. Enhanced forms of energy are put forth after the system becomes relaxed in the case of quantum dots, when electrons and holes recombine. The electrons are accepted back into the valence band, and their excitons, which are electron-hole pairs, are yielded. Before returning to the valence band, in levels above the conduction band, the electrons are deposited on the energy levels in an upper layer where they must emit energy in the form of photons during recombination, after reaching the valence band.

Due to quantum confinement in quantum dots (QDs), the photoluminescence (PL) emission wavelength is highly dependent on the size of the QDs. Smaller QDs contain a larger bandgap, which allows emission at shorter wavelengths, while larger QDs will emit at longer wavelengths. This phenomenon of size-dependent emission shifts helps in estimating the QD size and its distribution; hence, uniformity of the sample can be evaluated via PL spectroscopy (Fig. 7).

In addition, the degree of crystallinity can also be monitored by analyzing the shape, position, and intensity of the emission peak, along with the traps on the surface and the passivation layer on the surface. For instance, a broad or red-shifted emission peak suggests non-radiative recombination center surface defects, whereas a sharp, intense peak indicates strong surface passivation with monodisperse QDs.

Photoluminescence spectroscopy, along with time-resolved PL (TRPL), is often used to explore carrier dynamics, such as exciton lifetimes and radiative versus non-radiative decay pathways, which are crucial for optimizing QDs in light-emitting devices, photovoltaic cells, and biological imaging [34].

As stated previously, surface quality, the optical bandgap, and PL-quantum interdot transitions in quantum dots can be assessed using PL spectroscopy. Therefore, it is non-destructive and very sensitive. In addition, its capability to investigate photophysical processes makes it critical in nanomaterials research and optoelectronic device developments.

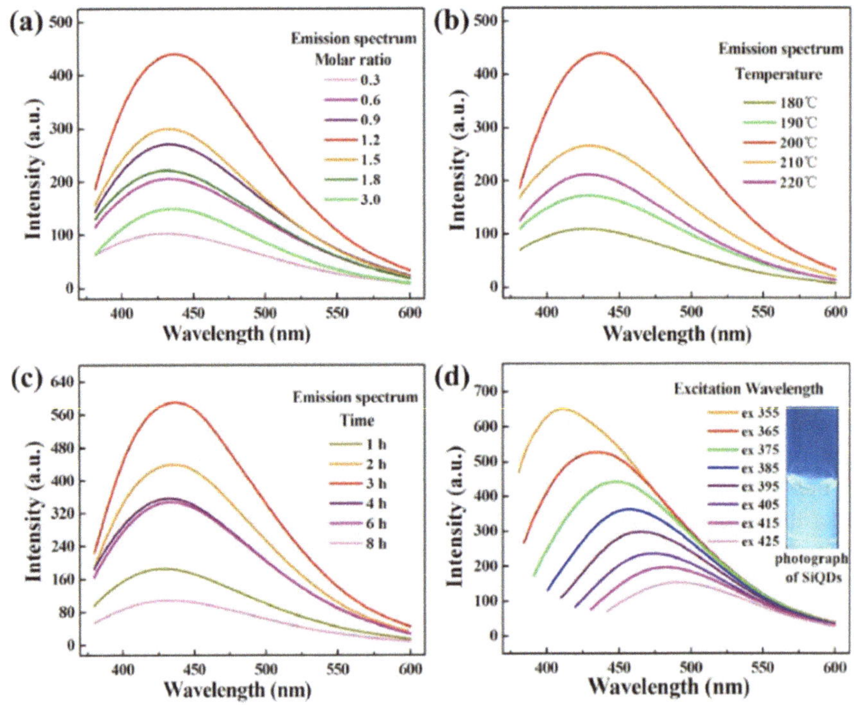

Fig. 7 (A) PL spectra of silicon quantum dots at varying **a** molar ratios, **b** temperatures, **c** reaction times ($\lambda_{ex} = 365$ nm), and **d** excitation wavelengths (355–425 nm). Reproduced from Ji et. al. [14] with permission under the terms and conditions of the Creative Commons Attribution 4.0 (CC BY) license

3.3 Raman Spectroscopy

Raman spectroscopy is a form of vibrational spectroscopy that informs the user about the molecular vibrations and the structure of the material, including its chemistry, crystallinity, and materials [20]. This form of spectroscopy is efficient in describing nanoscale materials, which consist of quantum dots (QDs), carbon nanostructures, and intricate molecular systems. This form describes the identity and energies of bonds in the molecules to the frequency of electromagnetic radiation. This technique identifies vibrational transitions in molecules. Light energy in Raman spectroscopy is always multiplied, which means that more energetic light is produced than consumed. A laser is commonly used to give monoenergetic light, which is required for Raman spectroscopy.

Nevertheless, only a tiny fraction of photons, approximately 1 in 10^6–10^8, interact with the molecular vibrational modes of the material and scatter inelastically. In this process, some would lose energy, while others would gain energy. This creates a phenomenon where the light that is scattered changes its wavelength, or frequency, which is referred to as the Raman shift [15]. The difference in energy between the photons striking and those that are scattered is equal to the vibrational energy of the molecules existing within the sample. There are three forms of Raman scattering (Fig. 8) [27].

- Stokes scattering: The scattered photon is of lesser energy than the incident photon (the electron moves to a higher vibrational level).
- Anti-Stokes scattering: The scattered photon is of greater energy compared to the incident photon (the electron starting from an excited vibrational state migrates down to a lower state).
- Rayleigh Scattering: This is an elastic scattering with no energy alteration.

With quantum dots, Raman spectroscopy is deftly applied to gauge lattice dynamics, phonon confinement, phonon modes, and crystalline structure, all of which impact the optoelectronic properties of these nanostructures. The Raman shift offers spectral identifiers for materials that reveal their composition, crystal phase, and surface phonon modes as they relate to the particle size, shape, and surface chemistry. Probing phonon confinement effects, size-dependent shifts, and broadening of Raman peaks, indirect but dependable data regarding the QD dimensions and structural order are provided. Multiple pieces of literature highlight the capabilities of Raman spectroscopy to capture doping and surface passivation in quantum core-shell quantum dot systems [30]. Besides, spatially resolved studies of QD thin films with detailed crystallinity and defect mapping are accessible with Raman mapping methods. Under ambient environmental conditions, other than with XRD, Raman Spectroscopy does not require extensive sample preparation, making it ideal for routine studies of surfaces and interfaces of QD-based nanocomposites. Raman spectroscopy enhances comprehension of the quality of quantum dots, their growth mechanisms, and their integration into optoelectronic devices by improving

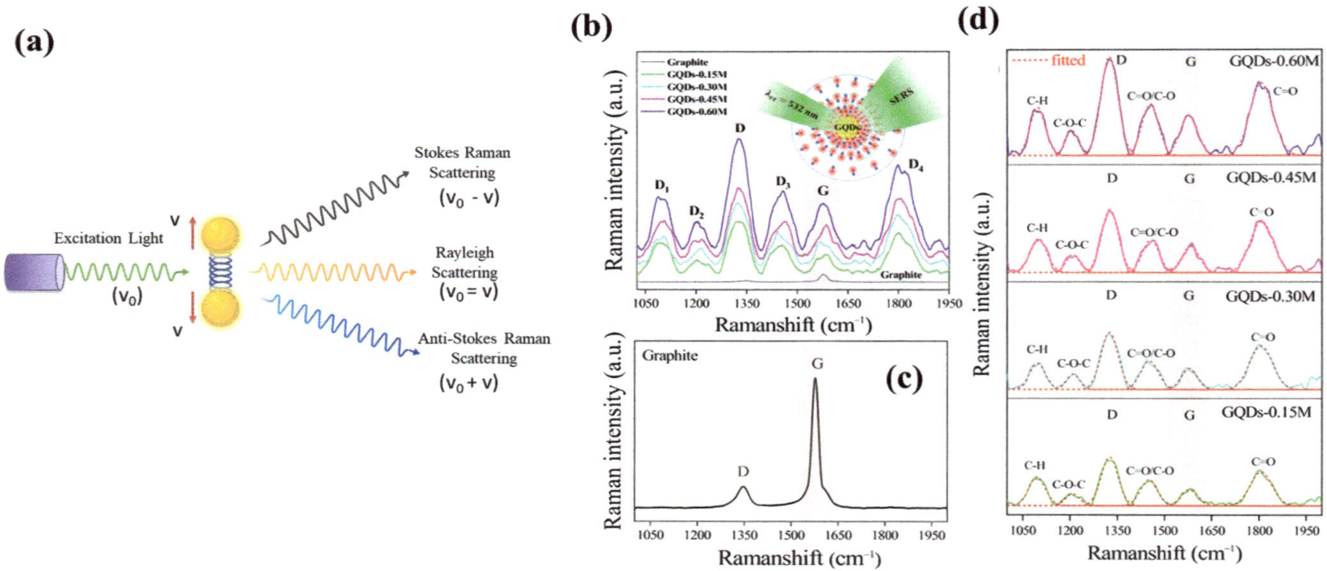

Fig. 8 **a** Types of Raman scattering when light interacts with a molecule. **b** Raman spectra of GQDs at various electrolyte concentrations with **c** Raman spectrum of the normal graphite for comparison, and **d** fitted Raman spectra of GQDs with a Gaussian function to obtain the peak positions. Reproduced from Panyathip et. al. [26] with permission under the terms and conditions of the Creative Commons Attribution 4.0 (CC BY) license

the understanding of the crystallinity, chemical bonding, and nanoscale stress/strain at which quantum dots are comprised.

Surface Enhanced Raman Scattering (SERS) has developed as an advanced application of Raman spectrometry, as it allows for the ultrasensitive detection of the individual molecular vibrations of a sample when a signal gain of up to 10^8 is applied. This signal gain is mainly due to the Localized Surface Plasmon Resonance (LSPR) produced by metallic nanostructures, usually gold (Au), silver (Ag), or copper (Cu), which amplify the electromagnetic field in the vicinity of the sample surface. In the last few years, there has been significant interest in SERS concerning the characterization and functional analysis of QD [7], particularly in hybrid systems in which QDs are integrated with plasmonic materials. Even though QDs in themselves demonstrate rather weak Raman signals, the integration with plasmonic nanostructures facilitates the enhancement of Raman signals to such a level that the study of surface ligands and interfacial interactions. Many investigations have used SERS to study the molecular and passivation coatings on QDs, which provide considerable information about their chemical stability and dynamics of ligand exchanges.

The enhancement mechanism within SERS-QD systems has been attributed to a combination of electromagnetic amplification by the metallic substrate and chemical enhancement from charge transfer between the QDs and surface-bound species. This combination enables SERS to be uniquely useful for non-destructive real-time probing of nanoscale surface and photochemistry, as well as structural change processes [7].

3.4 *Fourier Transform Infrared Spectroscopy*

Fourier Transform Infrared Spectroscopy (FTIR) is a commonly used method in material science and nanotechnology for determining functional groups, surface chemistries, and molecular interactions [6]. The principle upon which FTIR operates is that a molecule absorbs infrared (IR) radiation at certain frequencies, which represent the vibrating movements of the chemical bonds within the molecule. Each material has a unique fingerprint that can be designated using a diverse set of vibrational spectra, and this fingerprint can be obtained through selective absorption using IR radiation (Fig. 9) [10].

Unlike traditional dispersive IR techniques, where absorption is measured one wavelength at a time, FTIR gathers all wavelengths at once via a Michelson interferometer [17]. This optical component splits the IR beam and recombines it after reflection from a fixed and moving mirror, creating an interferogram which contains all spectral information. A Fourier Transform is performed to change the time-domain interferogram to a frequency-domain spectrum where absorbance (or transmittance) is displayed against wavenumber (usually 4000–400 cm^{-1}) on the axes.

Within the field of quantum dot (QD) nanostructures, FTIR spectroscopy is invaluable for characterizing surface chemistry. It is routinely employed for the detection and the bonding type characterization of organic ligands like thiols, phosphines,

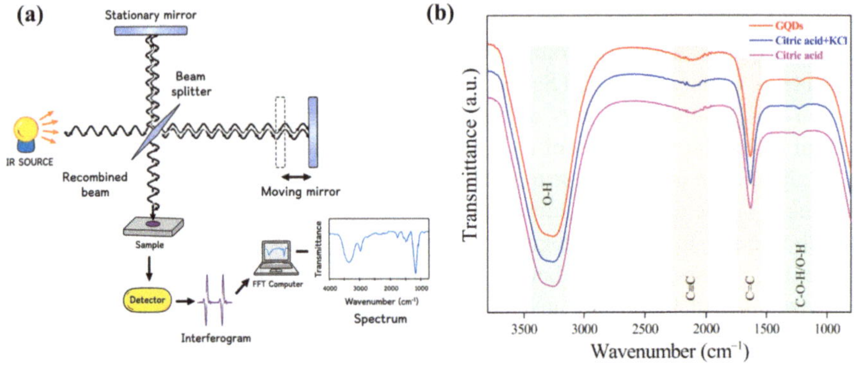

Fig. 9 **a** Schematic representation of an FTIR spectrometer. Reproduced from Faramarzi et. al. [10] with permission under the terms and conditions of the Creative Commons Attribution 4.0 (CC BY) license), **b** FTIR transmittance spectra of graphene quantum dots (GQDs) at different electrolyte solutions. Reproduced from Panyathip et. al. [26] with permission under the terms and conditions of the Creative Commons Attribution 4.0 (CC BY) license

carboxylic acids, or amines, which act to stabilize QDs during their synthesis and control their solubility and multifunctionality. Moreover, FTIR sheds some light on processes such as ligand exchange, surface oxidation, and other chemical alterations done on the surfaces of QDs for bioconjugation or integration into devices.

It is important to note that FTIR or spectrochemical analysis with infrared radiation has a wide range of uses with the non-destructive mode of evaluation that allows the processing of various forms of samples, such as films, powders, and even colloidal suspensions, with little to no preparation needed. Its use for the verification of functionalization and the preservation of the chemical properties of nanomaterials in fundamental science and applied science is exceedingly smooth.

Table 2 outlines important optical characterization methods, including UV-Vis spectroscopy, Photoluminescence (PL) spectroscopy, Raman spectroscopy, and Fourier Transform Infrared (FTIR) spectroscopy of quantum dots. UV-Vis spectroscopy reveals the bandgap as well as the size of the particles in question based on their behaviors toward absorption, while PL uncovers the emission properties and the surface quality through fluorescence. Raman spectroscopy provides data on vibrations, which enable the appraisal of the crystal structure and phonon, while FTIR identifies surface ligands and functional groups, which are crucial for the description of surface chemistry of QDs. All these techniques, in conjunction, deliver reasonably complete information about the quantum dots in question.

Table 2 Summary of optical techniques employed to characterize quantum dots (QDs)

Technique	Purpose	Key information provided
UV–vis spectroscopy	Determines optical absorption and bandgap	Bandgap, particle size estimation
Photoluminescence (PL) spectroscopy	Measures emission properties and quantum yield	Emission peak, size uniformity, surface defects
Raman spectroscopy	Probes vibrational modes and crystal structure	Crystallinity, phonon modes, lattice strain
FTIR spectroscopy	Identifies surface ligands and functional groups	Surface chemistry and molecular bonding

4 Composition Determination Techniques

Understanding the structure-property relationships of Quantum Dots (QDs) is quintessential for mastering their application, especially in intricate architectures such as core–shell or doped systems. For such purposes, **X-ray Photoelectron Spectroscopy (XPS)** and **Energy Dispersive X-ray Spectroscopy (EDX)** are the two most widely used techniques (Fig. 10).

Providing XPS data, inspection, and detailed quantitative evaluation of a sample is done on the order of parts per million in homogeneous bulk materials, which requires considerable surface cleaning. XPS is noted for its high sensitivity to surfaces. Furthermore, it provides detailed quantitative and chemical state information of the elements within a 10 nm thickness. The process of irradiating QDs with monochromatic X-rays ejects core-level electrons. These electrons are then analyzed for their kinetic energy to determine their binding energies for specific elements and their oxidation states. XPS has robustly been employed for the elucidation of ligand binding and surface compositions, as well as oxidation states for QDs. It has a unique part to play in estimating the chemical shifts and stoichiometry of core and shell materials in heterostructure QDs, which have complex surfaces being mastered in surface passivation and functionalization processes [11] (Fig. 11).

In comparison, EDX, which is frequently coupled with Scanning Electron Microscopy (SEM) or Transmission Electron Microscopy (TEM), offers semi-quantitative elemental appraisal for wider areas that cover the surface and bulk of the sample. Characteristic X-rays are emitted from the sample upon interaction with an electron beam and are later detected and analyzed to yield the constituent elements. EDX has considerable application in confirming key element constituents of QDs and in studying elemental distribution in core–shell, alloyed, or doped nanocrystal structures. Although EDX does not have the chemical state discrimination of XPS, it provides quick spatially resolved compositional mapping, which is advantageous with high-resolution electron microscopy.

XPS, together with EDX, provides a comprehensive composition analysis of quantum dots within a single robust framework. Surface chemistry and oxidation states are best examined with XPS, while elemental identification and distribution

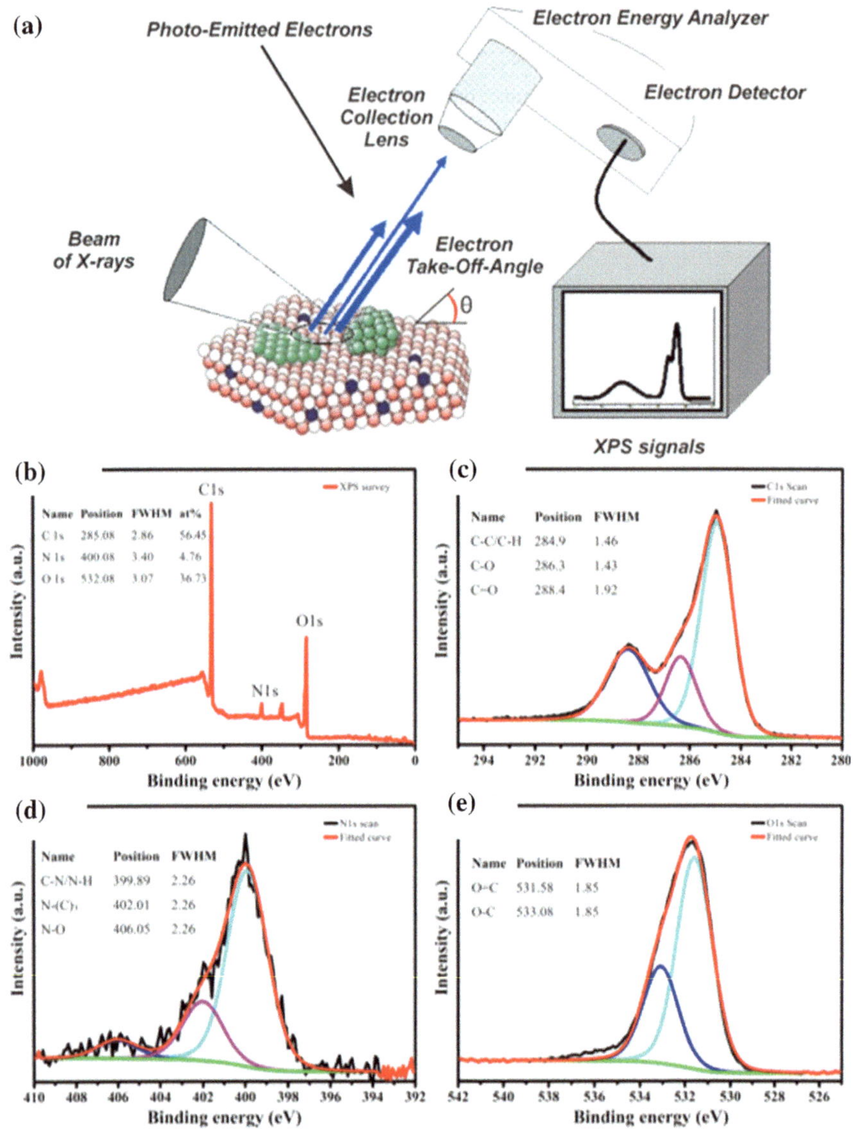

Fig. 10 **a** Schematic representation of an X-ray photoelectron spectroscopy instrument. Reproduced from Bagus et. al. [2] with permission under the terms and conditions of the Creative Commons Attribution 4.0 (CC BY) license), **b** X-ray photoelectron spectroscopy (XPS) survey spectrum of GQDs, **c** C1s, **d** N1s, and **e** O1s high resolution spectra of GQDs, showing oxygen and nitrogen functionalities. Reproduced from Abbas et. al. [1] with permission under the terms and conditions of the Creative Commons Attribution 4.0 (CC BY) license

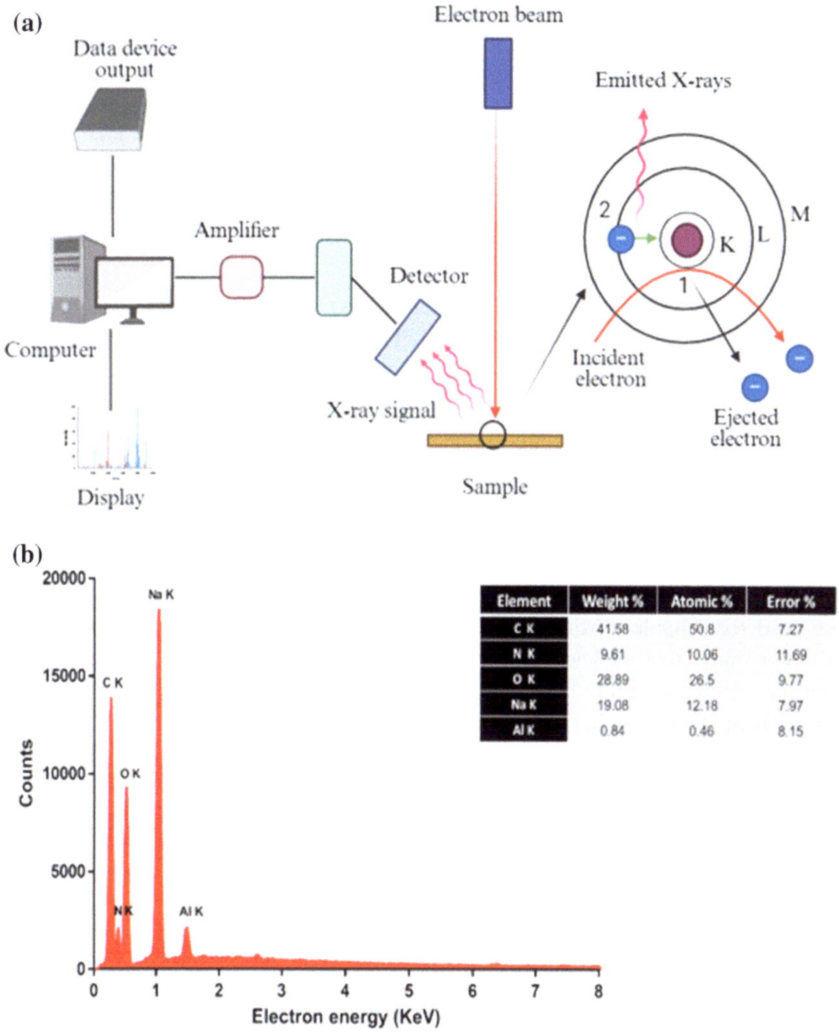

Element	Weight %	Atomic %	Error %
C K	41.58	50.8	7.27
N K	9.61	10.06	11.69
O K	28.89	26.5	9.77
Na K	19.08	12.18	7.97
Al K	0.84	0.46	8.15

Fig. 11 **a** Figure displaying the concepts of energy dispersive x-ray analysis (EDX). (1) An incoming electron stimulates interaction with an atom's inner electron shell, leading to the removal of one electron from the atom. (2) An outer shell subsequently moves down to fill the vacancy in the shell, releasing a photon with the energy equal to the binding energy corresponding to the vacancy. The emitted photon's energy will be characteristic to the element from which the electron was removed. **b** Energy-dispersive X-ray analysis of carbon quantum dots. Reproduced from Fawaz et. al. [11] with permission under the terms and conditions of the Creative Commons Attribution 4.0 (CC BY) license

evaluations are performed with EDX. Their combination enables assessment of material uniformity, doping concentrations, and fidelity to the synthesis, which has been found useful to inform the design and QD-based materials optimization for use in optoelectronics, photovoltaics, and nanomedicine [32].

5 Conclusion

An accurate understanding of quantum dots (QDs) stems from their thorough structural, optical, and compositional features. This chapter strived to explain the core measures of evaluation on QDs which include as their size and shape to spectroscopic methods (UV-Vis, PL, Raman, FTIR) for optical and vibrational property evaluation as well as elemental composition methods (XPS, EDX) for chemical composition assessment. Each method has its unique strengths and limitations, and their combined use provides a comprehensive understanding of QD Properties critical to customizing their functions in diverse and multi-layered applications. Integrating QDs into emerging technologies such as modern optoelectronics, biomedical imaging, sensing, energy harvesting, increases their importance, emphasizing the need for reliable advanced methods for characterization. Therefore, the need to foster innovative advancement in nanotechnology highlights the importance of appropriate tool selection based on the context of application within the system of materials in QD research.

References

1. Abbas A, Tabish TA, Bull SJ et al (2020) High yield synthesis of graphene quantum dots from biomass waste as a highly selective probe for Fe3+ sensing. Sci Rep 10:21262
2. Bagus PS, Freund HJ (2024) X-ray photoelectron spectroscopy as a useful tool to study surfaces and model systems for heterogeneous catalysts: a review and perspective. Surf Sci 745:122471
3. Basak M, Rahman ML, Ahmed MF, Biswas B, Sharmin N (2022) The use of X-ray diffraction peak profile analysis to determine the structural parameters of cobalt ferrite nanoparticles using Debye-Scherrer, Williamson-hall, Halder-Wagner and size-strain plot: different precipitating agent approach. J Alloys Compd 895:162694
4. Brar SK, Verma M (2011) Measurement of nanoparticles by light-scattering techniques. TrAC Trends Anal Chem 30:4–17
5. Chen X, Wang Q, Wang XJ et al (2021) Synthesis and performance of ZnO quantum dots water-based fluorescent ink for anti-counterfeiting applications. Sci Rep 11:5841
6. Dendisová M, Jeništová A, Parchaňská-Kokaislová A, Matějka P, Prokopec V, Švecová M (2018) The use of infrared spectroscopic techniques to characterize nanomaterials and nanostructures: a review. Anal Chim Acta 1031:1–14
7. Doğan İ, Gresback R, Nozaki T, van de Sanden MC (2016) Analysis of temporal evolution of quantum dot surface chemistry by surface-enhanced Raman scattering. Sci Rep 6:29508
8. Ebrahimi S, Souri D, Khezripour A (2022) Strain-size features of bare ZnSe: cu and ZnSe: cu@ ZnS core/shell quantum dots from extensive X-ray diffraction analysis, and their photoluminescence properties. J Lumin 244:118757

9. Edward JT (1970) Molecular volumes and the stokes-Einstein equation. J Chem Educ 47:261
10. Faramarzi B, Moggio M, Diano N et al (2023) A brief review of FT-IR spectroscopy studies of sphingolipids in human cells. Biophysica 3:158–180
11. Fawaz W, Hasian J, Alghoraibi I (2023) Synthesis and physicochemical characterization of carbon quantum dots produced from folic acid. Sci Rep 13:18641
12. Muniz FTL, Miranda MR, Morilla dos Santos C, Sasaki JM (2016) The Scherrer equation and the dynamical theory of X-ray diffraction. Found Crystallogr 72:385–390
13. García de Arquer FP, Talapin DV, Klimov VI, Arakawa Y, Bayer M, Sargent EH (2021) Semiconductor quantum dots: technological progress and future challenges. Science 373:eaaz8541
14. Gokulsaswath V, Suganya G, Monika S, Kalpana G (2024) Investigation of zirconium disulphide quantum dots for supercapacitor applications. Cryst Res Technol 59:2300111
15. Radamson HH, Sychugov I (2023) Scanning probe microscopies (SPMs) part I: atom force microscope (AFM). In: Analytical methods and instruments for micro-and nanomaterials. Springer, pp 243–282
16. Ji Y, Zhang H, Zhang C et al (2019) Fluorescent and mechanical properties of silicon quantum dots modified sodium alginate-carboxymethylcellulose sodium nanocomposite bio-polymer films. Polym (Basel) 11:1476
17. Keresztury G, Chalmers J, Griffith P (2002) Raman spectroscopy: theory. Handb Vib Spectrosc 1:71–87
18. Kongkanand A, Tvrdy K, Takechi K et al (2008) Quantum dot solar cells. Tuning photoresponse through size and shape control of CdSe-TiO2 architecture. J Am Chem Soc 130:4007–4015
19. Kunanta N, Artsang P, Sangtawesin S, Talataisong W, Meemon P (2023) Laboratory implementation of Fourier transform infrared (FTIR) spectrograph using a super-continuum laser. In: Journal of physics: conference series. IOP Publishing, p 012107
20. Kurouski D, Dazzi A, Zenobi R, Centrone A (2020) Infrared and Raman chemical imaging and spectroscopy at the nanoscale. Chem Soc Rev 49:3315–3347
21. Li G, Zhang H, Han Y (2023a) Applications of transmission electron microscopy in phase engineering of nanomaterials. Chem Rev 123:10728–10749
22. Li Q, Anpo M, You J, Yan T, Wang X (2023b) Photoluminescence (PL) spectroscopy. In: Handbook of advanced catalyst characterization. Springer, pp 295–321
23. Liu X, Hao J, Liu J, Tao H (2018) Green synthesis of carbon quantum dots from lignite coal and the application in Fe3+ detection. IOP Conf Ser Earth Environ Sci 113:012063
24. Liu Y, Sheng L, Abdullah M, Shen X, He Y, Liu J, Chen X (2024) High resolution observation of a noble-metal/ZnO-QDs/rGO ternary system through an ultra-thin SiN x window using in situ liquid cell scanning electron microscopy. New J Chem 48:14538–14547
25. Magonov SN, Whangbo M-H (2008) Surface analysis with STM and AFM: experimental and theoretical aspects of image analysis. Wiley & Sons
26. Mohammadi A, Belbasi M, Safari H et al (2025) Green synthesis of silver sulfide quantum dots using tarragon and evaluation of cytotoxic effect on HFF2 cells. Sci Rep 15:10217
27. Panyathip R, Sucharitakul S, Phaduangdhitidhada S et al (2021) Surface enhanced Raman scattering in graphene quantum dots grown via electrochemical process. Molecules 26:5484
28. Prabhash PG, Nair SS (2016) Synthesis of copper quantum dots by chemical reduction method and tailoring of its band gap. AIP Adv 6:055003
29. Sanmartín-Matalobos J, Bermejo-Barrera P, Aboal-Somoza M, Fondo M, García-Deibe AM, Corredoira-Vázquez J, Alves-Iglesias Y (2022) Semiconductor quantum dots as target analytes: properties, surface chemistry and detection. Nano 12:2501
30. Todescato F, Minotto A, Signorini R, Jasieniak JJ, Bozio R (2013) Investigation into the heterostructure interface of CdSe-based core–shell quantum dots using surface-enhanced Raman spectroscopy. ACS Nano 7:6649–6657
31. Wang Z, Xing X, Yang Y et al (2018) One-step hydrothermal synthesis of thioglycolic acid capped CdS quantum dots as fluorescence determination of cobalt ion. Sci Rep 8:8953
32. Weigert F, Müller A, Häusler I, Geißler D, Skroblin D, Krumrey M, Unger W, Radnik J, Resch-Genger U (2020) Combining HR-TEM and XPS to elucidate the core–shell structure of ultrabright CdSe/CdS semiconductor quantum dots. Sci Rep 10:20712

33. Zhang KY, Yu Q, Wei H, Liu S, Zhao Q, Huang W (2018) Long-lived emissive probes for time-resolved photoluminescence bioimaging and biosensing. Chem Rev 118:1770–1839
34. Zhang J, Guo Z, Sun T, Guo P, Liu X, Gao H, Dai S, Xiong L, Huang J (2024) Energy-efficient organic photoelectric synaptic transistors with environment-friendly CuInSe2 quantum dots for broadband neuromorphic computing. SmartMat 5:e1246

Chapter 4
Applications of Quantum Dots in Bioassay and Bioimaging

Srijita De, Saurabh Dubey, Tanusree Ghoshal, Rahul Deka, Nayanjyoti Kakati, Jiwajyoti Mahanta, Prathu Raja Parmar, Sanbed Das, Surjendu Maity, and Dipankar Bandyopadhyay

Abstract In the fight against the ever-increasing number and types of diseases in the modern world, timely diagnosis is perhaps one of the most important aspects to be considered. Understandably, a significant volume of research has been dedicated to assessing, labeling, and imaging at molecular and cellular levels for the prominent visualization of biological processes that can be harnessed toward the quick and sensitive detection of markers specific to different diseases. This has brought about the advent of bioassays and bioimaging techniques involving quantum dots (QDs), which have revolutionized healthcare research. As such, QDs have garnered significant attention as the labeling elements in biosensing and bioanalytical applications. Semiconductor QDs have distinct photophysical characteristics due to their size-dependent color tunability, including intense, broad absorption and bright emission with a narrow wavelength bandwidth. This chapter explores the utilization of QDs in observing biological responses, qualitatively and quantitatively, using bioassays and advanced bioimaging technologies. Different QD-based ultrasensitive bioassays are classified according to their signal response, which includes fluorescent and electrochemical outputs. QDs have also shown great promise in electrochemiluminescence (ECL) and photoelectrochemical (PEC) biolabeling. The synthesis and surface functionalization of QDs, exploring novel bioconjugation approaches for fundamental biomedical research, cellular imaging, and clinical diagnosis, are

S. De · S. Dubey · D. Bandyopadhyay (✉)
Centre for Nanotechnology, Indian Institute of Technology Guwahati, Assam, India
e-mail: dipban@iitg.ac.in

T. Ghoshal · N. Kakati · J. Mahanta · P. R. Parmar · D. Bandyopadhyay
Department of Chemical Engineering, Indian Institute of Technology Guwahati, Assam, India

R. Deka · D. Bandyopadhyay
Jyoti and Bhupat Mehta School of Health Sciences and Technology, Indian Institute of Technology Guwahati, Assam, India

S. Das
Department of Chemical Engineering, Indian Institute of Technology Bombay, Maharashtra, India

S. Maity
Duke University School of Medicine, Duke University, Durham, NC, USA

T. Bhuyan et al. (eds.), *Multifunctional Quantum Dots*, Materials Horizons: From Nature to Nanomaterials, https://doi.org/10.1007/978-981-92-0172-3_4

also elucidated. Additionally, the role of QDs as energy transfer (ET) donors and acceptors is explained within a bioanalytical context. Finally, a future perspective on the *in vivo* applications of biocompatible QDs, specifically in the human body, is discussed thoroughly for the state-of-the-art assessing-imaging methods and real-life clinical applications.

Keywords Quantum dots · Bioassay · Bioimaging · Biomarkers · Nanomaterials · Energy transfer · Bioconjugation

1 Introduction

Over the years, technology has become an indispensable part of modern human life, especially in the field of healthcare. The advancements in understanding diseases, recognizing the specific biomarkers associated with them, accuracy in diagnosis, labeling the biological process, monitoring the effect of external materials inside a body, and capturing the correct response for future analysis have engendered the leap biomedical science has been achieving recently. In this regard, visualization is one of the most necessary aspects of disease detection to understand a biological process at a molecular level [8]. Thus, of late, fluorescence-based technological imaging has garnered significant attention. For example, organic fluorophores [8, 13], organic dyes [43, 103], green fluorescent protein (GFP), and its various mutants [117] are conventionally used for visualizing complex processes in cells. Further, photoactivatable fluorescent proteins [43, 93, 103] and nucleotides have also been developed for high-resolution imaging of intracellular molecules [14, 97]. However, it is difficult to visualize fluorescent tags when the concentration of a biomolecule is very small and imaging is required for a longer time [84]. Additionally, the poor photostability of fluorescent proteins, narrow excitation bandwidth, and interferences in multiplexed experiments make them redundant in biological imaging. One of the main difficulties with fluorescent tags is photobleaching. Further, various photochemical reactions inside the cellular environment end up degrading and permanently altering the molecular structures of organic dyes due to excitation light absorption and converting them to non-fluorescent tags [8, 109]. Prior art suggests that the anti-fade reagents are developed to reduce photobleaching, but their toxicity in living cells makes them incompatible [8, 142].

The evolution of quantum dots (QDs) offers an efficient alternative to traditional fluorescent proteins and organic dyes [142]. They are nanocrystalline colloidal semiconductors possessing exclusive optical properties [37]. The QDs are bright and photochemically stable fluorophores with a wide excitation wavelength spectrum. Depending on the particle size, the absorption wavelength ranges from the ultraviolet to the infrared spectra [2, 8, 20, 84]. They typically consist of II-VI (ZnO, ZnS, CdS, CdSe, or CdTe), III-V (GaN, GaP, GaAs, InP, or InAs), and IV-VI binary and alloyed semiconductor materials with a 3D confinement (1–20 nm) [47, 142]. Other distinguished characteristics of QDs include negligible photobleaching, prominent optical response [32, 84, 127], better quantum yield, along with long fluorescence response [84], dimension-tunable emission spectra coupled with broad range of absorption spectra [18], quantum confinement effect [8], and concurrent excitation of multiple QDs using a single light source [46]. These features of quantum dots ensure reliability in long-term imaging studies, track disease progression and cell behavior, and render them crucial to the biomolecular imaging world as well as in medical science.

The QDs provide exceptional biocompatibility in bioimaging, enabling long-term live-cell imaging, in vivo studies, and high-resolution visualization [108]. QDs possess versatile optical properties apart from fluorescence, such as chemiluminescence (CL), electrochemiluminescence (ECL), phosphorescence, and photoluminescence (PL) [66, 67]. Their versatility extends to advanced bioanalytical techniques, such as Förster Resonance Energy Transfer (FRET) and Charge Transfer (CT) sensing, which enhance the precision of molecular interaction studies [48]. In multiplexed imaging, multiple molecules are labeled with different quantum dots. Their emission spectra are so sharp and narrow that QDs produce non-overlapping colors for imaging [2, 8]. Biocompatibility is a prerequisite for using QDs in biomedical and biological research. Despite initial concerns about biocompatibility and toxicity, advancements in surface functionalization, encapsulation, bioconjugation, and alternative materials, such as carbon and graphene QDs, have addressed these challenges [142]. Consequently, QDs are now pivotal in biological assays, including DNA sorting, immunoassays, and FRET-based assays, to name a few.

This chapter depicts the potential of quantum dots (QDs) in advancing bioassay and bioimaging technologies for healthcare diagnostics. The objectives focus on synthesizing biocompatible QDs using advanced functionalization techniques. The chapter will also discuss optimization techniques for bioconjugation strategies that minimize toxicity and improve stability. The study will summarize ultrasensitive QD-based bioassays for different materials. It will also explore QD applications in live-cell imaging, single-molecule tracking, and in vivo studies while developing multiplexed imaging protocols for simultaneous monitoring and visualization. Additionally, the chapter investigates the use of QDs in advanced bioanalytical techniques to understand molecular interaction, as represented in Fig. 4.1. Subsequently, the chapter will explain the limitations of QDs in biomolecular imaging, comparative analysis, and some of the alternatives for progression in this field. Eventually, the chapter will end with a summary of the entire discussion and shed light on future prospects regarding the implementation of QDs in the domain of biomedical processes and healthcare.

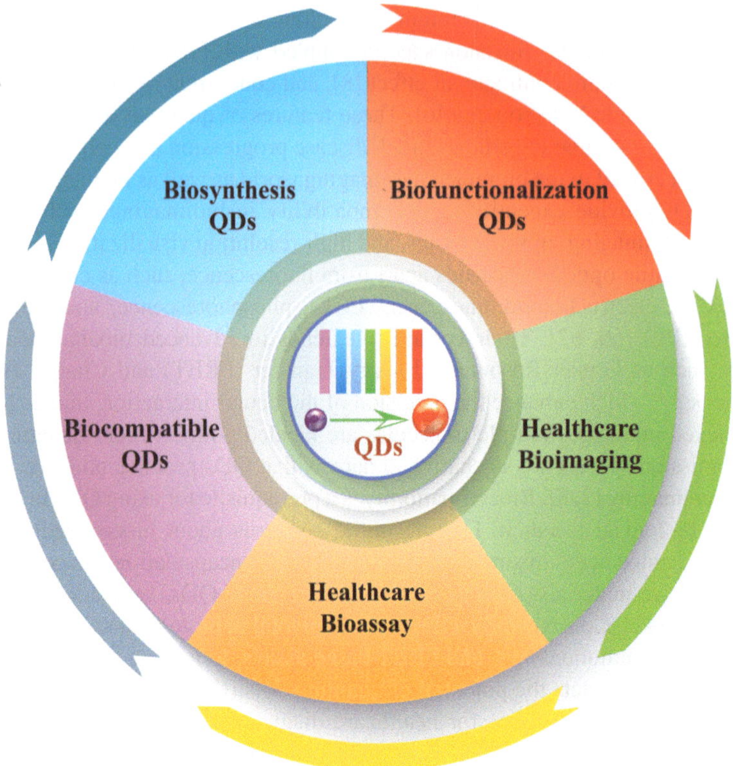

Fig. 4.1 Schematic of biosynthesis and biofunctionalization of biocompatible quantum dots and their applications in bioassay and bioimaging

2 Overview of Quantum Dots

2.1 Definition and Basic Properties

Quantum dots are ultra-small semiconducting nanocrystals manufactured from materials like silicon (Si), cadmium selenide (CdSe), or cadmium sulfide (CdS). Measuring between 2 and 10 nm, approximately the width of two to five silicon atoms, they are tiny, but their properties are anything but ordinary [33, 34]. Quantum dots absorb light and re-emit it in distinct colors dictated by their size and composition. These remarkable properties are due to their nanoscale size, which leads to the 'quantum confinement effect'. Quantum confinement in semiconductor nanocrystals arises from the spatial restriction of excitons—electron—hole pairs bound by Coulombic forces in semiconductors. When the size of a nanoparticle approaches the exciton Bohr radius, which is the average distance between the electron and hole in an excited state, the movement of these charge carriers becomes constrained, giving rise to quantum confinement effects. This phenomenon causes nanocrystals to absorb higher energy than their bulk counterparts, resulting in light emission at shorter wavelengths [8]. As a result, smaller quantum dots emit higher energy blue light, while larger ones generate lower energy red light [104], depicted in Fig. 4.2. It is analogous to striking bells of different sizes—a larger bell makes a deep, low-energy sound, while a smaller one rings with a piercing, high-energy tone.

This size-dependent emission makes quantum dots adaptable in applications that need fine color tweaking [9, 128]. They are perfect for a range of applications because of their tunability, including multiplexed imaging and particular bioimaging jobs that call for exact control over emission wavelengths. Their wide absorption spectra also contribute to their versatility, enabling the simultaneous excitation of several QDs of various sizes by a single light source. Moreover, their narrow emission peaks minimize spectral overlap, enhancing signal clarity and sensitivity, a crucial factor for multiplexed detection. Beyond their capacity to modify color, quantum dots also claim remarkable photostability and brightness. Unlike typical fluorescent dyes, quantum dots resist fading over time, even under prolonged exposure to light. QDs resist photobleaching and photodegradation. This makes them ideal for long-term imaging, time-lapse studies, and monitoring biological processes over

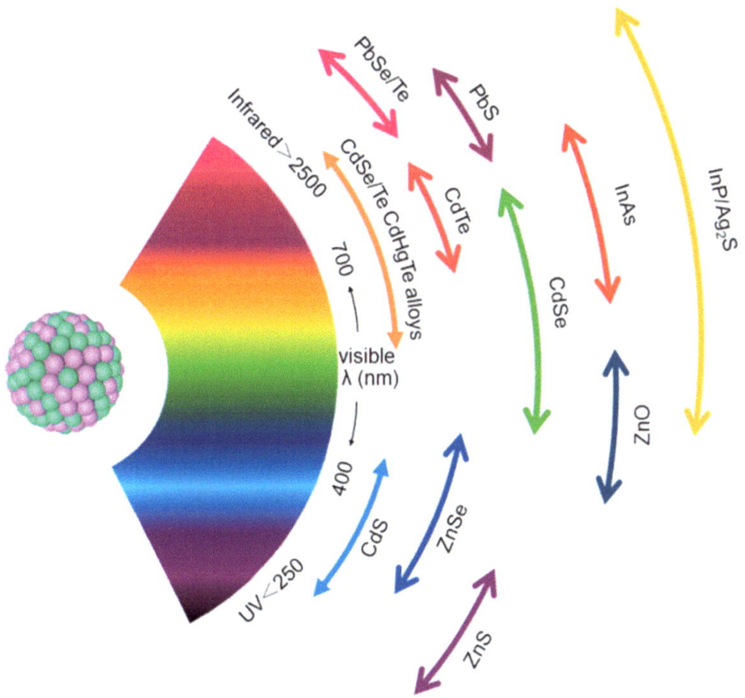

Fig. 4.2 Schematic of QD core materials with their emission wavelengths across the electromagnetic spectrum (Reprinted with permission from [95]). Copyright 2018 American Chemical Society

extended periods without the risk of signal loss. Their high quantum yields, which often exceed 80%, ensure that QDs generate bright, intense fluorescence signals even at low concentrations. Their brightness, coupled with photostability, allows stronger signals, promotes the high-resolution quality of images in medical imaging [56, 140], and ensures vibrant, long-lasting colors in TV displays and solid-state lighting. These unique qualities of size-dependent emission, photostability, and brightness have rendered quantum dots indispensable in cutting-edge technologies. From boosting the efficiency of solar panels to enabling high-resolution bioimaging and providing striking graphics on modern screens, quantum dots are proving that sometimes, the smallest things make the biggest impact [33, 36, 104]. Besides their optical characteristics, the electrical activity of QDs is greatly influenced by quantum confinement. By altering the size of the QDs, their electrical properties can be fine-tuned due to the discrete energy levels that come from their small size. Smaller QDs emit higher energy (blue-shifted) light, while larger ones emit lower energy (red-shifted) light, resulting in an adjustable bandgap. These special qualities increase their efficiency in Förster Resonance Energy Transfer (FRET) systems and other applications needing precise energy transfer [48].

2.2 *Synthesis Methods*

The QDs are mainly synthesized by the following top-down and bottom-up methods [87]

a. Top-down method

   - Wet chemical etching
   - Reactive ion etching
   - Focused ion beam

b. Bottom-up method

   - Wet chemical method
   - Vapor phase methods.

Figure 4.3 summarizes some of the quantum dot synthesis methods, including chemical, electrochemical, and biological synthesis.

2.2.1 Chemical Synthesis

Among the synthesis methods, the water-phase and organic-phase synthesis methods are regarded as the conventional procedures for preparing QDs with good physical, chemical, electrical, magnetic, optical, and morphological properties. The primary methods of QD synthesis by chemical and physical methods include [90, 95].

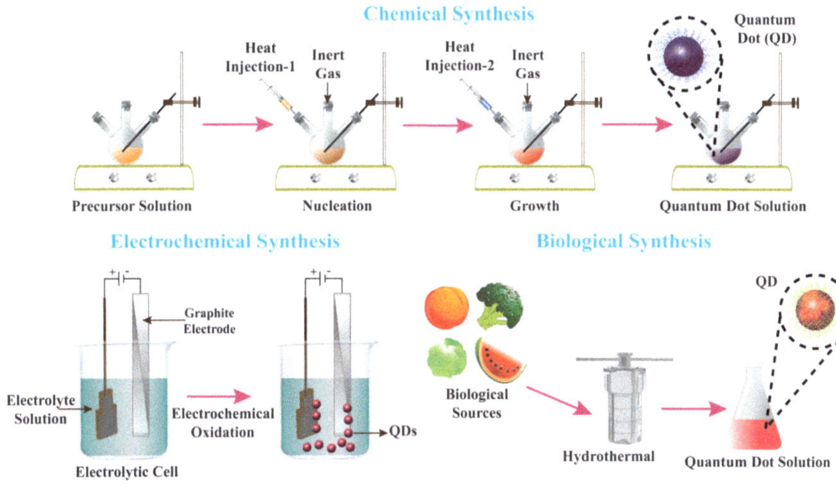

Fig. 4.3 Representation of the various mechanisms for quantum dot synthesis

a. Colloidal synthesis method where solution phase synthesis technique is utilized. Nucleation and development of QDs happen at controlled temperatures and conditions under the solvent phase.
b. The vapor-phase synthesis technique includes chemicals and physical vapor deposition techniques, where semiconducting material is deposited in the gas phase in a relatively high-temperature reaction environment.
c. The electrochemical synthesis method uses electric potential as the external field across the electrolytic reaction mixture, which reduces the metal ions, subsequently leading to nucleation and growth of the QDs.

A considerable body of literature exists on the production of these QDs. The work by Yakoubi and the team demonstrated the synthesis of CdZnS and Cu-doped CdZnS QDs using an aqueous-based method, as represented in Fig. 4.4a. QDs with enhanced properties such as high emissivity could be produced using 3-mercaptopropionic acid (MPA) or N-acetylcysteine (NAC) as a capping ligand [132]. Amirav and the team showed the production of semiconducting QDs using the spray-based method. In these methods, aqueous/organic semiconductor salt is dispersed using a thermospray nebulizer into small droplets. As these droplets move ahead, the saturation of the precursor leads to spontaneous salt condensation caused by solvent evaporation. This solvent vaporization produces one semiconductor drop from one droplet. This method has been applied to various groups of QDs, like CdS, ZnS, and MnS from II-VI, and PbS and MoS_2 from IV-VI group [3]. Another method by Lignos and co-workers synthesized monodispersed PbS and PbSe QDs for various experimental conditions. NIR fluorescence spectroscopy was utilized for real-time assessment and fine-tuning of the synthesized QDs [72]. Previous research showed vapor-phase synthesis of ZnO QDs on Si substrate. Scanning electron microscopy (SEM) and atomic force microscopy (AFM) confirmed the 6 nm height with 15 nm diameter

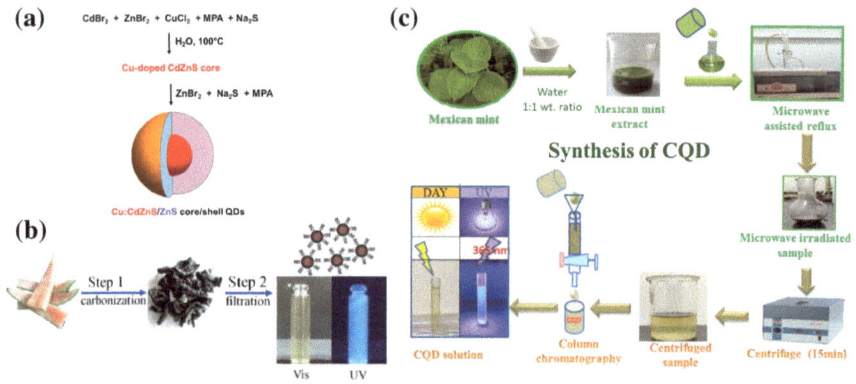

Fig. 4.4 Schematic illustration of **a** chemical synthesis of Cu-doped CdZnS/ZnS core–shell QDs (Reprinted from [132]). Copyright (2016), with permission from Elsevier. **b** Biosynthesis of carbon QDs from watermelon peel (Reprinted from [141]). Copyright (2012), with permission from Elsevier. **c** Carbon QDs biosynthesis from Mexican mint (Reprinted from [7]). Copyright (2021), with permission from Elsevier

ZnO QDs formation. Photoluminescence at room temperature suggests synthesized QDs exhibit UV emission at 3.32 eV. The ZnO QDs bandgap is determined to be 3.41 eV, indicating the quantum confinement effect [77]. Another work uses an electrochemical method for the synthesis to produce 2 to 3 nm of graphene and graphene oxide-based QDs. A weak acid and strong alkali combination was used as an electrolyte. Different alkali concentrations of various reaction mixtures in the microenvironment result in the production of oxygen-rich functional groups on graphene QDs, forming graphene oxide QDs. The work generated surface defects on the graphite rod after heating it at a high temperature [1].

- First, these defects provide more places for exfoliation and oxidation, facilitating electrochemical exfoliation.
- Second, variation in the NaOH concentration in the electrolyte produces GQDs/GOQDs with different mechanical and visual properties.

Similar procedures have been used to produce QDs, like tungsten disulfide and molybdenum disulfide [42, 118].

2.2.2 Biocompatible Quantum Dot Synthesis

With the advent of modern science and medical research, QDs are enormously utilized in healthcare applications. So, because of the toxicity issues, the synthesis of biocompatible QDs is preferred. For QD synthesis, living organisms (prokaryotes to eukaryotes) have already been explored to synthesize biocompatible QDs to be comfortably used in biological and biomedical applications. The mild synthesis condition of biosynthesis methods paves the way for better biostability and other properties of QDs. In this green synthesis route, additional encapsulation or cap exchange steps are unnecessary. So, various steps like water solubilization and biofunctionalization of chemically synthesized QDs are carried out.

There are three ways to make biocompatible quantum dots [108, 133, 134]:

a. QD synthesis in biological systems,
b. QD synthesis in biomimetic systems,
c. Biofunctionalization of chemically synthesized QDs.

The use of biosynthesis as the reaction platform has three important features,

a. Amino acids, proteins, and chelating peptides are naturally occurring binding and metal-reducing agents that may decompose metallic compounds, safeguard the products of their breakdown, and reduce metal ions.
b. Easy platform manipulation employs genetic engineering to express various proteins and modulate the QD evolution, size distribution, and emission of fluorescence.
c. It provides moderate reaction conditions, neglecting high temperature, severe deoxygenation, or harmful chemical solvents.

The synthesis of CdS QDs has been done through various H_2S-producing enzymatic reactions. The hydrolysis of thiophosphate by alkaline phosphatase, where the catalytic oxidation of 1-thio-β-D-glucose takes place by glucose oxidase, and the breakdown of S-adenosyl-L-homocysteine by S-adenosyl-L-mono cysteine hydrolase and methionine γ-lyase happens simultaneously [99–101] to synthesize CdS QDs. Another method for synthesizing CdS QDs involves acetylthiocholine hydrolysis to thiocholine using acetylcholinesterase. In this reaction, H_2S is also produced by the decomposition of sodium thiosulfate. Subsequently, a reaction between the Cd ion and this H_2S produces CdS QDs [101]. On top of that, these enzymatic reactions lead to small biomolecules (e.g., glutathione and phosphate) that can play the role of a stabilizer by acting as a surface ligand. Dameron and his team synthesize the CdS semiconductor crystallites using Candida glabrata yeasts and Schizosaccharomyces pombe with cadmium salts [28]. Similar work has been proposed by Li and the team, where they prepare CdSe QDs with a high quantum yield (4.7%) using genetically engineered yeast cells [71]. Another research team produces protein-capped CdTe QDs by culturing yeast cells with Cd and Te precursors [11, 12].

Some prokaryotes can also be used for QD synthesis in living organisms. A literature work by Sweeney and his team uses *E. coli* to synthesize QDs. The mixture of stationary phase E. coli, Cd, and sulfide precursor has been used to obtain polydisperse QDs with a size range of 2–5 nm. Because of the simple cellular metabolism and single-compartment structure, *E. coli* shows good options for precise size and crystalline tuning of QDs [113]. Genetic engineering developments now offer the production of specific proteins and peptides within *E. coli* cells, showing a new way of controlling the growth and properties of quantum dots (QDs). Chen and team use genetic recombination of *E. coli* cells to express phytochelatin synthase to synthesize CdS nanocrystals. This engineered biofactory of bacterium produces water-soluble CdS nanocrystals of tunable properties [20, 85]. Though in situ, real-time bioimaging is positively assisted by intracellularly synthesized QDs. However, the key limitation of this is subsequent isolation, which is complete and disruptive. So, recent literature by Bao and co-workers has shown an exclusive single-step extracellular synthesis method to make CdTe QDs in the Luria – Bertani medium [11, 12]. Very recently, Chen has demonstrated artificial protein condensates for the production of quantum dots in the bacterium *E. coli*. This is facilitated by overexpressing spider silk protein [22]. This overexpression served multiple roles,

a. **Binding Precursor Ions**: The increased presence of spider silk protein facilitates the binding of essential precursor ions, which are crucial for the subsequent stages of protein condensate formation.
b. **Attracting Components**: The abundant spider silk protein acts as a scaffold, attracting and concentrating other necessary components within the cellular environment.
c. **Inducing β-sheet Formation**: The increased concentration of spider silk protein triggers additional β-sheet structure formation within the protein. These β-sheets are critical for properly assembling and maturing the protein condensates.

This phenomenon allows the production and colocalization of fluorescent QDs to "light up" the condensates and reduce the cytotoxicity of the precursor heavy metal ions and resultant QDs [22].

Carbon quantum dots (CQDs) and graphene quantum dots (GQDs) are another set of biocompatible QDs. Completely nonvulnerable green methods can easily produce these. These methods require one eco-friendly material as a carbon source. Some of the methods reported for the green synthesis of these QDs are as follows:

a. Hydrothermal synthesis method,
b. Microwave synthesis method,
c. Carbonization synthesis method.

Previous research by Liu and the team reported the synthesis of nitrogen-doped carbon quantum dots (N-CQDs) using tomatoes as an eco-friendly raw material. Ethylenediamine and urea have been used as an N-dopant. This method produced N-CQD particles in nano/micro-sized [31]. In 2012, fluorescent water-soluble CQDs were formed from watermelon peel, involving a low-temperature carbonization step at 220 °C for 2 h and subsequent filtration [141], depicted in Fig. 4.4b. Figure 4.4c illustrates the formation of CQDs from the *Plectranthus amboinicus* (Mexican Mint) leaves using a microwave-assisted reflux method with a fluorescence quantum yield of 17% [7]. Another previous literature report uses waste cotton linter as an eco-friendly carbon source. They synthesized water-dispersed CQDs very rapidly using microwave-assisted hydrothermal processes. The synthesized CQDs exhibited a 10.14-nm size [35]. Lemon and onion juice have also been used as natural carbon sources for the synthesis of CQDs with a 6.15-nm size. That is later utilized to determine riboflavin in mineral/multivitamin supplements [88]. Another similar work reports pomelo peels as the carbon precursor that was used to synthesize CQDs at 200 °C using a hydrothermal route. The synthesized QD size is found to be 2–4 nm with a quantum yield of 6.9%. Later, these CQDs were utilized for Hg^{2+} ions in lake water samples [78]. Carbonization is another effective method for the synthesis of QDs. Here, the material is heated in the presence/absence of air at a high temperature. Murugan and team have synthesized CQDs using finger millet ragi (*Eleusine coracana*) as the source material by carbonization at 300 °C. Later, these CQDs are utilized to detect Cu^{2+} ions in actual water samples with a detection limit of 10 nM [89]. To add to that, sweet pepper has also been used as a source of carbon. The CQD synthesis method followed two steps:

a. Low-temperature carbonization of pepper at 180 °C for 5 h.
b. Separation of the product.

Later, hypochlorite is detected using these CQDs in tap water with a limit of detection of 0.05 mmol/L [567, 136]. Another biocompatible QD, graphene QDs or GQDs, is being used in healthcare applications [6, 57]. Guo and team synthesize N—(rhodamine B) lactam-ethylenediamine (RBD)-functionalized GQDs (RBD-GQDs). These QDs act as Fe^{3+} turn-on fluorescent nanosensors in living cells. The detection limit is found to be 0.02 μM. RBD-GQDs can easily penetrate the cancer stem cells (CSCs). As Fe^{3+} ions in these CSCs are very responsive toward RBD-GQDs, a strong orange-red fluorescence of Fe^{3+} binding to RBD-GQDs has been demonstrated as a marker for CSCs imaging [45].

2.2.3 Surface Functionalization and Bioconjugation

Sometimes, chemically synthesized QDs undergo further passivation and functionalization using various methodologies. The main problem with most of the chemically synthesized QDs is the presence of hydrophobic organic ligands on their surfaces. This limits their wide application in biological systems. The use of ligand exchange or amphiphilic molecules as surface attachments has been widely shown in literature as a solution. In one of the previous works of Chan and the team, mercaptoacetic acid was used as a permutable ligand to make biocompatible ZnS-CdSe QDs. Further, these QDs were utilized to recognize the receptors of HeLa cells [19]. Mattoussi and his team synthesizes a dithiolated ligand of dihydrolipoic acid (DHLA) to passivate the QDs to achieve the desired properties and functionalization [83]. Many other mono-thiolated compounds (e.g., mercaptocarbonic acids [50, 75, 120], cysteine [76, 115], artificially thiolated agents (e.g., cyclodextrin [91], silane [41], protein [124], and DNA [86]) have been utilized to solve this water solubility issue of QDs [142].

Covalent conjugation for bioconjugation of QDs with different proteins and peptides has been used since 1998. They contain many carboxyl groups and amino groups that can be easily used to anchor biomolecules on the surface of quantum dots. In order to promote a chemical reaction, carbodiimide reagents like 1-ethyl-3-(3-dimethylaminoopropyl) Carbodiimide (EDC) and dicyclohexylcarbodiimide (DCC) can be used in conjunction with N-hydroxysuccinimide (NHS). An amino and carboxylic acid group are joined in this process to form a stable amide bond [19]. Other than this, many other covalent conjugation methods, like amine and sulfhydryl chemistry, phenylboronic acid and diol chemistry, and aniline-catalyzed hydrazone chemistry have also been utilized in the literature [142]. Yan and his team utilize glucose oxidase (GOD) to conjugate it to Mn-ZnS QDs using EDC/NHS as coupling chemistry. GOD-Mn-ZnS QDs are widely used for most glucose-based bioassays as they do not interfere with autofluorescence or light scattering [126]. Amine- as well as sulfhydryl-reactive conjugation chemistry is used to implant amine and sulfhydryl-comprising antibodies on QDs [17]. The maleimide-modified QDs contain a covalent thioether link with the sulfhydryl groups of antibodies in sulfhydryl-reactive conjugation chemistry. These can directly be used to identify cellular membranes.

Electrostatic interaction is a noncovalent coupling method where two oppositely charged molecules interact with QDs and with each other to facilitate the conjugation process. These electrostatic interactions between biomolecules and QDs are nonspecific [142], and these do not have problems with QD aggregation, unlike covalent conjugation methods. Mattoussi and his coworker uses negatively charged lipoic acid-covered CdSe/ZnS QDs and positively charged maltose binding protein (MBP) to synthesize QD-MBP. The QD-MBP shows good optical properties, stability, and biological activity. Later, this was used to develop a fluorescence resonance Energy transfer (FRET)-based sensor [27]. Bhang and his team demonstrates the use of charged interaction between positively charged CdSe/CdS/ZnS QDs and negatively charged HA. This interaction leads to the formation of HA-QD conjugates. Later, these are applied to mark lymphatic vessels for several days. The surface passivation by dithiol and -NH_2 containing linker molecules at both ends facilitates the positively charged cloud around the synthesized CdSe/CdS/ZnS QDs [15].

3 Quantum Dots in Bioassay

3.1 QD-Based Energy Transfer and Fluorescent Assay

QD-DNA Combination: Two different quantum dots are used to develop a biosensing assay for the specific detection of DNA sequences in nucleic acid, following two different methods. The first method uses two types of QDs with separate emission peak wavelengths. DNA probes are biotinylated and coupled to two different QDs through biotin-streptavidin interaction. The two QD-labeled probes are hybridized at different binding sites of the same target strand of DNA. On the other hand, the second method uses one type of QD and two single-stranded DNA. One strand of DNA is biotinylated, and the other one is tagged with an organic fluorophore. In the final step, quantum dots were added to bind to the biotinylated sandwich structures, creating QD-sandwich nanoassemblies. In the second method, QDs were used as both nanoscaffolds and fluorescent tags, capturing multiple fluorescently tagged hybridization products, which amplifies the target signals in the bioassay [135].

QD-FRET Assay: Protease activity is observed based on the fluorescence and bioluminescence resonance energy transfer (FRET/BRET) of the QDs in a multiplexed and high-throughput protease assay [59]. In 2013, researchers developed a multiplexable Förster Resonance Energy Transfer (FRET) immunoassay using Tb donors and QD acceptors to determine prostate-specific antigen (PSA) in low-volume serum samples. They used two stable, biocompatible QDs with different colors to characterize assays using three types of antibodies [123]. Another research introduces a novel, single-step, and amplification-free FRET assay for detecting microRNAs, crucial disease development and progression biomarkers. By combining advanced biophotonic design with RNA hybridization and ligation techniques, this assay offers precise, multiplexed quantification of miRNAs with high sensitivity and specificity, enabling both large-scale analysis and potential point-of-care applications for medical diagnostics [55]. A single-step multiplexed nucleic acid hybridization assay was developed to quantify DNAs and miRNAs for low concentrations at a 'nanomolar scale' from the same amount of sample used for a single measurement. The study demonstrates the employment of single-time-gated (TG)-FRET-pair multiplexing arrangement for molecular (Tb-to-dye) and nanoparticle (Tb-to-quantum-dot) biosensing. This TG-FRET-based donor–acceptor conjugation is applicable for tuning the photoluminescence (PL) decay and measuring biological recognition. The superior photophysical properties of QDs enable higher-order color-multiplexing and facile conjugation of nucleic acids [96]. In the next study, the author demonstrated that ($\approx$ 6.5 kDa) histidine-tagged very small albumin-binding domain-derived affinity proteins (ADAPTs) can self-assemble to zwitterionic ligand-coated QDs, and these ADAPT-QD conjugates were applicable for the sensitive detection of the human epidermal growth factor receptor 2 (HER2) in serum-containing samples through time-gated Tb-to-QD-FRET-based immunoassays [129]. The same group later developed a DNA hybridization assay by attaching DNA to a QD via a peptide linker by Tb-to-QD-FRET and biomolecular modeling. They investigate the conformation of 31 nucleotides long DNA differentiating between single-stranded DNA (ssDNA) and double-stranded DNA (dsDNA), exploiting the PL effect of QD-DNA conjugates [44]. Hong and the team developed an enzyme-free hydrogel-based QD-DNA FRET sensing assay. Figure 4.5a shows a short single-stranded target DNA catalyzes the initiating cycle of DNA hairpin loops, which then carries a fluorescence quencher into 'Y-shaped DNA' (Y-DNA). These Y-DNAs are quickly self-assembled with ssDNA-functionalized QDs for the construction of QD-DNA hydrogels. This 3D QD-DNA hydrogel enhances the resonance energy transfer from donors to acceptors. As the individual target DNA molecule can open the DNA hairpin loops numerous times, this improves the sensitivity of the QD-DNA hydrogel assay, illustrated in Fig. 4.5b [49].

QD Electrochemical Sensing: Quantum dot-based electrochemical DNA bioassay has also been developed in prior literature. The graphite surface of the screen-printed electrode was immobilized with bismuth citrate for the detection of the C634R

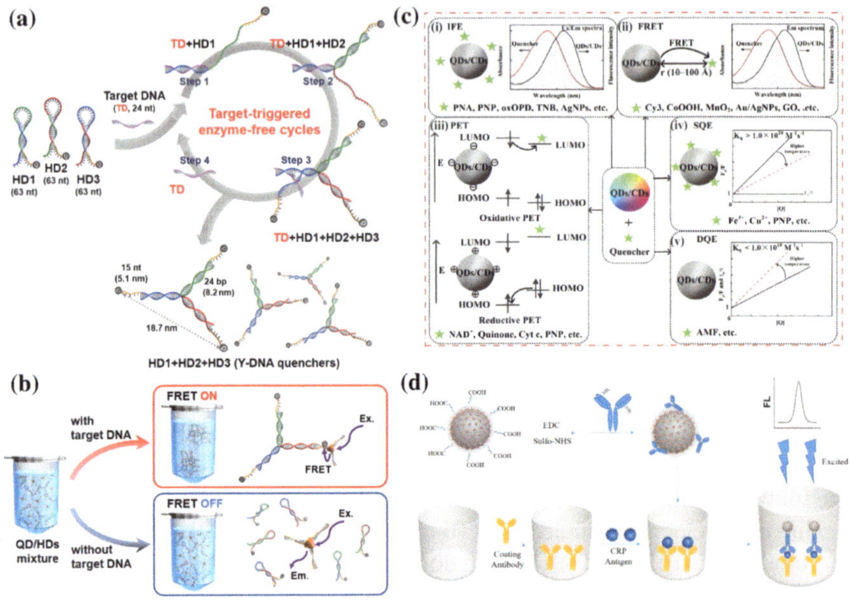

Fig. 4.5 **a** Mechanism and **b** principle of QD-DNA hydrogel-based FRET assay (Reprinted from [49]). Copyright (2021), with permission from Elsevier. **c** Mechanism of QD-based fluorescent assay for monitoring enzymatic activity (Reprinted from [116]). Copyright (2020), with permission from Elsevier. **d** Schematic of QD-based fluorescent immunoassay mechanism (Reprinted from [66]). Copyright (2019), with permission from Elsevier

mutation through hybridization. Subsequently, the surface was modified with biotin-tagged target oligonucleotide and complementary probe DNA with streptavidin-conjugated PbS QDs. The surface of the sensor was utilized as a voltammetric QD transducer, which was based on bismuth nanoparticles. The electrochemical transduction response involved the determination of the Pb(II) released upon acidic dissolution of the QDs, showing a linear relationship of the Pb(II) peak current with logarithmic concentrations of the target DNA [61].

QD Fluorescent Assay: QD-DNA conjugate was used in fluorescence in situ hybridization (FISH) assay to locate specific gene targets in metaphase chromosomes of model organisms for detection. The probe DNA conjugated to QD was hybridized with the target gene, giving a higher sensitivity and signal-to-noise ratio [10]. In 2020, Zhu and the team reviewed the design of different QDs/CDs-based fluorescent assays for monitoring enzymatic activity, namely, inner-filter effect (IFE), Förster Resonance Energy Transfer (FRET), photo-induced electron transfer (PET), static quenching effect (SQE), and dynamic quenching effect (DQE)-based fluorescent assays, as described in Fig. 4.5c. In the IFE mechanism, enzymatic products, such as peptide-modified Au nanoparticles (AuNPs) and thio-2-nitrobenzoic acid (TNB), were used as quenchers in enzymatic reactions, where the absorption range of quencher molecules overlaps with the emission or excitation wavelengths of QDs/

CDs. These absorption signals were converted into fluorescence signals, affecting the quenching efficiency [24, 73, 138]. This IFE mechanism is successfully utilized to monitor α-glucosidase (α-Glu) [53, 62], N-acetyl-β-D-glucosaminidase (NAG) [82], and α-L-fucosidase (Afu) [25] activity.

In the FRET-based sensing mechanism, nanomaterials (AuNPs, MnO_2) and fluorescent organic molecules (Cy3, rhodamine) were deployed as quenchers to facilitate energy transfer with QDs/CDs, which act as energy donors. The transfer of energy happens from the excited state energy donor to the energy acceptor in the ground state through long-range dipole–dipole molecular interactions [54, 111]. Metalloproteinase trypsin [106], matrix metalloproteinase (MT1-MMP) [26], α-Glu [64], and alkaline phosphatase (ALP) [74, 105] are some of the enzymes whose enzymatic activities were observed through FRET-based enzymatic fluorescent assay. Trypsin was also detected using PET-based enzymatic fluorescent assay, where graphene QDs (GQDs) are employed as electron donors and cytochrome c (Cyt c) is the quencher or electron acceptor [70]. The electron transfer mechanism in PET depends on the lowest unoccupied molecular orbitals (LUMO)-highest occupied molecular orbitals (HOMO) energy level difference between the electron donor and acceptor [79, 112]. In oxidative PET, the electron transfer happens at the LUMO energy level, whereas the HOMO level electron transfer takes place in reductive PET. SQE and DQE show contrasting characteristics. QD/CDs interact with a quencher (enzymatic product) to form a ground-state complex, which is non-fluorescent and falls back to the ground state in SQE without emission of photons. However, in DQE, QDs/CDs return to the ground state after a collision with the quencher, transferring electrons between them [73, 144]. β-galactosidase (β-Gal) is detected via host–guest recognition in SQE-based assay [114], and amifostine (AMF) and ALP were quantified by DQE-based enzymatic fluorescent assay [58, 116].

Another QD-DNA FRET-based bioassay was created to detect nerve agents, where QD-DNA bioconjugates form a covalent link to quencher molecules to form FRET donor–acceptor pairs for quantifying a metabolite, namely methylphosphonic acid (MePA), acting as a nerve agent [29]. Recently, a bipolar electrochemiluminescent (BPE-ECL)-based bioassay was developed for the detection of hepatitis C virus envelope protein E2 [39].

3.2 QD-Based Immunoassay

The incorporation of Quantum dots in fluorescence-linked immunosorbent assay (FLISA) has been receiving significant attention due to their high sensitivity and marvelous optical properties. Here, antigens or antibodies are integrated with QDs as targets to fabricate bioassays, where QDs are also employed as fluorescent probes to capture the change in fluorescence intensity. QD-FLISA consists of two methods: the sandwich immunoassay and the competition immunoassay. In the sandwich one, a microplate is used to capture a known antibody for binding a target antigen sample. The particular QD-antibody is then added to combine with the specific antigen. The antigen attaches with two units of antibody at the same time to build a sandwich structure. The microplate is coated with an antigen or antibody and follows the previously explained steps in competitive immunoassay, only differing in the competition between the test sample and the coating material. The sandwich immunoassay promotes a positive correlation, where fluorescence intensity increases with the concentration of antigen, whereas the competitive immunoassay portrays a negative correlation [80]. C-reactive protein (CRP) detection is performed via a sandwich immunoassay, explaining the mechanism in Fig. 4.5d [66], while Aflatoxin B1 (AFB1) and Zearalenone (ZEN) were detected simultaneously using a competitive immunoassay [68]. Other examples of biomolecules detected using QD-based fluorescence immunoassays include human alpha-fetoprotein (AFP) [137], *E. coli* [23], fumonisin mycotoxins [4], and SARS-CoV-2 [122], among many others [110].

4 Quantum Dots in Bioimaging

The striking properties of QDs are enhanced fluorescent ability, adaptable emission range, and sustained stability to photochemical changes, for which they act as the most viable agents for biolabeling and bioimaging alike. Recently, different bioimaging techniques using QDs display advancements in the biomedical domain, including in vitro and in vivo imaging, targeted cellular imaging, fluorescence lifetime imaging, near-infrared imaging, multimodal imaging, and single-particle tracking of biological processes [16, 119]. The uptake of QDs through the cellular environment is achieved mainly through energy-independent endocytosis, membrane receptor engagement, endocytic entry, mechanically enhanced cell delivery, and blended approaches. Through these approaches, the channelization of QDs into cellular organelles and their role in cellular bioimaging can be studied extensively [142]. Cellular bioimaging is important for various applications, such as targeted drug delivery or the identification of any abnormalities in the cell. Water-soluble perovskite $CsPbBr_3$ QDs were used to identify exosomes through the process of Single-molecule localization microscopy (SMLM). Also, the same QDs composition, if conjugated with biotin, nuclear membrane, or cell membrane, could be visualized in high resolution using SMLM [133, 134].

Multicolor imaging by quantum dots serves as a viable multiplex option. A study wherein CdSe QDs are synthesized in a water-dependent reaction with customizable emission ranges. Cellular uptake studies of the CdSe QDs are carried out in human colon cancer (HCT116) cell lines. The results show internalization into the cytoplasm, and the QDs also crossed the nuclear membrane and showed varied fluorescence intensities/colors. This could be a good tool for targeted drug delivery through bioimaging [107]. In a similar approach to multicolor bioimaging, PDMS-coated CdTe QDs are developed to identify many different cellular organelles, namely, lysosomes, mitochondria, nucleus, and actin, tested in vitro in the human hepatoblastoma (HEPG2) cell line. Specific targeting is facilitated by binding EDC-NHS with antibodies such as monoclonal (CD68), lysosomes, mitochondria, nuclear antigen, and smooth muscle actin antibodies. Different spectral fluorescence lifetimes are observed during excitation at a wavelength of 405 nm [92]. Lysosomes are an important class of cellular organelles that maintain cellular growth and development, metabolic rates, cellular survival, and other homeostasis inside the cell. The study mainly focused on monitoring cellular apoptosis, necrosis, and tracking lysosomal-endosomal proteins employing boron (B) and nitrogen (N) co-doped carbon QDs, as shown in Fig. 4.6a. The QDs present an emission range in the red wavelength 618 nm with a higher quantum yield of around 30% and sustained fluorescence capability with a dip of 3% from the total in a time range of 1 h [30]. Autofluorescence is an issue in imaging, particularly in live tissue imaging. The research and development of QDs showing absorption in the NIR region of 700 to 900 nm can mitigate this issue. Also, NIR waves can deeply penetrate living tissues without loss in fluorescent intensity. Intracellular signaling inside a human system greatly depends on calcium ions (Ca^{2+}), considered secondary messengers that play a pivotal role. A study has reported the identification of Ca^{2+} ions in the NIR range of 620–700 nm using NIR-water-soluble $AgInS_2$ quantum dots. These QDs are used for fluorescence imaging of zebrafish eggs and larvae. Figure 4.6b displays the bright-field and fluorescence images of zebrafish after 30 min treatment with $AgInS_2$ QDs and co-incubation of AgInS2 QDs with Ca^{2+}. The third column of the image captured the normal development of zebrafish eggs containing $AgInS_2$ QDs in laser scanning confocal microscopy (CLSM), rendering low cytotoxicity of QDs and their applicability in animal live imaging. The fluorescence yield factor was recorded to be in the range of 20–24% [125]. Similarly, the bioimaging of GSH using near-infrared (NIR)-emitting fluorescent Ag_2S QDs is carried out in a platform composed of CoOOH nanosheets. The negatively charged Ag_2S quantum dots (QDs) can attach to CoOOH nanosheets (NSs) through electrostatic forces, forming a nanoprobe. This interaction causes the fluorescence of Ag_2S QDs at 685 nm to be quenched due to the inner-filter effect (IFE). However, when GSH was introduced, it reduced CoOOH NSs to Co^{2+}, which restores the fluorescence of Ag_2S QDs. This QDs platform can act as a cost-effective testing bed for diseases wherein GSH is an important biomarker [21].

5 Discussion and Analysis

As noted in earlier sections of this chapter, quantum dots may be broadly classified into several categories based on their chemical composition, namely, Semiconductor, Carbon, and Perovskite quantum dots. This section is devoted to a brief discussion of the pros and cons and applications of each, as well as their future prospects.

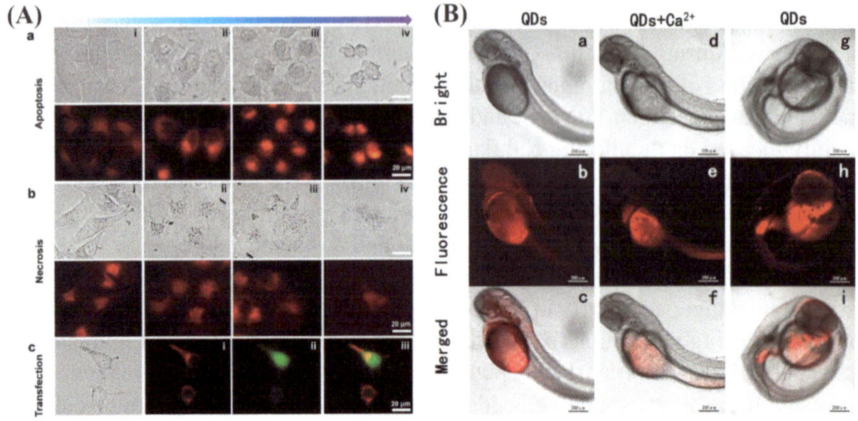

Fig. 4.6 **a** Optical microscopic images and corresponding fluorescent images of apoptosis **a**, necrosis **b** evolution of Hela cells, and tracking of enhanced green fluorescent protein (EGFP) plasmid transfection in 293 T cells by CQDs (Reprinted with permission from [30]). Copyright 2023 American Chemical Society. (B) CLSM images of zebrafish incubated with $AgInS_2$ QDs containing cultures, Ca^{2+} cultures, and zebrafish eggs incubated with $AgInS_2$ QDs containing culture medium, respectively (Reprinted from [125]). Copyright (2024), with permission from Elsevier

5.1 A Brief Comparative Study of Quantum Dots and Their Bioanalytical Applications

Ever since the discovery of quantum dots, a growing volume of research has been directed toward their use in bioanalytical assays and other diagnostic techniques. Quantum dots have tunable surface chemistry and optical properties, narrow emission peaks, broad excitation spectra, and features that give them an edge over organic fluorescent dyes, which are susceptible to photobleaching.

Zhu and the team have developed a fluorescent immunosorbent assay using red CdTe quantum dots with an emission at 655 nm to detect Cry1Ab protein in genetically modified maize [143]. The assay was highly specific toward the target protein and was found to be more sensitive than conventional immunosorbent assays. Similarly, water-soluble CdSe/ZnS core–shell QDs have been used as fluorescent probes to measure the levels of C-Reactive protein, a biomarker for infection. The novel assay proved to be more sensitive than standard commercial assays, with a much shorter reaction time [81]. Semiconductor quantum dots have also been deployed for in vivo and in vitro cellular and tissue imaging. For example, Zhang and team used PbS/CdS quantum dots coated with Polyethylene Glycol to study blood vessels in mice [139]. The authors were able to image femoral and tumor vasculature at varying depths.

However, despite these, semiconductor quantum dots are hindered by their own share of drawbacks, not least of which is their cytotoxicity, especially concerning in vivo studies. Heavy metals such as Cadmium are popular in the synthesis of quantum dots, and their cytotoxicity is well-documented. Zhang's research team has reported that roughly a quarter of the PbS/CdS QDs injected in the mice were retained within their bodies [139]. Thus, Hoshino and the team summarize that the use of such quantum dots in imaging studies is accompanied by the risk of the release of toxic elements into cells [51]. Environmental pollution due to the presence of heavy metals [38] and the cost of fabrication are also major drawbacks of semiconductor quantum dots.

This has opened up the avenue for Carbon Dots, which, as the name suggests, are quantum dots composed of carbon-based materials. They are a relatively recent entrant in the domain of quantum dots, having been discovered in 2004 by Xu and the team [131]. These nanostructures can be fabricated via cost-effective methods from a collection of precursors, ranging from graphene and its derivatives to plant and crop waste. Their chemical composition makes them biocompatible and non-cytotoxic, not to mention environmentally friendly. Like semiconductor QDs, carbon dots also offer tunable optical properties, making it easier to modify their surface chemistry for functionalization. Carbon dots have surfaced as a promising replacement to semiconductor QDs; carbon dots have begun to see increasing use in bioanalytical applications, including bioassays [116, 63, 65]. For example, a well plate-based sandwich immunoassay developed to detect human α-fetoprotein (AFP) by using fluorescent carbon dots made from citric acid via a single-step hydrothermal process [130]. The performance of the novel high-throughput immunoassay was comparable to standard benchmark assays using Horseradish Peroxidase and Fluorescein Isothiocyanate.

Similarly, Carbon dots have been deployed for the purpose of imaging intracellular organelles [40] and microorganisms [121], as well as the imaging of plants [69] and animal tissue [52, 65]. However, carbon dots are not without their drawbacks, such as limited photoluminescent quantum yield (PLQY) and lower photostability relative to semiconductor QDs. Perovskite quantum dots, yet another recent discovery reported by Protesescu and the team [94], offer higher PLQY, along with similarly tunable optical and surface properties. This makes them suitable as fluorescent probes for bioanalytical applications. For example, Kim's research team has developed uniformly sized cesium lead bromide ($CsPbBr_3$) quantum dots (QDs) to act as fluorescent probes in an immunoassay to detect human hepatitis B surface antigen, α-fetoprotein, and HIV antibody [60]. $CsPbBr_3$ QDs have also been used to develop a chemiluminescent assay to measure levels of the antibiotic cefazolin [102]. Ryu and the team have used $CsPbBr_3$ scintillators for the real-time X-ray imaging of malignant tumors grown in mice [98]. Despite their excellent optical properties, Perovskite QDs are also limited by their instability and the toxicity of their constituent heavy metals.

5.2 *Challenges and Future Outlook*

It has now been established that quantum dots can act as robust fluorescent probes in biological assays as well as imaging techniques. Their large-scale use, however, is hindered by the limitations that have been discussed previously. Semiconductor and

Perovskite QDs still largely rely on heavy metals like lead and cadmium, which are not only toxic to living tissue but also harm the environment upon release or disposal. Furthermore, Perovskite QDs are susceptible to structural damage, depending on their surrounding media. These challenges may be mitigated by using core–shell QDs wherein the material of the outer shell of the nanocrystals is non-toxic in nature while protecting the cores from heat- or media-induced damage. Attempts have been made to fabricate and use QDs of metals like copper and Indium as alternatives to cadmium and lead. Carbon dots are non-toxic, biocompatible, do not pollute the environment, and can be developed from inexpensive sources, unlike semiconductor and perovskite QDs. However, their PLQY is relatively low; improving it is a lucrative avenue of research.

Additional challenges that ought to be addressed include scalability and maintenance of uniformity in the size and surface properties of quantum dots. As the optical properties of QDs are governed by their size, surface defects, and chemistry, their uniformity should be ensured. Once that has been achieved, the question of translating the production of QDs from the lab scale to the industrial or commercial scale should be addressed.

6 Summary

The findings from recent advancements in quantum dot (QD) technology highlight their transformative potential in bioassays and bioimaging. It focuses on enhancing both diagnostic sensitivity and imaging precision. QDs have demonstrated extraordinary capabilities in multiplexed bioassays, offering the ability to track multiple targets simultaneously due to their size-dependent emission and broad absorption spectra. These properties enable QDs to outperform traditional organic fluorophores and provide sharper, brighter, and long-lasting signals critical for real-time, high-resolution bioimaging. In bioassay applications, QDs have been successfully employed for ultrasensitive detection in a variety of platforms, such as fluorescence, electrochemical, FRET, and photoelectrochemical assays. Functionalizing QDs with biomolecules such as antibodies or peptides has expanded their use in targeted imaging and diagnostics. For instance, QDs can be tailored to bind specifically to cancer cells or pathogens, enabling precise detection and even tracking of disease progression. In the case of bioimaging, QDs have enabled breakthroughs in cellular and molecular imaging, with the development of near-infrared (NIR) QDs, which facilitate deep-tissue penetration and minimize background autofluorescence. These advancements improve the resolution of in vivo imaging, allowing for more accurate monitoring of disease progression and treatment responses. QDs are integrated with cutting-edge technologies, for instance, microfluidics, lab-on-a-chip devices, and wearable sensors for advanced applications in point-of-care diagnostics. However, it's not all smooth sailing. Despite these significant advancements, challenges remain, particularly in addressing the toxicity and stability of QDs in complex biological systems. Nonetheless, continuous innovations in QD synthesis, advancements in biocompatibility, and their integration with data-driven technologies such as AI for image analysis hold great promise for the future of QD-based bioassays and bioimaging. These developments position QDs as pivotal tools in revolutionizing non-invasive, real-time diagnostics and therapeutic monitoring. In the future, QDs are expected to revolutionize the treatment of patients. They will allow for quick and easy medical tests and continuous tracking of their health. Moreover, QD-based imaging techniques could be employed in surgical procedures, offering real-time tumor margin visualization. After a thorough, comprehensive investigation, it can be proposed that QD technology is likely to become a cornerstone in biomedical research and precision medicine, with innovations in bioassays, bioimaging, and nanomedicine driving the next generation of diagnostic tools and targeted therapeutic strategies.

References

1. Ahirwar S, Mallick S, Bahadur D (2017) Electrochemical method to prepare graphene quantum dots and graphene oxide quantum dots. ACS Omega 2(11). https://doi.org/10.1021/acsomega.7b01539
2. Ali MK, Javaid S, Afzal H, Zafar I, Fayyaz K, Ain Q, Rather MA, Hossain MJ, Rashid S, Khan KA, Sharma R (2023) Exploring the multifunctional roles of quantum dots for unlocking the future of biology and medicine. Environ Res 232. https://doi.org/10.1016/j.envres.2023.116290
3. Amirav L, Amirav A, Lifshitz E (2005) A spray-based method for the production of semiconductor nanocrystals. J Phys Chem B 109(20). https://doi.org/10.1021/jp051337w
4. Anfossi L, Di Nardo F, Cavalera S, Giovannoli C, Spano G, Speranskaya ES, Goryacheva IY, Baggiani C (2018) A lateral flow immunoassay for straightforward determination of fumonisin mycotoxins based on the quenching of the fluorescence of CdSe/ZnS quantum dots by gold and silver nanoparticles. Microchim Acta 185(2). https://doi.org/10.1007/s00604-017-2642-0
5. Ansari L, Hallaj S, Hallaj T, Amjadi M (2021) Doped-carbon dots: recent advances in their biosensing, bioimaging and therapy applications. Coll Surf B Biointerf 203. https://doi.org/10.1016/j.colsurfb.2021.111743
6. Arab K, Jafari A, Shahi F (2024) The role of graphene quantum dots in cutting-edge medical therapies. Polym Adv Technol 35(9):e6571. https://doi.org/10.1002/PAT.6571
7. Architha N, Ragupathi M, Shobana C, Selvankumar T, Kumar P, Lee YS, Kalai Selvan R (2021) Microwave-assisted green synthesis of fluorescent carbon quantum dots from Mexican Mint extract for Fe3+ detection and bio-imaging applications. Environ Res 199. https://doi.org/10.1016/j.envres.2021.111263
8. Arya H, Kaul Z, Wadhwa R, Taira K, Hirano T, Kaul SC (2005) Quantum dots in bio-imaging: Revolution by the small. In: Biochemical and biophysical research communications, vol 329(4). https://doi.org/10.1016/j.bbrc.2005.02.043
9. Bagga A, Chattopadhyay PK, Ghosh S (2006) Origin of stokes shift in InAs and CdSe quantum dots: exchange splitting of excitonic states. Phys Rev B Condens Matter Mater Phys 74(3). https://doi.org/10.1103/PhysRevB.74.035341
10. Banerjee A, Pons T, Lequeux N, Dubertret B (2016) Quantum dots-DNA bioconjugates: synthesis to applications. Interf Focus 6(6). https://doi.org/10.1098/rsfs.2016.0064
11. Bao H, Hao N, Yang Y, Zhao D (2010a) Biosynthesis of biocompatible cadmium telluride quantum dots using yeast cells. Nano Res 3(7). https://doi.org/10.1007/s12274-010-0008-6

12. Bao H, Lu Z, Cui X, Qiao Y, Guo J, Anderson JM, Li CM (2010b) Extracellular microbial synthesis of biocompatible CdTe quantum dots. Acta Biomater 6(9). https://doi.org/10.1016/j.actbio.2010.03.030
13. Bates M, Huang B, Dempsey GT, Zhuang X (2007) Multicolor super-resolution imaging with photo-switchable fluorescent probes. Science 317(5845). https://doi.org/10.1126/science.1146598
14. Betzig E, Patterson GH, Sougrat R, Lindwasser OW, Olenych S, Bonifacino JS, Davidson MW, Lippincott-Schwartz J, Hess HF (2006) Imaging intracellular fluorescent proteins at nanometer resolution. Science 313(5793). https://doi.org/10.1126/science.1127344
15. Bhang SH, Won N, Lee TJ, Jin H, Nam J, Park J, Chung H, Park HS, Sung YE, Hahn SK, Kim BS, Kim S (2009) Hyaluronic acid-quantum dot conjugates for in vivo lymphatic vessel imaging. ACS Nano 3(6). https://doi.org/10.1021/nn900138d
16. Bilan R, Fleury F, Nabiev I, Sukhanova A (2015) Quantum dot surface chemistry and functionalization for cell targeting and imaging. Bioconj Chem 26(4). https://doi.org/10.1021/acs.bioconjchem.5b00069
17. Blanco-Canosa JB, Medintz IL, Farrel D, Mattoussi H, Dawson PE (2010) Rapid covalent ligation of fluorescent peptides to water solubilized quantum dots. J Am Chem Soc 132(29). https://doi.org/10.1021/ja910988d
18. Chan WCW, Maxwell DJ, Gao X, Bailey RE, Han M, Nie S (2002) Luminescent quantum dots for multiplexed biological detection and imaging. Curr Opin Biotechnol 13(1). https://doi.org/10.1016/S0958-1669(02)00282-3
19. Chan WCW, Nie S (1998) Quantum dot bioconjugates for ultrasensitive nonisotopic detection. Science 281(5385). https://doi.org/10.1126/science.281.5385.2016
20. Chen AY, Deng Z, Billings AN, Seker UOS, Lu MY, Citorik RJ, Zakeri B, Lu TK (2014) Synthesis and patterning of tunable multiscale materials with engineered cells. Nat Mater 13(5). https://doi.org/10.1038/nmat3912
21. Chen H, Tang Y, Gui J, Li P, Zhu X, Wu C, Yin P, Liu M, Zhang Y, Yao S (2023) Detection and bioimaging of glutathione in cells via a turn-on near-infrared fluorescent nanoprobe composed of Ag_2S quantum dots and CoOOH nanosheets. ACS Appl Nano Mater 6(3):1870–1879. https://doi.org/10.1021/ACSANM.2C04853
22. Chen MT, Hu CF, Huang HB, Qian ZG, Xia XX (2022) Spatially directed biosynthesis of quantum dots via spidroin templating in Escherichia coli. Angewandte Chem Int Edn 61(49). https://doi.org/10.1002/anie.202214177
23. Chen R, Huang X, Li J, Shan S, Lai W, Xiong Y (2016) A novel fluorescence immunoassay for the sensitive detection of Escherichia coli O157:H7 in milk based on catalase-mediated fluorescence quenching of CdTe quantum dots. Anal Chim Acta 947. https://doi.org/10.1016/j.aca.2016.10.017
24. Chen S, Yu YL, Wang JH (2018) Inner filter effect-based fluorescent sensing systems: a review. Anal Chim Acta 999:13–26. https://doi.org/10.1016/J.ACA.2017.10.026
25. Chen Z, Ren X, Meng X, Zhang Y, Chen D, Tang F (2012) Novel fluorescence method for detection of α-l-fucosidase based on CdTe quantum dots. Anal Chem 84(9):4077–4082. https://doi.org/10.1021/AC300166N
26. Chung EY, Ochs CJ, Wang Y, Lei L, Qin Q, Smith AM, Strongin AY, Kamm R, Qi YX, Lu S, Wang Y (2015) Activatable and cell-penetrable multiplex FRET nanosensor for profiling MT1-MMP activity in single cancer cells. Nano Lett 15(8):5025–5032. https://doi.org/10.1021/ACS.NANOLETT.5B01047
27. Clapp AR, Medintz IL, Mauro JM, Fisher BR, Bawendi MG, Mattoussi H (2004) Fluorescence resonance energy transfer between quantum dot donors and dye-labeled protein acceptors. J Am Chem Soc 126(1). https://doi.org/10.1021/ja037088b
28. Dameron CT, Reese RN, Mehra RK, Kortan AR, Carroll PJ, Steigerwald ML, Brus LE, Winge DR (1989) Biosynthesis of cadmium sulphide quantum semiconductor crystallites. Nature 338(6216). https://doi.org/10.1038/338596a0
29. Demers SME, Kuhne WW, Swindle AR, Dick DD, Coopersmith KJ (2023) Quantum dot-DNA FRET conjugates for direct analysis of methylphosphonic acid in complex media. ACS Omega 8(25). https://doi.org/10.1021/acsomega.3c02173

30. Deng Y, Long Y, Song A, Wang H, Xiang S, Qiu Y, Ge X, Golberg D, Weng Q (2023) Boron dopants in red-emitting B and N Co-doped carbon quantum dots enable targeted imaging of lysosomes. ACS Appl Mater Interfaces 15(13):17045–17053. https://doi.org/10.1021/ACSAMI.3C01705
31. Dhariwal J, Rao GK, Vaya D (2024) Recent advancements towards the green synthesis of carbon quantum dots as an innovative and eco-friendly solution for metal ion sensing and monitoring. RSC Sustainability 2(1):11–36. https://doi.org/10.1039/D3SU00375B
32. Dubertret B, Skourides P, Norris DJ, Noireaux V, Brivanlou AH, Libchaber A (2002) In vivo imaging of quantum dots encapsulated in phospholipid micelles. Science 298(5599). https://doi.org/10.1126/science.1077194
33. Efros AL (2019) Quantum dots realize their potential. Nature 575(7784). https://doi.org/10.1038/d41586-019-03607-z
34. El-Sayed MA (2004) Small is different: shape-, size-, and composition-dependent properties of some colloidal semiconductor nanocrystals. Acc Chem Res 37(5). https://doi.org/10.1021/ar020204f
35. Eskalen H, Uruş S, Cömertpay S, Kurt AH, Özgan Ş (2020) Microwave-assisted ultra-fast synthesis of carbon quantum dots from linter: fluorescence cancer imaging and human cell growth inhibition properties. Ind Crops Prod 147. https://doi.org/10.1016/j.indcrop.2020.112209
36. Feng D, Zhang G, Li Y, Feng D, Zhang G, Li Y (2024) Semiconductor quantum dots: synthesis, properties and applications. Nanomaterials 14(22):1825. https://doi.org/10.3390/NANO14221825
37. Frasco MF, Chaniotakis N (2009) Semiconductor quantum dots in chemical sensors and biosensors. Sensors 9(9). https://doi.org/10.3390/s90907266
38. Fthenakis VM (2004) Life cycle impact analysis of cadmium in CdTe PV production. Renew Sustain Energy Rev 8(4):303–334. https://doi.org/10.1016/j.rser.2003.12.001
39. Gao L, Huang Y, Zhang S, Chen Y, Yan S, Dai H, Zeng B (2024) Mxene quantum dots bipolar electrochemiluminescent platform for hepatitis C virus envelope protein E2 detection. Talanta 268. https://doi.org/10.1016/j.talanta.2023.125301
40. Geng X, Sun Y, Li Z, Yang R, Zhao Y, Guo Y, Xu J, Li F, Wang Y, Lu S, Qu L (2019) Retrosynthesis of tunable fluorescent carbon dots for precise long-term mitochondrial tracking. Small 15(48):1901517. https://doi.org/10.1002/smll.201901517
41. Gerion D, Pinaud F, Williams SC, Parak WJ, Zanchet D, Weiss S, Alivisatos AP (2001) Synthesis and properties of biocompatible water-soluble silica-coated CdSe/ZnS semiconductor quantum dots. J Phys Chem B 105(37). https://doi.org/10.1021/jp0105488
42. Gopalakrishnan D, Damien D, Li B, Gullappalli H, Pillai VK, Ajayan PM, Shaijumon MM (2015) Electrochemical synthesis of luminescent MoS2 quantum dots. Chem Commun 51(29):6293–6296. https://doi.org/10.1039/C4CC09826A
43. Griffin BA, Adams SR, Tsien RY (1998) Specific covalent labeling of recombinant protein molecules inside live cells. Science 281(5374). https://doi.org/10.1126/science.281.5374.269
44. Guo J, Qiu X, Mingoes C, Deschamps JR, Susumu K, Medintz IL, Hildebrandt N (2019) Conformational details of quantum dot-DNA resolved by förster resonance energy transfer lifetime nanoruler. ACS Nano 13(1). https://doi.org/10.1021/acsnano.8b07137
45. Guo R, Zhou S, Li Y, Li X, Fan L, Voelcker NH (2015) Rhodamine-functionalized graphene quantum dots for detection of Fe3+ in cancer stem cells. ACS Appl Mater and Interf 7(43). https://doi.org/10.1021/acsami.5b06523
46. Han M, Gao X, Su JZ, Nie S (2001) Quantum-dot-tagged microbeads for multiplexed optical coding of biomolecules. Nat Biotechnol 19(7). https://doi.org/10.1038/90228
47. Hildebrandt N (2011) Biofunctional quantum dots: controlled conjugation for multiplexed biosensors. ACS Nano 5(7). https://doi.org/10.1021/nn2023123
48. Hildebrandt N, Spillmann CM, Russ Algar W, Pons T, Stewart MH, Oh E, Susumu K, Díaz SA, Delehanty JB, Medintz IL (2017) Energy transfer with semiconductor quantum dot bioconjugates: a versatile platform for biosensing, energy harvesting, and other developing applications. Chem Rev 117(2). https://doi.org/10.1021/acs.chemrev.6b00030

49. Hong CA, Park JC, Na H, Jeon H, Nam YS (2021) Short DNA-catalyzed formation of quantum dot-DNA hydrogel for enzyme-free femtomolar specific DNA assay. Biosens Bioelectron 182. https://doi.org/10.1016/j.bios.2021.113110
50. Hong R, Fischer NO, Verma A, Goodman CM, Emrick T, Rotello VM (2004) Control of protein structure and function through surface recognition by tailored nanoparticle scaffolds. J Am Chem Soc 126(3). https://doi.org/10.1021/ja037470o
51. Hoshino A, Fujioka K, Oku T, Suga M, Sasaki YF, Ohta T, Yasuhara M, Suzuki K, Yamamoto K (2004) Physicochemical properties and cellular toxicity of nanocrystal quantum dots depend on their surface modification. Nano Lett 4(11):2163–2169. https://doi.org/10.1021/nl048715d
52. Hua X-W, Bao Y-W, Wu F-G (2018) Fluorescent carbon quantum dots with intrinsic nucleolus-targeting capability for nucleolus imaging and enhanced cytosolic and nuclear drug delivery. ACS Appl Mater Interfaces 10(13):10664–10677. https://doi.org/10.1021/acsami.7b19549
53. Huang S, Yang E, Yao J, Liu Y, Xiao Q (2018) Carbon dots doped with nitrogen and boron as ultrasensitive fluorescent probes for determination of α-glucosidase activity and its inhibitors in water samples and living cells. Microchim Acta 185(8):394. https://doi.org/10.1007/s00604-018-2932-1
54. Ji X, Wang W, Mattoussi H (2016) Controlling the spectroscopic properties of quantum dots via energy transfer and charge transfer interactions: concepts and applications. Nano Today 11(1):98–121. https://doi.org/10.1016/J.NANTOD.2015.09.004
55. Jin Z, Geißler D, Qiu X, Wegner KD, Hildebrandt N (2015) A rapid, amplification-free, and sensitive diagnostic assay for single-step multiplexed fluorescence detection of MicroRNA. Angewandte Chem Int Edn 54(34). https://doi.org/10.1002/anie.201504887
56. Kalsoom UE, Yi R, Qu J, Liu L (2021) Nonlinear optical properties of CdSe and CdTe core-shell quantum dots and their applications. Front Phys 9. https://doi.org/10.3389/fphy.2021.612070
57. Karami MH, Abdouss M, Rahdar A, Pandey S (2024) Graphene quantum dots: background, synthesis methods, and applications as nanocarrier in drug delivery and cancer treatment: an updated review. Inorgan Chem Commun 161. https://doi.org/10.1016/j.inoche.2024.112032
58. Ke CB, Lu TL, Chen JL (2020) Fluorometric determination of amifostine and alkaline phosphatase on amphiprotic molecularly imprinted silica crosslinked with binary functional silanes and carbon dots. Biosens Bioelectron 151:111965. https://doi.org/10.1016/J.BIOS.2019.111965
59. Kim GB, Kim YP (2012) Analysis of protease activity using quantum dots and resonance energy transfer. Theranostics 2(2). https://doi.org/10.7150/thno.3476
60. Kim H-R, Bong J-H, Park J-H, Song Z, Kang M-J, Son DH, Pyun J-C (2021) Cesium lead bromide (CsPbBr 3) perovskite quantum dot-based photosensor for chemiluminescence immunoassays. ACS Appl Mater Interfaces 13(25):29392–29405. https://doi.org/10.1021/acsami.1c08128
61. Kokkinos C, Prodromidis M, Economou A, Petrou P, Kakabakos S (2015) Quantum dot-based electrochemical DNA biosensor using a screen-printed graphite surface with embedded bismuth precursor. Electrochem Commun 60. https://doi.org/10.1016/j.elecom.2015.08.006
62. Kong W, Wu D, Xia L, Chen X, Li G, Qiu N, Chen G, Sun Z, You J, Wu Y (2017) Carbon dots for fluorescent detection of α-glucosidase activity using enzyme activated inner filter effect and its application to anti-diabetic drug discovery. Anal Chim Acta 973:91–99. https://doi.org/10.1016/J.ACA.2017.03.050
63. Li D, Jing P, Sun L, An Y, Shan X, Lu X, Zhou D, Han D, Shen D, Zhai Y, Qu S, Zbořil R, Rogach AL (2018) Near-infrared excitation/emission and multiphoton-induced fluorescence of carbon dots. Adv Mater 30(13):1705913. https://doi.org/10.1002/adma.201705913
64. Li G, Kong W, Zhao M, Lu S, Gong P, Chen G, Xia L, Wang H, You J, Wu Y (2016) A fluorescence resonance energy transfer (FRET) based "Turn-On" nanofluorescence sensor using a nitrogen-doped carbon dot-hexagonal cobalt oxyhydroxide nanosheet architecture and application to α-glucosidase inhibitor screening. Biosens Bioelectron 79:728–735. https://doi.org/10.1016/J.BIOS.2015.12.094

65. Li H, Yan X, Qiao S, Lu G, Su X (2018) Yellow-emissive carbon dot-based optical sensing platforms: cell imaging and analytical applications for biocatalytic reactions. ACS Appl Mater Interfaces 10(9):7737–7744. https://doi.org/10.1021/acsami.7b17619
66. Li J, Lv Y, Li N, Wu R, Xing M, Shen H, Li LS, Chen X (2019a) Robust synthesis of bright multiple quantum dot-embedded nanobeads and its application to quantitative immunoassay. Chem Eng J 361:499–507. https://doi.org/10.1016/J.CEJ.2018.12.068
67. Li M, Chen T, Gooding JJ, Liu J (2019b) Review of carbon and graphene quantum dots for sensing. ACS Sensors 4(7). https://doi.org/10.1021/acssensors.9b00514
68. Li R, Wen Y, Yang L, Liu A, Wang F, He P (2022) Dual quantum dot nanobeads-based fluorescence-linked immunosorbent assay for simultaneous detection of aflatoxin B1 and zearalenone in feedstuffs. Food Chem 366:130527. https://doi.org/10.1016/J.FOODCHEM.2021.130527
69. Li W, Zhang H, Zheng Y, Chen S, Liu Y, Zhuang J, Liu W-R, Lei B (2017) Multifunctional carbon dots for highly luminescent orange-emissive cellulose based composite phosphor construction and plant tissue imaging. Nanoscale 9(35):12976–12983. https://doi.org/10.1039/C7NR03217J
70. Li X, Zhu S, Xu B, Ma K, Zhang J, Yang B, Tian W (2013a) Self-assembled graphene quantum dots induced by cytochrome c: a novel biosensor for trypsin with remarkable fluorescence enhancement. Nanoscale 5(17):7776–7779. https://doi.org/10.1039/C3NR00006K
71. Li Y, Cui R, Zhang P, Chen BB, Tian ZQ, Li L, Hu B, Pang DW, Xie ZX (2013b) Mechanism-oriented controllability of intracellular quantum dots formation: the role of glutathione metabolic pathway. ACS Nano 7(3). https://doi.org/10.1021/nn305346a
72. Lignos I, Protesescu L, Stavrakis S, Piveteau L, Speirs MJ, Loi MA, Kovalenko MV, DeMello AJ (2014) Facile droplet-based microfluidic synthesis of monodisperse IV-VI semiconductor nanocrystals with coupled in-line NIR fluorescence detection. Chem Mater 26(9). https://doi.org/10.1021/cm500774p
73. Liu H, Ding J, Zhang K, Ding L (2019) Construction of biomass carbon dots based fluorescence sensors and their applications in chemical and biological analysis. TrAC, Trends Anal Chem 118:315–337. https://doi.org/10.1016/J.TRAC.2019.05.051
74. Liu JJ, Tang D, Chen Z, Yan X, Zhong Z, Kang L, Yao J (2017) Chemical redox modulated fluorescence of nitrogen-doped graphene quantum dots for probing the activity of alkaline phosphatase. Biosens Bioelectron 94:271–277. https://doi.org/10.1016/J.BIOS.2017.03.017
75. Liu L, Guo X, Li Y, Zhong X (2010) Bifunctional multidentate ligand modified highly stable water-soluble quantum dots. Inorgan Chem 49(8). https://doi.org/10.1021/ic902469d
76. Liu W, Hak SC, Zimmer JP, Tanaka E, Frangioni JV, Bawendi M (2007) Compact cysteine-coated CdSe(ZnCdS) quantum dots for in vivo applications. J Am Chem Soc 129(47). https://doi.org/10.1021/ja073790m
77. Lu JG, Ye ZZ, Huang JY, Zhu LP, Zhao BH, Wang ZL, Fujita S (2006) ZnO quantum dots synthesized by a vapor phase transport process. Appl Phys Lett 88(6). https://doi.org/10.1063/1.2172154
78. Lu W, Qin X, Liu S, Chang G, Zhang Y, Luo Y, Asiri AM, Al-Youbi AO, Sun X (2012) Economical, green synthesis of fluorescent carbon nanoparticles and their use as probes for sensitive and selective detection of mercury(II) ions. Anal Chem 84(12). https://doi.org/10.1021/ac3007939
79. Luo X, Han Y, Chen X, Tang W, Yue T, Li Z (2020) Carbon dots derived fluorescent nanosensors as versatile tools for food quality and safety assessment: a review. Trends Food Sci Technol 95:149–161. https://doi.org/10.1016/J.TIFS.2019.11.017
80. Lv Y, Fan J, Zhao M, Wu R, Li LS (2023) Recent advances in quantum dot-based fluorescence-linked immunosorbent assays. Nanoscale 15(12). https://doi.org/10.1039/d2nr07247e
81. Lv Y, Wu R, Feng K, Li J, Mao Q, Yuan H, Shen H, Chai X, Li LS (2017) Highly sensitive and accurate detection of C-reactive protein by CdSe/ZnS quantum dot-based fluorescence-linked immunosorbent assay. J Nanobiotechnol 15(1):35. https://doi.org/10.1186/s12951-017-0267-4

82. Ma J, Zhang H, Peng F, Yang X, Li ZL, Sun L, Jiang H (2020) Carbon dots as fluorescent nanoprobe for the determination of N-acetyl-β-d-glucosaminidase activity. Anal Chim Acta 1101:129–134. https://doi.org/10.1016/J.ACA.2019.12.018
83. Mattoussi H, Matthew Mauro J, Goldman ER, Anderson GP, Sundar VC, Mikulec FV, Bawendi MG (2000) Self-assembly of CdSe-ZnS quantum dot bioconjugates using an engineered recombinant protein. J Am Chem Soc 122(49). https://doi.org/10.1021/ja002535y
84. Medintz IL, Uyeda HT, Goldman ER, Mattoussi H (2005) Quantum dot bioconjugates for imaging, labelling and sensing. Nat Mater 4(6). https://doi.org/10.1038/nmat1390
85. Mi C, Wang Y, Zhang J, Huang H, Xu L, Wang S, Fang X, Fang J, Mao C, Xu S (2011) Biosynthesis and characterization of CdS quantum dots in genetically engineered Escherichia coli. J Biotechnol 153(3–4). https://doi.org/10.1016/j.jbiotec.2011.03.014
86. Mitchell GP, Mirkin CA, Letsinger RL (1999) Programmed assembly of DNA functionalized quantum dots [10]. J Am Chem Soc 121(35). https://doi.org/10.1021/ja991662v
87. Mohamed WAA, Abd El-Gawad H, Mekkey S, Galal H, Handal H, Mousa H, Labib A (2021) Quantum dots synthetization and future prospect applications. Nanotechnol Rev 10(1). https://doi.org/10.1515/ntrev-2021-0118
88. Monte-Filho SS, Andrade SIE, Lima MB, Araujo MCU (2019) Synthesis of highly fluorescent carbon dots from lemon and onion juices for determination of riboflavin in multivitamin/mineral supplements. J Pharmaceut Anal 9(3). https://doi.org/10.1016/j.jpha.2019.02.003
89. Murugan N, Prakash M, Jayakumar M, Sundaramurthy A, Sundramoorthy AK (2019) Green synthesis of fluorescent carbon quantum dots from Eleusine coracana and their application as a fluorescence 'turn-off' sensor probe for selective detection of Cu2+. Appl Surf Sci 476:468–480. https://doi.org/10.1016/J.APSUSC.2019.01.090
90. Owen J, Brus L (2017) Chemical synthesis and luminescence applications of colloidal semiconductor quantum dots. J Am Chem Soc 139(32). https://doi.org/10.1021/jacs.7b05267
91. Palaniappan K, Xue C, Arumugam G, Hackney SA, Liu J (2006) Water-soluble, cyclodextrin-modified CdSe-CdS core-shell structured quantum dots. Chem Mater 18(5). https://doi.org/10.1021/cm051602q
92. Pandey S, Bodas D (2023) Exploiting the UV excited size-dependent emission of PDMS-coated CdTe quantum dots for in vitro simultaneous multicolor imaging of HepG2 cellular organelles. Mater Adv 4(7):1694–1701. https://doi.org/10.1039/D2MA00964A
93. Patterson GH, Lippincott-Schwartz J (2002) A photoactivatable GFP for selective photolabeling of proteins and cells. Science 297(5588). https://doi.org/10.1126/science.1074952
94. Protesescu L, Yakunin S, Bodnarchuk MI, Krieg F, Caputo R, Hendon CH, Yang RX, Walsh A, Kovalenko MV (2015) Nanocrystals of cesium lead halide perovskites (CsPbX3, X = Cl, Br, and I): novel optoelectronic materials showing bright emission with wide color gamut. Nano Lett 15(6):3692–3696. https://doi.org/10.1021/nl5048779
95. Pu Y, Cai F, Wang D, Wang JX, Chen JF (2018) Colloidal synthesis of semiconductor quantum dots toward large-scale production: a review. Ind Eng Chem Res 57(6):1790–1802. https://doi.org/10.1021/ACS.IECR.7B04836
96. Qiu X, Guo J, Jin Z, Petreto A, Medintz IL, Hildebrandt N (2017) Multiplexed nucleic acid hybridization assays using single-FRET-pair distance-tuning. Small 13(25). https://doi.org/10.1002/smll.201700332
97. Rust MJ, Bates M, Zhuang X (2006) Sub-diffraction-limit imaging by stochastic optical reconstruction microscopy (STORM). Nat Methods 3(10). https://doi.org/10.1038/nmeth929
98. Ryu I, Ryu J-Y, Choe G, Kwon H, Park H, Cho Y-S, Du R, Yim S (2021) In vivo plain X-ray imaging of cancer using perovskite quantum dot scintillators. Adv Func Mater 31(34):2102334. https://doi.org/10.1002/adfm.202102334
99. Saa L, Mato JM, Pavlov V (2012a) Assays for methionine γ-lyase and S-adenosyl-l-homocysteine hydrolase based on enzymatic formation of CdS quantum dots in situ. Anal Chem 84(21):8961–8965. https://doi.org/10.1021/AC302770Q
100. Saa L, Pavlov V (2012b) Enzymatic growth of quantum dots: Applications to probe glucose oxidase and horseradish peroxidase and sense glucose. Small 8(22). https://doi.org/10.1002/smll.201201364

101. Saa L, Virel A, Sanchez-Lopez J, Pavlov V (2010) Analytical applications of enzymatic growth of quantum dots. Chem A Europ J 16(21). https://doi.org/10.1002/chem.200903373
102. Salari R, Amjadi M, Hallaj T (2023) Perovskite quantum dots as a chemiluminescence platform for highly sensitive assay of cefazolin. Spectrochim Acta Part A Mol Biomol Spectrosc 285:121845. https://doi.org/10.1016/j.saa.2022.121845
103. Sameiro M, Gonçalves T (2009) Fluorescent labeling of biomolecules with organic probes. Chem Rev 109(1). https://doi.org/10.1021/cr0783840
104. Sanderson K, Castelvecchi D (2023) Tiny "quantum dot" particles win chemistry Nobel. Nature 622(7982). https://doi.org/10.1038/d41586-023-03048-9
105. Shao K, Yang Y, Ye S, Gu D, Wang T, Teng Y, Shen Z, Pan Z (2020) Dual-colored carbon dots-based ratiometric fluorescent sensor for high-precision detection of alkaline phosphatase activity. Talanta 208:120460. https://doi.org/10.1016/J.TALANTA.2019.120460
106. Shi L, Rosenzweig N, Rosenzweig Z (2007) Luminescent quantum dots fluorescence resonance energy transfer-based probes for enzymatic activity and enzyme inhibitors. Anal Chem 79(1):208–214. https://doi.org/10.1021/AC0614644
107. Singh RK, Suryakanta U, Panigrahi B, Mishra S, Das S, Bhatta SR, Ghosh SK, Mandal D (2024) Facile approach to develop peptide-stabilized CdS/CdSe quantum dots for cellular imaging. Luminescence 39(12):e70048. https://doi.org/10.1002/BIO.70048
108. Smith AM, Duan H, Mohs AM, Nie S (2008) Bioconjugated quantum dots for in vivo molecular and cellular imaging. Adv Drug Deliv Rev 60(11). https://doi.org/10.1016/j.addr.2008.03.015
109. Song L, Hennink EJ, Young IT, Tanke HJ (1995) Photobleaching kinetics of fluorescein in quantitative fluorescence microscopy. Biophys J 68(6). https://doi.org/10.1016/S0006-3495(95)80442-X
110. Speranskaya ES, Drozd DD, Pidenko PS, Goryacheva IY (2020) Enzyme modulation of quantum dot luminescence: application in bioanalysis. TrAC Trends Anal Chem 127. https://doi.org/10.1016/j.trac.2020.115897
111. Stanisavljevic M, Krizkova S, Vaculovicova M, Kizek R, Adam V (2015) Quantum dots-fluorescence resonance energy transfer-based nanosensors and their application. Biosens Bioelectron 74:562–574. https://doi.org/10.1016/J.BIOS.2015.06.076
112. Sun X, Lei Y (2017) Fluorescent carbon dots and their sensing applications. TrAC, Trends Anal Chem 89:163–180. https://doi.org/10.1016/J.TRAC.2017.02.001
113. Sweeney RY, Mao C, Gao X, Burt JL, Belcher AM, Georgiou G, Iverson BL (2004) Bacterial biosynthesis of cadmium sulfide nanocrystals. Chem Biol 11(11). https://doi.org/10.1016/j.chembiol.2004.08.022
114. Tang C, Zhou J, Qian Z, Ma Y, Huang Y, Feng H (2017) A universal fluorometric assay strategy for glycosidases based on functional carbon quantum dots: β-galactosidase activity detection in vitro and in living cells. J Mater Chem B 5(10):1971–1979. https://doi.org/10.1039/c6tb03361j
115. Tohgha U, Varga K, Balaz M (2013) Achiral CdSe quantum dots exhibit optical activity in the visible region upon post-synthetic ligand exchange with D- or L-cysteine. Chem Commun 49(18). https://doi.org/10.1039/c3cc37987f
116. Tong X, Shi S, Tong C, Iftikhar A, Long R, Zhu Y (2020) Quantum/carbon dots-based fluorescent assays for enzyme activity. TrAC, Trends Anal Chem 131:116008. https://doi.org/10.1016/j.trac.2020.116008
117. Tsien RY (1998) The green fluorescent protein. Annu Rev Biochem 67. https://doi.org/10.1146/annurev.biochem.67.1.509
118. Valappil MO, Anil A, Shaijumon M, Pillai VK, Alwarappan S (2017) A single-step electrochemical synthesis of luminescent WS2 quantum dots. Chem A Europ J 23(38). https://doi.org/10.1002/chem.201701277
119. Vu TQ, Lam WY, Hatch EW, Lidke DS (2015) Quantum dots for quantitative imaging: from single molecules to tissue. Cell Tissue Res 360(1). https://doi.org/10.1007/s00441-014-2087-2

120. Wang Q, Xu Y, Zhao X, Chang Y, Liu Y, Jiang L, Sharma J, Seo DK, Yan H (2007) A facile one-step in situ functionalization of quantum dots with preserved photoluminescence for bioconjugation. J Am Chem Soc 129(20). https://doi.org/10.1021/ja071434y
121. Wang S, Zhang Y, Zhuo P, Hu Q, Chen Z, Zhou L (2020) Identification of eight pathogenic microorganisms by single concentration-dependent multicolor carbon dots. J Mater Chem B 8(27):5877–5882. https://doi.org/10.1039/D0TB00834F
122. Wang X, Shao S, Ye H, Li S, Gu B, Tang B (2023) Development of a quantum dot-based lateral flow immunoassay strip for rapid and sensitive detection of SARS-CoV-2 neutralizing antibodies. Sci Rep 13(1). https://doi.org/10.1038/s41598-023-49244-5
123. Wegner KD, Jin Z, Lindén S, Jennings TL, Hildebrandt N (2013) Quantum-dot-based förster resonance energy transfer immunoassay for sensitive clinical diagnostics of low-volume serum samples. ACS Nano 7(8). https://doi.org/10.1021/nn403253y
124. Willard DM, Carillo LL, Jung J, Van Orden A (2001) CdSe-ZnS quantum dots as resonance energy transfer donors in a model protein-protein binding assay. Nano Lett 1(9). https://doi.org/10.1021/nl015565n
125. Wu J, Li J, Cheng M, Li L, Yan R, Yue J (2024) Water-soluble near-infrared AgInS2 quantum dots for Ca2+ detection and bioimaging. Spectrochim Acta Part A Mol Biomol Spectrosc 322:124859. https://doi.org/10.1016/J.SAA.2024.124859
126. Wu P, He Y, Wang HF, Yan XP (2010) Conjugation of glucose oxidase onto Mn-doped Zns quantum dots for phosphorescent sensing of glucose in biological fluids. Anal Chem 82(4). https://doi.org/10.1021/ac902531g
127. Wu X, Liu H, Liu J, Haley KN, Treadway JA, Larson JP, Ge N, Peale F, Bruchez MP (2003) Immunofluorescent labeling of cancer marker Her2 and other cellular targets with semiconductor quantum dots. Nat Biotechnol 21(1). https://doi.org/10.1038/nbt764
128. Wu Y, Huang J, Zang J, Zhou J, Cheng C, Hu Z, Shan D, Yang W, Sychugov I, Sun L, Xu B (2024) Enlarging the Stokes shift of CuInS2 quantum dots using thiol–ene polymers for efficient large-area luminescent solar concentrators. Energy Environ Sci 17(17):6338–6349. https://doi.org/10.1039/D4EE02603A
129. Wu YT, Qiu X, Lindbo S, Susumu K, Medintz IL, Hober S, Hildebrandt N (2018) Quantum dot–based FRET immunoassay for HER2 using ultrasmall affinity proteins. Small 14(35). https://doi.org/10.1002/smll.201802266
130. Wu Y, Wei P, Pengpumkiat S, Schumacher EA, Remcho VT (2015) Development of a Carbon Dot (C-Dot)-linked immunosorbent assay for the detection of human α-fetoprotein. Anal Chem 87(16):8510–8516. https://doi.org/10.1021/acs.analchem.5b02019
131. Xu X, Ray R, Gu Y, Ploehn HJ, Gearheart L, Raker K, Scrivens WA (2004) Electrophoretic analysis and purification of fluorescent single-walled carbon nanotube fragments. J Am Chem Soc 126(40):12736–12737. https://doi.org/10.1021/ja040082h
132. Yakoubi A, Chaabane TB, Aboulaich A, Mahiou R, Balan L, Medjahdi G, Schneider R (2016) Aqueous synthesis of Cu-doped CdZnS quantum dots with controlled and efficient photoluminescence. J Lumin 175. https://doi.org/10.1016/j.jlumin.2016.02.035
133. Yang E, Zhang Y, Shen Y (2022a) Quantum dots for electrochemiluminescence bioanalysis—a review. Anal Chimica Acta 1209. https://doi.org/10.1016/j.aca.2021.339140
134. Yang Z, Dong Y, Zong S, Li L, Yang K, Wang Z, Zeng H, Cui Y (2022b) Water-dispersed CsPbBr 3 nanocrystals for single molecule localization microscopy with high location accuracy for targeted bioimaging. Nanoscale 14(17):6392–6401. https://doi.org/10.1039/D1NR08029F
135. Yeh HC, Ho YP, Wang TH (2005) Quantum dot-mediated biosensing assays for specific nucleic acid detection. Nanomed Nanotechnol Biol Med 1(2). https://doi.org/10.1016/j.nano.2005.03.004
136. Yin B, Deng J, Peng X, Long Q, Zhao J, Lu Q, Chen Q, Li H, Tang H, Zhang Y, Yao S (2013) Green synthesis of carbon dots with down- and up-conversion fluorescent properties for sensitive detection of hypochlorite with a dual-readout assay. Analyst 138(21). https://doi.org/10.1039/c3an01003a

137. Zhang B, Liang X, Hao L, Cheng J, Gong X, Liu X, Ma G, Chang J (2009) Quantum dots/particle-based immunofluorescence assay: synthesis, characterization and application. J Photochem Photobiol B Biol 94(1). https://doi.org/10.1016/j.jphotobiol.2008.09.008
138. Zhang J, Zhou R, Tang D, Hou X, Wu P (2019) Optically-active nanocrystals for inner filter effect-based fluorescence sensing: achieving better spectral overlap. TrAC, Trends Anal Chem 110:183–190. https://doi.org/10.1016/J.TRAC.2018.11.002
139. Zhang M, Yue J, Cui R, Ma Z, Wan H, Wang F, Zhu S, Zhou Y, Kuang Y, Zhong Y, Pang D-W, Dai H (2018) Bright quantum dots emitting at ~1,600 nm in the NIR-IIb window for deep tissue fluorescence imaging. Proc Natl Acad Sci 115(26):6590–6595. https://doi.org/10.1073/pnas.1806153115
140. Zhang S, Yong J, Cheng X, Chua JJ, Olivo M (2022) Characterization of quantum dots with hyperspectral fluorescence microscopy for multiplexed optical imaging of biomolecules. BioRxiv
141. Zhou J, Sheng Z, Han H, Zou M, Li C (2012) Facile synthesis of fluorescent carbon dots using watermelon peel as a carbon source. Mater Lett 66(1). https://doi.org/10.1016/j.matlet.2011.08.081
142. Zhou J, Yang Y, Zhang CY (2015) Toward biocompatible semiconductor quantum dots: from biosynthesis and bioconjugation to biomedical application. Chem Rev 115(21). https://doi.org/10.1021/acs.chemrev.5b00049
143. Zhu X, Chen L, Shen P, Jia J, Zhang D, Yang L (2011) High sensitive detection of Cry1Ab protein using a quantum dot-based fluorescence-linked immunosorbent assay. J Agric Food Chem 59(6):2184–2189. https://doi.org/10.1021/jf104140t
144. Zu F, Yan F, Bai Z, Xu J, Wang Y, Huang Y, Zhou X (2017) The quenching of the fluorescence of carbon dots: a review on mechanisms and applications. Microchim Acta 184(7):1899–1914. https://doi.org/10.1007/S00604-017-2318-9

Part II
Quantum Dots for Biological Applications

Chapter 5
Quantum Dots for Biosensors and Bio-diagnostic Applications

Subhash Chand, Ravi Kumar, and Kuldeep Kumar

Abstract Quantum dots (QDs) are nanoscale semiconductor materials that display distinctive optical and electrical characteristics, rendering them advantageous for bio-diagnostic tests and biosensing applications. Quantum dots are advantageous for bioremediation due to their high quantum yield, adjustable fluorescence, enhanced photostability and substantial surface-to-volume ratio, rendering them more sensitive and selective in detecting biomolecules. QD-based biosensors enable the detection of target compounds such as proteins, nucleic acids, enzymes and tiny molecules through mechanisms, including electrochemical interactions, quenching and fluorescence resonance energy transfer (FRET). The functionalization of biorecognition elements on quantum dots, including antibodies, aptamers and enzymes, improves the selectivity and applicability of biosensors in biosensing applications. Quantum dots incorporated into electrochemical and nanomechanical biosensors have realized substantial enhancements in signal transduction efficacy and detection limits. The domain of bio-diagnostics acquires novel imaging capabilities. Quantum dots enable real-time, high-resolution monitoring of biomolecules and cellular activity. Quantum dots exhibit potential for personalised medicine by facilitating cancer diagnosis, therapeutic development, pathogen identification and disease biomarker analysis; yet, their extensive application necessitates resolutions to cytotoxicity, scalability challenges and environmental impact concerns. Advancements in next-generation quantum dot-based biosensors and bio-diagnostic systems will rely on

S. Chand · K. Kumar (✉)
Department of Chemistry, Career Point University, Hamirpur, Himachal Pradesh, India
e-mail: kuldeep.sharma.753@gmail.com

S. Chand
Department of Chemistry, ABV Government Degree College Sunni, Shimla, Himachal Pradesh, India

R. Kumar
Department of Chemistry, School of Sciences, Baddi University of Emerging Sciences and Technology, Baddi, HP, India

K. Kumar
Centre for Nanoscience and Technology, Career Point University, Hamirpur, HP, India

T. Bhuyan et al. (eds.), *Multifunctional Quantum Dots*, Materials Horizons: From Nature to Nanomaterials, https://doi.org/10.1007/978-981-92-0172-3_5

further research into safe quantum dot formulations and interdisciplinary collaboration. Quantum dots undergo a comprehensive examination of their synthesis process and are subsequently utilized in functionalization and biological applications, highlighting their essential significance in health care and diagnostic advancements. This chapter highlights the importance of QDs in biosensors and bio-diagnostic applications due to their superior optical and electronic properties. Despite challenges like cytotoxicity, scalability and cost, interdisciplinary collaboration and innovative synthesis methods can overcome these.

Keywords Quantum dots · Electrochemical biosensors · Pathogen detection · Fluorescence · Biocompatibility

1 Introduction

Quantum dots (QDs) are nanometre-sized semiconductor crystals that display distinctive optical and electronic characteristics. It makes them significant in biosensing and bio-diagnostics applications as depicted in Fig. 1. The characteristics of QDs include elevated quantum yield, narrow and adjustable emission spectra and superior photostability, which are beneficial for the advancement of sensitive and selective biosensors [20, 52]. The quantum dots can be functionalized with diverse bioreceptors, which enhance their use in identifying certain biomolecules, thus improving the analytical efficacy of biosensors for detection limits, sensitivity and selectivity [72]. Their adaptable surface modification skills and unique photoelectrochemical properties enhance their potential as valuable components in biosensor development [48, 60]. Since their introduction, quantum dots have been vital in nanoscience and nanotechnology, especially in the advancement of biosensors and diagnostic instruments. Initially, quantum dots were predominantly utilized for their fluorescent characteristics in bioimaging and biosensing applications [32, 83]. In the last two decades, progress in quantum dot technology has resulted in the creation of more biocompatible and efficient quantum dot-based biosensors, enabling their application in both in vitro and in vivo settings [28]. The incorporation of quantum dots into biosensors has progressed to encompass diverse formats, including fluorescent, bioluminescent and chemiluminescent biosensors, each providing distinct benefits for the detection of biomolecules such as DNA, proteins, and enzymes [49].

Quantum dots have a vital role in enhancing bio-diagnostic technology through improved sensitivity and multiplexing capabilities. Their ability to function as energy transfer donors and acceptors in biosensing applications has transformed the advancement of novel biosensors, facilitating the precise identification of disease-associated proteins [10, 35]. The application of quantum dots in biosensors has resulted in substantial advancements in the identification of cancer biomarkers, inflammatory indicators and other essential biomolecules, thereby enabling early diagnosis and intervention for numerous diseases [2]. As QD technology advances, it possesses the

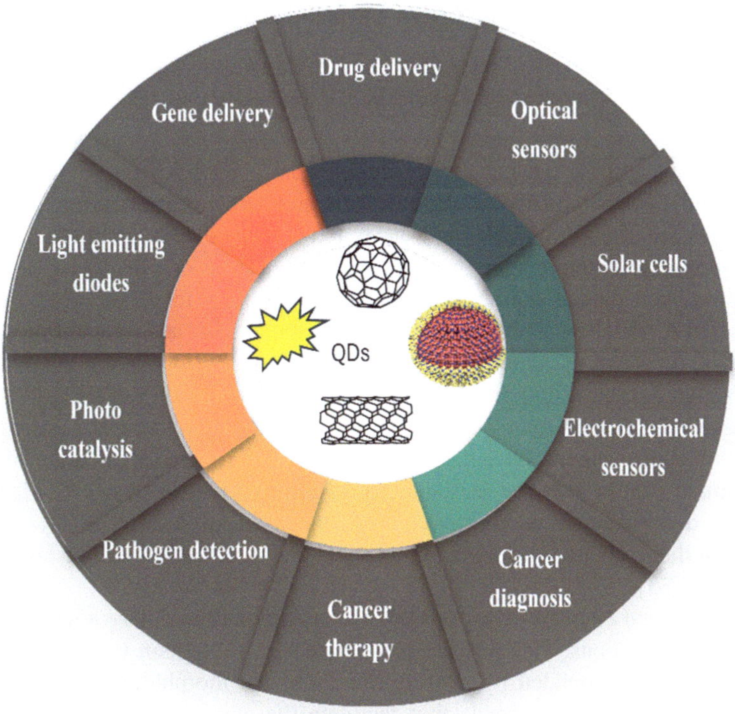

Fig. 1 Various applications of quantum dots

capacity to significantly enhance bio-diagnostic technologies, providing novel frameworks for identifying diseases and monitoring [69]. Recent research has concentrated on the creation of multifunctional quantum dots that work both as imaging agents and as nano scaffolds for diagnostic and therapeutic purposes. These multipurpose platforms hold significant potential for cancer diagnostics and regenerative medicine, serving dual purposes in imaging and therapeutic delivery [55].

This chapter explores the notable progress and uses of quantum dots in biosensors and bio-diagnostics, emphasizing their synthesis, functionalization and contribution to next-generation healthcare technology.

2 Fundamentals of Quantum Dots

2.1 Unique Optical and Electronic Properties of QDs

Quantum dots (QDs) are nanoscale semiconductor particles that exhibit unique optical and electronic properties due to their quantum confinement effects. These properties are highly tunable and have led to significant interest in various applications, including imaging, sensing and optoelectronics.

A prominent optical characteristic of quantum dots is their vibrant and intense fluorescence, which is highly adjustable over the visible and near-infrared spectrum. The tunability is attained by modifying the dimensions and composition of the quantum dots, enabling precise regulation of their emission wavelengths [26]. In contrast to traditional organic dyes, QDs may emit light in the near-infrared spectrum (>650 nm), which is advantageous for biological imaging applications because it minimizes light scattering and absorption in biological tissues [67]. Moreover, quantum dots demonstrate narrow emission spectra and elevated quantum yields, rendering them suitable for applications necessitating excellent colour fidelity and luminosity [33].

The electrical characteristics of QDs are equally extraordinary. Owing to their nanoscale dimensions, they possess distinct energy levels akin to those of atoms, which can be manipulated by altering their size and morphology [1]. This leads to size-dependent energy levels that facilitate the engineering of materials with certain electrical properties. So, they can be utilized to fabricate materials with adjustable light absorption and electronic transport characteristics, which are crucial for applications in photovoltaics and optoelectronic devices [72].

2.2 Synthesis Techniques of Quantum Dots

Quantum dots can be synthesized by various methods like microwave, hydrothermal, laser ablation, electrochemical oxidation, etc., as shown in Fig. 2 and are discussed below.

2.2.1 Colloidal Synthesis

In colloidal synthesis, metal precursors or organometallic compounds are dissolved in an appropriate solvent, subsequently undergoing a series of physical or chemical processes to produce quantum dots with tailored properties. This approach enables the efficient and economical fabrication of quantum dots with diversified properties. This method produces high-quality quantum dots, as demonstrated in the synthesis of CdSe quantum dots, which exhibit visible luminescence across a broad range of colours [40]. The ‘Hot Injection Method’ is a widely utilised approach

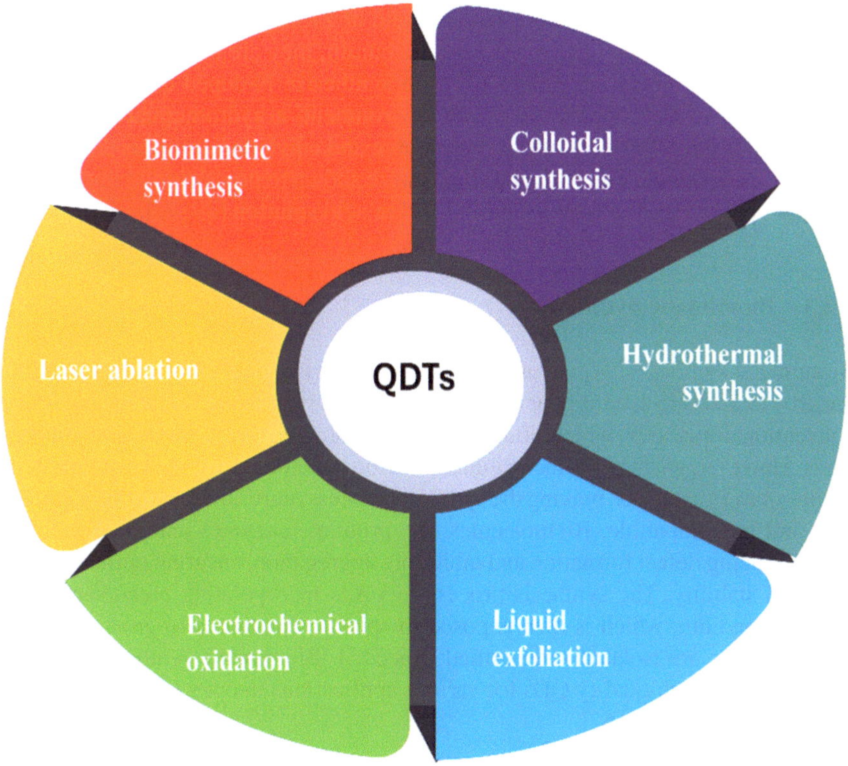

Fig. 2 Synthesis methods of quantum dots

in colloidal synthesis. A precursor solution containing the requisite ingredients is rapidly injected into a heated, high-temperature reaction mixture. Consequently, the precursors rapidly react to produce quantum dots, which promptly cool and solidify. The size and composition of colloidal quantum dots can be regulated by modifying the type, precursor concentration, reaction duration and temperature during synthesis [44].

2.2.2 Hydrothermal Method

Hydrothermal synthesis is a commonly employed method for the production of multifunctional quantum dots. The procedure entails a solution-based chemical reaction taking place within a sealed teflon-lined or steel vessel, typically known as an autoclave. This procedure is defined as a technique for synthesising single crystals, dependent on the solubility of minerals in high-temperature, high-pressure aqueous conditions. CdTe quantum dots are produced at comparatively low temperatures by controlling the concentration of CdTe precursors. The synthesised quantum dots show high luminescence with a quantum yield of 27.4%, and when they are conjugated with proteins such as bovine serum albumin, they are suitable for biological applications as fluorescent probes [42]. To synthesise N-doped quantum dots for the detection of Fe^{3+} ions in an acidic environment, a hydrothermal synthesis of microcrystalline cellulose as the carbon source and ethylenediamine as the nitrogen dopant was designed. The N-doped CQDs were shown to possess good fluorescence emission properties and high quantum yield in acid solution [79, 80].

2.2.3 Biomimetic Synthesis

Biomimetic synthesis of quantum dots (QDs) uses biological components to mimic the growth and self-assembly of QDs. This method offers numerous advantages over conventional methods, which often require high temperatures, chemicals and pressures. However, green biomimetic mineralization technology uses enzymes, proteins or microbes as catalysts, making the synthesis process more environmentally sustainable and biocompatible. Biomolecules like proteins, enzymes and DNA stabilize QDs, reducing defect formation and inhibiting aggregation, ensuring optimal dispersion and stability. The synthesis process involves incorporating precursors into a reaction medium, which is then exposed to specific conditions to grow QDs. The resulting QDs are isolated and purified if needed. Biomimetic synthesis is crucial for generating high-quality QDs for various applications in optoelectronics, energy, biology and medicine [62].

2.2.4 Other Methods

Laser ablation is a promising method to produce quantum dots, as it can produce high-quality materials. In this method, the target material is ablated by laser to produce quantum dots. For example, carbon quantum dots were synthesised using dual-beam pulsed laser ablation. It produces quantum dots suitable for bioimaging applications due to the homogeneity and high-quantum yield [15]. Similarly, researchers have produced black phosphorus quantum dots using pulsed laser ablation. The material is found to have a high quantum yield and stable photoluminescence. It makes material beneficial for cell bioimaging [68].

Quantum dots, produced through electrochemical etching, are widely used in various scientific disciplines due to their unique chemical and physical properties and cost-competitive fabrication processes. This process uses electrical and chemical reactions, with the anode and cathode being of the same metal. Silicon is a promising anode candidate for electrochemical etching-based synthesis of SiQDs, allowing rapid creation of SiQDs with a range of fluorescence light and narrow size dispersion. Photostable silicon quantum dots were produced by Chen et al. by using the electrochemical etching method followed by modifications by glutamate and polyvinyl pyrrolidone (PVP). Glutamate improves the fluorescence while PVP increases the dispersion of quantum dots [14]. The electrochemical method can be used to produce graphene quantum dots. Graphite materials are usually oxidised electrochemically and exfoliated to produce graphene quantum dots. Fu and team produced highly sensitive N-doped graphene quantum dots by this technique, which exhibited multicolour fluorescence. The material was found to be highly sensitive for detecting Fe^{3+} ions, which is significant for bioimaging applications [24].

Liquid exfoliation method is a prominent technique to synthesise quantum dots. In this method, bulk material is dispersed in a solvent to exfoliate it into quantum dots. For example, researchers have shown synthesis of tungsten disulphide quantum dots using this method. It is a green and sustainable approach to producing quantum dots useful for electrochemical sensing applications [27]. Similarly, highly photoluminescent boron nitride QDs were produced by using mechanochemical exfoliation in ethanol solvent for biosensing applications [7].

2.3 Surface Functionalization and Stabilization Methods

Surface functionalization is critical to enhance the properties and applications of quantum dots. Various techniques are used to tailor the surface of quantum dots to improve their stability, solubility and functionality for diverse applications. A notable method involves using proteins such as bovine serum albumin to functionalize the surface of quantum dots. This approach replaces hydrophobic ligands with multidentate ligands from bovine serum albumin. It significantly enhances their photoluminescence and stability under physiological conditions. This method has been successfully applied in the development of biosensors for detecting biomarkers like glycosylated haemoglobin (HbA1c) [41]. The use of different ligands can significantly alter the properties of quantum dots. For example, Cu_2S quantum are functionalized with stabilizers like mercaptoacetic acid, mercaptopropionic acid and glutathione. Among these, glutathione provides the best colloidal stability and lowest toxicity, making it suitable for bio-applications [78]. Similarly, CdSe/ZnS quantum dots are functionalized with carboxylate groups using heterobifunctional ligands to improve water solubility and maintain photoluminescence [64]. Microwave-assisted hydrosilylation method is used for silicon QDs, where molecules with various terminal functionalities are grafted onto the surface. This process improves their dispersibility, cell compatibility and photoluminescence properties. It makes them effective as fluorescent probes in biological applications [76].

Silica encapsulation is an effective method to stabilize QDs. It provides a protective layer that improves the stability and photoluminescence. For example, Li and team demonstrated the stabilization and development of nanobeads for biosensing applications [41]. Wang et al. [77] have demonstrated the use of poly-zwitterionic ligands to enhance the colloidal and photophysical stability of perovskite quantum dots. This method allows for easy phase transfer and preserves the photophysical characteristics of the QDs over extended periods. A combination of steric hindrance and charge stabilization can be used to stabilize quantum dots in various solvents. This hybrid approach has been used to transfer CdSe/CdS core/shell quantum dots between polar and nonpolar solvents while maintaining their photoluminescence [79, 80]. In order to reduce ion migration and crystal deformation, atomic layer deposition is used to coat quantum dots with inorganic layers, such as alumina. This method has demonstrated effective stabilization of quantum dots without compromising their photoluminescent quantum yield [81].

Park and team had elucidated the fluorescence characteristics of the quantum dots and the methodology for conjugating antibody quantum dots. The surface of the quantum dots was co-decorated with a zwitterionic ligand and a ligand containing a carboxylic acid. The carboxylic acid was employed for conjugation with the antibody by carbodiimide coupling (Fig. 3) [59].

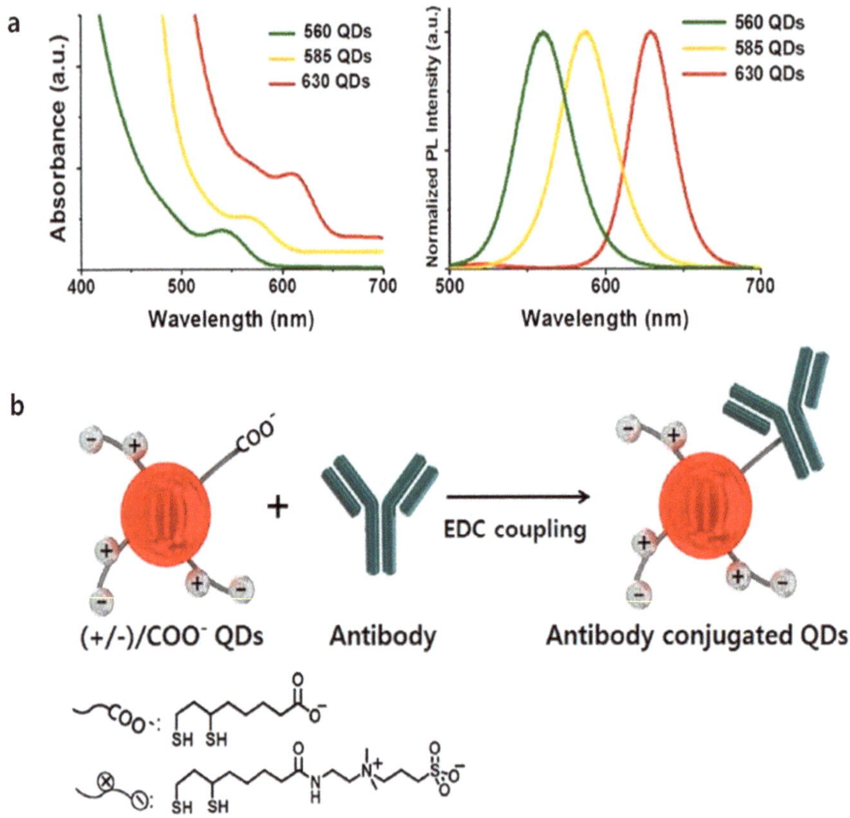

Fig. 3 **a** Absorption and fluorescence profiles of the quantum dots. **b** Scheme for antibody QD conjugation. 'Adapted with permission from Park et al. [59]. Copyright 2014 American Chemical Society'

3 Quantum Dots in Biosensing

3.1 Optical Biosensors

3.1.1 Mechanisms

Fluorescence resonance energy transfer (FRET)-based biosensors often incorporate quantum dots (QDs) because of their favourable photophysical characteristics, including their large absorption coefficients and sharp emission bands. FRET transfers energy from an excited donor (QD) to an acceptor molecule, observing a change in the donor's fluorescence intensity. Different biosensing processes, including gene sequencing and the identification of other biomolecules, apply this mechanism. Graphene quantum dots (GQDs) paired with AuNPs have been used to find the mecA gene sequence of *Staphylococcus aureus* using FRET. They were very good at quenching and could only find that sequence [70]. Similarly, QDs can be used in combination with graphene oxide (GO) for biomolecule sensing, where the interaction between QDs and GO leads to fluorescence quenching, which is reversed upon target recognition [18]. Raju et al. have reported sustainable N-doped carbon quantum dots (N-CQDs) and explored them as optical probes for the selective discrimination of H_2O_2 and aspartic acid in aqueous media based on simple photoluminescence 'turn on' and 'turn off' mechanisms [65].

Quenching mechanisms in QD-based biosensors involve the reduction of fluorescence intensity due to interactions with other molecules. This can happen through charge transfer and FRET, among other things, and the studies have designed it in different ways. For instance, fluorescent dyes or metal complexes can quench QDs through FRET or an electron exchange mechanism. Quenching pathways can be altered to construct multianalyte sensing interfaces, and research has shown that QDs, such as CdSe/ZnS, can be functionalized with peptides containing quenching moieties [3]. Furthermore, biosensing based on the plasmonic quenching, utilizing metal-QD nanohybrids, can also be used to modulate optical signals for biosensing, although it requires careful optimization to balance quenching and enhancement effects [31].

We can use colorimetric sensing with QDs to measure things by how their colour or fluorescence intensity changes when they interact with analytes. Researchers frequently combine this method with FRET or quenching mechanisms to enhance the sensitivity and selectivity of test. For example, a nanostructured biosensor that employs FRET quenching for glucose measurement in tears was designed to recover the emission of QDs where the presence of glucose restored the quenched emission of QDs, allowing for accurate detection [13]. Similarly, carbon quantum dots (CQDs) have been employed in sensing applications for the detection of various analytes by utilizing their fluorescence quenching and energy transfer properties [54]. Klostranec et al. [38] have shown the diagnostic scheme of quantum dots as shown in Fig. 4.

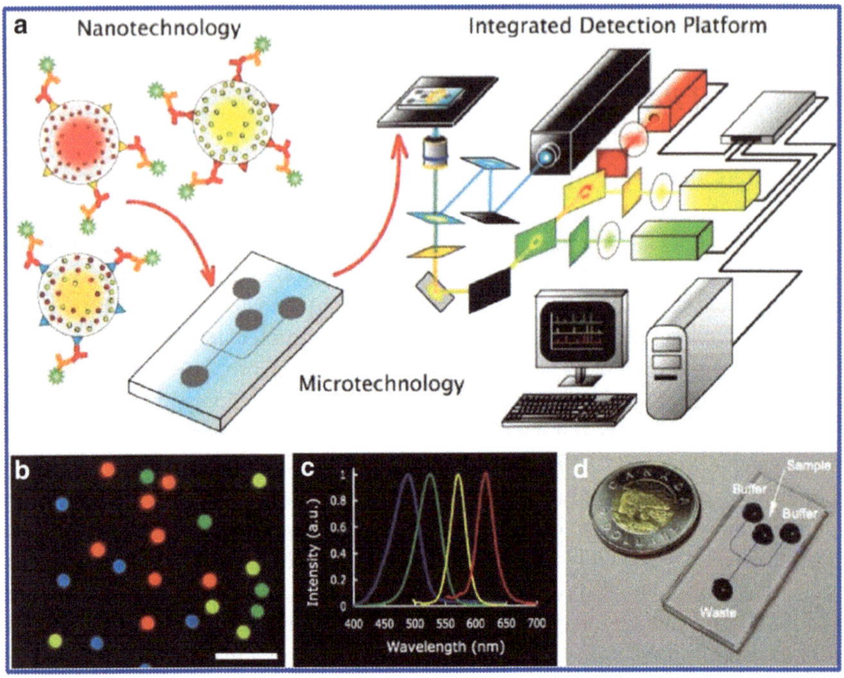

Fig. 4 Diagnostic scheme of quantum dots. Adapted with permission from Klostranec et al. [38]. Copyright 2007 American Chemical Society

3.1.2 Applications

Quantum dots (QDs) have attracted much attention in the field of optical biosensing because of their exceptional optical and electronic characteristics. These properties include high brightness, excellent photostability, wide absorption and narrow size-tunable emission. Such properties make them suitable for highly sensitive selective detection of biomolecules such as proteins, nucleic acids and small molecules.

Quantum dots find a clear application in facilitating the sensing of proteins and enzymes. Different biosensing techniques efficiently use these compounds for detection due to their high quantum yield and narrow emission spectra. For example, QDs directly attached to antibodies or aptamers are used to make the test more specific and to find low-molecular mass molecules in health and the environment. Furthermore, quantum dots (QDs) are utilized in immunoassays, including fluorescence-linked immunosorbent assays (FLISA) and fluorescence resonance energy transfer (FRET), for protein identification [19]. Xiao and his colleagues demonstrated the identification of human metaphase chromosomes by fluorescent in situ hybridisation (FISH) utilizing biotinylated DNA probes and QD-streptavidin conjugates, which is advantageous for investigative and therapeutic purposes [82]. Various ligands or peptide hormones can be conjugated to quantum dots (QDs) that interact with receptors in living cells, as demonstrated by Tomlinson et al., wherein Angiotensin II (Ang II) was linked to AMP-coated CdSe/ZnS core–shell nanocrystals via 1-[3-(Dimethylamino)propyl]-3-ethylcarbodiimide hydrochloride (EDC) coupling [73].

Quantum dots provide a flexible and very sensitive medium for nucleic acid detection, applicable in disease diagnostics and environmental monitoring. Their capacity for integration into many sensing methods, including FRET and nanoquencher systems, underscores their potential in enhancing nucleic acid detection technology. The continuous advancement of QD-based sensors ensures enhancements in sensitivity, selectivity and real-time analytical capabilities. Quantum dots are utilized in the detection of nucleic acids, including DNA and microRNAs. They possess a wide absorption spectrum, and their size tunable emission spectra can be precisely adjusted to accommodate multiplexed detection systems. Fluorescent biosensors utilizing quantum dots have been developed and synthesised for nucleic acid hybridisation assays, exhibiting great sensitivity and selectivity [48]. Graphene quantum dots are utilized for the detection of circulating cell-free nucleic acids. Due to their biocompatibility and photostability, these probes can be used for real-time detection in liquid biopsy-based techniques necessary for early diagnosis and treatment monitoring of diseases such as cancer [66]. In a study, researchers have explained the application of DNA-functionalized QDs for the identification of DNA and microRNA (miRNA). The QDs were conjugated to thiolated ssDNA for detection of nucleic acids with high sensitivity and specificity, which can differentiate between single nucleotide mismatches and can differentiate between different types of miRNA [71]. Researchers have developed nucleic acid/QD hybrid systems for optical sensing platforms. These systems employ different photophysical processes, including fluorescence and electron transfer quenching, to identify DNA and other biomolecules. The incorporation of these systems with electrodes enables the photocurrents as the transduction signal, which improves the sensitivity and general use of the sensors [23].

To improve the selectivity and sensitivity of the biosensors, quantum dots are immobilized or conjugated with biomolecules such as antibodies, aptamers or enzymes. These conjugates can detect small molecules through diverse assay techniques such as enzyme-linked methods and immunoassays [20]. Enzyme-linked assays used quantum dots for the detection of small molecules. These assays take advantage of the luminescent characteristic of QDs to make the detection step highly sensitive and selective. For example, QDs can be directly coated with enzymes or aptamers to produce biosensors that can quantitatively determine small molecules based on changes in emission intensity [4]. When aptamers, a type of short DNA or RNA molecule that have an affinity for target molecules, are linked to QDs, they create a highly sensitive biosensor. These aptamer-QD conjugates have been employed to detect numerous small molecules in biological and environmental matrices, recounting the flexibility and efficiency of QDs in biosensing systems [37]. Quantum dots have been employed in immunosensing applications to detect low-molecular weight analytes. By conjugating QDs with antibodies, researchers have developed immunoassays such as fluorescence-linked immunosorbent assays

(FLISA) and FRET-based methods, which are capable of detecting small analytes relevant to health, food safety and environmental monitoring [19].

3.2 *Electrochemical Biosensors*

QDs have become an important component when it comes to the design and fabrication of electrochemical biosensors because of their optical and electronic characteristics. Introducing QDs with electrodes improves the sensitivity and selectivity of the biosensors. For instance, a biosensor utilizing gold nanoparticles on titanium dioxide nanotubes fabricated with cadmium sulphide quantum dots exhibited a sixfold increase in sensitivity for cholesterol detection compared to pristine samples. This kind of structure promotes the rapid electron transfer and provides the voltage-tunable dual-analyte detection function, which makes it a potential enzyme-free biosensing platform [36]. Similarly, when incorporating the poly(propylene imine) dendrimer with a biosensor composed of 1-aminopropyltriethoxysilane-functionalized CdTe/CdSe/ZnSe QDs, charge transfer and analytical performance were improved for cholesterol detection [53]. Electrochemical biosensor was synthesized by researcher that can swiftly detect HTLV-1 DNA targets without the need for nucleic acid amplification [86].

Integrating QDs in glucose biosensors has been found to be quite effective. A novel electrochemiluminescent (ECL) bioassay based on CdTe QDs and AuNPs-modified glassy carbon electrodes exhibited excellent performance for glucose assay. The working principle of the biosensor is based on the quenching effect of ECL emission in the presence of added glucose, with a detection limit of 5.28 μmol L^{-1} [46]. Another strategy involved the fabrication of a highly sensitive biosensing interface based on CdTe QDs and carbon nanotubes. Such a configuration facilitated direct electron transfer between glucose oxidase and the electrode, which provided a remarkably sensitive response to glucose [47].

QDs integration on electrodes has also helped in the detection of cholesterol. An efficient mediator-free application of chitosan-capped CdS QDs-based biosensors was exploited for the detection of total cholesterol without the use of any mediators, proving the versatility of QDs in the field of clinical diagnosis [16]. Additionally, a biosensor of CdTe/CdSe/ZnSe QDs formed in poly(propylene imine) dendrimer detected cholesterol at 0.075 mM concentration and retained stability for one month. It indicates the robustness of quantum dot-based platforms [53]. Moreover, a biosensor employing a platinum/reduced graphene oxide/poly(3-aminobenzoic acid) film modified electrode displayed excellent linearity, high sensitivity for glucose and cholesterol, indicating the versatility of QD integrated system [63].

3.3 Nanomechanical Biosensors: Role of Quantum Dots in Enhancing Sensitivity and Precision

Quantum dots (QDs) are semiconductor crystalline materials with dimensions in the nanometre range and possess peculiar optical and electrical characteristics, which made them promising materials in biosensing. The properties include high brightness, photostability, broad absorption spectra and size-tunable narrow emission spectra, which make them suitable for increasing the sensitivity and accuracy of biosensors [48]. The large surface area of quantum dots available for binding of biomolecules significantly improves the sensitivity of biosensors. It makes it possible to detect analytes even at low concentrations. For example, ZnO quantum dots were used to improve the sensitivity of electrochemical biosensors by offering a stable platform for enzyme immobilization for detection of uric acid [5]. Similarly, to enhance photocurrent CdTe quantum dots were incorporated in photoelectrochemical biosensors for sensitive detection of microRNA-21 with a detection limit as low as 3.3 fM [84].

In addition, using quantum dots also increases the biosensors' accuracy because of their ability to reproduce consistent signals. For instance, graphene quantum dots that have notable photo-stability and biocompatibility play an important role in reproducibility and selectivity of biosensors [49]. Furthermore, incorporating of CdSe/ZnS QDs into porous silicon microcavity biosensors was demonstrated to enhance detection resolution due to enhanced Refractive Index which in turn enhances biomolecular recognition [85].

Due to the high surface area, QDs can be easily functionalized with a number of bioreceptors, hence improving biosensing properties. This functionalization capacity enables QDs to be tuned to specific receptors like DNA molecules, antioxidants, enzymes and other proteins, enhancing their use in biomedical science besides their use in disease diagnostics. The capacity to tune the surface of QDs with specific recognition molecules expands their usefulness in creating specific biosensors [22].

Recent development of utilizing QDs in biosensors is making new progress in improving their electronic characteristics and exploration of new synthesis techniques for modifying QDs surfaces for more efficiency. It is believed that the progress of QD-based biosensors can revolutionize several clinical diagnostics and bioanalysis by improving the rates of disease diagnosis at an early stage [20]. Summing up, quantum dots are documented to be indispensable for increasing the sensitivity and selectivity of nanomechanical biosensors. For this reason, they play a crucial role in creating novel biologically enhanced sensing platforms with significant implications for biomedical research and clinical diagnostics.

4 Quantum Dots in Bio-diagnostics

4.1 Imaging and Tracking Biomolecules

Quantum dots (QDs) are semiconductor nanoparticles that have recently found wide application in bioimaging and biomolecule tracking because of their fluorescent characteristics. These properties include size-tunable light emission, high brightness and photostability. It makes them more promising than traditional fluorescent dyes and proteins for in vitro and in vivo use [9]. They provide superior brightness and photostability relative to traditional organic dyes, essential for prolonged imaging applications. Their resistance to photobleaching aids in maintaining the strength of their signals for prolonged durations [17]. QDs can be conjugated with other imaging techniques such as magnetic or radioactive materials to develop multifunctional imaging probes. This integration increases the usefulness of QDs in different imaging techniques [8].

QDs are employed in single particle tracking methods to study the dynamics of biomolecules in living cells with high photostability and spatial precision. The refinement of QD surface coating and bioconjugation has improved their usability for live cell imaging [11]. Quantum dots conjugated with biomolecules like antibodies, peptides or small molecules can selectively label and image tumour biomarkers and vasculatures with high binding affinity and selectivity [17]. Quantum dots have been used in imaging microorganisms such as viral, bacterial and fungal microorganism because of their ability to couple with functional biomolecules [25]. Quantum dots have also been widely used in microbial imaging owing to their excellent photostability, long-term stability, and high sensitivity. Semiconductor QDs and carbon nanodots have been used to image different microorganisms including viruses, bacteria and fungi. This application expands the application of QD-based imaging techniques into realizing a new set of imaging protocols in microbial systems and in understanding microbial pathogenic processes [25]. A new application of QDs is laser desorption/ionization mass spectrometry imaging (LDI MSI) for the identification and visualization of small biomolecules. HGQDs, which are heteroatom-doped graphene quantum dots, were created to make it easier to find low-molecular weight compounds in tissues, such as metabolites. This has been applied effectively to address the problem of small biomolecule localisation in rat brain tissue as well as get valuable information about biochemical processes in the state of health and diseases [34].

4.2 Quantum Dots in Cancer Diagnosis and Therapy

Early cancer diagnosis is crucial for successful treatment. Traditional methods like CT scans, MRI and endoscopy have limitations like high costs, imprecision and limited resources. Nanotechnology, specifically quantum dots, offers a non-invasive, sensitive and biocompatible method for early cancer identification. Traditional treatments have side effects, making quantum dots a promising alternative and, rendering them valuable instruments in cancer diagnosis and therapy. Their elevated luminosity, stability and adjustable emission spectra provide sophisticated imaging and detection methodologies, potentially enhancing early cancer diagnosis and tailored treatments. QDs can also serve as a drug delivery system to deliver drugs intracellularly as shown in Fig. 5. Many tumours can be targeted via active or passive means. These two properties can be combined to develop a theranostic system that offers a dual advantage in diagnosis and therapy [39].

Quantum dots have been used in numerous diagnostic methods, like sentinel lymph-node mapping and tumour targeting in vivo. Altogether, their conjugation with biomolecules like peptides and antibodies permits the specification of tumour targeting, cancer cell metastasis detection and quantification of particular molecular markers [30]. Additionally, QDs are used in the monitoring of cancer-related microRNAs (miRNAs), which are overexpressed in cancer cells, providing insights into disease progression and treatment efficacy [51]. They are reported to show low cytotoxicity at the concentrations typically used for fluorescence imaging. The principal detection mechanism relies on the interaction between quantum dots and the surface groups of cancer cells. The quantum dots are internalized into the cells via endocytosis. Upon exposure to quantum dots, cancer cells typically accumulate in the nucleus or cytoplasm. Targeted staining with quantum dots for specific cellular imaging has been achieved by conjugating specialised ligands, including transferrin, folic acid and hyaluronic acid [61]. Researchers have shown synthesis of biosensor

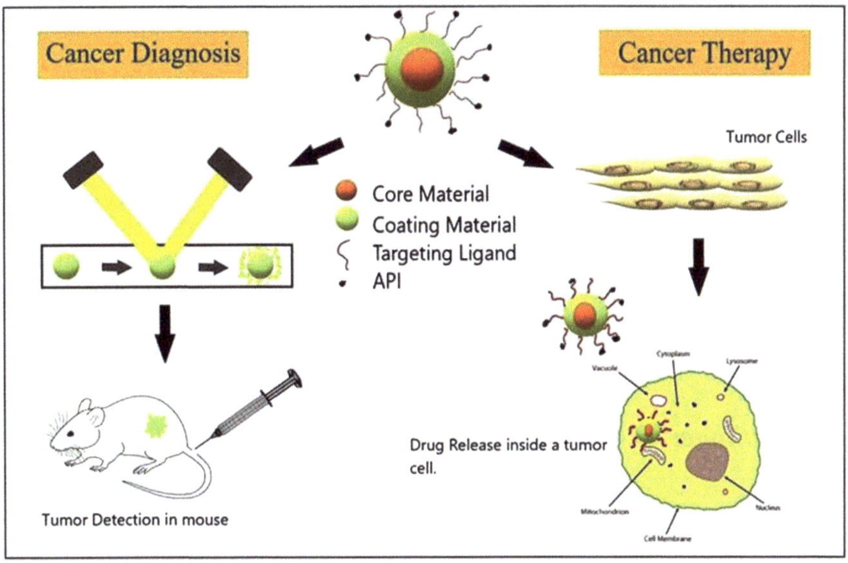

Fig. 5 Overview of the potential use of quantum dots in cancer diagnosis. Reused with permission from Kulkarni et al. [39]. Copyright 2019 Elsevier (Licence no. 5938071251490)

for detection of cancer biomarker carcinoembryonic antigen (CEA) in blood serum using CDs bio-functionalized with Horseradish Peroxidase (HRP)-conjugated CEA antibody as detection probe [12].

The folate receptor is overexpressed in cancer cells and tumours, including those of the breast, brain, cervix and ovary. Folic acid serves as an optimal ligand for the targeting and imaging of cancer cells. Researchers have synthesized luminous carbon quantum dots exhibiting a high quantum yield utilizing folic acid as a precursor. These quantum dots demonstrate exceptional characteristics, including great photostability, photoluminescent activity and favourable biocompatibility, enabling them to effectively penetrate cancer cells [45].

Nano-drug carriers for oncological therapy have gained attention due to their superior water solubility and biodegradability. Quantum dots can be integrated into these nanocarriers to improve drug delivery systems. Their fluorescence imaging capacity allows for monitoring drug response, distribution and activity through theranostic techniques. The removal of the linked drug moiety results in fluorescence recovery, which can be tracked over time. Quantum dots are linked through covalent or non-covalent bonds, offering enhanced control for selective release. For example, Gong et al. produced phosphorus and nitrogen-doped carbon quantum dots exhibiting improved drug loading capacity attributed to the presence of a central hollow cavity and several functional groups on the surface, including phosphoric, pyridinic, pyrrolic and hydroxyl [29].

Photoinduced therapy has garnered significant interest in recent years due to its non-invasiveness, excellent specificity and minimal injury to normal tissues [21]. This therapy generally comprises photothermal therapy (PTT) and photodynamic therapy (PDT). In PTT, near-infrared light is employed to produce heat that eradicates cancer cells by increasing tissue temperature. The photodynamic approach to cancer treatment involves the interaction between light and photosensitising drugs. This therapy involves the exposure of photosensitizers to light of a certain wavelength, resulting in the generation of reactive oxygen species that induce oxidative DNA damage and eliminate cancer cells [57]. The mechanism of DNA strand breakage and nucleobase damage caused by photoactivated quantum dots (QDs) involves the generation of reactive oxygen intermediates (ROIs), such as singlet oxygen and hydroxyl radicals. These ROIs initiate base modifications and hydrogen abstraction reactions, ultimately leading to strand scission. A schematic representation of these processes has been illustrated by Anas et al. [6] as shown in Fig. 6.

4.3 Pathogen Detection in Infectious Diseases

Quantum dots are promising materials for the detection of viral infections. Quantum dots can be used in the new generation of ultra-sensitive probes that combine biomarkers for multiplex detection of human viral infections. This kind of measurement enables multiple clinical factors to be measured, which saves both time and cost

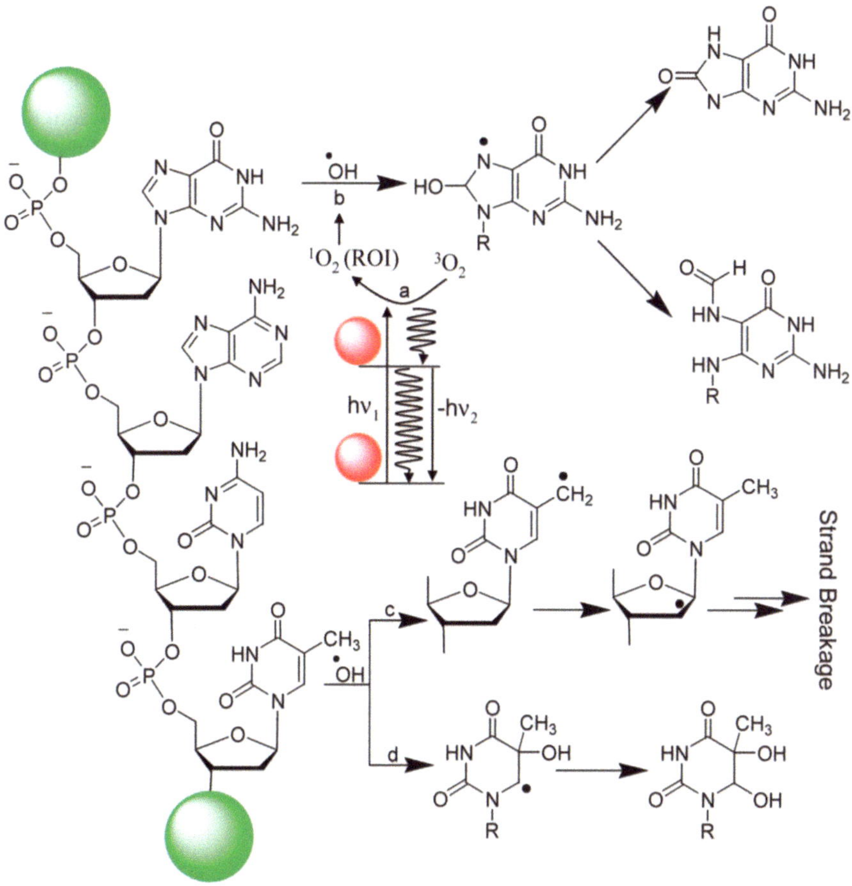

Fig. 6 Schematic presentation of **a** the photoinduced production of reactive oxygen intermediates (ROI) by quantum dots (QDs), **b**, **d** the nucleobase damage caused by ROI and **c** hydrogen abstraction leading to DNA strand breakage. Reproduced with permission from Anas et al. [6]. Copyright © 2008 American Chemical Society (Licence no. 6082600849271)

[56]. In a research, anodic electrochemiluminescence emitters utilizing selenium-based polymer dots are explored for the simultaneous detection of two pneumonia pathogens with CdS quantum dots [50].

Liang et al. developed a relative fluorescence biosensor utilizing carbon quantum dots (CQD) and cadmium telluride quantum dots (CdTe QDs) for the detection of HIV dsDNA in serum. Mercapto-propionic acid-coated CdTe quantum dots were initially conjugated with mitoxantrone. MTX is a synthetic anthraquinone compound that disrupts DNA, leading to red fluorescence quenching at 599 nm as a result of electron transfer between CdTe quantum dots and MTX. In this scenario, fluorescence emission at 435 nm is solely detectable due to the green quantum dots. Ultimately, the CdTe QDs disperse from the MTXssDNA complex as a result of electrical repulsion. Moreover, the presence of HIV-dsDNA facilitated the specific binding of MTX to dsDNA, leading to the dissociation of MTX from CdTe QDs [43]. Quantum dots facilitate simultaneous detection of many diseases. In research, quantum dots were utilized in an electrochemical biosensor to identify non-protein-coding RNA sequences from *Vibrio cholerae*, *Salmonella* sp. and *Shigella* sp. The functionalization of quantum dots with DNA probes tailored to each pathogen facilitates unique electrochemical signals for every target [75]. A highly sensitive immunosensor was developed by researchers using streptavidin-conjugated quantum dots to detect the non-structural protein-1, a dengue biomarker, at very low concentrations. It enables the identification of dengue infection at early stages. The fluorescence emission spectra reveal the linear response to the NS1 antigen concentrations, which demonstrates the potential of quantum dots for point-of-care applications [74]. A tunable biosensor utilizing localised surface plasmon resonance (LSPR) has been created for virus detection by regulating the distance between fluorescent CdZnSeS/ZnSeS quantum dots (QDs) and gold nanoparticles (AuNPs). The detection process entails the quenching of quantum dot fluorescence resulting from the attenuation of the localized surface plasmon resonance signal of gold nanoparticles in proximity to the quantum dots, as depicted in Fig. 7 [58].

5 Challenges of Using Quantum Dots in Biomedical and Bio-diagnostic Applications

5.1 Cytotoxicity

The cytotoxicity associated with quantum dots (QDs) presents a considerable obstacle in their application for biosensors and bio-diagnostic purposes. Quantum dots, often made from heavy metals such as cadmium, lead or selenium, have the potential to leach into biological systems, resulting in cellular toxicity. The mechanisms underlying cytotoxicity encompass the liberation of heavy metals, the induction of oxidative stress and the interactions associated with surface coatings. Factors that affect cytotoxicity encompass particle size, surface modifications, as well as concentration and dosing parameters. Strategies for mitigation encompass biocompatible surface coatings, core–shell architectures and non-toxic alternatives. The impact of cytotoxicity on biosensors and bio-diagnostics presents significant safety concerns that must be carefully considered.

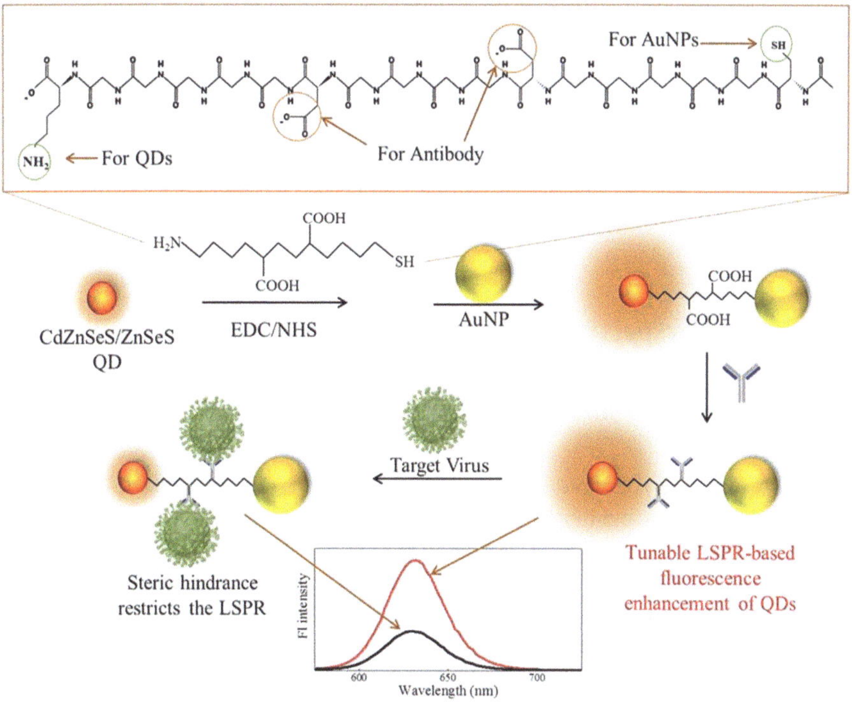

Fig. 7 Schematic diagram for the preparation of CdZnSeS/ZnSeS QD-peptide-AuNP nanocomposite and its detecting mechanism towards influenza virus. Reused with permission from Nasrin et al. [58]. Copyright 2020 Elsevier (Licence no. 5938110745462)

5.2 *Scalability*

The synthesis of quantum dots presents a notable challenge: achieving scalability. A variety of methods are used for the synthesis of quantum dots; however, attaining uniform quality and properties at an industrial scale presents significant challenges. The synthesis techniques require modification and diligent control to achieve consistency in morphology, size and surface functionalization. Furthermore, the integration of quantum dots for commercial applications presents challenges primarily due to the expenses associated with large-scale synthesis processes.

5.3 *Environmental and Biocompatibility Issues*

The degradation of quantum dots and the potential release of toxic components into the environment are growing environmental concerns. It is very important that quantum dots stay stable in biological and environmental settings, since breaking them down can cause harmful chemicals to leak out. We need to explore the development of environmentally friendly QDs to address these concerns. Biocompatibility is crucial for the safe use of quantum dots in biological applications. Quantum dots require surface functionalization to reduce toxicity and improve compatibility with biological systems. The way QDs interact with biological molecules and their ability to trigger bad immune responses are important issues that need to be looked into to make sure they can be used safely in biosensors and bio diagnostics.

6 Future Perspectives

6.1 Emerging Trends in Quantum Dot Technology

Quantum dot (QD) technology is improving as a result of material synthesis, surface modification and integration with other nanotechnologies. Carbon-based and certain lead-free perovskite QDs are being explored to reduce toxicity concerns. The combination of QDs with other nanostructures such as graphene, MOFs and nanowires is also useful, as it provides synergistic phenomena to improve the biosensor detection capabilities. Miniaturization and wearable technology are helping to develop portable QD-based diagnostic devices and endless point-of-care testing and continuous health monitoring.

6.2 Potential Impact on Personalized Medicine and Point-of-Care Diagnostics

Quantum dots have unique optical and electronic properties, making them a promising material for advancements in personalized medicine and point-of-care diagnostics. Due to bright, tunable fluorescence characteristics, they facilitate precise multiplexed detection of biomarkers. Biosensors incorporated with quantum dots can detect genetic, proteomic or even metabolic biomarkers and thereby give valuable insights into disease progression and the effectiveness of therapeutic interventions. Its detection abilities are relatively fast and highly accurate, which are critical for helping healthcare providers to make clinical decisions early. Scalability is a significant challenge in the production of quantum dots. There are numerous methods to produce quantum dots, but it is a significant challenge to attain uniform quality and properties in large-scale productions. In order to achieve uniformity in morphology, size and surface functionalization, modifications are required for precise and controlled synthesis methods. The cost associated with large-scale synthesis of quantum dots is also an obstacle to using quantum dots for bio-diagnostic and biosensing applications at commercial scale.

6.3 Opportunities for Commercialization and Interdisciplinary Research

The advancement of biosensors employing quantum dots (QDs) is growing due to the demand for sophisticated sensing technologies and robust healthcare systems. Significant potential exists in advancing scalable synthesis techniques and incorporating quantum dots into commercial biosensors and bio-diagnostic applications. Academia must collaborate with industries to ensure adherence to regulations and create material that prioritises real life use. Significant potential exists in advancing scalable production techniques and incorporating quantum dots into commercially feasible applications. To realize the complete capability of quantum dots in biosensing applications, interdisciplinary research is critical to tackle issues such as biocompatibility, stability, cytotoxicity and functional integration. Collaborations with data scientists and clinicians can enhance the development of intelligent diagnostic platforms.

7 Conclusion

Quantum dots are crucial for biosensing and bio-diagnostic applications due to their excellent optical and electronic properties. Their brightness, stability, adjustable fluorescence properties and versatility for surface modifications have led to advancements in various biosensor technologies. It has led to significant progress in health care and clinical diagnostics due to improved sensitivity, selectivity and precision in biomolecule detection by quantum dots. QDs have revolutionized imaging techniques, enhanced cancer diagnostics, pathogen detection and therapeutic monitoring in bio-diagnostics. Advancements like multiplexed imaging, targeted drug delivery and real-time biomolecular tracking have significantly impacted personalized medicine and point-of-care diagnostics. However, cytotoxicity, scalability, cost and environmental impact are challenges which researchers need to address. In order to address these challenges, researchers must collaborate with industries and look for more environmentally friendly and biocompatible designs. The combination of QDs with advanced technologies like wearable devices, lab-on-a-chip systems and AI-driven diagnostics is expected to transform health care and enhance global health outcomes.

References

1. Abdelsalam H, Zhang QF (2022) Properties and applications of quantum dots derived from two-dimensional materials. Adv Phys: X 7:2048966. https://doi.org/10.1080/23746149.2022.2048966

2. Ahmad A, Imran M, Ahsan H (2023) Biomarkers as biomedical bioindicators: approaches and techniques for the detection, analysis, and validation of novel biomarkers of diseases. Pharmaceutics 15:1630. https://doi.org/10.3390/pharmaceutics15061630
3. Algar WR, Khachatrian A, Melinger JS, Huston AL, Stewart MH, Susumu K, Blanco-Canosa JB, Oh E, Dawson PE, Medintz IL (2017) Concurrent modulation of quantum dot photoluminescence using a combination of charge transfer and Förster resonance energy transfer: competitive quenching and multiplexed biosensing modality. J Am Chem Soc 139:363–372. https://doi.org/10.1021/jacs.6b11042
4. Algar WR, Tavares AJ, Krull UJ (2010) Beyond labels: a review of the application of quantum dots as integrated components of assays, bioprobes, and biosensors utilizing optical transduction. Anal Chim Acta 673:1–25. https://doi.org/10.1016/j.aca.2010.05.026
5. Ali M, Shah I, Kim SW, Sajid M, Lim JH, Choi KH (2018) Quantitative detection of uric acid through ZnO quantum dots based highly sensitive electrochemical biosensor. Sens Actuators, A 283:282–290. https://doi.org/10.1016/j.sna.2018.10.009
6. Anas A, Akita H, Harashima H, Itoh T, Ishikawa M, Biju V (2008) Photosensitized breakage and damage of DNA by CdSe–ZnS quantum dots. J Phys Chem B 112:10005–10011. https://doi.org/10.1021/jp8018606
7. Angizi S, Hatamie A, Ghanbari H, Simchi A (2018) Mechanochemical green synthesis of exfoliated edge-functionalized boron nitride quantum dots: application to vitamin C sensing through hybridization with gold electrodes. ACS Appl Mater Interfaces 10:28819–28827. https://doi.org/10.1021/acsami.8b07332
8. Biju V, Itoh T, Ishikawa M (2010) Delivering quantum dots to cells: bioconjugated quantum dots for targeted and nonspecific extracellular and intracellular imaging. Chem Soc Rev 39:3031. https://doi.org/10.1039/b926512k
9. Bilan R, Nabiev I, Sukhanova A (2016) Quantum dot-based nanotools for bioimaging, diagnostics, and drug delivery. ChemBioChem 17:2103–2114. https://doi.org/10.1002/cbic.201600357
10. Borse S, Rafique R, Murthy ZVP, Park TJ, Kailasa SK (2022) Applications of upconversion nanoparticles in analytical and biomedical sciences: a review. Analyst 147:3155–3179. https://doi.org/10.1039/D1AN02170B
11. Chang Y, Pinaud F, Antelman J, Weiss S (2008) Tracking bio-molecules in live cells using quantum dots. J Biophotonics 1:287–298. https://doi.org/10.1002/jbio.200810029
12. Chellachamy Anbalagan A, Korram J, Doble M, Sawant SN (2024) Bio-functionalized carbon dots for signaling immuno-reaction of carcinoembryonic antigen in an electrochemical biosensor for cancer biomarker detection. Discover Nano 19:37. https://doi.org/10.1186/s11671-024-03980-3
13. Chen L, Tse WH, Chen Y, McDonald MW, Melling J, Zhang J (2017) Nanostructured biosensor for detecting glucose in tear by applying fluorescence resonance energy transfer quenching mechanism. Biosens Bioelectron 91:393–399. https://doi.org/10.1016/j.bios.2016.12.044
14. Chen Y, Sun L, Liao F, Dang Q, Shao M (2019) Fluorescent-stable and water-soluble two-component-modified silicon quantum dots and their application for bioimaging. J Lumin 215:116644. https://doi.org/10.1016/j.jlumin.2019.116644
15. Cui L, Ren X, Wang J, Sun M (2020) Synthesis of homogeneous carbon quantum dots by ultrafast dual-beam pulsed laser ablation for bioimaging. Mater Today Nano 12:100091. https://doi.org/10.1016/j.mtnano.2020.100091
16. Dhyani H, Ali MdA, Pal SP, Srivastava S, Solanki PR, Malhotra BD, Sen P (2015) Mediator-free biosensor using chitosan capped CdS quantum dots for detection of total cholesterol. RSC Adv 5:45928–45934. https://doi.org/10.1039/C5RA07012K
17. Díaz-González M, De La Escosura-Muñiz A, Fernandez-Argüelles MT, García Alonso FJ, Costa-Fernandez JM (2020) Quantum dot bioconjugates for diagnostic applications. Top Curr Chem (Z) 378:35. https://doi.org/10.1007/s41061-020-0296-6
18. Dong H, Gao W, Yan F, Ji H, Ju H (2010) Fluorescence resonance energy transfer between quantum dots and graphene oxide for sensing biomolecules. Anal Chem 82:5511–5517. https://doi.org/10.1021/ac100852z

19. Esteve-Turrillas FA, Abad-Fuentes A (2013) Applications of quantum dots as probes in immunosensing of small-sized analytes. Biosens Bioelectron 41:12–29. https://doi.org/10.1016/j.bios.2012.09.025
20. Farzin MA, Abdoos H (2021) A critical review on quantum dots: from synthesis toward applications in electrochemical biosensors for determination of disease-related biomolecules. Talanta 224:121828. https://doi.org/10.1016/j.talanta.2020.121828
21. Ferroni C, Del Rio A, Martini C, Manoni E, Varchi G (2019) Light-induced therapies for prostate cancer treatment. Front Chem 7:719. https://doi.org/10.3389/fchem.2019.00719
22. Frasco MF, Chaniotakis N (2009) Semiconductor quantum dots in chemical sensors and biosensors. Sensors 9:7266–7286. https://doi.org/10.3390/s90907266
23. Freeman R, Girsh J, Willner I (2013) Nucleic acid/quantum dots (QDs) hybrid systems for optical and photoelectrochemical sensing. ACS Appl Mater Interfaces 5:2815–2834. https://doi.org/10.1021/am303189h
24. Fu Y, Gao G, Zhi J (2019) Electrochemical synthesis of multicolor fluorescent N-doped graphene quantum dots as a ferric ion sensor and their application in bioimaging. J Mater Chem B 7:1494–1502. https://doi.org/10.1039/C8TB03103G
25. Gao G, Jiang Y-W, Sun W, Wu F-G (2018) Fluorescent quantum dots for microbial imaging. Chin Chem Lett 29:1475–1485. https://doi.org/10.1016/j.cclet.2018.07.004
26. García De Arquer FP, Talapin DV, Klimov VI, Arakawa Y, Bayer M, Sargent EH (2021) Semiconductor quantum dots: technological progress and future challenges. Science 373:eaaz8541. https://doi.org/10.1126/science.aaz8541
27. Garg M, Chatterjee M, Sharma AL, Singh S (2020) Label-free approach for electrochemical ferritin sensing using biosurfactant stabilized tungsten disulfide quantum dots. Biosens Bioelectron 151:111979. https://doi.org/10.1016/j.bios.2019.111979
28. Gidwani B, Sahu V, Shukla SS, Pandey R, Joshi V, Jain VK, Vyas A (2021) Quantum dots: prospectives, toxicity, advances and applications. J Drug Deliv Sci Technol 61:102308. https://doi.org/10.1016/j.jddst.2020.102308
29. Gong X, Zhang Q, Gao Y, Shuang S, Choi MMF, Dong C (2016) Phosphorus and nitrogen dual-doped hollow carbon dot as a nanocarrier for doxorubicin delivery and biological imaging. ACS Appl Mater Interfaces 8:11288–11297. https://doi.org/10.1021/acsami.6b01577
30. Guo W, Song X, Liu J, Liu W, Chu X, Lei Z (2024) Quantum dots as a potential multifunctional material for the enhancement of clinical diagnosis strategies and cancer treatments. Nanomaterials 14:1088. https://doi.org/10.3390/nano14131088
31. Hildebrandt N, Lim M, Kim N, Choi DY, Nam J-M (2023) Plasmonic quenching and enhancement: metal–quantum dot nanohybrids for fluorescence biosensing. Chem Commun 59:2352–2380. https://doi.org/10.1039/D2CC06178C
32. Jessy Mercy D, Girigoswami K, Girigoswami A (2024) A mini review on biosensor advancements-emphasis on quantum dots. Results Chem 7:101271. https://doi.org/10.1016/j.rechem.2023.101271
33. Jiang M, Sun Y, Chen M, Ji H, Liu Y, Qin R, Li X, Gao H, Zhang R, Zhang L (2024) Multicolor luminescence of carbon dots: from mechanisms to applications. Chem Eng J 496:153761. https://doi.org/10.1016/j.cej.2024.153761
34. Jin Z, Liu M, Huang X, Zhang X, Qu Z, Zhu J-J, Min Q (2022) Top-down rational engineering of heteroatom-doped graphene quantum dots for laser desorption/ionization mass spectrometry detection and imaging of small biomolecules. Anal Chem 94:7609–7618. https://doi.org/10.1021/acs.analchem.2c00802
35. Jokerst JV, Raamanathan A, Christodoulides N, Floriano PN, Pollard AA, Simmons GW, Wong J, Gage C, Furmaga WB, Redding SW, McDevitt JT (2009) Nano-bio-chips for high performance multiplexed protein detection: determinations of cancer biomarkers in serum and saliva using quantum dot bioconjugate labels. Biosens Bioelectron 24:3622–3629. https://doi.org/10.1016/j.bios.2009.05.026
36. Khaliq N, Rasheed MA, Khan M, Maqbool M, Ahmad M, Karim S, Nisar A, Schmuki P, Cho SO, Ali G (2021) Voltage-switchable biosensor with gold nanoparticles on TiO_2 nanotubes decorated with CdS quantum dots for the detection of cholesterol and H_2O_2. ACS Appl Mater Interfaces 13:3653–3668. https://doi.org/10.1021/acsami.0c19979

37. Kim D, Yoo S (2021) Aptamer-conjugated quantum dot optical biosensors: strategies and applications. Chemosensors 9:318. https://doi.org/10.3390/chemosensors9110318
38. Klostranec JM, Xiang Q, Farcas GA, Lee JA, Rhee A, Lafferty EI, Perrault SD, Kain KC, Chan WCW (2007) Convergence of quantum dot barcodes with microfluidics and signal processing for multiplexed high-throughput infectious disease diagnostics. Nano Lett 7:2812–2818. https://doi.org/10.1021/nl071415m
39. Kulkarni NS, Guererro Y, Gupta N, Muth A, Gupta V (2019) Exploring potential of quantum dots as dual modality for cancer therapy and diagnosis. J Drug Deliv Sci Technol 49:352–364. https://doi.org/10.1016/j.jddst.2018.12.010
40. Landry ML, Morrell TE, Karagounis TK, Hsia C-H, Wang C-Y (2014) Simple syntheses of CdSe quantum dots. J Chem Educ 91:274–279. https://doi.org/10.1021/ed300568e
41. Li J, Fan J, Wu R, Li N, Lv Y, Shen H, Li LS (2022) Biomolecular surface functionalization and stabilization method to fabricate quantum dots nanobeads for accurate biosensing detection. Langmuir 38:4969–4978. https://doi.org/10.1021/acs.langmuir.2c00392
42. Li M, Ge Y, Chen Q, Xu S, Wang N, Zhang X (2007) Hydrothermal synthesis of highly luminescent CdTe quantum dots by adjusting precursors' concentration and their conjunction with BSA as biological fluorescent probes. Talanta 72:89–94. https://doi.org/10.1016/j.talanta.2006.09.028
43. Liang S-S, Qi L, Zhang R-L, Jin M, Zhang Z-Q (2017) Ratiometric fluorescence biosensor based on CdTe quantum and carbon dots for double strand DNA detection. Sens Actuators, B Chem 244:585–590. https://doi.org/10.1016/j.snb.2017.01.032
44. Liu B, Altintas Y, Wang L, Shendre S, Sharma M, Sun H, Mutlugun E, Demir HV (2020) Record high external quantum efficiency of 19.2% achieved in light-emitting diodes of colloidal quantum wells enabled by hot-injection shell growth. Adv Mater 32:1905824. https://doi.org/10.1002/adma.201905824
45. Liu H, Li Z, Sun Y, Geng X, Hu Y, Meng H, Ge J, Qu L (2018) Synthesis of luminescent carbon dots with ultrahigh quantum yield and inherent folate receptor-positive cancer cell targetability. Sci Rep 8:1086. https://doi.org/10.1038/s41598-018-19373-3
46. Liu L, Ma Q, Li Y, Liu Z, Su X (2015) A novel signal-off electrochemiluminescence biosensor for the determination of glucose based on double nanoparticles. Biosens Bioelectron 63:519–524. https://doi.org/10.1016/j.bios.2014.07.087
47. Liu Q, Lu X, Li J, Yao X, Li J (2007) Direct electrochemistry of glucose oxidase and electrochemical biosensing of glucose on quantum dots/carbon nanotubes electrodes. Biosens Bioelectron 22:3203–3209. https://doi.org/10.1016/j.bios.2007.02.013
48. Ma F, Li C, Zhang C (2018) Development of quantum dot-based biosensors: principles and applications. J Mater Chem B 6:6173–6190. https://doi.org/10.1039/C8TB01869C
49. Mansuriya B, Altintas Z (2020) Applications of graphene quantum dots in biomedical sensors. Sensors 20:1072. https://doi.org/10.3390/s20041072
50. Mao Z, Dai C, Xu Y, Jia J, Ke L, Zhou Y (2024) Novel efficient selenium-based D–π–A NIR polymer dots anodic electrochemiluminescence emitter and its application in simultaneous detection of two pneumonia pathogens with CdS quantum dots. Anal Chem 96:12760–12766. https://doi.org/10.1021/acs.analchem.4c01873
51. Martins CSM, LaGrow AP, Prior JAV (2022) Quantum dots for cancer-related miRNA monitoring. ACS Sens 7:1269–1299. https://doi.org/10.1021/acssensors.2c00149
52. Mathur A, Gupta R, Kondal S, Wadhwa S, Pudake RN, Shivani, Kansal R, Pundir CS, Narang J (2018) A new tactics for the detection of S. aureus via paper based geno-interface incorporated with graphene nano dots and zeolites. Int J Biol Macromol 112:364–370. https://doi.org/10.1016/j.ijbiomac.2018.01.143
53. Mokwebo KV, Oluwafemi OS, Arotiba OA (2018) An electrochemical cholesterol biosensor based on a CdTe/CdSe/ZnSe quantum dots—poly(propylene imine) dendrimer nanocomposite immobilisation layer. Sensors 18:3368. https://doi.org/10.3390/s18103368
54. Molaei MJ (2020) Principles, mechanisms, and application of carbon quantum dots in sensors: a review. Anal Methods 12:1266–1287. https://doi.org/10.1039/C9AY02696G

55. Moradi M, Abdolhosseini M, Zarrabi A, Johari B (2019) A review on application of nanostructures and nano-objects with high potential for managing different aspects of bone malignancies. Nano-Struct Nano-Objects 19:100348. https://doi.org/10.1016/j.nanoso.2019.100348
56. Mousavi SM, Hashemi SA, Yari Kalashgrani M, Omidifar N, Lai CW, Vijayakameswara Rao N, Gholami A, Chiang W-H (2022) The pivotal role of quantum dots-based biomarkers integrated with ultra-sensitive probes for multiplex detection of human viral infections. Pharmaceuticals 15:880. https://doi.org/10.3390/ph15070880
57. Naik K, Chaudhary S, Ye L, Parmar AS (2022) A strategic review on carbon quantum dots for cancer-diagnostics and treatment. Front Bioeng Biotechnol 10:882100. https://doi.org/10.3389/fbioe.2022.882100
58. Nasrin F, Chowdhury AD, Takemura K, Kozaki I, Honda H, Adegoke O, Park EY (2020) Fluorometric virus detection platform using quantum dots-gold nanocomposites optimizing the linker length variation. Anal Chim Acta 1109:148–157. https://doi.org/10.1016/j.aca.2020.02.039
59. Park Y, Ryu Y-M, Jung Y, Wang T, Baek Y, Yoon Y, Bae SM, Park J, Hwang S, Kim J, Do E-J, Kim S-Y, Chung E, Kim KH, Kim S, Myung S-J (2014) Spraying quantum dot conjugates in the colon of live animals enabled rapid and multiplex cancer diagnosis using endoscopy. ACS Nano 8:8896–8910. https://doi.org/10.1021/nn5009269
60. Pechnikova NA, Domvri K, Porpodis K, Istomina MS, Iaremenko AV, Yaremenko AV (2024) Carbon quantum dots in biomedical applications: advances, challenges, and future prospects. Aggregate e707. https://doi.org/10.1002/agt2.707
61. Peng Z, Han X, Li S, Al-Youbi AO, Bashammakh AS, El-Shahawi MS, Leblanc RM (2017) Carbon dots: biomacromolecule interaction, bioimaging and nanomedicine. Coord Chem Rev 343:256–277. https://doi.org/10.1016/j.ccr.2017.06.001
62. Pérez-Donoso JM, Monrás JP, Bravo D, Aguirre A, Quest AF, Osorio-Román IO, Aroca RF, Chasteen TG, Vásquez CC (2012) Biomimetic, mild chemical synthesis of CdTe-GSH quantum dots with improved biocompatibility. PLoS ONE 7:e30741. https://doi.org/10.1371/journal.pone.0030741
63. Phetsang S, Jakmunee J, Mungkornasawakul P, Laocharoensuk R, Ounnunkad K (2019) Sensitive amperometric biosensors for detection of glucose and cholesterol using a platinum/reduced graphene oxide/poly(3-aminobenzoic acid) film-modified screen-printed carbon electrode. Bioelectrochemistry 127:125–135. https://doi.org/10.1016/j.bioelechem.2019.01.008
64. Pong B-K, Trout BL, Lee J-Y (2008) Modified ligand-exchange for efficient solubilization of CdSe/ZnS quantum dots in water: a procedure guided by computational studies. Langmuir 24:5270–5276. https://doi.org/10.1021/la703431j
65. Raju KS, Das GS, Tripathi KM (2024) Nitrogen-doped carbon quantum dots from biomass as a FRET-based sensing platform for the selective detection of H_2O_2 and aspartic acid. RSC Sustain 2:223–232. https://doi.org/10.1039/D3SU00343D
66. Ratre P, Jain B, Kumari R, Thareja S, Tiwari R, Srivastava RK, Goryacheva IY, Mishra PK (2022) Bioanalytical applications of graphene quantum dots for circulating cell-free nucleic acids: a review. ACS Omega 7:39586–39602. https://doi.org/10.1021/acsomega.2c05414
67. Reineck P, Gibson BC (2017) Near-infrared fluorescent nanomaterials for bioimaging and sensing. Adv Opt Mater 5:1600446. https://doi.org/10.1002/adom.201600446
68. Ren X, Zhang F, Zhang X (2018) Synthesis of black phosphorus quantum dots with high quantum yield by pulsed laser ablation for cell bioimaging. Chem Asian J 13:1842–1846. https://doi.org/10.1002/asia.201800482
69. Sharma A, Sharma N, Kumari A, Lee H-J, Kim T, Tripathi KM (2020) Nano-carbon based sensors for bacterial detection and discrimination in clinical diagnosis: a junction between material science and biology. Appl Mater Today 18:100467. https://doi.org/10.1016/j.apmt.2019.100467
70. Shi J, Chan C, Pang Y, Ye W, Tian F, Lyu J, Zhang Y, Yang M (2015) A fluorescence resonance energy transfer (FRET) biosensor based on graphene quantum dots (GQDs) and gold nanoparticles (AuNPs) for the detection of mecA gene sequence of Staphylococcus aureus. Biosens Bioelectron 67:595–600. https://doi.org/10.1016/j.bios.2014.09.059

71. Su S, Fan J, Xue B, Yuwen L, Liu X, Pan D, Fan C, Wang L (2014) DNA-conjugated quantum dot nanoprobe for high-sensitivity fluorescent detection of DNA and micro-RNA. ACS Appl Mater Interfaces 6:1152–1157. https://doi.org/10.1021/am404811j
72. Thakur A, Kumar A (2022) Recent advances on rapid detection and remediation of environmental pollutants utilizing nanomaterials-based (bio)sensors. Sci Total Environ 834:155219. https://doi.org/10.1016/j.scitotenv.2022.155219
73. Tomlinson ID, Mason JN, Blakely RD, Rosenthal SJ (2005) Peptide-conjugated quantum dots: imaging the angiotensin type 1 receptor in living cells. In: NanoBiotechnology protocols. Humana Press, New Jersey, pp 051–060. https://doi.org/10.1385/1-59259-901-X:051
74. Tran L, Park S (2021) Highly sensitive detection of dengue biomarker using streptavidin-conjugated quantum dots. Sci Rep 11:15196. https://doi.org/10.1038/s41598-021-94172-x
75. Vijian D, Chinni SV, Yin LS, Lertanantawong B, Surareungchai W (2016) Non-protein coding RNA-based genosensor with quantum dots as electrochemical labels for attomolar detection of multiple pathogens. Biosens Bioelectron 77:805–811. https://doi.org/10.1016/j.bios.2015.10.057
76. Wang J, Liu Y, Peng F, Chen C, He Y, Ma H, Cao L, Sun S (2012) A general route to efficient functionalization of silicon quantum dots for high-performance fluorescent probes. Small 8:2430–2435. https://doi.org/10.1002/smll.201102627
77. Wang S, Du L, Jin Z, Xin Y, Mattoussi H (2020) Enhanced stabilization and easy phase transfer of $CsPbBr_3$ perovskite quantum dots promoted by high-affinity polyzwitterionic ligands. J Am Chem Soc 142:12669–12680. https://doi.org/10.1021/jacs.0c03682
78. Wang Y, Liu L, Wang Q, Hu S, Zou P, Shi J, Zhang X (2016) Optimization of the aqueous synthesis of Cu_2S quantum dots with different surface ligands. Nanotechnology 27:015705. https://doi.org/10.1088/0957-4484/27/1/015705
79. Wu F, Zhang Y, Zhang Z, Li G, Li M, Lan X, Sun T, Jiang Y (2017) Hybrid colloidal stabilization mechanism toward improved photoluminescence and stability of CdSe/CdS core/shell quantum dots. Langmuir 33:7124–7129. https://doi.org/10.1021/acs.langmuir.7b01438
80. Wu P, Li W, Wu Q, Liu Y, Liu S (2017) Hydrothermal synthesis of nitrogen-doped carbon quantum dots from microcrystalline cellulose for the detection of Fe^{3+} ions in an acidic environment. RSC Adv 7:44144–44153. https://doi.org/10.1039/C7RA08400E
81. Xiang Q, Zhou B, Cao K, Wen Y, Li Y, Wang Z, Jiang C, Shan B, Chen R (2018) Bottom up stabilization of $CsPbBr_3$ quantum dots-silica sphere with selective surface passivation via atomic layer deposition. Chem Mater 30:8486–8494. https://doi.org/10.1021/acs.chemmater.8b03096
82. Xiao Y (2004) Semiconductor nanocrystal probes for human metaphase chromosomes. Nucleic Acids Res 32:28e–228. https://doi.org/10.1093/nar/gnh024
83. Yao J, Li P, Li L, Yang M (2018) Biochemistry and biomedicine of quantum dots: from biodetection to bioimaging, drug discovery, diagnostics, and therapy. Acta Biomater 74:36–55. https://doi.org/10.1016/j.actbio.2018.05.004
84. Yuan Y, Hu T, Zhong X, Zhu M, Chai Y, Yuan R (2020) Highly sensitive photoelectrochemical biosensor based on quantum dots sensitizing Bi_2Te_3 nanosheets and DNA-amplifying strategies. ACS Appl Mater Interfaces 12:22624–22629. https://doi.org/10.1021/acsami.0c04536
85. Zhou R, Jia Z, Lv X, Huang X (2019) The enhanced sensitivity of a porous silicon microcavity biosensor based on an angular spectrum using CdSe/ZnS quantum dots. Sensors 19:4872. https://doi.org/10.3390/s19224872
86. Zibadi F, Rezayi M, Kazemi Oskuee R, Bolourinezhad M, Darroudi M, Taghdisi SM, Radfar S (2024) Designing and evaluation of a novel electrochemical biosensor based on carbon quantum dots and gold core-shell to detect and measure human T-lymphotropic virus-1 (HTLV-1) in clinical samples. Microchem J 201:110563. https://doi.org/10.1016/j.microc.2024.110563

Chapter 6
Quantum Dots as Drug Delivery Systems

Simran Gupta, Ankita Mahadik, Mohammad Talha Nagori, Siddharth Ramesh Talla, and Govinda Kapusetti

Abstract Quantum dots (QDs) were discovered as highly flexible nanomaterials for drug delivery due to their distinctive optical, structural, and surface engineering characteristics. This chapter explains how QDs are incorporated into nanocarrier platforms, highlighting their utility in enhancing targeted therapy, controlled release, and real-time imaging of drug delivery agents. The tunable fluorescence, large surface-area-to-volume, and tunable surface interfaces allow them to be capable of dual functionality for diagnosis and therapy. Several design approaches are discussed in detail, such as covalent and non-covalent conjugation, ligand targeting, encapsulation, and stimuli-responsive coatings. Mechanisms of cellular uptake, release from endosomes, and mechanisms of biodistribution are critically discussed to present the capability of crossing biological barriers with QDs. The chapter also states some of the most important challenges, like toxicity, long-term retention, and regulatory constraints. With the continued development of cadmium-free and biodegradable QDs, these nanoplatforms can revolutionize the future of precision medicine with their more efficient, safer, and multimodal drug delivery options.

Keywords Nanotechnology · Nanocarriers · Drug delivery systems · Targeted therapy · Therapeutic efficacy · Biocompatibility

1 Introduction

Nanomedicine has emerged as a major frontier in biomedical science, with innovative approaches for diagnosing, treating, and monitoring disease through nanoscale interventions. At the core of this revolution lies the creation of advanced drug delivery systems that can maximize therapeutic efficacy while reducing systemic toxicity [13]. Among numerous nanocarriers, quantum dots (QDs)—nanoscale semiconductor crystals with exceptional optical and surface characteristics—have been

S. Gupta · A. Mahadik · M. Talha Nagori · S. Ramesh Talla · G. Kapusetti (✉)
Department of Medical Devices, National Institute of Pharmaceutical Education and Research, Kolkata, India
e-mail: govindphysics@gmail.com

T. Bhuyan et al. (eds.), *Multifunctional Quantum Dots*, Materials Horizons: From Nature to Nanomaterials, https://doi.org/10.1007/978-981-92-0172-3_6

proposed as multifunctional platforms with the potential to transform drug delivery. Their fluorescence tunability, stability, and ease of surface modification enable their specific drug conjugation, targeted delivery, and in vivo imaging, placing them at the interface of therapeutics and diagnostics [61]. This chapter presents an all-inclusive overview of incorporating quantum dots into drug delivery modes, reporting on their physicochemical properties, design principles, functional applications, and the pitfalls their biomedical translation is fraught with. Through a critical review of the recent developments and the upcoming trends, this chapter endeavors to demarcate the reach of QDs in remaking innovative, targeted, and responsive drug delivery systems within the general context of precision nanomedicine.

1.1 Overview of Nanomedicine and Drug Delivery

Nanomedicine is a novel paradigm in modern medicine with unparalleled potential to overcome conventional limits in disease diagnosis, treatment, and monitoring. The site of this revolution is the nanoscale drug delivery system technology capable of potentially optimizing the pharmacokinetics and biodistribution of drugs, reducing systemic toxicity, and enabling delivery at the cellular and molecular level [103]. In contrast to traditional formulations, nanocarriers can be designed to cross biological barriers, recognize specific stimuli, and engage with cell components directly, thus enhancing therapeutic effectiveness and reducing off-target toxicity. In the last two decades, liposomes, dendrimers, polymeric nanoparticles, and inorganic nanomaterials have been designed for their spatiotemporal drug delivery and release [102].

1.2 Emergence of Quantum Dots in Biomedical Applications

Among the broad nanomaterials, quantum dots (QDs) are semiconductor nanocrystals of 2–10 nm size that have been of significant interest due to their physicochemical features. They exhibit size-tunable fluorescence, excellent photostability, good quantum yield, and surface modifiability, all of which have extreme utility in various biomedical applications. Initially designed for optoelectronic devices, QDs have rapidly entered the life sciences, specifically imaging and diagnostics [88]. In addition, their prospect as multifunctional theranostic platforms, integrating diagnostic and therapeutic functions in a single system, has more recently emerged. For drug delivery, QDs can be designed to deliver therapeutic payload and targeting moieties together with real-time drug distribution and release monitoring through fluorescence imaging. Such characteristics position QDs at the interface of nanotechnology and precision medicine [118].

1.3 Scope and Objectives of the Chapter

This chapter discusses the evolving role of quantum dots as multimodal nanocarriers in drug delivery. The chapter tries to elucidate the mechanisms that govern QD-based delivery systems, highlight recent advances in their design and surface functionalization, and evaluate their performance in preclinical and clinical settings. Special attention is paid to applying QDs in targeted drug delivery, stimulus-induced release mechanisms, and theranostics [5]. Besides, the chapter critically assesses the challenges surrounding biocompatibility, toxicity, and regulatory obstacles that currently limit their clinical translation. Through such a detailed exploration, the chapter tries to provide an advanced understanding of how quantum dots are destined to transform the discipline of nanomedicine and open the gates to the future generation of innovative drug delivery systems [28].

2 Fundamentals of Quantum Dots (QDs)

Quantum dots are nanosized semiconductor particles with unique optical and electronic properties due to quantum confinement. Only by altering the size of QDs can several properties like photoluminescence, emission wavelength, solubility, and absorption be changed. This causes QDs to exhibit interesting behaviors. The high surface area allows the functionalization of QDs by various biomolecules, ligands, or drugs. Although QDs can be used in various fields, especially in the biomedical field for medical imaging, theranostics, and drug delivery.

2.1 Definition, Structure, and Composition

Tiny crystals were first discovered and synthesized by Alexei Ekimov in the 1980s. Later, the term "Quantum Dots" was coined by physicist Mark Reed in 1986 to describe those tiny crystals. They are also known as artificial atoms [26] and are artificial semiconductors, considered zero-dimensional, typically ranging from 2 to 10 nm in size, and have unique optical, electrical, and structural features. QDs have quantized energy levels. It requires more energy to excite the electrons in a small QD because of its greater bandgap. Smaller QDs release radiation with more energy, frequency, or shorter wavelengths because the frequency of light emitted is related to the energy. Larger QDs, on the other hand, emit photons with lower energy or frequency or longer wavelengths and have closely spaced energy levels. As QDs get bigger, they release photons with longer wavelengths and lower energy or frequency, and vice versa. When the QDs absorb a photon (light), the electrons in the QDs go from a lower to a higher energy state. When they return to their ground state, they emit energy known as photoluminescence. QDs can produce a distinctive light color, and the specific color will depend mainly on the size and material. The reason for this is something called quantum confinement and bandgap tuning. The property that distinguishes QDs from other bulk materials is quantum confinement. Quantum confinement is a phenomenon where a semiconductor nanoparticle's optical and electrical properties differ from its bulk counterparts due to reduced size. The surface-area-to-volume ratio and the surface-to-bulk atom ratio drastically increase when the particle size is reduced to a few nanometers. Due to this quantum confinement, QD shows size-dependent photoluminescence, with the larger dots (diameter 5–6 nm) emitting light at longer wavelengths with orange or red color and the smaller dots (diameter 2–3 nm) emitting radiation at a shorter wavelength with blue and green color [72]. The Bohr condition must be met to attain quantum confinement. This suggests the nanoparticle's size must be less than the corresponding exciton Bohr radius [2, 29].

The structure of QD is responsible for the electrical and optical properties required for a specific application. For example, spherical QDs differ from rod- or disk-shaped QDs in optical and electrical characteristics. In contrast to larger bulk materials, QDs have many different factors. Because of their small size, they have a high surface-to-volume ratio, which makes them highly reactive (or active). As shown in Fig. 1, the QDs are mainly composed of elements (such as Cd, Pb, Hg) from groups II–VI, III–IV, and IV–VI in the periodic table. The composition of QDs plays an essential role in their properties [61]. QDs are composed of a core made of a semiconductor material (CdSe, CdS, InP) alone or a core and a shell (e.g., ZnS) made of a separate material, which serves as a surface passivation layer and protects from oxidation. Moreover, the core and shell can be of different sizes and shapes, and their qualities can be adjusted to fit specific applications. The QD surface can be conjugated to different biomolecules or ligands, such as proteins or antibodies [15] (Fig. 1).

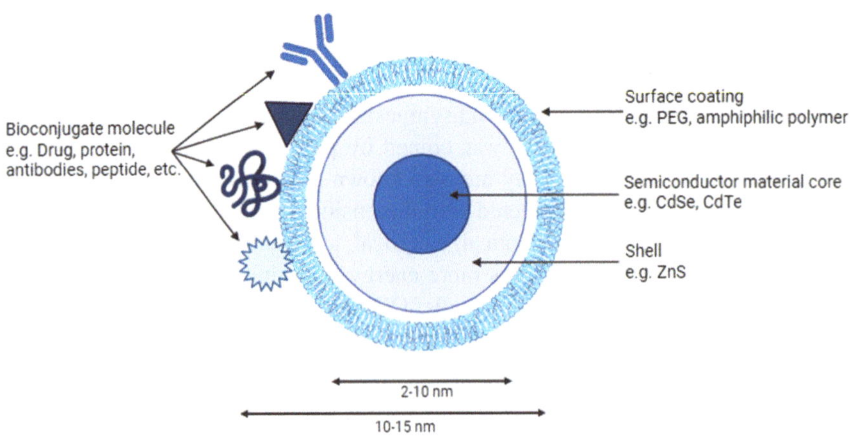

Fig. 1 The structural composition of a bioconjugated quantum dot, featuring a semiconductor core (e.g., CdSe), a protective shell (e.g., ZnS), surface coatings (e.g., PEG), and attached biomolecules (e.g., drugs, antibodies) for targeted drug delivery applications

2.2 Physicochemical and Optical Properties

Due to their unique optical and physicochemical characteristics, ease of surface functionalization, high photostability, biocompatibility, and water dispersity, QDs have become an attractive class of fluorescent nanomaterials. These features provide them an advantage over traditional fluorophores, including organic dye molecules, in detecting, diagnosing, and treating several illnesses, including cancer [45]. The optical properties are related to the small size, shape, composition, quantum confinement, and surface chemistry of the QDs. QDs are excellent choices for both in vitro and in vivo imaging applications due to their strong and tunable photoluminescence. The organic core of QDs is the source of their optical features. QDs, with their adjustable optical properties, have gained much attention as traditional organic label dyes lack the near-infrared (NIR) (>650 nm) emission possibility, which is highly desired for biomedical imaging due to its reduced light scattering, as well as inadequate tissue absorption. Advances in technology have improved the stability and reduced the toxicity of QDs, particularly graphene and carbon QDs, which show high photostability and low toxicity [61].

2.3 Types of Quantum Dots

Quantum dots can be classified into different types based on structure and composition. Includes core-type QDs, core–shell QDs, and alloyed QDs. Core-type QDs comprise a single semiconductor material, such as selenides, sulfides, cadmium, and zinc. Core–shell QDs comprise a semiconductor material core enclosed in a shell of another semiconductor material, which enhances optical properties and stability. Alloyed QDs are made from a core of two or more different materials, combining various materials to achieve tailored properties. QDs are of three main types: fluorescent, magnetic, and colloidal. Colloidal QDs are nanoscale semiconductor crystals used in biosensing, solar cells, photodetectors, displays, etc. Magnetic QDs derived from iron oxide nanoparticles have excellent superparamagnetic and fluorescent properties and are used in magnetic resonance imaging, theranostics, nanorobots, etc. [93]. Fluorescent QDs display unique electronic properties and are useful in imaging and cancer theranostics [90] and sensing because of their high quantum yield fluorescence and resistance to photobleaching. Nanocrystals composed of semiconductor materials such as CdSe, InP, PbS, or ZnS that exhibit quantum mechanical properties are called semiconductor QDs (SQDs). SQDs have special photophysical characteristics, like size-dependent optical characteristics, large quantum fluorescence yields, and superior durability against photobleaching. These characteristics make SQDs an attractive optical label for biological applications, such as cell sorting and tracing, *in vivo* imaging, multiplexed study of immunocomplexes or DNA hybridization processes, and biomedical diagnostics. To enhance stability and optical features, they frequently have a core–shell construction, in which a core semiconductor is encased in a shell made of another semiconductor material; for further applications, a capping outer layer like silica can provide a substantial surface area for covalently attaching to biorecognition molecules including peptides, antibodies, nucleic acids, and small-molecule ligands [80].

Carbon dots (CDs) are a subtype of QDs that can be considered the intersection of semiconductor material and carbon material. The three main categories of CDs are carbon QDs (CQDs), graphene QDs (GQDs), and carbonized polymer dots (CPDs) [95]. Carbon quantum dots (CQDs) are zero-dimensional fluorescent carbon-based nanomaterials typically less than 10 nm in size. CQDs are also called carbon nanolights since they have strong luminescence properties. The sp^2/sp^3 hybridized carbon atoms that comprise CQDs are frequently organized in an amorphous or partly crystalline form. Easy synthesis, high photostability, good water solubility, high photoresponse, minimal cytotoxicity, easy surface functionalization, good catalysis, and adjustable excitation-emission are some attractive characteristics of CQDs. CQDs can interact with therapeutic and targeted substances by covalent or non-covalent interaction. CQDs have broad applications in light-emitting devices, drug delivery, biological imaging, and photocatalysts. Top-down and bottom-up approaches can synthesize CQDs [104].

Graphene quantum dots, recognized as GQDs, contain sp2 or sp3 carbon atoms. GQDs are crystalline nanosized graphene fragments in the size range of typically below 20 nm. GQDs have restricted electronic transfer in all three spatial dimensions, which can produce a quantum size effect and exciton confinement. Compared to colloidal QDs, GQDs are significantly easier to handle, have extraordinary optoelectrical properties, are non-toxic, and have a more molecule-like character [12]. Due to their unique qualities, which include low cytotoxicity, high electrical conductivity, excellent water solubility, good biocompatibility, chemical stability, photoluminescence, low photobleaching, environmental friendliness, and optoelectronic features, GQDs are praised by numerous researchers. GQDs can absorb and emit light because they have distinct energy levels, unlike bulk graphene. This property makes them photoluminescent. Biological uses such as biosensing, medication delivery, and bioimaging make GQD extremely promising [40].

Perovskite QDs (PQDs) have drawn much interest recently because of their exceptional optical and electrical characteristics. PQDs are superior to traditional QDs due to their high photoluminescence quantum yields, higher color purity, and narrow emission bandwidths. They have applications in various optoelectronic devices, including solar cells and LEDs. Optoelectronics could undergo a significant transformation with the advancement of PQDs [2].

2.4 Synthesis Techniques and Surface Functionalization

Several types of techniques are used to synthesize QDs. These can be categorized into two approaches: top-down and bottom-up. The top-down method involves breaking bulk material into nanoscale particles through various methods such as molecular beam epitaxy (MBE), ion implantation, electron beam lithography, and X-ray lithography. In the bottom-up approach, atomic or molecular precursors are assembled to form QDs, and notable bottom-up techniques are wet chemical techniques and vapor-phase techniques. Wet chemical techniques include sol–gel, hot-solution decomposition, microemulsion, and electrochemical synthesis. These depend on particle development and nucleation, either homogeneous or heterogeneous. Vapor-phase techniques allow for a synthesis of QDs without patterning by employing atom-by-atom growth through MBE, sputtering, or gaseous aggregation [94]. Post-synthesis QDs require surface modification to enhance their stability, solubility, and functionality. Surface functionalization refers to altering the QD surface to tailor their chemical, physical, and biological properties for a particular application. Strategies include ligand exchange and attachment. For example, QDs can be loaded into liposomes, attached to the liposome surface, or integrated into a liposome phospholipid bilayer to produce luminous labels for both in vitro and in vivo imaging. Applying hydrophilic ligands, amphiphilic polymers, or silica to QD is another solubilization approach [16].

3 Rationale for Using QDs in Drug Delivery

The choice of an ideal nanocarrier is a cornerstone to the success of any drug delivery system, particularly if precision targeting, controlled release, and real-time therapeutic monitoring are required. Quantum dots (QDs), with their excellent physicochemical and optical properties, are a potential alternative to traditional nanocarriers such as liposomes, polymeric nanoparticles, and dendrimers [7]. The draw to using QDs in drug delivery systems also surpasses compositional variation since QDs hold possibilities as composite systems that combine diagnostics, treatment, and monitoring into a single nanoscale assembly, as illustrated in Fig. 2.

3.1 Unique Advantages Over Conventional Nanocarriers

Conventional nanocarriers have significantly enhanced drug delivery by enhancing solubility, circulation time, and bioavailability. They cannot achieve immediate feedback on drug distribution and treatment efficacy. Quantum dots accomplish most of these limitations through their inherent luminescence properties, creating a long-lasting optical signal for in vivo tracing [27]. Unlike organic dyes and conventional fluorophores, which suffer from photobleaching and large emission bandwidths, QDs have excellent photostability, narrow emission bandwidths, and size-tunable fluorescence. These characteristics enable simultaneous multi-color imaging and long-term biodistribution tracking without appreciable signal loss. Additionally, the surface chemistry of the QDs is such that it is quite possible to introduce a wide variety of

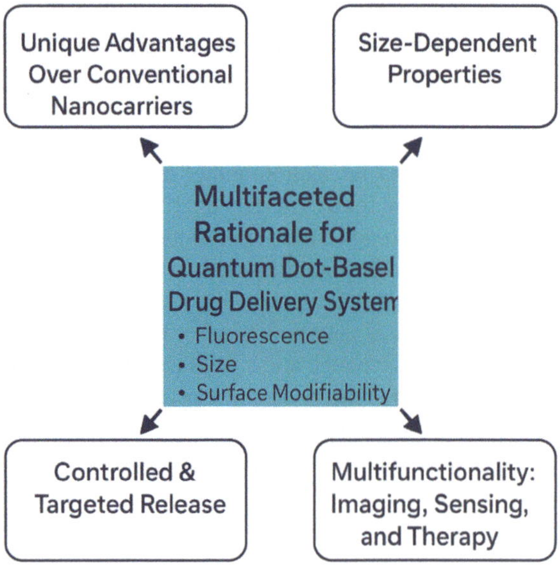

Fig. 2 Multifaceted rationale for using quantum dot-based drug delivery systems, highlighting their fluorescence, tunable size, and modifiable surfaces. Key advantages include size-dependent properties, multifunctionality (imaging, sensing, therapy), targeted release, and superiority over conventional nanocarriers

biomolecules, such as antibodies, peptides, and small-molecule ligands, providing targeted delivery to diseased tissue, such as tumors or inflammatory tissue [74].

3.2 *Size-Dependent Properties and Enhanced Permeability*

QD size, typically between 2 and 10 nm, determines their biological activity and use as drugs. Because of their small dimensions, they can penetrate effectively through barriers in the body, including the vascular endothelial and the tumor interstitial barriers, by taking advantage of the enhanced permeability and retention effect characteristic of most solid neoplasms [41]. This passive targeting functionality is complemented by the optical properties of QDs that are tunable in size, thereby enabling them to be optimally designed for imaging at specific wavelengths and tissue depths. Furthermore, the nanometric dimensions of QDs will allow them to be readily internalized in cells by endocytosis, whereby therapeutic cargo is provided to the intracellular environment through which most drugs exert their effect. The high surface-area-to-volume ratio also provides adequate areas for drug loading and surface manipulation, thus increasing their functionality as delivery vectors [117].

3.3 Controlled and Targeted Release Potential

One of the most significant benefits of applying QDs in drug delivery is that they can design controlled and stimulus-responsive drug release systems. Intelligent surface modification is applied to functionalize QDs with redox-cleavable linkers, pH-sensitive linkers, or enzyme-sensitive groups so that the drugs are selectively released depending on the target tissue microenvironment. In addition to being minimized, targeted delivery also enhances systemic toxicity and maximizes drug concentration at the target site [71]. Lastly, the surface of QDs is modular to enable multifunctional agent co-delivery, such as chemotherapeutic drugs, genes, or immunomodulators, in temporally synchronized patterns. Such intricate release profiles play a role in being tailored to specific therapeutic demands, particularly for cancer diseases, where spatial and temporal drug delivery is important [64].

3.4 Multifunctionality: Imaging, Sensing, and Therapy

One of the most creative aspects of QDs in drug delivery is their inherent multifunctionality. Theranostic therapy using QDs combines therapeutic and diagnostic functions into a single nanosystem. Their tunable and high fluorescence allows real-time imaging of drug biodistribution and cellular drug uptake, enables therapeutic process monitoring, and supports treatment decision-making [68]. Aside from imaging, QDs may be engineered as biosensors recognizing a molecular alteration in the local microenvironment of pH and enzymes, or to identify particular biomarkers. This sensing capability can trigger a feedback loop toward on-demand controlled drug delivery or real-time feedback for therapeutic effects. Unifying imaging, sensing, and therapy in QD-based systems is a pivotal advancement in precision medicine that promises more personal, adaptive, and optimal therapy [24].

4 Functionalization Strategies for Drug Loading

QDs have shown great potential in the biomedical field in two critical areas: drug delivery and imaging. Their large surface area for drug attachment, adjustable optical characteristics, and customizable solubility make them an attractive platform for several therapeutic applications. As nanotechnology continues to advance, the future of QDs seems bright. Enhancing the effectiveness of QDs in biomedical applications requires functionalization techniques for drug loading. Conjugating QDs with targeting ligands can be functionalized to target particular cells or tissues. Strategies include covalent and non-covalent conjugation, ligand attachment for targeted delivery, stimuli-responsive coatings, encapsulation, and surface adsorption approaches. The adaptability of QDs in the drug delivery system is facilitated by the distinct benefits and difficulties presented by each technique [92].

4.1 Covalent and Non-covalent Conjugation Techniques

Covalent conjugation involves the formation of a strong chemical bond between the drug molecule and surface ligand or functional group, ensuring a stable and durable interaction via amide, thioether, ester, or disulfide bonds. Covalent conjugation provides better control over the biomolecules' orientation on the QD surface, essential for maintaining their functionality and optimizing drug delivery efficiency. Covalent conjugation typically involves attaching QD-free carboxylic and amino groups to corresponding groups on drugs or other biomolecules. Examples include COOH-containing QDs, which enable easy conjugation with proteins, free amino groups, peptides, carbohydrate hydroxyl groups, and antibodies. Conjugating biomolecules with $-NH_2$-containing QDs is a standard method of adding the thiol group to the molecules [68]. Non-covalent conjugation is comparatively milder than covalent conjugation, reversible, and does not change the structural characteristics of nanomaterials. It makes it possible for different medications, ligands, proteins, antibodies, peptides, etc. to be physically adsorbed onto the QD surfaces surface via weak intermolecular forces like π–π stacking, hydrogen bonding, van der Waals, electrostatic, and hydrophobic interaction without changing their chemical makeup, maintaining the drug's structure and bioactivity. For example, aptamer-conjugated CQDs, which were used for the turn-on spectrofluorimetric detection of tumor cells, were produced by the non-covalent electrostatic interaction between the negatively charged DNA aptamers and the positively charged amine groups of CQDs [45]. A study was done for cancer treatment in which folic acid-carbon dots (FA–CDs) NPs were made to deliver doxorubicin (Dox) in a covalently and non-covalently bound form. FA was attached to the surface of CDs to target tumor cells with overexpressed folate receptors. The interaction between the drug Dox and FA-CD and CD composition determines Dox loading and release data. FA was covalently bonded to the amine group on the surface of CDs via an amide bond to increase the tumor cells' targeted specificity. FA is a water-soluble vitamin with a strong affinity for folate receptors (FRs) and is necessary for cell viability. Numerous breast, kidney, ovary, cervical, lung, colorectal, and brain tumor cells overexpress FR. FA enhances intracellular targeting efficiency, particularly against cancer cells that overexpress FR. As a result, non-covalent FA-CD-Dox showed the best result for the breast cancer cell line [23].

4.2 Ligand Attachment for Targeted Delivery

Ligand attachment on the QD surface significantly improves targeted drug delivery, especially in cancer therapy, by conjugating ligands to QD, which receptors on the surface of cancer cells can selectively recognize. It helps targeted drug delivery of drugs like doxorubicin and cisplatin, etc. by accelerating their nuclear accumulation and DNA cleavage activity, illustrated in Fig. 3. This can improve selectivity and specificity and minimize unwanted side effects. Targeting ligands such as peptides, folic acid, biotin, antibodies, and arginine glycine aspartic acid can be attached through covalent and non-covalent interactions by various techniques. For example, the targeting ligand riboflavin can be connected to GQD by π–π interaction using a pyrene linker to anticancer drug methyl 3,3-dimethyl-2-(3-methyl-2,3-dihydrobenzofuran-2-yl) butanoate (BFG). The drug is covalently attached to the GQD surface through a cleavable PEG linker, which is activated inside the body [36]. Another study was done by carefully designing a cleavable linker that can be activated explicitly inside cells. GQD is covalently attached to the tumor-targeting module biotin (vitamin K or vitamin B7), effectively detecting overexpressed biotin receptors on tumors. By utilizing the superior absorption capabilities of carbon nanomaterials, the drug employed in this investigation, Doxorubicin, was loaded onto the GQD surface via π–π interaction. Further, its built-in fluorescence makes monitoring the drug's release possible [37]. In another study, ligand-modified GQDs were used. The anticancer drug Dox was loaded onto folic acid (FA)-conjugated GQDs. The produced nano-assembly can effectively deliver the drug to the desired cells and distinguish tumor cells from healthy cells. Continuous monitoring of the cellular absorption of the Dox–GQD–FA nano-assembly and the resulting drug release is made possible by the intrinsic stable fluorescence of GQDs. HeLa cells selectively absorb the nano-assembly quickly via receptor-mediated endocytosis, although Dox release and accumulation take longer. According to *in vitro* toxicity results, the Dox–GQD–FA nano-assembly has a significant decrease in cytotoxicity to non-target cells while being able to target HeLa cells differently and effectively [99].

4.3 Stimuli-Responsive Coatings

Stimuli-responsive drug delivery systems (SRDDS) are mainly used for targeted drug delivery and minimizing side effects. The stimulus-responsive coating on QD enables QD to release therapeutic agents in response to stimuli, as shown in Fig. 3. There are two approaches for designing SRDDS: one involves exogenous stimuli, like temperature, ultrasound, light, magnetic, and electric, and endogenous stimuli, like pH, enzymes, and redox potential. Multi-stimuli-responsive DDS offers higher potential in cancer therapy than single-stimuli-responsive DDS [78]. A pH-responsive DDS for intracellular controlled drug release was investigated based on ZnO QDs. Hyaluronic acid (HA), as a targeting ligand, was attached to ZnO QDs to specifically attach to the

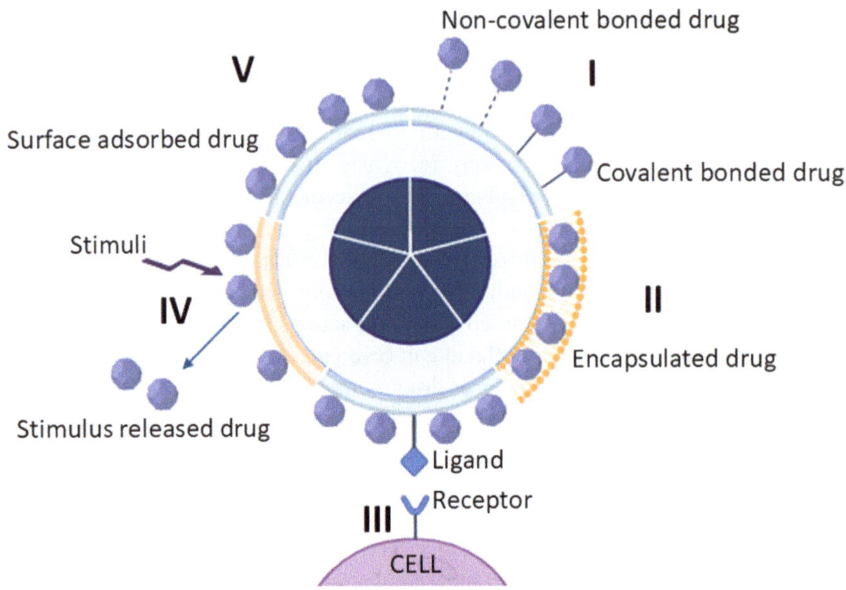

Fig. 3 Five distinct drug-loading and release strategies using quantum dots: non-covalent and covalent bonding (I), encapsulation (II), ligand–receptor-mediated interaction (III), stimulus-responsive release (IV), and surface adsorption (V). Ligand–receptor interactions enable targeted delivery to specific cells. External stimuli trigger controlled drug release, enhancing therapeutic precision. This multifunctional approach supports customizable and efficient nanomedicine platforms

glycoprotein CD44, which is overexpressed by cancer cells. The successful loading of doxorubicin (Dox) molecules onto PEG-functionalized ZnO QDs was achieved by covalent interactions and the creation of the metal–Dox complex. Following absorption by tumor cells, the pH-sensitive ZnO QDs are dispersed to Zn^{2+} in an acidic endosome or lysosome, causing the metal–drug complex to dissociate and a regulated release of Dox. A synergistic therapy was eventually achieved because Zn^{2+} and Dox both have anticancer effects [17]. Another study explored an innovative stimuli-responsive drug delivery system (DDS), which can release medication based on the target's MRSA (Methicillin-resistant Staphylococcus aureus) concentration. More release of drugs will result from a greater concentration of MRSA in the system, and vice versa. Novel carbon-dot-coated calcium alginate beads (CA-CDs), which have exceptional stability, have been successfully employed in novel DDS. To illustrate the phenomenon, allicin-containing garlic extract (GE) was used as a model drug system. With CA and CA-CD, GE loading was found to be 19 and 78%, respectively. The pH-dependent regulated drug release shown by CA-CD-GE leads to improved therapeutic efficacy. CA-CD-GE is a stimulus-responsive and regulated drug release system since it distributes the drugs according to the pathogen concentration (MRSA). Drug release, MRSA concentration, and medium pH are interconnected as cell division produces secondary metabolites, which lower the medium's pH. A decrease in pH triggers drug release from the beads. Additionally, MRSA cell death confirms the drug's effect [60].

4.4 Encapsulation

Developing novel DDS that can encapsulate medications inside nanocarriers and transport them exclusively to desired locations by active targeting is one effective way to prevent the adverse effects of conventional DDS. A study was done to improve the delivery in macrophages of an anti-TB medication compound, zinc–rifampicin complex (Zn–RIF), which was encapsulated into transferrin-conjugated silver quantum dots (Zn–RIF–Tf–QD). Transferrin encapsulation on silver quantum dot surfaces enhanced drug molecule binding efficiency, illustrated in Fig. 3. Silver QDs encapsulated in Zn–RIF and transferrin successfully target the macrophages, remain stable for up to 48 h, and significantly decrease the bacterial burden within the macrophages. It was found that the diameters of transferrin-conjugated QDs and silver QDs ranged from 5 to 20 nm. According to activity testing, Zn–RIF–Tf–QD has ten times the antibacterial activity of Zn–RIF, RIF, and Zn against Mycobacterium smegmatis and Mycobacterium Bovis-BCG [69]. In another study, Mitomycin C (MMC) drug-encapsulated manganese (Mn)-doped Mn–ZnS QDs conjugated with chitosan (CS)-based nanocarrier DDS were explored for bladder cancer treatment. MMC is physically entrapped within a polymeric matrix of CS-coated QDs. A UV–visible spectrophotometer has been used to extensively analyze the loading capacity, release properties, and drug encapsulation efficiency of this synthesized nanocarrier, MMC–CS–Mn–ZnS QDs, at a fixed wavelength of 358 nm. The developed Mitomycin C nanocarrier system synergistically anticipates the effective transport of chemotherapeutic medicines to the target tumor cells, allowing them to be sustained for a more extended period, etc. As a result, the created nanocarrier technology increases drug-loading effectiveness in inhibiting cancer from returning and spreading in non-muscle invasive bladder disorders [3].

4.5 Surface Adsorption Approaches

Surface adsorption is a non-covalent technique to physically attach the drug onto the QDs surface via weak, reversible interactions such as hydrogen bonding, π–π stacking, van der Waals, and hydrophobic interactions, which preserve the drug's and the QD's structural integrity and allow targeted drug delivery, as shown in Fig. 3. However, it has low stability compared to covalent interactions and has various applications in chemotherapy, theranostics, and gene delivery. In a study, GQD oxide was chosen as a carbon-based nanocarrier via theoretical approaches, such as molecular dynamics simulation. The loading and effectiveness of the hydrophobic anticancer drug docetaxel (DTX) onto the GQD oxide surfaces via non-covalent interaction, including π–π stacking and van der Waals forces, with and without a PEG-b-PLA (polyethylene glycol-block-poly lactic acid) copolymer acting as a surface modifier. It was found that in the DTX-GQD system, DTX was adsorbed onto the GQD from the beginning of the process. On the other hand, the copolymer created instability in the DTX-POL-GQD system, which initially inhibited the DTX from reaching the GQD oxide surface due to steric hindrance. After some time, the DTX solubility and adsorption were enhanced [109]. In another study, pH-sensitive QD-based DDS was used to deliver paclitaxel (PTX) for lung cancer treatment by using ZnO QDs. In this study, PTX for chemotherapy was delivered via the ZnO-adipic dihydrazide-heparin nano-complex, which was created by attaching heparin and adipic dihydrazide (ADH) to the surface of ZnO QDs. The Fourier transform infrared spectroscopy confirmed the PTX loading and surface modification. The ADH contains many amino groups that may react with free carboxylic groups to generate an amide bond. It is utilized to change the surface of ZnO QDs. Heparin (Hep) and PTX were physically linked to the ZnO surface following surface modification via non-covalent interaction, creating the ZnO-ADH-Hep-PTX nano-complex. ZnO QDs were bound to heparin via the amino groups of ADH. At pH below 5.5, ZnO QDs may disintegrate into Zn^{2+}. Researchers have discovered that Zn^{2+} may destroy tumor cells more preferentially while having minimal effect on healthy cells. To investigate how pH affects PTX release, ZnO-ADH-Hep-PTX was exposed to pH 5.0 (tumor microenvironment) and 7.4 (normal tissue). The findings indicated that a lower pH enhanced PTX release; at pH 5.0, 70% of PTX was released, presumably due to ZnO QD breakdown [101].

5 Mechanisms of Drug Delivery Using QDs

Quantum dots have been extensively studied in drug delivery as they have promising potential for efficient and targeted drug delivery. There is a need to study energy-dependent pathways, such as clathrin-mediated endocytosis and caveolae-mediated endocytosis, mediating the cellular uptake of the quantum dots. However, for the actual functionality of quantum dots, endosomal escape is a crucial step; hence, the methods of cytoplasmic targeting, such as electroporation, microporation, and cell-penetrating polydisulfide, have been explained. The biodistribution and pharmacokinetics parameters are also briefly described.

5.1 Cellular Uptake and Internalization Pathways

Cellular uptake and internalization are necessary to understand the biological fate of nanoparticles. Cell membranes are permeable, such that they only allow small molecules to pass across. Still, various mechanisms facilitate the uptake of nutrients and other micro-components to pass through the plasma membrane, which includes free diffusion of oxygen, carbon dioxide, and other nonpolar molecules driven by their concentration gradient. Other small molecules, including ions and amino acids, pass through the plasma membrane and utilize an active transport system facilitated by integral membrane protein pumps or ion channels. Additionally, nanoscale hydrophilic molecules are transported into the cell through the endocytosis pathway. It is a type of active transport in which cells absorb materials by encasing them in vacuoles or vesicles that are split off from the cytoplasmic membrane. The three known mechanisms are phagocytosis, pinocytosis, and caveolae-dependent or clathrin-mediated endocytosis [113]. There is potential to use QDs in biological applications, cellular uptake and internalization must be considered. The cell plasma membrane has several uptake mechanisms, and among all QDs, it uses clathrin-mediated endocytosis, micropinocytosis, and caveolae-mediated endocytosis [46]. Several factors, including size, surface charge, and surface functionality, affect the intracellular delivery of QDs [9]. The main transport of QDs in mammalian cells is by clathrin-mediated endocytosis. However, specific cells employ a combination of pathways for QD uptake. For example, Le et al. showed that both pathways, i.e., clathrin-mediated endocytosis and caveolae-mediated endocytosis, mediate the uptake of QDs [46]. Clathrin-mediated transport is the process by which cells take up material inside using clathrin-coated vesicles. The pathway is essential for neurotransmission, signal transduction, and regulation of plasma membrane activities in many eukaryotic cells [62]. For building clathrin protein into clathrin-coated vesicles, a range of accessory adaptor proteins are found in various cell types that bind to varied receptors [9]. Clathrin-mediated endocytosis proceeds with the formation of membrane invagination pit by (nucleation) vesicle budding. The involvement of proteins such as FCH Domain Only (FCHO) proteins and Epidermal Growth Factor Receptor Pathway Substrate 15 (ESPS15) ensures that clathrin is recruited to the right place, i.e., defines specific spots on the plasma membrane where clathrin-coated pits will form and play an essential role in vesicle formation. Adaptor protein 2 (AP2) mediates the selection of cargo. The clathrin triskelia are recruited to adaptor-rich sites. Then, the clathrin polymerization gives structural integrity to vesicle curvature. BAR domain proteins recruit dynamin (an enzyme) to the vesicle neck for further membrane fission. The role of dynamin is to release the vesicle by GTP hydrolysis. In the last step, the clathrin coat is disassembled by HSC70 and auxilin, which allows the vesicle to fuse with its target [62]. In 2011, Zhang et al. showed that QD655-COOH endocytosis was recognized by clathrin and scavenger receptors and regulated by F-actin and phospholipase C [110]. The caveolin-mediated transport pathway is primarily used for intracellular trafficking of molecules and nanoparticles across cell membranes via caveolae, which are small flask-shaped pits in the plasma membrane of the cholesterol-binding protein called caveolin. The pathway delivers the extracellular molecules into cellular interiors through specific ligands. During caveolin-mediated endocytosis, the extracellular ligand binds the lipid anchoring protein through signal transduction. Unlike clathrin-mediated endocytosis, the caveolin-mediated transport bypasses lysosomal fusion, which prevents the degradation of the extracellular ligand and maintains structural integrity. Hence, this pathway is efficient for cellular uptake and intracellular delivery of nanomaterials [57, 113].

5.2 Intracellular Trafficking and Drug Release

Quantum dots (QDs) intracellular trafficking is mainly through energy-dependent endocytosis, such as the clathrin-mediated, caveolae-mediated, lipid raft-related pathway, which varies according to the size, surface charge, and coating of QDs. After entering the cells, QDs are encapsulated in endosomes and transported to late endosomes or lysosomes, where they are generally sequestered. It has been reported that the intracellular accumulation of QDs could be classified into four visually distinct stages: initial attachment to the cell membrane surface, clustered formation in the cytoplasm, perinuclear localization, and, ultimately, forming multivesicular-like structures along the complex vesicular trafficking process. The narrow release of QDs from the vesicles into the cytosol represents an important obstacle to in-cell drug delivery. QD has a preferential endocytic pathway for cell entry, depending on various factors, including the QD surface coating, charge, and cell type. Of these factors that may influence the endocytotic ways, the QDs' size is the key. Clathrin-mediated endocytosis is the predominant route of QDs in mammalian cells and tissues. Particles of a large size might even be taken up through clathrin-dependent endocytosis in the case of 100 nm particles instead of 60–80 nm particles that might be taken up through the caveolae-mediated endocytosis pathway [111].

5.3 Endosomal Escape and Cytoplasmic Targeting

Endosomal escape is the process by which internalized nanoparticles, such as quantum dots (QDs), or therapeutic agents escape from the endosomal compartment of a cell to the cytosol. Following uptake into cells, most nanoparticles are sequestered in endosomes (for which maturation into late endosomes and lysosomes, organelles possessing degradative enzymes, can occur). For quantum dots or any drugs to work inside the cell, they must be released from these endosomes into the cytoplasm before they are degraded in the lysosome [19]. Cytoplasmic targeting of quantum dots (QDs) is essential because many biological and biomedical applications require the delivery of QDs directly into the cell cytosol rather than being trapped on the cell surface or inside endosomes. For cytoplasmic delivery of QDs, there are numerous methods, which include physical methods like electroporation and microporation and chemical methods such as cell-penetrating peptides (CPP) and cell-penetrating polydisulfide (CPD).

In the electroporation approach, electric voltages are introduced onto cells during electroporation, forming transient pores in the plasma membrane and enabling impermeable materials such as QDs to directly enter the cytosol from the extracellular medium. In this method of intracellular delivery of QDs, the cells are suspended in a low-conductivity medium with controlled electric pulses. An electric field on the membrane produces transmembrane potential, stretching and deforming, forming pores. Consequently, molecules or particles smaller than the size of the pores can enter the cell while the pores remain intact. Derfus et al. pioneered electroporation for intracellular QD delivery of QDs in living cells. They delivered CdSe/ZnS QDs, coated with polyethylene glycol (PEG), as conglomerates of up to 500 nm rather than single particles. Further, microporation is a physical approach for intracellular delivery of materials by membrane disruption at a localized site, which enables precise delivery of substances (e.g., peptides, proteins, and lipid vesicles) at desired locations within the cell (e.g., cytosol, nucleus, or other large organelles), ensuring efficient delivery with low cell mortality, enabling precise control over the amount and location of delivery. Traditionally, injection needles were used to deliver the substance into the cells under an inverted microscope. This technique is highly efficient for intracellularly delivering materials with different properties (size and charge). However, it does require highly skilled operators, and delivery is performed one cell at a time. Saurabh et al. made streptavidin-QDs (CdSe/CdTe core coated by a ZnS shell) loaded with biotin-Hapten. This molecular recognition moiety binds with high affinity to a genetically encoded protein, which is injected into the nanosystem using microinjection. CPD quantum dots have significant use in drug delivery; hence, it is essential to overcome barriers to the uptake of QDs into the cell and ensure QDs are available to cells freely in the cytoplasm. Till now, physical methods like microinjection, electroporation, or osmotic shock remain the efficient methods for cytoplasmic targeting of QDs. The cell-penetrating peptides (CPP) method has been developed, which is limited to being trapped in the plasma membrane. About 70 QDs per cell could be delivered efficiently utilizing the recently invented cell-penetrating poly(disulfide) technique, with 90% of QDs freely diffusing in the cytoplasm rather than being trapped in intracellular compartments. In this process, streptavidin-coated QDs are combined with biotinylated CPDs, which firmly attach to the QDs, forming the CPD-QD complex due to the strong binding affinity between streptavidin and biotin. It is important to note that this method cannot employ QDs alone or with non-biotinylated CPDs. Here, the mechanism involves CPDs interacting with thiol groups on the cell membrane surface, making a covalent disulfide exchange that enables CPD-QD to cross the cell membrane and enter the cytosol. After the complex uptake in the cytoplasm, CPDs are degraded by depolymerization, making this process minimal in toxicity and efficient for delivering the QDs into the cytoplasm [25].

5.4 *Biodistribution and Pharmacokinetics*

The liver and spleen, the RES's principal organs, eliminate most QD formulations from circulation in a comparatively short time (20 min) [79]. Using micropet imaging and gamma counting, the study tracked CdSe QDs in the living mice. Yaghini et al. [105] demonstrated this after administering indium-based cadmium-free quantum dots IV into rats and mice. The rapid distribution of QDs was found in multiple organs, including the liver, spleen, kidney, lung, brain, and intestine. High accumulation levels were consistently found in the liver and spleen, however, after distribution, QDs were eliminated very slowly, indicating long-term tissue retention [105].

Pharmacokinetics studies the absorption, distribution, metabolism, and excretion of QDs from living organisms after administration. The administration route of QDs affects distribution, metabolism, and excretion. Intravenous is a standard route for administration, and sometimes subcutaneous, intramuscular, and inhalation routes for administration of QDs. QDs are absorbed mainly via receptor-mediated endocytosis, and after intravenous injection, they tend to accumulate selectively in targeted tissues, mostly when conjugated with targeting ligands. After absorption, the QDs, to some extent, accumulate in the liver, brain, and kidneys. Metabolically, the QD core remains essentially unchanged, while the shell and capping materials undergo oxidation and photolysis. Size and physicochemical characteristics significantly affect excretion, the kidneys eliminate QDs less than 5 nm. It has been reported that within 5 days, silica-coated CdSe QDs are eliminated by urine and faeces [31].

6 Applications in Therapeutics

Quantum dots (QDs) have grown into multifunctional nanoplatforms with extensive therapeutic potential. Their distinctive physicochemical features—narrow size tunability, outstanding optical behavior, and surface functionalization capability—make them the best drug carrier candidates for targeted, controlled drug delivery, real-time imaging, and combined therapeutic-diagnostic (theragnostic) functions [32]. This section discusses the wide therapeutic area of QD-based systems, focusing on established and emerging clinical applicability and translational feasibility applications.

6.1 Cancer Drug Delivery

Cancer remains a global health problem, and there has never been a greater need for selective, effective, and non-toxic drug delivery systems. Quantum dots offer a multi-dimensional approach with twofold uses as drug carriers and imaging agents. Their nanometer dimension range (typically <10 nm) offers better permeation and retention in cancer by the EPR effect of enhanced permeability and retention. As demonstrated by Caratti and Trapani [10], ligand functionalization toward cell-type specificity for targeting the tumor cells (such as folate, HER2, transferrin, or RGD peptides) maximizes cell specificity and avoids off-target cytotoxicity. Several preclinical models have also demonstrated evidence for the efficacy of QD–drug conjugates (such as doxorubicin, paclitaxel, and siRNA) against targeted reduction in tumor burden in xenograft and orthotopic models. QDs enable controlled and targeted therapy and are also designed for stimuli-responsive release-pH, enzymatic activity, or redox conditions. Combinatorial therapy with QDs for photodynamic or photothermal ablation and chemotherapeutic drugs is also aggressively explored, with synergistic cancer-cell-killing ability [114]. Certain cadmium-free QDs (e.g., graphene QDs, silicon

QDs) are biosafe, hence, the field stands at the doorsteps of clinical translation. Apart from their use in delivery and diagnostics, QDs are incorporated into ultra-low concentration biomarker-sensing platforms to detect cancer early and enable lead-time diagnosis [76]. Bioorthogonal chemistry and click conjugation methods have also boosted QD bioconjugation stability and efficiency, with more systemic circulation and less premature clearance. Apart from that, hybrid platforms made up of QDs with polymers, liposomes, or inorganic nanoparticles are being designed to control drug release profiles and prolong circulation half-life. Other emerging approaches involve the use of non-toxic QDs to control immune checkpoints by co-delivering immunotherapeutic drugs or monitor the behavior of immune cells within the tumor microenvironment, which could be more efficient to overcome resistance to conventional treatments [70]. QD integration with the real-time biosensing and machine learning algorithms is being explored more comprehensively for therapeutic response prediction, minimal residual disease detection, and adaptively modulated therapy. With improvement in the regulatory framework and determination of the accuracy of the toxicity profiles, especially in cadmium-free systems, QD-based platforms are ready to bridge the bench-to-bedside gap, ushering a new era of precision oncology [73].

6.2 Miscellaneous Applications

Apart from the main uses, QDs are being molded into many treatment modalities. In eye treatments, drug-releasing QDs delivered by nanocarriers facilitate prolonged drug release for cures of retinal pathology such as age-related macular degeneration and diabetic retinopathy [59]. Cardiovascular uses involve drug release, controlling anti-thrombosis, and visualization of vulnerable plaques with fibrin- or integrin-specific QDs [65]. QD-based gene delivery systems are also designed to be engineered toward immunotherapy, i.e., cytokine delivery, immune checkpoint inhibitors, or mRNA vaccines. Due to their capability of protecting delicate biomolecules and increasing uptake, quantum dots are increasingly used in immunotherapy. Quantum dots possess a new role in immunomodulation by facilitating targeted cytokine and gene delivery and shielding therapeutic agents against degradation. In addition, [112] have demonstrated that CRISPR/Cas9 or antisense oligonucleotide modification of QDs allows for precise gene editing, for which initial success has already been demonstrated in animal models for monogenic disease treatment. The therapeutic potential of QDs is immense and expanding day by day. Despite ongoing research, concerns regarding biosafety, chronic toxicity, and large-scale clinical translation remain significant challenges. Nevertheless, their potential applications highlight the promise of a paradigm-shifting revolution in drug delivery as illustrated in Fig. 4 the diversified biomedical applications of functionalized quantum dots (QDs), a semiconductor core with therapeutic and targeting ligand coatings. Divided into four quadrants, it marks their applications in cancer drug delivery, antimicrobial/antiviral therapy, wound healing, and miscellaneous medical applications, including ocular

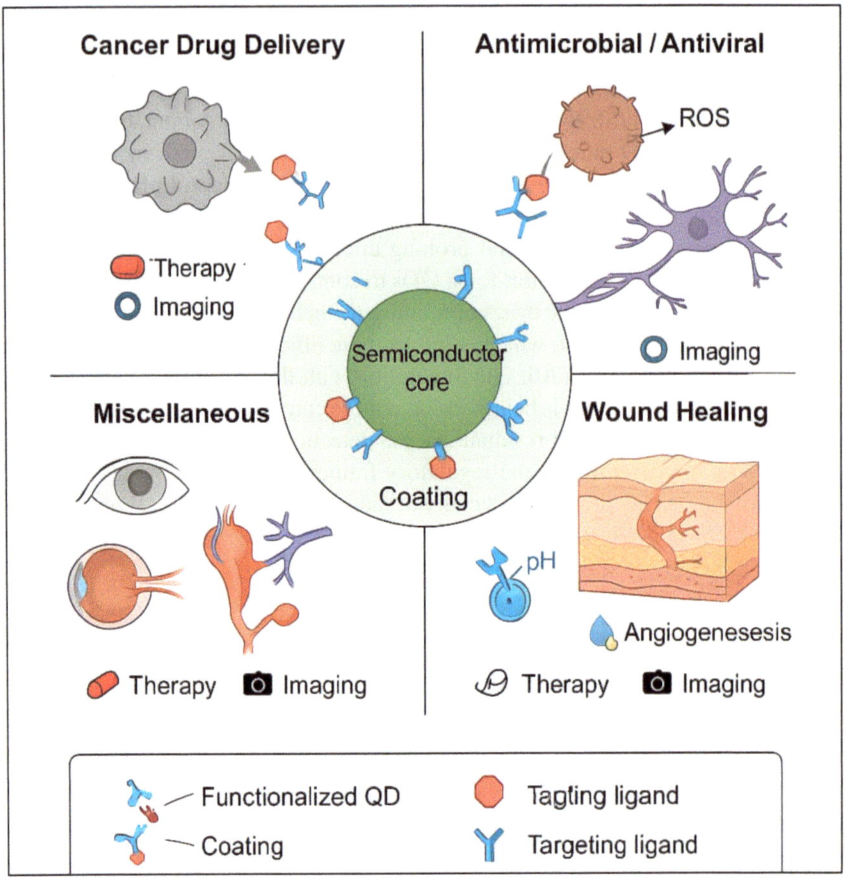

Fig. 4 Therapeutic applications of QD-based nanocarriers

and neural imaging. All four quadrants graphically demonstrate how QDs enable therapy and imaging, augmented with functionalities like pH sensitivity and reactive oxygen species (ROS) generation. The legend below describes symbols for targeting ligands, therapeutic agents, and functional coatings to emphasize quantum dots' customized and multifunctionality nature in precision medicine.

7 Toxicological and Biocompatibility Concerns

Quantum dots are substances with numerous applications in biology and medicine [86]. Their unique electrical and optical properties have sparked enormous interest throughout the globe. Quantum dots (QDs) have many special optical and electrical characteristics, but heavy metal-containing QDs (such as cadmium-based QDs) can be toxic over long-term exposure. According to a study, different physicochemical parameters, like composition, size, shape, dosage, surface chemistry, and exposure route, affect the detrimental effects of QDs [47]. Quantum dots might be detrimental through oxidative stress, and Cd^{2+} QDs can release toxic Cd^{2+} ions. Compared to bigger QDs, smaller ones often exhibit greater cellular absorption and toxicity [47]. Positively charged QDs are more harmful than neutral or even negatively charged ones. With all their important applications, there are concerns regarding the potential risks associated with quantum dots (QDs), particularly those negatively charged. After long-term exposure, the concern remains the same regarding the potential risk and the toxicity [98].

7.1 *In Vitro and in Vivo Toxicity Studies*

The quantum dots show toxic effects at the cellular scale. Multiple experiments have been conducted to evaluate the effects of QDs on different aspects of cell growth, viability, structure, and survival. The pharmacokinetics of the quantum dots in vivo, as quantum dots are non-toxic or less toxic if they stay for a shorter duration [51]. The physicochemical properties of QDs, like size, surface charge, and chemical composition, play pivotal roles in their toxicity profiles, in vitro and in vivo [83]. The pharmacokinetics of the quantum dots in vivo, as quantum dots are non-toxic or less toxic if they stay for a shorter duration [51]. The physicochemical properties of QDs, like size, surface charge, and chemical composition, play pivotal roles in their toxicity profiles, in vitro and in vivo [83]. An in vitro toxicity study is shown using various cell models. Research results showed that the toxicity of QDs is not only related to the physical and chemical properties, like preparation methods, surface properties, and particle size [43, 107].

The toxicity of quantum dots in vitro cell cultures must be highly valued, but in vivo animal models, the quantum dots are not toxic or very weak in short-term exposure. Figure 5 depicts the QDs, information regarding adsorption, distribution, metabolism, and excretion is essential for toxicological testing and evaluation. The toxicity of quantum dots is a complex process, including the interaction with the body, cells, and molecules. For testing the QD penetration through the plasma membrane and accumulation in cells [87] used QDs with the same chemical composition but different hydrodynamic diameters (HDDs) and surface charges: CdSe/CdS/ZnS (6 + 3 ML)–PEG–OH, CdSe/CdS (8 ML)–PEG–OH, CdSe/ZnS–PEG–OH, CdSe/ZnS–PEG–COOH, and CdSe/ZnS–PEG–NH_2. The study was performed on SK-BR-3 human breast cancer cells because tumor cells have a high rate of intracellular transport. The cells were studied after adding the quantum dots at 24, 48, and 72 h intervals. QD accumulation in cells is estimated by the integrated fluorescence intensity normalized to the number of cells. The study observed that positively charged QDs, despite their larger diameter, most effectively penetrated the cell membrane as early as 24 h of incubation. The reason behind this is that the negative charge of the membrane facilitated the penetration of the positively charged QDs. They could

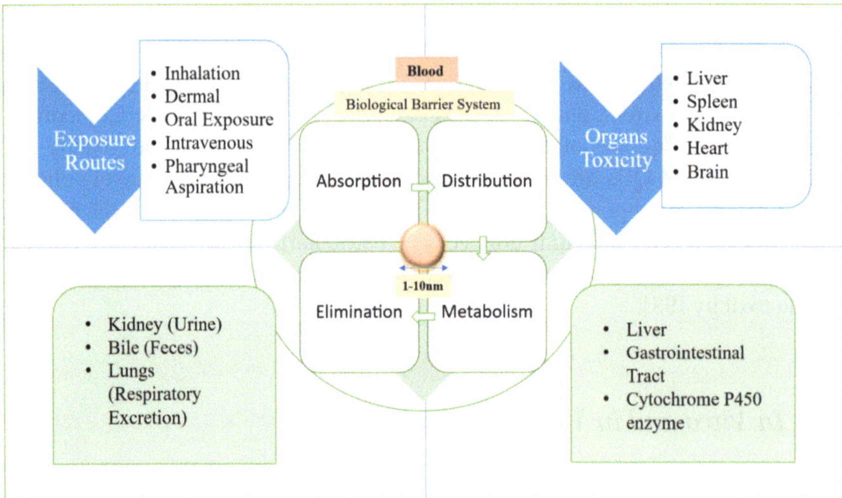

Fig. 5 The pharmacokinetics of the QDs in vivo

also effectively penetrate the nuclear membrane and interact with DNA's negatively charged sugar-phosphate backbone. After incubating for 48 and 72 h, the QDs with smaller hydrodynamic diameters (HDDs) penetrated cells more effectively. This confirms that positively charged QDs with small hydrodynamic diameters are the most toxic for cells in vitro [87]. The evaluation of the in vivo toxicity test of QDs depends on the cytotoxicity of QDs. The safety of QDs must be studied before QDs are applied in vivo. Because cytotoxicity research can pre-determine the potential toxicity of QDs and the existing mechanism of drug production before animal exposure, it is essential to study the cytotoxicity of QDs. Previous studies have found that QDs caused different degrees of cytotoxicity, such as decreasing cell viability, altering cell morphology [116], inducing autophagy [48, 97], causing cells to produce living oxygen substances, and changing the expression of genes [6].

Most QDs enter cells through endocytosis, which cells regulate in various ways, including G protein-coupled receptors. Although non-specific endocytosis can be prevented by comprehending the mechanism of QDs and cell reaction, cell endocytosis is an intricate mechanism controlled by several variables. The investigation into quantum dot endocytosis is still in its early stages, and more research is needed. The substance that enters the cell by endocytosis will be transported to different parts of the cell as necessary to perform its function. The present results show that QDs are transported to the lysosome mainly through the early and late endosomes after entering the cell. Since the lysosome is an acidic organelle, QDs are highly likely to degrade under low pH conditions, inducing cell organelle damage and functional damage. Unfortunately, experiments have proved that QDs caused a series of cellular organ pressures, such as cell membrane defects, nuclear shrinkage, lysosome swelling [53], mitochondrial lipid oxidation, and membrane potential decline

[30]. Although the cell has a specific self-repair function, when the organelle pressure exceeds a certain threshold, the cell will undergo irreversible death. The high concentration of QDs further activates the apoptosis pathway based on organelle damage, leading to cell apoptosis or inducing cells to trigger programmed death autophagy. At the body level, QDs may cause adverse effects on systems, such as the respiratory, cardiovascular, nervous, immune, and endocrine systems [117].

7.2 Long-Term Retention and Clearance

Targeted nanoparticles help to diagnose and cure human diseases, but their possible toxicity due to long-term retention remains a primary barrier. In several in vivo studies, the injection of cadmium-based quantum dots has produced morphological and functional problems in the liver. Both the accumulation of QDs in the liver and their breakdown and clearance over a longer duration can cause hepatic toxicity [50]. For example, a recent work showed that intravenous injection of CdSe/ZnS QDs in mice resulted in liver dysfunction and inflammation, and that the presence of cadmium and the production of ROS induced hepatotoxicity. The levels of liver enzymes increased following in vivo administration of cadmium-based QDs have been reported by several studies, suggesting impaired liver function/or hepatocellular injury [91]. In a study by [91], higher levels of creatinine (Cr) were observed after mice were injected with cadmium-based QDs, which later resulted in kidney injury. Hyperplasia and swelling of the glomerular capillary endothelium were identified in the H&E-stained sections of the kidney [91]. Yaghini et al. [106] found that CdSe/ZnS QDs did not induce appreciable toxicity in Sprague–Dawley rats throughout their study [106]. Behari et al. [14] observed significant toxicity of CdSe/ZnS QDs in Wistar rats [14], including higher liver enzymes and liver inflammation. In a study by [75], lung injury and inflammation were reported following the administration of CdSe/ZnS QDs in rats [75].

Organic quantum dot coating changes the kidney clearance of the quantum dots and prevents serum absorption. The quantum dots with negative surface charge are less prone to glomerular filtration than the positively charged surface quantum dots [22]. Moreover, in contrast to cationic QDs, those with stiffness and various quantum dot shapes can control the glomerular filtration rate. Additionally, certain surface and chemical modifications may decrease the metallic toxicity of quantum dots, PEGylation procedures are one such strategy, resulting in a prolonged shelf-life [22]. Further, it has been shown that QDs can be altered to increase their affinity for specific tissues, including malignancies, by incorporating alternative targeting ligands, polymers, or antibodies. Reports indicate that the kidneys readily eliminate QDs with an average hydrodynamic diameter of less than 5.5 nm through glomerular filtration. The smaller 8–10 nm diameter quantum dots can escape this issue. Zwitterionic or neutral organic QD coatings change renal QD clearance and affect serum protein adsorption. Furthermore, compared to positively charged QDs, those with neutral or negatively charged surfaces are less susceptible to glomerular filtration. Compared to positively charged QDs, those with neutral or negatively charged surfaces are less susceptible to glomerular filtration. The shape and stiffness of quantum dots can also be altered to affect the glomerular filtration rate. On the other hand, multiple surface alterations have also been developed to reduce the metallic QDs, which can have toxic effects over long-term exposure.

7.3 Surface Modification to Reduce Cytotoxicity

QDs frequently undergo surface alterations, including conjugating biomolecule recognition, leading to particular biological roles. It was observed that QD surface modification can alter toxicity [81]. The authors state that core–shell QDs are less toxic than bare QDs. To enhance the fluorescence performance of QDs, a semiconductor shell layer is typically wrapped around their surface. By blocking Cd^{2+} from the medium in the environment, lowering Cd^{2+} leakage and free ion production, and lowering toxicity, this package can enhance the fluorescence characteristics of QDs. The in vitro cytotoxicity studies showed that PEG-g-CS as a ligand displayed very low cytotoxicity [39]. The testing indicated that the properties of QDs and the release of Cd^{2+} result in developmental toxicity. Moreover, two types of CdSe/ZnS QDs exhibited developmental toxicity that affected heart development. In contrast, carboxyl QDs exhibited greater toxicity, possibly due to their higher affinity and release to embryos and larvae [115]. So, multiple factors of surface modification are responsible for QD toxicity, but their impact on toxicity varies.

One of the primary mechanisms triggered by the quantum dots is oxidative stress, as it produces ROS and antioxidant depletion on cellular interaction. Figure 6 depicts the core of QD degradation is with the release of heavy metal ions such as Cd^{2+} later producing reactive oxygen species (ROS) and resulting in apoptosis, and finally interacting with components in the surrounding medium. Different surface functional groups (e.g., COOH, NH2, or OH) also affect their biological behavior and toxic effects. Zheng et al. [115] utilized zebrafish embryos to compare the toxicity in vivo of CdSe/ZnS QDs modified with carboxyl (–COOH) and amino (–NH2) groups. Both CdSe/ZnS QDs reduced the survival rate, hatching rate, and embryonic movement of zebrafish, ultimately leading to pericardial oedema and cardiac dysfunction. Each property of QD toxicity contributes to charge > functionalization > and size. Specifically, positively charged QDs are more toxic than negative QDs; QDs with long ligands are more toxic than those functionalized with shorter chains; and the toxicity of QDs with small particle sizes is greater than that of larger particle sizes. It is challenging to assess the toxicity profile of quantum dots for several reasons, such as the variety in sizes, the inhibition of ion leaching, and aggregation. A surface coating, such as silica, can reduce the toxicity of QDs [11].

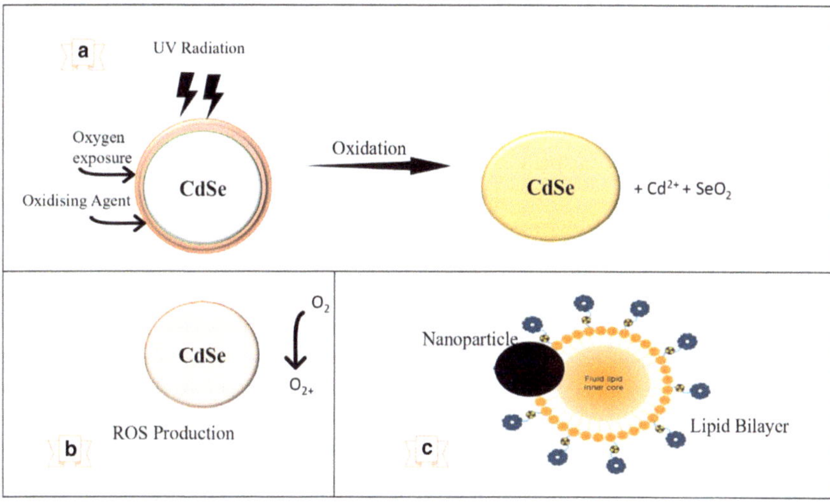

Fig. 6 Cytotoxic mechanism following three points: **A** The core of QDs' degradation with release of heavy metal ions such as Cd^{2+}. **B** Producing reactive oxygen species (ROS) and induce apoptosis. **C** Interacts with components in the surrounding medium

7.4 Regulatory and Ethical Considerations

Quantum dots' ethical and regulatory aspects pose ethical and legal challenges due to their unique characteristics and effectiveness in various industries, specifically healthcare. Addressing the ethical issues and legal frameworks surrounding their creation and important biological and medical applications is essential as their use grows. Despite having beneficial potential, addressing the ethical issues effectively is important to guarantee its safe and fair utility. Regulatory Considerations: In terms of safety assessments, several guidelines have been developed by regulatory bodies for assessing the efficacy of QDs, mainly focusing on their interactions with biological systems and potential ecotoxicological effects [84, 89]. During the standardization, there is a need for standardized characterization techniques to evaluate QDs effectively, which is essential for their approval in applications in different branches [84].

Ethical considerations and informed consent in biomedical applications where QDs are utilized for therapy and diagnosis, ethical considerations concerning patient self-reliance and informed consent are crucial [84]. Because QD-based therapies are novel and their long-term effects are erratic, it is imperative to get informed consent. Also, the potential ecological risks associated with QDs necessitate careful consideration of their lifecycle and disposal. Engaging the public in discussions about the implications of QD technology is essential to address societal concerns and foster trust [89].

8 Challenges and Limitations

A minimal number of clinical trials is a significant drawback. It can be attributed to the complex challenges associated with them from the pharmaceutical, industrial, technical, and biological aspects.

8.1 Scale-Up and Manufacturing Issues

Quantum dots encounter multiple industrial drawbacks upon shifting from the laboratory-scale to large-scale production. First, the various surface modifications essential for altering and advancing their performance make the mass production of QDs complicated and a multi-step process. Second, it is not easy to maintain uniform physicochemical properties of QDs on a large industrial scale. Third, contaminating residues in the solvents or constituent materials that are insignificant on a tiny scale are also magnified, significantly affecting the quality and functionality of the synthesized QDs. Peng et al. [70] came across alkyl phosphonic and phosphonic acid impurities in trioctylphosphine oxide (TOPO), which is used as a coordinating solvent, that play a drastic role in the performance of CdSe QDs upon scale-up, resulting in poor optical performance. Certain advantages and developments have taken place to resolve the above challenges, and synthesizing CdSe QDs faces difficulties upon scale-up [70].

Conversely, the particle size distribution and, in turn, the optical characteristics of the resulting QDs are influenced by the injection speed and stirring velocities. A one-pot method was developed to address this problem, which includes mixing the precursors at room temperature and heating each reaction component precisely. The novel approach was resilient and injection-free, showing superior scale-up and equivalent optical characteristics to QDs produced by the conventional method. Using a unique electrochemical process, [85] synthesized PbS QDs on an industrial scale. A specialized lithography track system eliminated metal cross-contamination, while nitrogen purging prevented oxidation hazards [85]. The low-temperature operation conditions maintained the device's performance without damage. The articulated method increased the quantum properties of PbS QDs by 50%. Fourth, the primary synthesis of QDs is associated with environmental risks because of potentially hazardous metals like Cd and Pb or certain solvents. In order to reduce the ecological hazards associated with heavy metals, cadmium-free QDs, such as carbon, graphene, and silicon QDs, have been created and have shown corresponding optical performance. Starting in October 2019, the European Commission decided to ban cadmium-containing quantum dots, dramatically affecting the QD market, with the overall sales of cadmium-free QDs dominating over 80% in 2018. Moreover, environment-friendly green synthetic methods have also emerged. An innovative and ecologically sound method for creating carbon QDs is hydrothermal synthesis. Hydrothermal synthesis is an economical technique and provides QDs with the right capabilities. Many environmentally conscious techniques have been developed, such as dry heating, microwave synthesis, and material recycling [1].

8.2 Stability and Shelf-Life of QD Formulations

The two primary obstacles are the stability and pharmacokinetics of QDs. The performance of QDs, including degradation, may hamper drug delivery vehicles. Strategies such as surface modification and encapsulation techniques are used to improve these parameters and the effectiveness of QDs [55]. By preventing biological components from breaking down QDs, these tactics hope to increase their bioavailability and lengthen their stay in circulation [96].

8.3 Immunogenicity and Off-Target Effects

Quantum dots containing heavy metals have raised concerns regarding toxicity, especially cadmium, immunogenicity, and off-target effects. It can affect different cell functions and lead to significant immunotoxicity [33]. In cellular impact, the cytotoxic effects of CdTe QDs have resulted in CD^{4+} and CD^{8+} T cells, affecting cell viability and apoptosis. Quantum dots may produce significant off-target effects, including inflammation and long-term tissue accumulation. Direct administration of quantum dots can induce cellular inflammation, as evidenced by the infiltration of inflammatory markers in vivo studies. On the other hand, quantum dots accumulate in long-term exposure and show chronic effects on the immune system [56]. Numerous applications in different fields make quantum dots more demanding. However, the synthesis and modification according to demand make them expensive. Specifically, metal cadmium-containing quantum dots require surface modification to reduce their toxicity and increase their effectiveness. The high cost is mainly due to the requirement for precision in different sizes and shapes, compatibility study, and the chemistry of the quantum dots [38].

8.4 Cost and Commercial Viability

A vast number of applications in various fields make quantum dots more demanding. However, their cost and economic viability are influenced by multiple factors. In addition to synthesis, surface changes needed to improve targeting and biocompatibility result in additional costs. These include conjugating QDs with biomolecules or coating them with various polymers, such as polyethylene glycol (PEG), raising the overall cost. The lack of large-scale production facilities for biomedical quantum dots limits cost reduction opportunities. With increased demand, manufacturing processes increase and become more streamlined, and economies of scale might help lower costs. Commercial viability and specific market growth related to the global quantum dots market are experiencing significant growth. The cost of quantum dots was estimated at USD 5.97 billion in 2023 and is expected to increase at a compound annual growth rate to USD 25.05 billion by 2032 with CAGR of 17.4%. Critical display technologies, solar cells, and biomedical imaging applications drive this growth. Biomedical applications of QDs have significant applications in targeted medication delivery, imaging, and diagnostics in the biomedical industry. They are perfect for monitoring biological processes because of their intense, steady fluorescence. However, studies into safer substitutes like carbon and silicon-based QDs have been spurred by worries about toxicity, primarily associated with heavy metal-containing QDs [1].

Limitations of quantum dots in the pharmaceutical field: The increase in production and application of QDs has led to a growing risk of unintentional occupational or environmental exposure to QDs. Some of the bioconjugation of quantum dots causes difficult delivery to the target cell. Large biomolecules such as proteins or antibodies can be conjugated with QDs to increase their hydrodynamic diameter significantly. This expansion may decrease the effectiveness of tissue penetration and hinder cellular absorption. Larger QDs may also accumulate in organs like the liver and spleen because they are less likely to be eliminated by the kidneys. Certain quantum dots with bioconjugation, especially with surface modification, can show cytotoxicity over a long exposure period. For example, carbon dots with carboxylic groups have shown delayed toxicological effects due to the aggregation over a long period and can cause toxicity to organs like the kidney [20]. Because of the accumulation of quantum dots, they tend to kill the live cell [21]. It is currently unknown whether quantum dots are metabolized and eliminated, which can cause toxicity in the body. Poor size control results from coating the core material with a shell [58]. Quantum dots deteriorate due to their surface and blinking characteristics [118]. The pharmaceutical industry has very little clinical data on quantum dots [8]. The processes of QD toxicity, bioaccumulation, organ-specific effects, and their implications on organisms in actual exposure situations are still poorly understood.

9 Future Perspectives and Emerging Trends

Quantum dot (QD) technology for cancer therapy is poised to break through critical barriers in safety, targeting, and treatment individualization in the future. Emerging designs in biodegradable and biocompatible QDs are poised to eliminate chronic toxicity challenges, enhancing their utility in clinical oncology. In addition to these advances, stimuli-responsive QD platforms are moving toward exploiting the unique aspects of the tumor microenvironment, such as low pH and enzymatic and redox gradients, to afford highly specific and tunable drug release to the appropriate target [67]. Besides, integrating QDs into the theranostics platform has tremendous promise for precision oncology, enabling simultaneous tumor imaging and targeted drug delivery, which can enable real-time monitoring of therapeutic response and personalized treatment modulation. Overall, these directions illustrate the revolutionary potential of QDs to transform drug delivery for cancer and increase patient therapy [82].

9.1 Next-Generation Biocompatible QDs for Safe Cancer Therapy

The development of quantum dots (QDs) as promising drug carriers for cancer therapy is founded on material synthesis, overcoming centuries-old issues of toxicity and biocompatibility. Traditional QDs, composed chiefly of heavy metals like cadmium, selenium, and lead, are of pivotal concern to safety due to their possibility of cytotoxicity, aside from time-dependent accumulation in essential organs, thereby restricting their future use in the clinic [100]. To overcome the disadvantages discussed above, emerging investigations have moved toward synthesizing future QDs using native biocompatible and biodegradable materials such as carbon, graphene, silicon, and indium phosphide. The resultant new QDs show relatively low toxicity and provide good photophysical properties, including high quantum yield (QY) of fluorescence and tunable emission spectra of great significance for applications in the diagnosis and therapy of cancer diseases [83]. Surface engineering has also played a leading role in increasing the biocompatibility of QDs. Hydrophilic polymer (e.g., polyethylene glycol) functionalization, peptides, and other biomolecules enhance solubility and biological milieu stability, decreasing nonspecific reticuloendothelial system uptake and increasing systemic circulation time. This leads to more effective targeted therapy of the tumor through enhanced permeability and retention (EPR) effect, as well as through ligand–receptor interaction [66]. Secondly, built-in biodegradability ensures they can be metabolized adequately by excreting or metabolizing once they have completed their therapeutic task, so chronic toxicity is not a concern. These advances indicate a radical change in the chemistry of toxic QDs concerning the new generation of biocompatible nanosystems, ultimately designed for successful and safe cancer therapy. Meeting the safety-performances trade-off, next-generation QDs are expected to bridge the gap between experimental nanomaterials and clinically applicable cancer therapy and finally realize targeted, minimally invasive, and personalized cancer therapy [54].

9.2 Stimuli-Responsive QD Platforms for Targeted Cancer Drug Delivery

Stimuli-responsive quantum dot (QD) systems are a next-generation approach to enhancing the efficacy and specificity of cancer chemotherapy drug delivery by utilizing the tumor microenvironment's unique biochemical and physiological features. In contrast to the conventional drug delivery systems, smart QDs are engineered to be responsive specifically to inner stimuli such as reduced pH, enhanced activity of specific enzymes, redox imbalance, and hypoxia inside tumor tissue [44]. In Fig. 7, the endogenous and exogenous stimuli have been illustrated. By attaching responsive coatings or linkers to these stimuli, either due to their degradation or a change in structure, stimulus-responsive QDs allow a controlled and locally targeted drug payload release inside the TME. Tumor-directed activation enables localized therapy and reduced systemic side effects and toxicity [42]. Alternatively, other bioreactors introduce exogenous stimuli such as light, temperature, or magnetic fields to trigger drug release temporally and spatially with high spatial resolution and thus to modulate treatment based on disease dynamically. Integrating stimulus-responsive function in the QD-based drug delivery systems will therefore be very attractive for achieving an efficient strategy to combat the deficiency of traditional chemotherapy, leading to the delivery of potent anticancer drugs with enhanced selectivity and therapeutic efficiency [18].

9.3 Theranostic QDs in Precision Oncology

Theranostic quantum dots (QDs) have presented an unprecedented opportunity in precision oncology owing to the possibility of cancer imaging, target therapy, and therapeutic response monitoring in one nanoscaled platform. Such multivalent nanostructures take advantage of the natural optical properties of QDs, i.e., broad absorption spectra, narrow, highly tunable emission bands, high quantum yield, and stability that are superior to those of conventional organic dyes in sensitivity and longevity [63]. Through surface functionalization with tumor-penetrating ligands (e.g., antibodies, peptides, aptamers) and chemotherapeutic agents, QDs can be engineered to actively target cancer cells with the capability to image and treat at both cellular and molecular levels in real time. Its dual modality function enables one to track biodistribution, intracellular delivery, and tumor response in real time, significantly increasing treatment specificity for cancer diagnosis and enabling timely therapeutic intervention [49]. A few recent advances in theranostics using QDs are the utilization of stimulus-sensitive linkers, allowing for on-demand drug delivery toward tumor-specific stimuli like acidic pH, excessive glutathione levels, or overexpressed enzymes like matrix metalloproteinases. This allows for selective spatiotemporal drug delivery with minimal systemic toxicity and maximum therapeutic effect [52]. Moreover, QDs can be incorporated into other imaging modes,

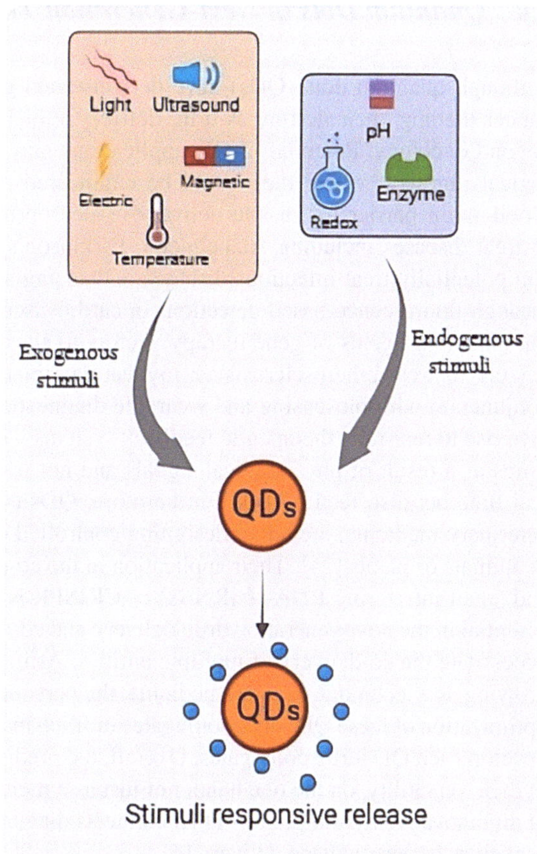

Fig. 7 Schematic diagram of stimuli-responsive nanomaterials. The figure illustrates the stimulus-responsive release of payloads from quantum dots (QDs). The figure highlights two types of stimuli: **a** Exogenous stimuli, such as temperature, electric, light, ultrasound, and magnetic field. **b** Endogenous stimuli—pH change, redox reaction, and enzymes. Both endogenous and exogenous stimuli interact with QDs. When exposed to the appropriate stimulus, QDs undergo the change that triggers the release of the payload. This approach enables precise and targeted delivery of drugs using quantum dots

such as MRI or PET, to construct multimodal platforms that provide complementary anatomical and functional data for enabling extensive tumor profiling. The novel construct also employs NIR emission to permit deeper tissue penetration and minimize auto-fluorescent background to improve imaging contrast in vivo [108]. Theranostic QDs represent the ultimate synergy between nanotechnology and personalized medicine. They may offer ultra-highly customized treatment strategies as a function of real-time tumor characteristics and patient-specific responses. Although there are challenges, such as long-term biocompatibility, clearance pathways, and regulatory issues, advancements in surface chemistry, biodegradable core material composition, and hybrid nanostructure relentlessly drive toward conquering these challenges. QDs play a central role in cancer treatment in theranostics, enabling earlier and more precise diagnosis, better stratification, and even targeted, finely tuned, patient-specific therapeutic intervention [82].

9.4 Quantum Dots in Next-Generation Therapeutics

Although quantum dots (QDs) have demonstrated great promise in oncology for cancer therapy, their destiny as drug delivery vehicles lies far into the future and beyond oncology. Potential of QD applications may lie in the treatment of neurological illnesses, where they could be engineered and programmed to cross the blood–brain barrier for in situ delivery of neuroprotective or gene editing drugs to treat diseases including Alzheimer's, Parkinson's, and glioblastoma [35]. QDs can potentially treat infectious diseases, enhancing specificity and rapid diagnosis through fluorescence-based detection. In cardiovascular diseases, QDs target anti-inflammatory agents or gene therapy vectors to sites of inflammation in the blood vessels to treat atherosclerosis or myocardial infarction [34, 77]. Moreover, in conjunction with biosensing and wearable diagnostic technologies, this could also give rise to real-time therapeutic feedback systems—miniaturized systems that give drugs as a result of physiological signals and not just measure patient response in real time but also feed it back. Furthermore, QDs can be used in agricultural and veterinary medicine, including designing controlled drug/vaccine delivery systems in animals or plants [24]. Their application in the co-delivery of conventional drugs and small interfering RNAs (siRNAs) or CRISPR is also one reason why they are essential in the next-generation drug delivery and co-delivery methodologies, which necessitate the co-delivery of multiple entities. Artificial Intelligence (AI) has been growing as a technique for nanomedicine, the personalized and scale-up design and optimization of these QD–drug conjugates in silico may change the way we currently develop such QD–drug conjugates. Overall, the prediction for QDs in the future lies in their versatility, on the one hand, not just as a mere drug carrier, but in the frame of an innovative hybrid platform that connects diagnostics and therapy, monitoring, as well as in personalized delivery [4].

9.5 Conclusion

The unforeseen development of nanotechnology has significantly influenced the new drug delivery scenario of the modern world, offering unparalleled opportunities to design more efficient, targeted, and less harmful drugs. Among the various nanocarriers investigated, quantum dots (QDs) are an up-and-coming category due to their attractive physicochemical characteristics, including size-tunable luminescence, high surface-area-to-volume ratio, and ease of surface modification. These features enable QDs to be multifunctional platforms, integrating diagnostics, imaging, and drug delivery with controlled release in a single construct. In this chapter, QD-based drug delivery systems' intrinsic properties, design strategy, and functional application have been elaborately detailed. Emphasis has been placed on their advantages over conventional delivery systems, most importantly enhancing intracellular delivery, enabling real-time biodistribution imaging, and enabling stimuli-responsive release profiles.

In addition, the merging of therapeutic and diagnostic capabilities, theranostics by another name, is leading QDs to the height of precision nanomedicine. Though these auspicious features are auspicious for clinical translation of QDs, their clinical translation is marred by multiple issues, such as heavy-metal-based core toxicity issues, long-term bioretention issues, and scalable synthesis and surface modification issues. New advancements in cadmium-free and biodegradable carbon-based QDs offer a potential avenue for clinical translation that circumvents critical safety and regulation issues. Future research will require these translational barriers to be overcome by integrating biocompatible, non-immunogenic QDs, globally recognized means for determining toxicity, and innovative manufacturing techniques. Multidisciplinary collaborative studies involving material science, molecular biology, pharmacology, and clinical medicine will have the onus of unlocking the full potential of QDs as future nanocarriers. Since this is the case, quantum-dot-system-based technology stands to revolutionize the drug delivery paradigm, redefining how therapies are built, delivered, and monitored in real-world clinical applications.

Acknowledgements The authors are grateful to the National Institute of Pharmaceutical Education and Research, Kolkata, Kolkata 700054, and the Department of Science and Technology for supporting the students through fellowships.

Competing Interests None declared.

Ethical Approval Not required.

Conflicts of Interest The authors declare that they have no known competing financial interests or personal relationships that could have appeared to influence the work reported in this paper.

References

1. Abdellatif AA, Younis MA, Alsharidah M, Al Rugaie O, Tawfeek HM (2022) Biomedical applications of quantum dots: overview, challenges, and clinical potential. Int J Nanomed 17:1951–1970. https://doi.org/10.2147/IJN.S357980
2. Agarwal K, Rai H, Mondal S (2023) Quantum dots: an overview of synthesis, properties, and applications. Mater Res Express 10(6). Institute of Physics. https://doi.org/10.1088/2053-1591/acda17
3. Aina F, Manan A, Yusof NA, Abdullah J, Mohammad F, Nurdin A, Yazan LS, Khiste SK, Al-Lohedan HA, Racoviceanu R, Alina M, My JA (2021) Pharmaceutics drug release profiles of Mitomycin C encapsulated quantum dots-chitosan nanocarrier system for the possible treatment of non-muscle invasive bladder cancer. https://doi.org/10.3390/pharmaceutics
4. Ali MK, Javaid S, Afzal H, Zafar I, Fayyaz K, Ain Q, Rather MA, Hossain MJ, Rashid S, Khan KA, Sharma R (2023) Exploring the multifunctional roles of quantum dots for unlocking the future of biology and medicine. Environ Res 232:116290. https://doi.org/10.1016/j.envres.2023.116290
5. Alipour B, Mortezazadeh T, Abdulsahib WK, Arzhang A, Malekzadeh R, Farhood B (2023) A systematic review of multimodal application of quantum dots in breast cancer diagnosis: effective parameters, status and future perspectives. J Drug Deliv Sci Technol 86:104682. https://doi.org/10.1016/j.jddst.2023.104682

6. Alothaid H, Al-Anazi MR, Al-Qahtani AA, Colak D, Yusuf A, Aldughaim MS, Mahzari AM, Habibullah MM, Alarifi S, Alkahtani S, Al-Qahtani AA (2023) Exposure to cadmium telluride quantum dots and gene expression profile of Huh-7 hepatocellular carcinoma cell line. Dose-Response 21(3). https://doi.org/10.1177/15593258231185457
7. Alshawwa SZ, Kassem AA, Farid RM, Mostafa SK, Labib GS (2022) Nanocarrier drug delivery systems: characterization, limitations, future perspectives and implementation of artificial intelligence. Pharmaceutics 14(4):883. https://doi.org/10.3390/pharmaceutics14040883
8. Amol Singh P, Bajwa N (2016) Nano technical trends for cancer treatment: quantum dots a smart drug delivery system. Int J Pharm Sci Res 7(4):1360. https://doi.org/10.13040/IJPSR.0975-8232.7(4).1360-74
9. Anas A, Okuda T, Kawashima N, Nakayama K, Itoh T, Ishikawa M, Biju V (2009) Clathrin-mediated endocytosis of quantum dot−peptide conjugates in living cells. ACS Nano 3(8):2419–2429. https://doi.org/10.1021/nn900663r
10. Caratti A, Trapani F (2024) Boosting comprehensive two-dimensional gas chromatography with artificial intelligence: computer vision helps to see what we smell. In: 18th international symposium on hyphenated techniques in chromatography and separation technology, leuven
11. Auclair J, Turcotte P, Gagnon C, Peyrot C, Wilkinson KJ, Gagné F (2019) The influence of surface coatings on the toxicity of silver nanoparticle in rainbow trout. Comp Biochem Physiol C: Toxicol Pharmacol 226:108623. https://doi.org/10.1016/j.cbpc.2019.108623
12. Bacon M, Bradley SJ, Nann T (2014) Graphene quantum dots. Particle Particle Syst Charact 31(4):415–428. Wiley-VCH Verlag. https://doi.org/10.1002/ppsc.201300252
13. Bansal M, Kumar A, Malinee M, Sharma TK (2020) Nanomedicine: diagnosis, treatment, and potential prospects, pp 297–331. https://doi.org/10.1007/978-3-030-29207-2_9
14. Behari J, Behari J, Jin T (2011) Bio-distribution and toxicity assessment of intravenously injected anti-HER2 antibody conjugated CdSe/ZnS quantum dots in Wistar rats. Int J Nanomed 463. https://doi.org/10.2147/ijn.s15124
15. Bilan R, Fleury F, Nabiev I, Sukhanova A (2015a) Quantum dot surface chemistry and functionalization for cell targeting and imaging. Bioconjugate Chem 26(4):609–624. American Chemical Society. https://doi.org/10.1021/acs.bioconjchem.5b00069
16. Bilan R, Fleury F, Nabiev I, Sukhanova A (2015b) Quantum dot surface chemistry and functionalization for cell targeting and imaging. http://pubs.acs.org
17. Cai X, Luo Y, Zhang W, Du D, Lin Y (2016) PH-sensitive ZnO quantum dots-doxorubicin nanoparticles for lung cancer targeted drug delivery. ACS Appl Mater Interfaces 8(34):22442–22450. https://doi.org/10.1021/acsami.6b04933
18. Cao Z, Zuo X, Liu X, Xu G, Yong K-T (2024) Recent progress in stimuli-responsive polymeric micelles for targeted delivery of functional nanoparticles. Adv Coll Interface Sci 330:103206. https://doi.org/10.1016/j.cis.2024.103206
19. Chatterjee S, Kon E, Sharma P, Peer D (2024) Endosomal escape: a bottleneck for LNP-mediated therapeutics. Proc Natl Acad Sci 121(11). https://doi.org/10.1073/pnas.2307800120
20. Chen N, He Y, Su Y, Li X, Huang Q, Wang H, Zhang X, Tai R, Fan C (2012) The cytotoxicity of cadmium-based quantum dots. Biomaterials 33(5):1238–1244. https://doi.org/10.1016/j.biomaterials.2011.10.070
21. Cho SJ, Maysinger D, Jain M, Röder B, Hackbarth S, Winnik FM (2007) Long-term exposure to CdTe quantum dots causes functional impairments in live cells. Langmuir 23(4):1974–1980. https://doi.org/10.1021/la060093j
22. Choi HS, Ipe BI, Misra P, Lee JH, Bawendi MG, Frangioni JV (2009) Tissue- and organ-selective biodistribution of nir fluorescent quantum dots. Nano Lett 9(6):2354–2359. https://doi.org/10.1021/nl900872r
23. Dada SN, Babanyinah GK, Tetteh MT, Palau VE, Walls ZF, Krishnan K, Croft Z, Khan AU, Liu G, Wiese TE, Glotser E, Mei H (2022) Covalent and noncovalent loading of doxorubicin by folic acid-carbon dot nanoparticles for cancer theranostics. ACS Omega 7(27):23322–23331. https://doi.org/10.1021/acsomega.2c01482

24. Das S, Mazumdar H, Khondakar KR, Mishra YK, Kaushik A (2024) Review—quantum biosensors: principles and applications in medical diagnostics. ECS Sens Plus 3(2):025001. https://doi.org/10.1149/2754-2726/ad47e2
25. Derivery E, Bartolami E, Matile S, Gonzalez-Gaitan M (2017) Efficient delivery of quantum dots into the cytosol of cells using cell-penetrating Poly(disulfide)s. J Am Chem Soc 139(30):10172–10175. https://doi.org/10.1021/jacs.7b02952
26. Deshmukh A, Bhaiswar J, Kapse V, Meghe D (2023) Quantum dots: a brief review. Quantum Dots Emerg Mater Versatile Appl 41–66. https://doi.org/10.1016/B978-0-323-85278-4.00001-5
27. Din FU, Aman W, Ullah I, Qureshi OS, Mustapha O, Shafique S, Zeb A (2017) Effective use of nanocarriers as drug delivery systems for the treatment of selected tumors. Int J Nanomed 12:7291–7309. https://doi.org/10.2147/IJN.S146315
28. Đorđević S, Gonzalez MM, Conejos-Sánchez I, Carreira B, Pozzi S, Acúrcio RC, Satchi-Fainaro R, Florindo HF, Vicent MJ (2022) Current hurdles to the translation of nanomedicines from bench to the clinic. Drug Deliv Transl Res 12(3):500–525. https://doi.org/10.1007/s13346-021-01024-2
29. Efros AL, Brus LE (2021) Nanocrystal quantum dots: from discovery to modern development. ACS Nano 15(4):6192–6210. American Chemical Society. https://doi.org/10.1021/acsnano.1c01399
30. Fang Y, Zhuo L, Yuan H, Zhao H, Zhang L (2023) Construction of graphene quantum dot-based dissolving microneedle patches for the treatment of bacterial keratitis. Int J Pharm 639:122945. https://doi.org/10.1016/j.ijpharm.2023.122945
31. Gidwani B, Sahu V, Shukla SS, Pandey R, Joshi V, Jain VK, Vyas A (2021) Quantum dots: prospectives, toxicity, advances and applications. J Drug Deliv Sci Technol 61. Editions de Sante. https://doi.org/10.1016/j.jddst.2020.102308
32. Guo W, Song X, Liu J, Liu W, Chu X, Lei Z (2024) Quantum dots as a potential multifunctional material for the enhancement of clinical diagnosis strategies and cancer treatments. Nanomaterials 14(13):1088. https://doi.org/10.3390/nano14131088
33. Hoshino A, Hanada S, Manabe N, Nakayama T, Yamamoto K (2009) Immune response induced by fluorescent nanocrystal quantum dots in vitro and in vivo. IEEE Trans Nanobiosci 8(1):51–57. https://doi.org/10.1109/TNB.2009.2016550
34. Hu Q, Fang Z, Ge J, Li H (2022) Nanotechnology for cardiovascular diseases. Innovation 3(2):100214. https://doi.org/10.1016/j.xinn.2022.100214
35. Hu Y, Wang X, Niu Y, He K, Tang M (2024) Application of quantum dots in brain diseases and their neurotoxic mechanism. Nanoscale Adv 6(15):3733–3746. https://doi.org/10.1039/D4NA00028E
36. Iannazzo D, Pistone A, Celesti C, Triolo C, Patané S, Giofré SV, Romeo R, Ziccarelli I, Mancuso R, Gabriele B, Visalli G, Facciolà A, Di Pietro A (2019) A smart nanovector for cancer targeted drug delivery based on graphene quantum dots. Nanomaterials 9(2). https://doi.org/10.3390/nano9020282
37. Iannazzo D, Pistone A, Salamò M, Galvagno S, Romeo R, Giofré SV, Branca C, Visalli G, Di Pietro A (2017) Graphene quantum dots for cancer targeted drug delivery. Int J Pharm 518(1–2):185–192. https://doi.org/10.1016/j.ijpharm.2016.12.060
38. Jean J, Xiao J, Nick R, Moody N, Nasilowski M, Bawendi M, Bulović V (2018) Synthesis cost dictates the commercial viability of lead sulfide and perovskite quantum dot photovoltaics. Energy Environ Sci 11(9):2295–2305. https://doi.org/10.1039/C8EE01348A
39. Jiang Z, Zhao C, Liu X (2014) Synthesis of poly(ethylene glycol)-graft-chitosan and using as ligand for fabrication of water-soluble quantum dots. Colloids Surf, B 115:260–266. https://doi.org/10.1016/j.colsurfb.2013.12.001
40. Kadyan P, Malik R, Bhatia S, Al Harrasi A, Mohan S, Yadav M, Dalal S, Ramniwas S, Kumar Kataria S, Arasu T (2023) Comprehensive review on synthesis, applications, and challenges of Graphene Quantum Dots (GQDs). J Nanomater 2023. Hindawi Limited. https://doi.org/10.1155/2023/2832964

41. Kambhampati P (2021) Nanoparticles, nanocrystals, and quantum dots: what are the implications of size in colloidal nanoscale materials? J Phys Chem Lett 12(20):4769–4779. https://doi.org/10.1021/acs.jpclett.1c00754
42. Kaushik N, Borkar SB, Nandanwar SK, Panda PK, Choi EH, Kaushik NK (2022) Nanocarrier cancer therapeutics with functional stimuli-responsive mechanisms. J Nanobiotechnol 20(1):152. https://doi.org/10.1186/s12951-022-01364-2
43. Kominkova M, Milosavljevic V, Vitek P, Polanska H, Cihalova K, Dostalova S, Hynstova V, Guran R, Kopel P, Richtera L, Masarik M, Brtnicky M, Kynicky J, Zitka O, Adam V (2017) Comparative study on toxicity of extracellularly biosynthesized and laboratory synthesized CdTe quantum dots. J Biotechnol 241:193–200. https://doi.org/10.1016/j.jbiotec.2016.10.024
44. Kumar A, Patel DD, Banerjee J, Sharma M, Bhatia D, Banerjee S, Yadav AK (2025) Next-generation cancer theragnostic: applications of carbon quantum dots. ChemNanoMat 11(5). https://doi.org/10.1002/cnma.202500061
45. Kundu S, Ghosh M, Sarkar N (2021) State of the art and perspectives on the biofunctionalization of fluorescent metal nanoclusters and carbon quantum dots for targeted imaging and drug delivery. Langmuir 37(31):9281–9301. American Chemical Society. https://doi.org/10.1021/acs.langmuir.1c00732
46. Le N, Zhang M, Kim K (2022) Quantum dots and their interaction with biological systems. Int J Mol Sci 23(18):10763. https://doi.org/10.3390/ijms231810763
47. Li JY, Huang PL (2018) Effects of CdTe quantum dots on the content of cytochrome P450 in rat liver]. Zhonghua Lao Dong Wei Sheng Zhi Ye Bing Za Zhi = Zhonghua Laodong Weisheng Zhiyebing Zazhi = Chin J Ind Hygiene Occup Dis 36(3):174–178. https://doi.org/10.3760/cma.j.issn.1001-9391.2018.03.004
48. Liang Q, Yu F, Cai H, Wu X, Ma M, Li Z, Tedesco AC, Zhu J, Xu Q, Bi H (2023) Photo-activated autophagy-associated tumour cell death by lysosome impairment based on manganese-doped graphene quantum dots. J Mater Chem B 11(11):2466–2477. https://doi.org/10.1039/D2TB02761E
49. Liang Z, Khawar MB, Liang J, Sun H (2021) Bio-conjugated quantum dots for cancer research: detection and imaging. Front Oncol 11. https://doi.org/10.3389/fonc.2021.749970
50. Lin C-H, Yang M-H, Chang LW, Yang C-S, Chang H, Chang W-H, Tsai M-H, Wang C-J, Lin P (2011) Cd/Se/Te-based quantum dot 705 modulated redox homeostasis with hepatotoxicity in mice. Nanotoxicology 5(4):650–663. https://doi.org/10.3109/17435390.2010.539712
51. Lin X, Chen T (2023) A review of in vivo toxicity of quantum dots in animal models. Int J Nanomed 18:8143–8168. https://doi.org/10.2147/IJN.S434842
52. Lin X, Huang Q (2023) Editorial: smart nanomaterials for biosensing and therapy applications. Front Bioeng Biotechnol 11. https://doi.org/10.3389/fbioe.2023.1137508
53. Liu J, Huang Z, Yin S, Zhou X, Jiang Y, Shao L (2023) The lysosome-mitochondrion crosstalk engaged in silver nanoparticles-disturbed mitochondrial homeostasis. Sci Total Environ 889:164078. https://doi.org/10.1016/j.scitotenv.2023.164078
54. Liu Q, Ding X, Pang Y, Cao Y, Lei J, Wu J, Zhang T (2022) New insights into the safety assessment of quantum dots: potential release pathways, environmental transformations, and health risks. Environ Sci Nano 9(9):3277–3311. https://doi.org/10.1039/D2EN00252C
55. Liu R, Fang F, Liu P, Duan X, Wang K, Sun XW (2024) Liquid-encapsulated quantum dot for enhanced UV and thermal stability of quantum dot color conversion films. Nano Res 17(11):10127–10133. https://doi.org/10.1007/s12274-024-6971-0
56. Ma N, Lu Y, Wang J, Liang X, Dong S, Zhao L (2023) Role of CdTe quantum dots on peripheral Immunocytes and selenoprotein P: immunotoxicity at the molecular and cellular levels. Toxicol Res 12(6):1041–1050. https://doi.org/10.1093/toxres/tfad095
57. Ma N, Ma C, Li C, Wang T, Tang Y, Wang H, Mou X, Chen Z, He N (2013) Influence of nanoparticle shape, size, and surface functionalization on cellular uptake. J Nanosci Nanotechnol 13(10):6485–6498. https://doi.org/10.1166/jnn.2013.7525
58. Ma X, Li Y, Hussain I, Shen R, Yang G, Zhang K (2020) Core–shell structured nanoenergetic materials: preparation and fundamental properties. Adv Mater 32(30). https://doi.org/10.1002/adma.202001291

59. Mahaling B, Low SWY, Ch S, Addi UR, Ahmad B, Connor TB, Mohan RR, Biswas S, Chaurasia SS (2023) Next-generation nanomedicine approaches for the management of retinal diseases. Pharmaceutics 15(7):2005. https://doi.org/10.3390/pharmaceutics15072005
60. Majumdar S, Krishnatreya G, Gogoi N, Thakur D, Chowdhury D (2016) Carbon-dot-coated alginate beads as a smart stimuli-responsive drug delivery system. ACS Appl Mater Interfaces 8(50):34179–34184. https://doi.org/10.1021/acsami.6b10914
61. Matea C, Mocan T, Tabaran F, Pop T, Mosteanu O, Puia C, Iancu C, Mocan L (2017) Quantum dots in imaging, drug delivery and sensor applications. Int J Nanomed 12:5421–5431. https://doi.org/10.2147/IJN.S138624
62. McMahon HT, Boucrot E (2011) Molecular mechanism and physiological functions of clathrin-mediated endocytosis. Nat Rev Mol Cell Biol 12(8):517–533. https://doi.org/10.1038/nrm3151
63. Mishra Y (2024) Quantum dots as emerging theranostics in cancer treatment. Indian J Pharm Sci 86(5). https://doi.org/10.36468/pharmaceutical-sciences.1424
64. Mitchell MJ, Billingsley MM, Haley RM, Wechsler ME, Peppas NA, Langer R (2021) Engineering precision nanoparticles for drug delivery. Nat Rev Drug Discovery 20(2):101–124. https://doi.org/10.1038/s41573-020-0090-8
65. Naylor-Adamson L, Price TW, Booth Z, Stasiuk GJ, Calaminus SDJ (2024) Quantum dot imaging agents: haematopoietic cell interactions and biocompatibility. Cells 13(4):354. https://doi.org/10.3390/cells13040354
66. Niveria K, Singh P, Yadav M, Verma AK (2023) Quantum Dot (QD)-induced toxicity and biocompatibility. In: Handbook of II–VI semiconductor-based sensors and radiation detectors. Springer International Publishing, pp 181–211. https://doi.org/10.1007/978-3-031-19531-0_8
67. Omidian H, Wilson RL, Cubeddu LX (2024) Quantum dot research in breast cancer: challenges and prospects. Materials 17(9):2152. https://doi.org/10.3390/ma17092152
68. Pareek A, Kumar D, Pareek A, Gupta MM (2025) Advancing cancer therapy with quantum dots and other nanostructures: a review of drug delivery innovations, applications, and challenges. Cancers 17(5). Multidisciplinary Digital Publishing Institute (MDPI). https://doi.org/10.3390/cancers17050878
69. Pati R, Sahu R, Panda J, Sonawane A (2016) Encapsulation of zinc-rifampicin complex into transferrin-conjugated silver quantum-dots improves its antimycobacterial activity and stability and facilitates drug delivery into macrophages. Sci Rep 6. https://doi.org/10.1038/srep24184
70. Peng L, Gao Z, Liang Y, Guo X, Zhang Q, Cui D (2025) Nanoparticle-based drug delivery systems: opportunities and challenges in the treatment of esophageal squamous cell carcinoma (ESCC). Nanoscale 17(14):8270–8288. https://doi.org/10.1039/D4NR05114A
71. Pham SH, Choi Y, Choi J (2020) Stimuli-responsive nanomaterials for application in antitumor therapy and drug delivery. Pharmaceutics 12(7):1–19. MDPI AG. https://doi.org/10.3390/pharmaceutics12070630
72. Ameta R, Jayesh (2022) Quantum Dots: fundamentals, synthesis and applications. Elsevier
73. Ratre P, Thareja S, Mishra PK (2024) Quantum imaging-based nanobiosensors: Pioneering point-of-care approach for early diagnosis of environmental-linked breast cancer. J Reprod Healthcare Med 5:9. https://doi.org/10.25259/JRHM_10_2024
74. Resch-Genger U, Grabolle M, Cavaliere-Jaricot S, Nitschke R, Nann T (2008) Quantum dots versus organic dyes as fluorescent labels. Nat Methods 5(9):763–775. https://doi.org/10.1038/nmeth.1248
75. Roberts JR, Antonini JM, Porter DW, Chapman RS, Scabilloni JF, Young S-H, Schwegler-Berry D, Castranova V, Mercer RR (2013) Lung toxicity and biodistribution of Cd/Se-ZnS quantum dots with different surface functional groups after pulmonary exposure in rats. Part Fibre Toxicol 10(1):5. https://doi.org/10.1186/1743-8977-10-5
76. Salaudeen HD, Akinniranye RD (2024) Precision nanotechnology for early cancer detection and biomarker identification. Int J Res Publ Rev 5(11):6313–6327. https://doi.org/10.55248/gengpi.5.1124.3404

77. Sandhya M, Senthilraja G, Priyadharshini E, Rani LU, Harideekshayini R, Nishanthi M, Anand T, Subramanian KS (2025) Early detection of plant diseases and their management using quantum dots: status and strategies. J Fluoresc. https://doi.org/10.1007/s10895-025-04344-5
78. Saravanakumar K, Hu X, Ali DM, Wang M-H (2019) Emerging strategies in stimuli-responsive nanocarriers as the drug delivery system for enhanced cancer therapy. Curr Pharm Des 25(24):2609–2625. https://doi.org/10.2174/1381612825666190709221141
79. Schipper ML, Iyer G, Koh AL, Cheng Z, Ebenstein Y, Aharoni A, Keren S, Bentolila LA, Li J, Rao J, Chen X, Banin U, Wu AM, Sinclair R, Weiss S, Gambhir SS (2009) Particle size, surface coating, and PEGylation influence the biodistribution of quantum dots in living mice. Small 5(1):126–134. https://doi.org/10.1002/smll.200800003
80. Shao L, Gao Y, Yan F (2011) Semiconductor quantum dots for biomedicial applications. Sensors 11(12):11736–11751. https://doi.org/10.3390/s111211736
81. Silva BF, Andreani T, Gavina A, Vieira MN, Pereira CM, Rocha-Santos T, Pereira R (2016) Toxicological impact of cadmium-based quantum dots towards aquatic biota: effect of natural sunlight exposure. Aquat Toxicol 176:197–207. https://doi.org/10.1016/j.aquatox.2016.05.001
82. Singh V (2024) Theranostics: integrated diagnostics and therapy using nanomedicine, pp 505–530. https://doi.org/10.1007/978-3-031-72467-1_17
83. Sobhanan J, Rival JV, Anas A, Sidharth Shibu E, Takano Y, Biju V (2023) Luminescent quantum dots: synthesis, optical properties, bioimaging and toxicity. Adv Drug Deliv Rev 197:114830. https://doi.org/10.1016/j.addr.2023.114830
84. Souto EB, Blanco-Llamero C, Krambeck K, Kiran NS, Yashaswini C, Postwala H, Severino P, Priefer R, Prajapati BG, Maheshwari R (2024) Regulatory insights into nanomedicine and gene vaccine innovation: safety assessment, challenges, and regulatory perspectives. Acta Biomater 180:1–17. https://doi.org/10.1016/j.actbio.2024.04.010
85. Steckel, Stephen J (2006) The synthesis of inorganic semiconductor nanocrystalline materials for the purpose of creating hybrid organic/inorganic light-emitting devices. Massachusetts Institute of Technology
86. Ganti SS, Mamidala VS, Subba SS, Chirla S, Tadikonda RR (2024) A review on quantum dots and their applications. Open Access Res J Biol Pharm 12(2):024–033. https://doi.org/10.53022/oarjbp.2024.12.2.0040
87. Sukhanova A, Bozrova S, Gerasimovich E, Baryshnikova M, Sokolova Z, Samokhvalov P, Guhrenz C, Gaponik N, Karaulov A, Nabiev I (2022) Dependence of quantum dot toxicity in vitro on their size, chemical composition, and surface charge. Nanomaterials 12(16):2734. https://doi.org/10.3390/nano12162734
88. Sumanth Kumar D, Jai Kumar B, Mahesh HM (2018) Quantum Nanostructures (QDs): an overview. In: Synthesis of inorganic nanomaterials. Elsevier, pp 59–88. https://doi.org/10.1016/B978-0-08-101975-7.00003-8
89. Swarnaltha Y, Siddharthan S (2022) Quantum dots: policy and ethics. In: Graphene, nanotubes and quantum dots-based nanotechnology. Elsevier, pp 887–899. https://doi.org/10.1016/B978-0-323-85457-3.00031-1
90. Tandale P, Choudhary N, Singh J, Sharma A, Shukla A, Sriram P, Soni U, Singla N, Barnwal RP, Singh G, Kaur IP, Suttee A (2021) Fluorescent quantum dots: an insight on synthesis and potential biological application as drug carrier in cancer. Biochem Biophys Rep 26:100962. https://doi.org/10.1016/j.bbrep.2021.100962
91. Tang Y, Han S, Liu H, Chen X, Huang L, Li X, Zhang J (2013) The role of surface chemistry in determining in vivo biodistribution and toxicity of CdSe/ZnS core–shell quantum dots. Biomaterials 34(34):8741–8755. https://doi.org/10.1016/j.biomaterials.2013.07.087
92. Tope S, Saudagar S, Kale N, Khambayat S, Bhise K (2014) Review: therapeutic application of quantum dots (QD). Pharma Innovat J 2(12). www.thepharmajournal.com
93. Tufani A, Qureshi A, Niazi JH (2021) Iron oxide nanoparticles based magnetic luminescent quantum dots (MQDs) synthesis and biomedical/biological applications: a review. Mater Sci Eng, C 118:111545. https://doi.org/10.1016/j.msec.2020.111545

94. Valizadeh A, Mikaeili H, Samiei M, Farkhani SM, Zarghami N, Kouhi M, Akbarzadeh A, Davaran S (2012) Quantum dots: synthesis, bioapplications, and toxicity. Nanoscale Res Lett 7(1):480. https://doi.org/10.1186/1556-276X-7-480
95. Wang H, Wang B, Bai J, Lu S (2024) Shiny quantum dots: synthesis, properties, and electroluminescent light-emitting diodes. J Mater Chem A 12(34):22417–22441. Royal Society of Chemistry. https://doi.org/10.1039/d4ta03218g
96. Wang K, Tao Y, Tang Z, Zhao H, Sun X, Rosei F, Liu D, Xiong Y (2025) Stability of photoelectrochemical cells based on colloidal quantum dots. Chem Soc Rev 54(7):3513–3534. https://doi.org/10.1039/D4CS00587B
97. Wang L, Lin M, Hou X, Dou L, Huang Z, Liu R, Zhang J, Cai C, Chen C, Liu Y, Wang D, Guo D, An R, Wei L, Yao Y, Zhang Y (2023) Black phosphorus quantum dots induce autophagy and apoptosis of human bronchial epithelial cells via endoplasmic reticulum stress. Chemosphere 327:138463. https://doi.org/10.1016/j.chemosphere.2023.138463
98. Wang X, He K, Hu Y, Tang M (2022) A review of pulmonary toxicity of different types of quantum dots in environmental and biological systems. Chem Biol Interact 368:110247. https://doi.org/10.1016/j.cbi.2022.110247
99. Wang X, Sun X, Lao J, He H, Cheng T, Wang M, Wang S, Huang F (2014) Multifunctional graphene quantum dots for simultaneous targeted cellular imaging and drug delivery. Colloids Surf, B 122:638–644. https://doi.org/10.1016/j.colsurfb.2014.07.043
100. Wang X, Wu T (2023) An update on the biological effects of quantum dots: from environmental fate to risk assessment based on multiple biological models. Sci Total Environ 879:163166. https://doi.org/10.1016/j.scitotenv.2023.163166
101. Xie C, Zhan Y, Wang P, Zhang B, Zhang Y (2020) Novel surface modification of ZnO QDs for paclitaxel-targeted drug delivery for lung cancer treatment. Dose-Response 18(2). https://doi.org/10.1177/1559325820926739
102. Yadav AP (2025) Current status and future prospect of receptor targeted nanocarrier delivery of cancer therapeutics. J Adv Sci Res
103. Yadav K (2024) Nanotechnology in diabetes management: revolutionizing treatment and diagnostics. J Mol Liq 414:126117. https://doi.org/10.1016/j.molliq.2024.126117
104. Yadav PK, Chandra S, Kumar V, Kumar D, Hasan SH (2023) Carbon quantum dots: synthesis, structure, properties, and catalytic applications for organic synthesis. Catalysts 13(2). MDPI. https://doi.org/10.3390/catal13020422
105. Yaghini E, Tacconi E, Pilling A, Rahman P, Broughton J, Naasani I, Keshtgar MRS, MacRobert AJ, Della Pasqua O (2021) Population pharmacokinetic modelling of indium-based quantum dot nanoparticles: preclinical in vivo studies. Eur J Pharm Sci 157:105639. https://doi.org/10.1016/j.ejps.2020.105639
106. Yaghini E, Turner H, Pilling A, Naasani I, MacRobert AJ (2018) In vivo biodistribution and toxicology studies of cadmium-free indium-based quantum dot nanoparticles in a rat model. Nanomed Nanotechnol Biol Med 14(8):2644–2655. https://doi.org/10.1016/j.nano.2018.07.009
107. Yang Y, Xia R, Zhang X, Wang X, Zhou Y, Wang H, Feng Y, Lv S, Ji S (2021) Effects of oral exposure to Mn-Doped ZnS quantum dots on intestinal tract and gut microbiota in mice. Front Physiol 12. https://doi.org/10.3389/fphys.2021.657266
108. Yin C, Hu P, Qin L, Wang Z, Zhao H (2024) The current status and future directions on nanoparticles for tumor molecular imaging. Int J Nanomed 19:9549–9574. https://doi.org/10.2147/IJN.S484206
109. Yoosefian M, Fouladi M, Atanase LI (2022) Molecular dynamics simulations of docetaxel adsorption on graphene quantum dots surface modified by PEG-b-PLA Copolymers. Nanomaterials 12(6). https://doi.org/10.3390/nano12060926
110. Zhang LW, Bäumer W, Monteiro-Riviere NA (2011) Cellular uptake mechanisms and toxicity of quantum dots in dendritic cells. Nanomedicine 6(5):777–791. https://doi.org/10.2217/nnm.11.73
111. Zhang M, Kim DS, Patel R, Wu Q, Kim K (2022) Intracellular trafficking and distribution of Cd and InP quantum dots in HeLa and ML-1 thyroid cancer cells. Nanomaterials 12(9):1517. https://doi.org/10.3390/nano12091517

112. Zhang Y, Wu Z-Y (2024) Gene therapy for monogenic disorders: challenges, strategies, and perspectives. J Genet Genomics 51(2):133–143. https://doi.org/10.1016/j.jgg.2023.08.001
113. Zhao F, Zhao Y, Liu Y, Chang X, Chen C, Zhao Y (2011) Cellular uptake, intracellular trafficking, and cytotoxicity of nanomaterials. Small 7(10):1322–1337. https://doi.org/10.1002/smll.201100001
114. Zhao X, Bai J, Yang W (2021) Stimuli-responsive nanocarriers for therapeutic applications in cancer. Cancer Biol Med 18(2):319–335. https://doi.org/10.20892/j.issn.2095-3941.2020.0496
115. Zheng N, Yan J, Qian W, Song C, Zuo Z, He C (2021) Comparison of developmental toxicity of different surface modified CdSe/ZnS QDs in zebrafish embryos. J Environ Sci 100:240–249. https://doi.org/10.1016/j.jes.2020.07.019
116. Zheng W, Xu Y-M, Wu D-D, Yao Y, Liang Z-L, Tan HW, Lau ATY (2018) Acute and chronic cadmium telluride quantum dots-exposed human bronchial epithelial cells: the effects of particle sizes on their cytotoxicity and carcinogenicity. Biochem Biophys Res Commun 495(1):899–903. https://doi.org/10.1016/j.bbrc.2017.11.074
117. Zhou J, Yang Y, Zhang C (2015) Toward biocompatible semiconductor quantum dots: from biosynthesis and bioconjugation to biomedical application. Chem Rev 115(21):11669–11717. https://doi.org/10.1021/acs.chemrev.5b00049
118. Zrazhevskiy P, Sena M, Gao X (2010) Designing multifunctional quantum dots for bioimaging, detection, and drug delivery. Chem Soc Rev 39(11):4326–4354. https://doi.org/10.1039/b915139g

Chapter 7
Quantum Dots in Agriculture and Food Industry

Sunny Kumar and Nilambar Bariha

Abstract Quantum dots (QDs) are nanometer-sized semiconductor particles with remarkable optical and electronic properties that provide photostability and versatile surface chemistry. QDs are being explored in agriculture for plant health monitoring and pest control. QDs can be used as sensors for early detection of plant pathogens or environmental stresses, leading to timely interventions, thereby mitigating crop losses. QDs serve as advanced delivery systems for agrochemicals. Their surface can be precisely engineered to achieve targeted binding and controlled release of fertilizers or pesticides in response to specific environmental cues such as soil pH and plant physiological signals. This targeted delivery approach minimizes off-target effects, reduces the overall environmental impact of agricultural inputs, and promotes sustainable crop protection. QDs are applied to food safety packaging materials and quality monitoring in the food industry, which help detect contaminants and spoilage. Their high sensitivity and specificity make them valuable in ensuring food safety standards. In food safety and nutrition, QDs have been incorporated into innovative packaging materials that visibly change fluorescence to indicate spoilage, contamination, or temperature abuse, thus enhancing supply chain transparency and consumer protection. Their high sensitivity allows the detection of trace levels of toxins, heavy metals, or residual pesticides that traditional colorimetric sensors might miss. Additionally, QDs can enhance the nutritional value of food by enabling the fortification of vitamins and minerals. The multifunctional QD platforms represent a transformative in agriculture technology, sustainable crop protection, and food safety monitoring for more resilient food systems.

Keywords Agriculture · Food safety · Quality monitoring · Nutrients enhancements · Pathogens

S. Kumar (✉)
School of Chemical and Biomolecular Engineering, Georgia Institute of Technology, Atlanta, GA, USA
e-mail: skumar710@gatech.edu; chemicalsunnyraj@gmail.com

N. Bariha
Department of Fire and Safety Engineering, Rajiv Gandhi Institute of Petroleum Technology, Sivasagar Campus, Sivasagar, Assam, India

T. Bhuyan et al. (eds.), *Multifunctional Quantum Dots*, Materials Horizons: From Nature to Nanomaterials, https://doi.org/10.1007/978-981-92-0172-3_7

1 Introduction

With the global population on the rise, the world is facing food scarcity and challenges to maintain a sustainable food supply chain in the field of agriculture. In a few years, quantum dots have been introduced, which are well known as members of nanomaterial, paved away to the emergence of new generation technologies poised to revolutionize horticulture. Quantum dots are core-type, core–shell, and alloyed type [51], which are composed of semiconductor quantum dots (CdSe, CdS, CdTe, and ZnS), carbon-based quantum dots, perovskite quantum dots, and silicon quantum dots. Carbon quantum dots were unexpectedly discovered during the purification process of carbon nanotubes [24]. It has the potential to be revolutionary in agriculture and horticulture research. Quantum dots have gained attention in agricultural research because of their unique properties, such as photoluminescence, high biocompatibility, efficient photo-induced electron transfer, low toxicity, and cost-effectiveness. These features make quantum dots an auspicious material for promoting plant growth and improving crop yield in sustainable agriculture [16].

Quantum dots (QDs) technologies play a crucial role in different fields due to their physicochemical properties and progressively provide their nutritional quality and safety. QDs are applied to food safety packaging materials and quality monitoring in the food industry, which helps detect contaminants and spoilage. Their high sensitivity and specificity make them valuable in ensuring food safety standards. Additionally, QDs can enhance the nutritional value of food by enabling the fortification of vitamins and minerals. Food quality and safety are important to ensure the best product quality to meet consumer demands and public health. Exploration of the QDs provides tremendous applications in food quality and safety in terms of minimizing the risk of foodborne illnesses, contamination, drug residue, and other potential hazards in food [96]. Quantum dots are being explored in agriculture for plant health monitoring and pest control. They can be used as sensors to detect pathogens or environmental stresses in plants, leading to timely interventions. During the last decades, the development of nanomaterials has rapidly increased and produced by the synthesized classical method. These quantum dots nanomaterials are toxic and not suitable for their synthesis procedure. The green synthesis of quantum dots offers a sustainable method using biomass and waste materials, with an emphasis on environmentally friendly production mechanisms. These quantum dots have diverse applications in energy generation, storage, healthcare, environmental remediation, and food and agriculture products [23, 60].

This book chapter discusses the applications of quantum dots in agriculture and the food industry. It begins with the fundamental optical and electronic properties of quantum dots. These properties are utilized for a wide range of applications, particularly in agriculture and the food industry. In the context of agriculture, we explore the monitoring and growth of plant health influenced by the Quantum dots. QD's ability to convert ultraviolet (UV) and blue light into red wavelength and efficient absorption by chlorophyll enhance photosynthesis and crop productivity. QD-based nano sensors enable real-time monitoring of plant physiology and crop yield. In plant health, quantum dots are also being explored to enrich the nutritional value of food. By enhancing light absorption and stress resilience in crops, QDs contribute indirectly to improved nutrient density. In the food industry, QDs play a pivotal role in smart packaging. QD's fluorescent properties are utilized to detect spoilage indicators, microbial contamination, and changes in environmental conditions. Overall, the multifunctional quantum dots enhanced plant health and yield to improved food safety and nutritional quality for future sustainable agriculture and food systems.

2 Quantum Dots Properties

Semiconductor quantum dots exhibit optical and electronic properties due to quantum confinement effects. When their size is reduced to below the exciton Bohr radius, the movement of electrons and holes is confined in all three spatial dimensions, resulting in discrete energy levels. Quantum dots can be classified based on the relative position of their valence and conduction bands. In one type, the core possesses a higher band gap than the shell. In another configuration, the core's conduction and valence bands are situated within the band gap of the shell material. The synthesis of QDs typically follows a bottom-up approach, involving the combination of metal or chalcogenide precursors with ligands that pass through the nucleation and growth stage. Within QDs, the excitons are confined in all three dimensions and are held together by strong Coulomb interactions. The quality and functional properties of the QD can be tuned during synthesis by optimizing parameters such as morphology, size, and crystal structure. The precursor reactivity, concentration, and reaction duration significantly influence the final structure and properties of QD [76, 80].

Chalcogenide based QDs encompass range of composition cadmium or lead chalcogenide QDs, silver chalcogenide, and silver indium chalcogenide QD. Within these, cadmium chalcogenide QDs (CdX, X = S, Se, Te) are commonly synthesized using the hot injection method [22]. Cadmium precursors used in this process include $CdMe_2$, CdO, $Cd(AcO)_2$, and $CdCO_3$. Chalcogen sources typically include sulfides, selenides, tellurides bound to ligands such as $P(C_8H_{17})_3$, PBu_3, and $(TMS)_2S$ [58].

Silicon quantum dots can use different methods laser pyrolysis, plasma-assisted decomposition, and microemulsion. SiH_4 precursors are exposed to high-temperature carbon dioxide laser pyrolysis, which forms Si QDs ranging 5–10 nm [47]. With the microemulsion method, $SiCl_4$ serves as the precursor, dissolved in hexane along with Dodecyl poly (ethylene glycol), and further mixture is then reacted with allylamine with UV light for 4 h. Silicon quantum dots are highly attractive due to their presumed biocompatibility, primarily attributed to their low toxicity to living organisms [7].

$CuInS_2$ quantum dots have attracted growing attention in photocatalysis applications due to their absorption, favorable alignment with sunlight, long-lived photogenerated electron–hole pairs, and environmentally friendly composition [71]. The colloidal synthesis of CuIn chalcogenide nanocrystals is more complex due to the mismatched reactivity of metal cations with S^{2-}, Se^{2-}, and Te^{2-} [91]. According to the Hard-Soft Acid–Base theory, Cu^+ is a soft Lewis acid, In^{3+} is a hard Lewis acid, and chalcogenide anions are soft Lewis bases, with their softness increasing in the order $S^{2-} < Se^{2-} < Te^{2-}$. To overcome these synthetic challenges, a variety of methods have been employed, including solvothermal, hot injection, hydrothermal, microwave-assisted, and heating-up techniques [35].

InAs quantum dots synthesis can be controlled with reaction conditions because of the covalent character of their bonds, elevated reaction temperature, and precursor reactivity. With under optimized conditions, achieving uniform particle size and size distribution remains a significant challenge. InAs was synthesized in 1995 through a room temperature reaction between In(III) halides and $As(SiMe_3)_3$ in toluene and InAs powder was collected after annealed at ~400 °C [83].

Highly monodisperse colloidal InAs quantum dots (QDs) with excellent optoelectronic properties are being actively developed [43]. Subsequently, Guzelian et al. developed a method to synthesize InAs QDs via a reaction between $InCl_3$ and $As(SiMe_3)_3$ at 240–265 °C, using trioctylphosphine as both the solvent and capping agent [32]. This approach produced particles ranging from 2.5 to 6 nm in size, with a size distribution of 10–15%, which were refined further through post-synthetic size-selective precipitation.

Carbon dots are small-sized carbon nanoparticles and can be broadly classified as aqueous or other suspensions, and their different types are shown in Fig. 1. The nanostructure of a "core–shell" consists of a nanoscale carbon core and on its functional groups. They can also be determined based on structural properties such as optical, electronic, and catalytic [17]. Carbon dots are derived from carbon-based sources such as graphene and carbon nanotubes. The unique properties opened up applications like bioimaging, sensing technologies, and electronic devices [31]. These multifunctional properties of quantum dots are valuable materials in scientific research and industry. The physicochemical properties of carbon dots have drawn attention for their unique optical and high compatibility, including their low cost and application in the sensing field [68].

At present, the detection principles involved in the field of nano-biological technologies, such as fluorescence quenching mechanisms, have a huge scope in the food area. The food safety ensures the best quality of product is reached to meet consumer demands and public health. So, the advanced technologies play an important role in minimizing the risk of foodborne illness, contamination, drug residue, and other potential hazards in food [96]. Maintaining nutrient levels in food remains a key challenge for the restricted substances [66]. Table 1 shows that the different categories of food can be detected based on their technologies [65, 74]. The importance of food quality and safety lies in ensuring the best product quality to meet consumer demands and public health. Advanced technologies play a crucial role in minimizing the risk of foodborne illnesses, contamination, drug residue, and other potential hazards in food.

The shape of quantum dots is generally spherical particles with a size of less than 10 nm, with unique optical and electrical properties [39]. They have many applications in biological imaging, such as fluorescence resonance energy transfer, which measures conformational changes and the interaction between different molecules. The quantum dots in food science are an important area of research in detecting and

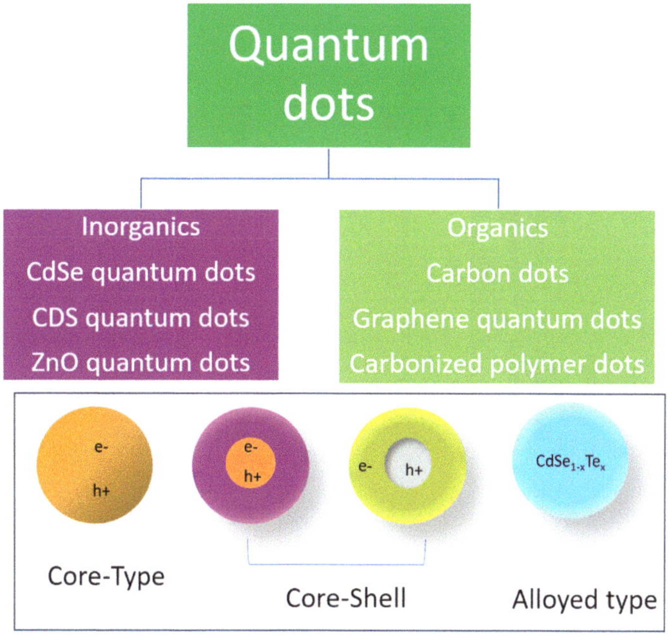

Fig. 1 Quantum dots classification and subdivision of metallic semiconductor QDs, core-type, core–shell, and alloyed

monitoring food pathogens of various sizes that can emit fluorescence signals during excited with a single wavelength of light [9].

Quantum dots have become increasingly attractive to researchers in the food industry due to their low toxicity, environmental friendliness, and excellent photostability. Food analysis plays a crucial role in ensuring both quality assessment and safety control. Quantum dot-based fluorescence sensors offer faster, more accurate, highly sensitive, and cost-effective methods for food testing. These sensors are widely used to detect a range of food contaminants, including additives, pathogens, heavy metals, nutrients, antibiotics, and pesticide residues [75].

3 Quantum Dots for Plant Health

Quantum dots (QDs) are widely applied in chemistry, biology, medicine, and agriculture. However, their rapid proliferation raises concerns about environmental dispersal and its potential effects on human health and environmental safety [61]. The quantum dots have versatile applications that vary from medicine to agriculture, offering immense potential for industrial advancements. Their properties, governed by size and surface modifications, are characterized by discrete energy levels and band gap

Table 1 Detection of different categories of food based on their detection technologies

Detecting object	Categories of food	Sources of carbon	Detecting principle	LOD Excitation wavelength	Excitation wavelength (nm)	Emission wavelength (nm)	References
Melamine	Milk	Histidine	Fluorescence quenching	36 nM	350	438	[18]
Ascorbic acid	Fresh fruits, vegetables, and fruit juices	Sodium citrate	Fluorescence recovery	42 nM	360	441	[48]
Tannic acid	Wine	6-Bromohexylboronic acid	Fluorescence quenching	0.018 mg/L	362	440	[1]
Tartrazine	Steamed buns, honey, candy	Aloe	Fluorescence quenching	73 nM	441	503	[85]
Pesticide residues	Apple	N-Methylethanolammonium thioglycolate	Fluorescence quenching	5 ppb	350	520	[45]
Aflatoxin B1	Peanut, corn	Pancreatin	Fluorescence recovery	5 pg/mL	365	425	[81]

variations. In agriculture, QDs are rapidly gaining prominence as growth enhancers and diagnostic tools for plant diseases [31].

Hybrid quantum dots present a novel opportunity to integrate nanotechnology into agriculture, enhancing crop management and environmental monitoring. By merging quantum dot properties with biological components, these nanoparticles enable real-time tracking of plant health, optimizing nutrient absorption, and assessing environmental conditions. They provide continuous insights that support data-driven decision-making, improve resource efficiency, and promote ecosystem sustainability. However, concerns regarding safety and cost must be addressed before widespread adoption. Despite these challenges, hybrid quantum dots have the potential to significantly boost crop yields and contribute to the development of smarter, more sustainable agricultural practices [29]. Li et al. [44] reported that 5 nm sized carbon dots (CDs) with varying oxygen contents influenced rice-plant growth, resulting in a 14.8% increase in total rice yield and improved resistance of rice plants to diseases [44].

Advancements in science and technology are driving innovation in the agricultural sector, offering novel approaches to address long-standing challenges. Nanotechnology, in particular, has enabled the development of more efficient and environmentally friendly nano formulations that support the goals of sustainable agriculture [25]. Figure 2 shows the schematic representation of the next-generation materials that have the potential to interact with plant physiology immediately upon entering the complex plant-soil environment, creating new opportunities to study and harness their effects [56].

The controlled release of active compounds requires a thorough understanding of both the beneficial and adverse interactions between these nanomaterials with plants. This knowledge can guide the development of safer and more effective nanomaterial-based agricultural solutions. While current findings suggest that environmental concentrations of nanomaterials are typically well below harmful levels, there remains a pressing need for comprehensive safety evaluations to understand their long-term impacts fully [6].

Nanosensors enable real-time monitoring of growth conditions, plant health, and disease detection with precision. Additionally, nanotechnology-driven approaches in agriculture enhance the quality of agricultural products [34] and reduce agriculture waste [14]. Crucially, nanotechnology in agriculture enhances upstream efficiency by maximizing water and energy utilization.

4 Quantum Dots Used as Sensors to Detect Pathogens in Plants

Quantum dots (QDs) have emerged as versatile nanomaterials in the development of biosensors for the detection of microorganisms, owing to their superior photoluminescence, high surface-to-volume ratio, and tunable optical properties. QDs-based

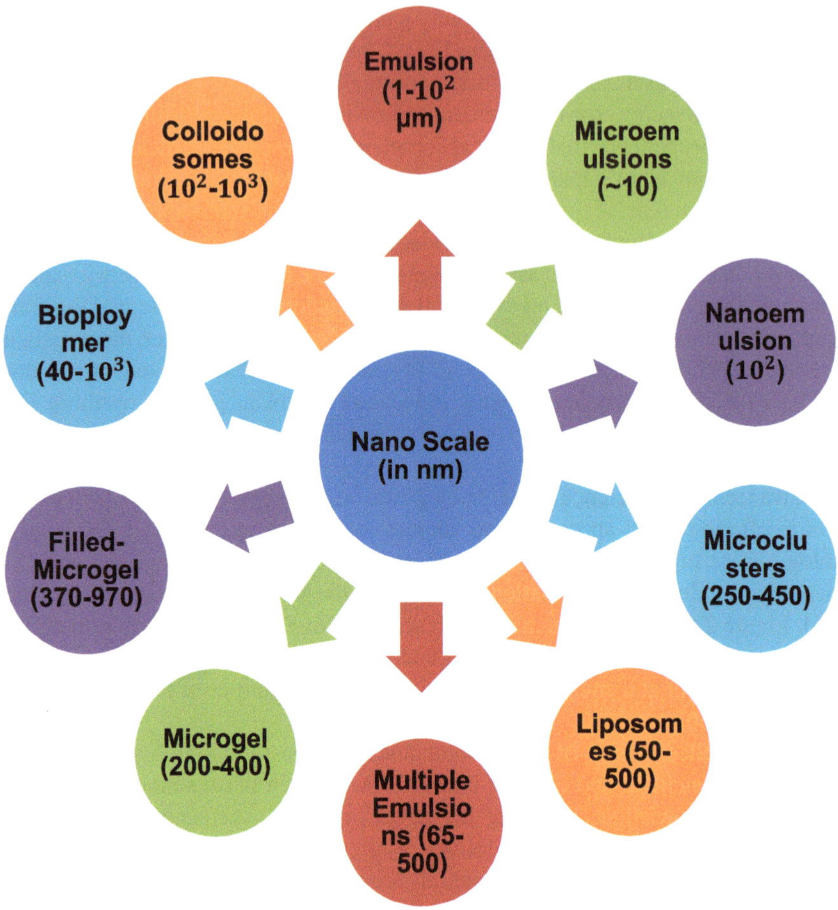

Fig. 2 Potential interaction of nanoparticle scale based on the plant system. Reproduced from Mittal et al. [56], under the terms of the Creative Commons Attribution License (CC BY 4.0)

biosensors are employed for the identification of a wide range of biological targets, including enzymes, antibodies, and microbial pathogens [41].

Quantum dots (QDs), with their high efficiency, stability, and biocompatibility, have advanced imaging and sensing, particularly in the near-infrared range. In plant pathology, they enable pathogen detection and tracking of plant–microbe interactions, offering promising tools for disease diagnosis and management [70]. Kashyap et al. [38] studied the quick detection of plant pathogens which is vital for crop protection and food security [38]. Traditional methods are slow and labor-intensive, while current immunological and molecular techniques have limitations. Integrating nanotechnology enables portable, sensitive devices using functionalized nanoparticles for efficient field diagnostics. Plant pathogens are cause of decreased crop production and can contribute to food shortages for both humans and livestock. While the primary method is still chemical control for managing foliar fungal diseases, its repeated use can lead to reduced pathogen susceptibility, environmental contamination, and increased economic burden on farmers. Effective disease prevention and management depend on the accurate, sensitive, and rapid detection of the causal pathogens, enabling the timely implementation of appropriate control strategies [21].

Plant pathogens significantly limit crop yields and threaten food security. Chemical controls have drawbacks, including resistance and environmental harm. Early, accurate detection is crucial, but conventional methods are often slow and impractical in the field, as shown in Table 2. Nanotechnology-based tools offer rapid, sensitive, and high-throughput alternatives for effective plant pathogen diagnosis [72].

Quantum dots are functionalized with specific biomolecules (antibodies, DNA, or aptamers) to target plant pathogens [72]. When quantum dots (QDs) encounter plant pathogens, their interaction generates observable signals, including changes or reductions in fluorescence (Fig. 3). Consequently, there is a pressing need to develop low-cost methods that enhance the accuracy and speed of plant pathogen diagnosis. Nanotechnology, including nanoparticles and quantum dots (QDs), has emerged as a powerful approach for the rapid and highly accurate detection of specific biological markers. Technologies such as biosensors, QDs, nanostructured

Table 2 The comparative study of plant pathogen detection techniques. Reproduced from Shivashakarappa et al. [72], under the Creative Commons Attribution-NonCommercial-No Derivatives 4.0 International License (CC BY-NC-ND 4.0)

Existing method	Detection limit	Time required for		Advantages
		N Assay	Results	
Conventional	10^7–10^8 cfu mL^{-1}	72 h	72 h	Cost-effective assay with precise high target
Nucleic acid-based	10^3–10^4 cfu mL^{-1}	1–3 h	6–12 h	Reliable and accessible technology, characterized by ease of use, high sensitivity, and accuracy
Immunological	10^5–10^6 cfu mL^{-1}	1–3 h	6–12 h	Affordable and easy to transport, they allow precise measurement and classification of pathogens by taxonomy
Nanotechnology	1 fmol^{-1}	30 min^{-1} h	30 min^{-1} h	Offers rapid detection, ease of use, affordability, high specificity and sensitivity, strong on-site applicability, and supports high-throughput analysis

platforms, nanoimaging, and nanopore DNA sequencing offer significant potential to improve the sensitivity, specificity, and speed of pathogen detection. These tools also enable high-throughput analysis and hold promise for effective monitoring and crop protection [42].

QDs are semiconducting fluorescent nanoparticles that are highly suitable for the detection and control of foodborne pathogens, owing to their unique size-dependent optical properties and exceptional photostability [57]. They are semiconductor nanocrystals characterized by unique photophysical properties, making them highly promising as optical nanoprobes in various biomedical and analytical applications [27]. Wang et al. [82] study the effectiveness as biosensors in imaging plants and detecting diseases on plants [82].

This interaction allows for exceptionally sensitive detection of pathogens at early stages of infection. Moreover, the availability of QDs in various colors enables the simultaneous identification of multiple pathogens. Consequently, QDs offer the potential for real-time monitoring of plant infections. Early identification significantly reduces crop losses, enhancing food security and promoting sustainable agriculture by minimizing reliance on harmful chemical inputs. Polyethyleneimine-coated MXene quantum dots show improvement to cotton tolerance to *Verticillium dahliae* by maintaining reactive oxygen species homeostasis, highlighting their therapeutic potential in plant protection [64]. Additionally, high production costs and the

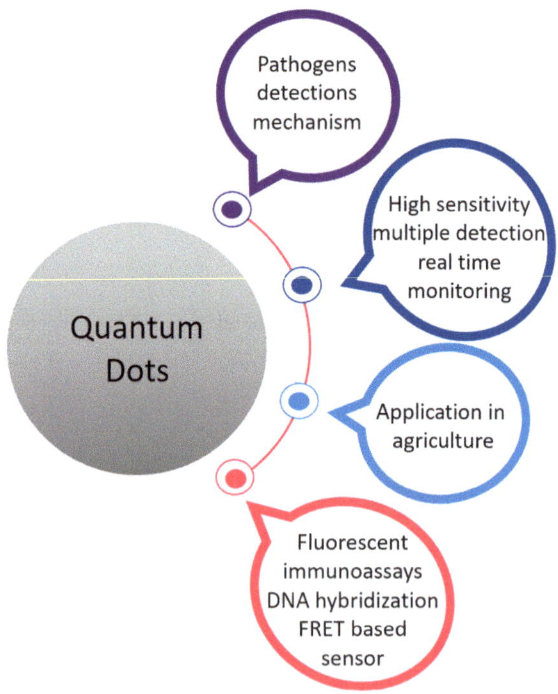

Fig. 3 Flowchart for the quantum dots sensors to detect plants

need for portable, robust detection systems must be addressed for sustainable plant disease management [54].

5 Quantum Dots for Nutrient Value for Food

Quantum dots (QDs) contribute indirectly to enhancing the nutritional value of food by improving crop growth, extending shelf life, and enabling more efficient food safety analysis. Although QDs are not directly consumed as a nutrient, their role in these processes supports a more nutritious and sustainable food system. Tiny quantum dots shed new light on nutrient flow, where quantum dots can be used to track nutrients moving through plants in real time [84]. QDs can bind with specific nutrients and glow under certain light conditions, which allows for visual monitoring of nutrient transport pathways, offering unprecedented insight into how plants absorb and distribute essential elements like nitrogen and phosphorus. Zinc quantum dot biochar (Zn-QDB) enhances antioxidants and nutrient absorption, which promote growth and yield [3]. The drought stress is reduced with the application of 0.8% QDB which improves the leaf water contents and nutrient absorption in rapeseed [77]. Drought stress generally hampers the absorption of needed nutrients by plants, affecting growth and yield.

Quantum dots enhance food nutrient value with precise analysis and improved agriculture practices that produce more nutritious food. Quantum dots enhance food nutrient analysis, where their physicochemical properties enable precise detection of vitamins, minerals, and contaminants. Sun et al. [78] examined the food toxicity, antibacterial properties, and applications of carbon quantum dots in detecting heavy metals, pathogens, and prohibited material, as well as their use in enhancing nutritional components and preserving the freshness of fruits and vegetables [78].

Carbon quantum dots enhances plant photosynthesis, which increases crop yield and nutrient absorption during quantum dots' interaction with plant tissue and chloroplasts, which leads to more nutritious food and better quality control [30]. Tan et al. [79] studied that the electron transport rate in both rice (+57.3%) and corn (+65.5%) was increased under carbon quantum dots treatment [79].

Cheng et al. [15] successfully synthesized the low-cost, biomass-derived carbon dots (B-CDs) [15]. These B-CDs, featuring a small particle size and blue fluorescence, were readily absorbed by crops, which significantly enhanced photosynthesis by improving light-harvesting efficiency. Various methods exist for the synthesis of CQDs, but increasing attention is being given to low-cost and eco-friendly approaches. Biomass-derived CQDs, produced through green synthesis, are regarded as one of the most promising fluorescent nanomaterials. This sustainable method typically involves converting waste materials into high-value CQDs, thereby minimizing resource waste and supporting environmentally sustainable development. Consequently, it has become an increasingly important raw material in CQD production. Table 3 shows the numerous types of biomass that have already been successfully used as effective precursors for CQD synthesis [20].

Table 3 Summary of precursors and applications of carbon quantum dots

Precursors	Quantum yield	Application	References
Alkali lignin glucose	0.176	Used in live imaging within biological systems	[90]
Garlic peels	0.262	Detection of europium ions	[10]
Salvia hispanica L. seeds	0.178	Radiolabeling for rat model	[53]
Citrus clementina peel	–	Specific sensing of ferric ions	[69]
Grape seed	0.275	Duel detection of Cu^{2+} and ascorbic acid	[46]
Waste acorn cups	0.0738	Extraction process	[86]
Expired passion fruit shell	0.018	–	[87]
Aloe barbadensis miller	0.31	Bioimaging in live cell environments	[50]
Cellulolytic enzyme lignin	0.0823	Quantification of trypsin	[89]
Biomass lignin	0.2368	Ultra-sensitive detection of Fe^{3+}	[95]
Corncob	0.5346	Detection of DNA and metal ions	[36]
Quinoa saponin powder	0.222	Sensing of Co^{2+}	[93]
Waste bones	–	Detection of detergent residue	[88]
Maple tree leaves	–	Cesium detection and glycerol oxidation electrocatalysis	[13]
Date palm leaves	–	Cellular level imaging	[4]
Water hyacinth	0.033	Fe^{3+} ion detection	[92]
Mopan persimmons	0.0839	Selective detection of Fe^{3+} ions	[49]

6 Quantum Dots in Food Safety Packaging Materials

Quantum dots (QDs) are emerging as promising candidates for innovative food packaging applications due to their multifunctional properties. QD polymer nanocomposites offer significant advancements in food packaging by enabling the detection and monitoring of foodborne pathogens, as well as the assessment of food quality through sensing and imaging technologies. They exhibit high sensitivity and rapid response in detecting contaminants, thereby enhancing food safety and consumer protection [19, 28]. The integration of QDs into polymer matrices enhances the functionality of packaging systems, particularly in monitoring applications. These materials may contribute significantly to improving food safety and quality. However, further research and development are required to fully understand their safety implications within the food industry, as well as their regulatory and commercial viability [55]. The responsible assessment and management of nanomaterials are essential to ensure food safety, which holds to promise for advancing food industry practices, benefiting consumers, and supporting public health outcomes [26]. QD-based techniques have been explored for the detection of acrylamide in potato chips, a compound commonly present in thermally processed foods such as fried snacks, vegetable crisps, breakfast cereals, and various baked and roasted products, including cakes and cookies. Acrylamide is a known carcinogen with high toxicity; because of that, the reliable detection of acrylamide in food products is critical for maintaining food safety standards and minimizing health risks to consumers [33].

They can be embedded in radio frequency identification technology into packaging materials to monitor food quality by detecting environmental changes like pH, temperature, and gas levels, providing real-time indicators of spoilage through visible color shifts or fluorescence [97].

QDs also serve as anti-counterfeiting tools (Fig. 4), enabling tamper-proof labeling via unique optical signatures [52]. Quantum dots can be used to track and trace advanced technologies and create unique, tamper-proof labels or invisible security codes on food packages. Their precise optical properties allow for customized patterns that are difficult to replicate, enhancing the safety and authenticity of food products.

Additionally, quantum dots with antimicrobial activity can extend shelf life by inhibiting microbial growth [8]. Antimicrobial properties of quantum dots are associated with packaging materials to inhibit the growth of bacteria, fungi, and other microorganisms [96]. This feature can significantly enhance the shelf life of perishable foods. Some quantum dots, such as carbon quantum dots (CQDs) and sulfur-based QDs, offer advantages like non-toxicity, sustainability, and low-cost synthesis, making them attractive for edible, biodegradable, and active packaging.

Their ability to block harmful UV rays further protects sensitive food items from degradation. Packaging materials embedded with QDs can offer superior UV protection, reducing the degradation of food products caused by exposure to UV light [62].

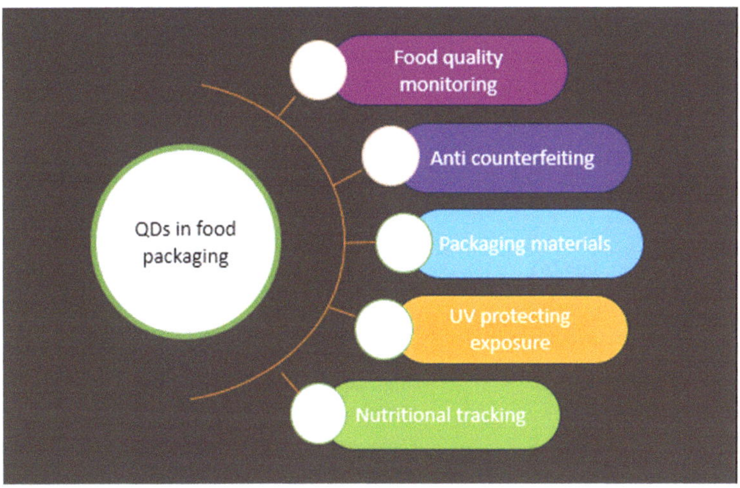

Fig. 4 Flowchart for the quantum dots in food safety packaging

This modified packaging is particularly beneficial for sensitive items like dairy, oils, and beverages [12].

Nonetheless, with innovations in synthesis and integration, QDs continue to gain interest in intelligent packaging, nutrient tracking, and enhanced food safety [37]. Functionalized QDs can be used to monitor nutrient levels in packaged foods or even fortify packaging with nanomaterials that help preserve nutrients during storage [5]. Quantum dots and their potential prospects and use of the quantum dots which are applied on food safety and quality enhancement are shown in Table 4.

Traditional QDs typically consist of heavy metal cores encapsulated by protective shells. However, growing concerns regarding their cytotoxicity have led to increased research efforts toward developing more biocompatible alternatives, such as carbon quantum dots, sulfur quantum dots [2].

CQDs have emerged as highly versatile tools across various research domains, including bioimaging, biosensing, drug delivery, solar energy, and food packaging. They have increasingly replaced traditional semiconductor quantum dots, primarily due to their non-toxic nature. Their incorporation into food packaging materials has shown significant promise, owing to their antioxidant, antimicrobial, and barrier properties. CQDs contribute to extending product shelf life, inhibiting microbial growth, enhancing mechanical strength, improving gas barrier performance, blocking harmful UV radiation, and ultimately reducing food spoilage and waste [73]. Currently, Sulfur quantum dots are emerging as promising functional nanomaterials

Table 4 Commercially available quantum dots for food industries

Companies/QDs	Core/shell composition	Typical application	References
Thermo Fisher Scientific Qdot™ Nanocrystals	CdSe/ZnS	Fluorescent labeling for plant imaging and pathogen-detection assays in agriculture	[63, 94]
Ocean NanoTech QDs	CdSe (various shells), InP, Ag_2S	Integration into fluorescence-based sensors for pesticide-residue detection and soil-nutrient monitoring	[59]
Sigma-Aldrich (Merck) CdSe/ZnS QDs	CdSe/ZnS	Lab-scale assays: nutrient tracking and microbe detection in food samples	[96]
Nanosys Heavy-Metal-Free Quantum Dots	InP/ZnS	Smart-label inks for intelligent packaging (freshness/temperature indicators)	[96]
PlasmaChem $CuInS_2$/ZnS QDs	$CuInS_2$/ZnS	Kits available for sensor development (e.g., rapid mycotoxin detection in stored grains)	[67]
Nanoco Technologies CFQD™ InP/ZnS QDs	InP/ZnS	Cadmium-free QDs used for detecting food toxins (e.g., aflatoxin B_1)	[11]

in diverse research fields, owing to their non-toxicity, intrinsic antibacterial activity, biocompatibility, and excellent photoluminescence properties [40].

7 Conclusion

This chapter has illuminated the transformative potential of quantum dots (QDs) within the agriculture and food sectors. Their unique physicochemical properties enable significant multifunctional capabilities. Specifically, we have discussed their efficacy in the early detection of plant pathogens and their application in real-time monitoring of food quality and safety through intelligent packaging. Furthermore, we have discussed how QDs facilitate precise analysis of nutrient content and bioavailability, supporting innovations in food processing and crop enhancement. These insights highlight the exciting opportunities for more innovative and responsive systems within the agri-food landscape. Continuing biocompatibility and environmental safety research will be crucial in realizing their full potential. Ultimately, the integration of QDs represents a promising step toward achieving sustainable agriculture and bolstering global food security in the future.

References

1. Ahmed GHG, Laíño RB, Calzón JAG, García MED (2015) Fluorescent carbon nanodots for sensitive and selective detection of tannic acid in wines. Talanta 132:252–257. https://doi.org/10.1016/j.talanta.2014.09.028
2. Alle M, Priyadarshi R, Purohit SD (2025) Functional nanomaterials and nanocomposites for biodegradable food packaging. Springer
3. Alotaibi NM, Alotibi MM, Younis U, Hussain GS, Dawar K, Hareem M, Ghoneim AM, Fahad S, Danish S (2024) Zn-quantum dot biochar regulates antioxidants and nutrient uptake to improve rapeseed growth and yield in drought stress. Plant Stress 11:100286. https://doi.org/10.1016/j.stress.2023.100286
4. Athinarayanan J, Periasamy VS, Al-Harbi LN, Alshatwi AA (2023) Phoenix dactylifera leaf-derived biocompatible carbon quantum dots: application in cell imaging. Biomass Convers Biorefinery 13:12989–12998. https://doi.org/10.1007/s13399-021-02159-5
5. Bakeshlouy Afshar M, Poursattar Marjani A, Gozali Balkanloo P (2024) Introducing graphene quantum dots in decomposable wheat starch-gelatin based nano-biofilms. Sci Rep 14:2069. https://doi.org/10.1038/s41598-024-52560-z
6. Batley GE, Kirby JK, McLaughlin MJ (2013) Fate and risks of nanomaterials in aquatic and terrestrial environments. Acc Chem Res 46:854–862. https://doi.org/10.1021/ar2003368
7. Beri D (2023) Silicon quantum dots: surface matter, what next? Mater. Adv. 4:3380–3398. https://doi.org/10.1039/D2MA00984F
8. Bijoy G, Sangeetha D (2024) Biomass derived carbon quantum dots as potential tools for sustainable environmental remediation and eco-friendly food packaging. J Environ Chem Eng 12:113727. https://doi.org/10.1016/j.jece.2024.113727
9. Bonilla JC, Bozkurt F, Ansari S, Sozer N, Kokini JL (2016) Applications of quantum dots in food science and biology. Trends Food Sci Technol 53:75–89. https://doi.org/10.1016/j.tifs.2016.04.006

10. Boruah A, Saikia M, Das T, Goswamee RL, Saikia BK (2020) Blue-emitting fluorescent carbon quantum dots from waste biomass sources and their application in fluoride ion detection in water. J Photochem Photobiol B 209:111940. https://doi.org/10.1016/j.jphotobiol.2020.111940
11. CFQD (n.d.) Nanoco. https://www.nanocotechnologies.com/products-markets/cfqd-quantum-dots/. Accessed 06 February 2025
12. Chelladurai D, Alaguthevar R, Murugesan B, Subburamu K, Khan A, Rhim J-W (2024) Carbon quantum dots: progress toward food safety and sustainability. Food Biosci 61:105016. https://doi.org/10.1016/j.fbio.2024.105016
13. Chellasamy G, Arumugasamy SK, Govindaraju S, Yun K (2022) Green synthesized carbon quantum dots from maple tree leaves for biosensing of cesium and electrocatalytic oxidation of glycerol. Chemosphere 287:131915. https://doi.org/10.1016/j.chemosphere.2021.131915
14. Chen H, Yada R (2011) Nanotechnologies in agriculture: new tools for sustainable development. Trends Food Sci Technol, Agri-Food Nano Appl.: Ensuring Soc. Benefits 22:585–594. https://doi.org/10.1016/j.tifs.2011.09.004
15. Cheng B, Yang Z, Chen F, Yue L, Cao X, Li J, Qian H-L, Yan X-P, Wang C, Wang Z (2023) Biomass-derived carbon dots with light conversion and nutrient provisioning capabilities facilitate plant photosynthesis. Sci Total Environ 901:165973. https://doi.org/10.1016/j.scitotenv.2023.165973
16. Chowmasundaram YA, Tan TL, Nulit R, Jusoh M, Rashid SA (2023) Recent developments, applications and challenges for carbon quantum dots as a photosynthesis enhancer in agriculture. RSC Adv 13:25093–25117. https://doi.org/10.1039/D3RA01217D
17. Cui L, Ren X, Sun M, Liu H, Xia L (2021) Carbon dots: synthesis, properties and applications. Nanomaterials 11:3419. https://doi.org/10.3390/nano11123419
18. Dai H, Shi Y, Wang Y, Sun Y, Hu J, Ni P, Li Z (2014) A carbon dot based biosensor for melamine detection by fluorescence resonance energy transfer. Sens Actuators B Chem 202:201–208. https://doi.org/10.1016/j.snb.2014.05.058
19. Das TK, Ganguly S (2023) Revolutionizing food safety with quantum dot-polymer nanocomposites: from monitoring to sensing applications. Foods 12:2195. https://doi.org/10.3390/foods12112195
20. Dong Z, Qi J, Yue L, Zhou H, Chen L, Gu J, He Y, Wu H (2024) Biomass-based carbon quantum dots and their agricultural applications. Plant Stress 11:100411. https://doi.org/10.1016/j.stress.2024.100411
21. Dyussembayev K, Sambasivam P, Bar I, Brownlie JC, Shiddiky MJA, Ford R (2021) Biosensor technologies for early detection and quantification of plant pathogens. Front Chem 9. https://doi.org/10.3389/fchem.2021.636245
22. Ebrahim F, Al-Hartomy O, Wageh S (2023) Cadmium-based quantum dots alloyed structures: synthesis, properties, and applications. Materials 16:5877. https://doi.org/10.3390/ma16175877
23. El-Azazy M, Osman AI, Nasr M, Ibrahim Y, Al-Hashimi N, Al-Saad K, Al-Ghouti MA, Shibl MF, Al-Muhtaseb AH, Rooney DW, El-Shafie AS (2024) The interface of machine learning and carbon quantum dots: from coordinated innovative synthesis to practical application in water control and electrochemistry. Coord Chem Rev 517:215976. https://doi.org/10.1016/j.ccr.2024.215976
24. Ezati P, Priyadarshi R, Rhim J-W (2022) Prospects of sustainable and renewable source-based carbon quantum dots for food packaging applications. Sustain Mater Technol 33:e00494. https://doi.org/10.1016/j.susmat.2022.e00494
25. Fraceto LF, Grillo R, de Medeiros GA, Scognamiglio V, Rea G, Bartolucci C (2016) Nanotechnology in agriculture: which innovation potential does it have? Front Environ Sci 4. https://doi.org/10.3389/fenvs.2016.00020
26. Fukuda K (2015) Food safety in a globalized world. Bull World Health Organ 93:212. https://doi.org/10.2471/BLT.15.154831
27. Gao G, Jiang Y-W, Sun W, Wu F-G (2018) Fluorescent quantum dots for microbial imaging. Chin Chem Lett 29:1475–1485

28. Ghasemi R, Mirahmadi-zare SZ, Allafchian A, Behmanesh M (2022) Fast fluorescent screening assay and dual electrochemical sensing of bacterial infection agent (*Streptococcus agalactiae)* based on a fluorescent-immune nanofibers. Sens Actuators B Chem 352:130968. https://doi.org/10.1016/j.snb.2021.130968
29. Gorla US, Gupta JK, Samathoti P, Sahithi A, Nuli MV, Kumar VR, Kumar RR (2024) Suitability of hybrid quantum dots as payload for plants. In: Thomas S, Das P, Ganguly S (eds) Quantum dots based nanocomposites: design, fabrication and emerging applications. Springer Nature Switzerland, Cham, pp 359–375. https://doi.org/10.1007/978-3-031-54779-9_18
30. Guirguis A, Yang W, Conlan XA, Kong L, Cahill DM, Wang Y (2023) Boosting plant photosynthesis with carbon dots: a critical review of performance and prospects. Small 19:2300671. https://doi.org/10.1002/smll.202300671
31. Gupta A, Mehta SK, Kunal K, Mukhopadhyay K, Singh S (2022) 10—Quantum dots as promising nanomaterials in agriculture. In: Ghosh S, Thongmee S, Kumar A (eds) Agricultural nanobiotechnology, Woodhead publishing series in food science, technology and nutrition. Woodhead Publishing, pp 243–296. https://doi.org/10.1016/B978-0-323-91908-1.00016-X
32. Hu C, Channa AI, Xia L, Li X, Li Z, Wang ZM, Tong X (2025) Colloidal InAs quantum dots: synthesis, properties, and optoelectronic devices. Adv Funct Mater 2500280. https://doi.org/10.1002/adfm.202500280
33. Hu Q, Xu X, Li Z, Zhang Y, Wang J, Fu Y, Li Y (2014) Detection of acrylamide in potato chips using a fluorescent sensing method based on acrylamide polymerization-induced distance increase between quantum dots. Biosens Bioelectron 54:64–71. https://doi.org/10.1016/j.bios.2013.10.046
34. Huang R, Vaze N, Soorneedi A, Moore MD, Luo Y, Poverenov E, Rodov V, Demokritou P (2021) A novel antimicrobial technology to enhance food safety and quality of leafy vegetables using engineered water nanostructures. Environ Sci Nano 8:514–526. https://doi.org/10.1039/D0EN00814A
35. Islas-Rodriguez N, Muñoz R, Rodriguez JA, Vazquez-Garcia RA, Reyes M (2023) Integration of ternary I-III-VI quantum dots in light-emitting diodes. Front Chem 11. https://doi.org/10.3389/fchem.2023.1106778
36. Jagannathan M, Dhinasekaran D, Soundharraj P, Rajendran S, Vo D-VN, Prakasarao A, Ganesan S (2021) Green synthesis of white light emitting carbon quantum dots: fabrication of white fluorescent film and optical sensor applications. J Hazard Mater 416:125091. https://doi.org/10.1016/j.jhazmat.2021.125091
37. Jia Z, Shi C, Yang X, Zhang J, Sun X, Guo Y, Ying X (2023) QD-based fluorescent nanosensors: production methods, optoelectronic properties, and recent food applications. Compr Rev Food Sci Food Saf 22:4644–4669. https://doi.org/10.1111/1541-4337.13236
38. Kashyap PL, Kumar S, Srivastava AK (2017) Nanodiagnostics for plant pathogens. Environ Chem Lett 15:7–13. https://doi.org/10.1007/s10311-016-0580-4
39. Khan A, Ezati P, Priyadarshi R, Zhang W, Roy S, Riahi Z, Rhim J-W (2024) Advances and prospects of sulfur quantum dots in food sensing applications. Sustain Mater Technol 41:e01105. https://doi.org/10.1016/j.susmat.2024.e01105
40. Khan A, Ezati P, Priyadarshi R, Zhang W, Roy S, Riahi Z, Rhim J-W (2024) Advances and prospects of sulfur quantum dots in food sensing applications. Sustain Mater Technol e01105
41. Khan MAU, Arshad H, Majid A (2022) QDs for sensing of microorganisms. In: Majid A, Arshad H, Khan MAU (eds) Quantum dots for plant systems. Springer International Publishing, Cham, pp 137–159. https://doi.org/10.1007/978-3-031-10216-5_6
42. Khiyami MA, Almoammar H, Awad YM, Alghuthaymi MA, Abd-Elsalam KA (2014) Plant pathogen nanodiagnostic techniques: forthcoming changes? Biotechnol Biotechnol Equip 28:775–785. https://doi.org/10.1080/13102818.2014.960739
43. Kim T, Park S, Jeong S (2021) Diffusion dynamics controlled colloidal synthesis of highly monodisperse InAs nanocrystals. Nat Commun 12:3013. https://doi.org/10.1038/s41467-021-23259-w
44. Li H, Huang J, Lu F, Liu Y, Song Y, Sun Y, Zhong J, Huang H, Wang Y, Li S, Lifshitz Y, Lee S-T, Kang Z (2018) Impacts of carbon dots on rice plants: boosting the growth and improving

the disease resistance. ACS Appl Bio Mater 1:663–672. https://doi.org/10.1021/acsabm.8b00345
45. Li H, Sun C, Vijayaraghavan R, Zhou F, Zhang X, MacFarlane DR (2016) Long lifetime photoluminescence in N, S co-doped carbon quantum dots from an ionic liquid and their applications in ultrasensitive detection of pesticides. Carbon 104:33–39. https://doi.org/10.1016/j.carbon.2016.03.040
46. Li J, Xu O, Zhu X (2021) A facile green and one-pot synthesis of grape seed-derived carbon quantum dots as a fluorescence probe for Cu(II) and ascorbic acid. RSC Adv 11:34107–34116. https://doi.org/10.1039/D1RA05656E
47. Li R, Xu J, Mu X, Zeng F (2025) A comprehensive review on the synthesis methods and applications of silicon quantum dots (SiQDs). Nanotechnol 7:100144. https://doi.org/10.1016/j.nxnano.2025.100144
48. Liu J, Chen Y, Wang W, Feng J, Liang M, Ma S, Chen X (2016) "Switch-On" fluorescent sensing of ascorbic acid in food samples based on carbon quantum dots–MnO_2 probe. J Agric Food Chem 64:371–380. https://doi.org/10.1021/acs.jafc.5b05726
49. Ma H, Guan L, Chen M, Zhang Y, Wu Y, Liu Z, Wang D, Wang F, Li X (2023) Synthesis and enhancement of carbon quantum dots from Mopan persimmons for Fe^{3+} sensing and anti-counterfeiting applications. Chem Eng J 453:139906. https://doi.org/10.1016/j.cej.2022.139906
50. Malavika JP, Shobana C, Ragupathi M, Kumar P, Lee YS, Govarthanan M, Selvan RK (2021) A sustainable green synthesis of functionalized biocompatible carbon quantum dots from *Aloe barbadensis* Miller and its multifunctional applications. Environ Res 200:111414. https://doi.org/10.1016/j.envres.2021.111414
51. Mallick T, Karmakar A, Sultana Z (2022) Quantum dots: potential cell imaging agent. In: Barik P, Mondal S (eds) Application of quantum dots in biology and medicine: recent advances. Springer Nature, Singapore, pp 191–207. https://doi.org/10.1007/978-981-19-3144-4_10
52. Maritano V, Barge P, Biglia A, Comba L, Ricauda Aimonino D, Tortia C, Gay P (2024) Anticounterfeiting and fraud mitigation solutions for high-value food products. J Food Prot 87:100251. https://doi.org/10.1016/j.jfp.2024.100251
53. Marouzi S, Darroudi M, Hekmat A, Sadri K, Kazemi Oskuee R (2021) One-pot hydrothermal synthesis of carbon quantum dots from *Salvia hispanica* L. seeds and investigation of their biodistribution, and cytotoxicity effects. J Environ Chem Eng 9:105461. https://doi.org/10.1016/j.jece.2021.105461
54. Masood HA, Qi Y, Zahid MK, Li Z, Ahmad S, Lv J-M, Shahid MS, Ali HE, Ondrasek G, Qi X (2024) Recent advances in nano-enabled immunomodulation for enhancing plant resilience against phytopathogens. Front Plant Sci 15. https://doi.org/10.3389/fpls.2024.1445786
55. McCarthy M, Brennan M, Kelly AL, Ritson C, De Boer M, Thompson N (2007) Who is at risk and what do they know? Segmenting a population on their food safety knowledge. Food Qual Prefer 18:205–217. https://doi.org/10.1016/j.foodqual.2005.10.002
56. Mittal D, Kaur G, Singh P, Yadav K, Ali SA (2020) Nanoparticle-based sustainable agriculture and food science: recent advances and future outlook. Front Nanotechnol 2. https://doi.org/10.3389/fnano.2020.579954
57. Mohamadi E, Moghaddasi M, Farahbakhsh A, Kazemi A (2017) A quantum-dot-based fluoro assay for detection of food-borne pathogens. J Photochem Photobiol B 174:291–297. https://doi.org/10.1016/j.jphotobiol.2017.08.005
58. Mussa Farkhani S, Valizadeh A (2014) Review: three synthesis methods of CdX (X = Se, S or Te) quantum dots. IET Nanobiotechnol 8:59–76. https://doi.org/10.1049/iet-nbt.2012.0028
59. Ocean NanoTech, LLC [WWW Document] (2025). https://oceannanotech.com/web/products.php?category2=Quantum_Dots&utm_source=chatgpt.com. Accessed 06 February 2025
60. Osman AI, Zhang Y, Farghali M, Rashwan AK, Eltaweil AS, Abd El-Monaem EM, Mohamed IMA, Badr MM, Ihara I, Rooney DW, Yap P-S (2024) Synthesis of green nanoparticles for energy, biomedical, environmental, agricultural, and food applications: a review. Environ Chem Lett 22:841–887. https://doi.org/10.1007/s10311-023-01682-3

61. Pagano L, Maestri E, White JC, Marmiroli N, Marmiroli M (2018) Quantum dots exposure in plants: minimizing the adverse response. Curr Opin Environ Sci Health, Nanomater Plants 6:71–76. https://doi.org/10.1016/j.coesh.2018.09.001
62. Patil AS, Waghmare RD, Pawar SP, Salunkhe ST, Kolekar GB, Sohn D, Gore AH (2020) Photophysical insights of highly transparent, flexible and re-emissive PVA@WTR-CDs composite thin films: a next generation food packaging material for UV blocking applications. J Photochem Photobiol Chem 400:112647. https://doi.org/10.1016/j.jphotochem.2020.112647
63. Qdot Nanocrystals—Section 6.6—US [WWW Document] (n.d.) https://www.thermofisher.com/us/en/home/references/molecular-probes-the-handbook/ultrasensitive-detection-technology/qdot-nanocrystal-technology.html. Accessed 06 February 2025
64. Qiu P, Li J, Zhang L, Chen K, Shao J, Zheng B, Yuan H, Qi J, Yue L, Hu Q, Ming Y, Liu S, Long L, Gu J, Zhang X, Lindsey K, Gao W, Wu H, Zhu L (2023) Polyethyleneimine-coated MXene quantum dots improve cotton tolerance to Verticillium dahliae by maintaining ROS homeostasis. Nat Commun 14:7392. https://doi.org/10.1038/s41467-023-43192-4
65. Qu J-H, Wei Q, Sun D-W (2018) Carbon dots: principles and their applications in food quality and safety detection. Crit Rev Food Sci Nutr
66. Qu J-H, Wei Q, Sun D-W (2018) Carbon dots: principles and their applications in food quality and safety detection. Crit Rev Food Sci Nutr 58:2466–2475. https://doi.org/10.1080/10408398.2018.1437712
67. Quantum Dot Kit [WWW Document] (n.d.) https://shop.plasmachem.com/90-quantum-dot-kit?utm_source=chatgpt.com. Accessed 06 February 2025
68. Kumar A, Kumar D, Saikia M (2023) A review on plant derived carbon quantum dots for bio-imaging. Mater Adv 4:3951–3966. https://doi.org/10.1039/D3MA00254C
69. Šafranko S, Stanković A, Hajra S, Kim H-J, Strelec I, Dutour-Sikirić M, Weber I, Bosnar MH, Grbčić P, Pavelić SK, Széchenyi A, Mishra YK, Jerković I, Jokić S (2021) Preparation of multifunctional N-doped carbon quantum dots from Citrus clementina peel: investigating targeted pharmacological activities and the potential application for Fe^{3+} sensing. Pharmaceuticals 14:857. https://doi.org/10.3390/ph14090857
70. Sandhya M, Senthilraja G, Priyadharshini E, Rani LU, Harideekshayini R, Nishanthi M, Anand T, Subramanian KS, Sandhya M, Senthilraja G, Priyadharshini E, Rani LU, Harideekshayini R, Nishanthi M, Anand T, Subramanian KS (2025) Early detection of plant diseases and their management using quantum dots: status and strategies. J Fluoresc
71. Shin SJ, Koo J-J, Lee J-K, Chung TD (2019) Unique luminescence of hexagonal dominant colloidal copper indium sulphide quantum dots in dispersed solutions. Sci Rep 9:20144. https://doi.org/10.1038/s41598-019-56762-8
72. Shivashakarappa K, Reddy V, Tupakula VK, Farnian A, Vuppula A, Gunnaiah R (2022) Nanotechnology for the detection of plant pathogens. Plant Nano Biol 2:100018. https://doi.org/10.1016/j.plana.2022.100018
73. Sistani S, Shekarchizadeh H (2022) Applications of carbon quantum dots in detection and packaging of foods. J Food Sci Technol Iran 19:193–209. https://doi.org/10.22034/FSCT.19.127.193
74. Sistani S, Shekarchizadeh H, Sistani S, Shekarchizadeh H (2022) Applications of quantum dots in the food industry. In: Quantum dots—recent advances, new perspectives and contemporary applications. IntechOpen. https://doi.org/10.5772/intechopen.107190
75. Sistani S, Shekarchizadeh H, Sistani S, Shekarchizadeh H (2022) Applications of quantum dots in the food industry. In: Quantum dots—recent advances, new perspectives and contemporary applications. IntechOpen. https://doi.org/10.5772/intechopen.107190
76. Sobhanan J, Rival JV, Anas A, Sidharth Shibu E, Takano Y, Biju V (2023) Luminescent quantum dots: synthesis, optical properties, bioimaging and toxicity. Adv Drug Deliv Rev 197:114830. https://doi.org/10.1016/j.addr.2023.114830
77. Sun M, Zhao C, Shang H, Hao Y, Han L, Qian K, White JC, Ma C, Xing B (2023) ZnO quantum dots outperform nanoscale and bulk particles for enhancing tomato (*Solanum lycopersicum*) growth and nutritional values. Sci Total Environ 857:159330. https://doi.org/10.1016/j.scitotenv.2022.159330

78. Sun Y, Zhang M, Bhandari B, Yang C (2022) Recent development of carbon quantum dots: biological toxicity, antibacterial properties and application in foods. Food Rev Int 38:1513–1532. https://doi.org/10.1080/87559129.2020.1818255
79. Tan TL, Zulkifli NA, Zaman ASK, binti Jusoh M, Yaapar MN, Rashid SA (2021) Impact of photoluminescent carbon quantum dots on photosynthesis efficiency of rice and corn crops. Plant Physiol Biochem 162:737–751. https://doi.org/10.1016/j.plaphy.2021.03.031
80. Vasudevan D, Gaddam RR, Trinchi A, Cole I (2015) Core–shell quantum dots: properties and applications. J Alloys Compd 636:395–404. https://doi.org/10.1016/j.jallcom.2015.02.102
81. Wang B, Chen Y, Wu Y, Weng B, Liu Y, Lu Z, Li CM, Yu C (2016) Aptamer induced assembly of fluorescent nitrogen-doped carbon dots on gold nanoparticles for sensitive detection of AFB1. Biosens Bioelectron 78:23–30. https://doi.org/10.1016/j.bios.2015.11.015
82. Wang H, Wang Y, Hou X, Xiong B (2020) Bioelectronic nose based on single-stranded DNA and single-walled carbon nanotube to identify a major plant volatile organic compound (p-ethylphenol) released by Phytophthora cactorum infected strawberries. Nanomaterials 10:479. https://doi.org/10.3390/nano10030479
83. Wells RL, Aubuchon SR, Kher SS, Lube MS, White PS (1995) Synthesis of nanocrystalline indium arsenide and indium phosphide from indium (III) halides and tris(trimethylsilyl)pnicogens. Synthesis, characterization, and decomposition behavior of I_3In. cntdot. $P(SiMe_3)_3$. Chem Mater 7:793–800
84. Whiteside Matthew (2008) Tiny quantum dots shed new light on nutrient flow [WWW Document]. https://www.science.org/content/article/tiny-quantum-dots-shed-new-light-nutrient-flow. Accessed 06 Feb 2025
85. Xu H, Yang X, Li G, Zhao C, Liao X (2015) Green synthesis of fluorescent carbon dots for selective detection of tartrazine in food samples. J Agric Food Chem 63:6707–6714. https://doi.org/10.1021/acs.jafc.5b02319
86. Xu N, Gao S, Xu C, Fang Y, Xu L, Zhang W (2021) Carbon quantum dots derived from waste acorn cups and its application as an ultraviolet absorbent for polyvinyl alcohol film. Appl Surf Sci 556:149774. https://doi.org/10.1016/j.apsusc.2021.149774
87. Yang H, Zhou B, Zhang Y, Liu H, Liu Y, He Y, Xia S (2021) Valorization of expired passion fruit shell by hydrothermal conversion into carbon quantum dot: physical and optical properties. Waste Biomass Valorization 12:2109–2117. https://doi.org/10.1007/s12649-020-01132-z
88. Ye H, Liu B, Wang J, Zhou C, Xiong Z, Zhao L (2022) A hydrothermal method to generate carbon quantum dots from waste bones and their detection of laundry powder. Molecules 27:6479. https://doi.org/10.3390/molecules27196479
89. Yin C, Chen L, Niu N (2021) Nitrogen-doped carbon quantum dots fabricated from cellulolytic enzyme lignin and its application to the determination of cytochrome c and trypsin. Anal Bioanal Chem 413:5239–5249. https://doi.org/10.1007/s00216-021-03496-0
90. Zhang B, Liu Y, Ren M, Li W, Zhang X, Vajtai R, Ajayan PM, Tour JM, Wang L (2019) Sustainable synthesis of bright green fluorescent nitrogen-doped carbon quantum dots from alkali lignin. ChemSusChem 12:4202–4210. https://doi.org/10.1002/cssc.201901693
91. Zhang J, Bifulco A, Amato P, Imparato C, Qi K (2023) Copper indium sulfide quantum dots in photocatalysis. J Colloid Interface Sci 638:193–219. https://doi.org/10.1016/j.jcis.2023.01.107
92. Zhao L, Bai T, Wei H, Gardea-Torresdey JL, Keller A, White JC (2022) Nanobiotechnology-based strategies for enhanced crop stress resilience. Nat Food 3:829–836. https://doi.org/10.1038/s43016-022-00596-7
93. Zhou C, Wu S, Qi S, Song W, Sun C (2021) Facile and high-yield synthesis of N-doped carbon quantum dots from biomass quinoa saponin for the detection of Co^{2+}. J Anal Methods Chem 2021:9732364. https://doi.org/10.1155/2021/9732364
94. Zhou J-W, Zou X-M, Song S-H, Chen G-H (2018) Quantum dots applied to methodology on detection of pesticide and veterinary drug residues. J Agric Food Chem 66:1307–1319. https://doi.org/10.1021/acs.jafc.7b05119
95. Zhu L, Shen D, Liu Q, Wu C, Gu S (2021) Sustainable synthesis of bright green fluorescent carbon quantum dots from lignin for highly sensitive detection of Fe^{3+} ions. Appl Surf Sci 565:150526. https://doi.org/10.1016/j.apsusc.2021.150526

96. Zou Y, Shi Y, Wang T, Ji S, Zhang X, Shen T, Huang X, Xiao J, Farag MA, Shi J, Zou X (2024) Quantum dots as advanced nanomaterials for food quality and safety applications: a comprehensive review and future perspectives. Compr Rev Food Sci Food Saf 23:e13339. https://doi.org/10.1111/1541-4337.13339
97. Zuo J, Feng J, Gameiro MG, Tian Y, Liang J, Wang Y, Ding J, He Q (2022) RFID-based sensing in smart packaging for food applications: a review. Futur Foods 6:100198. https://doi.org/10.1016/j.fufo.2022.100198

Chapter 8
Toxicity and Cytotoxicity of Quantum Dots

Anuja Vibhute and Arpita Pandey Tiwari

Abstract Quantum dots (QDs) are nanomaterials with remarkable optical and electronic properties, making them highly attractive for applications in biomedical, environmental, and industrial fields. However, their potential toxicity and cytotoxicity present significant challenges for their safe use, particularly in clinical settings. This chapter explores the factors influencing QD toxicity, including their synthesis methods, surface modifications, and interactions with biological systems. We discuss the impact of QDs on cellular processes, focusing on oxidative stress, DNA damage, and organ toxicity, with particular emphasis on heavy-metal-based QDs. Strategies to mitigate these risks, such as surface engineering, biocompatible coatings, and the development of greener alternatives like graphene and silicon QDs, are also examined. Additionally, the chapter highlights the importance of understanding the long-term effects of QDs, their biodegradation, and clearance mechanisms, as well as the need for standardized toxicity testing protocols. Addressing these challenges will be key to the successful integration of QDs in various applications.

Keywords Biocompatibility · Cytotoxicity · DNA damage · Oxidative stress · Quantum dots · Toxicity

1 Introduction

Quantum dots (QDs) are nanometer-sized particles or crystals made from semiconducting materials, typically ranging in size from 2 to 10 nanometres (containing around 10–50 atoms). In 1979, Alexei Ekimov initiated research on doped glasses at the S.I. Vavilov State Optical Institute, Russia. His goal was to elucidate the

A. Vibhute · A. Pandey Tiwari (✉)
Department of Medical Biotechnology and Stem Cell and Regenerative Medicine, Centre for Interdisciplinary Research, D. Y. Patil Education Society (Deemed to be University), Kolhapur, Maharashtra, India
e-mail: arpitaptiwari@gmail.com

A. Vibhute
e-mail: anujavibhute06@gmail.com

T. Bhuyan et al. (eds.), *Multifunctional Quantum Dots*, Materials Horizons: From Nature to Nanomaterials, https://doi.org/10.1007/978-981-92-0172-3_8

chemical composition and structural characteristics of colloidal particles in coloured glasses and to understand their growth mechanisms [12]. QDs are composed of just a few thousand atoms, featuring a semiconductor core made from materials such as cadmium selenium (CdSe), lead selenium (PbSe), or indium arsenide (InAs). These cores are surrounded by an outer shell, typically zinc sulphide (ZnS) or cadmium sulphide (CdS), which serves to reduce toxicity [18]. QDs boast a wide range of applications, including single-electron transistors, solar cells, LEDs, lasers, single-photon emitters, second-harmonic generation, quantum computing, cell biology research, advanced microscopy, and medical imaging [8].

Understanding the toxicity and cytotoxicity of QDs is paramount due to their growing usage in biomedical, environmental, and industrial sectors. The key reasons for this include ensuring safe biomedical applications, assessing environmental impact, adhering to regulatory standards, and protecting public health. Research into QD toxicity and cytotoxicity is essential for the responsible development of QD technology and its safe incorporation into diverse applications.

In biomedicine, understanding the toxicity of QDs is crucial for the safe application of these materials in imaging, drug delivery, and theranostic (combined therapeutic and diagnostic) uses. Investigating their interactions with cells, the induction of oxidative stress and long-term effects can provide valuable insights into their biocompatibility [27].

From an environmental perspective, studying the ecotoxicity and behaviour of QDs in ecosystems is vital to minimize ecological harm resulting from their production, use, and disposal. Research into safer material designs, such as cadmium-free QDs and surface modifications, can optimize their applications while reducing associated risks. Additionally, advanced models like 3D cell cultures and organ-on-a-chip systems offer innovative approaches for predictive toxicity studies. This research is essential for regulatory compliance, ethical use, and the sustainable development of QD technologies.

This chapter examines the toxicity and cytotoxicity of QDs, investigating their underlying mechanisms and evaluating their implications for both biomedical and environmental safety. It also outlines strategies to mitigate adverse effects, laying the groundwork for the safe and sustainable advancement of QD technologies.

2 Characteristics of Quantum Dots Relevant to Toxicity

2.1 Chemical Composition and Core-Shell Structures

QDs often consist of cores made from heavy metals such as cadmium selenide (CdSe), cadmium telluride (CdTe), or lead sulphide (PbS), which can release toxic metal ions like Cd^{2+} and Pb^{2+} upon degradation, posing serious risks to both biological systems and the environment. To address this, alternative QDs based on less toxic materials like zinc compounds (ZnS, ZnSe) or carbon dots have been developed, although these alternatives may sometimes exhibit inferior optical performance compared to their heavy metal counterparts.

To mitigate toxicity, QDs are typically encapsulated with a shell material such as zinc sulphide (ZnS), which enhances their stability, reduces ion leakage, and improves optical properties; however, the protective effect of the shell depends greatly on its durability, as degradation can expose the toxic core, leading to cytotoxicity. The release of heavy metal ions, particularly under acidic, oxidative, or photolytic conditions, significantly contributes to QD-induced toxicity, with the extent of ion leakage being largely influenced by the shell's composition and integrity.

Furthermore, the core-shell architecture influences the generation of reactive oxygen species (ROS), which are major contributors to oxidative stress and cellular damage; well-engineered core-shell systems can effectively minimize ROS generation and thereby reduce toxic effects [44]. QD cores are synthesized using a variety of metal complexes, including semiconductors, noble metals, and magnetic transition metals, with Group III–V QDs comprising materials like indium phosphate (InP), indium arsenide (InAs), gallium arsenide (GaAs), and gallium nitride (GaN), while Group II–VI QDs include cores such as zinc sulphide (ZnS), zinc selenide (ZnSe), cadmium selenide (CdSe), and cadmium telluride (CdTe) [17, 19].

Advancements in synthetic techniques have enabled the fabrication of more complex structures like CdTe/CdSe and CdSe/ZnTe, as well as hybrid QDs incorporating lead selenide (PbSe) [25]. Thus, the chemical composition and core-shell structure of QDs play a critical role in defining their toxicity, underscoring the importance of careful material design and functionalization to minimize adverse effects while preserving their desirable functional properties.

2.2 Surface Modifications and Functionalization

Surface modifications using biocompatible coatings, such as polymers or ligands, enhance the stability of QDs and minimize their direct interactions with cells, effectively reducing their toxicity. The small size of QDs enables them to traverse biological barriers, including cell membranes, and accumulate in tissues. However, their high surface area-to-volume ratio, while beneficial for certain applications, also increases their reactivity, potentially leading to elevated oxidative stress and cellular damage.

Positively charged QDs often demonstrate increased cytotoxicity because of their strong interactions with negatively charged cell membranes. Additionally, surface charge plays a critical role in determining the stability and dispersion of QDs in biological and environmental media [49]. Assigning biocompatible coatings or functional groups to the QD core-shell structure can impart specific bioactivity. Newly synthesized QDs are naturally hydrophobic and not biologically useful due to a hydrophobic cap formed during synthesis in organic solvents. To make them biologically compatible and active, QDs undergo 'functionalization' with secondary coatings, which enhances water solubility, core durability, and suspension characteristics, thereby providing the desired bioactivity [3].

For instance, coating QD cores with hydrophilic polyethylene glycol (PEG) groups enhances their biocompatibility. This modification allows the QDs to be further conjugated with bioactive molecules, enabling targeted interaction with specific biological processes or cellular characteristics. By attaching various molecular components to the QD core, these particles can be customized for particular diagnostic or therapeutic applications. Functionalization methods include electrostatic interactions, adsorption, multivalent chelation, and covalent bonding, which are critical factors in determining QD durability, stability, and reactivity in vivo. In scientific literature, these properties are often referred to as 'core-shell-conjugate' structures or vice versa [23].

The degradation of QD coatings can lead to unexpected or undesirable reactions in vivo. Additionally, certain coating materials, such as mercaptoacetic acid, have been shown to exhibit cytotoxicity. This underscores the importance of QD physicochemical properties in assessing their toxicity. The stability of the QD core-coating bioactive complexes plays a critical role in their safety, as instability may render QDs harmful. Since QDs are prone to degradation under photolytic and oxidative conditions, ensuring their stability will be a pivotal factor in the successful commercialization of QD-based products [35].

2.3 Size, Shape, and Physical Properties

Cellular uptake happens when cells interact with particles ranging from a few manometers to several hundred manometers. Cells possess highly regulated and precisely defined functions to manage the uptake and transport of biological components within this size range [62]. The particle size of QDs is a critical factor influencing their efficiency in cellular internalization and their cytotoxicity. Numerous studies have demonstrated these findings, highlighting, for instance, the differences in subcellular distribution and toxicity between red and green light-emitting QDs [45]. Smaller QDs are generally more cytotoxic than larger ones at equivalent concentrations. Lewinski and team observed that red QDs predominantly localized in the cytoplasm, whereas green QDs, being smaller, were found primarily in the nucleus. This suggests that the ability of smaller QDs to penetrate the nucleus increases the risk of DNA damage and the induction of apoptosis, accounting for their higher cytotoxicity [29].

The effect of particle shape on cellular uptake is equally important as particle size, although it is more complex and not always straightforward to study. This complexity arises because not all particles can be easily fabricated into various shapes. Research has demonstrated that cells can absorb colloidal particles of diverse shapes, including

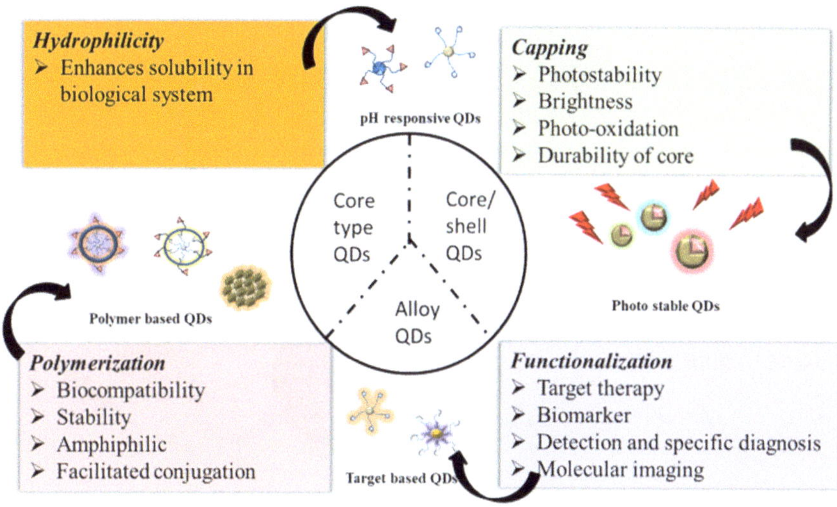

Fig. 1 Representation of QDs characteristics relevant to toxicity. Reprinted with permission [41]

spheres [16], rods [57], tubes [34], and sheets [36]. QDs can be manufactured into rod-like structures with different aspect ratios, in addition to spherical shapes. However, spherical QDs typically diffuse through the plasma membrane much more rapidly than rod-like QDs with a high aspect ratio [50]. Figure 1 represents characteristics of QDs including polymer based QDs, photostable QDs, target-based QDs, pH responsive QDs [41].

3 Mechanisms of Toxicity

The toxicity of QDs primarily stems from their core compositions, such as cadmium selenide (CdSe) and cadmium telluride (CdTe) [21, 47]. Figure 2 shows mechanism of toxicity. Internalized QDs experience endosomal acidity, leading to coating degradation and environment of endosomes, leading to the degradation of surface coatings and the subsequent release of toxic Cd^{2+}. These metal ions contribute to oxidative stress by promoting the generation of reactive oxygen species (ROS), particularly through interactions with mitochondria, ultimately affecting organelle function and compromising plasma membrane integrity. QDs may also translocate to the nucleus, where they interact with DNA and induce genotoxic effects. Additionally, the precipitation and accumulation of QDs on the plasma membrane can result in structural disruption and cellular stress [37].

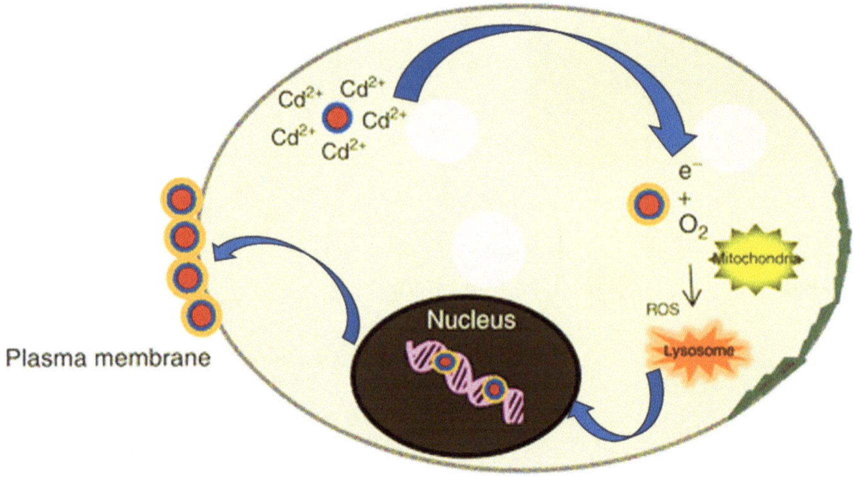

Fig. 2 Mechanism of QD toxicity. Reprinted with permission from Nadeau [37]

4 In Vitro Cytotoxicity of QDs

QDs have diverse biomedical applications, including bioimaging, targeted drug delivery, and photodynamic therapy. The suitability of QDs for these applications is determined by key factors such as (i) biocompatibility, (ii) cytotoxicity, and (iii) fluorescence properties. To evaluate QD cytotoxicity in vitro, researchers commonly employ cell lines such as human embryonic kidney (HEK 293) cells, Chinese hamster ovary (CHO) cells, human lung epithelial (A549) cells, and mouse fibroblast (NIH 3T3) cells. Depending on the specific research focus and desired tissue response, specialized neuronal cell lines like PC12 may also be used [13].

QDs are generally internalized into cells through endocytosis, with the specific pathway influenced by their surface modifications. Common endocytic routes include clathrin-mediated endocytosis, caveolin-mediated endocytosis, and macropinocytosis. After internalization, QDs localize in various subcellular compartments: (1) endosomes and lysosomes: These are the primary destinations immediately following endocytosis. (2) Nucleus: Localization occurs when QDs are equipped with nuclear-targeting ligands. (3) Cytoplasm: Achieved when QDs escape from endosomal compartments. The strong fluorescence properties of QDs enable real-time imaging and precise tracking of their intracellular distribution [13].

Mitigating in vitro toxicity of QDs is essential for their safe application in biomedical research. Employing biocompatible coatings, such as silica, PEG, or zwitterionic polymers, is an effective strategy. These coatings stabilize the QD core, prevent the release of toxic ions, and reduce interactions with cellular components, thereby minimizing oxidative stress and membrane disruption.

The choice of core materials is critical, with cadmium-free QDs like indium phosphide (InP) or graphene-based QDs emerging as safer alternatives, thereby reducing the risk of heavy metal ion release in cell culture environments. Another essential factor is dosing optimization, as high concentrations of QDs and prolonged exposure can increase cytotoxic effects, including apoptosis, necrosis, and the generation of ROS.

Moreover, the culture environment, including pH and oxidative conditions, must be carefully controlled because these factors affect QD stability and the extent of ion release. By addressing these aspects, the in vitro toxicity of QDs can be minimized, facilitating safer and more reliable use in cellular studies. When investigating QD cytotoxicity in vitro, comparing the sensitivities of different cell types reveals variations in their responses. These differences are often influenced by cellular characteristics such as metabolism, proliferation rate, and membrane composition. Notably, cancer cells and normal cells typically exhibit differing sensitivities to QDs, offering valuable insights for biomedical applications like targeted imaging and therapy.

- **Cancer Cells**: Cancer cells, distinguished by their high metabolic activity, rapid proliferation, and altered membrane properties, exhibit enhanced uptake of QDs, particularly those functionalized for targeted delivery. This increased internalization often results in elevated cytotoxic effects in cancer cells, driven by ROS generation, mitochondrial dysfunction, and DNA damage. The heightened vulnerability of cancer cells to QD-induced cytotoxicity presents a therapeutic advantage, allowing selective elimination of tumour cells while minimizing harm to normal cells [28].
- **Normal Cells**: Normal, non-cancerous cells typically show lower QD uptake due to less active endocytic pathways and a reduced number of surface receptors for targeted QDs. As a result, these cells generally exhibit lower sensitivity to QD-induced toxicity under similar conditions. However, normal cells can still experience adverse effects from QDs, especially at high concentrations or with prolonged exposure, which may cause oxidative stress, membrane damage, and apoptosis [58].
- **Cell-Type Specific Variations**: The extent of QD toxicity can vary significantly among different normal or cancer cell lines, influenced by factors such as membrane charge, receptor expression, and intracellular antioxidant capacity. For instance, neurons are particularly sensitive to QDs due to their limited ability to manage oxidative stress. Liver cells, such as HepG2, are highly susceptible because of their detoxification role and active endocytosis. Stem cells may exhibit variable responses depending on their differentiation state and metabolic activity. These variations underscore the importance of considering cell-specific characteristics when assessing the cytotoxicity of QDs in biomedical applications. Akhavan et al. [2] examined the potential genotoxic effects of graphene nanoplatelets and nanoribbons on human mesenchymal stem cells. Their findings indicate that these graphene-based materials are capable of penetrating the cells and triggering ROS generation. This oxidative stress leads to chromosomal abnormalities and DNA fragmentation, even at low exposure concentrations, highlighting their impact on genomic integrity. Furthermore, Mahto et al. investigated the cellular

compatibility of surface-functionalized CdSe/ZnS quantum dots with BALB/3T3 fibroblast cells [33].

- **Implications for In Vitro Studies**: Understanding these differential sensitivities is crucial for designing experiments and interpreting results. For therapeutic applications, QDs can be functionalized to preferentially target cancer cells, minimizing off-target effects on normal cells. Conversely, in non-therapeutic contexts, such as long-term imaging, minimizing QD cytotoxicity in normal cells becomes paramount [42]. This comparison underscores the importance of tailoring QD design and experimental parameters to the specific cell types under investigation to achieve desired outcomes while mitigating potential toxic effects.

5 In Vivo Toxicity of QDs

Several animal models have been utilized to study the in vivo toxicity of QDs, each offering unique insights into their potential risks and mechanisms of toxicity. Rodents, such as mice and rats, are commonly employed to investigate systemic toxicity, biodistribution, and long-term effects of QDs. Zebrafish (Danio rerio) serves as valuable models for assessing developmental toxicity, organ-specific effects, and interactions with the environment due to their transparent embryos and rapid development. Drosophila melanogaster (fruit flies) are used for genetic studies and to explore molecular pathways associated with QD toxicity. Additionally, xenograft models allow researchers to evaluate the behaviour of QDs in tumour environments, making them particularly useful for studying therapeutic applications. These diverse models collectively aid in understanding QD biocompatibility and hazards, paving the way for the development of safer and more effective QD-based technologies.

Understanding the distinction between acute and chronic toxicity, as well as short-term and long-term effects, is essential in evaluating the impact of QDs. Acute toxicity refers to the harmful effects that manifest shortly after a single or brief exposure to QDs. These effects are typically assessed by monitoring immediate symptoms, such as organ damage, cell death, or behavioural changes. In contrast, chronic toxicity arises from prolonged exposure to QDs and involves cumulative damage over time. Chronic toxicity studies focus on long-term consequences, including cancer development, reproductive issues, or persistent organ damage. This comprehensive understanding of toxicity is critical for ensuring the safe application of QDs in biomedical and other fields [11]. Short-term effects refer to the immediate or early-stage responses to QD exposure, typically observable within days or weeks. These effects may include inflammation, oxidative stress, or acute organ damage. In contrast, long-term effects develop over months or years of continuous or repeated exposure to QDs. Such effects can include persistent health issues, such as chronic inflammation, genetic mutations, or prolonged organ dysfunction. Both acute and chronic toxicity studies, along with assessments of short-term and long-term effects, are essential for gaining a comprehensive understanding of QD safety and the potential health risks they may pose [55].

Animal studies have demonstrated that QDs can accumulate in various organs due to their primary exposure routes, posing potential risks to vital organs. QDs primarily alter key biomarkers in target organs, particularly by inducing oxidative damage, and in severe cases, can lead to hyperemia, tissue necrosis, or even death. In addition to directly damaging target organs, QDs can also trigger the accumulation of immune cells, causing inflammatory reactions in other major organs. Whether aiming to mitigate the risk of human exposure to QDs in production and daily life or to enable their clinical application, conducting a thorough in vivo toxicity assessment of QDs is crucial (Fig. 3).

QDs tend to accumulate in organs linked to the reticulo-endothelial system, such as the liver, spleen, lungs, and kidneys, after crossing biological barriers through various exposure routes. This accumulation in the spleen, in particular, can trigger inflammatory reactions, leading to adverse effects and toxicity. Research has shown that smaller QDs exhibit heightened toxicity to the immune system, as their size allows them to more easily infiltrate immune cells in vivo [5]. Interestingly, surface coating has emerged as an effective strategy to mitigate the toxicity of QDs [46]. For instance, in a study involving adult Sprague Dawley rats, six weeks of oral exposure to silver nanoparticles (AgNPs) resulted in significant silver accumulation in the spleen and an increased immune response. However, when AgNPs were coated with polyethylene pyrrolidone (PVP-AgNPs), no harmful effects were observed in the treated animals, underscoring the potential of surface modifications to alter the

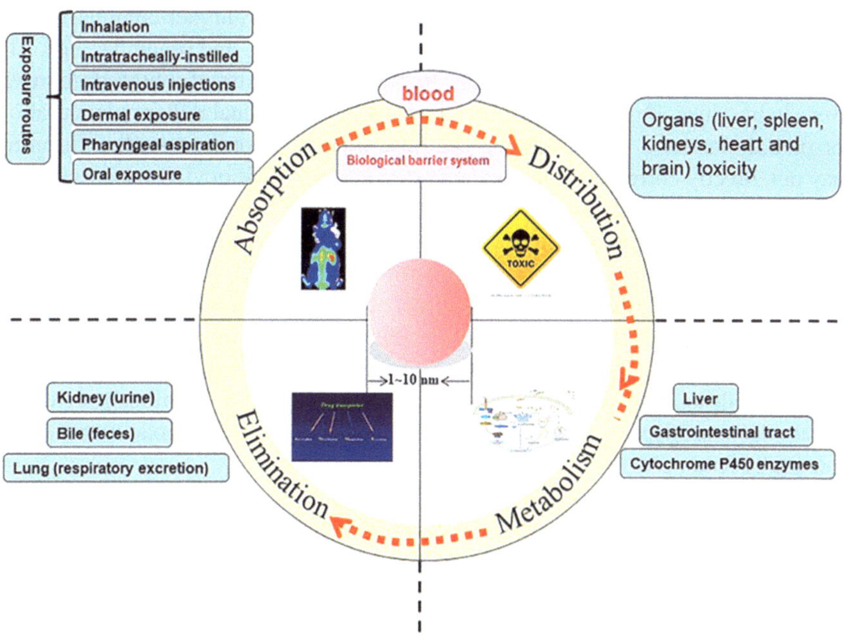

Fig. 3 Pharmacokinetics of the QDs in vivo. Reprinted with permission [32]

toxicological profile of QDs [4]. Cadmium ions (Cd^{2+}) are recognized as potent nephrotoxic agents, with irreversible effects on renal tubules, leading to proteinuria [31]. Prolonged exposure to Cd^{2+} can impair renal tubular function, ultimately causing glomerular damage and a decline in glomerular filtration rates. Due to its extended biological half-life, Cd^{2+} tends to accumulate predominantly in the kidneys and bones, posing long-term health risks [22]. This distribution is often accompanied by immune responses and tissue damage, as summarized in Table 1. Among these immune responses, inflammation emerges as a key reaction triggered by QDs.

6 Factors Influencing Toxicity

6.1 Concentration and Dose

At high concentrations, QDs can saturate cellular uptake mechanisms, leading to excessive accumulation within cells or tissues. This accumulation can cause oxidative stress, cellular damage, and toxicity. When QD concentrations exceed a certain threshold, the risk of cytotoxic effects such as mitochondrial dysfunction or apoptosis increases. Although lower concentrations might reduce direct cytotoxic effects, they can still induce sublethal stress, particularly with prolonged exposure [30]. The total administered dose of QDs, whether from a single exposure or repeated exposures, significantly influences the degree of toxicity. A higher cumulative dose raises the risk of adverse effects, such as systemic toxicity and organ damage. The delivery rate of QDs is also important. Rapid administration of a high dose might overwhelm physiological defence mechanisms, while gradual delivery could allow for partial clearance or adaptation. The dose used in laboratory studies (in vitro) may not directly correlate to biological systems (in vivo), as in vivo systems involve complex interactions, including immune responses and biodistribution [22].

At low concentrations or doses, QDs generally show minimal acute toxicity, though they may still pose risks of long-term effects, such as genotoxicity or subtle changes in immune responses. As the concentration or dose increases to moderate levels, oxidative stress becomes more prominent, leading to moderate inflammation, cellular stress responses, or disruption of normal cellular functions. At high concentrations or doses, toxicity is more severe, often resulting in significant cytotoxic effects, such as apoptosis or necrosis. In extreme cases, systemic toxicity can occur, potentially leading to organ dysfunction or failure [14].

A study administered single doses of CdSe QDs at 10, 20, 40, and 80 mg/kg intravenously to male Wistar rats, observing them for 30 days. Results indicated dose-dependent biochemical and oxidative damage, with notable bioaccumulation in the spleen, liver, and lungs. The highest accumulation was in the spleen (120 μg/g), followed by the liver (78 μg/g) and lungs (38 μg/g). Elevated white blood cell counts and decreased red blood cells suggested inflammatory responses [9]. Another study

Table 1 Summary of the adverse effects of quantum dots on liver, lung, kidney, brain, spleen in vivo

QDs duration	Animal model	Exposure route	Concentration toxicity	References
MPA-CdTe QDs: 5 nm	Male BALB/c mice	Mice were administered CdTe QDs intravenously at a dose of 5 mg/kg body weight and were monitored at time intervals of 2 h, 24 h, 3 days, and 1 week	Exposure to CdTe QDs induced oxidative stress in the liver, evidenced by a reduction in total glutathione levels, enhanced superoxide dismutase activity, and altered gene expression of several biomarkers associated with oxidative stress. Additionally, CdTe-QD treatment triggered apoptosis in liver cells and caused an increase in both the size and number of mitochondria within the hepatocytes of treated mice	[39]
TGA/ TGH-CdTe QDs: 3.8 nm	Female and male Kunming mice	CdTe QDs were administered through intravenous injection at doses of 0.2, 2, and 10 mg/kg, with observations conducted over a 7-day period	QDs primarily accumulated in the liver and kidneys. Histological analysis revealed tissue damage and liver enlargement, accompanied by elevated ALT expression in the medium- and high-dose groups 7 days after QD administration. Furthermore, QD exposure induced oxidative stress in the liver	[10]
Cd/se core–ZnS shell QDs stabilized with trioctylphosphine oxide (TOPO) and coated with poly (maleic anhydride-alt-1-octadecene) (PMAT)	Male mice from the C57BL/6J and A/J strains	The mice were co-exposed to house dust mite allergens and quantum dots QDs through oropharyngeal aspiration, with doses of 100 μg HDM and 140 ng cadmium equivalents of QDs	QDs appeared to amplify the impact of house dust mite allergens on lung permeability. Compared to mice exposed to HDM alone, co-exposure to HDM and QDs resulted in a significant increase in total protein levels in bronchoalveolar lavage fluid (BALF) in both C57BL/6J and A/J mice. Additionally, in A/J mice, co-exposure to HDM and QDs led to a marked elevation in BALF IL-33 levels, which was not observed with HDM exposure alone	[48]

(continued)

Table 1 (continued)

QDs duration	Animal model	Exposure route	Concentration toxicity	References
CdTe QDs with a size range of 2.7–5.6 nm	Healthy male Institute of Cancer Research (ICR) mice and male C57BL/6 mice	CdTe QD solutions were administered via intravenous injection at doses of 0.15, 1.5, and 15 μmol/kg (calculated based on the molar mass of cadmium). The mice were observed at 24 h, 7 days, 14 days, and 28 days post-injection. Additionally, a group of 4-week-old C57BL/6 mice was fed with 100 μM QDs, while a control group was fed with deionized water for one month	Cd-QD treatment resulted in adverse effects to mouse primary kidney cells. When cd QDs entered into cells, excessive ROS are generated, which can lead to cell apoptosis and inhibition on cellular viability	[60]
CdTe-MPA (mercaptopropionic acid) quantum dots QDs with a size of 15 nm	Adult BALB/c mice	CdTe-MPA QDs were administered via intravenous injection at doses of 8 or 16 nmol/kg body weight, with observations made at 24 h post-injection	QDs were detected in all tested organs, with the liver and kidneys being the primary sites of CdTe-QD accumulation. QD exposure led to renal dysfunction. In the high-dose group, uric acid levels decreased in a dose-dependent manner, while the levels of creatinine and blood urea nitrogen were significantly reduced	[20]
CdTe/ZnS QDs with an average size of 5.7 ± 0.8 nm and a hydrated particle size of 8.57 ± 2.60 nm	Institute of Cancer Research mice	CdTe/ZnS QDs were administered via intravenous injection at a concentration of 2.5 mM, with observations made at 3, 6, and 12 h post-injection	3 h after the intravenous injection of CdTe/ZnS QDs, the concentration of QDs in the mouse hippocampus peaked. These QDs were found to polarize microglia and induce damage to peripheral neurons. 6 h after injection, the QD concentration in the kidney increased rapidly, while levels in other organs decreased, suggesting that the kidney plays a key role in QD excretion	[15]

(continued)

Table 1 (continued)

QDs duration	Animal model	Exposure route	Concentration toxicity	References
CdTe-MPA QDs with an average size of 15.25 ± 0.34 nm	Male BALB/c mice	CdTe-MPA QDs were administered intravenously at a dose of 6 mg/kg body weight, with observations made up to one week post-administration	Following intravenous injection, CdTe QDs primarily accumulated in the liver, spleen, and kidneys. At higher doses, exposure to CdTe QDs led to mild dehydration, lethargy, ruffled fur, hunched posture, and body weight loss. Histological analysis of tissues revealed hepatic haemorrhage and necrotic areas in the spleen when the dose was 6 mg/kg or higher	[38]

demonstrated that while a single administration of QDs showed moderate species-specific toxicity, repeated administration over 14 days, even at low doses (0.5 mg/kg), led to delayed and irreversible organ damage, including hematotoxic effects and impairment of liver and kidney functions [26]. Research on zebrafish embryos exposed to differently functionalized CdTe QDs revealed that QDs with amino groups (QD-NH_3) were more toxic than those with carboxyl (QD-COOH) or polyethylene glycol (QD-PEG) groups. Unexpectedly, lower concentrations exhibited higher toxicity, attributed to changes in particle agglomeration affecting bioavailability [6]. These findings underscore the importance of considering QD concentration and exposure frequency when evaluating their toxicity.

6.2 Physicochemical Properties

The physicochemical properties of QDs, including solubility, stability, and aggregation, play a critical role in determining their toxicity. For biological and medical applications, QDs must be water-soluble. Poorly soluble QDs may aggregate in physiological environments, leading to localized toxicity or impaired biodistribution. Conversely, highly soluble QDs with proper surface coatings can reduce toxicity by improving dispersibility and limiting aggregation.

The solubility of the core material also affects the release of toxic ions (e.g., cadmium ions from CdSe QDs). Higher solubility can increase ion leaching, exacerbating cytotoxic and systemic toxic effects. Solubility may vary with pH, influencing QD behaviour in different biological compartments, such as acidic lysosomes versus neutral cytoplasm, potentially leading to differential toxicity [40]. Stable QDs maintain their structural integrity, preventing degradation into smaller, potentially more toxic components. In contrast, unstable QDs degrade more quickly, releasing toxic ions or harmful by products into biological systems. When QDs are exposed to light, they may undergo photodegradation, generating reactive oxygen species that contribute to oxidative stress, DNA damage, and cell death. QDs that lack oxidative stability are prone to oxidation, leading to the breakdown of their core material and increasing toxicity due to ion release or the formation of reactive intermediates [61].

QDs can aggregate due to interactions with proteins, salts, or other biomolecules in physiological fluids. Aggregation can change their size, shape, and surface properties, leading to two main issues: (i) reduced cellular uptake if the aggregates are too large and (ii) increased localized toxicity due to physical obstruction or high local concentrations. Surface functionalization with polymers (e.g., PEG) or biomolecules can reduce aggregation by enhancing colloidal stability. Without these coatings, QDs are more likely to aggregate, which increases their toxicity potential. Additionally, larger aggregates may bypass renal clearance, leading to prolonged retention in organs such as the liver, spleen, or kidneys, thereby exacerbating toxicity [54].

A study investigated the cytotoxicity of water-soluble core/shell QDs with varying sizes, surface charges, and chemical compositions. The findings revealed that smaller QDs exhibited higher in vitro cytotoxicity. Additionally, QDs with a low negative surface charge were more toxic than those with a higher negative or positive charge. Interestingly, the chemical composition of the QD core did not significantly affect cytotoxicity, underscoring the importance of size and surface charge in determining QD toxicity [49]. Research on CdSe/ZnS QDs with various surface modifications including poly(ethylene glycol)-carboxyl (anionic), mercaptopropionic acid (anionic), glutathione (zwitterionic), and cysteamine (cationic) revealed that toxicity in early-stage zebrafish was predominantly influenced by the type of surface modification. Notably, QDs modified with poly(ethylene glycol)-carboxyl groups exhibited the most severe toxicity, affecting mortality, hatching rates, heartbeat, and causing malformations [53]. These findings underscore the necessity of carefully considering surface charge, modification, and size when assessing the safety and designing applications of QDs.

6.3 Environmental and Biological Factors

The toxicity of QDs is profoundly influenced by both environmental and biological factors. Environmental parameters such as pH, temperature, ionic strength, light exposure, and oxidative conditions play a pivotal role in determining QD stability, solubility, and ion release. Acidic and oxidative environments, in particular, accelerate QD degradation, leading to toxicity. Similarly, biological factors, including cellular uptake, tissue accumulation, immune responses, protein corona formation, oxidative stress generation, and clearance mechanisms, govern QD behaviour within living systems. Sensitive organs like the liver and kidneys, combined with prolonged exposure, increase the risk of bioaccumulation and chronic effects [43].

In animal models, QDs have shown the potential for bioaccumulation and organ-specific toxicity. Studies have reported accumulation in organs such as the liver and kidneys, leading to concerns about long-term health effects. However, the extent of toxicity varies depending on factors like QD composition and exposure duration [53]. A study examined the occurrence, behaviour, and ecotoxicity of QDs in aquatic environments. The study highlighted that QDs can undergo dissolution, aggregation, and transformation, influenced by factors such as ultraviolet (UV) irradiation, the presence of natural organic matter, and temperature variations. These behaviours affect their stability, bioavailability, and potential toxicity to aquatic organisms [59]. The intricate interplay between these environmental and biological factors underscores the importance of controlled conditions and thoughtful design in QD applications to minimize toxicity.

To mitigate the toxicity of QDs in biomedical and environmental applications, it is crucial to understand the factors influencing their behaviour. Strategies such as surface modification, precise dosing, and regulation of environmental conditions can effectively minimize toxicological risks. Adopting eco-friendly alternatives, such as graphene quantum dots (GQDs) or silicon QDs, offers a sustainable solution to replace heavy-metal-based QDs, particularly in applications involving extended exposure. By integrating thoughtful design, enhancing biocompatibility, and maintaining environmental control, the potential toxicity of QDs can be substantially reduced. Tailoring these approaches to specific applications ensures the safe utilization of QDs across biomedical, environmental, and industrial domains [51].

7 Current Challenges and Knowledge Gaps

The scarcity of clinical trials involving QDs is primarily due to the multifaceted challenges they present across pharmaceutical, industrial, technical, and biological dimensions, as outlined in Fig. 4 and detailed below. One notable issue is their poor solubility in aqueous environments, particularly when organometallic synthesis is employed—a widely used and highly efficient method for QD production. Proposed solutions to address this limitation include replacing hydrophobic surface ligands with hydrophilic, thiol-based compounds, applying hydrophilic coatings, or utilizing aqueous-based synthesis techniques, as previously discussed [24].

Scaling up QD production from laboratory to industrial levels presents several challenges. Firstly, the extensive surface modifications required to optimize their performance make large-scale production highly complex and involve multiple steps. Secondly, achieving consistent physicochemical properties of QDs on a large scale is difficult [52]. Lastly, impurities present in the raw materials or solvents, which may have minimal impact in small-scale production, are proportionally magnified during scaling up, significantly affecting the quality and functionality of the QDs [56].

Beyond the pharmaceutical and industrial challenges, the in vivo use of QDs faces significant hurdles. QDs inherently lack organ specificity, interacting indiscriminately with tissues and cellular membranes. Furthermore, many metallic QDs

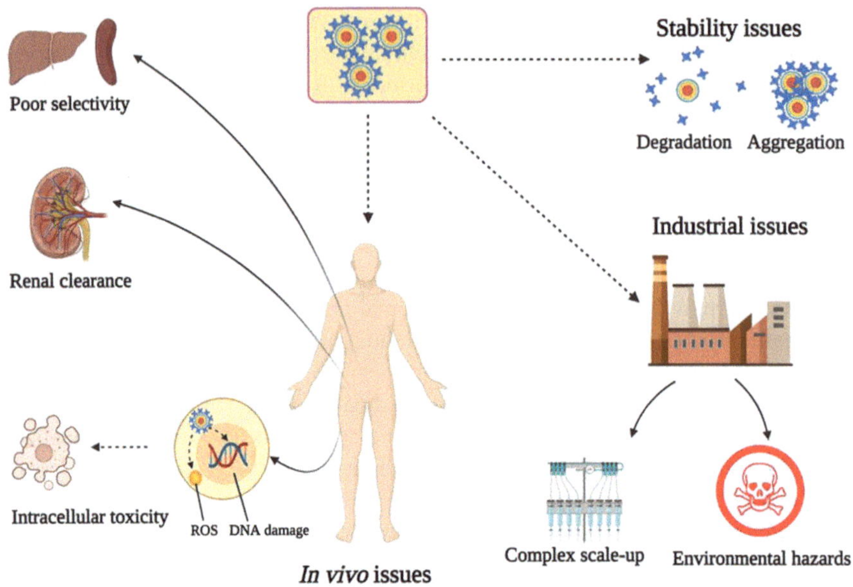

Fig. 4 Challenges in the clinical translation of QDs. Key issues include instability (degradation and aggregation), industrial limitations (scale-up complexity and heavy metal hazards), and in vivo concerns (rapid renal clearance, poor selectivity, and ROS- or DNA damage-induced toxicity). Reprinted with permission from Abdellatif et al. [1]

induce intracellular toxicity by causing oxidative damage to cellular components or interacting with DNA. The risks are exacerbated when heavy metals like lead and cadmium are used, as these elements tend to accumulate in bones and are not efficiently eliminated from the body [7].

Despite considerable progress, several critical gaps remain in understanding the toxicity of QDs, limiting their full potential. A key area of uncertainty involves the long-term effects of QDs in biological systems, including their bioaccumulation, biodegradation, and clearance mechanisms. Additionally, limited knowledge about the interactions of QDs with complex biological systems, such as the immune system and microbiome, poses challenges to their clinical application. The precise mechanisms of QD-induced toxicity, including their ability to generate ROS and induce cellular damage, are not yet fully elucidated. Environmental concerns also persist, particularly regarding the lifecycle of QDs, from their synthesis to disposal, with heavy-metal-based QDs warranting special attention. Furthermore, the absence of standardized toxicity testing protocols complicates the comparison of studies and the establishment of safety guidelines. Addressing these gaps will require collaborative efforts across nanotoxicology, materials science, and biomedical research to create safer QDs for various applications.

Addressing the challenges of toxicity assessment and developing environmentally sustainable QDs are critical steps for their broader adoption in future applications.

Ongoing research and innovation in this field are essential to pave the way for translating this remarkable technology into clinical settings, benefiting both patients and industries. With persistent efforts and advancements, this promising area of research is likely to achieve significant breakthroughs soon.

8 Conclusions

QDs offer tremendous potential in biomedical, environmental, and industrial applications, thanks to their exceptional optical and physicochemical properties. However, their toxicity and cytotoxicity remain significant hurdles that must be addressed to ensure their safe and effective use. A thorough understanding of the factors influencing QD toxicity—such as synthesis techniques, interactions with biological systems, and environmental conditions—is vital to mitigating these risks. Innovations in surface modification, the adoption of biocompatible materials, and the shift towards eco-friendly alternatives like graphene and silicon QDs provide promising strategies for reducing toxicity.

Moreover, in-depth research into the long-term effects of QDs, including their biodegradation, clearance, and interactions with complex biological systems, is crucial for advancing their clinical applications. Environmental concerns surrounding heavy-metal-based QDs underscore the importance of sustainable production, responsible use, and proper disposal. The development of standardized protocols for toxicity testing is equally essential for establishing consistent safety standards. By fostering interdisciplinary collaboration and driving continuous innovation, these challenges can be effectively addressed. This will pave the way for the safe and widespread use of QDs across various domains, ensuring their contribution to the advancement of nanotechnology while prioritizing human health and environmental safety.

References

1. Abdellatif AAH, Younis MA, Alsharidah M, Al Rugaie O, Tawfeek HM (2022) Biomedical applications of quantum dots: overview, challenges, and clinical potential. Int J Nanomedicine 17:1951–1970. https://doi.org/10.2147/IJN.S357980
2. Akhavan O, Ghaderi E, Emamy H, Akhavan F (2013) Genotoxicity of graphene nanoribbons in human mesenchymal stem cells. Carbon 54:419–431. https://doi.org/10.1016/j.carbon.2012.11.058
3. Baig MS, Suryawanshi RM, Zehravi M, Mahajan HS, Rana R, Banu A, Subramanian M, Kaundal AK, Puri S, Siddiqui FA, Sharma R (2023) Surface decorated quantum dots: synthesis, properties and role in herbal therapy. Front Cell Dev Biol 11:1139671. https://doi.org/10.3389/fcell.2023.1139671
4. Beus M, Pongrac IM, Capjak I, Ilić K, Vrček E, Ćurlin M, Milić M, Čermak AM, Pavičić I (2023) Particle surface functionalization affects mechanism of endocytosis and adverse effects

of silver nanoparticles in mammalian kidney cells. J Appl Toxicol 43:416–430. https://doi.org/10.1002/jat.4392
5. Bi J, Mo C, Li S, Huang M, Lin Y, Yuan P, Liu Z, Jia B, Xu S (2023) Immunotoxicity of metal and metal oxide nanoparticles: from toxic mechanisms to metabolism and outcomes. Biomater Sci 11:4151–4183. https://doi.org/10.1039/D3BM00271C
6. Bosch S, Botha TL, Wepener V (2023) Influence of different functionalized CdTe quantum dots on the accumulation of metals, developmental toxicity and respiration in different development stages of the zebrafish (*Danio rerio*). Front Toxicol 5:1176172. https://doi.org/10.3389/ftox.2023.1176172
7. Choi HS, Ipe BI, Misra P, Lee JH, Bawendi MG, Frangioni JV (2009) Tissue-and organ-selective biodistribution of NIR fluorescent quantum dots. Nano Lett 9:2354–2359. https://doi.org/10.1021/nl900872r
8. Cotta MA (2020) Quantum dots and their applications: what lies ahead? ACS Appl Nano Mater 3:4920–4924. https://doi.org/10.1021/acsanm.0c01386
9. Das K, Meena R, Gaharwar US, Priyadarshini E, Rawat K, Paulraj R, Mohanta YK, Saravanan M, Bohidar HB (2023) Bioaccumulation of CdSe quantum dots show biochemical and oxidative damage in wistar rats. Oxid Med Cell Longev 1:7707452. https://doi.org/10.1155/2023/7707452
10. Du Y, Zhong Y, Dong J, Qian C, Sun S, Gao L, Yang D (2019) The effect of PEG functionalization on the in vivo behavior and toxicity of CdTe quantum dots. RSC Adv 9:12218–12225. https://doi.org/10.1039/C9RA00022D
11. Geys J, Nemmar A, Verbeken E, Smolders E, Ratoi M, Hoylaerts MF, Nemery B, Hoet PH (2008) Acute toxicity and prothrombotic effects of quantum dots: impact of surface charge. Environ Health Perspect 116:1607–1613. https://doi.org/10.1289/ehp.1156
12. Gheshlaghi N, Pisheh HS, Karim MR, Ünlü H (2016) Interface strain effects on ZnSe/(CdSe) based type I and ZnSe/CdS type II core/shell quantum dots. Energy Proc 102:152–163. https://doi.org/10.1016/j.egypro.2016.11.330
13. Gidwani B, Sahu V, Shukla SS, Pandey R, Joshi V, Jain VK, Vyas A (2021) Quantum dots: Prospectives, toxicity, advances and applications. J Drug Delivery Sci Technol 61:102308. https://doi.org/10.1016/j.jddst.2020.102308
14. Hardman R (2006) A toxicologic review of quantum dots: toxicity depends on physicochemical and environmental factors. Environ Health Perspect 114:165–172. https://doi.org/10.1289/ehp.8284
15. He K, Liang X, Wei T, Liu N, Wang Y, Zou L, Bai C, Yao Y, Wu T, Kong L, Zhang T (2020) A metabolomics study: CdTe/ZnS quantum dots induce polarization in mice microglia. Chemosphere 246:125629. https://doi.org/10.1016/j.chemosphere.2019.125629
16. Heng BC, Zhao X, Tan EC, Khamis N, Assodani A, Xiong S, Ruedl C, Ng KW, Loo JS (2011) Evaluation of the cytotoxic and inflammatory potential of differentially shaped zinc oxide nanoparticles. Arch Toxicol 85:1517–1528. https://doi.org/10.1007/s00204-011-0722-1
17. Hines MA, Guyot-Sionnest P (1996) Synthesis and characterization of strongly luminescing ZnS-capped CdSe nanocrystals. J Phys Chem 100:468–471. https://doi.org/10.1021/jp9530562
18. Hossain N, Mobarak MH, Mimona MA, Islam MA, Hossain A, Zohura FT, Chowdhury MA (2023) Advances and significances of nanoparticles in semiconductor applications–a review. Results Eng 19:101347. https://doi.org/10.1016/j.rineng.2023.101347
19. Hu L, Zhang C, Zeng G, Chen G, Wan J, Guo Z, Wu H, Yu Z, Zhou Y, Liu J (2016) Metal-based quantum dots: synthesis, surface modification, transport and fate in aquatic environments and toxicity to microorganisms. RSC Adv 6:78595–78610. https://doi.org/10.1039/C6RA13016J
20. Jiang S, Lin Y, Yao H, Yang C, Zhang L, Luo B, Lei Z, Cao L, Lin N, Liu X, Lin Z (2018) The role of unfolded protein response and ER-phagy in quantum dots-induced nephrotoxicity: an in vitro and in vivo study. Arch Toxicol 92:1421–1434. https://doi.org/10.1007/s00204-018-2169-0
21. Jin Y, Zhao X (2008) Cytotoxicity of photoactive nanoparticles. In: Safety of nanoparticles: from manufacturing to medical applications, vol 20, pp 19–31. https://doi.org/10.1007/978-0-387-78608-7_2

22. Johns CE, Gattu M, Camilli S, Desaraju A, Kolliputi N, Galam L (2023) The cd/Zn axis: emerging concepts in cellular fate and cytotoxicity. Biomolecules 13:316. https://doi.org/10.3390/biom13020316
23. Karakoti AS, Shukla R, Shanker R, Singh S (2015) Surface functionalization of quantum dots for biological applications. Adv Colloid Interf Sci 215:28–45. https://doi.org/10.1016/j.cis.2014.11.004
24. Kargozar S, Hoseini SJ, Milan PB, Hooshmand S, Kim HW, Mozafari M (2020) Quantum dots: a review from concept to clinic. Biotechnol J 12:2000117. https://doi.org/10.1002/biot.202000117
25. Kim S, Fisher B, Eisler HJ, Bawendi M (2003) Type-II quantum dots: CdTe/CdSe (core/shell) and CdSe/ZnTe (core/shell) heterostructures. J Am Chem Soc 125:11466–11467. https://doi.org/10.1021/ja0361749
26. Kuznetsov D, Krylsky D, Dezhurov S, Grachev A, Neschisliaev V, Orlova E, Kuznetsova A (2023) Quantum dots are time bomb: multiscale toxicological study. Chem Biol Interact 374:110396. https://doi.org/10.1016/j.cbi.2023.110396
27. Le N, Kim K (2023) Current advances in the biomedical applications of quantum dots: promises and challenges. Int J Mol Sci 24:12682. https://doi.org/10.3390/ijms241612682
28. Le N, Zhang M, Kim K (2022) Quantum dots and their interaction with biological systems. Int J Mol Sci 23:10763. https://doi.org/10.3390/ijms231810763
29. Lewinski N, Colvin V, Drezek R (2008) Cytotoxicity of nanoparticles. Small 4:26–49. https://doi.org/10.1002/smll.200700595
30. Liang Y, Zhang T, Tang M (2022) Toxicity of quantum dots on target organs and immune system. J Appl Toxicol 42:17–40. https://doi.org/10.1002/jat.4180
31. Lin PI, Cardenas A, Rifas-Shiman SL, Zota AR, Hivert MF, Aris IM, Sanders AP (2023) Non-essential and essential trace element mixtures and kidney function in early pregnancy—a cross-sectional analysis in project viva. Environ Res 216:114846. https://doi.org/10.1016/j.envres.2022.114846
32. Lin X, Chen T (2023) A review of in vivo toxicity of quantum dots in animal models. Int J Nanomedicine 31:8143–8168. https://doi.org/10.2147/IJN.S434842
33. Mahto SK, Park C, Yoon TH, Rhee SW (2010) Assessment of cytocompatibility of surface-modified CdSe/ZnSe quantum dots for BALB/3T3 fibroblast cells. Toxicol Vitro 24:1070–1077. https://doi.org/10.1016/j.tiv.2010.03.017
34. Mao Z, Zhou X, Gao C (2013) Influence of structure and properties of colloidal biomaterials on cellular uptake and cell functions. Biomater Sci 9:896–911. https://doi.org/10.1039/C3BM00137G
35. Moon H, Lee C, Lee W, Kim J, Chae H (2019) Stability of quantum dots, quantum dot films, and quantum dot light-emitting diodes for display applications. Adv Mater 31:1804294. https://doi.org/10.1002/adma.201804294
36. Mu Q, Su G, Li L, Gilbertson BO, Yu LH, Zhang Q, Sun YP, Yan B (2012) Size-dependent cell uptake of protein-coated graphene oxide nanosheets. ACS Appl Mater Interfaces 4:2259–2266. https://doi.org/10.1021/am300253c
37. Nadeau J (2013) Quantum dot reactive oxygen species generation and toxicity in bacteria: mechanisms and experimental pitfalls. In: Quantum dot sensors: technology and commercial applications, vol 24, pp 1–6. https://doi.org/10.1201/b14595-7
38. Nguyen KC, Zhang Y, Todd J, Kittle K, Patry D, Caldwell D, Lalande M, Smith S, Parks D, Navarro M, Massarsky A (2019) Biodistribution and systemic effects in mice following intravenous administration of cadmium telluride quantum dot nanoparticles. Chem Res Toxicol 32:1491–1503. https://doi.org/10.1021/acs.chemrestox.8b00397
39. Nguyen KC, Zhang Y, Todd J, Kittle K, Lalande M, Smith S, Parks D, Navarro M, Tayabali AF, Willmore WG (2020) Hepatotoxicity of cadmium telluride quantum dots induced by mitochondrial dysfunction. Chem Res Toxicol 33:2286–2297. https://doi.org/10.1021/acs.chemrestox.9b00526
40. Nikazar S, Sivasankarapillai VS, Rahdar A, Gasmi S, Anumol PS, Shanavas MS (2020) Revisiting the cytotoxicity of quantum dots: an in-depth overview. Biophys Rev 12:703–718. https://doi.org/10.1007/s12551-020-00653-0

41. Niveria K, Singh P, Yadav M, Verma AK (2023) Quantum dot (QD)-induced toxicity and biocompatibility. In: Handbook of II–VI semiconductor-based sensors and radiation detectors: materials and technology, vol 1. Springer International Publishing, Cham, pp 181–211. https://doi.org/10.1007/978-3-031-19531-0_8
42. Patil J, Bhattacharya S (2024) Exploring the potential of quantum dots and plasmonic nanoparticles for imaging and phototherapy in colorectal neoplasia. Results Chem 31:101689. https://doi.org/10.1016/j.rechem.2024.101689
43. Reshma VG, Mohanan PV (2019) Quantum dots: applications and safety consequences. J Lumin 205:287–298. https://doi.org/10.1016/j.jlumin.2018.09.015
44. Reshma VG, Sabareeswaran A, Rajeev KS, Mohanan PV (2020) In vitro and in vivo toxicity analysis of zinc selenium/zinc sulfide (ZnSe/ZnS) quantum dots. Food Chem Toxicol 145:111718. https://doi.org/10.1016/j.fct.2020.111718
45. Rivera-Gil P, Jimenez De Aberasturi D, Wulf V, Pelaz B, Del Pino P, Zhao Y, De La Fuente JM, Ruiz De Larramendi I, Rojo T, Liang XJ, Parak WJ (2013) The challenge to relate the physicochemical properties of colloidal nanoparticles to their cytotoxicity. Acc Chem Res 46:743–749. https://doi.org/10.1021/ar300039j
46. Ryan J, Jacob P, Lee A, Gagnon Z, Pavel IE (2022) Biodistribution and toxicity of antimicrobial ionic silver (Ag^+) and silver nanoparticle ($AgNP^+$) species after oral exposure, in Sprague-Dawley rats. Food Chem Toxicol 166:113228. https://doi.org/10.1016/j.fct.2022.113228
47. Sahu SC, Casciano DA (eds) (2009) Nanotoxicity: from in vivo and in vitro models to health risks. Wiley
48. Scoville DK, Nolin JD, Ogden HL, An D, Afsharinejad Z, Johnson BW, Bammler TK, Gao X, Frevert CW, Altemeier WA, Hallstrand TS (2019) Quantum dots and mouse strain influence house dust mite-induced allergic airway disease. Toxicol Appl Pharmacol 368:55–62. https://doi.org/10.1016/j.taap.2019.01.018
49. Sukhanova A, Bozrova S, Gerasimovich E, Baryshnikova M, Sokolova Z, Samokhvalov P, Guhrenz C, Gaponik N, Karaulov A, Nabiev I (2022) Dependence of quantum dot toxicity in vitro on their size, chemical composition, and surface charge. Nano 12:2734. https://doi.org/10.3390/nano12162734
50. Tarantola M, Pietuch A, Schneider D, Rother J, Sunnick E, Rosman C, Pierrat S, Sönnichsen C, Wegener J, Janshoff A (2011) Toxicity of gold-nanoparticles: synergistic effects of shape and surface functionalization on micromotility of epithelial cells. Nanotoxicology 5:254–268. https://doi.org/10.3109/17435390.2010.528847
51. Wagner AM, Knipe JM, Orive G, Peppas NA (2019) Quantum dots in biomedical applications. Acta Biomater 94:44–63. https://doi.org/10.1016/j.actbio.2019.05.022
52. Wang L, Wang Y, Xu T, Liao H, Yao C, Liu Y, Li Z, Chen Z, Pan D, Sun L, Wu M (2014) Gram-scale synthesis of single-crystalline graphene quantum dots with superior optical properties. Nat Commun 5:5357. https://doi.org/10.1038/ncomms6357
53. Wang Y, Pang S, Chen Z, Wang J, Liu L, Zhang L, Wang F, Song M (2023) Surface modification determines the distribution and toxicity of quantum dots during the development of early staged zebrafish. Environ Sci Technol 57:10574–10581. https://doi.org/10.1021/acs.est.3c01949
54. Yang HL, Bai LF, Geng ZR, Chen H, Xu LT, Xie YC, Wang DJ, Gu HW, Wang XM (2023) Carbon quantum dots: preparation, optical properties, and biomedical applications. Mater Today Adv 18:100376. https://doi.org/10.1016/j.mtadv.2023.100376
55. Yao Y, Chen Z, Zhang T, Tang M (2022) Adverse reproductive and developmental consequences of quantum dots. Environ Res 213:113666. https://doi.org/10.1016/j.envres.2022.113666
56. Yuan Y, Riehle FS, Gu H, Thomann R, Urban G, Krüger M (2010) Critical parameters for the scale-up synthesis of quantum dots. J Nanosci Nanotechnol 10:6041–6045. https://doi.org/10.1166/jnn.2010.2564
57. Yue H, Wei W, Yue Z, Wang B, Luo N, Gao Y, Ma D, Ma G, Su Z (2012) The role of the lateral dimension of graphene oxide in the regulation of cellular responses. Biomaterials 33:4013–4021. https://doi.org/10.1016/j.biomaterials.2012.02.021
58. Zhang LW, Bäumer W, Monteiro-Riviere NA (2011) Cellular uptake mechanisms and toxicity of quantum dots in dendritic cells. Nanomedicine 6:777–791. https://doi.org/10.2217/nnm.11.73

59. Zhang Y, Qin C, Wang H, Hu X, Gao Y (2024) Occurrence, behavior, and ecotoxicity of quantum dots in aquatic environments. Environ Sci Nano 11:4044–4059. https://doi.org/10.1039/D4EN00542B
60. Zhao L, Zong W, Zhang H, Liu R (2019) Kidney toxicity and response of selenium containing protein-glutathione peroxidase (Gpx3) to CdTe QDs on different levels. Toxicol Sci 168:201–208. https://doi.org/10.1093/toxsci/kfy297
61. Zhu C, Chen Z, Gao S, Goh BL, Samsudin IB, Lwe KW, Wu Y, Wu C, Su X (2019) Recent advances in non-toxic quantum dots and their biomedical applications. Progr Nat Sci Mater Int 29:628–640. https://doi.org/10.1016/j.pnsc.2019.11.007
62. Zhu M, Nie G, Meng H, Xia T, Nel A, Zhao Y (2013) Physicochemical properties determine nanomaterial cellular uptake, transport, and fate. Acc Chem Res 46:622–631. https://doi.org/10.1021/ar300031y

Part III
Quantum Dots for Environmental Science

Chapter 9
Quantum Dots for Detection of Environmental Contaminants

Shweta Kaushal, Richa Sharma, Dimpal Chauhan, and Kuldeep Kumar

Abstract The growing global crisis due to toxic metal ions, dyes, polyaromatic hydrocarbons, excess use of pesticides, herbicides, and many more pollutants is one of the prominent concerns for human health, the environment, and the ecosystem. Therefore, it is the utmost requirement of society to develop an easy, simple, and sensitively accurate tool that responds quickly to the detection of contaminants. Quantum dots possess unique properties due to the presence of exciton pairs, making them reveal exceptional optical and electronic characteristics. This chapter aims to explore the distinct properties, types of quantum dots, and their application in monitoring or detecting contaminants. Special attention is given to the fluorescence, electrochemical, and colorimetric-based sensors for monitoring. The chapter also highlights the advantages like real-time monitoring, versatility, multifunctionality, and challenges of quantum dots-based sensors along with the recent advances in quantum dot technology for environmental applicability. Finally, this chapter addresses the future perspectives for developing technologically favorable quantum dots-based sensors for applicability in real-world problems.

Keywords Contaminants · Fluorescence · FRET · Sensors · Quantum dots

---

S. Kaushal · K. Kumar
Department of Chemistry, Career Point University, Hamirpur, Himachal Pradesh, India

S. Kaushal · K. Kumar (✉)
Centre for Nano-Science and Technology, Career Point University, Hamirpur, Himachal Pradesh, India
e-mail: kuldeep.sharma.753@gmail.com

R. Sharma
Department of Chemistry, S.D.W.G. Government College Beetan, Una, Himachal Pradesh, India

D. Chauhan
Department of Chemistry, Government College Jukhala, Bilaspur, Himachal Pradesh, India

T. Bhuyan et al. (eds.), *Multifunctional Quantum Dots*, Materials Horizons: From Nature to Nanomaterials, https://doi.org/10.1007/978-981-92-0172-3_9

1 Introduction

Regardless of the well-established financial standing of nations, it is widely acknowledged that contaminants pose a serious threat to the environment. Environmental contaminants are those toxic substances that pollute the environment, ecosystem, wildlife, and human health. In the past decades, advancements in technology, agriculture, and the medical sector have led to improvements in human lifestyles, but they are causing problems by developing new contaminants popularly known as emerging contaminants (ECs). The origin of these ECs can be from a natural source or a chemical route that risks the environment and human life [62]. The majority of these contaminants remained unidentified for the last 15–20 years due to their regular use in everyday life and industrial activities [84]. The ECs are classified under many categories like pharmaceuticals, personal care products (PCPs), industrial chemicals, disinfectant by-products, toxic algae, biocidal metabolites, and disruption devices [2]. The essential need for some of these substances in contemporary society increases the output of these products by up to 500 million tons [43]. Their extensive monitoring in the environment has risen to a critical problem. Therefore, various detection techniques have been developed to answer this requirement. However, nanotechnology has tremendous potential to tightly link the detection and remediation of contaminants with the help of certain materials possessing unique chemical and physical properties.

Many interesting materials have been discovered in the domains of science, serving as efficient tools for environmental contamination. Quantum Dots (QDs) are nano-sized semiconductor crystals extensively used for LEDs, photodetectors, photovoltaics, and biological applications. In the 1980s, Alexei Ekimov, a Russian physicist, discovered QDs with dimensions even smaller than exciton radius [40]. The electrons are free to move in any direction when present in bulk semiconductor material while this movement of electrons is ceased due to the quantum confinement effect of nanomaterials. The extensive demand for QDs is due to their discrete energy levels and three-dimensional electron confinement at electronic levels, enabling them to show unique optical features [45]. QDs are a subgroup of nanomaterials that can be prepared using a single element or compound material. Generally, Si, Ge, and compounds from Group II-VI, III-V have pioneer semiconductor materials. The characteristic properties of QDs are small size (typically 1 to 10 nm), exceptional fluorescent quantum yield, high photostability, size-dependent emission, and blinking [7]. QDs have been used as sensors for many years due to their distinct properties. Various studies were cited in the literature—carbon-based QDs were used for the detection of various species, including pollutants, pesticides, proteins, and metal ions [59]. Zero-dimensional graphene QDs were another excellent example of sensors used in the biomedical field for diagnostics and imaging-like purposes [53]. In addition, inorganic QDs, namely CsPbBr3 were used as sensing elements for insecticides like omethoate [33].

2 Properties of Quantum Dots

2.1 *Optical Properties*

The optical properties of QDs are due to their confined size and discrete energy levels, which allow them to exhibit exceptional absorption and emission spectra. For example, Ramvall et al. [72] worked on the GaN QDs synthesized using the metalorganic vapor phase. The change from 2D to 3D and red shift in the photoluminescence was observed upon deposition of Si on AlGaN due to compressed excitonic radius. Bhattacharjee et al. [9] tuned the optical properties of carbon QDs employing near-field effects. The change introduced in the local states of assembly due to the near-field environment creates tunable optical characteristics. The study by Mittelstädt et al. [57] presented the alloying of antimony in the QDs, allowing improved carrier dynamics due to the hetero-dimensional confinement of holes and electrons. Hu et al. [30] also reported the tuning of photoluminescence properties of traditional dots via varying the reagents and reaction conditions. The resulting product was versatile and could be applied in white light. In addition to this, Sengupta et al. [78] chemically coupled ZnSe and CdS QDs to change their electronic characteristics, impacting their emission wavelength. The asymmetric coupling of different materials provides data storage and computational device applications. The study by Kandi et al. [38] also emphasized the modification of CdS QDs with $BiOI/Bi_2MoO_6$ for remediating the environment from antibiotics. The modified nanocomposite exhibited enhanced degradation performance against common antibiotics. These studies highlighted the impact of various parameters that efficiently improve the photoluminescence, fluorescence, and emission-like optical properties of QDs.

2.2 *Surface Functionalization*

The surface functionalization of QDs is the modification of the surface of QDs by varying the synthesis parameters, reagents, solvents, and environmental conditions, which make them exhibit enhanced properties and applications. Bilan et al. [10] reviewed the performance of QDs in biomedical applications by comparing the traditional fluorescent dyes and proteins. Mohagheghpour et al. [58] synthesized Alendronate-functionalized graphene QDs with 57% quantum yield to detect As (III) ions in nanomolar levels. The quenching of fluorescence intensity of functionalized QDs due to metal-ligand interaction emphasized the sensitivity and selectivity of the sensor probe for efficient applicability. The alterations in brightness, stability, bleaching, and broad absorption with tuned emission wavelength after surface modifications with bioactive molecules make them susceptible to bioimaging, targeting, nanocarriers, and detection-like applications. Li et al. [47] presented the Si QDs as prominent sensors after modifying their surface chemistry with NH_2, OH, CH_3,

SH, and H. The deformations, as confirmed by density functional calculations, introduced in the structural and optical properties could be used for the formation of unique devices. The study by Chandra et al. [11] investigated the variation of photoluminescence intensity after adding divalent cations on the surface of carbon QDs. The biodegradable polymer used for the surface modification via microwave-assisted synthesis tuned the excitation and emission properties marking the product as an important probe for sensing and bioimaging. In another study by Qian et al. [70], the emission color of graphene QDs was changed using small organic molecules containing OH, NH_2, and SH groups. The amine group modification exhibited 17.6% emission efficiency of graphene QDs. The semiconductor QDs are also regarded as efficient tools in the field of biomedical and environmental applications. The impact of surface modifications including various mechanisms leads to improved stability, biocompatibility, particle size, and oxidation factors of QDs [28]. Therefore, systematic surface modification is one of the efficient approaches to altering the properties of QDs for desirable applications.

2.3 *Quantum Yield*

The characteristics associated with the quantum yield of QDs have also endowed these materials with one more useful merit for practical applications. The efficiency and sensitivity of QDs are crucial assets in sensing, targeting, and imaging. For instance, Qu et al. [71] worked on the simple hydrothermal-based synthesis of graphene QDs showing a quantum yield of 23.6%. The tendency of high-yielded graphene QDs to up-converse the light helped it to degrade the methyl orange dye up to 95% when associated with TiO_2 nanotubes. Additionally, Yang et al. [96] synthesized ZnO QDs using an ultrasonic microreactor method yielding 65% QDs which in turn impacted the photoluminescence properties, types, and the number of defects in ZnO QDs. Huang et al. [31] observed the change in emission color just by changing the temperature of the solvent during perovskite QDs synthesis. The increase of 93% emission quantum yield resulted in the variation of color during emission.

3 Types of Quantum Dots Used for Environmental Detection

The salient features of QDs marked their invention as potent candidates for research, and the pictorial representation of types of QDs is shown in Fig. 1. Generally, the QDs can be classified based on (a) structure and (b) constituent material for synthesis:

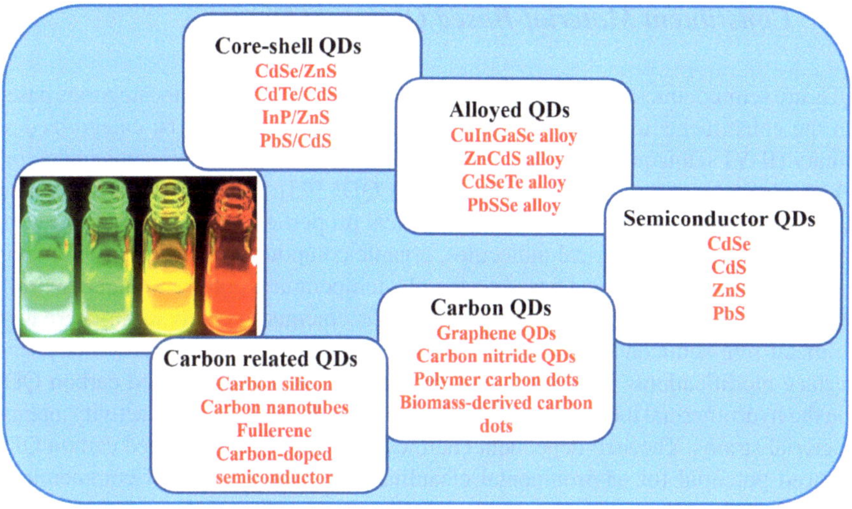

Fig. 1 Different types of QDs

3.1 Structure-Based Quantum Dots

The classification of QDs based on the structure can be further categorized as core-shell and alloyed QDs. The QDs are quantum-mechanically confined materials, so they need a strategy to sustain their confinement. One of the methods is to encapsulate the core with an organic surfactant or inorganic material to produce core-shell QDs. This increases the efficiency due to enhancement in electron-hole confinement and stabilizing the dangling bonds. The core-shell QDs were seen to improve the absorption coefficients and threshold energy-like characteristics due to the capping strategy [89]. The study by Selopal et al. [77] also emphasized the impact of core-shell QDs on the efficiency of solar cells. The reduction in surface defects of QDs after core encapsulation leads to engineering the band alignment and interface properties. Similarly, Nandan and Mehata [63] worked on tuning the wavefunction of excitons by increasing the thickness of the shell. As the thickness of the shell increases, the absorption shifts towards a lower wavelength, resulting in the blue shift. Additionally, Bailey and Nie [6] demonstrated the importance of CdSeTe alloyed QDs in advancing the electronic and optical properties for their usage as biomarkers or imaging. The quantum yield of 60% and red shift in absorption make them potent near-infrared fluorescence probes for deep tissue penetration-like purposes. Similarly, Yang et al. [95] designed AgAuSe colloidal alloyed QDs that showed emission in the second NIR region with a photoluminescence quantum yield of 65.3%. Suppressing vacancies and defects after alloying Au in QDs increases the lifetime and yield of alloyed QDs. Therefore, these studies underscore the crucial knowledge related to the motion of exciton pairs, alloying with noble metal atoms, and surface defects like parameters for synthesizing specific nanostructured semiconductor QDs.

3.2 Constituent Material-Based Quantum Dots

QDs are semiconductor materials that can be further divided into subcategories based on the constituents used for synthesis. Semiconductor QDs can be categorized as binary (II-VI semiconductors like CdSe and CdS), ternary (I-III-VI semiconductors like $CuInS_2$), and perovskite (like $CsPbBr_3$) QDs [93]. Semiconductor QDs have wide applications in sensors due to their optical properties, helping them to serve as sensing probes for biological molecules, organic compounds, ions, etc. The sensing arrays of QDs can be varied based on the pH, temperature, outer, and inner environment of the analyte [19]. Similarly, carbon QDs emerged recently as an alternative to metal-free semiconductor QDs due to their biocompatibility, less toxicity, and surface modifications. Travlou et al. [88] synthesized S and N-doped carbon QDs via the hydrothermal method, which exhibited excellent antibacterial activity against bacterial strains. The size-dependent chemical characteristics of doped carbon QDs showed potential for environmental cleanliness. The breakdown of commercially available and cheap starting material carbon fiber into graphene QDs of 1–4 nm in size showed process-based alterations and photoluminescence [67]. These graphene QDs with nanosecond lifetimes and luminescent stability will be applicable in biosensing. The study by Siddique et al. [82] worked on three variants of 0D carbon materials. The striking differences and similarities related to excitation-emission or photoluminescence properties of amorphous, graphene, and N-doped graphene QDs hint at important evidence related to the light emission mechanism and development of efficient carbon-based probes. Therefore, constituent-based synthesis of QDs will be an effective approach for detecting and eliminating environmental contaminants soon.

4 Detection of Contaminants Using Quantum Dots

4.1 Fluorescence-Based Detection

The detection of contaminants via sensitizing the luminescence of QDs after chemical or physical interactions with the target moiety is one of the methods of detection where QDs act as fluorescence probes. The mechanism of the fluorescence utilized detection is based on the energy flow among fluorophore QDs and analyte, which is popularly known as Forster Resonance Energy transfer (FRET) (Fig. 2). The efficiency of the FRET mechanism depends majorly on the distance, spatial arrangement, spectral overlap of the donor-acceptor pair, and fluorescence lifetime of QDs [64]. The study by Adegoke et al. [3] presented the use of CdSeTeS/ZnS alloyed QDs as a fluorescent sensor after their combination with graphene oxide. The developed nanocomposite showed a switch-off mode when alloyed QDs covalently interacted with the graphene oxide. However, the fluorescence intensity was enhanced after the adsorption of polyaromatic pollutants on the surface of graphene-incorporated QDs.

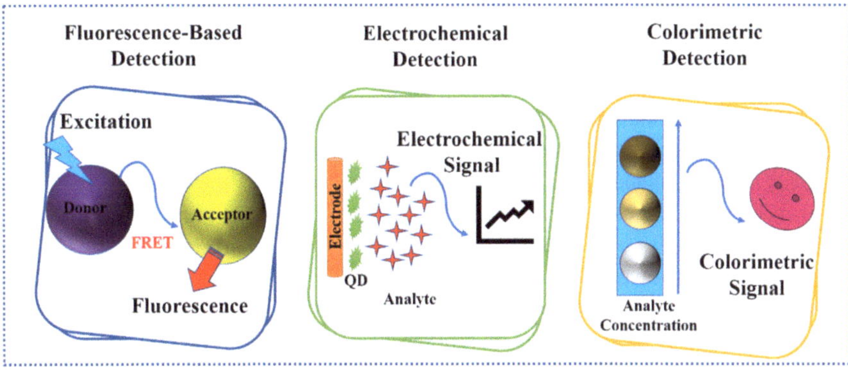

Fig. 2 Different methods of detection of contaminants using QDs

The pi-pi interactions constructed the switch-on mode of the probe and exhibited the affinity against four pollutants. Huang and Tong [32] also worked on the fabrication of a dual-functional fluorophore for the detection of tetracycline and Fe^{3+} ions. The green-colored CdTe QDs after modification with Eu^{3+} quenched the intensity which was restored by the adsorption of tetracycline. While the blue-colored boron-doped carbon QDs were synthesized with CdTe for fluorescence to Fe^{3+}. The synthesized fluorescent probe showed potential in detecting tetracycline and Fe^{3+} present in water. Kaur et al. [41] also synthesized the chitosan-capped ZnS QDs nanohybrid with nickel ferrites to detect pharmaceutical drugs. The practical applicability of the developed QDs sensor was evaluated on a real water sample, and the efficacy of the probe was affirmed. The recyclable stable nanohybrids offer a promising strategy for environmental remediation when used as fluorescent sensors. Therefore, strong luminescence, high surface area, emission/excitation wavelengths, and lifetime-like intrinsic properties of QDs led to the foundation of a fluorescence-based detection mechanism.

4.2 Electro-Chemical Detection

QDs also emerged as frontier electrochemical sensors due to their electronic properties, size tunability, chemically modified surface area, and interfacial properties. The utilization of the electron-hole pair generated after the excitation of QDs with the electrode surface helps in electrochemical sensing (Fig. 2). The mechanism is based on the injection of excited electrons from the conduction band into the electrode, leading to the generation of current [20]. The advantages of using QDs in electrochemical sensors are related to their role as pumps between the electrode and redox agent, blocking of excess substances to minus dark current, and easy coupling of QDs with chemical moieties [97]. In this context, Feminus and Deepa [17] synthesized CdSe QD/Meso-Silica/rGO composite for the electrochemical sensing of antioxidants. The difference in the electron transfer properties and surface area as compared to unmodified electrodes helped the composite to detect phenolic antioxidants at higher working potentials. The study by Wahab et al. [90] also emphasized the potential of Al_2O_3 QDs as detecting candidates for urea solution. The group synthesized QDs using the hydrothermal method and tested various attributes using different characterizations. The Al_2O_3 QDs enhanced the electrochemical impedance, showing their utilization as sensor material for tap water samples also. Bhanjana et al. [8] reported the fabrication of highly sensitive SnO_2 QDs serving as electrochemical sensors for cadmium ions. The Cd^{2+} ions were sensed via gold electrode modification with SnO_2 QDs as electron mediators. The highly sensitive mediator showed less than 2 secs response time to sense the ions. Therefore, the optical and electronic properties due to the 3D confinement effect of QDs made these semiconductor materials show potential in electrochemical sensing despite several challenges.

4.3 Colorimetric Detection

Among distinct sensing techniques for contaminants, colorimetric sensors gained interest due to simple, rapid, and on-site operation without the need for extra instruments or detectors. The real-time detection of contaminants with the naked eye is a simple and cost-effective approach that finds utility in practical applications (Fig. 2). Sawetwong et al. [76] synthesized Mn-ZnS QDs embedded on imprinted polymers for the detection of plant-killing herbicide popularly known as glyphosate. The binding selectivity of glyphosate on the cavities of the sensor mediates the color changes during the detection. The change from dark to light green color after oxidation of ABTS is dependent on the concentration of the analyte. Similarly, Han et al. [23] worked on the Eu^{3+}-doped CdTe QDs sensor, which exhibited dual-functional properties with multicolor variations. The sensor performed the detection of tetracycline residues in milk efficiently. On addition of the analyte to QDs, the output appeared in the form of a change in color from green to yellow, and lastly to red. This methodology showed applicability in food and environmental safety. The study by Arslan et al. [5] developed visible light-emitting Si QDs decorated on nanofibrous mats for colorimetric detection of trinitrotoluene. The surface functionalities of the Si QDs allowed the dressing of Nylon 6, 6 mats in a thin manner. The interactions between the aromatic ring of the analyte and the NH_2 group of Si QDs revealed selective detection in lower concentrations. The fluorescent properties of molecules like organic dyes, polymers, carbon dots, etc. can help to achieve colorimetric sensing. The on-off mode of the fluorescent entity-based sensors was advantageous due to stability and quantum yield [48].

5 Applications of Quantum Dots in Environmental Contaminant Detection

5.1 Heavy Metal Detection

Heavy metals (Pb^{2+}, Cd^{2+}, Hg^{2+}, As^{3+}, Fe^{2+}, Zn^{2+}, Al^{3+}, Co^{2+}, Ni^{2+}, etc.) are the metals that have a specific density of >5 g/cm3 and severely influence the environment and living creatures [34]. Industrial effluents, agriculture, mining, the use of insecticides or pesticides in soil, and many other human activities have posed a significant risk to the environment as they introduce toxic heavy metal ions into the soil, air, and water [100]. Mercury (Hg) is recognized by WHO as a dangerous pollutant due to its severe environmental and health impacts, including DNA damage and CNS impairment. Lead toxicity involves oxidative stress and ionic disruption, replacing essential cations and damaging cellular metabolism. Cadmium is a highly toxic metal that causes oxidative stress and disrupts enzymatic systems, leading to liver and kidney damage. Arsenic targets sulfhydryl groups in cells, impairing respiration, enzymes, and mitosis [34, 98]. Therefore, monitoring their levels over time is a crucial factor due to their non-biodegradable nature and toxicity even at trace levels. For example, ultra-trace levels of toxic metals such as Hg, As, etc. can cause catastrophic health effects, including impairment to the bones and other vital organs like the liver, kidney, etc., and the central nervous system [15]. These metal ions bind to the proteins, resulting in the malfunctioning of cells [34]. Detection of these heavy metal ions involves highly sensitive and selective technologies like atomic absorption spectroscopy, optical emission spectroscopy, mass spectrometry techniques, etc. However, they are quite expensive. In the field of heavy metal ion detection, sensing by nanomaterials has played a crucial role. Among nanomaterials, QDs are one of the most popular sensing materials due to their minuscule catalytic sizes that contribute to low detection limits, quick reaction times, excellent specificity, and anti-pathogenic properties. They display unique chemical and physical properties, such as large surface area, chemical affinity, and surface activity, which aid the sensing processes [54].

5.2 Detection of Pesticides, Herbicides, and Organic Pollutants

In recent years, QDs have been increasingly utilized for fluorescence-based sensing of pesticides; herbicides; and other organic pollutants like pharmaceuticals, industrial effluents, organic dyes, etc. These QDs are synthesized from organometallic precursors and are typically protected by a capping layer of organic ligands. The ligand's functional groups enable surface chemistry modifications, allowing QDs to be tailored for specific applications and coupled with receptors for selective target analyte detection [64]. C-dots synthesized from gelatin via a hydrothermal method were used as stabilizing agents in the development of AgNP-based colorimetric sensors for detecting different pesticides. These C-dots exhibited strong fluorescence (410 nm) and absorption (315 nm). The Ag/C-dot probe displayed color and absorbance changes upon pesticide addition, with detection efficiency in the order: imidacloprid > atrazine > chlorpyrifos [61]. Different findings by researchers have shown that imidacloprid binds strongly to CQDs through its -NO_2 and -NH groups, disrupting the self-complexation of the CQDs and causing them to transit into a more isolated state, which enhances their fluorescence. Similarly, thiabendazole, a benzimidazole fungicide, boosts the fluorescence of CQDs embedded in silica molecularly imprinted polymers [13]. Whereas studies by Mandal et al. [52] have shown that atrazine, chlorpyrifos, lindane, and tetradifon quench the fluorescence of CQDs. Different research groups have developed fluorescence sensors using functional materials based on QDs incorporated into silica nanoparticles to detect various pesticides like pyrethroids, Methomyl [46], parathion-methyl [94]. Mousa et al. developed a unique boron and nitrogen-doped carbon QDs-based fluorescent nano-sensor for the detection of antibiotics and phenol derivatives.

5.3 Water Quality Monitoring

In wastewater treatment, QDs are incredibly versatile. They are highly sensitive probes for monitoring water quality, strong photocatalytic agents for destroying organic pollutants, and excellent materials for the adsorption of contaminants [42] [50]. designed hexagonal boron nitride QDs as a nanozyme sensor to detect pathogens, leading to degradation of the water quality. The layer-by-layer assembly of synthesized QDs over the agarose polymer makes easy accessibility of QDs at the solid surface after the degradation of glycoside bonds of the polymer. The appearance of a blue color after the oxidation process signified the colorimetric detection of *Escherichia coli* in spiked water by a low-cost sensing strategy. The study by Pandit et al. [65] also worked on the problem caused by *E. coli* in contaminating potable water. The synthesis of biosensors based on the magnetic ions and Aptamer conjugated with CdTe-MPA QDs was applied for the on-spot detection of bacteria. The synthesized material was able to detect even small bacterial counts with high resolution when coupled with a photodiode and a convex lens. This approach to developing biosensors overcomes the limitations offered by conventional methods. Sharma et al. [81] developed ZnS QDs for the remediation of wastewater using a sustainable approach. The quantum-mechanically confined QDs possessed enhanced light absorption that works excellently by varying parameters, including pH and temperature. The modifications, like doping and heterojunction strategy, were also explored by the group for engineering the band gap and migration ability. Therefore, the time-consuming, cost-ineffective, and not able to detect on-spot like drawbacks of conventional methods can be effectively improved using QDs-based monitoring.

5.4 Air Quality Monitoring

QDs have emerged as promising technologies for air quality monitoring, offering several advantages such as high sensitivity, low cost, and ease of integration into real-time monitoring networks. These sensors can detect a variety of airborne pollutants, including particulate matter, volatile organic compounds (VOCs), and toxic gases like NO_2, SO_2, and O_3, making them highly valuable for assessing air pollution levels [79]. Mitri et al. [56] fabricated lead sulfide colloidal QDs for the detection of NO_2 due to the versatile functionalization of the synthesized gas sensor. The detection range of the sensor was up to 2.8 to 100 ppm. The extensive selectivity, repeatability, and response time of 12 secs with recovery within 26 min were achieved using colloidal QDs. QDs can detect VOCs through mechanisms such as photoluminescence quenching, energy transfer, or surface interactions. When VOC molecules interact with the QDs, changes in their emission intensity or wavelength occur, enabling the detection and quantification of pollutants. Ferrero et al. [18] developed a portable instrument for monitoring toxic VOCs in atmospheric conditions using photoluminescence changes in QDs embedded in a sol-gel matrix. This sol-gel material exhibited long-lived phosphorescence, quenched by trace levels of gaseous acetone. Researchers developed a highly efficient sensing system using N-doped graphene QDs to detect VOCs at room temperature. These QDs exhibited strong fluorescence, which was selectively quenched in the presence of specific VOCs, enabling their detection without the need for elevated temperatures or complex instrumentation [22]. Similarly, ZnS QDs can act as efficient acetone sensors. They have exceptional sensitivity, selectivity, and stability in detecting acetone even at low concentrations, making them ideal for monitoring health-hazardous gases in diverse environments [55].

6 Advantages and Challenges of Quantum Dot-Based Sensors

6.1 Advantages

6.1.1 High Sensitivity and Selectivity

QDs are renowned for their exceptional sensitivity and selectivity in diverse sensing applications. Their ability to facilitate single-molecule detection enhances their selectivity and sensitivity in analytical applications. The large surface area of QDs supports the attachment of various biomolecules like DNA, microRNAs, proteins, enzymes, small molecules, and viruses, amplifying their functionality in biosensing technologies. They are exceptionally sensitive in detecting various analytes, like heavy metal ions, herbicides, pesticides, pollutants, etc. [14, 37]. Moreover, they exhibit tunable optical properties which can be precisely controlled by adjusting their size and composition [12].

6.1.2 Versatility in Different Environmental Matrices (Air, Water, Soil)

The ability of QDs to function in diverse environmental matrices-soil, water, and air, makes them valuable tools for addressing pressing ecological challenges. In environmental monitoring. They have emerged as highly efficient photocatalysts due to their unique optical and electronic properties, including tunable band gaps, high surface-to-volume ratios, and excellent photostability. QDs have demonstrated exceptional potential in various environmental applications, including pollutant degradation, water purification, and toxic gas reduction, showcasing their versatility and effectiveness under diverse conditions [36].

Advances in their synthesis, such as greener and more sustainable methods, have further enhanced their functionality and reduced their environmental impact. By leveraging sunlight as a renewable energy source, QDs facilitate efficient water splitting, presenting a viable pathway for green hydrogen generation. Advances in synthesis techniques, including eco-friendly and cost-effective methods, have further improved their performance while addressing environmental concerns [73].

6.2 *Challenges*

6.2.1 Toxicity Concerns, Especially with Cadmium-Based QDs

Though QDs offer substantial benefits in various technological fields. However, their potential toxicity poses a significant challenge. Many QDs are composed of heavy metals like cadmium, which can leach into the environment and pose risks to human health and ecosystems. Additionally, their persistence in the environment and potential bioaccumulation raise concerns about long-term impacts. The release of toxic cadmium ions and subsequent effects on aquatic organisms such as bioaccumulation, oxidative stress, and disruption of vital biological processes, cannot be overlooked. Exposure to natural sunlight further exacerbates these toxic effects, as ultraviolet radiation can accelerate the degradation of QDs, increasing the release of harmful cadmium ions [83, 91]. Efforts to mitigate these concerns include the development of biocompatible coatings, controlled release mechanisms, and alternative synthesis methods using fewer toxic materials [92]. Additionally, rigorous toxicity evaluation protocols and lifecycle assessments are crucial for ensuring their responsible use [29].

6.2.2 Environmental Degradation and Stability Issues

While QDs offer promising applications across multiple fields, their environmental degradation and stability issues necessitate careful consideration. Continuous light exposure can lead to photobleaching, causing degradation of quantum dot properties, which affects their performance in display systems. QDs can degrade in the presence of reactive oxygen species (ROS) and biological fluids, leading to the release of toxic ions (e.g., cadmium or lead in CdSe or PbSe QDs), which may harm cells. Moreover, their tendency to aggregate, especially in solution or thin films, can lead to reduced quantum efficiency and less effective display performance. QDs, especially those based on materials like CdSe, can suffer from degradation due to environmental factors like oxygen, moisture, temperature fluctuations, and extreme pH conditions [60, 101].

6.2.3 Cost and Scalability for Widespread Application

The cost and scalability of QDs are pivotal factors influencing their widespread adoption across various industries, including displays, photovoltaics, and biomedical applications. Addressing these aspects requires a comprehensive understanding of the challenges inherent in current production methods and the exploration of innovative strategies to overcome them. Although, traditional colloidal synthesis methods are effective in producing high-quality QDs, these methods are labor-intensive and depend on expensive precursors and solvents, elevating production costs. The development of core-shell structures involves complex, multi-step processes that further increase production expenses [35]. Moreover, the use of rare or hazardous elements like cadmium, indium, or lead not only escalates costs due to their limited availability but also raises environmental and health concerns. Therefore, strategies must be adopted for reducing cost and improving scalability. Green, solvent-free, or aqueous-phase synthesis techniques must be developed which can reduce reliance on toxic and expensive chemicals. Moreover, one-pot synthetic approaches can be explored to streamline production, minimizing steps and associated costs. Transitioning to materials such as carbon dots, silicon QDs, or perovskites, which are more abundant and less hazardous, offers cost advantages [83, 85].

7 Recent Advances in Quantum Dot Technology for Environmental Applications

7.1 New Material Development

In particular, QD has been studied as a viable substitute for conventional organic dyes because of its ultra-small size, wide excitation wavelength, high photostability, and size-tunable fluorescence [44]. The need for radioactive reagents in vivo animal monitoring may be eliminated with the help of recently synthesized QDs with near-infrared emission [4, 16]. Gold nanoparticles serve dual purposes: they function as non-radiative quenchers and act as antennas to enhance the fluorescence of QDs due to their strong plasmonic properties. When positioned approximately 30 nm from the QD surface, gold nanoparticles enhance the excitation rates of QDs, thereby increasing fluorescence intensity. Additionally, QDs have been integrated with magnetic nanoparticles to achieve enhanced bio-detection capabilities [27].

7.2 Development of Non-toxic, Eco-Friendly Quantum Dots

The toxicity of QDs is largely influenced by their specific physicochemical characteristics and surrounding environmental factors. These include the size, charge, concentration, surface functional groups, and their stability under oxidative, photolytic, and mechanical conditions. Despite their remarkable electro-optical properties, Cd-based QDs have been found to exhibit toxic effects, including causing DNA damage that leads to cell death, as indicated by extensive research. The toxicity of QDs containing Cd and Pd has garnered significant attention. To mitigate concerns about heavy metal toxicity and the environmental impact of using and disposing of cadmium-based QDs, researchers have focused on developing less toxic or non-toxic alternatives. These alternatives are categorized into heavy metal-free QDs and metal-free QDs. Heavy metal-free QDs typically refer to ternary I-III-VI compounds, while metal-free QDs are primarily based on carbon, silicon, or biomolecules [21, 51]. The discussion of non-toxic eco-friendly QDs is given below.

Heavy Metal-free Ternary Rroup I-III-VI Quantum Dots–To overcome the negative effects of II-IV binary nanocrystals, non-toxic ternary group I-III-VI QDs have emerged as promising alternatives. These QDs are composed of elements from Group I (e.g., Cu, Ag), Group III (e.g., Al, Ga, In, Tl), and Group VI (e.g., S, Se, Te). Examples include $AgInS_2$, $CuInS_2$, and $ZnS–AgInS_2$. They offer enhanced control over bandgap energies, with optical properties that can be fine-tuned by adjusting their size and composition. For instance, $CuInS_2$ has a narrow bandgap energy of 1.45 eV, enabling emission in the near-infrared (NIR) region. Its emission peak can be tuned from 693 to 835 nm by leveraging the quantum size effect through reaction temperature adjustments. Additionally, $CuInS_2$ QDs exhibit a quantum confinement regime and a broad shoulder in the long-wavelength region. Coating these QDs with a ZnS inorganic shell enhances their quantum yield, extends emission lifetimes, reduces Cd^{2+} release, and protects the QDs from oxidation [21, 69].

Carbon-based Quantum Dots (CQDs) are quasi-spherical carbon nanoparticles, typically less than ~10 nm in size, composed of a mixture of graphite and turbostratic carbon in varying proportions. They exhibit an electronic energy spectrum similar to that of direct band-gap semiconductors and are characterized by low toxicity, excellent water solubility, chemical stability, and biocompatibility [91]. Typically, CQDs are synthesized using chemical precursors. However, growing interest has recently shifted toward producing CQDs from biomass waste materials, driven by economic and environmental considerations [49, 86]. Notably, thermochemical conversion of biomass waste—such as agricultural residues, food waste, and municipal solid waste—has been extensively studied as a sustainable method for synthesizing CQDs. Biomass-derived CQDs leverage renewable resources with high carbon content (45–55 wt %) as precursors, offering an eco-friendly alternative to traditional carbon-based materials produced from nonrenewable sources like coal and petroleum coke. Additionally, CQDs possess unique optoelectronic, chemical, and magnetic properties, enabling a wide range of potential applications in electronics, biomedicine, biosensors, bioimaging, and renewable energy [1].

Biomolecule-derived QDs (Bio-Dots) are QDs derived from biomolecules. Natural amino acids, with their diverse side chains containing various functional groups (R groups), serve as ideal precursors for the controlled synthesis of biomolecule-based carbon dots (biodots) with tailored properties [49]. Biodots exhibit distinctive physicochemical properties, such as their ultra-small size, which facilitates cellular uptake, tunable photoluminescence for bioimaging, outstanding aqueous solubility, and high chemical and photostability. Additionally, they demonstrate excellent singlet oxygen quantum yield and inherent biocompatibility, making them well-suited for photodynamic therapy. These attributes are crucial for theragnostic applications [99].

Silicon Quantum Dots (SiQDs) exhibit efficient and tunable optical properties due to quantum confinement, making them promising for applications in light-emitting diodes, lasers, and solar cells [21]. Several techniques have been established for synthesizing luminescent silicon dots. These include the etching of porous silicon, laser ablation, solution-phase approaches, low-pressure plasma synthesis, thermal treatment of silsesquioxane, and a two-step process involving laser pyrolysis of silane followed by acid etching [25]. In vivo studies have demonstrated that emissive SiQDs undergo rapid biodegradation, producing non-toxic silicic acid that is easily excreted through urine [66].

7.3 Advances in Surface Functionalization for Targeted Detection

Surface functionalization plays a crucial role in enhancing the separation of charge carriers. It reduces the recombination rates of electron-hole pairs by trapping them at specific surface sites. Various methods have been devised for surface modification and functionalization, such as π–π interactions, sol-gel techniques, coordination chemistry, and covalent bonding [68]. To use water-soluble QDs in bioimaging, detection, drug delivery, and related applications, they must be conjugated with functional biomolecules such as proteins, enzymes, antibodies, or nucleic acids. These biologically relevant molecules offer ample binding sites that enable QDs to achieve specific objectives, including targeted cell interaction, attachment of selective toxins, conjugation with FRET coupling molecules, or enhancing the overall cellular uptake of QDs [39].

7.4 Integration with Other Technologies

Alongside ongoing advancements in material quality and the performance of individual QD devices on silicon achieved through epitaxial growth, significant efforts are being directed toward the development of fully integrated photonic circuits. Currently, InP-based monolithic integration and silicon photonics represent the two primary technologies in photonic integration. QDs-based photonic integrated circuits (PICs) on silicon can leverage the progress made in both monolithic and heterogeneous integration approaches to enhance performance while lowering costs [80].

7.5 Combining Quantum Dots with Other Nanomaterials for Enhanced Performance

At present, the most effective way to stop heavy metal ions from leaching while preserving QD's optical qualities is to coat it with biocompatible materials. To reduce cytotoxicity and increase biocompatibility, QDs have been coated with various polymers and silica. Silica shell also protects the core materials of QDs from an oxidative environment, similar to the polymer coating layer. For in vitro cell imaging, QD nanocomposites tested can be safely adopted after conjugation with specific probes. The different conditions, like stability, blood circulation time, excretion pathways, and biodistribution, decide the in vivo fate of the nanocomposites. The silica and polymer-coated QDs have longer blood retention times due to the slow uptake of the nanocomposites. Recently, the biosensing capability, specificity, and sensitivity of QDs have been enhanced by their combination with other components.

Various QD biosensors based on electrochemistry, foster resonance energy transfer, and electrochemiluminescence are employed to enhance their biosensing. Further, to promote the multifunctionality of QD in diverse areas like drug delivery, multi-modal imaging, therapeutics, and diagnostics, etc., they are composited with other functional materials [75].

8 Future Perspectives

CQDs derived from biomass offer several advantages over traditional semiconductor QDs. Nanostructured carbon materials sourced from biomass have the potential to address economic, environmental, and social challenges. Various micro-structured carbon materials are prepared from specific types of biomass for diverse applications. However, during the commercialization of CQDs, careful planning is essential to ensure that the biomass used as a raw material does not compete with other critical uses. Additionally, the unique microstructure and chemical composition of each biomass type influence the structure, morphology, and composition of the resulting CQDs, making it challenging to standardize a single process. Selecting the appropriate biomass and optimizing the preparation method is crucial for producing CQDs suited to specific applications. Moreover, standardizing the reaction conditions for processes such as pyrolysis or hydrothermal carbonization (HTC) is necessary to achieve consistent and desirable CQD properties [87].

Significant progress has been made in enhancing CQDs' properties, including their optical characteristics and electron transfer capabilities. Techniques such as surface passivation, functionalization, and metal doping play crucial roles in improving these properties. Further research is necessary to develop methods for synthesizing CQDs with controllable shapes, sizes, and doping using metals or non-metals to achieve specific functionalities. Among CQDs' remarkable features, photoluminescence (PL) stands out, enabling its use in numerous applications. However, the underlying mechanisms of PL are not yet fully understood, warranting further investigation. Additionally, in immobilized electrode systems, exploring the interactions between nanomaterials and biomolecules on electrode or nanofilm surfaces is essential for advancing this field [86].

Understanding the relationship between reaction parameters in the synthesis of carbon dots (CDs) is crucial for refining production methods, uncovering unique properties, and advancing potential applications. Nonetheless, the experimental datasets associated with CD synthesis are vast and often noisy. Recently, machine learning (ML) has emerged as an effective tool for analyzing and screening high-performance materials [23]. Machine learning (ML), a branch of artificial intelligence, has gained significant attention for its effectiveness and versatility in processing large datasets efficiently across various domains [26]. Han et al. [24] introduced a machine learning (ML) guided method for synthesizing highly fluorescent carbon dots (CDs) using a hydrothermal process. They showed that integrating ML into the synthesis significantly improved the optical properties of the CDs, which displayed strong green fluorescence with a quantum yield of 39.3%. Additionally, the optimized CDs were effectively employed for the ultra-sensitive, rapid, and specific detection of iron ions, achieving a detection limit as low as 0.039 μM.

Strict guidelines and regulations are essential for ensuring that new drug products are safe, effective, and economically viable, particularly those categorized under the Generally Recognized as Safe (GRAS) framework. In the case of QDs-based biological and pharmaceutical applications, toxicity remains a significant concern. Despite this, various commercially available QDs are used in applications such as drug delivery, imaging, and biosensing. Research suggests that surface coating can effectively reduce toxicity, making QDs more suitable for biological applications and less likely to trigger immune responses. When considering the regulatory aspects of QDs, the type of product and its intended use serve as the primary criteria for government regulatory claims. To date, no comprehensive literature addresses the regulatory status of QDs. The primary focus remains on toxicity studies to ensure their safety for use.

QDs are recognized for their exceptional photophysical properties, which make them valuable in various biomedical applications. However, their use is significantly limited by concerns over toxicity. A major issue arises from the heavy metals used in their synthesis, particularly cadmium (Cd). The release of Cd ions from the QD core has been identified as a primary cause of cytotoxicity. Cadmium is classified as a hazardous material and is strictly regulated under current environmental policies. Similarly, other heavy metals like mercury (Hg) and lead (Pb) fall into the same category. To address these challenges, researchers have developed non-toxic alternatives, including carbon dots (C-dots), silicon quantum dots (SiQDs), graphene quantum dots (GQDs), as well as QDs based on Ag_2Se, Ag_2S, InP, ZnSe/ZnS, and $CuInS_2$/ZnS [74].

9 Conclusion

In conclusion, quantum dots emerge as new zero-dimensional nanomaterials that bridge the gap between available traditional methods and actual, practical on-site detection of contaminants. Despite all the challenges related to cost-effectiveness, stability, toxicity, and scalability, quantum dots attracted considerable attention in the research community. The effectiveness of this semiconductor material anticipates its use in vital applications. The integration of quantum dots with the leading nanotechnology can have an impact on developing materials for sustainably monitoring of environmental pollution.

References

1. Abbas A, Mariana LT, Phan AN (2018) Biomass-waste-derived graphene quantum dots and their applications. Carbon 140:77–99. https://doi.org/10.1016/j.carbon.2018.08.016
2. Abdulrazaq Y, Abdulsalam A, Rotimi AL, Abdulbasit AA, Clifford O, Abdulsalam OA, Racheal ON, Joy AA, Victor FO, Johannes ZM, Bilal M (2020) Classification, potential routes and risk of emerging pollutants/contaminant. In: Emerging contaminants, IntechOpen. https://doi.org/10.5772/intechopen.94447
3. Adegoke O, Montaseri H, Nsibande SA, Forbes PB (2017) Alloyed quaternary/binary core/shell quantum dot-graphene oxide nanocomposite: preparation, characterization and application as a fluorescence "switch ON" probe for environmental pollutants. J Alloys Compd 720:70–78. https://doi.org/10.1016/j.jallcom.2017.05.242
4. Allen PM, Liu W, Chauhan VP, Lee J, Ting AY, Fukumura D, Jain RK, Bawendi MG (2010) InAs (ZnCdS) quantum dots optimized for biological imaging in the near-infrared. J Am Chem Soc 132(2):470–471. https://doi.org/10.1021/ja908250r
5. Arslan O, Aytac Z, Uyar T (2017) Fluorescent Si QD decoration onto a flexible polymeric electrospun nanofibrous mat for the colorimetric sensing of TNT. J Mater Chem C 5(7):1816–1825. https://doi.org/10.1039/C6TC05521D
6. Bailey RE, Nie S (2003) Alloyed semiconductor quantum dots: tuning the optical properties without changing the particle size. J Am Chem Soc 125(23):7100–7106. https://doi.org/10.1021/ja035000o
7. Bera D, Qian L, Tseng TK, Holloway PH (2010) Quantum dots and their multimodal applications: a review. Materials 3(4):2260–2345. https://doi.org/10.3390/ma3042260
8. Bhanjana G, Dilbaghi N, Kumar R, Umar A, Kumar S (2015) SnO2 quantum dots as novel platform for electrochemical sensing of cadmium. Electrochim Acta 169:97–102. https://doi.org/10.1016/j.electacta.2015.04.045
9. Bhattacharjee L, Pal K, Jat SK, Koner AL, Ravichandran J, Bhattacharjee RR (2017) Tuning optical properties of printable carbon quantum dots using near-field environment. Carbon 125:409–418. https://doi.org/10.1016/j.carbon.2017.09.030
10. Bilan R, Fleury F, Nabiev I, Sukhanova A (2015) Quantum dot surface chemistry and functionalization for cell targeting and imaging. Bioconjug Chem 26(4):609–624. https://doi.org/10.1021/acs.bioconjchem.5b00069
11. Chandra S, Pathan SH, Mitra S, Modha BH, Goswami A, Pramanik P (2012) Tuning of photoluminescence on different surface functionalized carbon quantum dots. RSC Adv 2(9):3602–3606. https://doi.org/10.1039/C2RA00030J
12. Chen X, Liu Y, Ma Q (2018) Recent advances in quantum dot-based electrochemiluminescence sensors. J Mater Chem C 6(5):942–959. https://doi.org/10.1039/C7TC05474B
13. Chu HW, Unnikrishnan B, Anand A, Lin YW, Huang CC (2020) Carbon quantum dots for the detection of antibiotics and pesticides. J Food Drug Anal 28(4):539. https://doi.org/10.38212/2224-6614.1269
14. Cui L, He XP, Chen GR (2015) Recent progress in quantum dot based sensors. RSC Adv 5(34):26644–26653. https://doi.org/10.1039/C5RA01950H
15. Devi P, Rajput P, Thakur A, Kim KH, Kumar P (2019) Recent advances in carbon quantum dot-based sensing of heavy metals in water. TrAC Trends Anal Chem 114:171–195. https://doi.org/10.1016/j.trac.2019.03.003
16. Du Y, Xu B, Fu T, Cai M, Li F, Zhang Y, Wang Q (2010) Near-infrared photoluminescent Ag_2S quantum dots from a single source precursor. J Am Chem Soc 132(5):1470–1471. https://doi.org/10.1021/ja909490r
17. Feminus JJ, Deepa PN (2022) Fabrication and application of electrodeposited CdSe QD/Mesosilica/rGO electrode as an electrochemical sensor. Mater Chem Phys 289:126440. https://doi.org/10.1016/j.matchemphys.2022.126440

18. Ferrero FJ, Valledor M, Campo JC, López A, Llano-Suárez P, Fernández-Arguelles MT, Costa-Fernández JM, Soldado A (2020) Portable instrument for monitoring environmental toxins using immobilized quantum dots as the sensing material. Appl Sci 10(9):3246. https://doi.org/10.3390/app10093246
19. Frasco MF, Chaniotakis N (2009) Semiconductor quantum dots in chemical sensors and biosensors. Sensors 9(9):7266–7286. https://doi.org/10.3390/s90907266
20. Freeman R, Girsh J, Willner I (2013) Nucleic acid/quantum dots (QDs) hybrid systems for optical and photoelectrochemical sensing. ACS Appl Mater Interfaces 5(8):2815–2834. https://doi.org/10.1021/am303189h
21. Gao H, Teng CP, Huang D, Xu W, Zheng C, Chen Y, Liu M, Yang DP, Lin M, Li Z, Ye E (2017) Microwave assisted synthesis of luminescent carbonaceous nanoparticles from silk fibroin for bioimaging. Mater Sci Eng C 80:616–623. https://doi.org/10.1016/j.msec.2017.07.007
22. Gavgani JN, Dehsari HS, Hasani A, Mahyari M, Shalamzari EK, Salehi A, Taromi FA (2015) A room temperature volatile organic compound sensor with enhanced performance, fast response and recovery based on N-doped graphene quantum dots and poly (3, 4-ethylenedioxythiophene)–poly (styrenesulfonate) nanocomposite. RSC Adv 5(71):57559–57567. https://doi.org/10.1039/C5RA08158K
23. Han S, Yang L, Wen Z, Chu S, Wang M, Wang Z, Jiang C (2020a) A dual-response ratiometric fluorescent sensor by europium-doped CdTe quantum dots for visual and colorimetric detection of tetracycline. J Hazard Mater 398:122894. https://doi.org/10.1016/j.jhazmat.2020.122894
24. Han Y, Tang B, Wang L, Bao H, Lu Y, Guan C, Zhang L, Le M, Liu Z, Wu M (2020b) Machine-learning-driven synthesis of carbon dots with enhanced quantum yields. ACS Nano 14(11):14761–14768. https://doi.org/10.1021/acsnano.0c01899
25. He X, Ma N (2014) An overview of recent advances in quantum dots for biomedical applications. Colloids Surf B: Biointerfaces 124:118–131. https://doi.org/10.1016/j.colsurfb.2014.06.002
26. Hoffmann J, Bar-Sinai Y, Lee LM, Andrejevic J, Mishra S, Rubinstein SM, Rycroft CH (2019) Machine learning in a data-limited regime: augmenting experiments with synthetic data uncovers order in crumpled sheets. Sci Adv 5(4):peaau6792. https://doi.org/10.1126/sciadv.aau6792
27. Holzinger M, Le Goff A, Cosnier S (2017) Synergetic effects of combined nanomaterials for biosensing applications. Sensors 17(5):1010. https://doi.org/10.3390/s17051010
28. Hu L, Zhang C, Zeng G, Chen G, Wan J, Guo Z, Wu H, Yu Z, Zhou Y, Liu J (2016) Metal-based quantum dots: synthesis, surface modification, transport and fate in aquatic environments and toxicity to microorganisms. RSC Adv 6(82):78595–78610. https://doi.org/10.1039/C6RA13016J
29. Hu L, Zhong H, He Z (2021) Toxicity evaluation of cadmium-containing quantum dots: a review of optimizing physicochemical properties to diminish toxicity. Colloids Surf B: Biointerfaces 200:111609. https://doi.org/10.1016/j.colsurfb.2021.111609
30. Hu S, Trinchi A, Atkin P, Cole I (2015) Tunable photoluminescence across the entire visible spectrum from carbon dots excited by white light. Angew Chem Int Ed 54(10):2970–2974. https://doi.org/10.1002/anie.201411004
31. Huang H, Susha AS, Kershaw SV, Hung TF, Rogach AL (2015) Control of emission color of high quantum yield $CH_3NH_3PbBr_3$ perovskite quantum dots by precipitation temperature. Adv Sci 2(9). https://doi.org/10.1002/advs.201500194
32. Huang M, Tong C (2022) A dual-emission ratiometric fluorescence probe for highly selective and simultaneous detection of tetracycline and ferric ions in environmental water samples based on a boron-doped carbon quantum dot/CdTe–Eu^{3+} composite. Environ Sci Nano 9(5):1712–1723. https://doi.org/10.1039/D2EN00110A
33. Huang S, Guo M, Tan J, Geng Y, Wu J, Tang Y, Su C, Lin CC, Liang Y (2018) Novel fluorescence sensor based on all-inorganic perovskite quantum dots coated with molecularly imprinted polymers for highly selective and sensitive detection of omethoate. ACS Appl Mater Interfaces 10(45):39056–39063. https://doi.org/10.1021/acsami.8b14472

34. Jaishankar M, Tseten T, Anbalagan N, Mathew BB, Beeregowda KN (2014) Toxicity, mechanism and health effects of some heavy metals. Interdiscip Toxicol 7(2):60–72. https://doi.org/10.2478/intox-2014-0009
35. Jean J, Xiao J, Nick R, Moody N, Nasilowski M, Bawendi M, Bulović V (2018) Synthesis cost dictates the commercial viability of lead sulfide and perovskite quantum dot photovoltaics. Energy Environ Sci 11(9):2295–2305. https://doi.org/10.1039/C8EE01348A
36. Jouyandeh M, Khadem SSM, Habibzadeh S, Esmaeili A, Abida O, Vatanpour V, Rabiee N, Bagherzadeh M, Iravani S, Saeb MR, Varma RS (2021) Quantum dots for photocatalysis: synthesis and environmental applications. Green Chem 23(14):4931–4954. https://doi.org/10.1039/D1GC00639H
37. Kalkal A, Kadian S, Pradhan R, Manik G, Packirisamy G (2021) Recent advances in graphene quantum dot-based optical and electrochemical (bio) analytical sensors. Mater Adv 2(17):5513–5541. https://doi.org/10.1039/D1MA00251A
38. Kandi D, Behera A, Sahoo S, Parida K (2020) CdS QDs modified BiOI/Bi2MoO6 nanocomposite for degradation of quinolone and tetracycline types of antibiotics towards environmental remediation. Sep Purif Technol 253:117523. https://doi.org/10.1016/j.seppur.2020.117523
39. Karakoti AS, Shukla R, Shanker R, Singh S (2015) Surface functionalization of quantum dots for biological applications. Adv Colloid Interf Sci 215:28–45. https://doi.org/10.1016/j.cis.2014.11.004
40. Kargozar S, Hoseini SJ, Milan PB, Hooshmand S, Kim HW, Mozafari M (2020) Quantum dots: a review from concept to clinic. Biotechnol J 15(12):2000117. https://doi.org/10.1002/biot.202000117
41. Kaur J, Kaur M, Aggarwal D, Kumar V, Tikoo K, Kaushik A, Singhal S (2024) Unveiling the multifaceted applications of magnetically responsive chitosan capped ZnS QDs for sensing and annihilation of pharmaceutical drugs. Talanta 266:125084. https://doi.org/10.1016/j.talanta.2023.125084
42. Khan R, Shukla S, Kumar M, Barceló D, Zuorro A, Bhargava PC (2024) Progress and obstacles in employing carbon quantum dots for sustainable wastewater treatment. Environ Res 119671. https://doi.org/10.1016/j.envres.2024.119671
43. Khan S, Naushad M, Govarthanan M, Iqbal J, Alfadul SM (2022) Emerging contaminants of high concern for the environment: current trends and future research. Environ Res 207:112609. https://doi.org/10.1016/j.envres.2021.112609
44. Klostranec JM, Chan WC (2006) Quantum dots in biological and biomedical research: recent progress and present challenges. Adv Mater 18(15):1953–1964. https://doi.org/10.1002/adma.200500786
45. Kumar DS, Kumar BJ, Mahesh HM (2018) Quantum nanostructures (QDs): an overview. Synth Inorg Nanomater:59–88. https://doi.org/10.1016/B978-0-08-101975-7.00003-8
46. Li H, Qu F (2007) Synthesis of CdTe quantum dots in sol– gel-derived composite silica spheres coated with calix [4] arene as luminescent probes for pesticides. Chem Mater 19(17):4148–4154. https://doi.org/10.1021/cm0700089
47. Li QS, Zhang RQ, Lee ST, Niehaus TA, Frauenheim T (2008) Optimal surface functionalization of silicon quantum dots. J Chem Phys 128(24). https://doi.org/10.1063/1.2940735
48. Liu B, Zhuang J, Wei G (2020) Recent advances in the design of colorimetric sensors for environmental monitoring. Environ Sci Nano 7(8):2195–2213. https://doi.org/10.1039/D0EN00449A
49. Mahajan R, Bhadwal AS, Kumar N, Madhusudanan Ṁ, Pudake RN, Tripathi RM (2017) Green synthesis of highly stable carbon nanodots and their photocatalytic performance. IET Nanobiotechnol 11(4):360–364
50. Majumdar S, Gogoi D, Boruah PK, Thakur A, Sarmah P, Gogoi P, Sarkar S, Pachani P, Manna P, Saikia R, Chaturvedi V (2024) Hexagonal boron nitride quantum dots embedded on layer-by-layer films for peroxidase-assisted colorimetric detection of β-galactosidase producing pathogens. ACS Appl Mater Interfaces. https://doi.org/10.1021/acsami.4c01565

51. Mandal G, Darragh M, Wang YA, Heyes CD (2012) Cadmium-free quantum dots as time-gated bioimaging probes in highly autofluorescent human breast cancer cells. Chem Commun (Camb) 49(6):624. https://doi.org/10.1039/c2cc37529j
52. Mandal P, Sahoo D, Sarkar P, Chakraborty K, Das S (2019) Fluorescence turn-on and turn-off sensing of pesticides by carbon dot-based sensor. New J Chem 43(30):12137–12151. https://doi.org/10.1039/C9NJ03192H
53. Mansuriya BD, Altintas Z (2020) Applications of graphene quantum dots in biomedical sensors. Sensors 20(4):1072. https://doi.org/10.3390/s20041072
54. May BM, Bambo MF, Hosseini SS, Sidwaba U, Nxumalo EN, Mishra AK (2022) A review on I–III–VI ternary quantum dots for fluorescence detection of heavy metals ions in water: optical properties, synthesis and application. RSC Adv 12(18):11216–11232. https://doi.org/10.1039/D1RA08660J
55. Mishra RK, Choi GJ, Choi HJ, Gwag JS (2021) ZnS quantum dot based acetone sensor for monitoring health-hazardous gases in indoor/outdoor environment. Micromachines 12(6):598. https://doi.org/10.3390/mi12060598
56. Mitri F, De Iacovo A, De Luca M, Pecora A, Colace L (2020) Lead sulphide colloidal quantum dots for room temperature NO_2 gas sensors. Sci Rep 10(1):12556. https://doi.org/10.1038/s41598-020-69478-x
57. Mittelstädt A, Schliwa A, Klenovský P (2022) Modeling electronic and optical properties of III–V quantum dots—selected recent developments. Light Sci Appl 11(1):17. https://doi.org/10.1038/s41377-021-00700-9
58. Mohagheghpour E, Farzin L, Sadjadi S (2024) Alendronate-functionalized graphene quantum dots as an effective fluorescent sensing platform for arsenic ion detection. Biol Trace Elem Res 202(5):2391–2401. https://doi.org/10.1007/s12011-023-03819-5
59. Molaei MJ (2020) Principles, mechanisms, and application of carbon quantum dots in sensors: a review. Anal Methods 12(10):1266–1287. https://doi.org/10.1039/C9AY02696G
60. Moon H, Lee C, Lee W, Kim J, Chae H (2019) Stability of quantum dots, quantum dot films, and quantum dot light-emitting diodes for display applications. Adv Mater 31(34):1804294. https://doi.org/10.1002/adma.201804294
61. Murugesan P, Moses JA (2024) Carbon quantum dots stabilized silver nanoparticles-based colorimetric sensor for detection of pesticide residues. Mater Sci Eng B 304:117354. https://doi.org/10.1016/j.mseb.2024.117354
62. Naidu R, Espana VAA, Liu Y, Jit J (2016) Emerging contaminants in the environment: risk-based analysis for better management. Chemosphere 154:350–357. https://doi.org/10.1016/j.chemosphere.2016.03.068
63. Nandan Y, Mehata MS (2019) Wavefunction engineering of type-I/type-II excitons of CdSe/CdS core-shell quantum dots. Sci Rep 9(1):2. https://doi.org/10.1038/s41598-018-37676-3
64. Nsibande SA, Forbes PBC (2016) Fluorescence detection of pesticides using quantum dot materials–a review. Anal Chim Acta 945:9–22. https://doi.org/10.1016/j.aca.2016.10.002
65. Pandit C, Alajangi HK, Singh J, Khajuria A, Sharma A, Hassan MS, Parida M, Semwal AD, Gopalan N, Sharma RK, Suttee A (2022) Development of magnetic nanoparticle assisted aptamer-quantum dot based biosensor for the detection of Escherichia coli in water samples. Sci Total Environ 831:154857. https://doi.org/10.1016/j.scitotenv.2022.154857
66. Park JH, Gu L, Von Maltzahn G, Ruoslahti E, Bhatia SN, Sailor MJ (2009) Biodegradable luminescent porous silicon nanoparticles for in vivo applications. Nat Mater 8(4):331–336. https://doi.org/10.1038/nmat2398
67. Peng J, Gao W, Gupta BK, Liu Z, Romero-Aburto R, Ge L, Song L, Alemany LB, Zhan X, Gao G, Vithayathil SA (2012) Graphene quantum dots derived from carbon fibers. Nano Lett 12(2):844–849. https://doi.org/10.1021/nl2038979
68. Phang SJ, Tan LL (2019) Recent advances in carbon quantum dot (CQD)-based two dimensional materials for photocatalytic applications. Cat Sci Technol 9(21):5882–5905. https://doi.org/10.1039/C9CY01452G
69. Qi K, Wang Y, Wang R, Wu D, Li GD (2016) Facile synthesis of homogeneous $CuInS_2$ quantum dots with tunable near-infrared emission. J Mater Chem C 4(9):1895–1899. https://doi.org/10.1039/C5TC04232A

70. Qian Z, Ma J, Shan X, Shao L, Zhou J, Chen J, Feng H (2013) Surface functionalization of graphene quantum dots with small organic molecules from photoluminescence modulation to bioimaging applications: an experimental and theoretical investigation. RSC Adv 3(34):14571–14579. https://doi.org/10.1039/C3RA42066C
71. Qu A, Xie H, Xu X, Zhang Y, Wen S, Cui Y (2016) High quantum yield graphene quantum dots decorated TiO_2 nanotubes for enhancing photocatalytic activity. Appl Surf Sci 375:230–241. https://doi.org/10.1016/j.apsusc.2016.03.077
72. Ramvall P, Riblet P, Nomura S, Aoyagi Y, Tanaka S (2000) Optical properties of GaN quantum dots. J Appl Phys 87(8):3883–3890. https://doi.org/10.1063/1.372429
73. Rao VN, Reddy NL, Kumari MM, Cheralathan KK, Ravi P, Sathish M, Neppolian B, Reddy KR, Shetti NP, Prathap P, Aminabhavi TM (2019) Sustainable hydrogen production for the greener environment by quantum dots-based efficient photocatalysts: a review. J Environ Manag 248:109246. https://doi.org/10.1016/j.jenvman.2019.07.017
74. Reshma VG, Mohanan PV (2019) Quantum dots: applications and safety consequences. J Lumin 205:287–298. https://doi.org/10.1016/j.jlumin.2018.09.015
75. Lu ZS, Li CM (2011) Quantum dot-based nanocomposites for biomedical applications. Curr Med Chem 18(23):3516–3528. https://doi.org/10.2174/092986711796642634
76. Sawetwong P, Chairam S, Jarujamrus P, Amatatongchai M (2021) Enhanced selectivity and sensitivity for colorimetric determination of glyphosate using Mn–ZnS quantum dot embedded molecularly imprinted polymers combined with a 3D-microfluidic paper-based analytical device. Talanta 225:122077. https://doi.org/10.1016/j.talanta.2020.122077
77. Selopal GS, Zhao H, Wang ZM, Rosei F (2020) Core/shell quantum dots solar cells. Adv Funct Mater 30(13):1908762. https://doi.org/10.1002/adfm.201908762
78. Sengupta S, Ganguli N, Dasgupta I, Sarma DD, Acharya S (2011) Long-range visible fluorescence Tunability using component-modulated coupled quantum dots. Adv Mater 23(17):1998–2003. https://doi.org/10.1002/adma.201004126
79. Shahid I, Shahzad MI, Tutsak E, Mahfouz MM, Al Adba MS, Abbasi SA, Rathore HA, Asif Z, Chen Z (2024) Carbon based sensors for air quality monitoring networks; Middle East perspective. Front Chem 12:1391409. https://doi.org/10.3389/fchem.2024.1391409
80. Shang C, Wan Y, Selvidge J, Hughes E, Herrick R, Mukherjee K, Duan J, Grillot F, Chow WW, Bowers JE (2021) Perspectives on advances in quantum dot lasers and integration with Si photonic integrated circuits. ACS photonics 8(9):2555–2566. https://doi.org/10.1021/acsphotonics.1c00707
81. Sharma K, Raizada P, Hasija V, Singh P, Bajpai A, Nguyen VH, Rangabhashiyam S, Kumar P, Nadda AK, Kim SY, Varma RS (2021) ZnS-based quantum dots as photocatalysts for water purification. J Water Process Eng 43:102217. https://doi.org/10.1016/j.jwpe.2021.102217
82. Siddique AB, Hossain SM, Pramanick AK, Ray M (2021) Excitation dependence and independence of photoluminescence in carbon dots and graphene quantum dots: insights into the mechanism of emission. Nanoscale 13(39):16662–16671. https://doi.org/10.1039/D1NR04301C
83. Silva BF, Andreani T, Gavina A, Vieira MN, Pereira CM, Rocha-Santos T, Pereira R (2016) Toxicological impact of cadmium-based quantum dots towards aquatic biota: effect of natural sunlight exposure. Aquat Toxicol 176:197–207. https://doi.org/10.1016/j.aquatox.2016.05.001
84. Stefanakis AI, Becker JA (2016) A review of emerging contaminants in water: classification, sources, and potential risks. Impact Water Pollut Hum Health Environ Sustain 55–80. https://doi.org/10.4018/978-1-4666-9559-7.ch003
85. Terada S, Xin Y, Saitow KI (2020) Cost-effective synthesis of silicon quantum dots. Chem Mater 32(19):8382–8392. https://doi.org/10.1021/acs.chemmater.0c02320
86. Thangaraj B, Solomon PR, Ranganathan S (2019) Synthesis of carbon quantum dots with special reference to biomass as a source-a review. Curr Pharm Des 25(13):1455–1476. https://doi.org/10.2174/1381612825666190618154518
87. Thangaraj B, Solomon PR, Chuangchote S, Wongyao N, Surareungchai W (2021) Biomass-derived carbon quantum dots–a review. Part 1: preparation and characterization. Chem Bio Eng Rev 8(4):265–301. https://doi.org/10.1002/cben.202000029

88. Travlou NA, Giannakoudakis DA, Algarra M, Labella AM, Rodríguez-Castellón E, Bandosz TJ (2018) S-and N-doped carbon quantum dots: surface chemistry dependent antibacterial activity. Carbon 135:104–111. https://doi.org/10.1016/j.carbon.2018.04.018
89. Vasudevan D, Gaddam RR, Trinchi A, Cole I (2015) Core–shell quantum dots: properties and applications. J Alloys Compd 636:395–404. https://doi.org/10.1016/j.jallcom.2015.02.102
90. Wahab R, Khan F, Alam M, Ahmad J, Al-Khedhairy AA (2023) Aluminum oxide quantum dots (Al2O3): an immediate sensing aptitude for the detection of urea. Inorg Chem Commun 147:110238. https://doi.org/10.1016/j.inoche.2022.110238
91. Wang J, Liu R, Liu B (2016a) Cadmium-containing quantum dots: current perspectives on their application as nanomedicine and toxicity concerns. Mini Rev Med Chem 16(11):905–916
92. Wang S, Cole IS, Li Q (2016b) The toxicity of graphene quantum dots. RSC Adv 6(92):89867–89878. https://doi.org/10.1039/C6RA16516H
93. Wu HL, Li XB, Tung CH, Wu LZ (2019) Semiconductor quantum dots: an emerging candidate for CO_2 photoreduction. Adv Mater 31(36):1900709. https://doi.org/10.1002/adma.201900709
94. Yan X, Li H, Han X, Su X (2015) A ratiometric fluorescent quantum dots based biosensor for organophosphorus pesticides detection by inner-filter effect. Biosens Bioelectron 74:277–283. https://doi.org/10.1016/j.bios.2015.06.020
95. Yang H, Li R, Zhang Y, Yu M, Wang Z, Liu X, You W, Tu D, Sun Z, Zhang R, Chen X (2021) Colloidal alloyed quantum dots with enhanced photoluminescence quantum yield in the NIR-II window. J Am Chem Soc 143(6):2601–2607. https://doi.org/10.1021/jacs.0c13071
96. Yang W, Yang H, Ding W, Zhang B, Zhang L, Wang L, Yu M, Zhang Q (2016) High quantum yield ZnO quantum dots synthesizing via an ultrasonication microreactor method. Ultrason Sonochem 33:106–117. https://doi.org/10.1016/j.ultsonch.2016.04.020
97. Yue Z, Lisdat F, Parak WJ, Hickey SG, Tu L, Sabir N, Dorfs D, Bigall NC (2013) Quantum-dot-based photoelectrochemical sensors for chemical and biological detection. ACS Appl Mater Interfaces 5(8):2800–2814. https://doi.org/10.1021/am3028662
98. Zhang Y, Xu C, Fu Z, Shu Y, Zhang J, Lu C, Mo X (2018) Associations between total mercury and methyl mercury exposure and cardiovascular risk factors in US adolescents. Environ Sci Pollut Res 25:6265–6272. https://doi.org/10.1007/s11356-017-0905-2
99. Zheng XT, Ananthanarayanan A, Luo KQ, Chen P (2015) Glowing graphene quantum dots and carbon dots: properties, syntheses, and biological applications. Small 11(14):1620–1636. https://doi.org/10.1002/smll.201402648
100. Zhong WL, Yang JY (2024) Fluorescent carbon quantum dots for heavy metal sensing. Sci Total Environ 957:177473. https://doi.org/10.1016/j.scitotenv.2024.177473
101. Zhu ZJ, Yeh YC, Tang R, Yan B, Tamayo J, Vachet RW, Rotello VM (2011) Stability of quantum dots in live cells. Nat Chem 3(12):963–968. https://doi.org/10.1038/nchem.1177

Chapter 10
Quantum Dots as Environmental Remediation Agents

Akansha Gupta, Shivani Yadav, and Harsh Pipil

Abstract The development and synthesis of zero-dimensional quantum dots (QD) particles have profound applications in major engineering fields. Environmental remediation, pollution control, biosensors, bio-imaging and probe selection are a few areas where QD particles have proven their capabilities. Light-driven nanocomposites and nano-photocatalysts have been employed in the past to remove notorious dyes such as methylene blue, malachite green and other xenobiotics from water bodies and other ecosystem spheres. QD particles generally range from 1 to 10 nm in diameter, arranged in a confined boundary. They exhibit semiconductor properties through two mechanisms, one by behaving as electron sinks in N-doped chips and as preventers of electron–hole recombination. QD applications are also found in photocatalytic activity because of the high surface area and convenient regeneration after every cycle. Along with this, doping for other heteroatoms such as N and S has increased in optical properties and surface stability. Excessive industrialisation and resource extraction due to anthropogenic activities are creating a burden on the environment. Seven billion people rely on natural resources such as clean water and air, which are deteriorating with time because of a mixture of smoke and discarded effluents. To potentially augment and decrease such an extreme level of contamination, remediation of environmental components is necessary. Quantum dots serve as potential remediation agents for removing contamination by coagulation, precipitation and sorption techniques. This chapter elaborately discusses the role of QDs in environmental remediation.

Keywords Quantum dots · Environment · Solar cells · Photocatalysis · Remediation · Sorption

A. Gupta (✉)
Department of Biotechnology, IILM University, Greater Noida, Uttar Pradesh, India
e-mail: akansha.gupta@iilm.edu; gupta.akansha145@gmail.com

S. Yadav
Department of Environmental Engineering, Delhi Technological University, New Delhi, India

H. Pipil
Manav Rachna Centre for Advanced Water Technology and Management (MRCAWTM), Manav Rachna International Institute of Research and Studies, Faridabad, Haryana, India

T. Bhuyan et al. (eds.), *Multifunctional Quantum Dots*, Materials Horizons: From Nature to Nanomaterials, https://doi.org/10.1007/978-981-92-0172-3_10

1 Introduction

Current trends in environment domains are focusing on the principle of sustainability, society and a zero-waste approach. These three notable pillars of environmental science strengthen different perspectives of waste management, remedial and mitigation measures along with escalating virtuous science practice in energy generation and sequestration [35]. Nanotechnology has been a fascinating field because of its existence in the world, but with zero naked-eye visibility [34]. In common terms, nanotechnology can easily be defined as 'technology or science working at the nanoscale'. And this involves working, modifying and stimulating particles at a diameter of 1–100 nm. Nanotechnology in this era has broadened its horizon in automatic machinery, AI imaging and sensory experimentation. Richard Feynman in 1959 at a conference of the American Physical Society said that 'There is enough room at the bottom', which opened doors for innovations in all fields of science. The intervention of human beings in nanotechnology has impacted lifestyle, amended pharmaceuticals, drug delivery systems, risk assessment and characterization has thereby enhanced biodiversity and the existence of nature more distinctly [7, 31, 52].

Quantum dots (QDs) are one of the most noticed, engineered arrays of particles. They have emerged and potentialized in major domains of environmental safety and applications. The total cost of the QD nanoscale economy is estimated to cross 10 trillion USD by the end of 2030. QD materials are often carbon-based particles with a size ranging <10 nm, confined within a certain area dimension and defined energy state [22]. They are chemically inert particles with little cytotoxicity, great electrical conductivity and luminescent properties. Engineered QDs of two types—carbon-based QD (CQD) and graphene-based QD (GQD)—are varied based on functional group and side-chain moieties on their surfaces. Hybridization arrangement in GQD's is moreover sp^2, where small graphene particles are embedded in 2-D nanocrystal dimensions, whereas CQD's show an amorphous structure with sp^3 arrangement of atoms in layers. An adjustable energy band gap in GQD, along with the incorporation of oxygen and hydrogen atoms in monolayers, enhances the edge and conjugate effects in the modification of chemical characteristics and optoelectronic capabilities. In terms of core chemistry, CQD has a more complex design with a core carbogenic carbon coupled with carbon sheets; on the other hand, GQD contains spherical nanosheets of sp^2 hybridized carbon nanosheets, moreover mimicking the structural composition of carbon dots (CDs) [21]. Semiconductor-engineered QDs are generally synthesized from natural raw material sources, giving them both organic and inorganic existence. Carbon-based (CQD) nanoparticle approach has always been a stimulating field of research, where the major focus and challenges were to quantify energy particles for applicative purposes. It is physically, chemically and biologically significant, such as improvement in surface texture, rheology, enzymatic and adsorption techniques, etc. CQD's have excellent aqueous and non-aqueous surface passivation properties such that in the presence of aerobic conditions, a shift in band

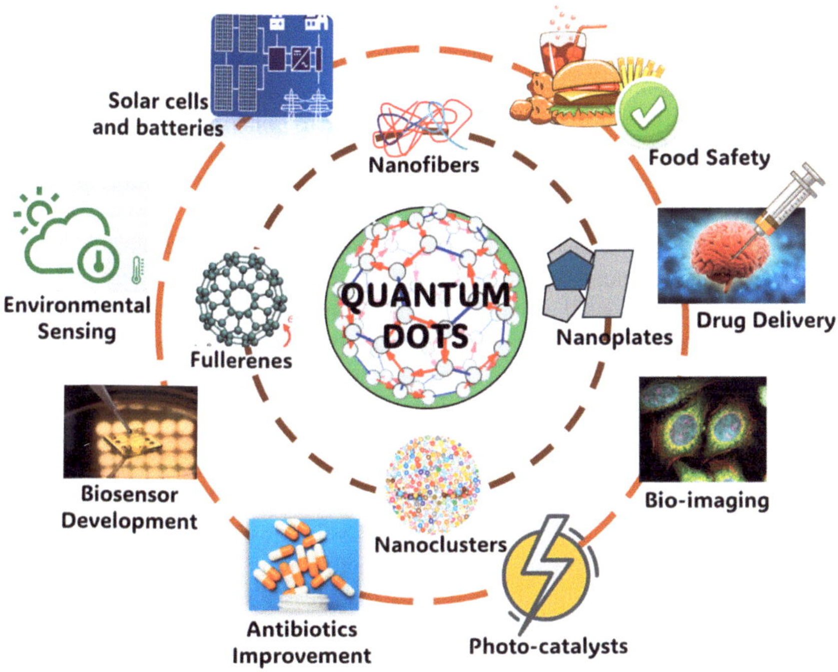

Fig. 1 Illustration depicting potential applications of QDs

gap of CQD's carbon atoms, enabling it for practical applications in sensor manufacturing. Quantum properties such as optical resolution, zero band gap distance, photoluminescence, absorption and adsorption spectral studies are widely acknowledged during their manufacturing and applications [37, 47], widely studied applications of QD are briefly illustrated in Fig. 1. For the above context, various applications of QD about environmental remediation techniques and protocols are discussed in this chapter.

2 Applications of QD for Environmental Remediation

QD demonstrates great applications in environmental remediation by its zero-dimensional fluorescent properties and biocompatibility in the ecosystem. The QD particles synthesized by various methods were studied for the presence of heavy metal analysis, UV spectrum visualization, pH sensors, etc. For pollution studies QDs work on two principles, either the response is attenuated by the interaction between environmental contaminants or by the surface sensor present on QD particles. For the latter, a generated florescence signal as response is received by the molecules present on the QD surface that are selectable (and aligned) towards target (environmental contaminant) recognition. The effect generated as response is suppressed by quenching mechanism and the florescence which was generated gets diminished. Such quenching methods are based on other approaches like static, dynamic, inner-filter, resonance-energy transfer, photo-induced and surface energy transfer quenching methods. Static quenching method applies electrostatic interaction between QD particles and quencher, whereas in contrast dynamic no direct interaction of QD and quencher is required.

QD in environmental studies is used for the detection of contaminants by the ion–molecule detection method [39, 65]. General contaminants are anionic in nature, and QD surfaces are coated with the relevant cations. This anion-cation combination thus quenched fluorescence during the detection experiment in wastewater. Unique and small molecule detection in the case of pesticide and fertilizer study conducted by Yang et al. [62] shows the quenching effect of QD particles by tetracycline compounds detected by the process of amine-aldehyde condensation. In a similar study, glyphosate contamination was detected by studying green-yellow florescence quenching aided by Foster resonance energy transfer (FRET) mechanism in real water sample [57].

Detection of pathogenic bacteria by slight modification on the surface of QD is a direct detection method, which is possible due to increased surface area, charge, water solubility and biocompatibility. An accuracy of 91.2% was achieved for detecting over six pathogenic bacteria in the study of Zhang et al. [66]. The quenching mechanism involved in detection was proposed by Watson and Crick, such that there exists an on–off switch between 5-carboxyfluorescein-labelled cDNA. The labelled cDNA hybridizes with the middle aptamer sequence in the absence of target molecules, producing a quenching effect.

2.1 *Polymer Nanocomposites*

Polymer-based bioactive QD is widely applied in biomedical fields for biosensor development, drug delivery systems and modern medicine. Zero-dimensional QDs are joined together in the form of a polymer with exciting affinities for binding with biomolecules. These polymers prevent light scattering for longer wavelengths, and thus, optical losses are reduced. In the solution blending method for synthesis, a nanofiller and a compatible polymer are solubilized in a solvent to achieve a uniform nano-dispersed polymerized matrix. Such solutions are stable even if the solvent vaporizes with time, depicting high shelf-life properties. As compared to organic dyes, the spectral range of semiconductor QD particles is much narrower, which provides a suitable choice for tuning reaction temperature and time. During the study of cell morphology, QD particles are generally excited by one photon or one wavelength. Electromagnetic spectrum studies suggest that during excitation, the absorption band is generally towards the blue end of the wavelength, but the response can be collected in a multi-range segment, thus giving a multi-imaging picture of the complete surface. QDs are replacing conventional chromophores during biological dyeing and biosensor development because of the constant excitation for a longer period on a time scale. The coating surface of QD particles generally consists of S-based thiols or hydroxyl groups because of commercial cut cost and biological relevance [4]. Polymeric coating on QD surface provides a multi-range functional group choice along with multivalent adsorbing interactions, a great barrier to prevent leakage of heavy metals, ionic strength and stability. Coating of ligands on QD surface is an important aspect, as with ligand interaction, QD surfaces can be lost in colloidal stability, disruption of surface functional groups, aggregation and finally, complete loss or precipitation of QD particles in the solvent can take place. Bigger functional groups or monodentate ligands of sulphur like S-carboxylic acids, thiols, di-sulphide bridges, denatured amine acids or proteins, bovine albumin serum (BSA), etc., find a major segment of functional groups, and these groups act as free or loosely bound ligand in the solvent solutions are usually work on the principle of displacement or replacement at the time of mechanism [15, 63]. Scientists have cultured techniques for multi-dentate ligands for QD surface coatings comprising polymeric units of cysteine amino acids along with short or medium-chain phenylalanine units, along with hydroxy sulphur bonded cyclic-alanine. Other polymeric molecules, such as peptide polymeric surface chains, are introduced after denaturing the peptide chain containing di-sulphide bridges, PEG chains and thiolactic acid. These functional groups provide consistent bio-functionality during aqueous conditions by reacting with free thiol groups and amine moieties, and doping with PEG molecules provides stability for H bonding in water [30].

2.2 *Role of QD for the Removal of Antibiotics Present in Different Domains*

Antibiotics rank high in the pharmaceutical research segment, with over 10,000 new biomolecules newly discovered each year by various pharma companies [59]. After administering antibiotics, over 50% of them find their way to environmental domains, thus contributing towards great contamination of water bodies, grasslands and turning up in biodiversity flora and fauna cycles by food or water infection [16]. Molecular complexity of pharma molecules poses a great difficulty towards their remediation from the environment, being rich in functional group presence, and undoubtedly, they enter different domains of the environment, harming several components of the ecosystem. Causing organ failure, blood infection and dystrophy ailments in ruminants and rupture of soil microbiota [38].

Researchers have developed a synergistic property of combining carbon-based QD for the photo-catalysing of antibiotics. These particles, in synergy, work as a promising action towards the amalgamation of semiconductor materials in a tightly bound composite. Facilitation of carbon QD in the remediation of such xenobiotic compounds, in the process of photocatalysis, is very important to study and craft QD particles by a bottom-up approach. Such an approach explains the construction and compacting up of the atoms in the QD segment until it reaches a specific nano-crystalline structure by application of physical or chemical forces.

They also augment towards free mobilization, transport and diffusion of charged and vacancies in the photocatalytic process, thus enabling a more efficient process of heterogeneous photocatalysis. Such synergistic carbon QD is now being synthesized by blending natural raw materials like fruit peels and oil-seed cakes. Usually, the structure of carbon QD is doped with nitrogen atoms, keeping the entire composition in a heterogeneous format for a convenient facilitation of electrons during the catalysis process. In the study with N-doped QD particles, it was found that improved adsorption and electronic conductivity, strong blue emission in N-doped QD because of the presence of citrazinic acid residues [3, 46]. Traditional methods for remediation fade as the xenobiotic properties of environmental contaminants become recalcitrant, thus inexhaustible approach of QD with promising photochemical and catalytic properties booms towards a reliable alternate for eradication of such molecules.

2.3 Application of QD for Colorimetric Sensing

Colorimetric sensing aided by QD is budding research for the detection of anions, cations, zwitter ions, etc., in environmental radiation and the detection of contaminants and pollutants. Fabrication and personalized development of sensors aids towards rapid detection and naked-eye result analysis by the colour detection, generated as a response to the presence/absence of contaminants. Peroxidase sensors mediate a blue colour signal during the detection of tetramethylbenzidine (TMB) in water samples. In the study of Zahra et al. [64], silver nanoparticle-doped QD particles synthesized from a raw material of *Lycii fructus* show a colour change from yellow to red during the detection of phoxim (an insecticide and potential acetylcholine inhibitor). Similarly, dual or multi-mode QD sensors were developed to efficiently detect multiple contaminants in a single sample. Such as in the study of Chen et al. [12] multifunctional sensor detects three contaminants: 2,5-diaminobenzenesulfonic acid, hydrochloride and 4-aminophenylbrononic acid in a single sample. Dual sensors, such as colorimetric and fluorescence, were also developed for detection of d-block metal ions, such Fe^{3+}, Cd^{2+} etc. [30]. Colorimetric QD sensors show high efficiency in pollution detection as compared to the fluorescence quenching method, as it does not require an external disturbance, and also, no adsorption mechanism is needed. Also, colorimetric detection eases the process of detection as chroma production provides an unaided visual appearance. Fluorescent materials involve the incorporation of expensive nano-enzymes and materials, which elevate the overall cost [20].

2.4 *Graphene QD Particles*

In some recent studies, graphene has created tremendous applications due to its versatile hybridization between carbon atoms and bulk structure. Some of the major bond interactions, such as intervalent covalent bonding, further supported by weak Vander wall interactions in bi-layer diatomic models are unique features of graphene [2]. Its 2-D structure makes it a distinctively outstanding molecule in optical, electrical, thermal, magnetic and transparent resonance, offering a pool of applications. On the other hand, zero-dimensional materials such as QD particles outsmart 2-D graphene sheets in terms of better solubility, optical and electrical properties. Blending and shrinking of such dimensional particles offer a unique combination of improvement, photovoltaics, optical fluorescence, semiconductor bio-imaging and sensing. Synthesis of graphene (2-D) blended quantum particles often made possible by hydrothermal cutting, electrical scissoring, ultrasonic treatment, milling, incarnation, extrusion etc.

Organic synthesis of GQD is assisted by hydrothermal process, by combing heteroatoms (synthesized organically) with precursor molecules at nanoscale and doping these atoms after combination step, with suitable stabilizing elements such as nitrogen (N) to form N-doped GQD's. These N-doped quantum particles show an increment in florescent quantum yield as compared to other quantum particles, doped with crystalline S, using sulphuric acid, citric acid, N-acetyl cysteine, TiO_2 etc. [18, 48]. Transition in surface morphology by doping (N) enhances electrochemical performances, cyclic stability, surface area and a popular choice as environmental remediating agent.

2.5 *QD-Based Photocatalysis*

Photocatalysis is a natural mechanism facilitating photosynthesis in plants daily and this involves transfer of photon energy packet to chemical bonds, resulting in breaking and generation of new chemical bonds. It is also a type of solar conversion mechanism. Usually, the catalysis mechanism takes advantage of an excessive surface area. QD blended semiconductor cells prevent rapid recombination of electron–hole pairs by facilitating the transfer of photoelectrons generated from the surface and behaving as electron sinks [53]. Traps and defects on the surface can act as barriers for smooth charge transfer and thus cause recombination of electrons and holes, which will hamper the photocatalysis process. Such a process can be conquered by using QD conductive surfaces or chips in solar cells, which will prevent recombination at junction's edges. QD particles themselves can generate electron–hole pairs without the use of semiconductors. In a study of Cu doped QD semiconductor, Cu-holding the maximum elemental composition shows an increment of 2.5 times photocatalytic activity as compared to QD alone. The study was conducted for the degradation of 1,4-DHP (Dihydropyridines) (aqueous solution) [24]. The mechanism of action

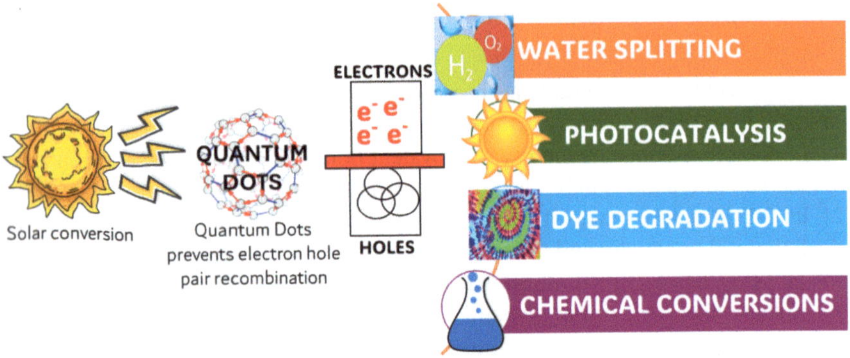

Fig. 2 Illustration of the basic mechanism for photocatalysis by QD

involves excitation of Cu-QD to Cu-QD*, followed by electron transfer from 1,4-DHP (electron donor) to Cu-QD− (radical) and 1,4-DHP+ (cation). The 1,4-DHP+ releases an electron, and this electron is transferred to O_2 (electron acceptor) and forms O_2 (radical). The final product is hence aromatized and degraded. O_2 (radical) also accepts electrons from Cu-QD− (radical), which releases Cu-QD back to a ground state, ready for the next cycle of photocatalysis. By-products of the reaction, i.e. O_2 (radical, anion), H+ decompose to form peroxide, which further decomposes to produce water and CO_2 in the presence of QD [9]. The above process is well illustrated in Fig. 2.

The application of QDs has widely increased in optical sensing and photocatalysis, thereby outsmarting conventionally used TiO_2 in semiconductor chips. TiO_2 chips were used because of their thermal and chemical stability, but suffered a major drawback because of their narrow absorption range in spectra (major absorption confined to UV wavelengths as compared to the complete solar spectra) [8]. Element doping and coupling by QD particles can enhance catalytic activity by shuttling electrons transfer by QD surfaces and facilitating charge separations. In the study, experimenting with methyl-blue degradation by N-doped TiO_2/QD particles reveals that hole stays longer on TiO_2 chips, increasing photocatalytic activity. Increased QD particles on the TiO_2 surface can hamper photodegradation, to prevent the ratio needs to be maintained carefully. Mechanism for photodegradation of methyl blue molecules (MBM) is such that MBM molecules presented on the hybrid surface are excited under visible light and convert to MBM* (excitation state), excited electrons in the MBM* are then transferred to TiO_2 conduction band that lies just above the MBM (normal state), the conduction band collects generated electrons, and MBM° (radical) is produced. In N-doped semiconductor band, MBM+ (radical) is separated from the hole and this separation further increases the photocatalyst activity [1, 25]. Figure 3 illustrates the above mechanism.

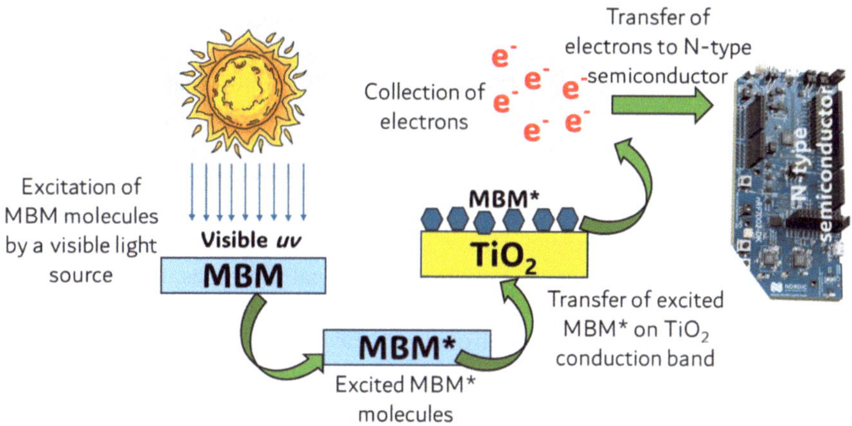

Fig. 3 Photodegradation of methyl blue molecules (MBM)

2.6 QD-Based Photodegradation

Photodegradation differs from photocatalysis such that photodegradation involves the breakdown of products in the presence of a visible spectrum light source or the sun, whereas photocatalysis is a process of photodegradation that just catalysis some specific process for the removal or breakdown of molecules. To harness the full solar spectrum for absorption, nanocomposites (hybrid structures for QD and TiO_2 and SiO_2) are applied. As TiO_2 is an affordable material for solar harnessing but faces a drawback because of a huge band gap [67]. In photodegradation of dyes, a nanocomposite hybrid of TiO_2-QD and SiO_2-QD results in 100% dye degradation in 25 and 15 min, whereas the same material without QD particles only resulted in less than 10% degradation. In the QD and TiO_2-mediated system, QD particles absorb light in the visible spectrum and emit in the UV spectra (320–425 nm wavelengths). This emitted light further excites the TiO_2 particles, generating new electron–hole pairs, which would further combine with suitable electron acceptors. This whole process is known as upconversion in a semiconductor [6].

For degradation of xenobiotic compounds and pharma molecules, QD particles can be modified for bandgap reductors, such as for diclofenac, photocatalytic activity can be enhanced by doping of porous graphite carbon nitride (g-C_3N_4). QD particles engineer the band composition and charge separation. The mechanism for diclofenac degradation is such that under the visible spectra, electrons of diclofenac are excited and transfer to QD-g-C_3N_4 conduction band. The energy for QD-g-C_3N_4 is lower in this case because of modification by QD, g-C_3N_4 acts as an intermediate for conduction of electrons from diclofenac to the conduction band. Application of only 0.6% of QD to g-C_3N_4 could increase photocatalytic activity by 3–4 times. This union also acts as an electron sink, preventing recombination [5, 49].

N-doping plays a vital role in the increase of photocatalytic activity aided by QD particles, such as in the case of numerous carbon nanostructures and nanotubes.

Spatially located doping system has made possible n-doped and p-doped semiconductor nanotubes in a single diode. In nanocomposites for hydroxyapatite, covalent bonding between QD and hydroxyapatite prevents the recombination of electron–hole and smooth conduct of charge from the surface, another it also inhibits the aggregation of QD particles in the system [13].

Commercial exploitation of QDs for environmental remediation is achieved by carbonization of waste biomass for the development of handy nanocomposite kits for the detection of heavy metal (Cd) in an aquatic ecosystem. These particles are exploited for exceptional optical and physicochemical properties, biodegradability [41]. Wastewater discharged from industries contains a large amount of volatile organic compounds (VOC) along with oxidized heavy metals and polycyclic aromatic compounds. Photocatalysis of wastewater contaminants by GQDs and modified/doped ZnO and SeO particles shows extraordinary removal of contaminants and pollutants. Photodegradation of contaminants in water is also a sustainable practice of utilizing natural light sources for degradation, and it also leverages renewable energy practices.

3 Conclusion and Prospects

QD particles have excelled in almost every field of engineering, outdoing the conventionally used molecules. These zero-dimensional particles have excellent absorption power, ranging across almost the entire wavelength of the light spectrum. Along with this optical illusion, great surface area and blending properties have made them a versatile substance for energy and remediation experiments. Environmental restoration is a major challenge for the twenty-first century, and xenobiotic compounds and obsolete pharma molecules with long persistence in the ecosystem interfere with the life cycle and metabolic functions of different flora and fauna. QD particles depict low toxicity, biocompatibility and physicochemical stability, along with drug and gene delivery, making them suitable for various applications. QD-based nanocomposites have been weaponized for environmental remediation by integration of intrinsic sensor and bio-imaging technology, along with withstanding external conditions such as light intensity, probe detection, environmental background. QD absorbents have overcome several obstacles, such as recycling, and enhanced and multiple-use adsorbing properties. The mechanism of degradation and absorption by easy excitation of contaminant electrons has made it very convenient for multiple applications. Theoretical simulations and predictive modelling have promoted the application of polymer cells, membrane tools for large-scale production in wastewater treatment plants, sewage and desalination operations. Effective cost synthesis and raw material availability for manufacturing QD particles have enhanced their practical application in anti-microbial domains without compromising on purity, homogeneity and selectivity.

Research findings	Challenges
In the study of García de Arquer et al. [19] on semiconductor QDs	QD surfaces and capping molecules are generally tailored to fix in a proper assembly unit inside a semiconductor solid to enable proper transduction and circulation of optical and energy stimuli. This involves critical research in bridging the gap between QD particles and dopants for semiconductor matrix. Along with enhancement of metastable structure, and reduction in dependency of heavy metals such as Cd, Hg. As per the life-cycle studies QD potent towards toxicity in certain domains of environment. Reproducibility of QD particles holds a big challenge in the same
In the study of Pareek et al. [42] on advanced cancer therapy with QD nanostructures	Doping of metal ions on QD surfaces involves two processes, adsorption of dopant ions on the surface and other is incorporation of dopants into the QD lattice structure. The former holds limited applications as colloidal synthesis of adsorbing dopants is energetically favourable but offer limited success
In the study of Pechnikova et al. [43] on CQD on biomedical applications and environmental remediation	CQDs are widely employed in biomedical research for targeted drug delivery systems. In environmental remediation, they can provide a perfect stimulus response in terms of increased pH, temp or enzymatic activity. Whereas a major drawback holds for crossing membrane barriers in case of blood-carriers system, brain-cell barriers during cancer therapies

Despite several advantages of QD, such as optical properties, and various applications in different fields of science and research. Several disadvantages, such as the low yield production of CQDs from renewable resources, are quite low, and the process of yielding turns out to be expensive considering the yield efficiency. In a top-down synthesis of CQDs, breaking large-sized C compounds into smaller nano-sized C particles results in a loss or low photoluminescence of generated articles and a very low yield of functional CQDs. Challenges will be addressed by focusing on future research methodology and conceptualizing a zero-waste approach during the synthesis of QD particles.

References

1. Abd Rani U, Ng LY, Ng CY, Mahmoudi E, Ng YS, Mohammad AW (2021) Sustainable production of nitrogen-doped carbon quantum dots for photocatalytic degradation of methylene blue and malachite green. J Water Process Eng 40:101816
2. Arab K, Jafari A, Shahi F (2024) The role of graphene quantum dots in cutting-edge medical therapies. Polym Adv Technol 35(9):e6571
3. Atulbhai SV, Swapna B, Kailasa SK (2023) Microwave synthesis of blue emissive carbon dots from 5-sulpho anthranilic acid and 1,5-diphenyl carbazide for sensing of levocetirizine and niflumic acid. Spectrochim Acta Part A Mol Biomol Spectrosc 287:122098

4. Babu AK, Raja MMM, Zehravi M, Mohammad BD, Anees MI, Prasad C et al (2023) An overview of polymer surface coated synthetic quantum dots as therapeutics and sensors applications. In: Progress in biophysics and molecular biology
5. Balakumar V, Ramalingam M, Sekar K, Chuaicham C, Sasaki K (2021) Fabrication and characterization of carbon quantum dots decorated hollow porous graphitic carbon nitride through polyaniline for photocatalysis. Chem Eng J 426:131739
6. Basavaraj N, Sekar A, Yadav R (2021) Review on green carbon dot-based materials for the photocatalytic degradation of dyes: fundamentals and future perspective. Mater Adv 2(23):7559–7582
7. Bijoy G, Sangeetha D (2024) Biomass derived carbon quantum dots as potential tools for sustainable environmental remediation and eco-friendly food packaging. J Environ Chem Eng 113727
8. Bokare A, Chinnusamy S, Erogbogbo F (2021) TiO_2-graphene quantum dots nanocomposites for photocatalysis in energy and biomedical applications. Catalysts 11(3):319
9. Chand P, Narula K, Vs R, Sharma S, Kumari S, Mondal N et al (2024) Mechanistic insights into cellular and molecular targets of zinc oxide quantum dots (ZnO QDs) in fungal pathogen, Candida albicans: one drug multi-targeted therapeutic approach. ACS Infect Dis
10. Chauhan VK, Shukla SK (2023) Performance analysis of prism shaped solar still using black phosphorus quantum dot material and lauric acid in composite climate: an experimental investigation. Sol Energy 253:85–99
11. Chauhan A, Kumar A, Deolia VK (2024) Zinc oxide doped with graphene quantum dots as improved electron transport layers for planner perovskite solar cells. Int J Mod Phys B 2550058
12. Chen J, Jia D, Johansson EM, Hagfeldt A, Zhang X (2021) Emerging perovskite quantum dot solar cells: feasible approaches to boost performance. Energy Environ Sci 14(1):224–261
13. Cheng C, Liang Q, Yan M, Liu Z, He Q, Wu T et al (2022) Advances in preparation, mechanism and applications of graphene quantum dots/semiconductor composite photocatalysts: a review. J Hazard Mater 424:127721
14. Cui P, Zhang J (2023) Photovoltaic performance of heteroatom-doped boron nitride quantum dots in quantum dot photovoltaic cells. Catal Commun 174:106590
15. Das S, Mondal S, Ghosh D (2024) Carbon quantum dots in bioimaging and biomedicines. Front Bioeng Biotechnol 11:1333752
16. Ding G, Yan F, Yang S, Fu Y (2024) Recent advances in carbon quantum dots for antibiotics detection. Rev Inorg Chem
17. Duan L, Hu L, Guan X, Lin CH, Chu D, Huang S et al (2021) Quantum dots for photovoltaics: a tale of two materials. Adv Energy Mater 11(20):2100354
18. Fu M, Chen W, Lei Y, Yu H, Lin Y, Terrones M (2023) Biomimetic construction of ferrite quantum dot/graphene heterostructure for enhancing ion/charge transfer in supercapacitors. Adv Mater 35(21):2300940
19. García de Arquer FP, Talapin DV, Klimov VI, Arakawa Y, Bayer M, Sargent EH (2021) Semiconductor quantum dots: technological progress and future challenges. Science 373(6555):eaaz8541
20. Guo Y, Li X, Shen P, Li X, Cheng Y, Chu K (2024) Dendritic-like MXene quantum dots@CuNi as an efficient peroxidase candidate for colorimetric determination of glyphosate. J Colloid Interface Sci 661:533–543
21. Hassan MS, Khan R, Amna T, Yang J, Lee IH, Sun MY et al (2016) The influence of synthesis method on size and toxicity of CeO_2 quantum dots: potential in the environmental remediation. Ceram Int 42(1):576–582
22. Hebbar A, Selvaraj R, Vinayagam R, Varadavenkatesan T, Kumar PS, Duc PA, Rangasamy G (2023) A critical review on the environmental applications of carbon dots. Chemosphere 313:137308
23. Hu L, Zhao Q, Huang S, Zheng J, Guan X, Patterson R et al (2021) Flexible and efficient perovskite quantum dot solar cells via hybrid interfacial architecture. Nat Commun 12(1):466
24. Iqbal T, Ashraf M, Afsheen S, Masood A, Qureshi MT, Obediat ST et al (2022) Copper sulfide (CuS) doped with carbon quantum dots (CQD) as an efficient photo catalyst. Opt Mater 125:112116

25. Jin Y, Tang W, Wang J, Ren F, Chen Z, Sun Z, Ren PG (2023) Construction of biomass derived carbon quantum dots modified TiO_2 photocatalysts with superior photocatalytic activity for methylene blue degradation. J Alloy Compd 932:167627
26. Jo HG, Kim KH, Ahn HJ (2021) Nitrogen-doped carbon quantum dots decorated on platinum catalysts for improved oxygen reduction reaction. Appl Surf Sci 554:149594
27. Kasirajan K, Karunakaran M, Choi HK (2024) Synthesis of environmentally-friendly carbon quantum dots from orange juice for selective detection of Fe^{3+} ions, antibacterial activity, and bio-imaging applications. J Environ Chem Eng 12(5):113535
28. Kim MJ, Istianah N, So BR, Kang HJ, Woo MJ, Park SJ et al (2022) Preparation of high-solid microfibrillated cellulose from Gelidium amansii and characterization of its physiochemical and biological properties. J Microbiol Biotechnol 32(12):1589
29. Kolaprath MKA, Benny L, Varghese A (2023) A facile, green synthesis of carbon quantum dots from Polyalthia longifolia and its application for the selective detection of cadmium. Dyes Pigm 210:111048
30. Kumar YA, Koyyada G, Ramachandran T, Kim JH, Hegazy HH, Singh S, Moniruzzaman M (2023) Recent advancement in quantum dot-based materials for energy storage applications: a review. Dalton Trans 52(25):8580–8600
31. Kumari K, Ahmaruzzaman M (2023) SnO_2 quantum dots (QDs): synthesis and potential applications in energy storage and environmental remediation. Mater Res Bull 112446
32. Latif Z, Shahid K, Anwer H, Shahid R, Ali M, Lee KH, Alshareef M (2024) Carbon quantum dots (CQDs) modified polymers: a mini review of non-optical applications. Nanoscale
33. Lin X, Chen T (2023) A review of in vivo toxicity of quantum dots in animal models. Int J Nanomed 8143–8168
34. Long C, Jiang Z, Shangguan J, Qing T, Zhang P, Feng B (2021) Applications of carbon dots in environmental pollution control: a review. Chem Eng J 406:126848
35. Mehta A, Mishra A, Basu S, Shetti NP, Reddy KR, Saleh TA, Aminabhavi TM (2019) Band gap tuning and surface modification of carbon dots for sustainable environmental remediation and photocatalytic hydrogen production—a review. J Environ Manag 250:109486
36. Molaei MJ (2020) The optical properties and solar energy conversion applications of carbon quantum dots: a review. Sol Energy 196:549–566
37. Montazer M, Pebdeni AB, Sheikholeslami MN, Abkenar SD, Firoozbakhtian A, Hosseini M, Dragoi EN (2023) Synthesis of cuttlebone/carbon quantum dots/nickel oxide nanocomposite for visible light photodegradation of malachite green used for environmental remediation. Chemosphere 333:138880
38. Mousa MA, Abdelrahman HH, Fahmy MA, Ebrahim DG, Moustafa AH (2023) Pure and doped carbon quantum dots as fluorescent probes for the detection of phenol compounds and antibiotics in aquariums. Sci Rep 13(1):12863
39. Nazibudin NA, Zainuddin MF, Abdullah CAC (2023) Hydrothermal synthesis of carbon quantum dots: an updated review. J Adv Res Fluid Mech Therm Sci 101(1):192–206
40. Pastuszak J, Węgierek P (2022) Photovoltaic cell generations and current research directions for their development. Materials 15(16):5542
41. Patil AB, Chaudhary PL, Adhyapak PV (2024) Carbon dots–cadmium sulfide quantum dots nanocomposite for 'on–off' fluorescence sensing of chromium (VI) ions. RSC Adv 14(18):12923–12934
42. Pareek A, Kumar D, Pareek A, Gupta MM (2025) Advancing cancer therapy with quantum dots and other nanostructures: a review of drug delivery innovations, applications, and challenges. Cancers 17(5):878
43. Pechnikova NA, Domvri K, Porpodis K, Istomina MS, Iaremenko AV, Yaremenko AV (2025) Carbon quantum dots in biomedical applications: advances, challenges, and future prospects. Aggregate 6(3):e707
44. Prabhu SA, Kavithayeni V, Suganthy R, Geetha K (2021) Graphene quantum dots synthesis and energy application: a review. Carbon Lett 31(1):1–12
45. Ratre P, Jain B, Kumari R, Thareja S, Tiwari R, Srivastava RK et al (2022) Bioanalytical applications of graphene quantum dots for circulating cell-free nucleic acids: a review. ACS Omega 7(44):39586–39602

46. Ren Y, Fan Z (2023) Synthesis of molecularly imprinted polymers based on nitrogen-doped carbon dots for specific detection of chlortetracycline by reversed phase microemulsion method. Talanta 265:124898
47. Seng RX, Tan LL, Lee WC, Ong WJ, Chai SP (2020) Nitrogen-doped carbon quantum dots-decorated 2D graphitic carbon nitride as a promising photocatalyst for environmental remediation: a study on the importance of hybridization approach. J Environ Manag 255:109936
48. Sengupta S, Pal S, Pal A, Maity S, Sarkar K, Das M (2023) A review on synthesis, toxicity profile and biomedical applications of graphene quantum dots (GQDs). Inorganica Chimica Acta 121677
49. Shi J, Luo Y, Yang T, Wang H, Ju C, Pu K et al (2022) Enhanced nonsacrificial photocatalytic generation of hydrogen peroxide under visible light using modified graphitic carbon nitride with doped phosphorus and loaded carbon quantum dots: constructing electron transfer channel. J Colloid Interface Sci 628:259–272
50. Si QS, Guo WQ, Wang HZ, Liu BH, Ren NQ (2020) Carbon quantum dots-based semiconductor preparation methods, applications and mechanisms in environmental contamination. Chin Chem Lett 31(10):2556–2566
51. Sikiru S, Oladosu TL, Kolawole SY, Mubarak LA, Soleimani H, Afolabi LO, Toyin AOO (2023) Advance and prospect of carbon quantum dots synthesis for energy conversion and storage application: a comprehensive review. J Energy Storage 60:106556
52. Soni H, Bhattu M, Priya SD, Kaur M, Verma M, Singh J (2024) Recent advances in waste-derived carbon dots and their nanocomposites for environmental remediation and biological applications. Environ Res 251:118560
53. Sun P, Xing Z, Li Z, Zhou W (2023) Recent advances in quantum dots photocatalysts. Chem Eng J 458:141399
54. Thoma T, Ma-Hock L, Schneider S, Honarvar N, Treumann S, Groeters S et al (2024) Toxicological inhalation studies in rats to substantiate grouping of zinc oxide nanoforms. Part Fibre Toxicol 21(1):24
55. Vercelli B (2021) The role of carbon quantum dots in organic photovoltaics: a short overview. Coatings 11(2):232
56. Vyas Y, Chundawat P, Dharmendra D, Punjabi PB, Ameta C (2021) Review on hydrogen production photocatalytically using carbon quantum dots: future fuel. Int J Hydrog Energy 46(75):37208–37241
57. Wang X, Yang Y, Huo D, Ji Z, Ma Y, Yang M et al (2020) A turn-on fluorescent nanoprobe based on N-doped silicon quantum dots for rapid determination of glyphosate. Microchim Acta 187:1–9
58. Wang S, Chen H, Xie H, Wei L, Xu L, Zhang L et al (2021) A novel thioctic acid-carbon dots fluorescence sensor for the detection of Hg^{2+} and thiophanate methyl via S-Hg affinity. Food Chem 346:128923
59. Wang J, Deng L, Chen M, Che Y, Li L, Zhu L et al (2024) Phytogenic feed additives as natural antibiotic alternatives in animal health and production: a review of the literature of the last decade. Anim Nutr
60. Wen F, Li P, Yan H, Su W (2023) Turmeric carbon quantum dots enhanced chitosan nanocomposite films based on photodynamic inactivation technology for antibacterial food packaging. Carbohyd Polym 311:120784
61. Yadav A, Chichkov NB, Avrutin EA, Gorodetsky A, Rafailov EU (2023) Edge emitting mode-locked quantum dot lasers. Prog Quantum Electron 87:100451
62. Yang M, Lian Z, Si C, Jan F, Li B (2022) Revealing the intrinsic relation between heteroatom dopants and graphene quantum dots as a bi-functional ORR/OER catalyst. Mol Catal 518:112109
63. Yempally S, Cabibihan JJ, Ponnamma D (2024) A facile approach to develop polyvinyl alcohol-based bio-triboelectric nanogenerator containing graphene-doped zinc oxide quantum dots. Energ Technol 12(3):2300992
64. Zahra F, Qureshi Z, Ali MN (2024) Carbon quantum dots, its synthesis and evaluation of its cytotoxicity. In: Emerging sustainable nanomaterials for biomedical applications. Springer Nature Switzerland, Cham, pp 349–383

65. Zhang J, Zhang S, Zhang Y, Al-Hartomy OA, Wageh S, Al-Sehemi AG et al (2023) Colloidal quantum dots: synthesis, composition, structure, and emerging optoelectronic applications. Laser Photonics Rev 17(3):2200551
66. Zhang Y, Liu Y, Yang Y, Li L, Tao X, Song E (2023) Rapid detection of pathogenic bacteria based on a universal dual-recognition FRET sensing system constructed with aptamer-quantum dots and lectin-gold nanoparticles. Chin Chem Lett 34(8):108102
67. Zhang H, Sun S, Ding H, Chen Y (2023) Quantum dots TiO_2 loaded amorphous SiO_2 composite photocatalysts: significant performance enhancement and the effect of SiO_2 surface hydroxyl groups. J Alloy Compd 960:170700
68. Zhichao Y, Zhenguo T, Nan Z, Zeyue C, Linhe H, Gang L (2022) TiO_2 nanotubes modified by CdS quantum dots and performance of photocatalytic water splitting for hydrogen. J Petrochem Univ 35(4):38

Chapter 11
Fate and Bioaccumulation of Quantum Dots

Aishwarya Joji Mathew and T. P. Vinod

Abstract Quantum Dots (QDs), nanoscale emissive particles with unique optoelectronic properties, are used as functional components in various technological applications, including imaging, sensing, drug delivery, electronic devices, and environmental monitoring. However, their widespread use raises concerns about their environmental impact, particularly regarding their fate and bioaccumulation. This chapter discusses different aspects of the fate and bioaccumulation of QDs within ecological systems. The environmental fate of QDs depends on their transport, transformation, partitioning, and degradation processes. Factors such as particle size, surface chemistry, and environmental conditions significantly influence their behavior and persistence in air, water, and soil. Bioaccumulation refers to the uptake, accumulation, and potential toxic effects of QDs in the environment. This process leads to the accumulation of QDs in tissues, with implications for the well-being of organisms and ecological balance. A deep understanding of the complex interactions between these nanoparticles and the ecosystem is crucial for accurately evaluating environmental issues and developing successful mitigation plans. To enhance our understanding of the environmental impact of these engineered nanomaterials, it is essential to gain detailed insight into the behavior, transformation, and fate of QDs. This chapter includes discussions on their modes of action, ecotoxicity, trophic transfer, and bioaccumulation across various trophic levels. Future research directions are also suggested to address existing knowledge gaps and to support the safe application and disposal of QDs.

Keywords Quantum dots · Bioaccumulation · Fate · Environment · Toxicity

A. J. Mathew · T. P. Vinod (✉)
Department of Chemistry, CHRIST University, Bangalore, India
e-mail: vinod.tp@christuniversity.in

T. P. Vinod
Centre for Renewable Energy and Environmental Sustainability, CHRIST University, Bangalore, India

T. Bhuyan et al. (eds.), *Multifunctional Quantum Dots*, Materials Horizons: From Nature to Nanomaterials, https://doi.org/10.1007/978-981-92-0172-3_11

1 Introduction

Quantum Dots (QDs) are nanoscale semiconductor particles that possess unique optical and electronic properties due to their size and quantum mechanical effects. QDs typically range from 1 to 10 nm in diameter [1]. At this scale, their properties are influenced by quantum confinement, where the electronic and optical characteristics are determined by their size, shape, and material composition [2, 3]. Quantum confinement leads to discrete energy levels, meaning that the electronic states of QDs are quantized. This results in size-dependent properties, such as color and fluorescence, which can be tuned by changing the size of the dots (Fig. 11.1) [4]. QDs exhibit strong and tunable photoluminescence, making them highly efficient light emitters [5]. When excited by light, they can emit photons at specific wavelengths, which depend on their size and material composition [6].

QDs can be synthesized using physical, chemical, and biological methods. Chemical methods for the preparation of QDs include solvothermal, hydrothermal, and colloidal synthesis. Physical methods such as electron-beam lithography, which creates QDs with defined sizes and pattern, are less commonly used for large-scale production. Biosynthetic approaches using microorganisms or plant extracts have been explored as environmentally friendly alternatives for producing QDs [7]. QDs are widely used in displays, such as Quantum Dot LED (QLED) TVs, where they enhance color accuracy and brightness [8]. In biomedical applications, QDs serve as

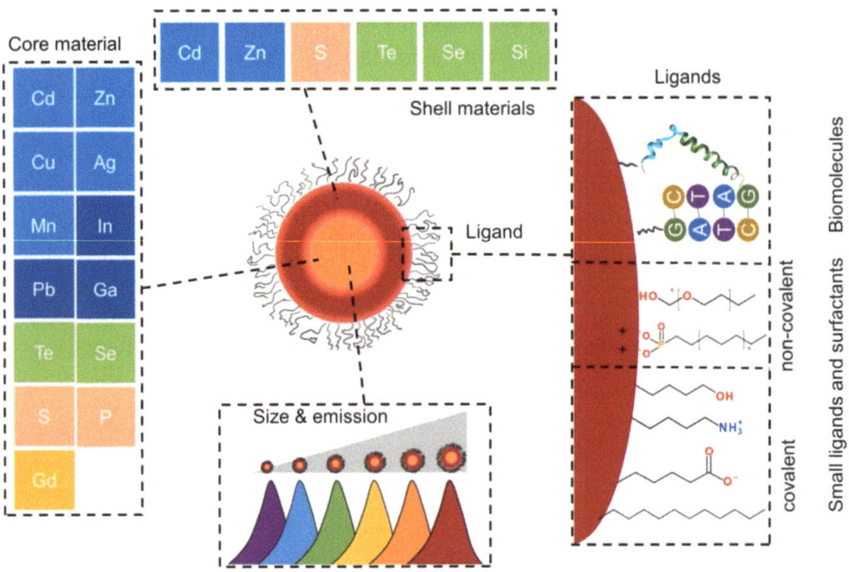

Fig. 11.1 Overview of Semiconductor Quantum Dots (SQDs) composition and properties. An example of core–shell SQD components with their surface ligands. Reproduced with permission [4], Copyright 2023, Elsevier

fluorescent labels for imaging, drug delivery systems, and as agents in photo-thermal therapy due to their size and photo-physical properties [9]. QDs can be incorporated into solar cells to improve efficiency by enhancing light absorption and charge separation. QDs are utilized in a range of sensing applications, such as environmental monitoring and biomolecule detection, because of their sensitivity and adjustable fluorescence [10], 11].

The stability of QDs is a critical factor that influences their performance and suitability for various applications. Surface defects can introduce energy states that affect the stability of QDs. Passivation techniques, which involve coating the surface with stabilizing materials, can help reduce these defects [12]. QDs are known for their photoluminescence properties, but prolonged exposure to light can lead to photobleaching, where the intensity of emitted light decreases over time. Stability can be enhanced through appropriate surface treatments and encapsulation. The stability of QDs is a multifaceted issue that impacts their functionality and safety across various applications [13]. By understanding the factors that influence stability and employing strategies to enhance it, we can optimize the performance of QDs while minimizing potential environmental risks [14].

The potential toxicity of QDs, particularly those containing heavy metals like cadmium, raises concerns about their impact on human health and ecosystems [15]. The release of toxic metal ions upon degradation or dissolution is a significant risk [16]. The smaller size and enhanced surface area of QDs can lead to physical interactions with cells. Their uptake can result in oxidative stress, inflammation, and apoptosis (programmed cell death) due to their ability to generate Reactive Oxygen Species (ROS) [16]. When QDs are thrown into the ecosystem, they can accumulate in aquatic and terrestrial ecosystems, posing risks to wildlife [17, 18. Their toxicity can affect various organisms, including microorganisms, plants, and animals, potentially disrupting food chains and ecosystems. Understanding the toxicity of QDs is essential for risk assessments that inform regulatory decisions regarding their production, use, and disposal. Research efforts focus on developing less toxic QDs and improving their safety profiles through better synthesis methods, surface modifications, and encapsulation techniques [19].

QDs play a significant role in environmental science, offering innovative solutions for monitoring, remediation, and sustainable practices [20, 21]. However, their environmental impacts and potential toxicity must be carefully assessed to ensure safe application and management [22]. In this chapter, we aim to offer a comprehensive overview of the fate and bioaccumulation of QDs. Additionally, we emphasize the stability, toxicity of QDs with their release, uptake, transport, and impact on aquatic and terrestrial organisms. We have also discussed certain models predicting the bioaccumulation of QDs in the ecosystem.

2 Fate of QDs in the Environment

The word "Biological fate" can be used to describe a number of potential outcomes that can include (a) possible pathways for human exposure to QDs; (b) the likelihood that QDs will break down into their constituent molecules; (c) the propensity of QDs to agglomerate, which could influence their biological environment; (d) Whether QDs will collect in cells or tissues after exposure, or if they will be expelled; and (e) what occurs to QDs after they are excreted or released into the environment.

2.1 Environmental Release Pathways

QDs can enter different environmental components through either point sources or area sources. Point sources include locations where nanoparticles are manufactured, such as landfills, wastewater treatment plants, and waste incineration facilities. Area sources involve the release of nanomaterials (such as metallic, metal oxide, polymer, and carbon-based nanoparticles) during the use of these nanomaterials. For instance, nanomaterials may be released when washing cloth with nanosilver or when rain washes off paints containing TiO_2 from building exteriors [5]. Nanomaterials are employed in agriculture to boost crop yields by enhancing the growth of plants or protecting crops from pests and organisms. For example, TiO_2 nanoparticles may be transported into the environment when applied to plant leaves to promote growth, while Cu nanoparticles can be released through the use of pesticides containing these nanoparticles [23]. Another possible source of nanomaterials is waste that consists of used nanomaterials. When these used nanogoods are processed in waste incineration facilities, nanoparticles may be discharged into the atmosphere. However, because of its high dispersion capacity and the short half-life (around 10 days) of particles smaller than 100 nm in the atmosphere, it is thought that the atmosphere is less exposed to these toxins than other environmental constituents. Furthermore, there is a chance that recycling nanoproducts will spread nanoparticles to secondary goods, like when textile nanoparticles find their way into paper goods. In this way, nanoparticles could contaminate environmental components and endanger human health and living things [24].

Identifying the possible exposure sources that could result in human uptake or absorption is crucial while dealing with the toxicological and biological destiny of QDs. Humans may come into contact with potentially dangerous substances in several ways: If the compounds are airborne, (i) they can be inhaled; (ii) absorbed through the skin or eyes; (iii) consumed; or (iv) injected (Fig. 11.2) [25]. According to Maynard et al., mass concentration, particle number, surface area, and the fundamental physicochemical properties of QDs should be taken into account in this context [26]. There are very few possible hazards associated with breathing in QDs. The production and handling of QDs probably have the greatest potential for aerosolization. However, aerosol formulations may also be used in future clinical

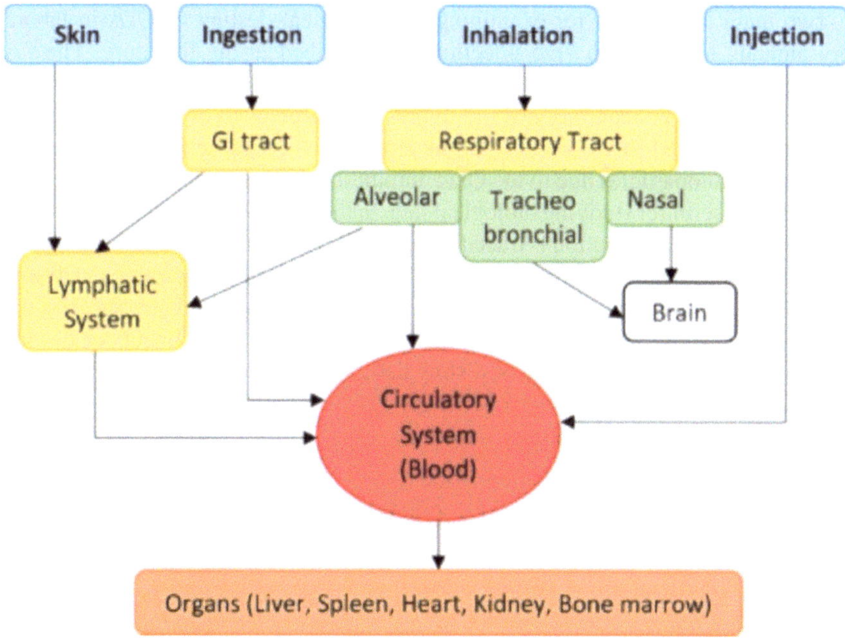

Fig. 11.2 Schematic illustration showing the environment release pathways for the nanomaterials. Reproduced with permission [25], Copyright 2022, Elsevier

uses of QDs. Aggressive and dangerous substances are rapidly added to massive volumes of fast-boiling organic solvents at high temperatures in classic QD production. Therefore, regulators will need to give careful thought to workplace exposure to aerosolized QDs [27].

Humans may be exposed locally or systemically to potentially hazardous substances through the skin and eyes. QDs have been shown to enter undamaged skin by Ryman-Rasmussen et al. [28]. Their findings indicate that QDs may penetrate the skin in a single day and have a variety of core/shell sizes, shapes, and surface coatings with different charges. The toxicity of QDs through their ingestion has to be addressed next. Regulators will need to assess the potential for accidental ingestion of QDs by individuals working in manufacturing, research laboratories, or diagnostic facilities. Additionally, there is a risk of ingestion from consuming food or water contaminated with nanoparticles [29]. According to Wang et al., when QDs were exposed to low pH levels, their structures broke down, releasing free cadmium ions, which increased the toxicity of QDs to the human small intestine [30]. The other way through which the environment is exposed to QDs toxicity is through direct injection (intravenous, subcutaneous, etc.) into humans and animals. In contrast to intravenous injections, which are more frequently utilized for other clinical applications, QDs would be delivered directly into tumor tissues. Depending on the injection method, QDs have different outcomes.

Two main approaches for examining the biological destiny of QDs in laboratory animals following in vivo injection are described in the literature. The first technique tracks the biodistribution of QDs by utilizing their fluorescence characteristics. Researchers can monitor the fluorescent particles to determine where the QDs are distributed in the body. This method offers a significant advantage over many other nanoparticles due to the QDs' luminescent properties, which make them particularly useful for studying biological fate. In the second approach, QDs are radiolabeled and their distribution is traced using these radioactive forms. The advantage of radiolabeling is the ability to gather quantitative information regarding the destiny of the QDs. Nevertheless, this technique is unable to distinguish between QDs that are active and those that have decomposed into their basic components or become inactive. Functional particles can be distinguished since only intact QD particles will continue to glow [19].

2.2 Transport Mechanisms in Air, Water, and Soil

When released into the environment, QDs can either undergo physicochemical reactions or stay in their original environmental compartment unaltered [1]. Nanomaterials typically enter the atmosphere by human activity, different types of atmospheric deposition (both wet and dry), or interactions between the atmosphere and the ground surface (both land and water). Additionally, atmospheric circulation can carry and disseminate nanomaterials in the air across long distances [31]. Nevertheless, there is no solid proof or data regarding the presence or concentration of QDs in the atmosphere, and they cannot move through the atmosphere easily. However, it is known that QDs can travel through soil and water (Fig. 11.3) [4]. The migration of QDs is restricted by their ability to penetrate soil particle gaps when they reach the soil from sources such as waste dumps or the breakdown of QD solar panels [32]. In deteriorated soils or with chelators present, this migration is accelerated. Furthermore, QDs may undergo partial breakdown and change as a result of soil aging, which would decrease their stability [33]. Additionally, QDs in the soil can be absorbed and collected by terrestrial creatures, such as plants and animals, or they can infiltrate into groundwater or surface or subsurface runoff to reach aquatic bodies. The behavior of QDs in an aquatic environment can be complicated. They can either agglomerate into the sediment or stay steadily distributed and floating in the water. Additionally, physical, chemical, and biological activities can change or degrade QDs in water [34].

QDs can become airborne through aerosolization processes, where they are emitted into the air as particles or droplets [35]. This typically occurs during manufacturing or handling. Once airborne, QDs are subjected to Brownian motion, which causes random movement due to collisions with air molecules. This affects their dispersion and concentration in the atmosphere. QDs can settle out of the air through gravitational settling or deposition on surfaces, such as soil, water bodies, or vegetation, influenced by particle size and density. Larger particles tend to settle faster

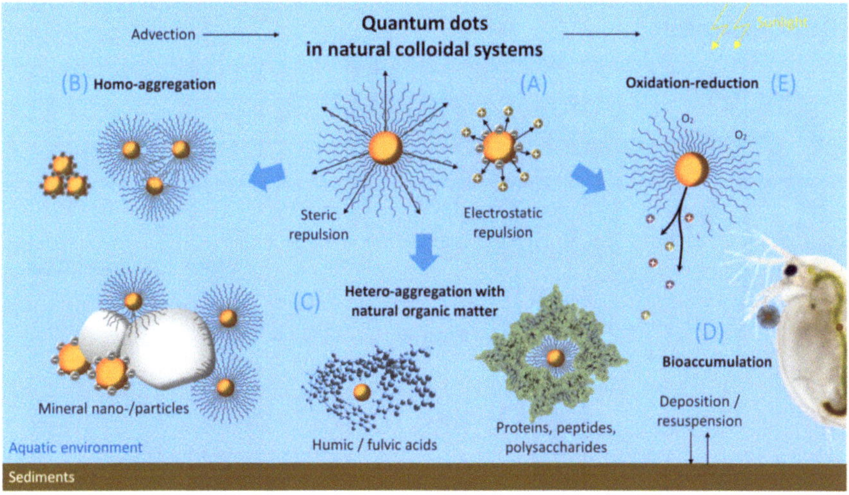

Fig. 11.3 Possible transformation processes of Engineered Nanoparticles (ENPs) in water for both electrostatically and sterically stabilized ENPs. (A) Ion pairing occurs between charged QDs and dissolved ions. (B) Colloidal instability arises from homo-aggregation. (C) Hetero-aggregation occurs with DOM or through attachment to nanoscale particles. (D) Bioaccumulation within organisms. (E) Dissolution through redox reactions or photochemical processes. Reproduced with permission [4], Copyright 2023, Elsevier

due to gravity [36]. Smaller QDs may remain suspended longer, while larger ones are more prone to settling. Humidity, temperature, and wind velocity can influence the transport and deposition patterns of airborne QDs. In aquatic environments, QDs can disperse evenly or aggregate, depending on factors such as ionic strength, pH, and the presence of organic matter. Aggregation can lead to sedimentation, affecting their bioavailability. QDs may move through water via diffusion, driven by concentration gradients. This process is influenced by the size and surface properties of the QDs. Water movements, such as currents and turbulence, can transport QDs over longer distances [37]. This process is critical in rivers, lakes, and coastal areas. The surface chemistry of QDs, including the type of coatings and ligands, can alter their stability, solubility, and interactions with water molecules and other particles. Factors like temperature, salinity, and the presence of organic or inorganic substances in the water can significantly affect QD behavior [38]. Some studies have shown that QDs can accumulate in the leaf epidermis and vascular tissue following application to the leaf surface. Additionally, many researchers have noted that QDs can be transported to other plant tissues. They suggested that QDs might also be absorbed through the stomata pathway (Fig. 11.4) [39]. QDs can migrate through soil profiles via leaching, where water percolates through the soil and carries QDs downwards. This process is influenced by soil moisture content and permeability. QDs can adsorb to soil particles, which can immobilize them and reduce their mobility [40]. The extent of adsorption depends on the QDs' surface characteristics and the soil's texture and composition. In some cases, QDs may behave as colloids, remaining suspended in soil moisture

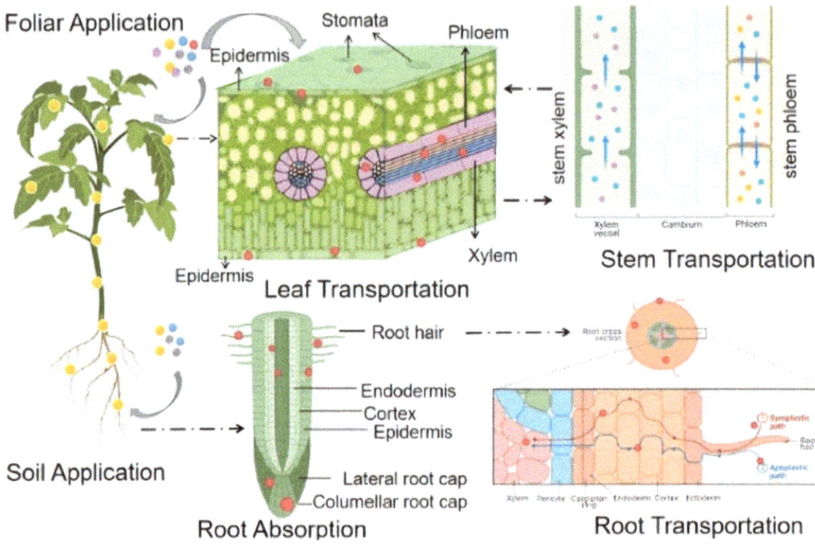

Fig. 11.4 Schematic illustration of how nanoparticles are taken up and transported within plants via foliar application or root exposure. Reproduced with permission [39], Copyright 2023, MDPI

and moving with water through the soil matrix. The type of soil (sand, silt, clay) affects the adsorption and retention of QDs. Clay soils may retain more QDs due to higher surface area and charge. These factors influence the chemical interactions between QDs and soil particles, impacting their mobility and bioavailability [41].

The transport mechanisms of QDs in air, water, and soil are complex and influenced by various physical, chemical, and environmental factors. Comprehending these dynamics is crucial for forecasting the environmental destiny of QDs and evaluating their possible advantages and disadvantages in diverse applications. Continued research in this area is necessary to develop strategies for managing QDs effectively in both industrial and environmental contexts.

2.3 *Transformation and Metabolism Processes*

QDs can go through a variety of changes regardless of whether they are moved across various environmental compartments, adsorbed onto soil, or disseminated in the water (Fig. 11.5) [42]. These modifications could involve interactions with macromolecules and physicochemical processes [43]. The impact of pH on the behavior of QDs is one of the primary elements influencing their metamorphosis. Variations in the dispersion medium's pH can affect the optical characteristics, stability, and aggregation of QDs during industrial manufacturing and their subsequent release into diverse environments. In low pH conditions, QDs tend to lose luminosity, whereas in high pH situations, they generally show stronger luminescence [44]. The nature of QDs is also influenced by illumination. The structure of QDs can change when exposed to light, such as UV or natural light, which can affect the release of core ions and the generation of Reactive Oxygen Species (ROS). Their toxicity, degradation, and luminescence therefore impact the environment [45]. Additionally, when QDs are exposed to air, UV light can degrade them and liberate free metal ions [42]. This makes the QDs more poisonous and compromises their stability. QDs can undergo photodissociation when exposed to natural light. According to studies, when exposed to natural sunlight, CDs, which are made of carbon, form optically inactive carbon nanoparticles. These subsequently experience indirect photolysis as a result of interactions with dissolved oxygen, metal ions, nitrates, and other aqueous substances. Some carbon atoms undergo oxidation to produce nitro compounds, while others undergo conversion to volatile forms like CO_2 or acetone. The remaining atoms undergo mineralization until the CDs are fully mineralized [46].

Another main factor that affects the behavior of QDs was found to be the influence of Natural Organic Matter (NOM) with the QDs. The transition of QDs is also influenced by the presence of NOM. NOM is a decisive factor in how nanoparticles cluster in soil and aggregate in water [47]. Research indicates that NOM primarily enhances the spatial potential resistance among nanoparticles, hence augmenting their stability in water. Additionally, these compounds boost nanoparticle stability by

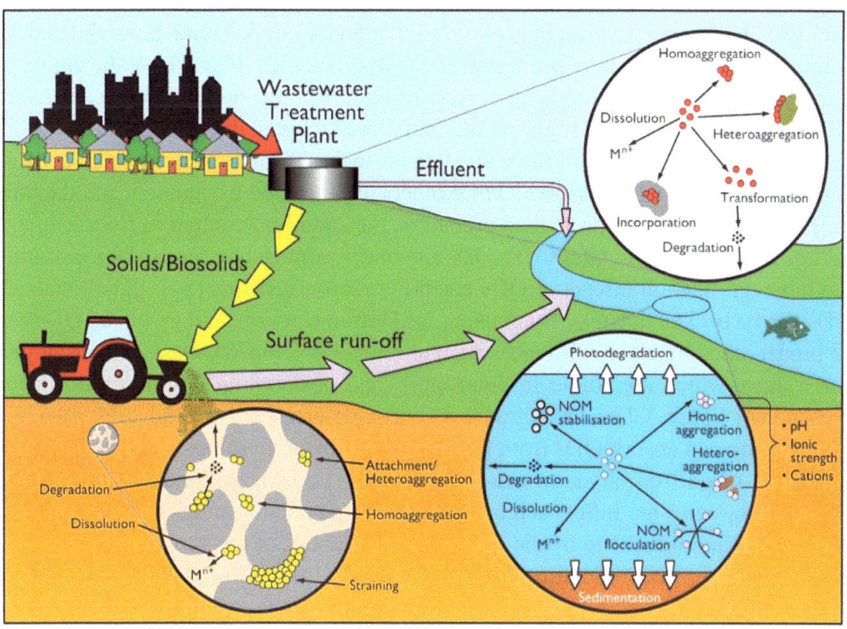

Fig. 11.5 Pathways and transformations of nanomaterials in the environment. Reproduced with permission [42], Copyright 2012, ACS

enhancing electrostatic repulsion, which decreases the rate of nanoparticle coalescence. For QDs, which are luminescent nanomaterials, NOM adsorbs onto their surface when they come into contact. Their movement in the water and the intensity of their luminescence are both impacted by this adsorption, which is mostly caused by hydrophobic interactions [48]. According to the studies, ionic strength also influences how QDs behave. Since Na^+ and Ca^{2+} are the main cations in natural water, investigations looking at how ionic strength affects QDs frequently utilize NaCl and $CaCl_2$. Studies that mimic real water conditions have demonstrated that the concentration of NaCl has a major impact on the deposition and transport behavior of graphene oxide QDs [49]. Higher ionic strength leads to increased deposition rates, likely due to the ionic bridge effect caused by higher NaCl concentrations. In general, the firmness and agglomeration of QDs in water are largely influenced by ionic strength. However, findings on this effect are inconsistent, and the exact mechanism remains unclear. The functional groups present on QDs might be a contributing factor to these variations [34].

When QDs enter or remain in the environment, they can disintegrate in living organisms, be absorbed by biological agents in different environmental compartments, or proceed through biomagnification and nutrient transfer [50]. Studies have shown that plants can absorb QDs from the soil along with the absorption of water or nutrients. The biological distribution of QDs entering plants varies in different cases [51]. On the surface of taproots, lateral roots, and fibrous roots, Polyethylene Glycol-(PEG-) COOH–CdS/ZnS QDs and Mercaptopropionic Acid-(MPA-) CdS/ZnS QDs have been seen to aggregate; this aggregation behavior is modulated by conjugated C–C bonds between lignin and QDs-MPA [52]. Plants have the ability to transmit QDs across the food chain. According to studies, rooted plants can absorb QDs from the soil through water intake, and these QDs can then be transmitted from plants to insects. QDs can infiltrate algal plants in freshwater settings by pinocytosis; some of these particles dissolve, while others are ingested by mussels and daphnia [53]. When ingested by protozoan predators, the cadmium content in *Pseudomonas aeruginosa* that accumulates CdSe QDs can rise by almost five times, demonstrating the biomagnification effects of QDs eaten by microorganisms [54]. Functionalized QDs can be broken down by fungi using the Fenton reaction, and graphene QDs can be biodegraded by other substances such as human myeloperoxidase and eosinophil peroxidase, which are found in neutrophils and eosinophils. Furthermore, in the presence of lipase, CDs can degrade via peroxide-catalyzed degradation. Fish and zooplankton can absorb QDs from freshwater protozoa. In saltwater, QDs have also been seen to go from algae to marine life [55]. The behavior and destiny of QDs in aquatic systems are influenced by a number of physicochemical changes on their surface that are influenced by the abiotic characteristics of the surrounding environment, including pH, ionic strength, and visible and ultraviolet light. Aggregation, which can be either homo- or hetero-agglomeration, is the main transformation of QDs in water. The frequency of collisions and the equilibrium of repulsive and attractive forces, such as electrostatic repulsions and van der Waals forces, at the QD surface have an impact on these processes. Compared to freshwater, QDs have a tendency to settle and aggregate more readily in seawater [56]. The kind of nanoparticle, its coatings, ligands, concentration, size, aggregation state, and environmental parameters like pH and salinity all affect how easily QDs dissolve.

2.4 Degradation and Persistence

Degradation of QDs includes the potential for QDs to either change through aggregation into more complex structures or decompose into their constituent molecules [57]. The total toxicity of the QDs was raised by their breakdown and the consequent release of free Cd (II) ions. The chemical stability of radiolabeled CdTe QDs in water was recently assessed by Woodward et al. They discovered that uncapped CdTe core QDs were released into the environment in three days after being suspended in an aqueous buffer. ZnS shell with CdTe cores, on the other hand, showed noticeably more stability and could stay in neutral buffer for almost a month. According to this work, ZnS capping significantly increases the stability of QDs in water [12]. In the investigation of the effects of pH levels (pH 4.8 versus pH 7.4) on the stability of CdSeS/SiO_2 QDs, Chen et al. found that the QDs maintained their integrity in both acidic and alkaline buffers for up to five days. They found that the CdSeS/SiO_2 QDs did not exhibit any discernible leaching of free cadmium ions, suggesting that these dots were extremely firm [58]. Therefore, one element that seems to affect QDs stability is their chemical makeup. CdSe QDs degraded and released free cadmium ions when exposed to air and UV light as studied by Derfus et al. When the QDs were exposed to these conditions, this deterioration made them more hazardous. Although instability caused by air and UV radiation might not be very important for QDs, it might be a big consideration when thinking about the possible long-term impacts of QDs discharged into the environment [45]. The carbon dots also have a propensity to group to form high-order structures. Red CdTe nanoparticles tended to clump together over time, according to Zhang et al.'s recent assessment of the durability of CdTe nanoparticles in cell culture settings [14]. Additionally, they observed that the QDs primarily aggregated outside of the cells. When CdTe QDs were dissolved in aquarium water, another group observed that they aggregated. Different QDs' propensity to self-aggregate is probably influenced by their surface chemistry, and aggregated QDs may have quite different effects on the environment and human health than non-aggregated particles. The tendency of QDs to aggregate has not been well studied, thus future studies will need to examine the effects of aggregation on QD stability and biological effects. This information will be essential to comprehend the biological and environmental fates of QDs [59].

QDs can undergo oxidation when exposed to oxygen, leading to changes in their chemical structure and properties. For example, the oxidation of cadmium selenide (CdSe) QDs can result in the formation of cadmium oxide and selenium oxides, which can release toxic metal ions into the environment. Exposure to light, particularly Ultraviolet (UV) radiation, can degrade QDs. The dissolution of the QDs composition and the release of heavy metal ions can result from the energy of UV light breaking chemical bonds [60] QDs can react with other environmental components, such as dissolved organic matter, pH changes, and salts, which can lead to destabilization and degradation. Certain microorganisms have been shown to interact with and degrade QDs. For instance, bacteria may produce enzymes or metabolic by-products that can alter the QDs, potentially facilitating their breakdown. The surface coatings of QDs can significantly affect their stability and degradation rates. More stable surface coatings can protect QDs from degradation, while less stable ones can lead to faster breakdown in environmental conditions [61]. QDs can accumulate in various environmental compartments, including water, soil, and sediments. Their persistence in these compartments can lead to long-term exposure for organisms, raising concerns about bioaccumulation and biomagnification in food webs. The core material of QDs (e.g., CdSe, PbS) plays a crucial role in their persistence. Heavy metal-based QDs are more likely to persist due to the stability of their metal cores. When QDs aggregate and settle, they can persist in sediments for extended periods, potentially affecting benthic organisms and the overall sediment ecosystem [62]. Their interaction with sediment particles can also impact their availability for degradation. The degradation and persistence of QDs in the environment are influenced by a complex interplay of chemical, physical, and biological factors. While QDs can degrade through various mechanisms, their persistence can lead to ecological risks, particularly related to bioaccumulation and the release of toxic by-products.

3 Bioaccumulation of QDs in the Environment

3.1 Uptake Pathways, Distribution, and Mechanism of Bioaccumulation

Nanomaterials have the potential to bioaccumulate at many stages of the food chain in addition to having direct harmful impacts on living beings [63]. The accumulation of these materials in the tissues of living organisms has been demonstrated by numerous research [22]. According to Hu et al., *E. fetida* may be harmed by the bioaccumulation of TiO_2 and ZnO nanoparticles in soil at concentrations higher than 1 g kg^{-1} [64]. Finding the precise areas of the body where nanomaterials accumulate should be a goal of research on their bioaccumulation. According to the studies, the characteristics of the nanomaterials can affect how much bioaccumulation occurs. Coutris et al.'s investigation into the bioaccumulation of cobalt and silver nanoparticles particularly in *E. fetida* lends credence to this. The worms consumed 69% of the Co nanoparticles from their meal after four weeks of exposure, whereas only 0.4% of the Ag Np was accumulated [65]. Their results showed that both the solubility of the nanoparticles and the source of the zinc have an impact on the degree of bioaccumulation of these compounds. The sluggish excretion of nanoparticles from *D. magna* was linked to the significant bioaccumulation of $nTiO_2$ observed by Zhu et al. [66]. Malnutrition resulted from delayed excretion, which made it harder to absorb meals.

A secondary source of contamination could be nanomaterials that build up within living things. These nanoparticles may pose a concern to the environment even after excretion or the organism's death. This issue has not yet been thoroughly explored in research. Additionally, it's crucial to recognize that nanomaterials within an organism can undergo various transformations, such as reacting with cellular components [67]. Bioaccumulation could also be linked to the risk of increasing nanoparticle concentrations as they move up through successive levels of the food chain, a phenomenon known as biomagnification. When Holbrook et al. studied the flow of QDs from bacteria (*E. coli*) to ciliates (*Tetrahymena pyriformis*) to rotifers (*Brachionus calyciflorus*), they did not find any indications of biomagnification [7]. Through trophic transfer from *Nicotiana tabacum* L. cv Xanthi (the producer) to the tobacco hornworm *Manduca sexta* (the consumer), Judy et al. observed a rise in the concentration of Au nanoparticles (with diameters of 5, 10, and 15 nm) [68]. The size of the Au nanoparticles had a substantial impact on the degree of bioaccumulation of the particles in the tobacco hornworm tissues; the highest bioaccumulation was seen for Au nanoparticles with a diameter of 10 nm, followed by 15 nm and then 5 nm.

Understanding the toxicities associated with different modes of nanoparticle exposure, especially compared to their dissolved metal ions, is an emerging focus in ecotoxicology. Brian and his colleagues examined the bioaccumulation, bioavailability, and toxicity of carboxyl-functionalized CdSe/Zn QDs in the amphipod *Leptocheirus plumulosus* when these amphipods are exposed to similar concentrations of cadmium (Fig. 11.6) [69]. The exposure resulted in a greater accumulation of QDs than dissolved cadmium. Notably, exposure to these nanoparticles through algae led to significant mortality in the amphipods. A strong correlation was found between cadmium and selenium levels in amphipods exposed to QDs in water, with these metals being spatially concentrated. On the other hand, the metals were less correlated and more distributed when the amphipods were exposed to QDs through algae, suggesting that QD dissolution could result in higher metal ion toxicity. This study suggests that QDs may aid in trophic transmission because they accumulate more than dissolved ions. QDs are toxic and bioavailable when ingested with algae, but not in the absence of algae [69]. The growing use of man-made nanoparticles in consumer items necessitates a thorough assessment of their ecological effects and trophic transfer. Govind and the group established a laboratory-scale aquatic microbial food chain with the ciliated protozoan *Paramecium caudatum* acting as the predator and *Escherichia coli* acting as the prey to investigate the impacts of Cadmium Telluride Quantum Dots (CdTe QDs) (Fig. 11.7) [70]. In comparison to the control, they discovered that exposure to CdTe QDs caused a decrease in Paramecium's bacterivorous ability as well as a roughly 12-h delay in doubling time. Fluorescence-based stoichiometric analysis indicated that QDs were bioaccumulated after its exposure. There was a significant increase in cellular concentration after 1 day compared to 1 h. Furthermore, *E. coli* accumulated Cadmium during its exposure to QDs, which transferred 1.4 times more cadmium to its predator, Paramecium [70].

QDs can enter the environment through industrial processes, accidental spills, or improper disposal of nanomaterial-containing products [71]. They can be aerosolized and deposited onto surfaces, including water bodies and soil. Smaller QDs can

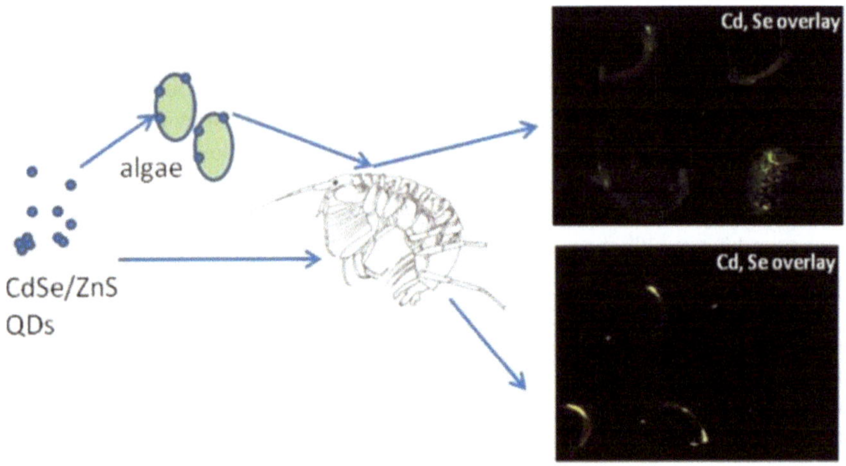

Fig. 11.6 Diagram showing bioaccumulation of carboxyl-functionalized CdSe/Zn QDs in the amphipod *Leptocheirus plumulosus*. Reproduced with permission [69], Copyright 2012, ACS

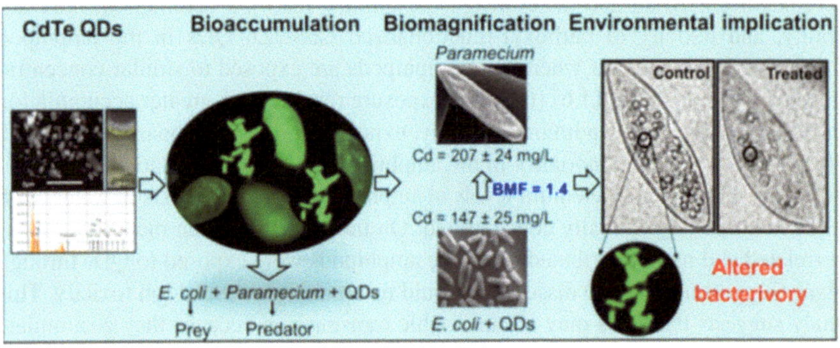

Fig. 11.7 Diagram illustrates a laboratory-scale aquatic microbial food chain model used to investigate the impacts of CdTe QDs with the ciliated protozoan *Paramecium caudatum* acting as the predator and *Escherichia coli* acting as the prey. Reproduced with permission [70], Copyright 2017, ACS

passively diffuse through biological membranes based on concentration gradients. Many organisms uptake QDs through endocytosis, a process where the cell membrane engulfs particles, forming vesicles. This is particularly common in microorganisms and some larger organisms. In some cases, QDs can interact with specific transport proteins, facilitating their entry into cells [72]. QDs may distribute unevenly within tissues due to their size, surface charge, and functionalization. For example, they may accumulate more in the liver or kidneys, which are responsible for detoxification and excretion. Within cells, QDs can localize in different organelles, including lysosomes, mitochondria, or the cytoplasm. This localization affects their interaction with cellular components and potential toxicity. The functionalization of QDs (e.g., with ligands or polymers) significantly influences their hydrophilicity/hydrophobicity, stability in biological systems, and interactions with cellular structures, impacting their distribution [18]. QDs can accumulate in organisms as long as their concentration in the environment is higher than in the organism. The bioaccumulation occurs when uptake exceeds excretion. Bioaccumulation can be exacerbated through trophic transfer, where predators consume prey-containing QDs, leading to higher concentrations in higher trophic levels (biomagnification). In some cases, QDs may dissolve in biological environments, releasing toxic metal ions (e.g., cadmium or lead) [73]. This release can lead to additional toxicity and contribute to the overall bioaccumulation of hazardous materials within the organism. QDs' physicochemical characteristics and environmental circumstances are just two of the many variables that affect their complicated absorption, dispersion, and bioaccumulation processes in the environment [74].

3.2 Impact on the Aquatic and Terrestrial Organisms

A major concern regarding the impact of QDs on organisms is their ability to penetrate into different types of cells. QDs can enter cells through various mechanisms. It has been observed that prokaryotic organisms have a limited capacity to absorb QDs, which might offer them some protection against the toxic effects of certain QDs. In contrast, eukaryotic organisms are more susceptible to QDs infiltration. For these organisms, QDs can enter cells through diffusion across cell membranes, endocytosis, or adhesion [75]. Plants are primarily exposed to nanomaterials through the stomata on their leaves and their root system. Additionally, nanoparticles can enter plants through areas where the plant organs (such as roots, stems, or leaves) have been physically damaged. The size of the holes in plant and fungal cell walls, which vary in size from 5 to 20 nm, determines the nanoparticles' capacity to enter the cells. However, according to Navarro et al., when nanomaterials interact with cell wall components, they can cause bigger pores to develop, which increases the cell wall's permeability [76]. The presence of nanomaterials can cause a variety of reactions in organisms, from changes in individual cells and organs to changes at the DNA level and within cell organelles, even leading to illness or the death of the entire organism [22].

The ecotoxicity of QDs across various trophic levels (including microorganisms, aquatic vertebrates and invertebrates) is strongly influenced by their physical and chemical characteristics, environmental conditions, concentration, exposure duration, and the species affected. Additionally, exposure to UV light can enhance their toxicity. Although the concentration of QDs among various trophic levels is not well-documented, their potential eco-toxicological impact is a growing concern [78]. The release of poisonous metals like cadmium from QDs, their change in the ecosystem (such as photo-activation and the production of ROS when exposed to ultraviolet or visible light), and their distinct nano-specific and bio-functional features are the reasons for this. The toxic effects of QDs to mammalian cells have been discussed in the works by Hardman and Pelley et al. To further assess environmental health, more data is required about their behavior, destiny, and eco-toxic effects in aquatic habitats across different trophic levels [19].

Numerous factors influence the behavior and destiny of nanomaterials in aquatic settings. Diffusion, oxidation, settling, re-suspension, stabilization by NOM, absorption onto suspended particulate matter, interactions with organisms, surface coating changes, aggregation or agglomeration, disaggregation or de-agglomeration, and biological transformations are some of these processes [79]. Dwivedi et al. categorized the processes that impact the behavior and destiny of QDs in the environment as physiologically induced reactions, macromolecular interactions, and physicochemical transformation [80]. The dissolution of QDs depends on factors such as the type of nanoparticle, its coatings and ligands, aggregate state, concentration, size, and environmental conditions like pH and salinity.

Kloepfer et al. found that adenosine monophosphate-coated CdSe/ZnS QDs caused QDs uptake in gram-positive *Bacillus subtilis* and *E. coli.* [81]. A simple diagram demonstrating the MoA and toxicity of Cd-based QDs to *E. coli* is presented in Fig. 11.8. These variables were associated with binding to the bacterial surface, changes in the activity of antioxidant enzymes, and a reduction in the antioxidant response [77]. QDs can have various effects on algae, including growth inhibition, modifications in enzyme activities, reduced photosynthetic activity, alterations in the antioxidant system through changes in gene expression, etc. They can also induce oxidative damage and alter the proteomic profile. Worms et al. observed that the freshwater microalgae *Chlamydomonas reinhardtii* exhibited interactions and adsorption of these nanoparticles on their surface upon exposure to carboxyl-coated CdSe/ZnS QDs [82]. A general scheme showing the MoA and in vivo toxicity of Cd-based QDs in the marine mussel *M. galloprovincialis* is shown in Fig. 11.9. This scheme includes DNA damage, immunological response, intracellular and extracellular ROS generation, and changes in antioxidant enzyme activity [77]. Fishes were employed to evaluate QD toxicity in the context of nano ecotoxicology. The rainbow trout (*O. mykiss*) and zebrafish (*D. rerio*) are the two most studied fish species. The mode of action of QDs and their dissolved counterpart, shown in (Fig. 11.10), were investigated using the zebra fish liver cell line [77].

The environmental fate of organisms is altered by their toxicokinetics, uptake, and biotransformation of QDs [83]. Experiments using reconstructed food chains were used to assess trophic transfer and biomagnification of QDs. In these investigations, higher trophic level animals were fed living organisms exposed to QDs. In rebuilt food chains encompassing both freshwater and marine species, trophic transmission of QDs was found; however, only weak biomagnification and detoxifying activities were seen [84].

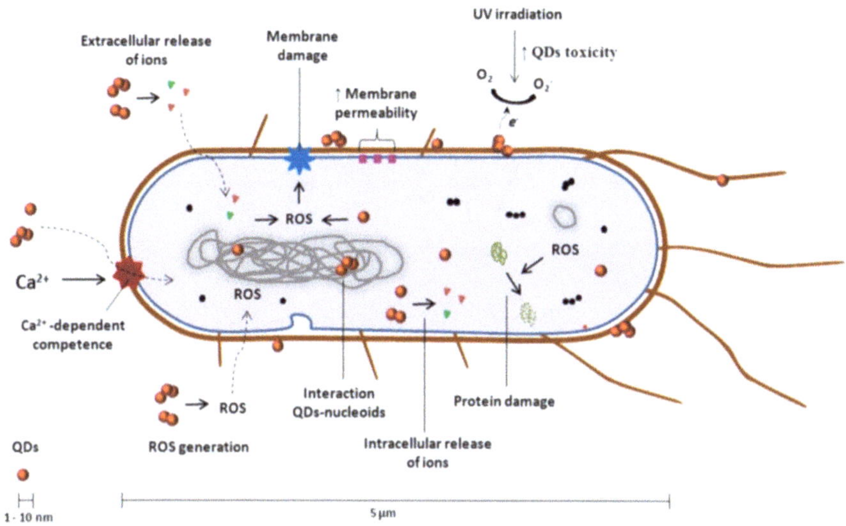

Fig. 11.8 Diagram illustrating the Mechanism of Action (MoA) of QDs in bacteria, using the gram-negative bacterium *Escherichia coli* as a model system. Reproduced with permission [77], Copyright 2017, Elsevier

3.3 Predicting Bioaccumulation Using Models

Understanding how organisms bioaccumulate nanomaterials is crucial for assessing their potential ecotoxicity. However, experimentally determining bioaccumulation is quite challenging, which has led to the development of predictive approaches using models for various nanomaterials (Fig. 11.11) [9]. The Quantitative Structure–Activity Relationship (QSAR) models, accumulation factor models, Biotic Ligand Model (BLM), and partition coefficients are examples of traditional modeling methods that were initially used to evaluate nanomaterials to provide quantitative estimates in a resource-efficient manner. However, because nanomaterials have different uptake behaviors, these approaches, which are based on the uptake patterns of these materials, may produce inaccurate results [85].

The concepts and modeling techniques of the free ion activity model and the fish gill complex model were combined to generate Biotic Ligand Models (BLMs). Based on the idea that mortality happens when the concentration of the metal-biotic ligand complex reaches a threshold level, BLMs are commonly used to predict metal toxicity when analyzing the bioavailability and toxicity of metal ions in water (Fig. 11.12) [9]. The model assesses the activity of free metal ions in the water, which is a crucial determinant of their biological effects. BLMs account for the strength of the interaction between the metal ions and the biotic ligands, which can include proteins, enzymes, or other cellular components [86]. The formation of metal–ligand complexes is critical to understanding how metals impact organisms. The model predicts that toxicity occurs when the concentration of these complexes exceeds a

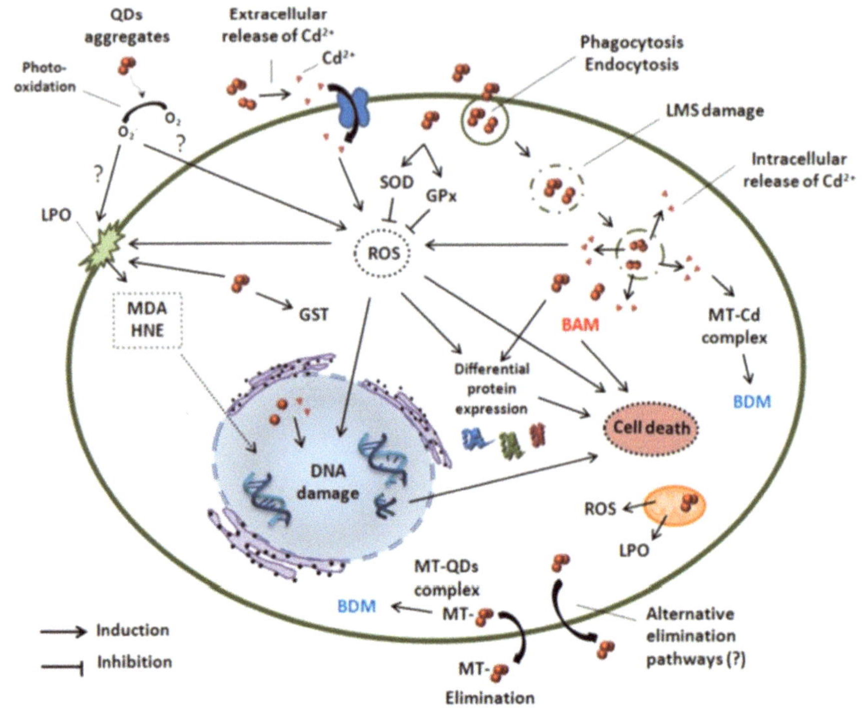

Fig. 11.9 Overview showing the mechanism of action and toxicity of cadmium-based QDs in the marine mussel *Mytilus galloprovincialis*. Reproduced with permission [77], Copyright 2017, Elsevier

certain threshold [87]. The surface coatings of QDs can influence their interactions with biological ligands. BLMs can incorporate information about these coatings to better predict how QDs bind to organisms. The physical characteristics of QDs, such as size and tendency to aggregate, affect their bioavailability. BLMs can integrate these factors to assess how they impact uptake and toxicity. Many QDs contain toxic metal ions (e.g., cadmium or lead) that may be released into the surrounding environment [88]. BLMs can help predict the bioavailability of these ions once they dissociate from the QD matrix. BLMs need to be calibrated with empirical data specific to the organism and QD type being studied. This calibration may include determining binding constants and other interaction parameters. The model evaluates the exposure levels of organisms to QDs in their environment, including concentration in water and potential ingestion through food [89].

Energy-based models, physiological models, and compartmental models are a few popular toxicokinetic models. Physiological and compartmental models are the most commonly used in studies of biological accumulation of these materials [90]. Compartmental models usually function as frameworks with one or two compartments. Physiological models, also known as (PBPK) Physiologically Based Pharmacokinetic Models, use information from physiological, anatomical, and biochemical

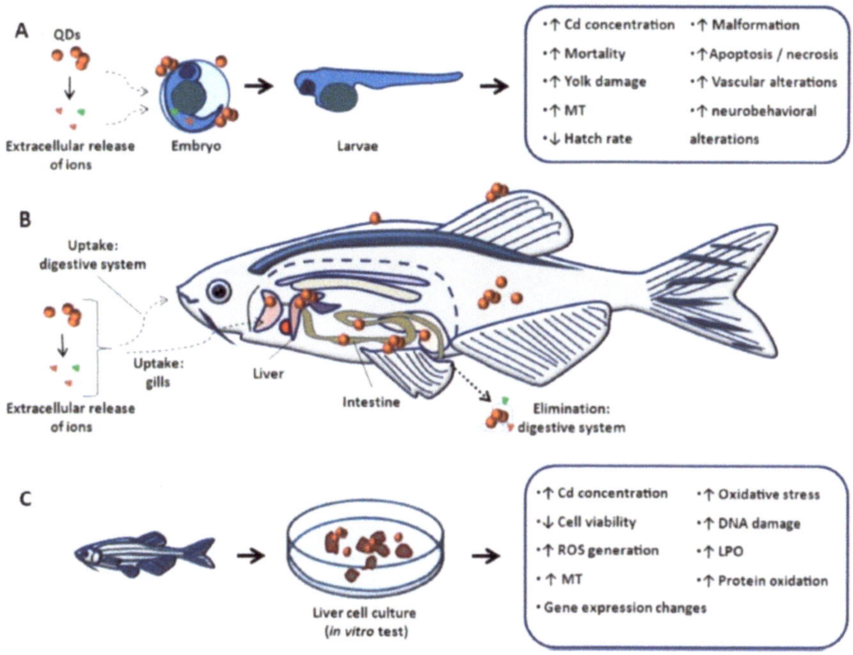

Fig. 11.10 Diagram illustrates many methods used to investigate the toxicity of QDs in zebra fish (*Danio rerio*) as the model organism. Among the techniques are (A) evaluating embryotoxicity, (B) exposing adult fish to QDs through intraperitoneal injection or waterborne and feeding channels, and (C) in vitro evaluation of cytotoxicity. Reproduced with permission [77], Copyright 2017, Elsevier

principles to model blood flow in different organs and tissues while adhering to the mass balance principle. This allows them to simulate how drugs metabolize in the body (Fig. 11.13) [9]. Each compartment represents different tissues or organs, such as the liver, kidneys, and lungs, which have distinct rates of uptake and elimination for QDs. The model incorporates parameters like absorption rates, distribution coefficients, elimination rates, and tissue partitioning coefficients specific to QDs. PBPK models can simulate various exposure routes to QDs, including oral, inhalation, or dermal exposure. This flexibility allows for comprehensive assessments of how QDs enter organisms [91]. The model helps predict the bioavailability of QDs based on their size, surface modifications, and environmental conditions. PBPK models allow researchers to predict how QDs distribute among different tissues, reflecting their accumulation in organs such as the liver, spleen, and kidneys. The model can simulate how QDs accumulate over time, taking into account both uptake from exposure and elimination through metabolism or excretion. PBPK models provide valuable insights for risk assessments by predicting potential bioaccumulation levels of QDs in various organisms and their implications for higher trophic levels in food webs [92]. Understanding the bioaccumulation dynamics of QDs through PBPK modeling can inform regulatory decisions regarding their safe use and environmental monitoring.

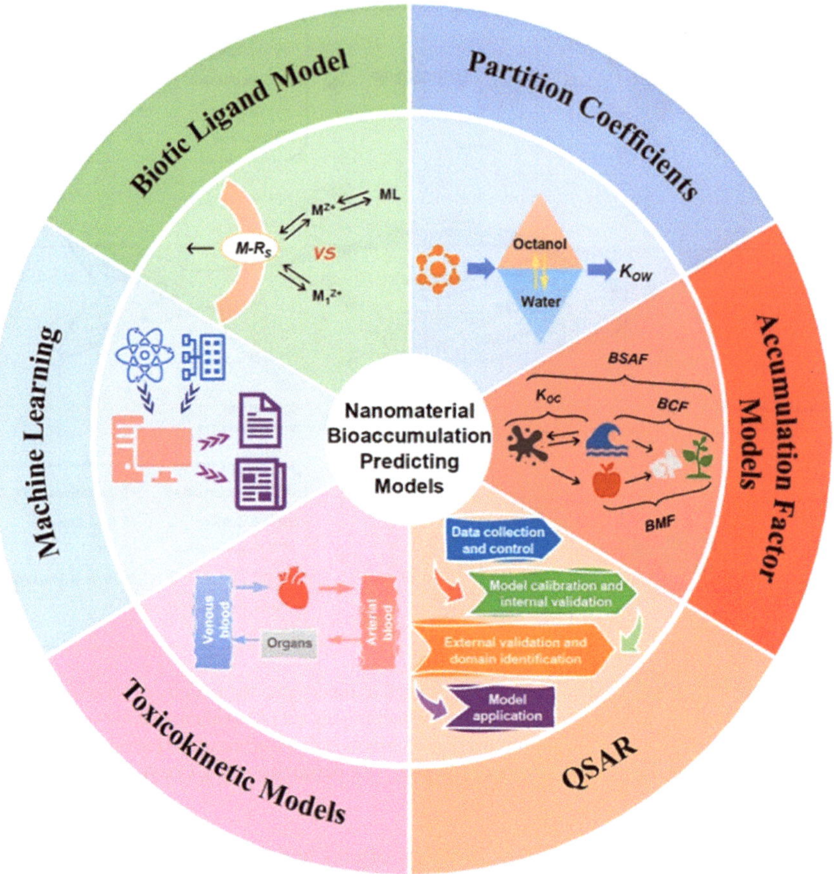

Fig. 11.11 Different models for foreseeing the bioaccumulation of nanomaterials in the ecosystem. Reproduced with permission [9], Copyright 2024, ACS

Nevertheless, framing models for the development of these materials in living things still presents many difficulties. Significant work is required to investigate underlying mechanisms and enhance these models' interpretability. Clarifying the relationships between different features and identifying internal mechanisms in complex systems can be facilitated by the creation of novel Machine Learning (ML) frameworks or the merging of ML models with conventional techniques [9]. Additionally, the majority of current models for these materials are mostly focused on aquatic creatures, with very few addressing land-dwelling species, and more complex models require extra factors. There is currently no proof that aquatic organisms can benefit from the use of PBPK models. Furthermore, little is known about the biomagnification of nanomaterials in the food chain. However, developing models for nanomaterials can help us better understand how they accumulate biodynamically, how they are exposed in the environment, and what transformation products they produce.

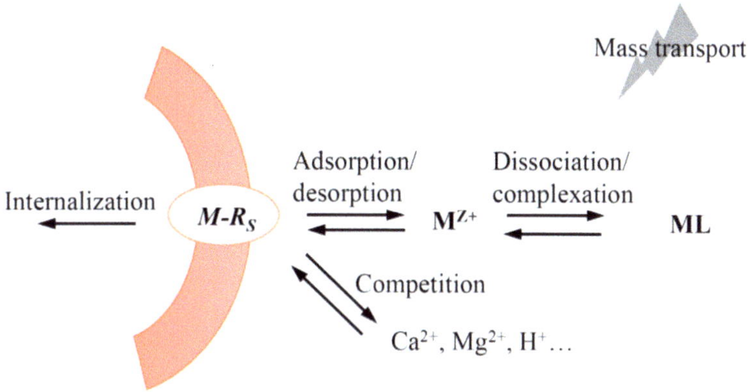

Fig. 11.12 Illustration of a typical physicochemical process showing how an aquatic organism absorbs a trace element. M and L represent the metal and ligand in solution, while RS indicates a sensitive site on the organism's surface (such as an ion channel, transporter, or carrier) within the BLM model. Reproduced with permission [9], Copyright 2024, ACS

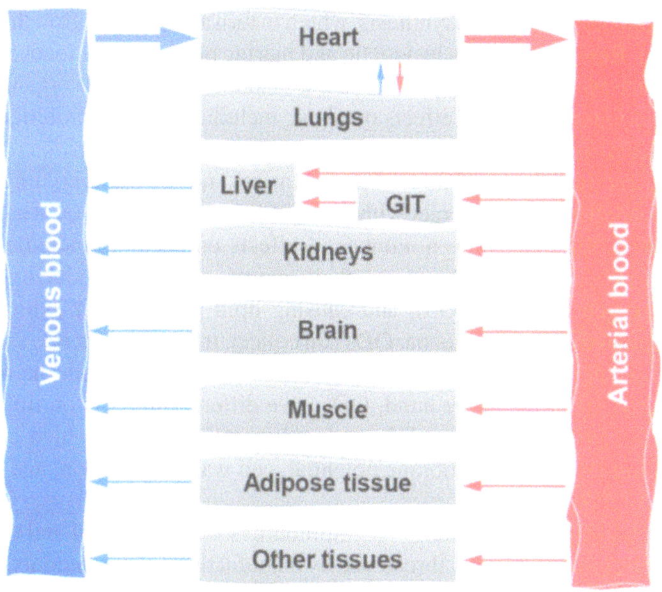

Fig. 11.13 Fundamental, whole-body physiologically based toxicokinetic model illustrating the processes and pathways involved in drug absorption, distribution, metabolism, and excretion. Reproduced with permission [9], Copyright 2024, ACS

3.4 *Potential Health Risks*

Recent investigations into the environmental fate of QDs have produced some fascinating results. Even for goods that have been reviewed, there is currently insufficient information available to conduct a thorough environmental risk assessment due to factors such as food chain transit, bioaccumulation, biomagnification, and persistence in natural habitats.

Understanding how different QDs affect ecosystem health is becoming a pressing concern because of their increasing production and their release into the environment. Evaluation of QDs' effects on various ecosystems, such as animals, plants, and microbes, is essential. For the purpose of doing risk assessments pertaining to ecosystem conservation, this information is crucial [93]. Plants can absorb QDs from the soil. One important way that QDs enter the food chain is by the uptake of QDs by plant roots. By measuring the fluorescence in different plant parts or by examining the distribution of components, one can evaluate the distribution of QDs within plants. Studies have found that when soil plants are exposed to heavy metal QDs, there is typically a reduction in chlorophyll levels. This decline in chlorophyll negatively impacts plant photosynthesis, which in turn affects plant growth and development [94]. Research on QDs in aquatic and marine plants mostly focuses on algae. Algae can act as carriers for toxic metals, transporting them into organisms. QDs can cause various detrimental effects on algae, including growth inhibition, reduced photosynthesis, oxidative stress, and diminished metabolism [95].

Researchers have explored how microorganisms such as bacteria, fungi, viruses, and certain small protozoa interact with QDs [96]. Functionalized, QDs can be taken up by bacteria. QDs can have a number of effects on bacteria, including slowing down their pace of growth, increasing their rate of mortality, producing too many Reactive Oxygen Species (ROS), and causing lipid peroxidation, which leads to oxidative damage [97]. In protozoa, QDs can collect, transmit nutrients, and undergo biomagnification; the degree of biomagnification is correlated with the organism's rate of clearance. On the other hand, QDs have different impacts on fungi than on protozoa. The primary surface of fungi is where QDs adsorb, which can lead to changes in mitochondrial membrane potential, cell wall degradation, and decreased cell viability [98].

The regular operation of microbial communities in both natural and constructed environments may be seriously threatened by manufactured nanomaterials, according to recent findings. Changes in the makeup of microbial communities have been the focus of the majority of studies on the effects of artificial nanomaterials on these populations. For this purpose, common analytical techniques include quantitative real-time PCR, Phospholipid Fatty Acid (PLFA) analysis, functional Multi-Substrate-Induced Respiration (MSIR) analysis, PCR-Denaturing Gradient Gel Electrophoresis (PCR-DGGE) [98], Terminal-Restriction Fragment Length Polymorphism (T-RFLP) and Catalyzed Reporter Deposition (CARDFISH) [99]. The most popular technique for community study is still PCR-DGGE, although there is a growing movement to examine how man-made nanomaterials affect microbial communities employing sequencing and omics technologies [100]. The impact of heavy metals emitted by QDs, including lead (Pb), zinc (Zn), cadmium (Cd), copper (Cu), and arsenic (As), on microbial populations must be taken into account. It has been demonstrated that metal pollution reduces microbial diversity, activity, and biomass [101]. Another important but little-studied topic is how QDs affect microbial ecological processes such as nitrogen fixation, denitrification, and methanogenesis. Although there is a dearth of information, QDs do have an impact on the ecological processes of microbes. For instance, the effect of QDs on carbon fixation has not yet been investigated, highlighting the necessity for further comprehensive and in-depth research.

Through eating or nutrient transmission, QDs can build up in mammals in terrestrial settings and may become hazardous [102]. Oral delivery of CdTe QDs at medium and low levels alters the structure of intestinal microbiota, while oral administration of large doses damages intestinal tissues. Additionally, it was noted that the degree of toxicity of QDs is correlated with the functionalization of their surface and can have a major effect on eggs and larvae [103]. These problems might be caused by QDs interacting with particular motor neurons that control contractions during egg-laying [16]. In wild-type nematodes, prolonged exposure to QDs can also result in reproductive issues such as dysregulated germ cell proliferation and differentiation, aberrant meiosis, decreased fertilized egg production and release, and decreased brood size and egg production rates, which can upset oogenesis [104].

4 Regulatory Measures and Risk Assessments

As the use of QDs expands across various industries, understanding their environmental fate and potential for bioaccumulation becomes crucial. Regulatory frameworks and risk assessment protocols are necessary to ensure the safe use of QDs while mitigating their potential risks [105]. Various international organizations, such as the Organization for Economic Co-operation and Development (OECD) and the World Health Organization (WHO), provide guidelines for the assessment of nanomaterials, including QDs. These guidelines emphasize the need for comprehensive safety assessments that consider unique properties of nanomaterials. Countries may have specific regulations governing the production, use, and disposal of QDs. In the U.S., for example, the Environmental Protection Agency (EPA) and the Food and Drug Administration (FDA) play critical roles in regulating nanomaterials based on their intended use, assessing both environmental and health impacts.

It is highly essential to identify the potential hazards associated with QDs, including their chemical composition, toxicity, and ecological impacts. This includes assessing the risks posed by heavy metals or other harmful components. Understanding the environmental fate of QDs is crucial [93]. This includes studying their degradation, persistence, transport mechanisms, and interactions with biological and abiotic components in ecosystems. Risk assessments must include studies on bioaccumulation potential, where the accumulation of QDs in organisms is measured over time [106]. This involves both laboratory experiments and field studies to assess how QDs transfer through food webs. Conducting toxicity tests on various organisms (e.g., fish, invertebrates, plants) helps determine the effects of QDs on different trophic levels [100]. Standardized tests can provide data on acute and chronic toxicity, genotoxicity, and potential long-term effects. Regulatory frameworks and risk assessment processes for QDs are vital for understanding their fate, bioaccumulation potential, and overall ecological impact. By following structured assessment protocols, regulatory bodies can ensure that the benefits of QDs are realized while minimizing potential risks to human health and the environment [98].

5 Conclusions and Future Prospects

QDs exhibit varying degrees of stability and persistence in the environment, influenced by their chemical composition and surface modifications. Understanding these factors is crucial for predicting their behavior in different ecosystems. QDs can accumulate in living organisms, especially those containing toxic heavy metals like cadmium and lead. This bioaccumulation poses significant risks to higher trophic levels and can impact food chains, raising concerns for both ecological and human health. Future studies should focus on the long-term environmental fate of QDs, including degradation and transformation in various ecosystems. Developing non-toxic or biodegradable QDs with improved surface coatings can enhance safety while maintaining functionality. Implementing robust environmental monitoring will help assess the real-world impacts of QDs on ecosystems. Investigating the effects of QDs on various species and trophic levels will provide deeper insights into their ecological impacts. Engaging stakeholders and fostering collaboration between researchers, regulators, and the public will support informed decision-making and responsible use of QDs. In summary, while QDs hold significant potential for various applications, understanding their fate and bioaccumulation is essential for minimizing risks and ensuring environmental safety. Continued research and proactive regulatory measures will be vital in navigating these challenges.

Acknowledgements Authors are thankful to the Centre for Research Projects (CRP), Christ University, for the Seed Money Grant SMSS-2334 and to Vision Group on Science and Technology, Govt. of Karnataka for K-FIST-L1 grant (GRD No. 1143).

References

1. Navarro DA, Banerjee S, Watson DF, Aga DS (2011) Differences in soil mobility and degradability between water-dispersible CdSe and CdSe/ZnS quantum dots. Environ Sci Technol 45(15):6343–6349. https://doi.org/10.1021/es201010f
2. John VL, Fasila PM, Chaithra KP, Vinod TP (2022) Carbon dots derived from frankincense soot for ratiometric and colorimetric detection of lead (II). Nanotechnology 33(49):495706. https://doi.org/10.1088/1361-6528/ac8e76

3. Sherka GT, Berry HD (2024) Insight into impact of size and shape on optoelectronic properties of InX (X = As, Sb, and P) semiconductor nanoparticles: a theoretical study. Front Phys 12. https://doi.org/10.3389/fphy.2024.1447997/full
4. Pavlicek A, et al (2023) The use and detection of quantum dots as nanotracers in environmental fate studies of engineered nanoparticles. Environ Pollut 317:120461. https://linkinghub.elsevier.com/retrieve/pii/S026974912201675X
5. Benn TM, Westerhoff P (2008) Nanopartle silver released into water from commercially available sock fabrics. Environ Sci Technol 42(11):4133–4139. https://doi.org/10.1021/es7032718.
6. John VL, Joy F, et al (2022) Amine functionalized carbon quantum dots from paper precursors for selective binding and fluorescent labelling applications. J Coll Interface Sci 617:730–744. https://linkinghub.elsevier.com/retrieve/pii/S0021979722004532
7. Holbrook RD, Murphy KE, Morrow JB, Cole KD (2008) Trophic transfer of nanoparticles in a simplified invertebrate food web. Nat Nanotechnol 3(6):352–355. https://www.nature.com/articles/nnano.2008.110
8. Chen Z, et al (2024) Color revolution: prospects and challenges of quantum-dot light-emitting diode display technologies. Small Methods 8(2). https://doi.org/10.1002/smtd.202300359
9. Zeng H et al (2024) Predicting bioaccumulation of nanomaterials: modeling approaches with challenges. Environ Health 2(4):189–201
10. Ayiloor Rajesh G, et al (2022) Carbon dots from natural sources for biomedical applications. Part Part Syst Charact 39(9). https://doi.org/10.1002/ppsc.202200017
11. Wang X, Wu T (2023) An update on the biological effects of quantum dots: from environmental fate to risk assessment based on multiple biological models. Sci Total Environ 879:163166. https://linkinghub.elsevier.com/retrieve/pii/S0048969723017850
12. Woodward JD, Kennel SJ, Mirzadeh S, Dai S, Wall JS, Richey T, Avenell J, Rondinone AJ (2007) In vivo SPECT/CT imaging and biodistribution using radioactive Cd 125 m Te/ZnS nanoparticles. Nanotechnology 18(17):175103. https://doi.org/10.1088/0957-4484/18/17/175103
13. John VL, et al (2023) Mulberry leaves (Morus Rubra)-derived blue-emissive carbon dots fed to silkworms to produce augmented silk applicable for the ratiometric detection of dopamine. Macromol Biosci 23(9). https://doi.org/10.1002/mabi.202300081
14. Zhang Y, Chen W, Zhang J, Liu J, Chen G, Pope C (2007) In vitro and in vivo toxicity of CdTe nanoparticles. J Nanosci Nanotechnol 7(2):497–503. https://doi.org/10.1166/jnn.2007.125
15. Ahmed HE, Soylak M (2024) Exploring the potential of carbon quantum dots (CQDs) as an advanced nanomaterial for effective sensing and extraction of toxic pollutants. TrAC Trends Anal Chem 180:117939. https://linkinghub.elsevier.com/retrieve/pii/S0165993624004229
16. Qu M, Qiu Y, Lv R, Yue Y, Liu R, Yang F, Wang D, Li Y (2019) Exposure to MPA-capped CdTe quantum dots causes reproductive toxicity effects by affecting oogenesis in nematode caenorhabditis elegans. Ecotoxicol Environ Saf 173:54–62. https://linkinghub.elsevier.com/retrieve/pii/S0147651319301708
17. Satish S, et al (2022) Risk assessment of selected pharmaceuticals on wildlife with nanomaterials based aptasensors. Sci Total Environ 836:155622. https://linkinghub.elsevier.com/retrieve/pii/S0048969722027188
18. Stewart DTR et al (2013) Quantum dots exhibit less bioaccumulation than free cadmium and selenium in the earthworm Eisenia Andrei. Environ Toxicol Chem 32(6):1288–1294
19. Pelley JL, Daar AS, Saner MA (2009) State of academic knowledge on toxicity and biological fate of quantum dots. Toxicol Sci 112(2):276–296. https://academic.oup.com/toxsci/article/1714008/State
20. Kumar A, et al (2023) Revolutionizing agriculture: a comprehensive review of nanotechnology applications. Int J Environ Clim Change 13(11):3586–3603. https://journalijecc.com/index.php/IJECC/article/view/3534
21. Raj A, Gauba P, Bhatt E (2024) Advanced nanoparticles for environmental remediation of emerging pollutants: a review. Soil Sediment Contam Int J 1–30. https://doi.org/10.1080/15320383.2024.2422387

22. Nowack B, Bucheli TD (2007) Occurrence, behavior and effects of nanoparticles in the environment. Environ Pollut 150(1):5–22. https://linkinghub.elsevier.com/retrieve/pii/S0269749107002734
23. Khot LR, et al (2012) Applications of nanomaterials in agricultural production and crop protection: a review. Crop Protect 35:64–70. https://linkinghub.elsevier.com/retrieve/pii/S0261219412000099
24. Gottschalk F, Sonderer T, Scholz RW, Nowack B (2009) Modeled environmental concentrations of engineered nanomaterials (TiO 2, ZnO, Ag, CNT, Fullerenes) for different regions. Environ Sci Technol 43(24):9216–9222. https://doi.org/10.1021/es9015553
25. Bakar NFA, et al (2022) Environmental impact of quantum dots. In: Graphene, nanotubes and quantum dots-based nanotechnology. Elsevier, pp 837–867. https://linkinghub.elsevier.com/retrieve/pii/B9780323854573000116
26. Maynard AD, Aitken RJ (2007) Assessing exposure to airborne nanomaterials : current abilities and future requirements. 1(March):26–41
27. Li L, Qian H, Ren J (2005) Rapid synthesis of highly luminescent CdTe nanocrystals in the aqueous phase by microwave irradiation with controllable temperature. 528–530
28. Ryman-Rasmussen JP, Riviere JE, Monteiro-Riviere NA (2006) Penetration of intact skin by quantum dots with diverse physicochemical properties. 91(1):159–165
29. Singh G, Thakur N, Kumar R (2024) Nanoparticles in drinking water: assessing health risks and regulatory challenges. Sci Total Environ 949:174940. https://linkinghub.elsevier.com/retrieve/pii/S0048969724050903
30. Wang L, Nagesha DK, Selvarasah S, Dokmeci MR, Carrier RL (2012) Toxicity of CdSe nanoparticles in Caco-2 cell cultures. 15:1–15
31. Roy D, Maitra S, De D, Mukherjee K (2014) Analysis of effect of temperature variation on computational faithfulness of a QCA XOR gate. In: International conference on electronics, communication and instrumentation (ICECI). IEEE, pp 1–4. http://ieeexplore.ieee.org/document/6767385/
32. Liu X, Sathishkumar K, Zhang H, Saxena KK, Zhang F, Naraginti S, Rajendiran R, Rajasekar A, Guo X (2024) Frontiers in environmental cleanup: recent advances in remediation of emerging pollutants from soil and water. J Hazard Mater Adv 16:100461. https://linkinghub.elsevier.com/retrieve/pii/S2772416624000627
33. Zhang P, et al (2020) Nanomaterial transformation in the soil–plant system: implications for food safety and application in agriculture. Small 16(21). https://doi.org/10.1002/smll.202000705
34. Zhang Y, Chen Y, Westerhoff P, Crittenden JC (2008) Stability and removal of water soluble CdTe quantum dots in water. Environ Sci Technol 42(1):321–325. https://doi.org/10.1021/es0714991
35. Memon R, Niazi JH, Qureshi A (2024) Biosensors for detection of airborne pathogenic fungal spores: a review. Nanoscale 16(33):15419–15445. https://xlink.rsc.org/?DOI=D4NR01175A
36. Xu L, Chen C (2012) Physiological behavior of quantum dots. Wiley Interdiscipl Rev Nanomed Nanobiotechnol 4(6):620–637
37. Liu Q et al (2022a) New insights into the safety assessment of quantum dots: potential release pathways, environmental transformations, and health risks. Environ Sci Nano 9(9):3277–3311. https://doi.org/10.1039/D2EN00252C
38. Amina et al (2024) A review on quantum dots: revolutionizing waste water cleaning. Chem Africa 7(9):5009–5023. https://doi.org/10.1007/s42250-024-01091-1
39. Wang X, Xie H, Wang P, Yin H (2023) Nanoparticles in Plants: Uptake, Transport and Physiological Activity in Leaf and Root." *Materials* 16(8): 3097. https://www.mdpi.com/1996-1944/16/8/3097.
40. Le N, Zhang M, Kim K (2022) Quantum dots and their interaction with biological systems. Int J Mol Sci 23(18):10763. https://www.mdpi.com/1422-0067/23/18/10763
41. Libralato G, et al (2017) Toxicity effects of functionalized quantum dots, gold and polystyrene nanoparticles on target aquatic biological models: a review. Molecules 22(9)

42. Batley GE, Kirby JK, McLaughlin MJ (2013) Fate and risks of nanomaterials in aquatic and terrestrial environments. Acc Chem Res 46(3):854–862. https://doi.org/10.1021/ar2003368.
43. Mei BC, et al (2008) Modular Poly(Ethylene Glycol) ligands for biocompatible semiconductor and gold nanocrystals with extended PH and ionic stability. J Mater Chem 18(41):4949. https://xlink.rsc.org/?DOI=b810488c
44. Medintz IL, Uyeda HT, Goldman ER, Mattoussi H (2005) Quantum dot bioconjugates for imaging, labelling and sensing. Nat Mater 4(6):435–446. https://www.nature.com/articles/nmat1390
45. Derfus AM, Chan WC, Bhatia SN (2004) Probing the cytotoxicity of semiconductor quantum dots. Nano Lett 4(1):11–18. https://doi.org/10.1021/nl0347334
46. Frank BP, et al (2020) Photochemical transformations of carbon dots in aqueous environments. Environ Sci Technol 54(7):4160–4170. https://doi.org/10.1021/acs.est.9b07437
47. Manciulea A, Baker A, Lead JR (2009) A fluorescence quenching study of the interaction of suwannee river fulvic acid with iron oxide nanoparticles. Chemosphere 76(8):1023–1027. https://linkinghub.elsevier.com/retrieve/pii/S0045653509005839
48. Jahan S, Alias YB, Bakar AF, Yusoff IB (2019) Transport and retention behavior of carbonaceous colloids in natural aqueous medium: impact of water chemistry. Chemosphere 217:213–222. https://linkinghub.elsevier.com/retrieve/pii/S004565351832112X
49. Lanphere JD, et al (2014) Stability and transport of graphene oxide nanoparticles in groundwater and surface water. Environ Eng Sci 31(7):350–359. https://doi.org/10.1089/ees.2013.0392
50. Al Salim N, et al (2011) Quantum dot uptake in the New Zealand environment. Mater Sci For 700:199–202. https://www.scientific.net/MSF.700.199
51. Whiteside MD, Treseder KK, Atsatt PR (2009) The brighter side of soils: quantum dots track organic nitrogen through fungi and plants. Ecology 90(1):100–108. https://doi.org/10.1890/07-2115.1
52. Sun H, Wang M, Lei C, Li R (2021) Cell wall: an important medium regulating the aggregation of quantum dots in maize (Zea Mays L.) seedlings. J Hazard Mater 403:123960. https://linkinghub.elsevier.com/retrieve/pii/S0304389420319506
53. Wang Y, Miao AJ, Luo J, Wei ZB, Zhu JJ, Yang LY (2013) Bioaccumulation of CdTe quantum dots in a freshwater alga ochromonas danica : a kinetics study. Environ Sci Technol 47(18):10601–10610. https://doi.org/10.1021/es4017188
54. Peyrot C, Gagnon C, Gagné F, Willkinson KJ, Turcotte P, Sauvé S (2009) Effects of cadmium telluride quantum dots on cadmium bioaccumulation and metallothionein production to the freshwater mussel, Elliptio Complanata. Comp Biochem Physiol Part C Toxicol Pharmacol 150(2):246–251. https://linkinghub.elsevier.com/retrieve/pii/S1532045609001136
55. Lewinski NA, Zhu H, Ouyang CR, Conner GP, Wagner DS, Colvin VL, Drezek RA (2011) Trophic transfer of amphiphilic polymer coated CdSe/ZnS quantum dots to danio rerio. Nanoscale 3(8):3080. https://xlink.rsc.org/?DOI=c1nr10319a
56. Grasso D, et al (2002) A review of non-DLVO interactions in environmental colloidal systems. Rev Environ Sci Bio/Technol 1(1):17–38. https://doi.org/10.1023/A:1015146710500
57. Dkhar DS, et al (2024) Versatile approaches of quantum dots in biosensing and imaging. WIREs Nanomed Nanobiotechnol 16(5). https://doi.org/10.1002/wnan.1998
58. Chen Z, et al (2008) Bio-distribution and metabolic paths of silica coated CdSeS quantum dots. Toxicol Appl Pharmacol 230(3):364–371. https://linkinghub.elsevier.com/retrieve/pii/S0041008X08001385
59. Wang L, et al (2008) Toxicity of CdSe nanoparticles in Caco-2 cell cultures. J Nanobiotechnol 6(1):11. https://doi.org/10.1186/1477-3155-6-11
60. Rocha TL, Mestre NC, Sabóia-Morais SMT, Bebianno MJ (2017a) Environmental behaviour and ecotoxicity of quantum dots at various trophic levels: a review. Environ Int 98:1–17. https://doi.org/10.1016/j.envint.2016.09.021
61. Hardman R (2006) A toxicologic review of quantum dots: toxicity depends on physicochemical and environmental factors. Environ Health Perspect 114(2):165–172

62. Chen LQ, Ding CZ, Ling J (2017) Intensive epidermal adsorption and specific venous deposition of carboxyl quantum dots in zebrafish early-life stages. Chemosphere 184:44–52. https://doi.org/10.1016/j.chemosphere.2017.05.173
63. Yasin MU, et al (2024) The synergistic potential of biochar and nanoparticles in phytoremediation and enhancing cadmium tolerance in plants. Chemosphere 354:141672. https://linkinghub.elsevier.com/retrieve/pii/S0045653524005654
64. Hu CW, et al (2010) Toxicological effects of TiO_2 and ZnO nanoparticles in soil on earthworm Eisenia Fetida. Soil Biol Biochem 42(4):586–591. https://linkinghub.elsevier.com/retrieve/pii/S0038071709004702
65. Pipan-Tkalec Ž, Drobne D, Jemec A, Romih T, Zidar P, Bele M (2010) Zinc bioaccumulation in a terrestrial invertebrate fed a diet treated with particulate ZnO or $ZnCl_2$ solution. Toxicology 269(2–3):198–203. https://linkinghub.elsevier.com/retrieve/pii/S0300483X09004326
66. Zhu X, Wang J, Zhang X, Chang Y, Chen Y (2010) Trophic transfer of TiO_2 nanoparticles from daphnia to zebrafish in a simplified freshwater food chain. Chemosphere 79(9):928–933. https://linkinghub.elsevier.com/retrieve/pii/S004565351000305X
67. Petersen EJ, Pinto RA, Zhang L, Huang Q, Landrum PF, Weber Jr WJ (2011) Effects of polyethyleneimine-mediated functionalization of multi-walled carbon nanotubes on earthworm bioaccumulation and sorption by soils. Environ Sci Technol 45(8):3718–3724. https://doi.org/10.1021/es103004r
68. Judy JD, Unrine JM, Bertsch PM (2011) Evidence for biomagnification of gold nanoparticles within a terrestrial food chain. Environ Sci Technol 45(2):776–781. https://doi.org/10.1021/es103031a
69. Jackson BP, Bugge D, Ranville JF, Chen CY (2012) Bioavailability, toxicity, and bioaccumulation of quantum dot nanoparticles to the amphipod leptocheirus plumulosus. Environ Sci Technol 46(10):5550–5556
70. Gupta GS et al (2017) Laboratory scale microbial food chain to study bioaccumulation, biomagnification, and ecotoxicity of cadmium telluride quantum dots. Environ Sci Technol 51(3):1695–1706
71. Abbas Q, et al (2024) Nanomaterials in environment: release, compartmental transformation, and nanotoxicity. In: Handbook of nanomaterials, vol 2. Elsevier, pp 745–789. https://linkinghub.elsevier.com/retrieve/pii/B9780323955133000101
72. Kermanizadeh A et al (2015) Nanomaterial translocation-the biokinetics, tissue accumulation, toxicity and fate of materials in secondary organs-a review. Crit Rev Toxicol 45(10):837–872
73. Lin CH, et al (2009) The chemical fate of the Cd/Se/Te-based quantum dot 705 in the biological system: toxicity implications. Nanotechnology 20(21)
74. Domingos RF, Simon DF, Hauser C, Wilkinson KJ (2011) Bioaccumulation and effects of CdTe/CdS quantum dots on chlamydomonas reinhardtii—nanoparticles or the free ions? Environ Sci Technol 45(18):7664–7669
75. Bhatt I, Tripathi BN (2011) Interaction of engineered nanoparticles with various components of the environment and possible strategies for their risk assessment. Chemosphere 82(3):308–317. https://linkinghub.elsevier.com/retrieve/pii/S0045653510011434
76. Navarro E, et al (2008) Environmental behavior and ecotoxicity of engineered nanoparticles to algae, plants, and fungi. Ecotoxicology 17(5):372–386. https://doi.org/10.1007/s10646-008-0214-0
77. Rocha TL, Mestre NC, Sabóia-Morais SM, Bebianno MJ (2017b) Environmental behaviour and ecotoxicity of quantum dots at various trophic levels: a review. Environ Int 98:1–17. https://linkinghub.elsevier.com/retrieve/pii/S016041201630486X
78. Aye M, et al (2013) Genotoxic and mutagenic effects of lipid-coated CdSe/ZnS quantum dots. Mutat Res/Genet Toxicol Environ Mutagen 750(1–2):129–138. https://linkinghub.elsevier.com/retrieve/pii/S1383571812003440.
79. Fabrega J, et al (2011) Silver nanoparticles: behaviour and effects in the aquatic environment. Environ Int 37(2):517–531. https://linkinghub.elsevier.com/retrieve/pii/S016041201000228X

80. Dwivedi AD, et al (2015) Fate of engineered nanoparticles: implications in the environment. Coord Chem Rev 287:64–78. https://linkinghub.elsevier.com/retrieve/pii/S0010854514003579
81. Kloepfer JA, et al (2003) Quantum dots as strain- and metabolism-specific microbiological labels. Appl Environ Microbiol 69(7):4205–4213. https://doi.org/10.1128/AEM.69.7.4205-4213.2003
82. Worms IA, Boltzman J, Garcia M, Slaveykova VI (2012) Cell-wall-dependent effect of carboxyl-CdSe/ZnS quantum dots on lead and copper availability to green microalgae. Environ Pollut 167:27–33. https://linkinghub.elsevier.com/retrieve/pii/S026974911200142X
83. Zhang Y, Qin C, Wang H, Hu X, Gao Y (2024) Occurrence, behavior, and ecotoxicity of quantum dots in aquatic environments. Environ Sci Nano 11(10):4044–4059. https://xlink.rsc.org/?DOI=D4EN00542B
84. Buffet PE, Poirier L, Zalouk-Vergnoux A, Lopes C, Amiard JC, Gaudin P, Risso-de Faverney C, Guibbolini M, Gilliland D, Perrein-Ettajani H, Valsami-Jones E (2014) Biochemical and behavioural responses of the marine polychaete hediste diversicolor to cadmium sulfide quantum dots (CdS QDs): waterborne and dietary exposure. Chemosphere 100:63–70. https://linkinghub.elsevier.com/retrieve/pii/S004565351400006X
85. Gottschalk F, Sun TY, Nowack B (2013) Environmental concentrations of engineered nanomaterials: review of modeling and analytical studies. Environ Pollut 181:287–300. https://linkinghub.elsevier.com/retrieve/pii/S0269749113003205
86. Smith KS, Balistrieri LS, Todd AS (2015) Using biotic ligand models to predict metal toxicity in mineralized systems. Appl Geochem 57:55–72. https://linkinghub.elsevier.com/retrieve/pii/S0883292714001607
87. Niyogi S, Wood CM (2004) Biotic ligand model, a flexible tool for developing site-specific water quality guidelines for metals. Environ Sci Technol 38(23):6177–6192. https://doi.org/10.1021/es0496524
88. Pagenkopf GK (1983) Gill surface interaction model for trace-metal toxicity to fishes: role of complexation, PH, and water hardness. Environ Sci Technol 17(6):342–347. https://doi.org/10.1021/es00112a007
89. Paquin PR, et al (2002) The biotic ligand model: a historical overview. Comp Biochem Physiol Part C Toxicol Pharmacol 133(1–2):3–35. https://linkinghub.elsevier.com/retrieve/pii/S1532045602001126
90. Li M, Al-Jamal KT, Kostarelos K, Reineke J (2010) Physiologically based pharmacokinetic modeling of nanoparticles. ACS Nano 4(11):6303–6317. https://doi.org/10.1021/nn1018818
91. Lankveld DPK, et al (2010) The kinetics of the tissue distribution of silver nanoparticles of different sizes. Biomaterials 31(32):8350–8361. https://linkinghub.elsevier.com/retrieve/pii/S0142961210008884
92. Grech A, et al (2017) Toxicokinetic models and related tools in environmental risk assessment of chemicals. Sci Total Environ 578:1–15. https://linkinghub.elsevier.com/retrieve/pii/S0048969716323324
93. Chen H, Gong Y, Han R (2014) Cadmium telluride quantum dots (CdTe-QDs) and enhanced ultraviolet-B (UV-B) radiation trigger antioxidant enzyme metabolism and programmed cell death in wheat seedlings ed Jin-Song Zhang. PLoS ONE 9(10):e110400. https://doi.org/10.1371/journal.pone.0110400
94. Xiao A, et al (2016) Carbon and metal quantum dots toxicity on the microalgae chlorella pyrenoidosa. Ecotoxicology and Environmental Safety 133:211–217. https://linkinghub.elsevier.com/retrieve/pii/S0147651316302895
95. Kang W, et al (2019) Study of the persistence of the phytotoxicity induced by graphene oxide quantum dots and of the specific molecular mechanisms by integrating omics and regular analyses. Environ Sci Technol 53(7):3791–3801. https://doi.org/10.1021/acs.est.8b06023
96. Abishagu A, et al (2024) Metal nanoparticles and their toxicity impacts on microorganisms. Biologia 79(9):2843–2862. https://doi.org/10.1007/s11756-024-01760-z.
97. Xiao Q, et al (2013) Evaluate the potential toxicity of quantum dots on bacterial metabolism by microcalorimetry. Thermochimica Acta 552:98–105. https://linkinghub.elsevier.com/retrieve/pii/S0040603112005448

98. Lu Z, et al (2008) Mechanism of antimicrobial activity of CdTe quantum dots. Langmuir 24(10):5445–5452. https://doi.org/10.1021/la704075r
99. Birnstiel S, Sebastián M, Romera-Castillo C (2022) Structure and activity of marine bacterial communities responding to plastic leachates. Sci Total Environ 834:155264. https://linkinghub.elsevier.com/retrieve/pii/S0048969722023579
100. Nogueira V, et al (2012) Impact of organic and inorganic nanomaterials in the soil microbial community structure. Sci Total Environ 424:344–350. https://linkinghub.elsevier.com/retrieve/pii/S0048969712002513
101. Chen MY, Fang Z, Xu LX, Zhou D, Yang XJ, Zhu HJ, Yong YC (2021) Enhancement of photo-driven biomethanation under visible light by nano-engineering of rhodopseudomonas palustris. Bioresourc Bioproc 8(1):30. https://doi.org/10.1186/s40643-021-00383-5.
102. Li X, et al (2020) Gut microbiota and lipid metabolism alterations in mice induced by oral cadmium telluride quantum dots. J Appl Toxicol 40(8):1131–1140. https://doi.org/10.1002/jat.3972
103. Živilė C, Nijolė K, Ričardas R, Vytautas K (2015) Toxicity of quantum dots and cadmium to rainbow trout (oncorhynchus mykiss) in early ontogenesis. Front Mar Sci 2. https://doi.org/10.3389/conf.FMARS.2015.03.00195
104. Contreras EQ, et al (2013) Toxicity of quantum dots and cadmium salt to caenorhabditis elegans after multigenerational exposure. Environ Sci Technol 47(2):1148–1154. https://doi.org/10.1021/es3036785
105. Liu Q, Ding X, Pang Y, Cao Y, Lei J, Wu J, Zhang T (2022b) New insights into the safety assessment of quantum dots: potential release pathways, environmental transformations, and health risks. Environ Sci Nano 9(9):3277–3311. https://xlink.rsc.org/?DOI=D2EN00252C
106. Rani R, Kumar K (2024) Environmental and health risk assessment of emerging organic pollutants. In: Biotechnological advances in agriculture, healthcare, environment and industry. CRC Press, Boca Raton, pp 183–98. https://doi.org/10.1201/9781032642567-16

Part IV
Quantum Dots for Energy Applications

Chapter 12
Quantum Dot-Based Materials for High-Performance Batteries and Supercapacitors

Ranjit Kate, Ganesh Lonkar, P. Periasamy, Bharat Kale, and Murthy Chavali

Abstract Quantum Dots (QDs)—semiconductor nanoparticles, have garnered significant interest in energy storage applications due to their unique properties, such as high surface-to-volume ratio, tunable electronic characteristics, and quantum size effect. These features make QDs a promising solution for addressing the growing need for energy conversion and storage, especially in light of global challenges such as climate change and the depletion of fossil fuels. This chapter explores recent advancements in applying QDs for lithium-ion, sodium-ion, potassium-ion, and Li sulfur batteries, highlighting their role in improving charge storage capacity, cycling stability, and rate performance. The application of QDs in Supercapacitors (SCs) and other battery chemistries is also explored. This chapter focuses on presenting and discussing recent advances and progress in application of QDs and related composite materials for energy storage technologies.

Keywords Quantum dots · Energy storage · Quantum size effect · Batteries · Supercapacitors

1 Introduction

The growing environmental issues related to fossil fuel consumption and the long-term sustainability of human society highlight the urgent need for efficient and clean energy conversion and storage solutions [40]. To address this, the use of renewable energy sources like wind, tidal, and solar power has been strongly promoted [42, 74,

R. Kate · G. Lonkar · B. Kale · M. Chavali (✉)
Department of Research and Development, MIT World Peace University (MIT-WPU), Pune, Maharashtra, India
e-mail: ChavaliM@gmail.com

P. Periasamy
Department of Physics, Nehru Institute of Engineering and Technology, Coimbatore, Tamil Nadu, India

T. Bhuyan et al. (eds.), *Multifunctional Quantum Dots*, Materials Horizons: From Nature to Nanomaterials, https://doi.org/10.1007/978-981-92-0172-3_12

88]. However, due to the intermittent nature of these energy sources, the electricity generated requires the integration of advanced storage technologies [46, 47]. As a result, batteries, supercapacitors, and their hybrid combinations are considered the most efficient and versatile technologies for energy storage applications [11, 12, 72, 81]. Among the materials used in these technologies, existing active materials, such as transition metal oxides, phosphates, carbon-based materials, composites, etc. have demonstrated significant advantages in terms of energy density, power density, and cycling stability [28, 30, 57, 72]. However, materials used in both of these technologies often face issues like slow charge transport, limited surface area for reactions, and structural degradation after cycling. To mitigate these issues, QDs have emerged as promising additives or standalone materials in energy storage systems.

QDs, as 0-D nanomaterials offer unique benefits due to their nanoscale dimensions and exceptional physicochemical properties [53, 69], their high surface-to-volume ratio increases the active area for electrochemical reactions, while the quantum confinement effect enables tunable electronic properties for optimizing charge transport and energy band alignment with electrode materials [70]. Furthermore, their incorporation can boost the conductivity of electrode materials, supporting higher power density and faster charge/discharge rates [48]. Figure 1 highlights the unique properties of QDs making them versatile for energy applications. To obtain QDs, various synthesis methods have been developed, which can be classified into top-down and bottom-up strategies. Top-down methods, with simpler procedures, are more suitable for large-scale production, while bottom-up methods offer high atomic utilization and controllable size, structure, and surface modifications, enabling core–shell construction [77]. The synthesis method allows control over size, structure, functional groups, and heteroatom properties, leading to more active sites, enhanced tunability, and better hybridization with other nanomaterials [57].

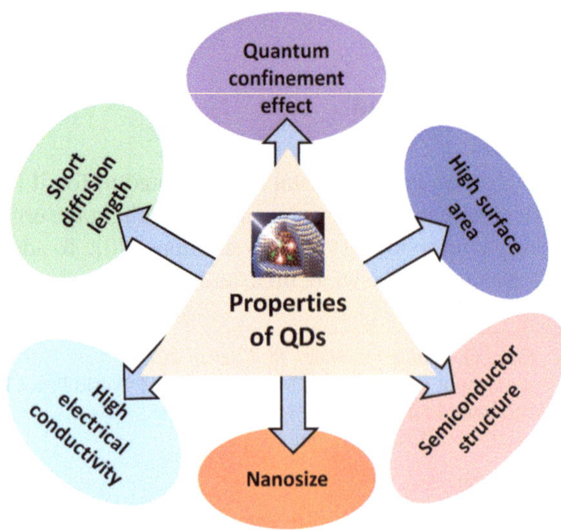

Fig. 1 Fundamental properties of Quantum Dots (QDs) and their role in enhancing energy storage performance

The overall performance of electrochemical energy storage devices is strongly influenced by the electrode materials employed. In this regard, the rapid advancement of active electrode materials has led to the incorporation of QDs in energy storage devices, such as rechargeable batteries and SCs, resulting in enhanced performance. In batteries, QDs contribute to improve cycling stability and capacity retention, thanks to their ultrafine particle size, which allows for better ionic conductivity and faster charge/discharge rates [48]. Unlike conventional active materials, QDs exhibit superior surface properties, reduced ion diffusion distance, excellent dispersibility, and improved structural integrity against volume changes during cycling. Furthermore, when used as electrodes in SCs, QDs offer high electrical conductivity, excellent stability, ultrahigh energy density, and superior specific capacitance [7].

In summary, this chapter explores the electrochemical applications of QDs in energy storage. It highlights the role of QDs in improving charge storage, capacity retention, and cycling stability. By understanding the relationship between QD properties and electrochemical performance, researchers can unlock new possibilities for advanced energy storage solutions.

2 Applications of QDs in Energy Storage

The applications of QDs in energy storage systems have gained significant attention due to their unique nanoscale properties. In batteries such as lithium-ion (Li-ion), sodium-ion (Na-ion), potassium-ion (K-ion), Li-sulfur (Li-S), and other battery chemistries, QDs improve performance by enhancing charge transport, increasing conductivity, and strengthening the structural stability of electrodes. Their small size and large surface area create more active sites for ion storage, boosting energy and power density. In SCs, QDs enhance charge storage by providing excellent electrical conductivity, faster ion diffusion, and greater stability during cycling. They also help to prevent electrode degradation and optimize ion transport, leading to better overall electrochemical performance. This section highlights how integrating QDs into electrode materials enhances their performance, enabling more efficient energy storage technologies.

2.1 Lithium-Ion Batteries (LIBs)

A battery consists of three primary components: a cathode, an anode, and an electrolyte (Fig. 2). While the cathode determines the battery's capacity and voltage, the anode plays a crucial role in energy density and stability. In Metal-Ion Batteries (MIBs), the anode is often a bottleneck due to challenges like volume expansion and poor conductivity. Recent studies in LIBs highlight that QD-based anode materials can overcome these limitations, offering high surface area, improved ion diffusion, and enhanced electrochemical properties.

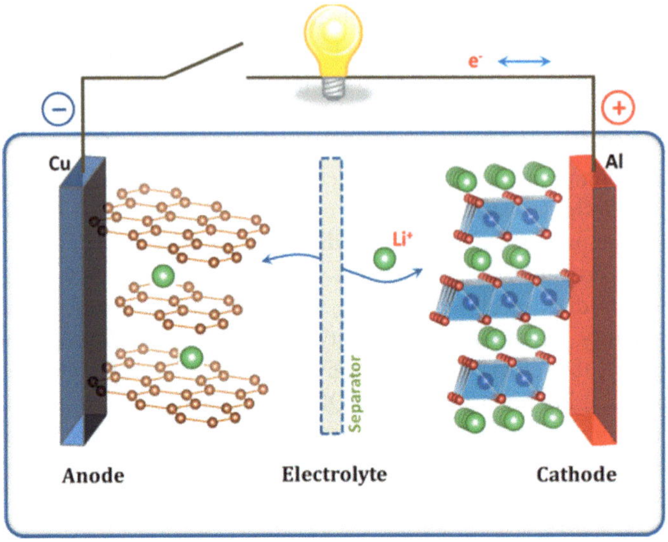

Fig. 2 Schematic representation of Lithium-Ion Battery (LIB) [15]. Reproduced with permission [15]. Copyright 2013, American Chemical Society

Metal Oxide (MO) and sulfide QDs-based nanocomposites have emerged as promising anode materials for LIBs. Yang et al. [71] prepared hierarchically porous carbon-wrapped ZnO QDs via MOF pyrolysis, demonstrating excellent performance as LIB anodes. This approach highlights the potential of cost-effective ZnO-based anodes. Zhang et al. [79] fabricated ZnO@ZnO QDs/C nanorod arrays on carbon cloth using an in situ ion exchange method (Fig. 3a), achieving excellent rate capability and cycling performance due to enhanced Li-ion diffusion, electron efficiency, and structural stability. Mo et al. [39] developed TiO_2 QDs/graphene composites as an anode material in LIBs as shown in Fig. 3b, c. The excellent electrochemical performance (Fig. 3d) is due to the graphene substrate, the even spread of TiO_2 QDs, and the combined effect of both materials. The graphene surface prevents the TiO_2 QDs from aggregating and helps maintain structural stability during cycling. Similarly, Huang et al. [21] synthesized TiO_2 QDs decorated Si nanocage composites as an anode material for LIBs. The interfacial engineering of TiO_2 QDs on Si nanocages enhances ion storage by providing structural stability, improved conductivity, and a stable SEI film. The resulting Si-0.04TiO_2 electrode achieves excellent cycling stability (616.3 mAh g^{-1} over 500 cycles at 1.0 A g^{-1}) and high capacity (513.6 mAh g^{-1} at 10.0 A g^{-1} after 150 cycles), offering a viable approach for addressing volume expansion in anode materials. Confining other metal oxide QDs within carbon materials can significantly enhance battery performance. For example, Zhao et al. [87] developed an in situ synthesis method (Fig. 3e, f) for well-dispersed SnO_2 QDs@GO with high mass loading, which exhibited excellent performance as an anode for LIBs. The SnO_2 QDs@GO anode achieved a capacity of 1121 mAh

g^{-1} at 100 mA g^{-1} and maintained 86% capacity after 2000 cycles (Fig. 3g), demonstrating the effective synergy between the components. The dispersion of SnO_2 QDs on GO enhances lithium storage and, with GO's conductivity, ensures excellent cycling stability. Additionally, other metal oxide QDs such as CoO [43], Co_3O_4 [27], CuO [66], Fe_3O_4@C [89], MnO [82], V_2O_5 [17] are also employed as anode materials for LIBs.

In addition to MO, metal chalcogenide QDs are also widely explored as anode materials to enhance the performance of LIBs. Xu et al. [67] synthesized ultrasmall MoS_2 QDs to enhance the electrochemical properties of $Li_4Ti_5O_{12}$ (LTO) by anchoring the MoS_2 QDs onto LTO nanosheets (Fig. 3h). As an anode in LIBs, the composite maintained 160 mAhg^{-1} after 1000 cycles at 10 C (Fig. 3i), showing significant potential for high-performance LIBs. Liao et al. [33] synthesized CoS_2–SnO_2@rGO QDs dots via hydrothermal methods, achieving high specific capacity and excellent cycling stability. The composite maintained 521.4 mAh g^{-1} after 300 cycles with 98.9% Coulomb Efficiency (CE) (Fig. 3j), with DFT calculations suggesting favorable Li ion diffusion paths. This work highlights the potential of dual transition metal heterojunctions for enhancing lithium-ion battery performance. Huang et al. [65] developed a method to embed ultrafine $SnSe_2$ QDs in 3D rGO, resulting in a composite with superior capacity (778.5 mAh g^{-1} at 50 mA g^{-1}), high-rate capability (324.5 mAh g^{-1} at 5000 mA g^{-1}), and long cycle life. Further, Zhu et al. [93] synthesized WTe_2 QDs embedded in NiSe/C nanowires (Fig. 3k), demonstrating excellent storage performance (~620 mAh g^{-1} at 1 A g^{-1} over 500 cycles) (Fig. 3i) and impressive rate capability as an anode material. Wang et al. [60] synthesized a novel binder-free anode, 3DOP Ge@N-C, featuring Ge QDs (2–5 nm) encapsulated in a 3D N-doped carbon framework Fig. 3m. This structure prevents Ge aggregation, accommodates volume expansion, and ensures excellent storage performance with a high capacity of 1160 mAh g^{-1} at 1 A g^{-1} and superior cycling stability (90% retention after 1200 cycles at 5 A g^{-1}).

The development of nonmetal QDs as an anode for LIBs is also highly promising. Meng et al. [72] designed black phosphorus (BP) QDs/Ti_3C_2 nanosheet anodes, achieving high capacity (910 mAh g^{-1} at 100 mA g^{-1}), excellent cycling stability (100% retention after 2400 cycles). Similarly, Zhang et al. [84] prepared Red Phosphorus (RP) QDs/$Ti_3C_2T_x$ composites with enhanced Li-ion transport, demonstrating superior electrochemical performance. Further, Wei et al. [62] developed a Ge/CNFs nanocomposite that effectively reduces capacity fading. The composite showed a high reversible capacity of 1204 mAh g^{-1} at 200 mA g^{-1}, 87.1% capacity retention after 100 cycles, demonstrating its potential for high-performance LIBs.

2.2 *Sodium-Ion Batteries (SIBs)*

Although LIBs offer high energy density and long lifespan, limited and uneven distribution of lithium restricts large-scale use. SIBs are emerging as a promising alternative due to sodium's abundance, low cost, and suitable redox potential. However, the

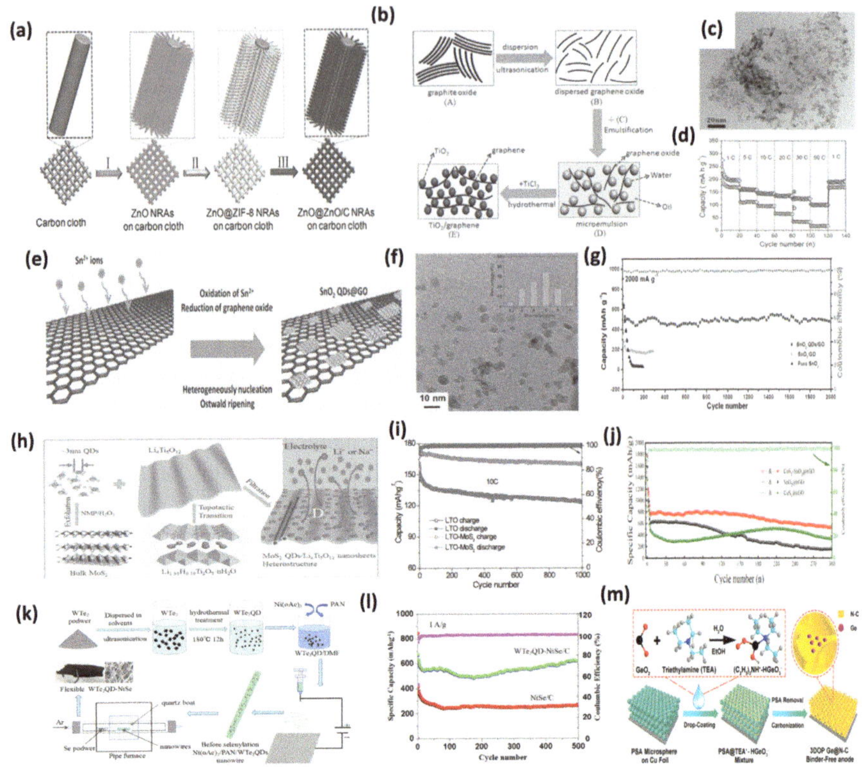

Fig. 3 **a** Synthesis process of ZnO@ZnO QDs/C Nanorod Arrays (NRAs) [79], **b** preparation technique of the TiO_2-QDs/graphene composite [39], **c** TEM image of the TiO_2-QDs/graphene [39], **d** comparison of the specific capacity between TiO_2-QDs/graphene and TiO_2-QDs electrodes [39], **e** synthesis procedure of SnO_2 QDs@Graphene Oxide (GO). [87], **f** HRTEM picture of SnO_2 QDs@GO [87], **g** high rate cycling performance and CE of different samples [87], **h** schematic synthesis of MoS_2 QDs/LTO composite [67], **i** cycling performance and the corresponding CE versus cycle number of the LTO-MoS_2-2 and LTO nanosheets at a rate of 10 C [67], **j** cyclability of CoS_2-SnO_2@rGO, SnO_2@rGO, and CoS_2@rGO samples at current density of 1 A g^{-1} [33], **k** schematic representation of the synthesis process of WTe_2QD-doped NiSe/C nanowires [93], **l** cycling performance of NiSe/C nanowires and WTe_2 QD-doped NiSe/C nanowires at the current density of 1 A g^{-1} [93], **m** scheme of the preparation process of 3DOP Ge@N-C composites [60]. **a** Reproduced with permission [79]. Copyright ©2015, Wiley-VCH. **b–d** Reproduced with permission [39]. Copyright © 2014, Wiley-VCH. **e–g** Reproduced with permission [87]. Copyright © 2016, Wiley-VCH. **h, i** Reproduced with permission [67]. Copyright © 2016, Wiley-VCH. **j** Reproduced with permission [33]. Copyright © 2023, Elsevier. **k, l** Reproduced with permission [93]. Copyright © [57], Elsevier. **m** Reproduced with permission [60]. Copyright © [60], Wiley-VCH

larger ionic radius of sodium ions makes finding suitable materials challenging. Researchers are exploring various types of electrode materials for SIBs. In this context, QDs are gaining attention for their ability to enhance the electrochemical performance of SIBs. When integrated with active materials, QDs can significantly improve charge and discharge kinetics by providing a high surface area for sodium ion adsorption and facilitating faster ion diffusion.

Liu et al. [36] synthesized 3D–0D graphene–Fe_3O_4 QD hybrids using a one-pot hydrothermal approach (Fig. 4a). Fe_3O_4 QD was uniformly loaded on 3D graphene (Fig. 4b), enhancing the electrode–electrolyte interface, preventing QD aggregation, and mitigating volume expansion. These hybrids demonstrated significantly enhanced cycling performance and higher capacities compared to pristine Fe_3O_4 at both low and high current densities. Further, Kong et al. [29] synthesized Fe_3O_4 QDs/MoS_2 nanosheet arrays on Graphite Paper (GP) as a high-performance anode for SIBs, offering excellent reversible capacity and rate performance. Yu et al. [75] synthesized a TiO_2 QDs/rGO composite as an anode for SIBs. Its porous structure ensures efficient electron/ion migration and mechanical stability, addressing the low conductivity and volume change issues of bulk MO in SIBs. Wang et al. [58] synthesized 2-CuO@C using a novel electrospinning method (Fig. 4c), embedding CuO QDs (~2 nm) in carbon nanofibers. The 2-CuO@C electrode demonstrated excellent capacity retention, high reversible capacity, and superior long-term cycling stability compared to bulk CuO (Fig. 4d). A capacity of 250 mAh g^{-1} was achieved even at a high current density of 5000 mA g^{-1}. Moreover, Wang et al. [59] developed a scalable 0D/3D SnO_2 QDs/graphene aerogel composite with superior anode performance for SIBs compared to standalone SnO_2.

Metal Sulfide (MS)-based QDs also hold significant potential as anode materials for SIBs. Hu et al. [20] synthesized an inverse opal-structured nanohybrid of carbon-confined transition MS QDs for SIB anodes (Fig. 4e). The 3D architecture features uniform QD dispersion in a conductive carbon matrix, encapsulating the QDs. The 3DOM Co_9S_8-QDs@NC exhibits minimal degradation over 2000 cycles, maintaining a specific capacity of 345 mAh g^{-1} with a CE over 99% (Fig. 4f). Additionally, full SIBs with 3D-ordered macroporous Co_9S_8-QDs@NC as the anode and $Na_3V_2(PO_4)_3$ as the cathode were fabricated and tested, demonstrating a discharge capacity of 314 mAh g^{-1}, a high-capacity retention of 91.8%, and a CE over 99% (Fig. 4g). Similarly, Guo et al. [16] synthesized a 2D heterostructure of CoS_x QDs (CoS and Co_9S_8) on N, S co-doped carbon nanosheets (CoS_x@NSC), enhancing conductivity and active sites. A CoS_x @NSC//$Na_{0.44}MnO_2$ full cell delivers a specific capacity of 414 mAh g^{-1} at 0.2 A g^{-1} (based on CoS_x @NSC mass) (Fig. 4h). Ding et al. [10] developed a 0D/2D SnS_2 QDs/N-doped $Ti_3C_2T_x$ MXene heterostructure using a solvent spatial confinement strategy. The uniformly distributed 3 nm SnS_2 QDs on the MXene matrix enhance Na^+ transport, achieving a high specific capacity of 763.2 mAh g^{-1} at 100 mA g^{-1} and excellent cycling stability of 345.3 mAh g^{-1} after 600 cycles. Superior electrochemical performance attributed to the strategic distribution of SnS_2 QDs onto $Ti_3C_2T_x$ MXene nanosheets which prevents restacking of the MXene layers and mitigates volume expansion during cycling, thus improving the structural integrity of the electrode (Fig. 4i). In addition to the above QDs, SnS [56], ZnS [83], and nonmetallic RP [78] QDs are also regarded as promising anode materials for SIBs.

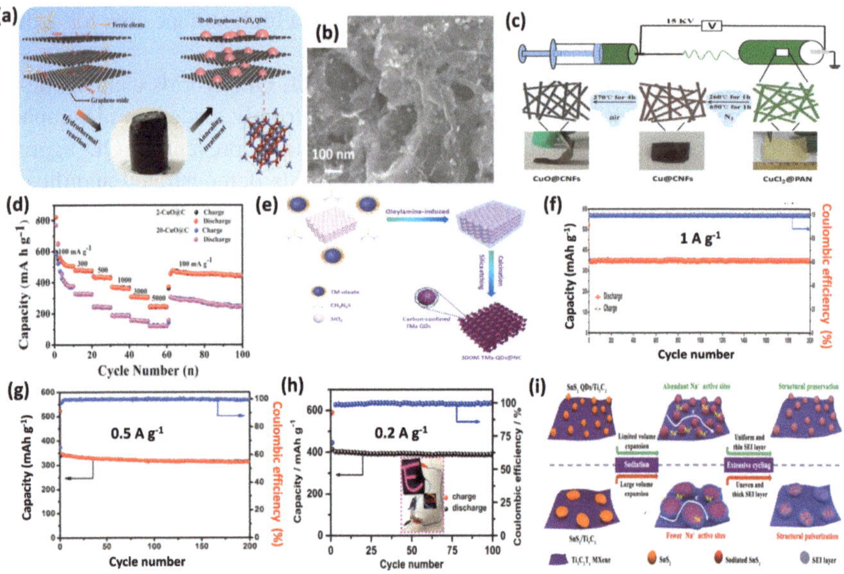

Fig. 4 **a** Synthesis method of graphene/Fe_3O_4 QDs [40]; **b** SEM picture of the graphene/Fe_3O_4 hybrid [36]; **c** schematic illustration of the formation process for CuO@C nanofibers [58], **d** rate performance of 2-CuO@C and 20-CuO@C at different current densities [58], **e** schematic illustration of the formation process of 3DOM TMs-QDs@NC [20], **f** cycling performance of 3DOM Co_9S_8-QDs@NC for 2000 cycles at the 1 A g^{-1} current density [20], **g** cycling performance of the full battery at 0.5 A g^{-1} [20], **h** cycle performance of the $Na_{0.44}MnO_2$//CoSx@NSC-1 full cell at a current density of 0.2 A g^{-1} [16], **i** schematic Illustration of SnS_2 QDs/Ti_3C_2 (upper) and SnS_2/Ti_3C_2 electrodes (lower) as anodes for SIBs before and after cycling [10]. **a**, **b** Reproduced with permission [36]. Copyright © 2016, American Chemical Society. **c**, **d** Reproduced with permission [58]. Copyright © 2016, Wiley-VCH. **e–g** Reproduced with permission [20]. Copyright © [72], Wiley-VCH. **h** Reproduced with permission [13]. Copyright © 2017, Wiley-VCH. **i** Reproduced with permission [10]. Copyright © [69], American Chemical Society

2.3 *Potassium-Ion Batteries (PIBs)*

PIBs are considered promising and appealing energy storage devices because of their cost-effectiveness and the abundance of potassium resources. However, their performance is limited by the large ionic radius of potassium ions (1.38 Å). Downsizing active materials to the QD scale provides an effective strategy to address these electrode-related challenges. Despite this potential, the field is still in its early stages of development, with only a limited number of QD-based anode materials reported for PIBs. Gao et al [13] synthesized CoS QDs nanoclusters uniformly anchored on graphene nanosheets via a facile two-step hydrothermal strategy. The prepared anode demonstrated an excellent capacity of 310.8 mAh g^{-1} after 100 cycles at 500 mA g^{-1} and superior high-rate performance in PIBs. Qi et al. [44] reported morphology-controllable ZnS QDs on graphene, which reduce the anode's volume effect and shorten the K-ion migration path. Liu et al. [57] developed $Co_{0.85}Se$ QDs encapsulated within mesoporous carbon via a direct hydrothermal method (Fig. 5a). As depicted in Fig. 5b, the $Co_{0.85}Se$ QDs/C20 electrode maintains a reversible capacity of 220 mAh g^{-1} even at an ultrahigh current density of 2000 mA g^{-1}. Hussain et al. [69] reported Co_3Se_4 QDs wrapped in N-doped carbon as an anode for PIBs, showing excellent discharge capacity and cyclic performance. DFT calculations revealed that the QDs had a lower K-ion transfer barrier and improved intercalation compared to bulk materials. Further, Cao et al. [3] reported the synthesis of $Cu_{12}Sb_4S_{13}$ QDs combined with few-layered Ti_3C_2 nanosheets (CAS-Ti_3C_2) as an advanced anode for PIBs (Fig. 5c). The Ti-S bond in the composite enables a strong synergistic effect, shortening the K^+ diffusion path and accommodating volumetric changes during cycling. The CAS-Ti_3C_2 anode delivers a high reversible capacity of 496.7 mAh g^{-1} at 100 mA g^{-1} after 200 cycles (Fig. 5d), outperforming most transition-metal disulfide QD-based anodes for PIBs. The superior performance is attributed to enhanced K^+ diffusion kinetics and a unique transformation-storage mechanism.

2.4 *Lithium-Sulfur Batteries (LSBs)*

LSBs utilize elemental sulfur as the cathode and lithium metal as the anode, offering a high theoretical capacity (1675 mAh g^{-1}) and energy density (2600 Wh kg^{-1}) [37]. Sulfur's low cost, non-toxicity, and abundance make it appealing [76]. However, the commercialization of LSBs faces major challenges, such as the shuttle effect from dissolving and migrating lithium polysulfides and the poor electronic conductivity of sulfur and its discharge products (Li_2S). Furthermore, the significant volume expansion during cycling makes it even harder to maintain the battery's stability and lifespan. The integration of QDs into Li-S cathodes significantly enhances electrochemical performance by mitigating the aforementioned challenges.

Jian et al. [23] developed ZnO QD-modified reduced graphene oxide (rGO@ZnO QDs) and incorporated it into Li-S cathodes. The resulting ZnO QDs/rGO/S cathode exhibited enhanced reaction kinetics, reduced surface impedance, and efficient polysulfide adsorption. It delivered an initial discharge capacity of 998.8 mAh g^{-1}, retaining 73.3% after 400 cycles at 1 C, and showed excellent rate capability up to 4C. Li et al. [60] synthesized multi-functionalized $NiFe_2O_4$ QDs as additives to reduce carbon content in sulfur cathodes for Li-S batteries (Fig. 5e). The QDs improved electrolyte wettability, adsorption, and electrical conductivity, reducing carbon content from ~26 to ~5%. They also enhanced polysulfide chemisorption, Li_2Sn phase conversion, and charge transfer. The resulting hybrid cathodes showed excellent rate performance, with only ~0.08% capacity fade over 500 cycles at 0.2 A g^{-1} and CE above 96%, even in $LiNO_3$-free electrolytes.

To mitigate the polysulfide shuttle reaction, several MS QDs were also incorporated to improve the conductivity of sulfur/lithium sulfides. Cai et al. [2] prepared a CdS-QDs-loaded CNT structure for sulfur infiltration in LSBs (Fig. 5f–g). This structure effectively adsorbed polysulfides while enabling fast Li-ion transfer, with the

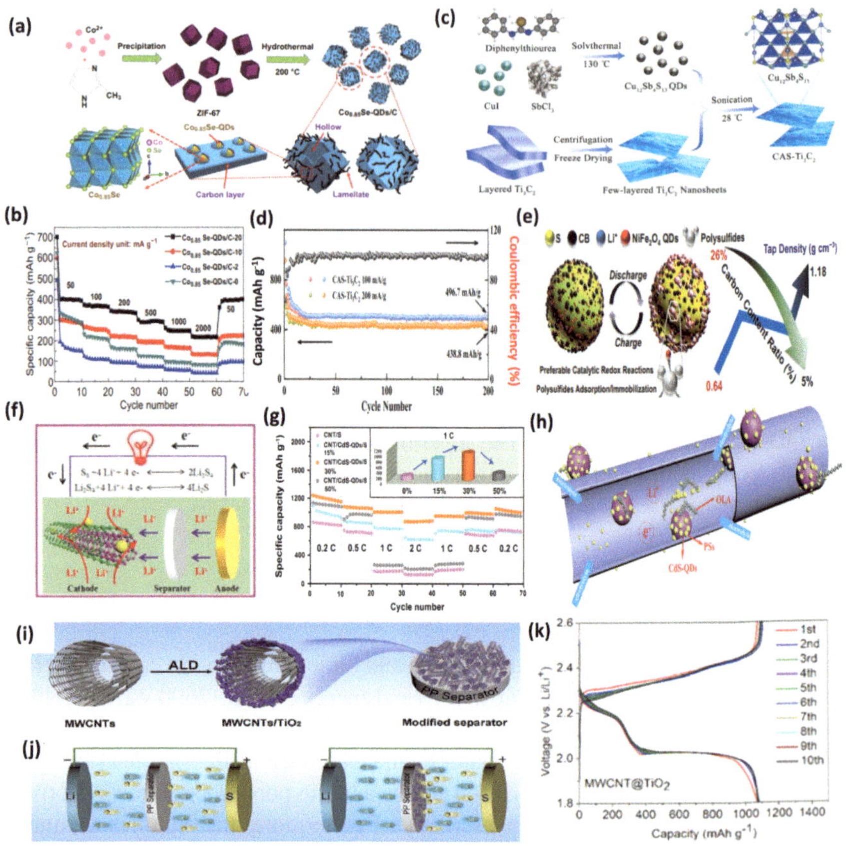

Fig. 5 **a** Preparation process for the $Co_{0.85}Se$-QDs/C composite [57], **b** rate capability under different current densities of the as-prepared $Co_{0.85}Se$-QDs/C samples [57], **c** scheme for the synthesis of hierarchical CAS-Ti_3C_2 composite [3], **d** cycling performance of CAS-Ti_3C_2 composites, $Cu_{12}Sb_4S_{13}$ QDs, and Ti_3C_2 nanosheets under 100 and 200 mA g^{-1} [3], **e** general diagram illustrating the configuration of devices (−)Li//S@CB ⊆ QDs(+) and the reaction mechanism of $NiFe_2O_4$ QDs [60], **f** structure of the CNT/CdS-QDs in LSB [2], **g** rate performance of CNT/S, CNT/CdS-QDs/S 15%, CNT/CdS-QDs/S 30%, and CNT/CdS-QDs/S 50% at different current densities [2], **h** schematic diagram of the effects of CdS-QDs on the electrochemical performance of LSBs [2], **i** schematic illustration of the fabrication method of MWCNTs@TiO_2 QDs [9], **j** schematic representation of LSB utilizing a commercial PP separator and the MWCNTs@TiO_2 QD-coated separator [9], **k** charge/discharge curves MWCNTs@TiO_2 QD [9]. **a, b** Reproduced from Liu [57]. Licensed under a Creative Commons Attribution 4.0 International License (http://creativecommons.org/licenses/by/4.0/). **c, d** Reproduced with permission [3]. Copyright © 2022, Wiley-VCH. **e** Reproduced from Li et al. [60]. Licensed under a Creative Commons Attribution 4.0 International License (http://creativecommons.org/licenses/by/4.0/). **f–h** Reproduced with permission [2]. Copyright © [57], Royal Society of Chemistry. **i–k** Reproduced with permission [9]. Copyright © [72], Elsevier

specific capacity gradually decreasing from ≈1238 mAh g^{-1} at 0.2 C to ≈918 mAh g^{-1} at 2 C, demonstrating excellent rate performance. Figure 5h shows the synergic effects of the hybrid sulfur hosts on the electrochemical performance. Shen et al. [50] fabricated a ZnS QD@rGO nanosheet to enhance interfacial electronic interactions and accelerate polysulfide transformation. DFT calculations and XPS results showed that the electronic interaction improved as the size of the QDs decreased. Additionally, other types of QDs such as MoS_2 [61], CuS [32], etc. have also been employed for LSBs. Some QDs are also used in functional separators to suppress lithium polysulfide shuttling and improve cathode conversion kinetics. Ding et al. [9] developed a separator with TiO_2 QDs decorated on multi-walled carbon tubes to mitigate the shuttle effect (Fig. 5i, j) and reduce electrode expansion during the reaction. This modification enhances the stability of LSBs (Fig. 5k) through strong interactions that restrain the shuttle effect. Furthermore, Liu et al. [37] introduced CoAl-LDH QDs on nitrogen-doped graphene as a functional separator for LSBs, offering active sites to capture lithium polysulfides and accelerate conversion reaction kinetics. LSB with this novel separator exhibited an excellent discharge capacity at 0.1 C (980.97 mAh g^{-1}). In addition to the QDs discussed above, various other QD-based nanocomposites, such as metal phosphide [69], metal telluride [69], and phosphorus [72], are also employed in LSBs. All the above findings highlight that electrical conductivity, suppression of the shuttling effect, and catalytic activity are key to optimizing sulfur-based cathodes. QDs, with their adjustable electronic structures, strong chemical polarity, and quantum-confined active sites, enhance conductivity and boost catalytic activity, making them ideal for high-performance sulfur cathodes.

2.5 *Other Emerging Batteries*

The role of QDs in systems such as LIBs, SIBs, PIBs, and LSBs is emerging and under active investigation. They are also being explored for other advanced battery technologies, including MIBs (e.g., zinc-ion and aluminum-ion) and metal-air batteries. These emerging batteries aim to address the limitations of current technologies by offering safer, cost-effective, and higher capacity solutions. Deng et al. [8] synthesized MOF-derived $ZnMn_2O_4$ (ZMO) QD@C using an in situ electrochemical induction strategy, where ZMO QDs are uniformly embedded in a porous carbon framework. Zn/ZMO QD@C batteries exhibit excellent cycling stability, maintaining 86.4% capacity retention after 1500 cycles at 1 A g^{-1}. Zhang et al. [81] reported novel V_2O_5/Carbon QD (CQD) hybrid nanobelts synthesized via a one-step hydrothermal method, with in situ partial reduction generating V^{4+}. As a Zinc-Ion Battery (ZIB) cathode, it achieved a high capacity of 460 mAh g^{-1}. Wu et al. [63] demonstrated that integrating CQDs with VO_2 significantly enhances electron/ion transport and structural integrity. The VO_2/CQDs composite shows superior electrochemical performance, including higher capacity (427 vs. 373 mAh g^{-1}), specific energy (333 vs. 287 Wh kg^{-1}), and rate capability (72% vs. 50% at 0.2 to 8 A

g^{-1}) compared to the unmodified sample. It also exhibits stable cycling over 2000 cycles. Sobhanmanesh et al. [51] synthesized an rGO/SnO_2 QDs composite cathode via microwave synthesis for an aluminum-ion battery. The rGO/SnO_2 QDs cathode showed outstanding electrochemical performance, including a discharge capacity of 231 mAh g^{-1} at 2 A g^{-1} over 300 cycles.

Metal-air batteries, which use O_2 from the air to convert it into electricity and produce water, offer a pollution-free solution. These batteries feature a metal anode, an air cathode, and an electrolyte, with a high theoretical energy density. Despite their advantages, metal-air batteries face several issues, including slow Oxygen Reduction/Oxygen Evolution Reaction (ORR/OER) kinetics, high overpotential, and poor stability of the air cathode Xu et al. [68]. To address these issues, efficient and stable cathode materials with superior ORR and OER are needed. Semiconducting QDs are promising because they provide more adsorption and reaction sites, enhance oxygen reaction kinetics, and significantly reduce overpotential [52, 55, 68). In particular, Li-air batteries are particularly notable for their ultrahigh energy density. Efficient catalysts for both the oxygen evolution and reduction reactions are crucial for improving the performance of Li-O_2 batteries. Sun et al. [52] developed MoS_2 QDs with sulfur vacancies using a novel top-down method involving lithium bis(trifluoromethyl sulfonyl)imide as a chemical clipper. These Vs-MoS_2 QDs exhibited outstanding ORR/OER activities in a Li-O_2 battery. Notably, the QDs maintained stable current density for 230 cycles, which was about nine times higher than that of bulk MoS_2. The enhanced performance is attributed to the QDs' ability to improve the conformal growth of Li_2O_2 on the cathode. Further, to solve the overpotential issue, Veeramani et al. [55] developed a photocathode nanocomposite (CdSe/ZnS QD@CNT) for photoelectrochemical Li-O_2 batteries. The QDs and carbon nanotubes facilitate efficient electron–hole pair transport, O_2 diffusion, and Li^+ conduction. With a charging voltage of 2.65–4.0 V, the QDs enable effective Li_2O_2 oxidation, achieving reduced overpotential and long-term stability over 100 cycles. These results highlight the potential of QDs in enhancing Li-O_2 battery performance. Cheng et al. [5] reported a BP QDs/δ-MnO_2 nanobelts cathode on carbon cloth, offering abundant active sites for ORR/OER and efficient reactant transport. The coupling of δ- MnO_2 and BP QDs promoted Li_2O_2 sheet growth, enabling high capacity and long cycle life. Although Li-O_2 batteries have drawn significant attention for their ultrahigh energy density, other metal-air batteries like Zn-air, Al-air, and Mg-air also hold considerable promise. Huang et al. [22] developed Mn_3O_4 QDs on N-doped multiwall CNTs as a cost-effective ORR catalyst, enabling a four-electron pathway. A Zn–air battery with this cathode achieved 745 mAh g^{-1} at 10 mA cm^{-2} and maintained voltage stability over 105 h of discharge. Liu et al. [38] synthesized bimetal oxide ($ZnCo_2O_4$) QDs combined with N-doped CNTs as bifunctional OER/ORR electrocatalysts for Zn-air batteries. The excellent performance was observed which is attributed to high conductivity and the increased specific surface area of $ZnCo_2O_4$ QDs, enhancing active species diffusion. Kim et al. (n.d.) used plasma engineering to anchor Fe QDs onto Nitrogen-doped Graphene (NG) and $Ti_3C_2T_x$ MXene composite for Al-air batteries. The negatively charged surface, induced by fluorine and hydroxyl groups, repelled Cl^- ions, protecting Fe

QD active sites. This design achieved high ORR catalytic activity, delivering a power density of 53.6 mW/cm^2 and a rate performance of 0.8 V at 50 mA/cm^2. Xu et al. [68] developed high-activity bifunctional ORR/OER electrocatalysts by integrating $NiFe_2O_4$ QDs with CNTs. The catalysts featured a large specific surface area (584 m^2 g^{-1}) and excellent conductivity, making them suitable cathodes for Zn–air and Mg–air batteries. $NiFe_2O_4$ QDs/CNTs demonstrated outstanding OER/ORR performance with a low potential difference (ΔE) of 0.9 V, highlighting the role of QDs in enhancing adsorption–desorption and active site availability to improve OER/ORR kinetics.

A summary of the electrochemical performance of QDs and derivatives in battery applications is given in Table 1.

QDs Quantum Dots; LIBs Lithium-Ion Batteries; SIBs Sodium-Ion Batteries; PIBs Potassium-Ion Batteries; LSBs Lithium-Sulfur Batteries.

2.6 *Supercapacitors*

SCs, characterized by their high power density ($\approx$10 kW kg^{-1}), fast charge/discharge rates, and long cycle life (>100,000 cycles), are broadly categorized into three types: Electrochemical Double-Layer Capacitors (EDLCs), pseudocapacitors, and hybrid SCs [14, 25]. A supercapacitor typically consists of three main components: the electrode, electrolyte, and separator as shown in Fig. 6. Among these, the electrode material is the most critical factor influencing the performance of SCs. Each type of SC uses different ways to store charge, allowing QDs to improve their performance in a specific and unique way.

2.6.1 QDs in Symmetric SCs

EDLCs store energy through the physical adsorption of ions on high surface area electrode materials, enabling rapid charge and discharge cycles with high power density. Due to these advantages, EDLCs are receiving increased attention, with current research focusing on developing low-cost bilayer electrode materials that achieve even higher energy and power densities. Carbonaceous materials are commonly used as electrode materials in EDLCs due to their high surface area, which boosts the specific capacitance (Csp). However, the low energy density of EDLCs limits their practical applications. Incorporating QDs and their composites into carbon electrodes helps to increase the specific surface area, thereby improving the SC's performance.

Li et al. [31] developed micro SCs by incorporating Carbon QDs (CQDs) as nanofillers into a graphene fiber (Fig. 7a). The interaction between CQDs and graphene enhanced the mechanical strength and electrical stability, forming bridges through hydrogen bonding. The CQDs/graphene fiber device exhibited a Csp of 607 mF/cm^2 and maintained cyclic stability for over 10,000 cycles. Chang et al.

Table 1 Summary of the electrochemical performance of QDs and derivatives in battery applications

QD	Sample	Discharge capacity (mAh/g)	Cycling stability (No. of cycles) (discharge capacity) (current density)	Refs.
LIBs				
ZnO	ZnO@ZnO QDs/C NRAs	785 at 500 mA g^{-1}	100, 699, 500 mA g^{-1}	[79]
SnO_2	SnO_2 QDs@GO	1121 at 100 mA g^{-1}	2000, 477, 2000 mA g^{-1}	[87]
SnO_2	SnO_2-CNT	1841.6 at 100 mA g^{-1}	100, 648, 500 mA g^{-1}	Lu et al. (2016)
TiO_2	Si-$_{0.04}TiO_2$	2006.1 at 0.5 A g^{-1}	500, 618.5, 1 A g^{-1}	[21]
TiO_2	TiO_2-QDs/GNs	272 at 1C	100, 190, 1C	[39]
TiO_2	TiO_2@C	632.4 at 25 mA g^{-1}	200, 523, 100 mA g^{-1}	[54]
Graphene	PF-GQD@SiNP	4066 at 50 mA g^{-1}	100, 3068, 100 mA g^{-1}	[35]
Graphene	Trp-GQD@MGA-n/Si	1427 at 100 mA g^{-1}	100, 1331, 100 mA g^{-1}	[34]
Carbon	C-VOCQD	420 at 0.3C	500, 173, 60C	[1]
Co_3O_4	Co_3O_4QD/Graphene	1618 at 0.1C	90, 1785, 0.1C	[92]
CoS_2	CoS_2NP@G-CoS_2QD	827 at 1 A g^{-1}	300, 831, 1 A g^{-1}	[18]
MoS_2	LTO/MoS_2	91 at 5C	200, 101, 2C	[67]
Germanium	Ge/GN sponge	2192 at 100 mA g^{-1}	50, 975, 100 mA g^{-1}	[45]
Sn	Sn QDs@CNFs	1176 at 0.1 A g^{-1}	500, 685, 0.2 A g^{-1}	[80]
SIBs				
Fe_3O_4	3D-0D graphene Fe_3O_4	525 at 30 mA g^{-1}	200, 312, 50 mA g^{-1}	[36]
Fe_3O_4	Fe_3O_4@MoS_2-GP	468 at 100 mA g^{-1}	300, 388, 100 mA g^{-1}	[29]
CuO	2-CuO@C	528 at 100 mA g^{-1}	500, 401, 100 mA g^{-1}	[58]
MoS_2	LTO/MoS_2	118 at 2C	200, 101, 2C	[67]
Co_9S_8	3DOM Co_9S_8-QDs@NC	727 at 0.1 A g^{-1}	2000, 345, 1 A g^{-1}	[20]
FeS_2	FeS_2 QDs/FGS	315 at 100 mA g^{-1}	100, 284 at 0.5 A g^{-1}	[49]
Carbon	C-VOCQD	328 at 0.3C	100, 132, 60C	[1]
Carbon	3D PCFs	130 at 5 A g^{-1}	5000, 137, 5 A g^{-1}	[19]

(continued)

Table 1 (continued)

QD	Sample	Discharge capacity (mAh/g)	Cycling stability (No. of cycles) (discharge capacity) (current density)	Refs.
PIBs				
CoS	CoS@G-25	799 at 0.1C	100, 310.8, 500 mA g^{-1}	[13]
$Co_{0.85}Se$	$Co_{0.85}Se$-QDs/C	402 at 50 mA g^{-1}	500, 228, 1 A g^{-1}	[57]
Co_3Se_4	Co_3Se_4 QD	380 at 50 mA g^{-1}	10,000, 243, 1 A g^{-1}	[69]
$Cu_{12}Sb_4S_{13}$	CAS-Ti_3C_2	163.3 at 5 mA g^{-1}	1800, 175.6, 1 A g^{-1}	[3]
LSBs				
ZnO	rGO@ZnO QDs	430.2 at 4C	200, 776.1, 0.2C	[23]
$NiFe_2O_4$	S@CB $\subseteq$ QDs	526 at 5 A g^{-1}	500, 910, 0.2 A g^{-1}	[60]
MoS_2	rGO- MoS_2 QDs/S	293 at 5C	100, 612, 0.5C	[61]
CdS	CNT/CdS-QDs	918.1 at 2C	150, 820.6, 0.5C	[2]
CuS	CuS-n/CA-S	1318 at 0.2C	500, 840, 0.5C	[32]
MoP	MPQ@G/S	681.2 at 2C	100, 1184.5, 0.2C	[69]
BP	PCNF/S/BP QD	910 at 1C	200, 1072, 0.1C	[72]
Other batteries				
$ZnMn_2O_4$	ZMOQD@C	318.4 at 100 mA g^{-1}	1500, 143.9, 1 A g^{-1}	[8]
V_2O_5	V_2O_5/C QDs	456 at 0.1 A g^{-1}	1500, 175, 4 A g^{-1}	[81]
Mn_3O_4	Mn_3O_4 QDs/ N-p-MCNTs	745 at 10 mA cm^{-2}	–	[22]
SnO_2	RGO/SnO_2 Q Ds	124.6 at 4 A g^{-1}	250,000, 163, 2.5 A g^{-1}	[51]
MoS_2	MoS_2 QD	–	230, ≈1000, 500 A g^{-1}	[52]

[4] synthesized 2D Porous Carbon Nanosheets (PCNs) from industrial sulfonated pitch via self-assembly and activation. Composed of graphene QD, the PCNs feature a 20 nm thickness, 2429 m^2/g surface area, 0.562 g/cm^3 packing density, 89 S/m conductivity, and excellent flexibility. In pure $EMIMBF_4$ electrolyte, the PCNs exhibit 90.1 F/cm^3 capacitance, 71% rate retention, 88% stability over 5000 cycles (Fig. 7b), and a maximum energy density of 39 Wh/L, making them ideal for high-performance EDLC electrodes. Oloore et al. [41] fabricated a thin-film supercapacitor using CdS-QDs and organometallic halide perovskite as active electrodes. The device demonstrated excellent cycling stability over 4000 cycles, large electrolyte-contact surface areas, low charge-transfer resistance, and short relaxation time, enhancing

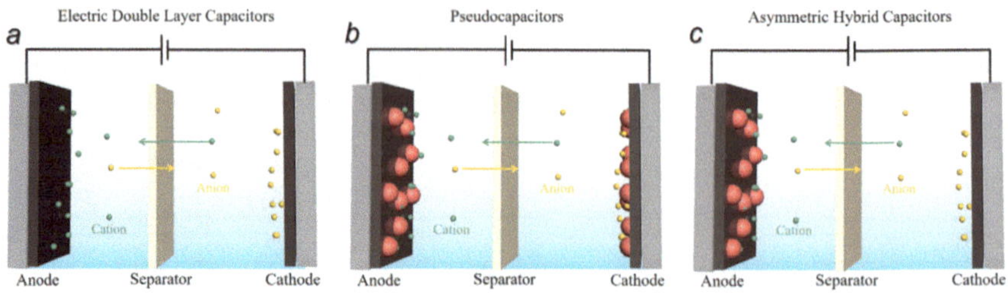

Fig. 6 Schematic representation of various types of SCs: **a** EDLCs with porous carbon symmetric electrodes on both sides; **b** pseudocapacitors utilizing QD-based materials as symmetric electrodes on both sides; **c** asymmetric SC featuring a porous carbon cathode and a QD-based anode (battery-type) [77]. Reproduced with permission [77]. Copyright © 2023, Wiley-VCH

overall performance. Yu et al. [69] developed 3D carbon composite electrodes for SCs by coating Carbon Dots (CDs) onto lignin-based Activated Carbons (ACs) (Fig. 7c). The CDs improved electrolyte affinity, increased capacitance from 125.8 to 301.7 F/g, and enhanced ion diffusion by 25.5 times. The electrodes showed excellent stability with 100% efficiency after 3000 cycles, highlighting their potential for advanced energy storage. Yin et al. [73] synthesized WS_2 QDs through a simple process and enhanced their pseudocapacitance properties via surface modification with 1,2-ethanedithiol. The modified QDs achieved a high specific capacitance of 457 F/g, excellent cycling stability over 8000 cycles, and a CE of 81%. Cong et al. [6] synthesized nonstoichiometric WO_3 QDs ($WO_{3\text{-}x}$, $x \approx 0.06$) with a size of 1.6 nm (Fig. 7d) for pseudocapacitors. Ligand exchange from octylamine to pyridine improved conductivity and hydrophilicity, enabling efficient ion transport, short diffusion paths, and enhanced electrochemical performance (Fig. 7e, f).

2.6.2 QDs in Asymmetric SCs

An Asymmetric SC (ASC) combines two different electrode materials, typically a battery-type faradaic electrode (pseudocapacitive) and a capacitive electrode (EDLC). This design combines the high energy density of the faradaic material with the fast charge–discharge capability of the capacitive material, achieving a wider voltage window, improved energy density, and enhanced cycling stability compared to symmetric SCs. However, synthesizing anode materials with high Csp remains a significant challenge. Fabricating ultrafine QDs in carbon substrates and developing ASCs effectively address existing challenges. Xu et al. [70] developed a binder-free C-$NiCo_2O_4$/C@Ni-Co-S nanoarray electrode supported on Ni foam (Fig. 8a), integrating 0D carbon QDs and a 2D carbon layer. The fabricated ASC device exhibited exceptional electrochemical performance (Fig. 8b, c). Its high energy density (95 Wh kg^{-1} at 0.9 kW kg^{-1}) and ability to power LEDs for 25 min highlight its potential for high-performance SCs (Fig. 8d), supported by its unique structural design and enhanced charge-transfer kinetics. Zhang et al. [85] synthesized a precursor

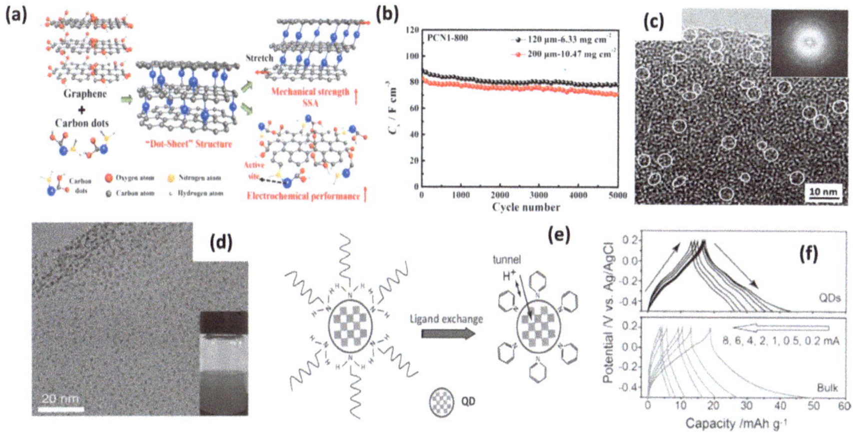

Fig. 7 **a** Schematic illustration of the dot-sheet porous structure formed between CDs and graphene [31], **b** cycling stability at 1 A g^{-1} of thin PCN1-800 electrode of 120 μm and thick PCN1-800 electrode of 200 μm [4], **c** CDs consisted of nanocrystallites [69], **d** TEM picture of as-obtained WO_3 QDs (average sizes: 1.6 nm), (inset of d) a photo of the QD dispersed in hexane [6], **e** replacement of ligand with conductive pyridine [6], **f** galvanostatic charge/discharge curves for QDs and bulk materials under different currents [6]. **a** Reproduced with permission [31]. Copyright © [72], Royal Society of Chemistry. **b** Reproduced with permission [4]. Copyright © [60], Elsevier. **c** Reproduced from Yu et al. [69]. Copyright © [69]. Published by American Chemical Society. This publication is licensed under CC-BY-NC-ND 4.0. **d–f** Reproduced with permission [6]. Copyright © 2014, Wiley-VCH

using ZIF-8 and CNTs to fabricate ZnO QDs/C/CNTs and N-doped carbon/CNTs via a controlled pyrolysis reaction (Fig. 8e). These materials were used as electrodes in SCs, demonstrating high capacitance. A solid-state ASC was then constructed using ZnO QDs/carbon/CNTs, PVA gel polymers, and porous N-doped carbon/CNTs (Fig. 8f), achieving a working potential of 1.7 V, with an energy density of 23.6 Wh kg^{-1} and a power density of 16.9 kW kg^{-1} (Fig. 8g). Zheng et al. [90] synthesized Ni_3S_4 QDs@GO/CNTs via a one-step method, achieving a Csp of 1200 F g^{-1} (4.5 times higher than Ni_3S_4 QDs) and improved rate capability (22–82% at 3 A g^{-1}). The design enhances conductivity, prevents QD agglomeration, and shows excellent potential for SCs. Xia et al. [64] developed a hybrid material by uniformly loading Fe_2O_3 QDs onto Functionalized Graphene Nanosheets (FGS) using a simple and efficient method (Fig. 8h). A 2 V ASC using Fe_2O_3 QDs/FGS anode and MnO_2/FGS cathode achieved 50.7 Wh kg^{-1} at 100 W kg^{-1} with excellent stability and power performance (Fig. 128i). Jing et al. [24] developed a NiO-dots/Gh composite via a simple alternating voltage electrochemical method using graphite and Ni flake electrodes. NiO QDs were uniformly dispersed on few-layer graphene, with oxygen functional groups forming Ni-O-C bonds to enhance conductivity. The composite exhibited a high Csp of 1181.1 F g^{-1} at 2.1 A g^{-1} and 66.2% retention at 42 A g^{-1}. When used in an ASC with AC as the negative electrode, it achieved an energy density of 27.3 Wh kg^{-1} at 1562.6 W kg^{-1} and 87.2% capacity retention after 5000 cycles (Fig. 8j).

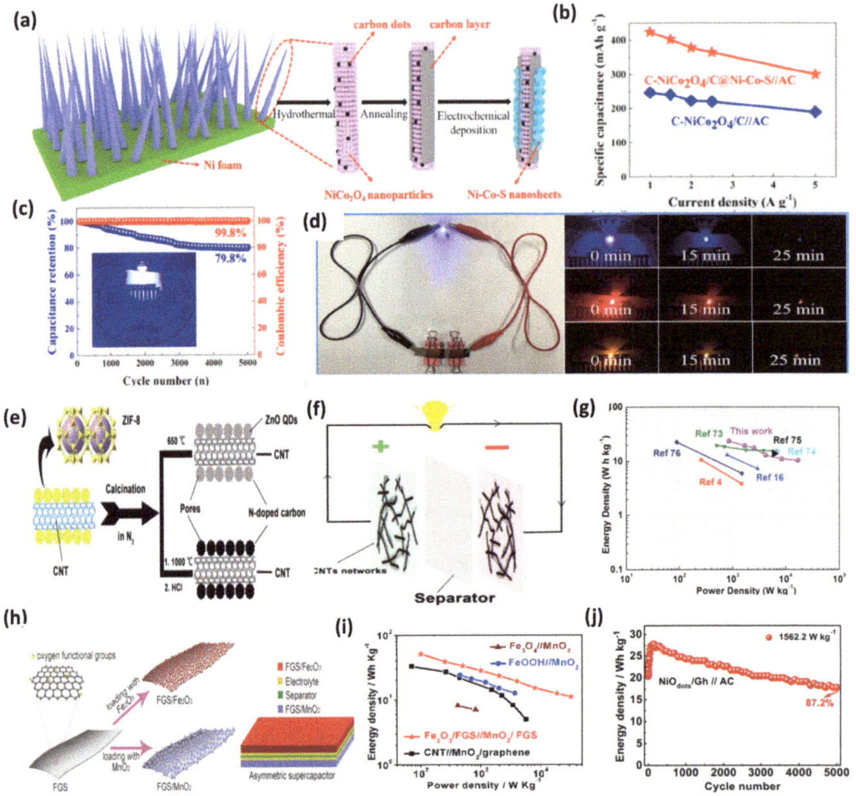

Fig. 8 **a** Illustration of the synthesis process of the C-$NiCo_2O_4$/C@Ni-Co-S nanoarrays [70], **b** Csp as a function of the current density of the C-$NiCo_2O_4$/C//AC and the C-$NiCo_2O_4$/C@Ni-Co-S//AC ASC [70], **c** capacitance retention and CE over 5000 charge–discharge cycles [70], **d** digital image of a blue, red, and yellow LED lighted by two series connection cells ASC [70], **e** schematic illustrations depicting the synthesis of ZnO QDs/carbon/CNTs and porous N-doped carbon/CNTs [85], **f** illustration of the assembled ASC [85], **g** ragone plots of the work and other all-solid-state ASC [85], **h** schematic illustration of the composite electrode design and the construction of ASC [64], **i** ragone plots of the Fe_2O_3/FGS//MnO_2/FGS ASC compared with reported data from the literature [64], **j** cyclability of NiO-dots/Gh//AC asymmetric capacitor at the power density of 1562.2 W kg^{-1} [24]. **a–d** Reproduced with permission [70]. Copyright © [69], American Chemical Society. **e–g** Reproduced with permission [85]. Copyright © 2016, Royal Society of Chemistry. **h**, **i** Reproduced with permission [64]. Copyright © 2015, Wiley-VCH. **j** Reproduced with permission [24]. Copyright © 2015, Elsevier

2.6.3 QDs in Hybrid SCs

Hybrid SCs (HSCs) have emerged as a promising energy storage technology, combining the high power density and long cycle life of traditional SCs with the enhanced energy density of batteries. By integrating electrostatic charge storage and faradaic redox mechanisms, HSCs achieve a unique balance between energy and power performance. Liu et al. [36] reported a one-pot hydrothermal reaction followed by calcination to synthesize Nb_2O_5 QDs embedded in ZIF-8-derived nitrogen-doped porous carbon (Nb_2O_5 QDs-NC). As a LIB anode, it exhibits excellent rate performance and cycling stability. Furthermore, an HSC device with an Nb_2O_5 QDs-NC anode and activated carbon cathode achieved remarkable electrochemical performance, including 76.9 W h kg^{-1} energy density, 11,250 W kg^{-1} power density, and long-term stability. Khalafallah et al. [26] synthesized Graphene QDs (GQDs)-decorated CuO nanoframeworks (GQDs/CuO) for high-performance SCs. Using a Cu-MOF template with confined porosity and Cu^{2+} as central atoms, CuO was prepared via in situ growth and annealing, followed by hydrothermal treatment. The GQDs enhanced conductivity and redox kinetics, achieving 729 F g^{-1} at 1 A g^{-1}, 82.2% retention after 3000 cycles for the GQDs/CuO, and 32.2 W h kg^{-1} energy density at 748.9 W kg^{-1} over 8000 cycles for the fabricated ASC.

3 Conclusion

In nutshell, QDs exhibit significant promise for transforming energy storage technologies, owing to their distinctive physicochemical characteristics, such as tunable size, large surface area, and superior electrical conductivity. Their incorporation into LIBs, SIBs, PIBs, and LSBs has led to a significant improvement in capacity, cycling stability, and rate performance by facilitating efficient ion diffusion and electron transport. Additionally, QDs have demonstrated significant promise in SCs, covering applications in symmetric, asymmetric, and hybrid configurations. These findings underscore the adaptability of QDs in tackling key challenges, including energy density, power density, and device longevity, across both batteries and SCs. Future research should prioritize scalable and cost-efficient synthesis methods, effective integration techniques, and improved long-term stability to fully exploit the potential of QDs for real-world energy storage solutions.

References

1. Balogun MS, Luo Y, Lyu F, Wang F, Yang H, Li H, Liang C, Huang M, Huang Y, Tong YX (2016) Carbon quantum dot surface-engineered VO_2 interwoven nanowires: a flexible cathode material for lithium and sodium ion batteries. ACS Appl Mater Interfaces 8(15):9733–9744. https://doi.org/10.1021/acsami.6b01305

2. Cai D, Wang L, Li L, Zhang Y, Li J, Chen D, Tu H, Han W (2019) Self-assembled CdS quantum dots in carbon nanotubes: induced polysulfide trapping and redox kinetics enhancement for improved lithium-sulfur battery performance. J Mater Chem A 7(2):806–815. https://doi.org/10.1039/c8ta09906e
3. Cao Y, Zhang Y, Chen H, Qin S, Zhang L, Guo S, Yang H (2022) Cu12Sb4S13 quantum dots/ few-layered Ti_3C_2 nanosheets with enhanced K+ diffusion dynamics for efficient potassium ion storage. Adv Funct Mater 32(6). https://doi.org/10.1002/adfm.202108574
4. Chang P, Yang F, Xie Q, Li T, Dong J (2020) 2D porous carbon nanosheet from sulfonated pitch-based graphene quantum dots for high volumetric performance EDLCs. J Power Sources 479. https://doi.org/10.1016/j.jpowsour.2020.228825
5. Cheng H, Xie J, Cao G, Lu Y, Zheng D, Jin Y, Wang K, Zhao X (2019) Realizing discrete growth of thin Li_2O_2 sheets on black phosphorus quantum dots-decorated δ-MnO2catalyst for long-life lithium–oxygen cells. Energy Storage Mater 23:684–692. https://doi.org/10.1016/j.ensm.2019.02.028
6. Cong S, Tian Y, Li Q, Zhao Z, Geng F (2014) Single-crystalline tungsten oxide quantum dots for fast pseudocapacitor and electrochromic applications. Adv Mater 26(25):4260–4267. https://doi.org/10.1002/adma.201400447
7. Das HT, Barai P, Dutta S, Das N, Das P, Roy M, Alauddin M, Barai HR (2022) Polymer composites with quantum dots as potential electrode materials for supercapacitors application: a review. In: Polymers, vol 14, no 5. MDPI. https://doi.org/10.3390/polym14051053
8. Deng S, Tie Z, Yue F, Cao H, Yao M, Niu Z (2022) Rational design of $ZnMn_2O_4$ quantum dots in a carbon framework for durable aqueous zinc-ion batteries. Angewandte Chemie – Int Edn 61(12). https://doi.org/10.1002/anie.202115877
9. Ding H, Zhang Q, Liu Z, Wang J, Ma R, Fan L, Wang T, Zhao J, Ge J, Lu X, Yu X, Lu B (2018) TiO_2 quantum dots decorated multi-walled carbon nanotubes as the multifunctional separator for highly stable lithium sulfur batteries. Electrochim Acta 284:314–320. https://doi.org/10.1016/j.electacta.2018.07.167
10. Ding J, Tang C, Zhu G, Sun W, Du A, He F, Wu M, Zhang H (2021) Integrating SnS_2. ACS Appl Energy Mater 4(1):846–854. https://doi.org/10.1021/acsaem.0c02730
11. Dubal DP, Ayyad O, Ruiz V, Gómez-Romero P (2015) Hybrid energy storage: the merging of battery and supercapacitor chemistries. In: Chemical society reviews, vol 44, no 7, pp 1777–1790. Royal Society of Chemistry. https://doi.org/10.1039/c4cs00266k
12. Fu W, Turcheniuk K, Naumov O, Mysyk R, Wang F, Liu M, Kim D, Ren X, Magasinski A, Yu M, Feng X, Wang ZL, Yushin G (2021) Materials and technologies for multifunctional, flexible or integrated supercapacitors and batteries. In: Materials today, vol 48, pp 176–197. Elsevier B.V. https://doi.org/10.1016/j.mattod.2021.01.026
13. Gao H, Zhou T, Zheng Y, Zhang Q, Liu Y, Chen J, Liu H, Guo Z (2017) CoS quantum dot nanoclusters for high-energy potassium-ion batteries. Adv Funct Mater 27(43). https://doi.org/10.1002/adfm.201702634
14. González A, Goikolea E, Barrena JA, Mysyk R (2016) Review on supercapacitors: technologies and materials. In: Renewable and sustainable energy reviews, vol 58, pp 1189–1206. Elsevier Ltd. https://doi.org/10.1016/j.rser.2015.12.249
15. Goodenough JB, Park KS (2013) The Li-ion rechargeable battery: a perspective. J Am Chem Soc 135(4). https://doi.org/10.1021/ja3091438
16. Guo Q, Ma Y, Chen T, Xia Q, Yang M, Xia H, Yu Y (2017) Cobalt sulfide quantum dot embedded N/S-doped carbon nanosheets with superior reversibility and rate capability for sodium-ion batteries. ACS Nano 11(12):12658–12667. https://doi.org/10.1021/acsnano.7b07132
17. Han C, Yan M, Mai L, Tian X, Xu L, Xu X, An Q, Zhao Y, Ma X, Xie J (2013) V_2O_5 quantum dots/graphene hybrid nanocomposite with stable cyclability for advanced lithium batteries. Nano Energy 2(5):916–922. https://doi.org/10.1016/j.nanoen.2013.03.012
18. He J, Chen Y, Li P, Fu F, Wang Z, Zhang W (2015) Self-assembled CoS2 nanoparticles wrapped by CoS2-quantum-dots-anchored graphene nanosheets as superior-capability anode for lithium-ion batteries. Electrochim Acta 182:424–429. https://doi.org/10.1016/j.electacta.2015.09.131

19. Hou H, Banks CE, Jing M, Zhang Y, Ji X (2015) Carbon quantum dots and their derivative 3D porous carbon frameworks for sodium-ion batteries with ultralong cycle life. Adv Mater 27(47):7861–7866. https://doi.org/10.1002/adma.201503816
20. Hu X, Jia J, Wang G, Chen J, Zhan H, Wen Z (2018) Reliable and general route to inverse opal structured nanohybrids of carbon-confined transition metal sulfides quantum dots for high-performance sodium storage. Adv Energy Mater 8(25). https://doi.org/10.1002/aenm.201801452
21. Huang Y, Lv Y, Zou Y, Chen N, Ao Z (2023) TiO2 quantum dots decorated Si nanocage for enhanced lithium ion batteries. J Electroanal Chem 930. https://doi.org/10.1016/j.jelechem.2022.117128
22. Huang Z, Qin X, Gu X, Li G, Mu Y, Wang N, Ithisuphalap K, Wang H, Guo Z, Shi Z, Wu G, Shao M (2018) Mn_3O_4 quantum dots supported on nitrogen-doped partially exfoliated multiwall carbon nanotubes as oxygen reduction electrocatalysts for high-performance Zn-air batteries. ACS Appl Mater Interfaces 10(28):23900–23909. https://doi.org/10.1021/acsami.8b06984
23. Jian Z, Zhang S, Guan X, Li J, Li H, Wang W, Xing Y, Xu H (2020) ZnO quantum dot-modified rGO with enhanced electrochemical performance for lithium-sulfur batteries. RSC Adv 10(54):32966–32975. https://doi.org/10.1039/d0ra04986g
24. Jing M, Wang C, Hou H, Wu Z, Zhu Y, Yang Y, Jia X, Zhang Y, Ji X (2015) Ultrafine nickel oxide quantum dots enbedded with few-layer exfoliative graphene for an asymmetric supercapacitor: enhanced capacitances by alternating voltage. J Power Sources 298:241–248. https://doi.org/10.1016/j.jpowsour.2015.08.039
25. Kate RS, Khalate SA, Deokate RJ (2018) Overview of nanostructured metal oxides and pure nickel oxide (NiO) electrodes for supercapacitors: a review. J Alloy Compd 734. https://doi.org/10.1016/j.jallcom.2017.10.262
26. Khalafallah D, Miao J, Zhi M, Hong Z (2021) Structuring graphene quantum dots anchored CuO for high-performance hybrid supercapacitors. J Taiwan Inst Chem Eng 122:168–175. https://doi.org/10.1016/j.jtice.2021.04.037
27. Kim KH, Song YJ, Ahn HJ (2023) Quantum dot-derived carbon nanopocket-confined Co_3O_4 within mesoporous carbon nanofiber for Cu-free anode of flexible Li-ion batteries. Appl Surf Sci 637. https://doi.org/10.1016/j.apsusc.2023.157905
28. Kim SW, Seo DH, Ma X, Ceder G, Kang K (2012) Electrode materials for rechargeable sodium-ion batteries: potential alternatives to current lithium-ion batteries. In: Advanced energy materials, vol 2, no 7, pp 710–721. https://doi.org/10.1002/aenm.201200026
29. Kong D, Cheng C, Wang Y, Huang Z, Liu B, Von Lim Y, Ge Q, Yang HY (2017) Fe_3O_4 quantum dot decorated MoS_2 nanosheet arrays on graphite paper as free-standing sodium-ion battery anodes. J Mater Chem A 5(19):9122–9131. https://doi.org/10.1039/c7ta01172e
30. Lamba P, Singh P, Singh P, Singh P, Bharti Kumar A, Gupta M, Kumar Y (2022) Recent advancements in supercapacitors based on different electrode materials: classifications, synthesis methods and comparative performance. J Energy Storage 48. https://doi.org/10.1016/j.est.2021.103871
31. Li Q, Cheng H, Wu X, Wang CF, Wu G, Chen S (2018) Enriched carbon dots/graphene microfibers towards high-performance micro-supercapacitors. J Mater Chem A 6(29):14112–14119. https://doi.org/10.1039/c8ta02124d
32. Li X, Hu K, Tang R, Zhao K, Ding Y (2016) CuS quantum dot modified carbon aerogel as an immobilizer for lithium polysulfides for high-performance lithium-sulfur batteries. RSC Adv 6(75):71319–71327. https://doi.org/10.1039/c6ra11990e
33. Liao SY, Chen J, Cui SF, Shang JQ, Li YZ, Cheng WX, Liu YD, Cui TT, Shu XG, Min YG (2023) CoS2 enhanced SnO2@rGO heterostructure quantum dots for advanced lithium-ion battery anode. J Power Sources 553. https://doi.org/10.1016/j.jpowsour.2022.232265
34. Lijuan K, Ruiyi L, Yongqiang Y, Zaijun L (2016) Multi-faceted design of a silicon anode for high performance lithium ion batteries using silicon nanoparticles encapsulated by a multiple graphene aerogel electrode material and a tryptophan-functionalized graphene quantum dot-sodium alginate binder. RSC Adv 6(80):76344–76354. https://doi.org/10.1039/c6ra15257k

35. Lijuan K, Yongqiang Y, Ruiyi L, Zaijun L (2016) Phenylalanine-functionalized graphene quantum dot-silicon nanoparticle composite as an anode material for lithium ion batteries with largely enhanced electrochemical performance. Electrochim Acta 198:144–155. https://doi.org/10.1016/j.electacta.2016.03.034
36. Liu H, Jia M, Zhu Q, Cao B, Chen R, Wang Y, Wu F, Xu B (2016) 3D–0D graphene-Fe_3O_4 quantum dot hybrids as high-performance anode materials for sodium-ion batteries. ACS Appl Mater Interfaces 8(40):26878–26885. https://doi.org/10.1021/acsami.6b09496
37. Liu Q, Han X, Park H, Kim J, Xiong P, Yuan H, Yeon JS, Kang Y, Park JM, Dou Q, Kim BK, Park HS (2021) Layered double hydroxide quantum dots for use in a bifunctional separator of lithium-sulfur batteries. ACS Appl Mater Interfaces 13(15):17978–17987. https://doi.org/10.1021/acsami.1c00974
38. Liu Z-Q, Cheng H, Li N, Yi Ma T, Su Y-Z (n.d.) $ZnCo_2O_4$ quantum dots anchored on nitrogen doped carbon nanotubes as reversible oxygen reduction/evolution electrocatalysts. https://doi.org/10.1002/adma
39. Mo R, Lei Z, Sun K, Rooney D (2014) Facile synthesis of anatase TiO2 quantum-dot/graphene-nanosheet composites with enhanced electrochemical performance for lithium-ion batteries. Adv Mater 26(13):2084–2088. https://doi.org/10.1002/adma.201304338
40. Ni Q, Bai Y, Wu F, Wu C (2017) Polyanion-type electrode materials for sodium-ion batteries. Adv Sci 4(3):1600275. https://doi.org/10.1002/advs.201600275
41. Oloore LE, Gondal MA, Popoola IK, Popoola AJ (2020) Cadmium sulfide quantum dots-organometallic halide perovskite bilayer electrode structures for supercapacitor applications. ChemElectroChem 7(2):486–492. https://doi.org/10.1002/celc.201901969
42. Panwar NL, Kaushik SC, Kothari S (2011) Role of renewable energy sources in environmental protection: a review. In: Renewable and sustainable energy reviews, vol 15, no 3, pp 1513–1524. https://doi.org/10.1016/j.rser.2010.11.037
43. Peng C, Chen B, Qin Y, Yang S, Li C, Zuo Y, Liu S, Yang J (2012) Facile ultrasonic synthesis of coo quantum dot/graphene nanosheet composites with high lithium storage capacity. ACS Nano 6(2):1074–1081. https://doi.org/10.1021/nn202888d
44. Qi Y, Yang Y, Hou Q, Zhang K, Zhao H, Su H, Zhou L, Liu X, Shen C, Xie K (2021) Uniform-dispersed ZnS quantum dots loading on graphene as a promising anode for potassium-ion batteries. Chin Chem Lett 32(3):1117–1120. https://doi.org/10.1016/j.cclet.2020.08.030
45. Qin J, Wang X, Cao M, Hu C (2014) Germanium quantum dots embedded in N-doping graphene matrix with sponge-like architecture for enhanced performance in lithium-ion batteries. Chem Eur J 20(31):9675–9682. https://doi.org/10.1002/chem.201402151
46. Raza SS, Janajreh I, Ghenai C (2014) Sustainability index approach as a selection criteria for energy storage system of an intermittent renewable energy source. Appl Energy 136:909–920. https://doi.org/10.1016/j.apenergy.2014.04.080
47. Sánchez A, Zhang Q, Martín M, Vega P (2022) Towards a new renewable power system using energy storage: an economic and social analysis. Energy Convers Manage 252. https://doi.org/10.1016/j.enconman.2021.115056
48. Shaker M, Ng S, Sadeghi Ghazvini AA, Javanmardi S, Gaho MA, Jin Z, Ge Q (2024) Carbon/graphene quantum dots as electrolyte additives for batteries and supercapacitors: a review. J Energy Storage 85. Elsevier Ltd. https://doi.org/10.1016/j.est.2024.111040
49. Shao Y, Yue J, Sun S, Xia H (2017) Facile synthesis of FeS_2 quantum-dots/functionalized graphene-sheet composites as advanced anode material for sodium-ion batteries. Chin J Chem 35(1):73–78. https://doi.org/10.1002/cjoc.201600637
50. Shen C, Zhang K, You Y, Wang H, Ning R, Qi Y, Li N, Ding C, Xie K, Wei B (2020) Inducing rapid polysulfide transformation through enhanced interfacial electronic interaction for lithium-sulfur batteries. Nanoscale 12(26):13980–13986. https://doi.org/10.1039/d0nr02429e
51. Sobhanmanesh MB, Ghaffarinejad A, Milani Hosseini SMR, Daneshtalab R (2024) Tin oxide quantum dots embedded between graphene sheets to improve non-aqueous aluminum ion battery performance. J Energy Storage 77. https://doi.org/10.1016/j.est.2023.109911
52. Sun Z, He J, Yuan M, Lin L, Zhang Z, Kang Z, Liao Q, Li H, Sun G, Yang X, Long R, Zhang Y (2019) Li^+-clipping for edge S-vacancy MoS_2 quantum dots as an efficient bifunctional

electrocatalyst enabling discharge growth of amorphous Li_2O_2 film. Nano Energy 65. https://doi.org/10.1016/j.nanoen.2019.103996

53. Talapin DV, Lee JS, Kovalenko MV, Shevchenko EV (2010) Prospects of colloidal nanocrystals for electronic and optoelectronic applications. Chem Rev 110(1):389–458. https://doi.org/10.1021/cr900137k
54. Tang Y, Liu L, Wang X, Jia D, Xia W, Zhao Z, Qiu J (2016) TiO2 quantum dots embedded in bamboo-like porous carbon nanotubes as ultra high power and long life anodes for lithium ion batteries. J Power Sources 319:227–234. https://doi.org/10.1016/j.jpowsour.2016.04.033
55. Veeramani V, Chen YH, Wang HC, Hung TF, Chang WS, Wei DH, Hu SF, Liu RS (2018) CdSe/ZnS QD@CNT nanocomposite photocathode for improvement on charge overpotential in photoelectrochemical Li-O2 batteries. Chem Eng J 349:235–240. https://doi.org/10.1016/j.cej.2018.05.012
56. Veerasubramani GK, Park MS, Choi JY, Kim DW (2020) Ultrasmall SnS quantum dots anchored onto nitrogen-enriched carbon nanospheres as an advanced anode material for sodium-ion batteries. ACS Appl Mater Interfaces 12(6):7114–7124. https://doi.org/10.1021/acsami.9b18997
57. Wang T, Chen HC, Yu F, Zhao XS, Wang H (2019) Boosting the cycling stability of transition metal compounds-based supercapacitors. In: Energy storage materials, vol 16, pp 545–573. Elsevier B.V. https://doi.org/10.1016/j.ensm.2018.09.007
58. Wang X, Liu Y, Wang Y, Jiao L (2016) CuO quantum dots embedded in carbon nanofibers as binder-free anode for sodium ion batteries with enhanced properties. Small 12(35):4865–4872. https://doi.org/10.1002/smll.201601474
59. Wang Y, Jin Y, Zhao C, Duan Y, He X, Jia M (2017) SnO_2 quantum dots/graphene aerogel composite as high-performance anode material for sodium ion batteries. Mater Lett 191:169–172. https://doi.org/10.1016/j.matlet.2016.12.072
60. Wang Y, Luo S, Chen M, Wu L (2020) Uniformly confined germanium quantum dots in 3D ordered porous carbon framework for high-performance Li-ion battery. Adv Funct Mater 30(16). https://doi.org/10.1002/adfm.202000373
61. Wei H, Ding Y, Li H, Zhang Q, Hu N, Wei L, Yang Z (2019) MoS2 quantum dots decorated reduced graphene oxide as a sulfur host for advanced lithium-sulfur batteries. Electrochim Acta 327. https://doi.org/10.1016/j.electacta.2019.134994
62. Wei W, Wang H, Tian A, Wang K, Wang J, Qu P, Zhang S, Guo L (2018) Confined metal Ge quantum dots in carbon nanofibers for stable rechargeable batteries. Nanoscale 10(15):6872–6877. https://doi.org/10.1039/c8nr00153g
63. Wu TH, Chen JA, Su JH, Ting YH (2023) Decorating carbon quantum dots onto VO2 nanorods to boost electron and ion transport kinetics for high–performance zinc–ion batteries. J Energy Storage 74. https://doi.org/10.1016/j.est.2023.109340
64. Xia H, Hong C, Li B, Zhao B, Lin Z, Zheng M, Savilov SV, Aldoshin SM (2015) Facile synthesis of hematite quantum-dot/functionalized graphene-sheet composites as advanced anode materials for asymmetric supercapacitors. Adv Func Mater 25(4):627–635. https://doi.org/10.1002/adfm.201403554
65. Xiang Huang Z, Liu B, Kong D, Wang Y, Ying Yang H (2018) SnSe2 Quantum Dot/rGO composite as high performing lithium anode. Energy Storage Mater 10:92–101. https://doi.org/10.1016/j.ensm.2017.08.008
66. Xie Q, Zhang S, Huang W, Zhao P (2023) CuO quantum dots embedded Cu3(BTC)2/CuO sugar gourd-like nanoarrays on copper foil as free-standing anodes for lithium-ion batteries with boosted performance. J Electroanal Chem 946. https://doi.org/10.1016/j.jelechem.2023.117714
67. Xu G, Yang L, Wei X, Ding J, Zhong J, Chu PK (2016) MoS2-quantum-dot-interspersed $Li_4Ti_5O_{12}$ nanosheets with enhanced performance for Li- and Na-Ion batteries. Adv Func Mater 26(19):3349–3358. https://doi.org/10.1002/adfm.201505435
68. Xu N, Zhang Y, Zhang T, Liu Y, Qiao J (2019) Efficient quantum dots anchored nanocomposite for highly active ORR/OER electrocatalyst of advanced metal-air batteries. Nano Energy 57:176–185. https://doi.org/10.1016/j.nanoen.2018.12.017

69. Xu R, Du L, Adekoya D, Zhang G, Zhang S, Sun S, Lei Y (2021) Well-defined nanostructures for electrochemical energy conversion and storage. In: Advanced energy materials, vol 11, no 15. John Wiley and Sons Inc. https://doi.org/10.1002/aenm.202001537
70. Xu Z, Li X, Sun S, Wei Z, Li H, Yin S (2021) Multidimensional carbon-modified NiCo2O4/Ni-Co-S nanocomposite electrode material for high-energy asymmetric supercapacitors. Energy Fuels 35(11):9692–9704. https://doi.org/10.1021/acs.energyfuels.1c01075
71. Yang SJ, Nam S, Kim T, Im JH, Jung H, Kang JH, Wi S, Park B, Park CR (2013) Preparation and exceptional lithium anodic performance of porous carbon-coated ZnO quantum dots derived from a metal-organic framework. J Am Chem Soc 135(20):7394–7397. https://doi.org/10.1021/ja311550t
72. Yang Y, Bremner S, Menictas C, Kay M (2018) Battery energy storage system size determination in renewable energy systems: a review. In: Renewable and sustainable energy reviews, vol 91, pp 109–125. Elsevier Ltd. https://doi.org/10.1016/j.rser.2018.03.047
73. Yin W, He D, Bai X, Yu WW (2019) Synthesis of tungsten disulfide quantum dots for high-performance supercapacitor electrodes. J Alloy Compd 786:764–769. https://doi.org/10.1016/j.jallcom.2019.02.030
74. Youm I, Sarr J, Sall M, Kane MM (n.d.) Renewable energy activities in Senegal: a review. www.elsevier.com/locate/rser
75. Yu J, Huang H, Bian F, Liang C, Xia Y, Zhang J, Gan Y, Zhang W (2018) Supercritical CO_2-fluid-assisted synthesis of TiO_2 quantum dots/reduced graphene oxide composites for outstanding sodium storage capability. ACS Appl Energy Mater 1(12):7213–7219. https://doi.org/10.1021/acsaem.8b01668
76. Yu J, Xiao J, Li A, Yang Z, Zeng L, Zhang Q, Zhu Y, Guo L (2020) Enhanced multiple anchoring and catalytic conversion of polysulfides by amorphous MoS3 nanoboxes for high-performance Li-S batteries. Angewandte Chemie – Int Edn 59(31):13071–13078. https://doi.org/10.1002/anie.202004914
77. Yu Y, Ma T, Huang H (2023) Semiconducting quantum dots for energy conversion and storage. In: Advanced functional materials, vol 33, no 16. John Wiley and Sons Inc. https://doi.org/10.1002/adfm.202213770
78. Zeng G, Hu X, Zhou B, Chen J, Cao C, Wen Z (2017) Engineering graphene with red phosphorus quantum dots for superior hybrid anodes of sodium-ion batteries. Nanoscale 9(38):14722–14729. https://doi.org/10.1039/c7nr05470j
79. Zhang G, Hou S, Zhang H, Zeng W, Yan F, Li CC, Duan H (2015) High-performance and ultra-stable lithium-ion batteries based on MOF-derived ZnO@ZnO quantum dots/C core-shell nanorod arrays on a carbon cloth anode. Adv Mater 27(14):2400–2405. https://doi.org/10.1002/adma.201405222
80. Zhang G, Zhu J, Zeng W, Hou S, Gong F, Li F, Li CC, Duan H (2014) Tin quantum dots embedded in nitrogen-doped carbon nanofibers as excellent anode for lithium-ion batteries. Nano Energy 9:61–70. https://doi.org/10.1016/j.nanoen.2014.06.030
81. Zhang J, Wei S, Wang H, Liu H, Zhang Y, Liu S, Wang Z, Lu X (2021) Carbon quantum dots promote coupled valence engineering of v2o5 nanobelts for high-performance aqueous zinc-ion batteries. Chemsuschem 14(9):2076–2083. https://doi.org/10.1002/cssc.202100223
82. Zhang L, Xia G, Huang Y, Wei C, Yu Y, Sun D, Yu X (2018) MnO quantum dots embedded in carbon nanotubes as excellent anode for lithium-ion batteries. Energy Storage Mater 10:160–167. https://doi.org/10.1016/j.ensm.2017.01.007
83. Zhang R, Xu J, Jia M, Pan E, Zhou C, Jia M (2019) Ultrafine ZnS quantum dots decorated reduced graphene oxide composites derived from ZIF-8/graphene oxide hybrids as anode for sodium-ion batteries. J Alloy Compd 781:450–459. https://doi.org/10.1016/j.jallcom.2018.11.122
84. Zhang S, Li X-Y, Yang W, Tian H, Han Z, Ying H, Wang G, Han W-Q (n.d.) Supporting information a novel synthesis of red phosphorus Nanodots/Ti_3 C_2 T x MXenes from low-cost Ti_3 SiC_2 MAX phases for superior lithium and sodium ion batteries
85. Zhang Y, Lin B, Wang J, Tian J, Sun Y, Zhang X, Yang H (2016) All-solid-state asymmetric supercapacitors based on ZnO quantum dots/carbon/CNT and porous N-doped carbon/CNT

electrodes derived from a single ZIF-8/CNT template. J Mater Chem A 4(26):10282–10293. https://doi.org/10.1039/c6ta03633c
86. Zhang Z, Ding T, Zhou Q, Sun Y, Qu M, Zeng Z, Ju Y, Li L, Wang K, Chi F (2021) A review of technologies and applications on versatile energy storage systems. In: Renewable and sustainable energy reviews, vol 148. Elsevier Ltd. https://doi.org/10.1016/j.rser.2021.111263
87. Zhao K, Zhang L, Xia R, Dong Y, Xu W, Niu C, He L, Yan M, Qu L, Mai L (2016) SnO2 quantum Dots@Graphene oxide as a high-rate and long-life anode material for lithium-ion batteries. Small 12(5):588–594. https://doi.org/10.1002/smll.201502183
88. Zhao N, You F (2020) Can renewable generation, energy storage and energy efficient technologies enable carbon neutral energy transition? Appl Energy 279. https://doi.org/10.1016/j.apenergy.2020.115889
89. Zhao S, Xie D, Yu X, Su Q, Zhang J, Du G (2015) Facile synthesis of Fe3O4@C quantum dots/graphene nanocomposite with enhanced lithium-storage performance. Mater Lett 142:287–290. https://doi.org/10.1016/j.matlet.2014.12.051
90. Zheng J, Lian X, Wu M, Zheng F, Gao Y, Niu H (2021) One-step preparation of Ni3S4 quantum dots composite graphene/carbon nanotube conductive network for asymmetric supercapacitor. J Alloy Compd 859. https://doi.org/10.1016/j.jallcom.2020.158247
91. Zhou L, Zhang K, Hu Z, Tao Z, Mai L, Kang YM, Chou SL, Chen J (2018) Recent developments on and prospects for electrode materials with hierarchical structures for lithium-ion batteries. In: Advanced energy materials, vol 8, no 6. Wiley-VCH Verlag. https://doi.org/10.1002/aenm.201701415
92. Zhou X, Shi J, Liu Y, Su Q, Zhang J, Du G (2014) Microwave irradiation synthesis of Co3O4 quantum dots/graphene composite as anode materials for Li-ion battery. Electrochim Acta 143:175–179. https://doi.org/10.1016/j.electacta.2014.08.023
93. Zhu X, Li S, Li J, Ali RN, Naz H, Liu P, Feng C, Xiang B (2019) Free-standing WTe2QD-doped NiSe/C nanowires for highly reversible lithium storage. Electrochim Acta 295:22–28. https://doi.org/10.1016/j.electacta.2018.10.128

Chapter 13
Building Energy Devices from Quantum Dots

P. Periasamy, V. P. Devarajan, Santanu Das, and Sudhagar Pitchaimuthu

Abstract Quantum Dots (QDs), nanoscale semiconductor particles with size-dependent optical and electronic properties, have emerged as a transformative technology in the field of energy devices. The properties, including tunable band gaps and high quantum yields, stem from quantum confinement effects, enabling the quantum dots to outperform traditional bulk materials in various applications. The synthesis, characterization, and integration of quantum dots in the context of energy devices such as Photovoltaics (PVs), Light-Emitting Diodes (LEDs), and thermoelectric systems are examined in depth in this chapter. The first part of the chapter provides a detailed overview of quantum dot synthesizing methods using chemical and physical techniques, coupled with a description of their structural, optical, and electronic properties. QDs also provide significant advantages in the field of photovoltaics, including multiple exciton generation and the absorption of broad-spectrum sunlight, providing improved efficiency in solar cells. We explore their integration into LEDs and demonstrate how tunable emission spectra and superior color purity enable the development of high-performance, low-energy displays and lighting. QDs enhance energy conversion efficiency in thermoelectric devices, they also help to improve the Seebeck coefficient and reduce thermal conductivity. While the applications shown here are very promising, this chapter also discusses the fundamental

P. Periasamy (✉)
Department of Physics, Nehru Institute of Engineering and Technology, Coimbatore, Tamil Nadu, India
e-mail: periasamy.nano@gmail.com

V. P. Devarajan
Department of Physics and Centre for Research and Development, K.S. Rangasamy College of Technology, Tiruchengode, Namakkal, Tamil Nadu, India

S. Das
Department of Ceramic Engineering, Indian Institute of Technology (Banaras Hindu University), Varanasi, India

S. Pitchaimuthu
Research Centre for Carbon Solutions, School of Engineering and Physical Sciences, Institute of Mechanical, Processing and Energy Engineering, Heriot-Watt University, Edinburgh, UK

T. Bhuyan et al. (eds.), *Multifunctional Quantum Dots*, Materials Horizons: From Nature to Nanomaterials, https://doi.org/10.1007/978-981-92-0172-3_13

difficulties of QD-based devices, including their stability and scalability, environmental issues, standards, and regulatory issues. This chapter highlights the transformative potential of quantum dots to enable a sustainable and energy-efficient future by reviewing recent advancements with the emergence of new trends that quantum dots offer.

Keywords Quantum dot · Quantum confinement · Energy storage · Photocatalytic · CO_2 conversion

1 Introduction to Quantum Dots (QDs)

Quantum Dots (QDs) are nanometer-sized semiconductor particles that possess unique electronic and optical properties due to their extremely small size. Quantum dots are typically 2–10 nm in diameter, exhibiting quantum mechanical effects, particularly quantum confinement [54]. Sometimes, these particles are referred to as "artificial atoms" because they often exhibit discrete and quantized energy levels (Fig. 1). However, size is the most significant difference between quantum dots and bulk semiconductor materials. Quantum dots are small in size and can exhibit novel properties that are different from those of their bulk counterparts, particularly in their electronic structure and optical properties [66]. Quantum dots are characterized by their ability to confine electrons and holes in three spatial dimensions [65]. This confinement results in the quantization of energy levels, which explains the unique light properties of quantum dots in controlling photoluminescence (emission wavelength) through a simple change in size. One of the hallmarks of quantum dots, which make them a powerful tool in technology, is their tunability. This is a key feature that makes quantum dots valuable in various technological areas, including optoelectronics, photovoltaics, and biological imaging.

The properties of quantum dots are unique, enabling them to find applications in various fields of science and technology. Optical and electronic properties are size-dependent and have quantum confinement. Quantum confinement in quantum dots is probably the most essential property. It describes the dimensions of a semiconductor crystal that shrunk down to be smaller than the exciton Bohr radius (the distance that an electron–hole pair remains bound in a semiconductor technology) [15, 64]. In this regime, the charge carriers (electrons and holes) are confined in all three spatial dimensions and therefore occupy discrete, quantized energy levels rather than the continuous energy bands found in bulk materials. These confinements significantly impact the quantum dot's electronic properties in multiple ways. Quantum dots, unlike bulk semiconductors, contain discrete energy levels rather than energy bands [55]. The reason for this is that when electrons are confined in a space, such as those in an atom, they're unable to move. Thus, the energy levels of quantum dots can be considered as quantified. As the size of the quantum dot decreases, the energy difference between the Highest Occupied Molecular Orbital (HOMO) and the Lowest Unoccupied Molecular Orbital (LUMO) increases. Such a phenomenon is known as

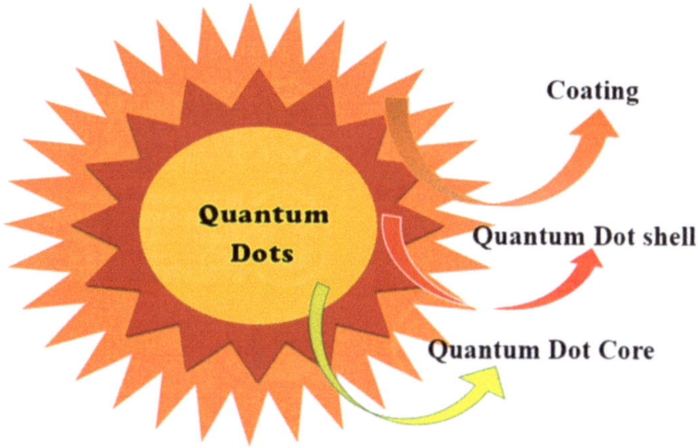

Fig. 1 Nanoscale semiconductor particle with a core, shell, and surface ligands

the "quantum size effect," as the confinement energy increases when the particle size decreases, thereby widening the bandgap of the quantum dot [2].

2 Unique Properties of Quantum Dots

The basic optical and electronic properties of quantum dots are strongly dependent on size. This size dependency directly results from quantum confinement and manifests in several ways (Fig. 2). In addition to quantum confinement and size-dependent properties, the unique characteristics of quantum dots make them attractive for applications in various areas. The surface chemistry of quantum dots also plays a prominent role in their properties. Due to their large surface-to-volume ratio, the surface states of Quantum Dots (QDs) significantly impact their physical properties. Since QDs are frequently coated with organic molecules called ligands to stabilize these surface states, ligands can passivate surface defects and prevent aggregation [34]. Additionally, the use of flexible ligands and surface passivation strategies enables further tuning of the optical and electronic properties of QDs, making them suitable for a wide range of applications. Quantum dots degrade more slowly in the face of light exposure, i.e., they have higher photostability compared to organic dyes. Due to this property, they are well suited for long-term imaging and display applications that require stable performance over time. Due to their poor optical properties and stability, quantum dots are often engineered as core/shell structures. The structures consist of a coating of different semiconductor materials with varying band gaps surrounding a core quantum dot. Further improvement in their optical performance is achieved by reducing nonradiative recombination and increasing their quantum yield through the use of core/shell design.

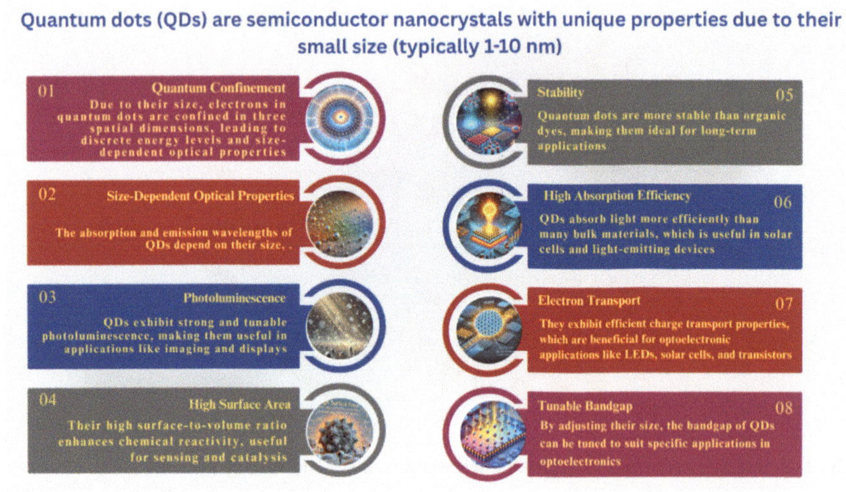

Fig. 2 Properties of quantum dots (tunable optical properties and strong quantum confinement)

3 Historical Development and Discovery of Quantum Dots

Several key discoveries and milestones in the journey of Quantum Dots (QDs) from their theoretical prediction to real-world applications are recognized. The historical development of quantum dots, including significant advancements by researchers and breakthroughs in our understanding and utilization of quantum dot nanostructures, is listed in Fig. 3. The quantum mechanical principles for understanding quantum confinement were laid down by Werner Heisenberg and Enrico Fermi in 1931 [57]. A famous lecture given by Richard Feynman in 1959 at Caltech, titled "There's Plenty of Room at the Bottom," was about manipulating matter at an atomic scale. It predicted the development of nanotechnology, which, by extension, led to the creation of quantum dots. The concept of quantum confinement emerged in semiconductor research in the 1970s. During this period, the concept of electrons being trapped within semiconductor materials and having their energy quantized was theorized [57]. In 1981, Alexei Ekimov, a Russian physicist, gave the first experimental evidence of quantum confinement in semiconductor crystals. In glass doped with semiconductor nanocrystals (CuCl), Ekimov observed size-dependent shifts in the optical absorption spectra, providing a direct demonstration of quantum size effects. In a colloidal solution of semiconductor nanocrystals (CdS) reported in the late 1970s, Louis E. Brus of Bell Laboratories observed quantum size effects. In addition to his contribution to the phenomenon of quantum confinement, Brus' work was instrumental in establishing semiconductor nanocrystals as a new material and setting them as a unique discipline of their own. Several researchers have begun to recognize the potential of quantum dots for various applications in optoelectronics and photonics [57].

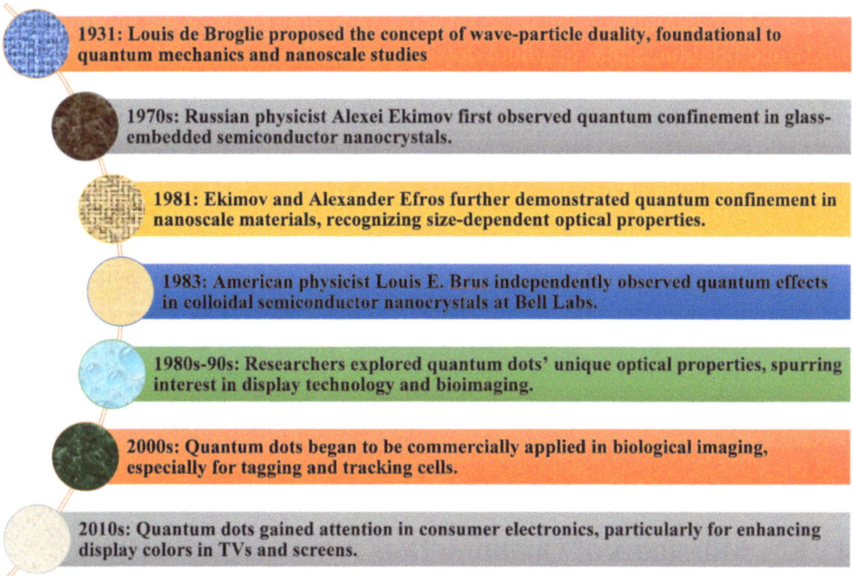

Fig. 3 Historical development of quantum dots

Through the development of a reproducible method for synthesizing monodisperse and high-quality quantum dots of well-defined sizes, Murray, Norris, and Bawendi synthesized quantum dots exhibiting numerous novel optical and spectroscopic properties [50]. Using the "hot injection" technique, they grew CdSe quantum dots with controlled growth and significantly improved optical properties. A core/shell quantum dot (CdSe/ZnS) synthesizing method was developed, which has improved photostability and quantum yield. Core/shell structures can provide a solution to the problem of surface defects and nonradiative recombination, which is a crucial step [60]. While quantum dots began to find use in commercial products, primarily in display technologies and biological imaging, they still required modifications to enhance their characteristics. For example, companies such as QD Vision and Nanosys have begun exploring quantum dot-enhanced displays, and quantum dots have emerged as fluorescent probes in biological and medical imaging. Research into Quantum Dot Solar Cells (QDSCs) began by incorporating quantum dots into solar cells. The potential for creating Multiple Exciton Generation (MEG) in Quantum Dots (QDs) offers novel opportunities for enhancing the efficiency of photovoltaic devices.

The major breakthrough was witnessed by the development of perovskite quantum dots, a new class of quantum dots that exhibit improved optical and electronic properties. The luminescence of perovskite quantum dots was higher, and their processing was more straightforward than that of traditional quantum dots, enabling new applications in optoelectronic devices. Then, quantum dots began to be incorporated into mass-produced consumer electronics, such as high-definition TVs. Samsung and

LG released quantum dot-enhanced displays, providing proof of the superior color purity and brightness that quantum dots can offer. Heavy metal-free quantum dots (InP, ZnSe) were synthesized to address environmental and health concerns associated with cadmium-based quantum dots [14]. This was also crucial for QDs to enter into the market, as consumer products often require new technologies. As new materials and composite structures evolve, so do quantum dots. Novel device architectures with improved performance are obtained through the integration of quantum dots with other nanomaterials, particularly carbon nanotubes and graphene. Next-generation quantum dot devices with unprecedented efficiencies are being ushered in by research into Multi-Exciton Generation (MEG) and quantum dot superlattices [21]. Now, the properties of quantum mechanics in quantum dots are being investigated in the field of quantum computing and communication. The developments demonstrated the versatility of quantum dots and their potential to revolutionize several areas of technology.

4 Key Milestones of Quantum Dots

Comprehending quantum confinement originated from the early work of Alexei Ekimov on quantum size effects in semiconductor nanocrystals (Fig. 4) [42]. Ekimov is sometimes regarded as a pioneer in the field of quantum dot research. In the 1980s, he was the first to demonstrate that the optical characteristics of quantum dots are size-dependent [42, 57]. However, Brus' independent discovery of quantum size effects in colloidal semiconductor nanocrystals reached a key milestone. While at Bell Labs, he also laid the foundations for future nanotechnology research and contributed to the development of quantum dots as a subject of study [77]. When he invented the "hot injection" method to produce high-quality quantum dots, he revolutionized the area. This synthesis method enabled the precise control of the size and new form of quantum dots, resulting in improved optical characteristics and a wider range of applications [77]. Victor Klimov is renowned for his groundbreaking research on MEG in quantum dots. He demonstrated that QDs can generate numerous electron–hole pairs per high-energy photon [18]. This finding opens new avenues for utilizing quantum dots in solar cells to enhance the performance of photovoltaic systems. Quantum dots have found applications in biological imaging, and Alivisatos has had a significant impact on the development and utilization of quantum dots. Beyond optoelectronics, however, there are many potential applications of Quantum Dots (QDs) [2], this research, which utilizes QDs as fluorescent probes, has improved medical diagnostics and bioimaging [26]. Stephen E. Braslavsky made a significant contribution to the advancement of core/shell quantum dot structures. He worked on fundamental issues of the stability of quantum dots and the functionality of their photoluminescence. His work has been instrumental in delivering quantum dots to individuals [33].

Fig. 4 Key milestones of quantum dots

5 Types of Quantum Dots

Quantum Dots (QDs) are categorized into different types based on their composition, structure, and synthesis methods (Fig. 5). To select an appropriate quantum dot for a specific application, it is essential to understand the types of quantum dots being referred to. Three types of quantum dots are categorized as core/shell quantum dots, perovskite quantum dots, and colloidal quantum dots. Different types of QDs are utilized depending on their function, with distinct properties for various technological applications [2].

Quantum dots have a core/shell structure, meaning they are semiconductors surrounded by another semiconductor, both of which have different band gaps. Typically, the core of the core/shell QD is composed of a material with a smaller bandgap, i.e., cadmium selenide (CdSe), and the shell consists of a material with a larger bandgap, i.e., zinc sulfide (ZnS). To improve the optical properties and stability of quantum dots, the core/shell structure was endorsed [17]. The shell material passivates the core surface, limiting nonradiative electron–hole pair recombination rates. The increase of the surface complex enhances the quantum yield and photoluminescence efficiency of quantum dots. The core/shell quantum dot is more robust in diverse environments because the shell provides a protective barrier that enhances the chemical and thermal stability of the core. The toxicity of some core materials (e.g., cadmium-based cores) may also be mitigated by encapsulating them in a less toxic shell material [28]. Wide applications have been reported in various fields that utilize core/shell quantum dots due to their high photoluminescence efficiency and

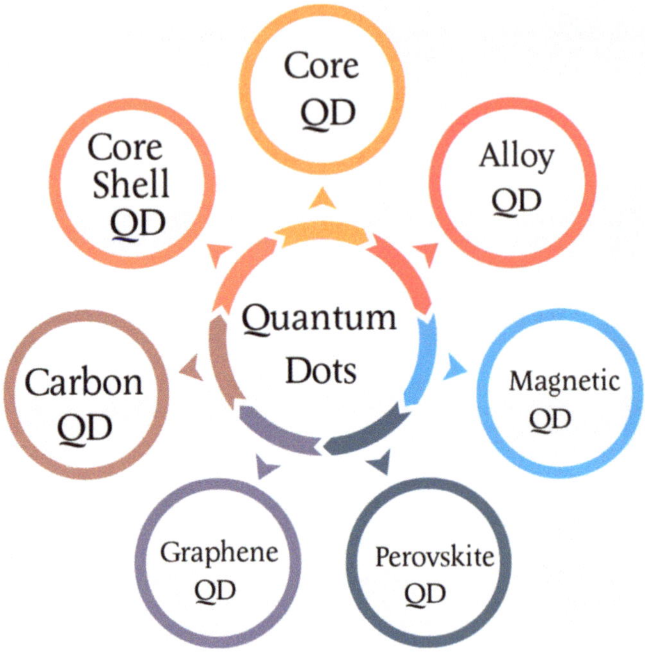

Fig. 5 Types of quantum dots used in various fields

stability. QLEDs utilize core/shell Quantum Dots (QDs) to produce vivid colors and high brightness. As fluorescent probes for biological imaging, core/shell QDs offer benefits including bright, stable emission and resistance to photobleaching. Core/shell Quantum Dots (QDs) are utilized in photovoltaic devices to enhance light absorption and charge separation efficiency [17].

A newer class of QDs called perovskite quantum dots are closely related to perovskite materials such as ABX_3 where A is a monovalent cation (e.g., methylammonium, cesium), B is a divalent metal cation (e.g., lead, tin), and X halide anion (e.g., chlorine, bromine, iodine). Perovskite Quantum Dots (QDs) differ from traditional semiconductor QDs in that they have a crystalline structure analogous to perovskite minerals, giving them their special optoelectronic properties. The light emission properties of perovskite QDs are exceptional, with high quantum yields, and they are highly efficient light emitters. Suitable for applications where high color purity is required, such as displays and lasers, they have broad absorption and narrow emission peaks [1]. The halide composition or the QD size can be easily controlled to tune the emission wavelength of the perovskite QDs across the visible spectrum. Simple and low-cost methods, such as solution processing, can be employed to synthesize perovskite Quantum Dots (QDs), which are particularly suitable for large-scale production. Perovskite quantum dots are promising for a large array of optoelectronic applications, including Perovskite QD-LEDs have brightness and efficiency on a par with conventional organic LEDs. Next-generation solar cells use perovskite QDs due to their extremely high-power conversion efficiencies and better light-harvesting capability. Perovskite QDs exhibit narrow emission linewidths attractive for low-threshold lasers. However, as with many other perovskite-based devices, perovskite QDs have their challenges, specifically, they lack stability in the presence of moisture and oxygen. There is continuing research into making improvements to their environmental stability and developing lead-free perovskite QDs to remove toxicity concerns [11].

6 Chemical Synthesis Techniques for Quantum Dots

The chemical synthesis of Quantum Dots (QDs) employs chemical methods to grow high-quality nanocrystals with precise control over their size, shape, and composition. Colloidal synthesis is one of the most popular techniques because it is simple, scalable, and provides high quantum yield and narrow size distribution of quantum dots [6, 11]. Here, precursors are dissolved in a hot solvent that contains coordinating ligands, which regulate the nucleation and growth of the quantum dots. Two main approaches have been applied in colloidal synthesis: the hot injection method, in which precursors are rapidly injected into solvent to generate the nucleation, and the heat-up method, by which the solution comprising all precursors is gradually heated, with simultaneous nucleation and growth. The versatility of colloidal synthesis allows the production of a host of quantum dot materials including core/shell and alloyed structures that are finding application in displays, solar cells, and biological imaging. A second chemical method is Chemical Vapor Deposition (CVD), which is utilized frequently to deposit quantum dots directly onto substrates. This encompasses the formation of quantum dots by the reaction of gaseous precursors on heated substrate surfaces under well-controlled conditions, resulting in quantum dots of high purity and uniformity. And we have variants of CVD (Metal–Organic CVD (MOCVD) and Plasma-Enhanced CVD (PECVD)) technique, which gives us more control in the deposition process, where we can use it for the fabrication of devices like LED lasers and solar cells [6, 11].

A particularly important technique in chemical synthesis also includes hydrothermal and solvothermal synthesis, which involve reactions occurring in heated, sealed autoclaves at high temperatures and pressures [40]. These methods are particularly suitable for the synthesis of quantum dots made of a material that is otherwise insoluble or lacks reactivity. Because the nucleation and growth are uniform in such a high-pressure and high-temperature environment, the produced quantum dots have narrow size distributions and high crystallinity. The hydrothermal synthesis used water as a solvent, while the solvothermal synthesis employed non-aqueous solvents, providing flexibility to obtain various types of quantum dots, including oxides, sulfides, and perovskites [51]. Microemulsion synthesis utilizes microemulsions, nanoscale droplets stabilized by surfactants, as confined environments for control of quantum dot growth within these environments. The nanoscale reaction environment provided by microemulsion droplets enables this method to exert precise control over the size and shape of the quantum dots. The core MCM-41 (Mobil Composition of Matter No. 41) nanoreactor is particularly advantageous for producing uniform and monodisperse quantum dots for use in drug delivery, sensors, and optoelectronic devices [75].

Template-assisted synthesis is a method that utilizes a physical or chemical template to direct the growth of quantum dots to a specific shape and size. The technique enables precise control of quantum dot morphology by confining nucleation and growth within the template's structure. By making use of templates, porous membranes, or nanostructured surfaces, quantum dots with specific geometries, e.g., nanorods or nanowires, can be generated with this technique and be useful in catalysis, photonic crystals, or advanced optoelectronic devices [32]. Third, the synthesis of quantum dots via sol–gel techniques is a wet chemical synthesis which forms quantum dots via the hydrolysis and condensation of metal precursors, typically used for forming oxide quantum dots of titanium dioxide (TiO_2) and silicon dioxide (SiO_2). The simplicity and low cost of this method make it easy to manufacture quantum dots in controlled size and shape on a large scale. Quantum dots of various compositions, sol–gel synthesized, are widely used in photocatalysis, sensors, and bioimaging, where oxide materials are preferred for their stability and biocompatibility. The different chemical synthesis techniques she employed for each type of quantum dot have distinct advantages for different applications, and together these techniques have played a vital role in the advancement of quantum dot technology in many fields [20].

7 Physical Synthesis Techniques for Quantum Dots

Physical synthesis techniques of Quantum Dots (QDs) make use of physical processes, e.g., evaporation, condensation, and ablation, to manipulate nanocrystals of fixed size and properties. Most importantly, these methods are particularly advantageous to high-purity and well-defined structure and composition quantum dots, thus suitable for those applications that demand high control of material characteristics [41]. Quantum dots and thin films are grown one layer at a time on a substrate using the highly controlled Molecular Beam Epitaxy (MBE) technique. Therefore, this method ensures an Ultra-High Vacuum (UHV) ambient, which ensures a contaminant-free atmospheric environment that is very critical for synthesizing the high-purity nanostructures. In many rooms, a heated substrate is exposed to atomic or molecular beams of precursor materials generated by evaporating source materials [41]. Once nucleation occurs, the atoms or molecules condense on the substrate and form the quantum dots. It allows for atomic level control of growth and enables synthesis of highly crystalline quantum dots with minimal defects. MBE is especially well suited to building quantum dot superlattices, quantum wells, and core/shell structures where size and shape, and maybe to some extent, composition, can be selected with precision, and it is particularly desirable for high-performance optoelectronic devices, such as lasers and infrared detectors.

Laser ablation instead creates quantum dots by vaporizing a target material using a high-powered laser, followed by condensation. The performance of this technique is carried out in a free-from-contamination-controlled ambiance, like a vacuum or noble gas ambiance. Rapid vaporization of material occurs when the laser pulse strikes the target material, and it creates a high-temperature plasma plume [36]. The material is vaporized and then cools, and that vapor condenses onto nanoparticles, it can also condense onto quantum dots. We can control the size and composition of the quantum dots by adjusting the laser parameters (pulse duration, energy, repetition rate) and ambient conditions (such as pressure and temperature). Laser ablation is a general method to synthesize quantum dots from a variety of materials, such as metals, semiconductors, and oxides, without the use of chemical precursors. This technique can be useful in applications that demand highly pure quantum dots with defined composition, like photocatalysis, sensors, and biomedical imaging [36].

Mechanical Milling, or Ball Milling, is a mechanical synthesis method where bulk materials in high-energy ball mills are mechanically grounded into nanoscale particles. Bulk materials and milling media (e.g., steel balls) are placed within a rotating cylindrical chamber in which mechanical forces fracture the material to reduce the material to smaller particles, including quantum dots. Quantum dot size can be controlled by milling time, speed, and the size of the milling media [62]. The second method to produce quantum dots, mechanical milling, is a relatively simple and cost-effective method of obtaining quantum dots from a huge array of materials, including oxides, metals, and semiconductors. However, the technique typically results in a broad size distribution and can introduce impurities due to media wear in the milling process, thereby limiting its application in high-purity and uniformity-demand areas [62]. Oxide-based quantum dots for photocatalysis and battery applications are usually produced by mechanical milling.

The formation of quantum dots by sputtering atoms from a target material onto a substrate is a Physical Vapor Deposition (PVD) process known as Sputtering. This involves bombarding the target material with high-energy ions (mostly from a plasma source) to eject atoms from the target material surface [23]. Once ejected, these atoms travel through the vacuum chamber and condense onto a substrate, forming quantum dots. By adjusting the sputtering parameters, such as ion energy, target-substrate distance, and deposition time, the size and distribution of the quantum dots can be controlled. This means that sputtering is an effective method for producing uniform quantum dot films on large-area substrates, which are useful for optoelectronic devices such as solar cells and LEDs. This method is also used to sequentially deposit sputtered materials to create multi-layered quantum dot structures or nanocomposites [23]. The unique advantages of physical synthesis techniques lie in their ability to manufacture high-purity quantum dots with well-defined structural and compositional properties. Molecular Beam Epitaxy (MBE) offers atomic level control and is highly suitable for fabricating high-performance optoelectronic devices [23].

8 Quantum Dots in Energy Applications

In recent years, Quantum Dots (QDs) have emerged as a revolutionary material for energy applications due to their unique optical and electronic properties, including tunable band gaps, high absorption coefficients, and Multiple Exciton Generation (MEG). The excellent properties of quantum dots are perfectly suited for various energy-related technologies, including solar cells, LEDs, photocatalysis, and energy storage [74]. The integration of quantum dots into these applications has resulted in significant improvements in efficiency, performance, and cost-effectiveness, making them essential for the advancement of next-generation energy devices. These applications offer various benefits over common materials, including the opportunity to increase efficiency, fine-tuning, and the potential for lower production costs.

9 Quantum Dots in Solar Energy Conversion

In the realm of solar energy conversion, quantum dots have revolutionized the field and Quantum Dot Solar Cells (QDSCs) are the prime examples. Quantum dots, unlike traditional silicon-based solar cells, can have size-tunable band gaps, giving the solar cell an absorption spectrum that can be tailored (Fig. 6). The tunability of band gap of QDs allows us for the development of multi-junction solar cells, composed of layers of quantum dots of different sizes which absorb different parts of the solar spectrum, significantly increasing the total efficiency of the cell [5]. These solar cells utilize colloidal quantum dots, examples of which include lead sulfide (PbS) and cadmium selenide (CdSe), due to their high absorption coefficients and tunable band gaps. Colloidal QDSCs have been demonstrated by researchers to achieve Power Conversion Efficiencies (PCEs) over 15%, positioning them as a potentially competitive replacement for traditional silicon-based solar cells [72]. For instance, Canadian researchers at the University of Toronto recently demonstrated PbS quantum dot solar cells with a 12% power conversion efficiency. With MEG, quantum dots have the potential to generate multiple electron–hole pairs (excitons) from a single high-energy photon [7]. Lead selenide (PbSe) and indium phosphide (InP) quantum dots are being researched as materials to use in MEG devices. Modifying the quantum dot size controls which wavelengths of light they will absorb, providing a desirable match to the solar spectrum and a potential improvement through the multi-junction solar cells. The possibility for quantum dots is MEG; however, it could enable us to go beyond the efficiency limits of conventional photovoltaic materials. Solution synthesis and low-cost processing techniques, such as spin coating or printing, are exploited to realize quantum dots compatible with flexible and lightweight solar cells.

Researchers [19], on the other hand, focused on high-efficiency solar cells and photodetectors via a quantization of graphene particles. The researchers used

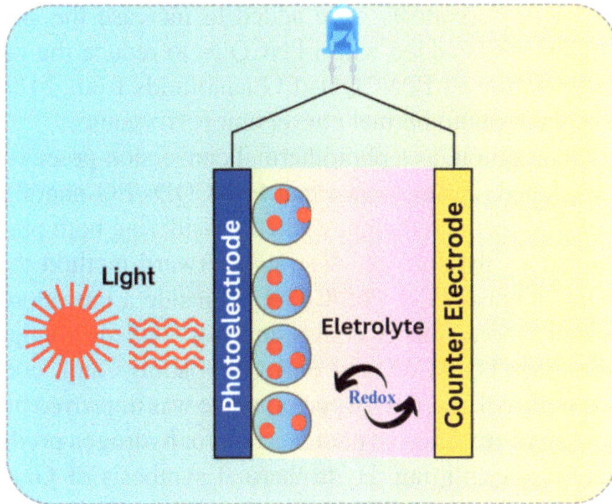

Fig. 6 Schematic illustration of the structure of a quantum dot-sensitized solar cell

Poly(3,4-ethylenedioxythiophene) (PEDOT) to explore the use of the PSS interlayer to modify and improve the photoelectrical performance of the rGO QDspSi heterojunction. The results showed that the presence of the PEDOT in the interlayer of photovoltaic characteristics was improved by the inclusion of the PSS layer to ~0.807 Vs, ~25.76 mA/cm^2, ~42.49%, ~8.83%. The modified device also demonstrated superior photodetection capabilities such as high responsivity, external quantum efficiency, detectivity, and quick on/off switching. These findings demonstrate the suitability of these devices for efficient optoelectronics and solar cell applications.

Eco-friendly colloidal core/shell Quantum Dots (QDs) in commercial solar technologies were studied [72]. To tailor the electronic band structure and charge carrier extraction of $AgInS_2$ (AIS)/ZnS QDs, Mn alloying and Cu doping were used by the researchers. Broadening of the band gap by the Mn alloying facilitated delocalization of photogenerated electrons into the shell. Cu doping offered Cu related states that captured photogenerated holes and accelerated electron transfer. The QDs act as light harvesters in a photoelectrochemical device for light-driven hydrogen evolution and have proved that high efficiency solar hydrogen can be produced.

Core/shell colloidal Quantum Dots (QDs) open new avenues for the synthesis and manipulation of electronic and optical properties suitable for a variety of optoelectronic devices. Multifunctional $AgGaS_2$ (AGS)/CdSeS core/shell QDs were grown using a two-step synthetic method [39]. Visible light absorption, emission redshift, and exciton lifetime extension were observed for the as-synthesized QDs. Photoelectrochemical cells and photodetectors for efficient solar energy conversion and light detection were fabricated using them. Versatile and high-performance optoelectronic applications are promising through the rational design of core/shell architecture. Novel N-doped Carbon Quantum Dots (FPNCQDs), synthesized using ammonium bifluoride (NH_4F) and perylene derivatives, were developed to enhance heat collection efficiency in Direct Absorption Solar Collectors (DASCs) [43]. NH_4F and perylene derivatives were added to increase the sp^2 conjugated structure and the graphite was added with FPNCQDs to reduce the band gap. It was shown that the temperature of FPNCQDs/EG nanofluids from 24 to 65.7 °C in 60 min indicated excellent photothermal conversion performance.

In addition, as a photothermal conversion process model for nanofluids was also developed, it was shown that FPNCQDs/EG nanofluids at 10 ppm had the largest solar collection efficiency when considering both photothermal conversion and heat loss from the surface. A straightforward method to enhance the loadings [35] of nanocomposites of PbS/CdS QDs inside a mesoporous electrode to realize highly efficient solar energy conversion was studied. Manipulation of TiO_2 surface charge increases both CdS and PbS QD loadings and, correspondingly, light-harvesting capability. Thus, device performance was improved by 33% for conversion efficiency and an unprecedented photocurrent for hydrogen production from photoelectrochemical water splitting. Hydrothermal synthesis of Graphene Quantum Dots (GQDs) and nitrogen-doped GQDs has been reported [46]. Both QDs show strong ultraviolet photoabsorption and excitation-dependent emission behavior. GQD- and N-GQD-sensitized ZnO nanorods were investigated for the photoelectrochemical properties. Photoconversion efficiency and IPCE of the GQD-modified nanorod photoanodes were considerably superior, and these devices showed superior photoelectrochemical properties compared to their N-GQD-sensitized architecture.

In another study, a modified synthetic method has been developed to improve the performance of colloidal $CuInS_2$ (CIS) Quantum Dot-Sensitized Solar Cells (QDSCs) [70]. The Cu/In non-stoichiometric ratios of CIS QDs have been carefully controlled, resulting in increased short-circuit photocurrent density and open-circuit voltage. A new record of 8.54% PCE has been achieved when the Cu/In precursor molar ratio is 1/4, in which CIS QDs can reduce surface defect states, effectively inhibiting charge recombination at TiO_2/CIS/electrolyte interfaces. The interfacial electron recombination mechanism of the solar cells was proposed, reducing photogenerated carrier loss and improving interfacial carrier collection. The study examines the photoelectrochemical properties of Ag-ZnO nanorods, which show superior photoconversion efficiency, photoresponsive properties, and incident-photon-to-current conversion efficiency compared to pristine ZnO nanorods [45]. S and N co-doped Graphene Quantum Dots (S, N-GQDs) were synthesized and used as a green photosensitizer to improve these properties. S, N-GQD-sensitized nanorods showed higher photoconversion efficiency and short-circuit photocurrent.

The study presents a highly stable, non-toxic bismuth trisulfide (Bi_2S_3) QDSSC with an iodide/triiodide redox electrolyte [61]. The DIPSILAR process (Dipping in the cationic solution, then rinsing and drying, and again dipping in anionic solution, followed by rinsing and drying is termed as DIPSILAR) was used to fabricate Bi_2S_3 quantum dots on the FTO/TiO_2 P90/TiO_2 P25 electrode. The quantum dots were found to be uniformly formed. A maximum power conversion efficiency of 0.26% was achieved from the best QDSSC with an open-circuit voltage of 482.9 mV and a short-circuit current density of 1.05 mA cm^{-2}. However, SILAR fabrication of Bi_2S_3 via a one-step method can dramatically cut costs compared with traditional light-harvesting materials. The interfacial structure of "giant" PbS/CdS Quantum

Dots (QDs) was modified by modulating the Cd:S molar ratio during growth. This control leads to hole transfer and determines the transition from double- to single-color emission. These QDs are optically active in the near infrared and may absorb solar energy. Under 1 Sun illumination, a photoelectrochemical cell can be used to offer a saturated photocurrent density of 5.3 mA/cm^2 with periodic illumination of a hybrid, TiO_2 mesoporous framework-based architecture. It also demonstrates good stability, 78% of the initial current density is retained after 2 h irradiation at 1 Sun [29].

A gradient band structure electron transport layer that improves the stability and performance of perovskite solar cells was investigated [22]. The GQDs and Au QDs, together with tin oxide (SnO_2) colloidal particles, have been successfully prepared by this method. With a conventional SnO_2 ETL, the cells reached a power conversion efficiency of 19.6%, while cells employing the new ETL produced an efficiency of 22.2%. The SnO_2 defects were compensated by the structure, which enhanced the charge carrier transport and separation ability and improved the electrical properties of SnO_2. This method also holds promise for other photovoltaic conversion devices. Different concentrations of lead sulfide quantum dots were used to create organic solar cells, resulting in slight changes in the Mott–Schottky analysis within the bulk heterojunction layer. A maximum power conversion efficiency of 5.15% was also achieved using quantum dots lead sulfide doped 0.4 wt% [16].

10 Quantum Dots in Energy Storage

Quantum dots have potential applications in energy storage technologies, like batteries and supercapacitors. These are next-generation energy storage devices because the high surface area and tunable electronic properties of quantum dots allow energy to be stored and released efficiently [27]. The transitional metal oxides or sulfide quantum dots, such as manganese oxide quantum dots (MnO_2) or molybdenum sulfide quantum dots (MoS_2), have been investigated as electrodes for lithium-ion batteries [9]. Their unique nanostructure and high surface area lead to high-specific capacities and excellent cycle stability for these materials. The high electrical conductivity and large surface area of quantum dots are also being considered for application in supercapacitors to enhance their charge storage capacity and power density [67]. A recent example is the application of carbon-based quantum dots as conductive additives in supercapacitor electrodes, resulting in improved performance. With their high surface area and controlled electronic characteristics, the unique properties of quantum dots enable higher energy and power densities in energy storage devices compared to conventional materials [73]. In battery electrodes, quantum dots provide chemical and mechanical stability superior to current technologies, resulting in longer cycle life and improved performance at higher usage levels.

The size-tunable properties of these materials enable precise control of the absorption and emission spectra, leading to higher efficiency in devices such as solar cells and LEDs [47]. High surface area-to-volume ratio and improved charge separation properties of quantum dots increase the efficiency of light-driven chemical reactions in photocatalysis [47]. The versatile and promising class of materials known as quantum dots is showing significant promise of advancing energy technologies by allowing us to achieve greater efficiencies at lower performance and cost than with more traditional materials. While research surrounding stability, toxicity, and scalability still has a way to go, quantum dots are set to be a major part of the future in sustainable energy solutions [80].

SCs are highly desirable due to their high-power density, safety, fast-charging time, low maintenance, and an extremely long-cycling lifespan. Aqueous Zinc-ion Batteries (AZIB) are promising because of their safety, cost-effectiveness, and competitive capacity. However, dendritic formation of zinc and undesirable side reactions limits their practical viability. To improve electrochemical performance, Carbon Quantum Dots (CQDs) can be used to modify inexpensive, tenacious, and hydrophilic Filter Paper (FP) as a separator. The FP-CQDs separator enables the dendrite-free and corrosion-resistant operation and possesses high ionic conductivity and zinc-ion transference number. In $Zn//V_2O_5$ full batteries, the composite separator displays superior capacity retention than FP separator alone, with excellent cycling stability as well [69]. Conductivity, hydrophobicity, and surface redox reactivity of MXene, a material with excellent energy storage properties, have been studied. The big surface area and high crystallinity of MOFs in an application regarding electrochemical application were investigated, but no good charging and discharging rate and specific capacity could be achieved. A new strategy was proposed such as a composite of Graphene Quantum Dots (GQDs), N-MXene, and iron-iridium MOF (Ir@Fe-MOF). This strategy lowers MXene sheet restacking through constraining the ion/electron transportation paths and heightening electroactive sites to achieve a highly specific capacity. Negligible reduction in charge and discharge potential was observed in the initial 5 cycles, but after that, it starts to increase until 30 cycles, owing to a slight increase in the charge-transfer resistance of the Ir@Fe-MOF/N-MXene/GQDs due to degradation of the Fe oxide [79].

A team of researchers focused on the development of a high-power-density polymer film capacitor for application with daily electronics and pulse electrical supplies [76]. Using thermal injection, the researchers synthesized perovskite Quantum Dots (QDs) and prepared a low QD polyetherimide (PEI) composite. Under high temperatures, deep trapping sites for charge carriers were created by QDs that reduced conduction loss. The behavior of a composite configuration that combines a dielectric elastomer with a piezoelectric polymer was demonstrated for improved energy capacity and dielectric reliability, with an energy density of 7.2 J/cm^3 and 90% efficiency at 350 MV/m. Capture of charge carriers by robust electrostatic interaction was contributed by the difference in trap energy between QDs and PEI. In this study, the authors use Cobalt-coordinated Carbon Quantum Dots (CoCQDs) for the enhanced energy storage performance of nickel–cobalt layered double hydroxides (NiCo-LDH) [12]. A composite electrode CoC@LDH was derived by co-depositing CoCQDs with nickel and cobalt ions. The improved conductivity and structure stability of the NiCo-LDH improved cycle stability and rate performance.

The 1.0CoC@LDH electrode demonstrated a high specific capacitance of 1867 F g^{-1} at 1 A g^{-1} and a maximum value of about 84.6% capacity retention after 6000 cycles [12].

Low-cost synthesizing process, outstanding optical properties, low toxicity, and high biocompatibility make Carbon Quantum Dots (CQDs) emerge as promising materials in the areas of optoelectronics, energy storage, and conversions [4]. Here, they have been noticed for their very specific electrical properties and the ability to host functional groups on electrode surfaces. The biological synthesis of CQDs from green algae *Halimeda opuntia* in the second study has shown that these CQDs have a specific capacitance of 311 F g^{-1}, which can be applied in energy storage and conversion. *Halimeda opuntia-based* biomass sources to synthesize CQDs were eco-friendly and addressed global waste management challenges. The battery-type electrode material consisted of graphene quantum dots synthesized from citric acid and ferrocenylbutanol. CNT and Fe_3O_4 were used to create different compositions of nanocomposites. The resulting nanocomposite exhibited ideal battery-type capacity, high energy and power density, and excellent cycle life. Along with superior electrochemical performance, the material was shown to improve the ion transfer rate between electrode/electrolyte surfaces, suggesting the possibility of using it in energy storage devices. Potential applications of the nanocomposite based on the superiority of its properties were indicated [53].

Nanocomposite polymer electrolyte films made of Polyethylene Oxide (PEO) and Carboxymethyl Cellulose (CMC) blends with gold quantum dots (AuQDs, CuNPs (copper nanoparticles) as hybrid nanofiller were investigated [59]. The composites revealed both lower degrees of crystallinity with rising AuQDs/CuNPs content, lower optical energy gap, and higher electrical conductivity. Higher values of dielectric properties and higher dielectric constant are observed at lower frequency regions. At lower frequencies, the Nyquist diagram appears as a semicircle, and at higher frequencies, a straight line. Such nanocomposite electrolyte films may be appropriate for capacitors and for other flexible energy storage devices. Improved capacitive and conductive performance of monoclinic amorphous $Ni_{1-x}Mn_xWO_4$ compounds have been synthesized using a hydrothermal technique [10]. Their structural, microstructural, and optical properties were investigated via various characterization techniques. The incorporation of Mn-ion in the $NiWO_4$ lattice decreases the sample's particle size to 4.5 nm and increases its bandgap energy significantly [10]. Cyclic voltammetry, galvanostatic charge–discharge characteristics, and electrochemical impedance spectroscopy were employed to reveal the electrochemical performance of these materials. Specifically, the addition of 2% Mn to $NiWO_4$ increases specific surface area and improves specific capacitance. The mechanism of charge storage for improving conductivity and electrochemical performance is revealed. Mn-doped $NiWO_4$ electrode material was examined by fabricating an asymmetric supercapacitor device with superior cyclic stability up to 5000 cycles [10].

A novel structure was proposed to improve the stability and flexibility of amorphous WO_3 (a-WO_3) in Electrochromic (EC) energy-storage devices. The structure introduces P-doped Carbon Quantum Dots (P-CQDs) into a-WO_3 through a low-temperature process, forming strong hydrogen bonds with a-WO_3. The functional groups and graphitic carbon of P-CQDs improve Li-ion wettability and electrical conductivity, enhancing electrochemical activity and kinetics. P-CQD graft-functionalized a-WO_3 electrodes show excellent EC energy-storage performances, with enhanced long-term cycle stability and flexibility. This heterostructure-inducing quantum dot graft functionalization with a-WO_3 is considered a breakthrough in the realization of high-performance flexible EC energy-storage devices [31]. Flexible energy-storage devices are becoming increasingly popular due to their portability and practicality. However, their capacitance decay and short cycle life limit their development [13]. A simple electrodeposition method has been developed to improve the electrochemical activity and structural stability of nickel cobalt-layered double hydroxides (NiCoLDH) nanosheet arrays on Carbon Cloth (CC) by introducing Carbon Quantum Dots (CQDs). The CQD-modified NiCoLDH nanosheet arrays on CC electrode (CQD/NiCoLDH@CC) not only expose more active sites but also enhance their structure and cycle life. The assembled flexible aqueous asymmetric CQD/NiCoLDH-3@CC//activated CC supercapacitor shows a high energy density and power density.

For applications in Lithium-ion Batteries (LIBs) and Supercapacitors (SCs), a new electrode material was fabricated by growing V_2O_3 Quantum Dots (QDs) on Mesoporous Carbon (MC) nanosheets. On account of their high intrinsic conductivity and open tunnel-like lattice structure, the material allows electron transfer and ion intercalation [25]. The composite architectures offer more electrochemical active sites, shorten ion/electron diffusion paths, and reduce volume swings during charging/discharging. Large capacity/capacitance and excellent cycling stability as well as high-rate performance of the composite are demonstrated. A good structure design for advanced metal oxide-based nanostructured electrode materials was suggested by this work.

11 Quantum Dots in Water Splitting and CO_2 Reduction

Two main photocatalytic applications of the quantum dots have been studied including water splitting for the H_2 fuel production and CO_2 reduction for value-added chemicals and fuels production [71] (Fig. 7). Photocatalytic water splitting is a process in which water molecules (H_2O) are split to hydrogen (H_2) and oxygen (O_2) using solar energy. Hydrogen presents an appealing clean and renewable energy carrier as a fossil fuel alternative [3]. For water splitting applications, quantum dots of titanium dioxide (TiO_2), cadmium sulfide (CdS), and lead sulfide (PbS) have been used as photocatalysts. Sunlight was absorbed by the electrode and electron–hole pairs are generated, which involved in the redox reactions to split water into H_2 and O_2 [78]. While traditional photocatalysts preferentially absorb ultraviolet light, quantum dots are engineered to absorb visible light, so that the efficiency of solar energy utilization is significantly increased. As an example, visible light illumination of water upon CdS quantum dots has been shown to be an efficient method of

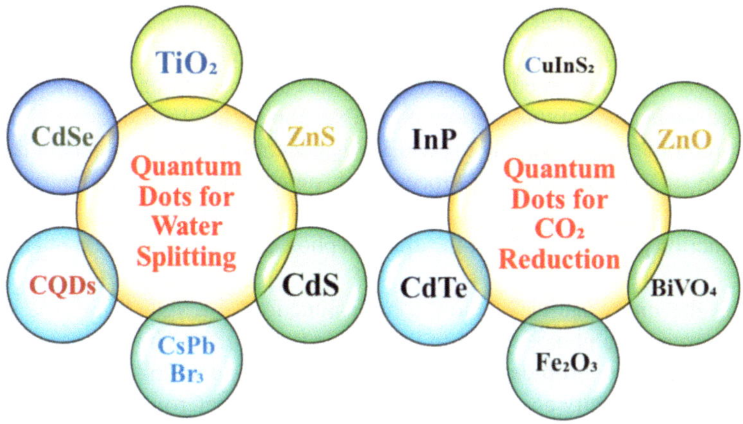

Fig. 7 Quantum dots for water splitting and CO_2 reduction

water splitting [8]. Z-scheme photocatalytic system employing quantum dots also serves to improve the photocatalytic efficiency of the water splitting process [52]. In such systems quantum dots can stand as photosensitizers that absorb light and pass electrons to another semiconductor that conducts the overall water splitting process. However, CdS quantum dots with co-catalysts platinum (Pt) or nickel (Ni) have shown high rates of hydrogen production under visible light. Other recent studies have examined the use of ternary quantum dots, including zinc-cadmium-sulfide (ZnCdS) to once again boost water splitting efficiency through optimization of bandgap and surface properties [52].

Hedau and team have developed a bifunctional Carbon Quantum Dot (CQD)-embedded Metal–Organic Framework (MOF) nanohybrid from natural resources for electrochemical water splitting and CO_2 reduction [24]. Fabrication of the nanohybrid, with key parameters like synthesis temperature, deposition time, pH, electrolyte type, and scan rate, was studied. Finally, an excellent hydrogen and oxygen evolution reaction was demonstrated with overpotentials below 50 mV to achieve a current density of 50 mA/cm^2. The electrode shows robustness and high chemical stability even after 4000 cycles or more than 17 h. The potential of the carbon nanohybrid for green energy applications is extended using a stable CQD@MOF/NF electrode that demonstrates an electrochemical CO_2 reduction with a specific current density for 12 h. The enhancement of the bismuth vanadate ($BiVO_4$) charge diffusion length and water oxidation kinetics for more effective photoelectrochemical catalytic performance was studied [38]. The addition of co-catalysts such as Nickel Molybdenum oxide (NM) and Carbon Quantum Dot (CQD) aid in initiation of more water oxidation reactions. $BiVO_4$/NM/CQD with hole scavenger yields the highest photocurrent densities, while the $BiVO_4$ electrode gives a density of 1.2 mA/cm^2. In addition, the study demonstrates excellent long-term stability with 90% photocurrent retention during illumination for 24 h. To this end, this study developed a high-temperature solid synthesis route to make highly stable SiO_2 glass-coated $CsPbI_3$

Quantum Dots (QDs) that are bright and stable under molten salt conditions [63]. Solvothermal synthesis confirmed the photogenerated electron transition into these QDs, which were implanted in g-C_3N_4 nanosheets. Charge transfer efficiency was improved further by polymerization of g-C_3N_4. Type II heterostructure was formed, which promoted photocatalytic H_2 and H_2O_2 generation. The mechanism revealed a two-step single-electron redox process with superoxide radicals as intermediates. In doing so, this study presents a new strategy for the preparation of g-C_3N_4-based heterojunction photocatalysts for the efficient photocatalytic production of H_2O_2 and H_2.

Using the impregnation method, a heterojunction of LNO@CdSe was synthesized for promising photocatalyst in the visible region [37]. The photoanodes were fabricated using the drop-casting method, demonstrating a 4.62-fold enhancement in photocurrent density compared to bare LNO. In the heterojunction, visible light in the near infrared region was absorbed, reducing the recombination of electron–hole pairs. This, in turn, hints that this material could be another possible green hydrogen and oxygen source when fed with solar energy. Although transition metal phosphates are good electrocatalysts for electrocatalytic water splitting, often they exhibit high overpotentials and slow HER and OER kinetics [44]. A two-step electrodeposition method was used to prepare hybrid catalysts consisting of Pt quantum dots-doped NiP (NiP-Pt) nano-embroidery spheres and $Co(OH)_2$ nanosheets. The excellent performance of the catalyst relies on the synergistic interactions between NiP and Pt quantum dots as well as the strong electronic interactions at the interface with $Co(OH)_2$. At an overpotential of 186 mV and a Tafel slope of 53.049 mV dec^{-1}, the catalyst shows good performance for OER at 10 mA cm^{-2}. Efficient photocatalysts for aqueous contamination destruction are green, in situ, one-pot, and economically synthesized from biocompatible Carbon Quantum Dots (CQDs) derived from garden thyme. Pure as well as co-doped CQDs with fluorine and boron were synthesized with a favorable quantum yield of 40%. In 10% B–10% F co-doped CQDs, removal rates of organic pollutant Rhodamine B (RhB) were investigated, and it gave the best performance with 10% F–10% B co-doped CQDs at 400 °C for 5 h. Upon exposure to visible light radiation, RhB oxidation on F, B co-doped CQDs catalysts was dominated by hydroxyl radicals as the reactive species [48]. This study reports a one-step method for preparing nickel oxide quantum dots on ultrathin amorphous carbon nanosheets (NiO dots/a-C). Researchers have shown that these nanosheets exhibit uniform size distribution and good activity and stability for Oxygen Evolution Reaction (OER) in 0.1 M KOH using rotating disk electrodes [30].

The cobalt ferrite, graphitic carbon nitride, and N-doped graphene quantum dots were prepared using hydrothermal method for a ternary composite [58]. The composite showed low overpotential for hydrogen evolution reactions at 287 mV and 445 mV for oxygen evolution reactions at 10 mA cm^{-2}. Long-term stability tests showed that the cell potential required for the electrode to achieve 10 mA cm^{-2} in total water splitting was as low as 2.0 V. Characterization methods and electrocatalytic processes of the composite were studied in an alkaline medium, so that the purity of the composite could be established. Direct and solid-state dispersion methods were used to prepare Graphene Quantum Dots (GQDs)/ZSM-5 type cobaltosilicate and GQDs/ZSM-5 type chromosilicate composites [49]. The photocatalytic activity of the GQDs/chromosilicate composite was better than that of the composites, while the GQDs/chromosilicate composite featured a superior performance. The rate constants for the composites ranged from 0.295 to 0.581 g mg^{-1} min^{-1} and followed the pseudo-second order kinetic model. Moreover, the composites displayed lower overpotentials for the HER and the OER than pure GDQs and cobaltosilicate. It is expected that these new GQDs-based composite materials will enable new doors in materials science and the electrocatalytic-related processes.

A direct Z-Scheme system of solar-driven water splitting and pollution treatment inspired by artificial photosynthesis has been developed by researchers [68]. As co-catalysts, CdS (1D nanorod)/WO_3 (2D nanosheet) heterostructures are used in the system. Authors have demonstrated a two-step pathway to form an effective photocatalyst, NiS-CdS/WO_3 (NCW), that possesses a high solar spectrum utilization efficiency and efficient photogenerated carrier transfer. High apparent quantum yield (AQY of 24.83% at 420 nm) and good long-term stability were demonstrated by NCW10's outstanding photocatalytic activity. This modified method optimizes the structure and properties of the photocatalyst, providing guidelines for the selection and development of solar conversion systems. Cu_2S/TiO_2 nanocomposites with varying concentrations of TiO_2 Quantum Dots (QDs) have been synthesized using a two-stage hydrothermal wet impregnation method. The formation of Cu_2S and TiO_2 in chalcocite and anatase phases is confirmed by X-ray diffraction analysis. TiO_2 quantum dots form a composite whose structure consists of nano-rod-like and nanosheet-like structures. The chemical state analysis of the nanocomposite indicates the presence of Cu^+, S^{2-}, Ni^{2+}, and O^{2-}. The excellent stability and durability of this photocatalyst make it suitable for large-scale applications related to the H_2 evolution system, with an H_2 evolution rate of approximately 45.6 mmol h^{-1} g^{-1} cat [56].

A promising approach to converting greenhouse gases into renewable fuels and chemicals is the photocatalytic reduction of carbon dioxide (CO_2) to value-added chemicals, such as methane (CH_4), methanol (CH_3OH), and formic acid (HCOOH). Photocatalysis for CO_2 reduction has been applied using quantum dots, such as cadmium telluride (CdTe), indium phosphide (InP), and Metal–Organic Framework (MOF)-derived quantum dots as photocatalysts. Such materials can harvest sunlight and produce high-energy electrons to convert CO_2 molecules into hydrocarbons and alcohols. Engineered quantum dots can be designed with specific electronic properties that enhance the selectivity and efficiency of the CO_2 reduction reaction. For instance, CdTe quantum dots exhibit high selectivity in producing methane, and InP quantum dots demonstrate high yields and efficiency in producing methanol. It is now found that hybrid quantum dot systems, CdTe quantum dots in conjunction with co-catalysts such as copper (Cu) or cobalt (Co), can achieve high CO_2 reduction rates under solar illumination. Surface engineering and quantum dot heterostructures have also been utilized for further developing these photocatalytic systems, thereby enhancing both selectivity and efficiency. Despite the promising potential of quantum dots in photocatalysis, several challenges must be addressed to realize their capabilities in practical applications fully.

12 Conclusion and Future Perspective

Due to their unprecedented ability to enhance efficiency across various energy devices, Quantum Dots (QDs) are likely to have a profound impact on the future of energy applications. In solar energy, QDs are capable of enhancing solar cell efficiency by providing adjustable bandgaps for wider light capture. QDs are also driving advancements in energy storage, including supercapacitors and batteries, by increasing charge-storage capacity and improving charge–discharge cycles. Furthermore, QDs are being researched for use in Light-Emitting Devices (QLEDs), which can provide electric displays with low power consumption. Their unique properties enable them to be used in photocatalysis for hydrogen production, as well as in environmental remediation. Quantum dots harness light and are on the cusp of revolutionizing the energy sector in key dimensions like sustainability, efficiency, and scalability. Research focuses on developing non-toxic, environmentally friendly Quantum Dots (QDs), as well as methods that enhance cost and stability. Additionally, QDs hold great promise in thermoelectric energy conversion technologies, waste heat recovery, and carbon sequestration. The incorporation of quantum dots into smart grids and automated energy monitoring systems may improve the instantaneous use of energy.

References

1. Aftab S, Koyyada G, Mukhtar M, Hegazy HH, Kim JH (2024) From energy to light: advancements in perovskite quantum dots for optoelectronics. J Mater Chem C. https://doi.org/10.1039/D4TC02230K
2. Agarwal K, Rai H, Mondal S (2023) Quantum dots: an overview of synthesis, properties, and applications. Mater Res Express 10(6):062001. https://doi.org/10.1088/2053-1591/ACDA17
3. Aldosari OF, Hussain I (2024) Unlocking the potential of TiO_2-based photocatalysts for green hydrogen energy through water-splitting: recent advances, future perspectives and techno feasibility assessment. Int J Hydrog Energy 59:958–981. https://doi.org/10.1016/J.IJHYDENE.2024.01.306
4. Al-Ghamdi SA, Darwish AAA, Hamdalla TA, Pasha A, Elnair ME, Al-Atawi A, Khasim S (2023) Biological synthesis of novel carbon quantum dots using Halimeda opuntia green algae with improved optical properties and electrochemical performance for possible energy storage applications. Int J Electrochem Sci 18(5):100102. https://doi.org/10.1016/J.IJOES.2023.100102
5. Anwar H, Arif I, Javeed U, Mushtaq H, Ali K, Sharma SK (2020) Quantum dot solar cells. In: Solar cells: from materials to device technology, pp 235–258. https://doi.org/10.1007/978-3-030-36354-3_9

6. Arshad H, Majid A, Khan MAU (2022) Quantum dots: synthesis, properties, and applications. In: Nanotechnology in the life sciences, pp 11–45. https://doi.org/10.1007/978-3-031-10216-5_2
7. Becker-Koch D, Albaladejo-Siguan M, Lami V, Paulus F, Xiang H, Chen Z, Vaynzof Y (2019) Ligand dependent oxidation dictates the performance evolution of high efficiency PbS quantum dot solar cells. Sustain Energy Fuels 4(1):108–115. https://doi.org/10.1039/C9SE00602H
8. Belman A, Jayarama A, Rao AS, Nagarkar SS, Dutta A, Duttagupta SP, Prabhu SS, Pinto R (2024) Challenges in photocatalytic hydrogen evolution: importance of photocatalysts and photocatalytic reactors. Int J Hydrog Energy 81:1442–1466. https://doi.org/10.1016/J.IJHYDENE.2024.07.262
9. Bera S, Mondal TK, Su YK, Saha SK (2024) Sulfur-tuned MoS_2 quantum dot decorated $Ti_3C_2T_x$ (MXene) electrode materials for high performance supercapacitor applications. J Alloy Compd 985:174010. https://doi.org/10.1016/J.JALLCOM.2024.174010
10. Chatterjee M, Saha S, Chatterjee T, Das S, Pradhan SK (2022) Mn-doped $NiWO_4$ quantum dots with superior electrochemical and conductivity performance for energy storage application. J Energy Storage 56:105946. https://doi.org/10.1016/J.EST.2022.105946
11. Chen D, Chen X (2019) Luminescent perovskite quantum dots: synthesis, microstructures, optical properties and applications. J Mater Chem C 7(6):1413–1446. https://doi.org/10.1039/C8TC05545A
12. Chen Q, Pan H, Chen Z, Jiang X, Li Y, Tian W, Liu H, Zhu S (2024) Cobalt coordinated carbon quantum dots boosting the performance of NiCo-LDH for energy storage. J Colloid Interface Sci 655:110–119. https://doi.org/10.1016/J.JCIS.2023.10.114
13. Chen W, Quan H, Chen D (2023) Carbon quantum dots boosted structure stability of nickel cobalt layered double hydroxides nanosheets electrodeposited on carbon cloth for energy storage. Surf Interfaces 36:102498. https://doi.org/10.1016/J.SURFIN.2022.102498
14. Cheng H, Ding S, Hao M, Wang L, Steele JA (2024) Perovskite quantum dots: what's next? Next Energy 4:100152. https://doi.org/10.1016/J.NXENER.2024.100152
15. Cotta MA (2020) Quantum dots and their applications: what lies ahead? ACS Appl Nano Mater 3(6):4920–4924. https://doi.org/10.1021/ACSANM.0C01386/ASSET/IMAGES/LARGE/AN0C01386_0001.JPEG
16. Cruz-Gómez J, de Moure-Flores FJ, Mayén-Hernández SA, Quiñones-Galván JG, Chettiar ADR, Pérez-García CE, Santos-Cruz J (2024) Lead sulfide quantum dots in an organic solar cell active layer. Mendeleev Commun 34(5):664–666. https://doi.org/10.1016/J.MENCOM.2024.09.012
17. Dabbousi BO, Rodriguez-Viejo J, Mikulec FV, Heine JR, Mattoussi H, Ober R, Jensen KF, Bawendi MG (1997) (CdSe)ZnS Core–shell quantum dots: synthesis and characterization of a size series of highly luminescent nanocrystallites. J Phys Chem B 101(46):9463–9475. https://doi.org/10.1021/JP971091Y
18. Ellingson RJ, Beard MC, Johnson JC, Yu P, Micic OI, Nozik AJ, Shabaev A, Efros AL (2005) Highly efficient multiple exciton generation in colloidal PbSe and PbS quantum dots. Nano Lett 5(5):865–871. https://doi.org/10.1021/NL0502672/SUPPL_FILE/NL0502672SI20050210_053049.PDF
19. El-Mahalawy AM, Abbas W, Mostafa O, Zidan NA, Issa HH, Fedawy M, Wassel AR (2024) Integrative role of PEDOT:PSS in adjusting the photoresponse efficiency of novel reduced graphene quantum dots/silicon heterojunction for optoelectronics and solar energy conversion applications. Surf Interfaces 46:103946. https://doi.org/10.1016/J.SURFIN.2024.103946
20. Flores-Gómez J, Mota-Macías S, Guerrero-Jiménez JP, Romero-Arellano VH, Morales-Rivera J (2024) Sol–gel synthesis of TiO_2 with pectin and their efficiency in solar cells sensitized by quantum dots. Gels 10(7):470. https://doi.org/10.3390/GELS10070470
21. Goodwin H, Jellicoe TC, Davis NJLK, Böhm ML (2018) Multiple exciton generation in quantum dot-based solar cells. Nanophotonics 7(1):111–126. https://doi.org/10.1515/nanoph-2017-0034
22. Guo Y, Li X, Chen B, Tang Y, Wang J, Lu H, Guo C, Tong SY (2024) Fabrication of gradient band tin oxide electron transport layer using self-separated dual-quantum dots for perovskite solar cells. Mater Today Energy 46:101708. https://doi.org/10.1016/J.MTENER.2024.101708

23. Ha JM, Lee SH, Park D, Yoon YJ, Yang IM, Seo J, Hwang YS, Lee CY, Suk JK, Park JK, Yeo S (2024) Synthesis mechanism from graphene quantum dots to carbon nanotubes by ion-sputtering assisted chemical vapor deposition. Discover Nano 19(1):1–9. https://doi.org/10.1186/S11671-024-04027-3/FIGURES/6
24. Hedau B, Park SJ, Kang BC, Ha TJ (2024) Bifunctional carbon quantum dot-embedded metal-organic framework nanohybrid as a highly efficient electrocatalyst for water splitting and CO_2 reduction. Carbon 216:118527. https://doi.org/10.1016/J.CARBON.2023.118527
25. Hei J, Cheng L, Fu Y, Du W, Qian Y, Li J, Yin Y, Wang N, Su L, Wang L (2023) Uniformly confined V_2O_3 quantum dots embedded in biomass derived mesoporous carbon toward fast and stable energy storage. Ceram Int 49(10):16002–16010. https://doi.org/10.1016/J.CERAMINT.2023.01.197
26. Hoang VC, Dave K, Gomes VG (2019) Carbon quantum dot-based composites for energy storage and electrocatalysis: mechanism, applications and future prospects. Nano Energy 66:104093. https://doi.org/10.1016/J.NANOEN.2019.104093
27. Ishmael AM, Abdel-Salam AI, Khalil MMH, Mansour AS (2024) Quantum dots composites for energy storage applications. In: Engineering materials, Part F2942, pp 429–454. https://doi.org/10.1007/978-3-031-54779-9_22
28. Jain K, Kishor S, Singh KS, Odelius M, Ramaniah LM (2021) Composition related tunability of "Green" core/shell quantum dots for photovoltaic applications from first principles. J Phys Chem C 125(49):27046–27057. https://doi.org/10.1021/ACS.JPCC.1C05230/ASSET/IMAGES/LARGE/JP1C05230_0007.JPEG
29. Jin L, Sirigu G, Tong X, Camellini A, Parisini A, Nicotra G, Spinella C, Zhao H, Sun S, Morandi V, Zavelani-Rossi M, Rosei F, Vomiero A (2016) Engineering interfacial structure in "Giant" PbS/CdS quantum dots for photoelectrochemical solar energy conversion. Nano Energy 30:531–541. https://doi.org/10.1016/J.NANOEN.2016.10.029
30. Jin B, Wang Q, Sainio J, Saveleva VA, Jiang H, Shi J, Ali B, Kallio AJ, Huotari S, Sundholm D, Han N, Kallio T (2024) Amorphous carbon modulated-quantum dots NiO for efficient oxygen evolution in anion exchange membrane water electrolyzer. Appl Catal B: Environ Energy 358:124437. https://doi.org/10.1016/J.APCATB.2024.124437
31. Jo MH, Kim KH, Ahn HJ (2022) P-doped carbon quantum dot graft-functionalized amorphous WO_3 for stable and flexible electrochromic energy-storage devices. Chem Eng J 445:136826. https://doi.org/10.1016/J.CEJ.2022.136826
32. Kang WS, Oh T, Nam GH, Kim HS, Kim KS, Park SH, Kim JH, Lee JH (2020) Template-assisted electrochemical synthesis of CdSe quantum dots—polypyrrole composite nanorods. Appl Sci 10(17):5966. https://doi.org/10.3390/APP10175966
33. Khan ZU, Khan LU, Brito HF, Gidlund M, Malta OL, Di Mascio P (2023) Colloidal quantum dots as an emerging vast platform and versatile sensitizer for singlet molecular oxygen generation. ACS Omega 8(38):34328–34353. https://doi.org/10.1021/ACSOMEGA.3C03962/ASSET/IMAGES/LARGE/AO3C03962_0019.JPEG
34. Kilina SV, Tamukong PK, Kilin DS (2016) Surface chemistry of semiconducting quantum dots: theoretical perspectives. Acc Chem Res 49(10):2127–2135. https://doi.org/10.1021/ACS.ACCOUNTS.6B00196/ASSET/IMAGES/MEDIUM/AR-2016-00196Z_0006.GIF
35. Kim JY, Jang YJ, Park J, Kim J, Kang JS, Chung DY, Sung YE, Lee C, Lee JS, Ko MJ (2018) Highly loaded PbS/Mn-doped CdS quantum dots for dual application in solar-to-electrical and solar-to-chemical energy conversion. Appl Catal B 227:409–417. https://doi.org/10.1016/J.APCATB.2018.01.041
36. Kim M, Osone S, Kim T, Higashi H, Seto T (2017) Synthesis of nanoparticles by laser ablation: a review. KONA Powder Part J 34(34):80–90. https://doi.org/10.14356/KONA.2017009
37. Kuchipudi A, Nandigana P, Prasanna M, Anupriya, Panda SK, Sreedhar G (2023) Enhanced photoelectrocatalytic performance of $LaNiO_3$ sensitized with CdSe quantum dots for photo-electrochemical water splitting. Mater Chem Phys 307:128123. https://doi.org/10.1016/J.MATCHEMPHYS.2023.128123

38. Kuo TR, Chiou YT, Huang HW, Kongvarhodom C, Saukani M, Yougbaré S, Chen HM, Lin LY (2024) Heterojunction and co-catalyst engineering of bismuth vanadate by nickel molybdenum oxide and carbon quantum dot decorations for photoelectrochemical water splitting. Int J Hydrog Energy 51:1099–1108. https://doi.org/10.1016/J.IJHYDENE.2023.10.344
39. Li X, Tong X, Yue S, Liu C, Channa AI, You Y, Wang R, Long Z, Zhang Z, Zhao Z, Liu XF, Wang ZM (2021) Rational design of colloidal $AgGaS_2$/CdSeS core/shell quantum dots for solar energy conversion and light detection. Nano Energy 89:106392. https://doi.org/10.1016/J.NANOEN.2021.106392
40. Li J, Wu Q, Wu J (2015) Synthesis of nanoparticles via solvothermal and hydrothermal methods. In: Handbook of nanoparticles, pp 1–28. https://doi.org/10.1007/978-3-319-13188-7_17-1
41. Li X, Xu Q, Zhang Z (2023) Molecular beam epitaxy growth of quantum wires and quantum dots. Nanomaterials 13(6):960. https://doi.org/10.3390/NANO13060960
42. Lims SC, Tran NA, Dao VD, Pham PV (2025) The world of quantum dot-shaped nanoparticles: Nobel prize in chemistry 2023: advancements and prospectives. Coord Chem Rev 528:216423. https://doi.org/10.1016/J.CCR.2024.216423
43. Liu Y, Dong C, Peng C, Zhang T, Zhang L (2024) Nitrogen-doped carbon quantum dots enable efficient photothermal conversion for direct absorption solar collectors. Sol Energy Mater Sol Cells 278:113178. https://doi.org/10.1016/J.SOLMAT.2024.113178
44. Liu Z, Guo Z, He H, Zhang Y, Ruan M, Liu Z (2022) Interface regulation of Pt quantum dots doped nickel phosphide and cobalt hydroxide to promote electrocatalytic overall water splitting. Int J Hydrog Energy 47(97):40986–40998. https://doi.org/10.1016/J.IJHYDENE.2022.09.182
45. Majumder T, Dhar S, Debnath K, Mondal SP (2017) Role of S, N co-doped graphene quantum dots as a green photosensitizer with Ag-doped ZnO nanorods for improved electrochemical solar energy conversion. Mater Res Bull 93:214–222. https://doi.org/10.1016/J.MATERRESBULL.2017.05.004
46. Majumder T, Mondal SP (2016) Advantages of nitrogen-doped graphene quantum dots as a green sensitizer with ZnO nanorod based photoanodes for solar energy conversion. J Electroanal Chem 769:48–52. https://doi.org/10.1016/J.JELECHEM.2016.03.018
47. Mimona MA, Rimon MIH, Zohura FT, Sony JM, Rim SI, Arup MMR, Mobarak MH (2025) Quantum dot nanomaterials: empowering advances in optoelectronic devices. Chem Eng J Adv 21:100704. https://doi.org/10.1016/J.CEJA.2025.100704
48. Mirzaeifard Z, Shariatinia Z (2024) Economical, one-pot, and green synthesis of plant-based carbon quantum dots for efficient visible-light photocatalytic dye degradation and water purification. J Taiwan Inst Chem Eng 163:105655. https://doi.org/10.1016/J.JTICE.2024.105655
49. Mokhtari A, Yavari A, Khatamian M, Sadeghi E, Peighambardoust NS, Aydemir U (2023) Facile synthesis of graphene quantum dots/ZSM-5 type metalosilicate composites and evaluating their performance in photocatalytic degradation of methylene blue and electrochemical water splitting. Adv Powder Technol 34(1):103892. https://doi.org/10.1016/J.APT.2022.103892
50. Murray CB, Norris DJ, Bawendi MG (1993) Synthesis and characterization of nearly monodisperse CdE (E = S, Se, Te) semiconductor nanocrystallites. J Am Chem Soc 115(19):8706–8715. https://doi.org/10.1021/JA00072A025/ASSET/JA00072A025.FP.PNG_V03
51. Nammahachak N, Aup-Ngoen KK, Asanithi P, Horpratum M, Chuangchote S, Ratanaphan S, Surareungchai W (2022) Hydrothermal synthesis of carbon quantum dots with size tunability via heterogeneous nucleation. RSC Adv 12(49):31729–31733. https://doi.org/10.1039/D2RA05989D
52. Ng BJ, Putri LK, Kong XY, Teh YW, Pasbakhsh P, Chai SP (2020) Z-Scheme photocatalytic systems for solar water splitting. Adv Sci 7(7):1903171. https://doi.org/10.1002/ADVS.201903171
53. Payami E, Teimuri-Mofrad R (2023) Development of ternary nanocomposite based on ferrocenyl-modified graphene quantum dots for high-performance energy storage applications. J Energy Storage 72:108346. https://doi.org/10.1016/J.EST.2023.108346

54. Poole CP, Owens FJ (2003) Introduction to nanotechnology, p 388. https://www.wiley.com/en-in/Introduction+to+Nanotechnology-p-9780471079354
55. Prakash R, Das S, Maiti P (2023) Non-toxic $CuInS_2$ quantum dot sensitized solar cell with functionalized thermoplast polyurethane gel electrolytes. Polymer 269:125708. https://doi.org/10.1016/J.POLYMER.2023.125708
56. Rao VN, Ravi P, Sathish M, Sakar M, Yang BL, Yang JM, Kumari MM, Shankar MV (2022) Titanate quantum dots-sensitized Cu_2S nanocomposites for superficial H_2 production via photocatalytic water splitting. Int J Hydrog Energy 47(95):40379–40390. https://doi.org/10.1016/J.IJHYDENE.2022.05.091
57. Reddy KTK, Madhavi Latha B, Gupta JK, Chaitanya A, Srinivasa Babu P, Vamseekrishna G, Manikanta Y, Sahithi A (2024) Discovery and history of quantum dots. In: Engineering materials, Part F2942, pp 13–34. https://doi.org/10.1007/978-3-031-54779-9_2
58. Reghunath BS, Rajasekaran S, Sunaja Devi KR, Pinheiro D, Jadan Resnik Jaleel UC (2023) N-doped graphene quantum dots incorporated cobalt ferrite/graphitic carbon nitride ternary composite for electrochemical overall water splitting. Int J Hydrog Energy 48(8):2906–2919. https://doi.org/10.1016/J.IJHYDENE.2022.10.169
59. Saeed A, Abolaban F, Al-Mhyawi SR, Albaidani K, Al Garni SE, Al-Marhaby FA, Alwafi R, Djouider F, Qahtan TF, Asnag GM (2023) Improving the polyethylene oxide/carboxymethyl cellulose blend's optical and electrical/dielectric performance by incorporating gold quantum dots and copper nanoparticles: nanocomposites for energy storage applications. J Mater Res Technol 24:8241–8251. https://doi.org/10.1016/J.JMRT.2023.05.073
60. Salaheldin AM, Walter J, Herre P, Levchuk I, Jabbari Y, Kolle JM, Brabec CJ, Peukert W, Segets D (2017) Automated synthesis of quantum dot nanocrystals by hot injection: mixing induced self-focusing. Chem Eng J 320:232–243. https://doi.org/10.1016/J.CEJ.2017.02.154
61. Senevirathne KMBB, Kandanapitiye MS, Udukala DN, Farhath MM, Fouad D, Kumar YA, Jaseetharan T (2024) Non-toxic Bi_2S_3 quantum dot-sensitized TiO_2 electrodes for cost-efficient solar power conversion. Electrochim Acta 507:145121. https://doi.org/10.1016/J.ELECTACTA.2024.145121
62. Sherif El-Eskandarany M, Al-Hazza A, Al-Hajji LA, Ali N, Al-Duweesh AA, Banyan M, Al-Ajmi F (2021) Mechanical milling: a superior nanotechnological tool for fabrication of nanocrystalline and nanocomposite materials. Nanomaterials 11(10):2484. https://doi.org/10.3390/NANO11102484
63. Shi W, Zhang X, Tian Y, Yang P (2024) Aqueous glass-coated $CsPbI_3$ quantum dots implanted in g-C_3N_4 nanosheets for efficient photocatalytic water splitting and H_2O_2 generation. J Environ Chem Eng 12(5):114018. https://doi.org/10.1016/J.JECE.2024.114018
64. Shiral Fernando KA, Sahu S, Liu Y, Lewis WK, Guliants EA, Jafariyan A, Wang P, Bunker CE, Sun Y-P (2015) Carbon quantum dots and applications in photocatalytic energy conversion. ACS Appl Mater Interfaces 7(16):8363–8376. https://doi.org/10.1021/acsami.5b00448
65. Shivaji K, Sridharan K, Kirubakaran DD, Velusamy J, Emadian SS, Krishnamurthy S, Devadoss A, Nagarajan S, Das S, Pitchaimuthu S (2023) Biofunctionalized CdS quantum dots: a case study on nanomaterial toxicity in the photocatalytic wastewater treatment process. ACS Omega 8(22):19413–19424. https://doi.org/10.1021/acsomega.3c00496
66. Singh S, Dhawan A, Karhana S, Bhat M, Dinda AK (2020) Quantum dots: an emerging tool for point-of-care testing. Micromachines 11(12):1–23. https://doi.org/10.3390/MI11121058
67. Wang Z, Che H, Lu W, Chao Y, Wang L, Liang B, Liu J, Xu Q, Cui X (2023) Application of inorganic quantum dots in advanced lithium-sulfur batteries. Adv Sci 10(19):2301355. https://doi.org/10.1002/ADVS.202301355
68. Wang S, Jiao Y, Yin J, Liu Y, Li X, Yang W, Han S, Zhao B, Wu F, Jiang J, Zhang H (2022) Innovation synthesis of NiS quantum dots modified CdS/WO_3 heterostructures as high-efficiency bifunctional photocatalysts for construction of visible light driven Z-scheme water-splitting and Cr (VI) degradation. Appl Surf Sci 602:154226. https://doi.org/10.1016/J.APSUSC.2022.154226

69. Wang T, Liu P, Chen X, Guo Y, He C, Guo J, Liu W, Gao S, Lv Y, Liu K (2025) Enhancing aqueous zinc-ion energy storage performance with ion-mediating carbon quantum dots-modified separators regulating zinc deposition behavior. J Colloid Interface Sci 678:301–310. https://doi.org/10.1016/J.JCIS.2024.08.175
70. Wang G, Wei H, Shi J, Xu Y, Wu H, Luo Y, Li D, Meng Q (2017) Significantly enhanced energy conversion efficiency of $CuInS_2$ quantum dot sensitized solar cells by controlling surface defects. Nano Energy 35:17–25. https://doi.org/10.1016/J.NANOEN.2017.03.008
71. Wei Z, Yan J, Zhang Y, Fang W, Wenfeng S (2024) Photocatalytic H_2O splitting and CO_2 reduction with internal electric field modulation. Chem Eng J 483:149267. https://doi.org/10.1016/J.CEJ.2024.149267
72. Xia L, Tong X, Li X, Imran Channa A, You Y, Long Z, Vomiero A, Wang ZM (2022) Synergistic tailoring of band structure and charge carrier extraction in "green" core/shell quantum dots for highly efficient solar energy conversion. Chem Eng J 442:136214. https://doi.org/10.1016/J.CEJ.2022.136214
73. Xiao J, Momen R, Liu C (2021) Application of carbon quantum dots in supercapacitors: a mini review. Electrochem Commun 132:107143. https://doi.org/10.1016/J.ELECOM.2021.107143
74. Xu Q, Niu Y, Li J, Yang Z, Gao J, Ding L, Ni H, Zhu P, Liu Y, Tang Y, Lv ZP, Peng B, Hu TS, Zhou H, Xu C (2022) Recent progress of quantum dots for energy storage applications. Carbon Neutrality 1(1):1–18. https://doi.org/10.1007/S43979-022-00002-Y
75. Yang L, Zhu J, Xiao D (2012) Microemulsion-mediated hydrothermal synthesis of ZnSe and Fe-doped ZnSe quantum dots with different luminescence characteristics. RSC Adv 2(21):8179–8188. https://doi.org/10.1039/C2RA21401F
76. Ye H, Gao W, Xu L (2024) High-temperature energy storage capability of polyetherimide composite incorporated with perovskite quantum dots. Colloids Surf, A 687:133479. https://doi.org/10.1016/J.COLSURFA.2024.133479
77. Yu K, Schanze KS (2023) Commemorating the Nobel prize in chemistry 2023 for the discovery and synthesis of quantum dots. ACS Cent Sci 9(11):1989. https://doi.org/10.1021/ACSCENTSCI.3C01296
78. Yusuf M, Rosha P, Qureshi F, Ali FM, Ibrahim H (2025) Recent avenues in the photocatalytic splitting of water for eco-friendly hydrogen production. Sustain Mater Technol 43:e01332. https://doi.org/10.1016/J.SUSMAT.2025.E01332
79. Zaka A, Iqbal MW, Alqorashi AK, Almutairi BS, Alrobei H (2024) Iron-iridium metal organic framework/nitrogen doped MXene/graphene quantum dots: a leading composite for electrolytic energy storage and hydrogen evolution reaction. Diam Relat Mater 149:111653. https://doi.org/10.1016/J.DIAMOND.2024.111653
80. Zhao H, Rosei F (2017) Colloidal quantum dots for solar technologies. Chem 3(2):229–258. https://doi.org/10.1016/J.CHEMPR.2017.07.007

Chapter 14
Quantum Dots in Photovoltaic (PV) Technology

Honey Mittal and Manika Khanuja

Abstract Quantum dots have revolutionized the landscape of photovoltaic research owing to their size-tunable bandgap, strong light absorption, and potential for multiple exciton generation. As the demand for efficient, low-cost, and stable solar energy materials grows, quantum dot-based photovoltaic systems present a promising solution. The first report on the efficiency of quantum dot sensitized solar cells was 0.12%. As of today, the efficiency is reported to be 18.1%, and further, researchers are working to improve the efficiency. This chapter provides a comprehensive overview of the fundamental working principles of quantum dots in photovoltaics, focusing on their synthesis, surface engineering, and integration strategies. It investigates the structural engineering of quantum dots, including size, composition, and morphology, and their effect on the solar energy conversion. The working principles of quantum dot photovoltaics are discussed in detail, highlighting photon absorption, exciton generation, charge separation, and transport processes. The individual components of quantum solar cells are also discussed in detail, including electrodes, transport layers, and quantum dot absorber layers. Recent technological developments are examined to present the progress in device efficiency, design innovation, and functional integration. The chapter also examines the challenge of long-term stability, presenting strategies to mitigate degradation caused by environmental and photochemical factors. The chapter concludes by discussing emerging trends, including flexible and wearable devices, and the potential role of machine learning in optimizing quantum dot properties for photovoltaic applications. Overall, this work aims to serve as a bridge between nanomaterial science and energy research, guiding future innovations toward sustainable and efficient solar energy technologies.

Keywords Quantum dot solar cell · Wearable photovoltaics · AI in photovoltaics · Bandgap engineering · Multiple exciton generation · Surface ligand engineering · Encapsulation

---

H. Mittal
Department of Physics and Environmental Sciences, Sharda School of Engineering and Science, Sharda University, Greater Noida, India

M. Khanuja (✉)
Centre for Nanoscience and Nanotechnology, New Delhi, India
e-mail: manikakhanuja@gmail.com

T. Bhuyan et al. (eds.), *Multifunctional Quantum Dots*, Materials Horizons: From Nature to Nanomaterials, https://doi.org/10.1007/978-981-92-0172-3_14

1 Introduction

As the global demand for clean and renewable energy continues to rise, solar energy has emerged as one of the most promising and sustainable alternatives to fossil fuels [43]. However, the efficiency, cost-effectiveness, and scalability of traditional photovoltaic technologies remain key challenges to widespread adoption. The silicon-based solar cells are dominant in the commercial sector; however, they have limitations such as high manufacturing costs, rigidity, and low-power conversion efficiencies under practical conditions [47]. To overcome these constraints, researchers have turned to novel materials and nanoscale engineering approaches to enhance solar energy efficiency. In this field, the quantum dots have attracted significant attention as next-generation materials for photovoltaic applications [44].

The quantum dots are semiconductor nanocrystals with tunable optical and electronic properties owing to quantum confinement effects that arise when their size approaches the exciton Bohr radius [13, 32]. These nanoscale materials exhibit discrete energy levels, allowing for precise control over bandgap energies by simply adjusting their size, shape, or composition [26]. This tunability allows quantum dots to absorb a broader range of the solar spectrum compared to bulk materials, including the near-infrared region, which is particularly beneficial for increasing solar cell efficiency [57]. Additionally, quantum dots exhibit high absorption coefficients, long carrier lifetimes, and the potential for multiple exciton generation (MEG), where a single high-energy photon can generate more than one electron–hole pair, thereby further enhancing photovoltaic performance [29]. In bulk materials, excess photon energy is mostly lost as heat. MEG in quantum dots redirects this energy into usable charge carriers [6].

The integration of quantum dots into photovoltaic architectures presents several advantages, including low-cost, solution-based methods such as spin-coating, printing, or dip-coating, making them attractive for flexible and lightweight solar cells [9]. Quantum dots also enable the development of third-generation photovoltaic devices such as quantum dot-sensitized solar cells, quantum dot heterojunction solar cells, and tandem solar cells, offering paths to surpass the Shockley–Queisser limit associated with single-junction devices [11]. Furthermore, their compatibility with various substrates and potential for hot-carrier extraction provide exciting opportunities for innovative solar cell designs.

One of the most promising and distinctive features of quantum dots that makes them highly attractive for photovoltaic applications is multiple exciton generation (MEG) [5]. MEG is a quantum phenomenon wherein the absorption of a single high-energy photon can produce two or more electron–hole pairs (excitons) rather than just one, as occurs in conventional semiconductors. As shown in Fig. 1, a high-energy photon (hν) generates a primary electron–hole pair, and the excess energy facilitates the formation of a second e^-–h^+ pair via impact ionization. This unique capability has the potential to significantly surpass the Shockley–Queisser limit (~33%) of single-junction solar cells, thereby enhancing photovoltaic efficiency. Quantum dots exhibit three-dimensional quantum confinement, which modifies the density of states

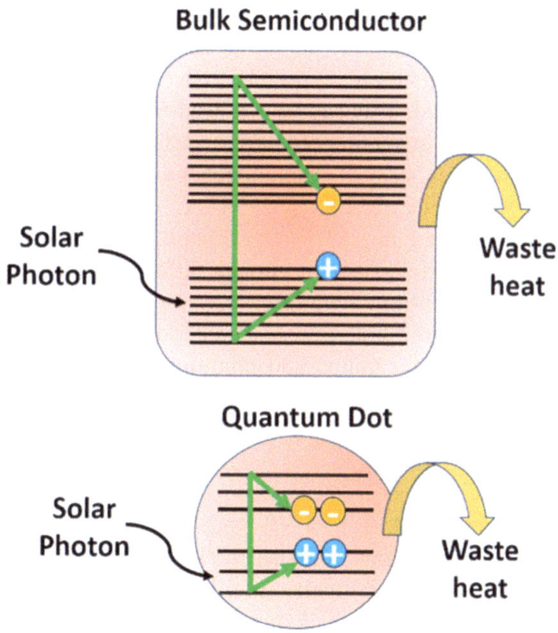

Fig. 1 Schematic illustration of multiple exciton generation (MEG) in a quantum dot

and enhances the Coulomb interaction between carriers. This enhanced interaction increases the probability of impact ionization, the key mechanism behind MEG. Unlike bulk semiconductors, where impact ionization is inefficient due to rapid carrier cooling, quantum dots slow down the cooling process due to phonon bottlenecks and discrete energy levels, thereby favoring MEG over thermalization [25].

2 Principle and Working of Quantum Dots

Quantum dots, owing to their unique optoelectronic properties, such as a size-tunable bandgap, strong light absorption, and multiple exciton generation, have emerged as promising materials for next-generation photovoltaic systems [7]. Their operation as photovoltaic materials is fundamentally governed by their ability to absorb solar photons, generate charge carriers (electrons and holes), and facilitate their separation and transport to external circuits for power extraction.

The operation of quantum dot-based photovoltaics begins with the absorption of incident solar photons. Due to quantum confinement, quantum dots exhibit size-dependent bandgaps, allowing the absorption spectrum to be tuned across the visible and near-infrared regions by controlling particle size, shape, and composition [33]. When a photon with energy equal to or greater than the bandgap is absorbed, an electron is excited from the valence band to the conduction band, leaving behind a hole, thus forming an exciton. In some cases, if the absorbed photon has energy significantly higher than the bandgap, multiple excitons may be generated via MEG, thereby improving the photon-to-charge conversion efficiency [46]. For the photogenerated exciton to contribute to the electric current, the electron–hole pair must be dissociated. This can be achieved through an energy offset at the quantum dot-electrode or quantum dot-matrix interface [44]. In typical quantum dot photovoltaic architectures, such as quantum dot-sensitized solar cells or quantum dot heterojunction solar cells, the quantum dots are interfaced with materials like TiO_2, ZnO, or conductive polymers, which facilitate exciton dissociation. Once separated, electrons and holes are transported through different pathways. The electrons are transferred to an electron-accepting layer (e.g., metal oxide or cathode), whereas, the holes are transferred to a hole-transport material (HTM) or directly to the anode. The efficient charge separation and minimized recombination are essential for achieving high photovoltaic performance. After charge separation and transport, electrons and holes reach their respective electrodes (typically transparent conductive oxides like ITO/FTO as the anode, and metal contacts as the cathode). The external circuit allows the flow of electrons, completing the electric circuit and enabling power delivery to an external load.

Figure 2 illustrates the structural and operational principles of a quantum dot solar cell. In the device architecture, light enters through a transparent glass substrate coated with ITO, which functions as the front electrode. Beneath this lies the quantum dot film, the active layer responsible for absorbing sunlight and generating charge carriers. A metal electrode forms the back contact, enabling charge extraction. Figure 2b depicts the energy band diagram of the system. Upon illumination, photons excite electrons in the quantum dots from the valence band to the conduction band, leaving behind holes. The energy difference between the quasi-Fermi levels of electrons and holes defines the open-circuit voltage (V_{OC}) of the solar cell. The ITO electrode collects the photogenerated electrons, while the metal electrode collects holes, creating a directional flow of current through an external circuit. This process effectively converts solar energy into electrical power, with the quantum dots offering tunable bandgaps and enhanced absorption properties that make them promising materials for next-generation photovoltaics.

3 Components of Quantum Dot Photovoltaic Systems

A quantum dot photovoltaic device consists of several critical components, each tailored to optimize light harvesting, charge separation, transport, and collection. The architecture and material selection significantly influence the device performance, stability, and fabrication cost. The essential components of a quantum dot photovoltaic system are:

i. **Quantum Dot Absorber Layer**: The quantum dots serve as the light-absorbing material responsible for harvesting photons and generating excitons (electron–hole pairs). Due to their size-tunable bandgap arising from quantum confinement

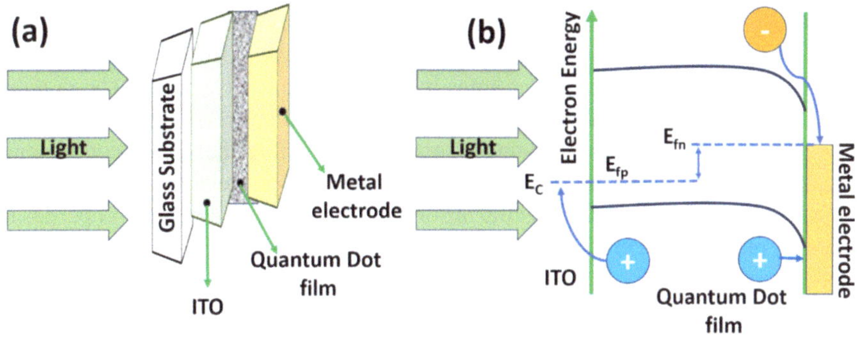

Fig. 2 **a** Schematic structure of a quantum dot solar cell consisting of a glass substrate, ITO electrode, quantum dot active film, and metal electrode. **b** Energy band diagram showing light-induced electron excitation and charge separations

effects, quantum dots can be engineered to absorb a broad range of the solar spectrum, including visible and near-infrared light.

ii. **Electron Transport Layer (ETL)**: The electron transport layer facilitates the selective extraction and transport of electrons from the quantum dot absorber to the cathode while blocking holes, thereby preventing recombination.

iii. **Hole-Transport Layer (HTL)**: The hole-transport layer extracts and transports holes from the quantum dot layer to the anode while blocking electrons. The HTL plays a crucial role in enhancing charge separation and minimizing recombination losses.

iv. **Substrate and Transparent Conductive Electrode**: The substrate provides mechanical support and serves as the bottom electrode that collects and transports one type of charge carrier (typically holes). It should be transparent to allow sunlight to reach the active layer.

v. **Back Electrode (Counter Electrode)**: The back electrode collects the electrons (or holes, depending on the architecture) and completes the external circuit. It must have high electrical conductivity and an appropriate work function to facilitate charge transfer.

vi. **Interfacial Layers and Passivation Layers**: The interfacial and passivation layers improve charge selectivity, reduce recombination, and enhance device stability. These layers also help in aligning energy levels between different materials and can act as buffer layers.

Figure 3 represents the schematic of a quantum dot solar cell, highlighting its layered structure and key functional components. At the base lies a transparent conducting electrode, typically made of materials like indium tin oxide (ITO), which allows light to pass through while serving as the front contact for collecting electrons. On top of this is a blocking layer that prevents charge recombination at the electrode interface. The core photoactive region consists of a quantum dot layer composed of semiconducting nanocrystals such as PbS, PbSe, CdS, or CdSe, which absorb sunlight and generate electron–hole pairs [42]. Above the quantum dots is a thin film of TiO_2, which functions as the electron transport layer, efficiently channeling electrons toward the external circuit. A hole transport layer is added on top that enables the movement of photogenerated holes toward the counter electrode. The counter electrode, typically a metal or conductive carbon-based layer, completes the circuit by collecting holes and maintaining charge balance. Altogether, these components work synergistically to harvest sunlight and convert it into electrical energy, with quantum dots offering broad spectral absorption and tunable electronic properties to enhance photovoltaic efficiency.

Fig. 3 Schematic representation of a quantum dot solar cell

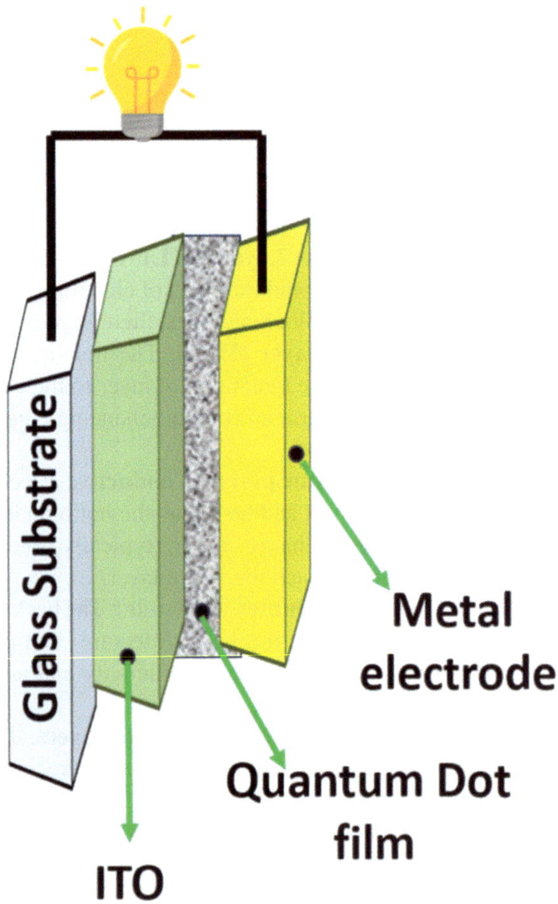

4 Performance Parameters

There are critical parameters of the photovoltaic system through which the working performance is analyzed, such as power conversion efficiency, fill factor, and open-circuit voltage. Understanding these critical performance metrics in quantum dot-based solar cells is essential for evaluating their efficiency, diagnosing losses, and designing strategies for improvement. Power conversion efficiency represents the fraction of incident solar energy that is converted into usable electrical energy. This value determines how effectively the solar cell converts sunlight into electricity. It directly reflects how well the quantum dot and associated device layers manage light absorption, exciton dissociation, charge transport, and collection. It is the most comprehensive performance metric and combines all other parameters:

$$\text{PCE} = \frac{J_{\text{SC}} \cdot V_{\text{OC}} \cdot \text{FF}}{P_{\text{in}}},$$

where PCE, J_{SC}, V_{OC}, FF, and P_{in} represent the power conversion efficiency, short-circuit current density (mA/cm^2), open-circuit voltage (V), fill factor, and incident solar power (typically 100 mW/cm^2 under AM 1.5G).

FF is a measure of the electrical quality of the solar cell and indicates how efficiently it operates near its maximum power point. High FF values approach the ideal rectangular I–V curve, showing minimal resistive and recombination losses. Low FF reduces the overall PCE even if J_{SC} and V_{OC} are relatively high. Improving FF typically requires optimizing transport layers, minimizing recombination, and controlling film morphology. FF is the ratio of the maximum obtainable power to the product of J_{SC} and V_{OC}. It reflects how ideal the I–V curve is:

$$\text{FF} = \frac{P_{\text{max}}}{J_{\text{SC}} V_{\text{OC}}}.$$

The maximum theoretical efficiency for conventional single-junction solar cells is ~33% for a semiconductor with a bandgap of 1.34 eV under AM1.5G illumination. This limit assumes one exciton per photon and perfect collection efficiency.

Quantum dots offer the possibility to exceed the Shockley–Queisser limit through MEG, where one high-energy photon produces multiple excitons. Theoretically, efficiencies of up to 42–44% under concentrated sunlight can be achieved using ideal MEG conditions [44]. For example, Ellingson et al. reported MEG in colloidal PbSe quantum dots, demonstrating quantum yields exceeding unity for photon energies $>3E_g$ [12]. Beard et al. demonstrated MEG efficiencies exceeding 100% in PbSe quantum dots, laying the foundation for potential increases in J_{SC} [52]. The graph, as shown in Fig. 4, illustrates key operational parameters, including short-circuit current (I_{SC}), open-circuit voltage (V_{OC}), and the maximum power point defined by (I_{MP}, V_{MP}). The fill factor (FF) is represented as the ratio of the green area (actual maximum power output) to the blue area (theoretical maximum power). This relationship determines the overall power conversion efficiency of the device.

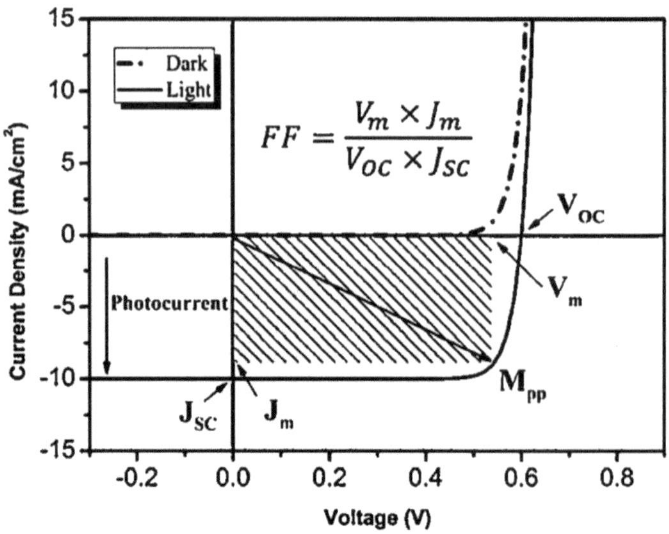

Fig. 4 Current–voltage (I–V) characteristics of a quantum dot solar cell [22]

5 Efficiency Enhancement Strategies

The researchers are focusing on structural engineering to tailor the physicochemical properties of quantum dots to enhance their photovoltaic performance. The optical absorption, charge carrier dynamics, bandgap energy, and interfacial charge transfer characteristics of quantum dots depend on the parameters such as size, shape, composition, and surface chemistry. These engineered properties have been strategically harnessed to improve light harvesting, reduce charge recombination, and enhance overall power conversion efficiency (PCE) in quantum dot-based solar cells.

i. **Bandgap Tunability**: The quantum dots have size-dependent bandgap due to the quantum confinement effect. The smaller quantum dots exhibit wider bandgaps, while larger quantum dots have smaller band gap that absorbs in the infrared region. For example, Bo Hou et al. demonstrated that tuning the size of PbS quantum dots allowed the control of their bandgap from 0.8 to 1.37 eV, optimizing them for solar cells where multiple quantum dot layers absorb distinct parts of the solar spectrum, thereby enhancing overall photovoltaic efficiency [15]. Along with size, the optical absorption also depends on the shape of the quantum dots, such as spherical, rod-like, or tetrapod, which influence the charge dynamics. The rod-shaped quantum dots provide directionality to charge separation and therefore have low electron–hole recombination, which results in high photovoltaic efficiency. For example, Huynh et al. have reported that the CdSe nanorods integrated into bulk heterojunction solar cells exhibited higher open-circuit voltage and improved charge transport compared to spherical CdSe quantum dots [18]. As shown in the Fig. 5, reducing the thickness of the quantum layer results in high power conversion efficiency, which is attributed to the reduction in bandgap as the doping concentration of the quantum dot film increases. The bandgap of the quantum dot can be varied by altering the size of the quantum dot. As shown in Fig. 5, upon increasing the size of the bandgap from 916 to 1536 nm, the absorption wavelength increased from 900 to 1500 nm, which implies a decrease in bandgap. The photo conversion efficiency also increases with an increase in the size of the quantum dots [39].

ii. **Core-Shell Structures and Heterostructuring**: Engineering quantum dots into core–shell heterostructures is a widely adopted strategy to passivate surface traps and enhance carrier separation. For instance, Talapin et al. show that CdSe/ZnS core–shell quantum dots exhibit improved photostability and quantum yield compared to bare CdSe. The improved efficiency is attributed to the reduced surface recombination losses in quantum dot-sensitized solar cells [48]. Similarly, N. J. Simi et al. have reported that $CdSe/Cu_2Se$ core–shell quantum dots have been employed to create favorable band alignments that facilitate electron injection into the electron transport layer and suppress recombination, leading to a reported efficiency improvement in quantum dot solar cells [45]. Figure 6 shows the HRTEM image of the ZnS/CdS hybrid quantum dot. The enhancement

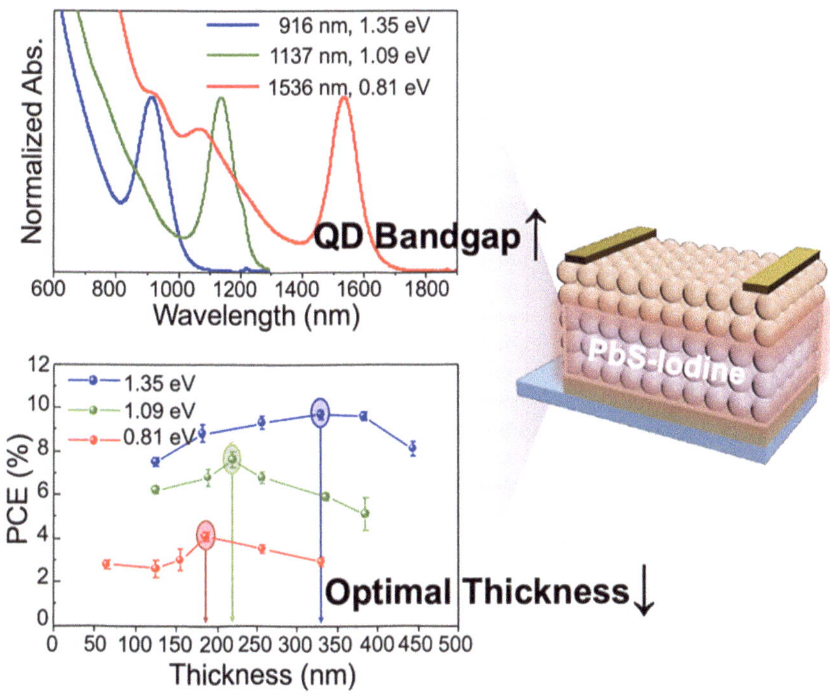

Fig. 5 Variation of bandgap with quantum dot size and effect of thickness of quantum dot on the power conversion efficiency [30]. (Reproduced with permission from C. Lim, Materials Today Energy; published by Elsevier, 2023)

of the efficiency of hybrid quantum dots is explained due to the carbon-centered radicals as the key active intermediates.

iii. **Doping and Alloying**: Incorporating heteroatoms via doping or alloying can significantly influence the electronic properties of quantum dots. For example, Ag-doped CdS quantum dots demonstrated enhanced charge carrier mobility and reduced recombination due to the introduction of shallow trap states and improved crystallinity [51]. Alloyed quantum dots such as $CdSe_xTe_{1-x}$ allow for precise tuning of the conduction and valence band positions, facilitating better band alignment with surrounding layers in the device structure [20]. The comparison between an undoped quantum dots with a Cu-doped quantum dot is shown in Fig. 7, highlighting the changes in electronic transitions resulting from doping. The transition in the undoped quantum dots (Fig. 7a) shows the relaxation of an excited electron from a higher energy state ($1P_e$) to a lower state ($1S_e$), followed by recombination with a hole. In the pristine quantum dot, relaxation occurs via intrinsic energy states without the involvement of intermediate levels. However, a new dopant-induced intragap state (due to Cu doping) enables an additional recombination pathway as shown in Fig. 7b. In this mechanism, an excited electron can recombine with a localized hole at the Cu site, represented

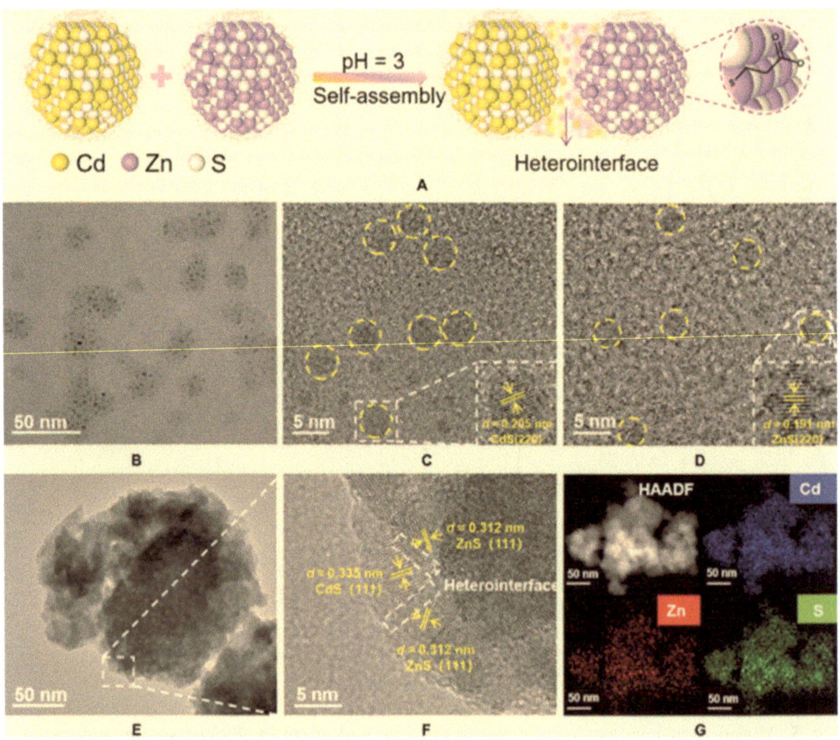

Fig. 6 HRTEM image of ZnS/CdS hybrid quantum dots [61]

as Cu*. This pathway results in modified emission properties, such as longer lifetimes or red-shifted photoluminescence, and potentially enhanced charge separation efficiency.

iv. **Surface Ligand Engineering**: The surface chemistry of quantum dots, governed largely by organic ligands, critically affects charge transport and energy alignment at interfaces. Long-chain insulating ligands typically used in synthesis hinder charge mobility, necessitating post-synthetic ligand exchange with shorter, more conductive ligands. Jingjing Xue et al. reported that exchanging ligands on quantum dots significantly improved the inter-dot electronic coupling, leading to enhanced carrier transport and a marked increase in PCE [56]. In colloidal quantum dots, organic ligands are initially used during synthesis to stabilize the nanocrystals and control their size and shape. However, these long-chain insulating ligands hinder charge transport in solid-state films, leading to poor device performance [19]. To overcome this, surface ligand exchange is

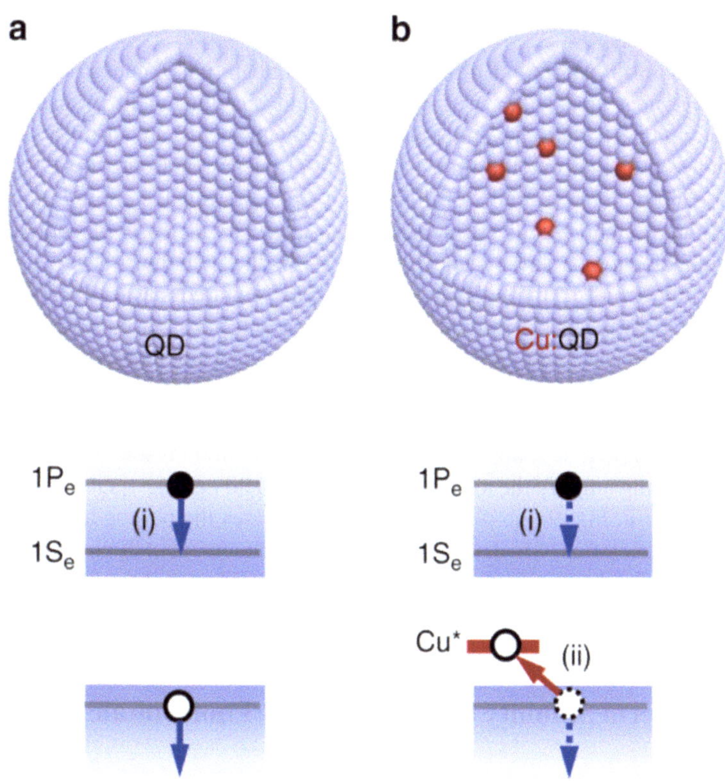

Fig. 7 Schematic representation of **a** undoped quantum dot and **b** Cu-doped quantum dot, illustrating their respective energy relaxation pathways [52]. (Reproduced with permission from L. Wang, Nature Communications; published by Springer Nature, 2019)

performed using shorter or more conductive molecules that can improve electronic coupling between adjacent quantum dots, reduce interparticle spacing, and enhance charge mobility. Moreover, the nature and functionality of the ligands play a crucial role in passivating surface trap states, modulating energy band alignment, and even inducing controlled doping (n-type or p-type) in quantum dot layers. By carefully selecting ligands with desired electronic and chemical properties, it is possible to tailor the optoelectronic behavior of quantum dots, suppress recombination losses, and create efficient p-n junctions, ultimately resulting in significantly improved solar cell efficiencies [55]. Figure 8 illustrates how ligand engineering, especially through custom surface modification (CSM), enhances quantum dot performance over conventional ligand exchange methods. The CSM approach results in a more ordered ligand shell, better surface coverage, and incorporation of functional atoms, significantly improving optoelectronic properties. Energy alignment is crucial in designing heterojunctions for solar cells. The graph shows how different ligands shift the valence band maximum (solid line) and conduction band minimum (dashed line) of carbon quantum dots.

v. **Photon and Plasmonic Enhancement**: The integration of nanostructured light management layers improves absorption and quantum yield in quantum dot photovoltaics. Photonic and plasmonic enhancements are critical strategies employed to improve the light-harvesting efficiency and overall performance of quantum dot solar cells [60]. Photonic strategies revolve around the manipulation of light within the solar cell structure to increase the optical path length of photons and enhance absorption in the quantum dot layer. This is typically achieved through photonic crystals, textured surfaces, or dielectric interference layers. By engineering light trapping through constructive interference and waveguiding, photonic structures can ensure that more photons are absorbed by the quantum dot layer, thereby generating more excitons [36]. Plasmonic enhancement involves integrating metallic nanoparticles into the solar cell. These particles support localized surface plasmon resonances, which can dramatically amplify the local electromagnetic field near the quantum dots [14]. As shown in Fig. 9, the plasmonic effect increases the photon absorption and exciton generation, which is visible through an increment in fluorescence. Additionally, plasmonic nanoparticles can scatter incident light, increase its path length, and further enhance light absorption.

6 Long-Term Stability of Quantum Dot Photovoltaic System

One of the major barriers to the commercialization of quantum dot photovoltaic technology is the challenge of long-term operational stability. While quantum dots offer tunable optical properties, solution-processability, and potential for low-cost manufacturing, their intrinsic and extrinsic stability under real-world conditions

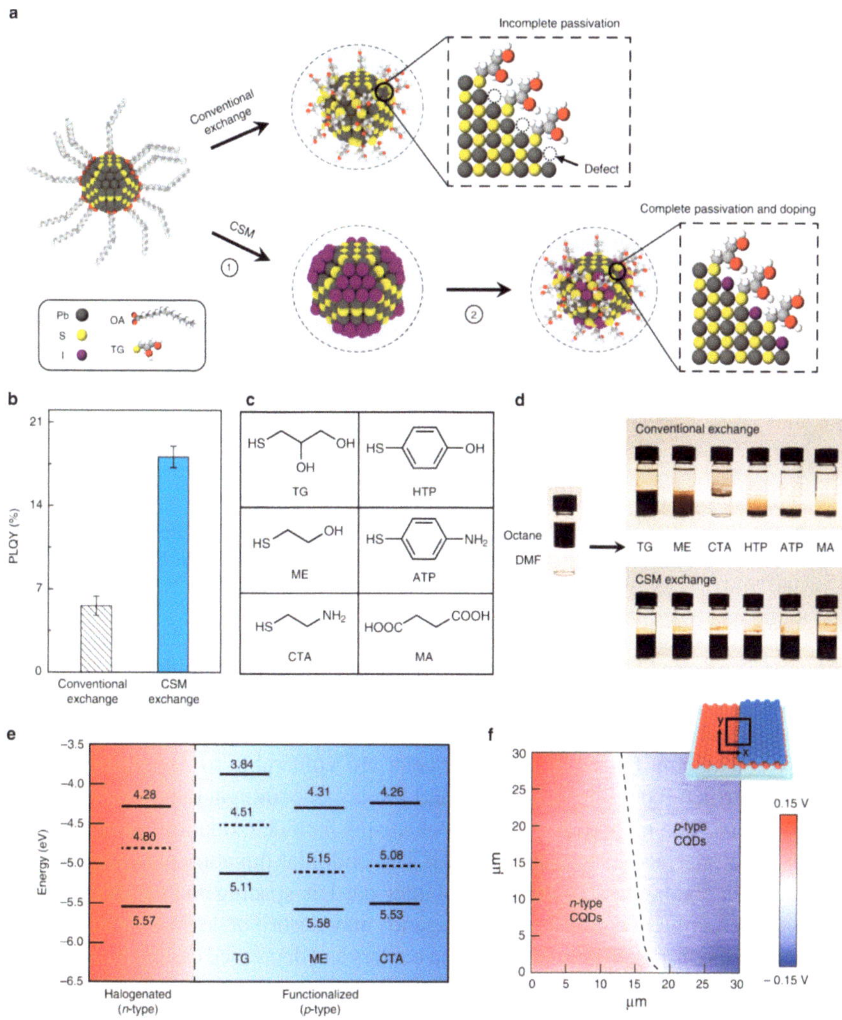

Fig. 8 Surface ligand engineering in colloidal quantum dots. **a** Comparison between conventional ligand exchange and custom surface modification. **b** PLQY improvement with CSM. **c** Structures of ligands used for tuning QD surface properties. **d** Solubility and dispersion differences after ligand exchange. **e** Energy level tuning by different ligand classes (n-type and p-type). **f** Built-in potential profile confirming p–n junction formation using engineered carbon quantum dot layers [8]. (Reproduced with permission from M. J. Choi, Nature Communications; published by Springer Nature, 2020)

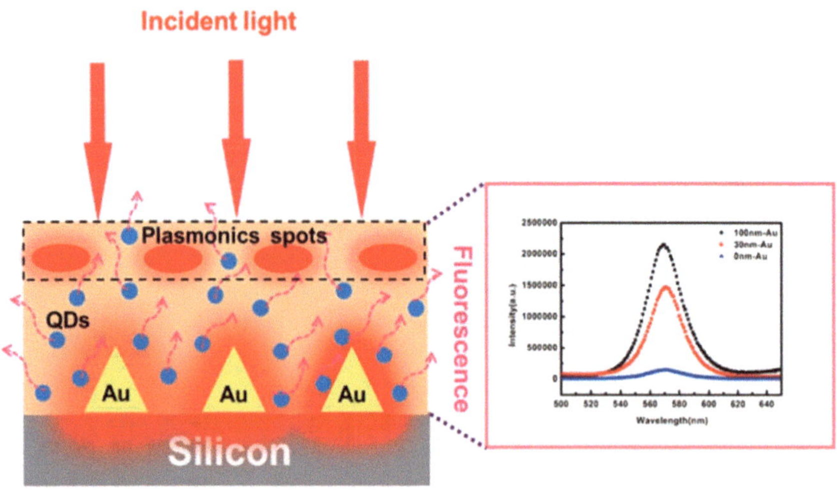

Fig. 9 Schematic illustration of the plasmonic effect in quantum dots and fluorescence plot [54]. (Reproduced with permission from Y. Wu, Solar Energy Materials and Solar Cells; published by Elsevier, 2014)

remains a concern [1]. Device degradation can occur due to multiple factors such as moisture, oxygen, light, heat, and chemical instability of surface ligands. Ensuring long-term stability is crucial for achieving reliable and economically viable solar energy conversion. Quantum dot solar cells are vulnerable to degradation due to their nanostructured, colloidal nature, and the organic–inorganic interfaces present in the device. The main degradation mechanisms include oxidation, thermal degradation, and interface instability [10, 53]. The colloidal quantum dots are typically capped with organic ligands that control inter-particle spacing and passivate surface traps. Under ambient conditions, these ligands may desorb or degrade, leading to the formation of non-radiative recombination centers and decreased carrier mobility. The quantum dots, such as PbS, CdSe, and $CsPbBr_3$, can undergo photooxidation under continuous light exposure, especially in the presence of oxygen and humidity [24, 31]. This process degrades the quantum dot crystal structure and reduces absorption and photoconductivity.

High temperatures accelerate ligand loss, increase trap density, and can lead to interfacial delamination or ion migration, particularly in perovskite-based quantum dot hybrids. Quantum dots, particularly perovskite quantum dots, are prone to decomposition and phase transitions when subjected to elevated temperatures [53]. This degradation affects the structural integrity, optoelectronic properties, and overall device performance. Figure 10 illustrates a comparative study of the thermal stability of perovskite quantum dots and their bulk film counterparts using in situ temperature-dependent X-ray diffraction (XRD) analysis. The graph highlights the enhanced thermal resilience of quantum dots over bulk films. The degradation at the interfaces between quantum dots and charge transport layers (ETLs/HTLs) can hinder charge extraction, increase recombination, and lower fill factor (FF) and open-circuit voltage (V_{OC}) over time.

Multiple approaches have been employed to improve the environmental and operational stability of quantum dot photovoltaic devices [10]. The replacement of long-chain organic ligands with short-chain, inorganic, or cross-linkable ligands enhances film conductivity, structural robustness, and chemical resistance [58, 63]. For example, Saif M. H. Qaid et al. reported improved stability after adding the polymer coating [37]. The encapsulation of quantum dots within wider-bandgap shells provides a protective barrier against moisture, oxidation, and UV damage while passivating surface traps [49]. Applying barrier coatings such as ALD-deposited Al_2O_3, glass, or polymer-based encapsulants can significantly reduce permeability to water and oxygen, thus extending device life [41]. Tzu Yi Lee et al. reported that quantum dot photovoltaics using atomic layer deposition (ALD) of Al_2O_3 as

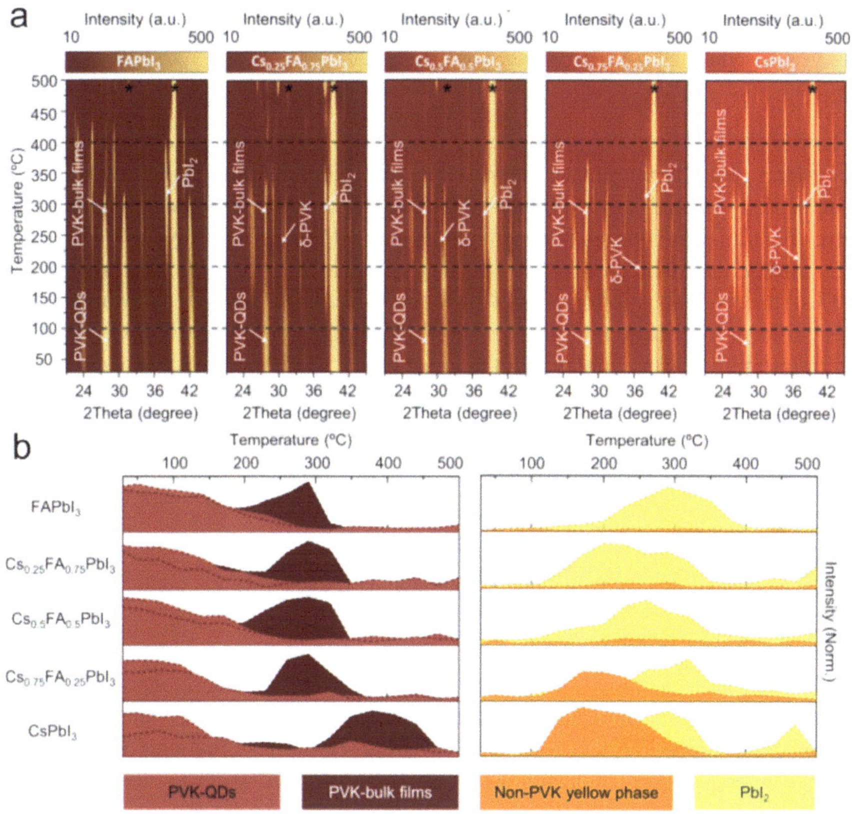

Fig. 10 **a** Temperature-dependent XRD patterns showing phase evolution in different A-site cation compositions (FAPbI3–CsPbI3), **b** phase intensity distribution with temperature for the α-phase (left) and degradation products (right) [40]. (Reproduced with permission from J. E. Rodigues, Journal of Materials Chemistry C; published by RSC, 2023)

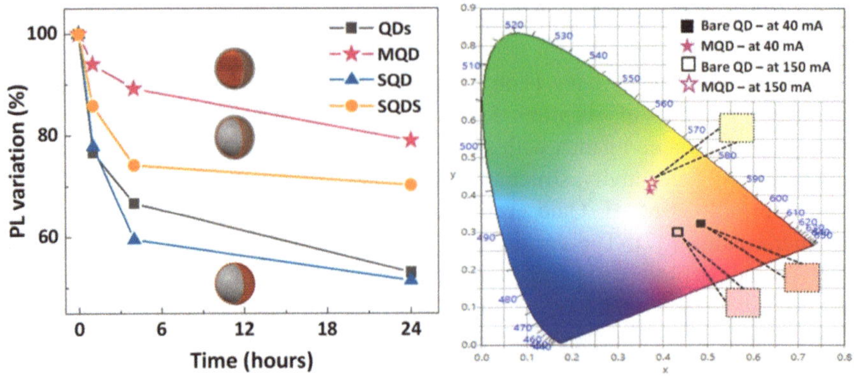

Fig. 11 Role of encapsulation in enhancing QD stability. (Left) Photoluminescence (PL) variation over 24 h for bare QDs, shell QDs (SQD), dual-protected shell QDs (SQDS), and multi-encapsulated QDs (MQDs). (Right) CIE color coordinates showing reduced emission color drift for MQDs compared to bare QDs at different current biases [35]. (Reproduced with permission from H. T. Nguyen, Materials Today Nano; published by Elsevier, 2024)

a barrier showed minimal performance loss [28]. Encapsulation not only improves long-term photoluminescence stability but also helps maintain color purity under electrical stress, both critical for reliable device operation. As shown in Fig. 11, the bare quantum dots undergo fast degradation with PL dropping below 70% after 24 h. The multi-encapsulated quantum dot shows the least PL variation, retaining nearly 90% of the original emission after 24 h. The shift in color coordinates (toward yellow) is significant for bare quantum dots, especially at higher current (150 mA), indicating color instability due to quantum dot degradation. The substitution of organic ligands and layers with inorganic materials, such as metal oxides and halide ligands, improves thermal and chemical stability [21]. All-inorganic quantum dot photovoltaics often show superior performance under elevated temperatures and long-term testing [17].

7 Developments in Quantum Dot Photovoltaic

Over the past two decades, quantum dot photovoltaics have evolved from laboratory-scale concepts to a promising class of next-generation solar technologies. This progress is driven by innovations in quantum dot synthesis, surface chemistry, device architecture, and photophysical understanding. These developments have significantly enhanced power conversion efficiency, broadened spectral absorption, and improved long-term stability. The evolution of quantum dot photovoltaic device configurations has contributed significantly to performance gains. Early quantum dot photovoltaic cells used simple metal–semiconductor contacts. Nowadays, the photovoltaic designs use structured heterojunctions such as Schottky junctions and p-n heterojunctions [50]. The quantum dots act as absorbers with metal oxides and improve the performance parameters of the photovoltaic system [4]. Additionally, the combination of quantum dots with conducting polymers or perovskites enhances carrier separation and light harvesting [4, 34]. The techniques, such as graded Bandgap and Tandem Architectures, have also surpassed the Shockley–Queisser limit [2]. Layering quantum dots of varying sizes allows harvesting across the solar spectrum. Low-temperature fabrication enables roll-to-roll processing, opening the path to wearable and lightweight PV applications.

One of the most exciting frontiers in the development of quantum dot photovoltaics is their integration into flexible and wearable electronics [62]. Traditional rigid solar panels are unsuitable for curved or portable surfaces, limiting their applicability in emerging technologies and self-powered wearable devices. Quantum dots offer a compelling route toward lightweight, low-cost, and deformable energy harvesting systems. Recent studies have demonstrated significant progress in this domain. For instance, Long Hu et al. reported highly flexible $CsPbI_3$ quantum dot solar cells with a certified power conversion efficiency of 15.1%, fabricated on PET substrates with a hybrid organic–inorganic architecture [16]. Similarly, Caofeng Pan et al. developed wearable quantum dot solar cells based on CdS/CdSe quantum dots [38]. These devices exhibited excellent operational stability under repeated flexing and ambient exposure. By enabling conformal energy harvesting systems that adapt to the human body or mobile surfaces, quantum dot-based flexible photovoltaics are paving the way for next-generation personal and ubiquitous energy solutions. Figure 12 illustrates the integration of quantum dot-based optoelectronic components into flexible and wearable devices for real-time photoplethysmography and health monitoring, emphasizing their potential extension into wearable photovoltaic platforms.

Artificial intelligence (AI) and machine learning (ML) are revolutionizing the development of quantum dot photovoltaics by enabling rapid screening of materials, predictive modeling, and optimization of fabrication processes. Traditional trial-and-error methods are often slow and resource-intensive, especially given the vast compositional and processing parameter space associated with quantum dot solar cells. AI-driven approaches, in contrast, allow researchers to efficiently navigate this complexity by identifying key correlations between quantum dot composition, device architecture, and performance metrics. Machine learning algorithms, such as neural networks and decision trees, are trained using experimental and simulated datasets to predict material properties and device outcomes. As shown in Fig. 13, these models can then guide the design of new quantum dots with tailored bandgaps, improved stability, and optimal charge transport characteristics [27]. For instance, neural networks have been used to predict the optimal ligand combinations for enhancing QD surface passivation and charge extraction, while decision tree models have assisted in optimizing annealing conditions and layer thicknesses to maximize efficiency.

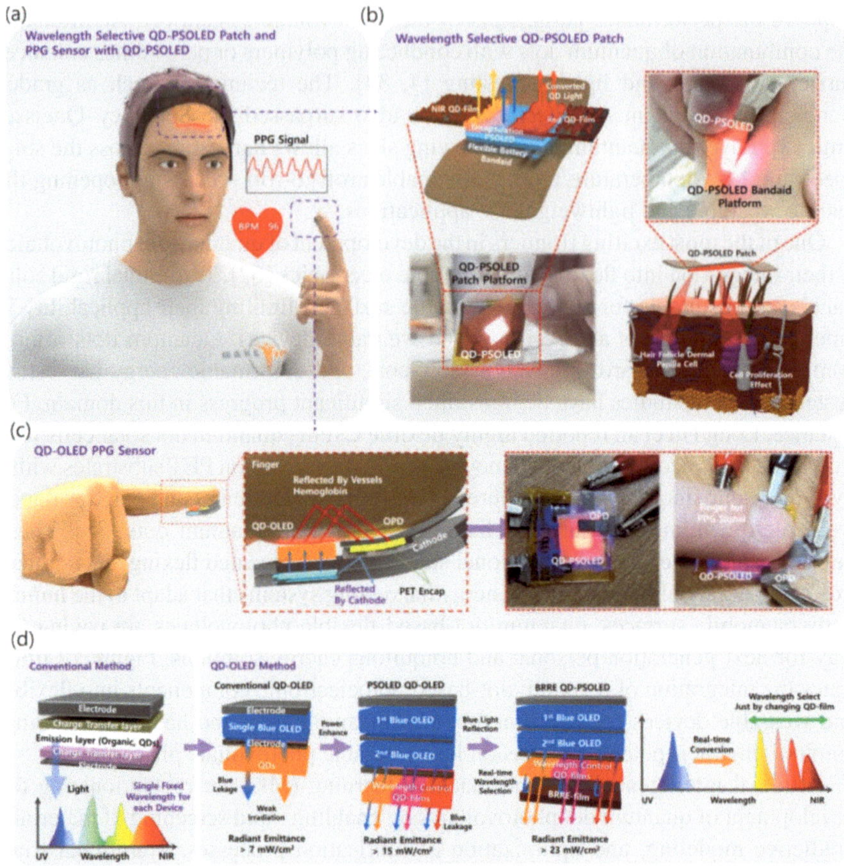

Fig. 12 Schematic overview of flexible and wearable quantum dot-based photonic systems for health monitoring applications: **a** Wearable QD-PSOLED patch and sensor, **b** structure and function of the QD-PSOLED patch, **c** QD-OLED-based PPG sensor, and **d** comparison of conventional versus QD-based light emission strategies [23]. Reproduced with permission from Y. W. Kim, Chemical Engineering Journal; published by Elsevier, 2024

8 Future Challenges and Perspectives

Despite the remarkable progress in quantum dot photovoltaic technologies over the past two decades, the pathway to their commercial realization remains hindered with several critical challenges. While quantum dots offer unique advantages such as bandgap tunability, solution-processability, and potential for multiple exciton generation, these strengths have yet to be fully translated into practical, market-ready solar devices. Overcoming both intrinsic material limitations and extrinsic device-level barriers will be key to unlocking their true potential.

One of the most pressing challenges is the long-term operational stability of quantum dots. Most quantum dots, especially those based on lead chalcogenides or

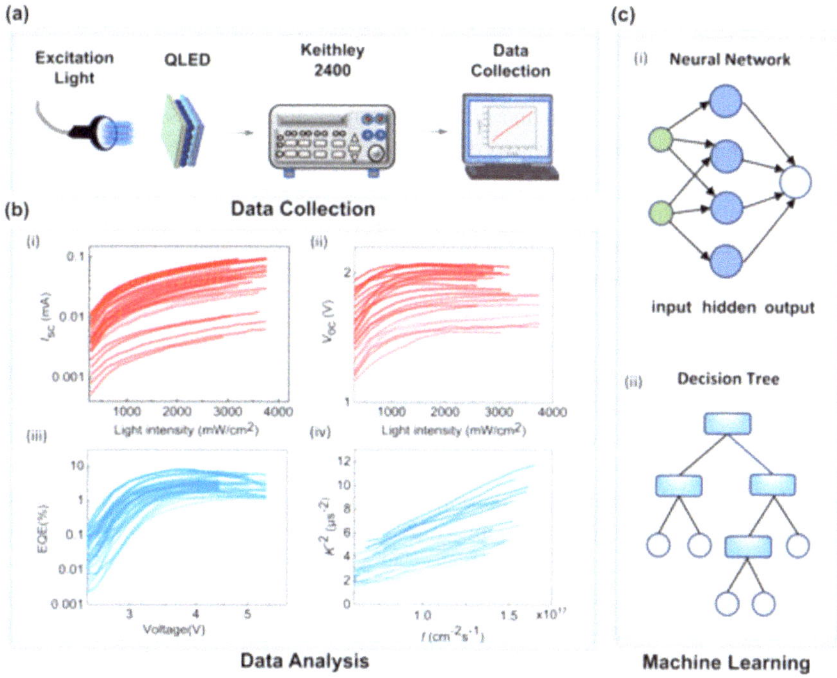

Fig. 13 **a** Experimental setup for data collection, **b** examples of performance metrics collected under varying light intensities, **c** application of machine learning models (i) neural networks and (ii) decision trees [59]. (Reproduced with permission from M. Yang, ACS Photonics; published by ACS, 2024)

perovskite materials, are highly sensitive to environmental factors such as moisture, oxygen, UV light, and temperature fluctuations. Degradation pathways ranging from ligand desorption and surface oxidation to ion migration and interfacial delamination significantly shorten the lifetime of quantum dot solar cells. While surface passivation strategies and encapsulation layers have shown promise in laboratory-scale studies, real-world durability under outdoor conditions (over 20–25 years) remains largely unproven. Developing robust, self-healing, and scalable encapsulation strategies will be critical to address these stability concerns.

Another major bottleneck lies in the toxicity and environmental impact of commonly used quantum dots, such as those based on cadmium or lead [3]. Although these materials offer excellent optoelectronic properties, their use raises significant environmental and regulatory concerns. As such, there is a growing push toward non-toxic, earth-abundant alternatives such as $CuInS_2$, ZnSe, or $AgBiS_2$. However, these "green" quantum dots often suffer from lower quantum yields, poor charge mobility, and synthetic complexity, which limit their photovoltaic performance. Future research must therefore focus on optimizing these non-toxic systems to compete with or surpass the efficiency and stability of their heavy-metal counterparts.

Furthermore, device architecture and charge transport optimization continue to present significant challenges. While ligand exchange techniques and hybrid heterojunction designs have improved charge carrier extraction and mobility, issues such as trap-assisted recombination and inefficient carrier diffusion persist. More refined control over interfacial energetics, defect passivation, and graded bandgap engineering will be needed to reduce energy losses and enhance device efficiency. From a theoretical standpoint, there remains untapped potential in leveraging quantum mechanical effects unique to quantum dots, particularly multiple exciton generation, hot-carrier extraction, and carrier multiplication. While these phenomena have been observed under specific conditions, integrating them into working devices with meaningful efficiency gains has proven elusive. Advances in quantum dot synthesis, ultrafast spectroscopy, and nanostructured device integration will be essential to harness these effects reliably. Looking forward, the integration of artificial intelligence (AI) and machine learning (ML) into materials discovery, process optimization, and device simulation offers an exciting avenue for accelerating quantum dot photovoltaic research. Data-driven approaches can help identify new quantum dot compositions, predict long-term stability, and fine-tune processing parameters with unprecedented speed and accuracy. Thus, AI-ML and simulation frameworks are anticipated to play a crucial role in shaping the future of quantum dots for photovoltaic applications and accelerating their transition from laboratory experiments to industrial applications. These approaches are expected to facilitate the development of environmentally benign quantum dots, scalable fabrication strategies, and intelligent photovoltaic systems for next-generation solar energy technologies.

References

1. Albaladejo-Siguan M, Baird EC, Becker-Koch D, Li Y, Rogach AL, Vaynzof Y (2021) Stability of quantum dot solar cells: a matter of (life)time. Adv Energy Mater 11(12):2003457. https://doi.org/10.1002/aenm.202003457
2. Almomani MS, Ahmed NM, Rashid M, Ibnaouf KH, Aldaghri OA, Madkhali N, Cabrera H (2022) Performance improvement of graded bandgap solar cell via optimization of energy levels alignment in Si quantum dot, TiO_2 nanoparticles, and Porous Si. Photonics 9(11):843. https://doi.org/10.3390/photonics9110843
3. Ashraf W, Sharma D, Khan A, Khanuja M (2025) Green carbon dots as adsorbents for removal of toxic chemicals. In: Green carbon dots: sustainable analytical approaches, vol 1506, pp 141–160. https://doi.org/10.1021/bk-2025-1506.ch006
4. Ballabio M, Cánovas E (2022) Electron transfer at quantum dot-metal oxide interfaces for solar energy conversion. ACS Nanosci Au 2(5):367–395. https://doi.org/10.1021/acsnanoscienceau.2c00015
5. Beard MC (2011) Multiple exciton generation in semiconductor quantum dots. J Phys Chem Lett 2(11):1282–1288. https://doi.org/10.1021/jz200166y
6. Beard MC, Midgett AG, Hanna MC, Luther JM, Hughes BK, Nozik AJ (2010) Comparing multiple exciton generation in quantum dots to impact ionization in bulk semiconductors: implications for enhancement of solar energy conversion. Nano Lett 10(8):3019–3027. https://doi.org/10.1021/nl101490z

7. Chistyakov AA, Zvaigzne MA, Nikitenko VR, Tameev AR, Martynov IL, Prezhdo OV (2017) Optoelectronic properties of semiconductor quantum dot solids for photovoltaic applications. J Phys Chem Lett 8(17):4129–4139. https://doi.org/10.1021/acs.jpclett.7b00671
8. Choi MJ, García de Arquer FP, Proppe AH, Seifitokaldani A, Choi J, Kim J, Baek SW et al (2020) Cascade surface modification of colloidal quantum dot inks enables efficient bulk homojunction photovoltaics. Nat Commun 11(1):103. https://doi.org/10.1038/s41467-019-13437-2
9. Choi MK, Yang J, Hyeon T, Kim DH (2018) Flexible quantum dot light-emitting diodes for next-generation displays. NPJ Flexible Electron 2(1):1–14. https://doi.org/10.1038/s41528-018-0023-3
10. Chuang C-H, Brown PR, Bulović V, Bawendi MG (2014) Improved performance and stability in quantum dot solar cells through band alignment engineering. Nat Mater 13(8):796–801. https://doi.org/10.1038/nmat3984.Improved
11. Ehrler B, Alarcón-Lladó E, Tabernig SW, Veeken T, Garnett EC, Polman A (2020) Photovoltaics reaching for the shockley-queisser limit. ACS Energy Lett 5(9):3029–3033. https://doi.org/10.1021/acsenergylett.0c01790
12. Ellingson RJ, Beard MC, Johnson JC, Pingrong Y, Micic OI, Nozik AJ, Shabaev A, Efros AL (2005) Highly efficient multiple exciton generation in colloidal PbSe and PbS quantum dots. Nano Lett 5(5):865–871. https://doi.org/10.1021/nl0502672
13. Fatima T, Khanuja M (2025) Novel Ternary Z scheme carbon quantum dots (CQDs) decorated WS_2/PANI ((CQDs@WS_2/PANI): 0D:2D:1D) nanocomposite for the photocatalytic degradation and electrochemical detection of pharmaceutical drugs. Nano Mater Sci 7:259–275. https://doi.org/10.1016/j.nanoms.2024.04.005
14. Guo R, Derom S, Väkeväinen AI, van Dijk-Moes RJA, Liljeroth P, Vanmaekelbergh D, Törmä P (2015) Controlling quantum dot emission by plasmonic nanoarrays. Opt Express 23(22):28206. https://doi.org/10.1364/oe.23.028206
15. Hou B, Cho Y, Kim BS, Hong J, Park JB, Ahn SJ, Sohn JI, Cha S, Kim JM (2016) Highly monodispersed PbS quantum dots for outstanding cascaded-junction solar cells. ACS Energy Lett 1(4):834–839. https://doi.org/10.1021/acsenergylett.6b00294
16. Hu L, Zhao Q, Huang S, Zheng J, Guan X, Patterson R, Kim J et al (2021) Flexible and efficient perovskite quantum dot solar cells via hybrid interfacial architecture. Nat Commun 12(1):466. https://doi.org/10.1038/s41467-020-20749-1
17. Huang CY, Li H, Ye W, Lin CH, Guan X, Long H, Kim J, Zhu X, Zeng H, Tom W (2023) Inorganic halide perovskite quantum dots: a versatile nanomaterial platform for electronic applications. Nano-Micro Lett 15:16. https://doi.org/10.1007/s40820-022-00983-6
18. Huynh WU, Dittmer JJ, Paul Alivisatos A (2002) Hybrid nanorod-polymer solar cells. Science 295(5564):2425–2427. https://doi.org/10.1126/science.1069156
19. Jia D, Chen J, Qiu J, Ma H, Mei Y, Liu J, Zhang X (2022) Tailoring solvent-mediated ligand exchange for $CsPbI_3$ perovskite quantum dot solar cells with efficiency exceeding 16.5%. Joule 6(7):1632–1653. https://doi.org/10.1016/j.joule.2022.05.007
20. Kamat PV (2013) Quantum dot solar cells. the next big thing in photovoltaics. J Phys Chem Lett 4(6):908–918. https://doi.org/10.1021/jz400052e
21. Khan J, Ullah I, Yuan J (2022) $CsPbI_3$ perovskite quantum dot solar cells: opportunities, progress and challenges. Mater Adv 3(4):1931–1952. https://doi.org/10.1039/d1ma01075a
22. Kim S, Hoang VQ, Bark CW (2021) Silicon-based technologies for flexible photovoltaic (Pv) devices: from basic mechanism to manufacturing technologies. Nanomaterials 11(11):2944. https://doi.org/10.3390/nano11112944
23. Kim YW, Kwon JH, Choi HR, Choi JG, Kwon OK, Kim KH, Cho EH et al (2024) Wearable quantum dots organic light-emitting diodes patch for high-power near infra-red photomedicene with real-time wavelength control. Chem Eng J 499(July):156121. https://doi.org/10.1016/j.cej.2024.156121
24. Kolobkova E, Nikonorov N (2017) CdS(Se) and PbS(Se) quantum dots with high room temperature quantum efficiency in the fluorine-phosphate glasses. In: Quantum-dot based light-emitting diodes. Intech. https://doi.org/10.5772/intechopen.68459

25. Kumar S, Saha K (2020) Application of MEG in increasing external quantum efficiency over 100%. Int J Innov Res Phys 1(3):57–59. https://doi.org/10.15864/ijiip.1307
26. Lee HK, Park T, Yoo H (2024) Device applications enabled by bandgap engineering through quantum dot tuning: a review. Materials 17(21):1–20. https://doi.org/10.3390/ma17215335
27. Lee HJ, Chiu A, Lin Y, Chintapalli S, Kamal S, Ji E, Thon SM (2024) Predicting PbS colloidal quantum dot solar cell parameters using neural networks trained on experimental data. Adv Intell Syst 7:2400310. https://doi.org/10.1002/aisy.202400310
28. Lee TY, Hsieh TH, Miao WC, Singh KJ, Li Y, Chang Ching T, Chen FC, Wen Chung L, Kuo HC (2022) High-reliability perovskite quantum dots using atomic layer deposition passivation for novel photonic applications. Nanomaterials 12(23):4140. https://doi.org/10.3390/nano12 234140
29. Li M, Begum R, Jianhui F, Qiang X, Koh TM, Veldhuis SA, Grätzel M, Mathews N, Mhaisalkar S, Sum TC (2018) Low threshold and efficient multiple exciton generation in halide perovskite nanocrystals. Nat Commun 9(1):4197. https://doi.org/10.1038/s41467-018-06596-1
30. Lim C, Choi M, Kim T, Shin D, Song JH, Jeong S (2023) Effect of bandgap variation on photovoltaic properties of lead sulfide quantum dot solar cell. Mater Today Energy 36:101357. https://doi.org/10.1016/j.mtener.2023.101357
31. Liu H, Dong Q, Lopez R (2018) The effect of light intensity, temperature, and oxygen pressure on the photo-oxidation rate of bare PbS quantum dots. Nanomaterials 8(5):341. https://doi.org/10.3390/nano8050341
32. Mimona MA, Rimon MIH, Zohura FT, Sony JM, Rim SI, Arup MMR, Mobarak MH (2025) Quantum dot nanomaterials: empowering advances in optoelectronic devices. Chem Eng J Adv 21:100704. https://doi.org/10.1016/j.ceja.2025.100704
33. Moreels I, Lambert K, meets D, De Muynck D (2009) Size-dependent optical properties of colloidal PbS quantum dots. ACS Nano 3(10):3023–30. http://pubs.acs.org/doi/abs/10.1021/nn900863a
34. Nguyen DT, Sharma S, Chen SA, Komarov PV, Ivanov VA, Khokhlov AR (2021) Polymer-quantum dot composite hybrid solar cells with a Bi-continuous network morphology using the block copolymer poly(3-hexylthiophene)-b-polystyrene or its blend with poly(3-hexylthiophene) as a donor. Mater Adv 2(3):1016–1023. https://doi.org/10.1039/d0ma00 770f
35. Nguyen HT, Tran TT, Kumar M, Song J, Choi SH, Jeong HJ, Song YH, Yun JH (2024) Enhancing thermal consistency and stability of quantum dots through encapsulation: a study on microscale-structured hybrid for WLEDs. Mater Today Nano 27:100504. https://doi.org/10.1016/j.mtnano.2024.100504
36. Prabhakaran P, Kim WJ, Lee K-S, Prasad PN (2012) Quantum dots (QDs) for photonic applications. Opt Mater Express 2(5):578. https://doi.org/10.1364/ome.2.000578
37. Qaid SMH, Ghaithan HM, Alharbi KK, Al-Asbahi BA, Aldwayyan AS (2021) Enhancement of light amplification of $CsPbBr_3$ perovskite quantum dot films via surface encapsulation by PMMA polymer. Polymers 13(15):2574. https://doi.org/10.3390/polym13152574
38. Que M, Guo W, Zhang X, Li X, Hua Q, Dong L, Pan C (2014) Flexible Quantum dot-sensitized solar cells employing CoS nanorod arrays/graphite paper as effective counter electrodes. J Mater Chem A 2(33):13661–13666. https://doi.org/10.1039/c4ta02052a
39. Rehman MA, Roy SB, Akhtar I, Bhopal MF, Choi W, Nazir G, Khan MF et al (2019) Thickness-dependent efficiency of directly grown graphene based solar cells. Carbon 148:187–195. https://doi.org/10.1016/j.carbon.2019.03.079
40. Rodrigues JE, Rosa AD, López-Sánchez J, Sebastiani-Tofano E, Nemes NM, Martínez JL, Alonso JA, Mathon O (2022) EXAFS evidence for the spin-phonon coupling in the monoclinic $PrNiO_3$ nickelate perovskite. J Mater Chem C 11(2):462–471. https://doi.org/10.1039/d2tc03 063b
41. Roelofs KE, Brennan TP, Dominguez JC, Bailie CD, Margulis GY, Hoke ET, McGehee MD, Bent SF (2013) Effect of Al_2O_3 recombination barrier layers deposited by atomic layer deposition in solid-state CdS quantum dot-sensitized solar cells. J Phys Chem C 117(11):5584–5592. https://doi.org/10.1021/jp311846r

42. Sayevich V, Kim WD, Robinson ZL, Kozlov OV, Jung H, Klimov VI (2025) Inverted CdSe/PbSe core/shell quantum dots with electrically accessible photocarriers. ACS Energy Lett 10:1062–1071. https://doi.org/10.1021/acsenergylett.4c03502
43. Şen Z (2004) Solar energy in progress and future research trends. Prog Energy Combust Sci 30(4):367–416. https://doi.org/10.1016/j.pecs.2004.02.004
44. Shilpa G, Mohan Kumar P, Kishore Kumar D, Deepthi PR, Sadhu V, Sukhdev A, Kakarla RR (2023) Recent advances in the development of high efficiency quantum dot sensitized solar cells (QDSSCs): a review. Mater Sci Energy Technol 6:533–546. https://doi.org/10.1016/j.mset.2023.05.001
45. Simi NJ, Bharathi Bernadsha S, Thomas A, Ison VV (2021) Quantum dot sensitized solar cells using type-II CdSe–Cu_2Se core-shell QDs. Results Opt 4:100088. https://doi.org/10.1016/j.rio.2021.100088
46. Smith C, Binks D (2014) Multiple exciton generation in colloidal nanocrystals. Nanomaterials 1:19–45. https://doi.org/10.3390/nano4010019
47. Smith DD, Cousins P, Westerberg S, De Jesus-Tabajonda R, Aniero G, Shen YC (2014) Toward the practical limits of silicon solar cells. IEEE J Photovolt 4(6):1465–1469. https://doi.org/10.1109/JPHOTOV.2014.2350695
48. Talapin DV, Lee JS, Kovalenko MV, Shevchenko EV (2010) Prospects of colloidal nanocrystals for electronic and optoelectronic applications. Chem Rev 110(1):389–458. https://doi.org/10.1021/cr900137k
49. Tu S, Chen M, Limin W (2021) Dual-encapsulation for highly stable all-inorganic perovskite quantum dots for long-term storage and reuse in white light-emitting diodes. Chem Eng J 412:128688. https://doi.org/10.1016/j.cej.2021.128688
50. Urban JJ, Milliron DJ (2010) Heterojunction solar cells based on colloidal quantum dots. Colloidal Quantum Dot Optoelectron Photovolt 9780521198:233–255. https://doi.org/10.1017/CBO9781139022750.010
51. Veerathangam K, Pandian MS, Ramasamy P (2017) Photovoltaic performance of Ag-doped CdS quantum dots for solar cell application. Mater Res Bull 94:371–377. https://doi.org/10.1016/j.materresbull.2017.06.024
52. Wang L, Chen Z, Liang G, Li Y, Lai R, Ding T, Kaifeng W (2019) Observation of a phonon bottleneck in copper-doped colloidal quantum dots. Nat Commun 10(1):4532. https://doi.org/10.1038/s41467-019-12558-y
53. Wang S, Zhao Q, Hazarika A, Li S, Yue W, Zhai Y, Chen X, Luther JM, Li G (2023) Thermal tolerance of perovskite quantum dots dependent on a-site cation and surface ligand. Nat Commun 14(1):2216. https://doi.org/10.1038/s41467-023-37943-6
54. Wu Y, Ren S, Xu X, Liu L, Wang H, Yu J (2014) Engineered fluorescence of quantum dots via plasmonic nanostructures. Sol Energy Mater Sol Cells 126:113–119. https://doi.org/10.1016/j.solmat.2014.03.050
55. Xu F, Gerlein LF, Ma X, Haughn CR, Doty MF, Cloutier SG (2015) Impact of different surface ligands on the optical properties of PbS quantum dot solids. Materials 8(4):1858–1870. https://doi.org/10.3390/ma8041858
56. Xue J, Lee JW, Dai Z, Wang R, Nuryyeva S, Liao ME, Chang SY et al (2018) Surface ligand management for stable $FAPbI_3$ perovskite quantum dot solar cells. Joule 2(9):1866–1878. https://doi.org/10.1016/j.joule.2018.07.018
57. Xue X, Hao Q, Chen M (2024) Very long wave infrared quantum dot photodetector up to 18 Mm. Light: Sci Appl 13(1):7397. https://doi.org/10.1038/s41377-024-01436-y
58. Yang J, Hahm D, Kim K, Rhee S, Lee M, Kim S, Chang JH et al (2020) High-resolution patterning of colloidal quantum dots via non-destructive, light-driven ligand crosslinking. Nat Commun 11(1):2874. https://doi.org/10.1038/s41467-020-16652-4
59. Yang M, Bao H, Xiangmin H, Sun S, Li M, Yan Y, Hou W et al (2024) Machine learning correlating photovoltaics and electroluminescence of quantum dot light-emitting diodes. ACS Photonics 11(5):2131–2137. https://doi.org/10.1021/acsphotonics.4c00413
60. Zaini MS, Ying J, Liew C, Ainliah S, Ahmad A, Mohmad AR, Kamarudin MA (2020) Quantum confinement effect and photoenhancement of photoluminescence of PbS and PbS/MnS quantum dots. Appl Sci 10:6282

61. Zhang LX, Qi MU, Tang ZR, Xu YZ (2023) Heterostructure-engineered semiconductor quantum dots toward photocatalyzed-redox cooperative coupling reaction. Research 6:0073. https://doi.org/10.34133/research.0073
62. Zhang X, Öberg VA, Juan D, Liu J, Johansson EMJ (2018) Extremely lightweight and ultra-flexible infrared light-converting quantum dot solar cells with high power-per-weight output using a solution-processed bending durable silver nanowire-based electrode. Energy Environ Sci 11(2):354–364. https://doi.org/10.1039/c7ee02772a
63. Zhang Y, Zhang Z, Liu Y, Gao H, Mao Y (2020) Short-chain ligands capped $CuInSe_2$ quantum dots as hole transport material for inverted perovskite solar cells. Mater Sci Semicond Process 120:105267. https://doi.org/10.1016/j.mssp.2020.105267

Chapter 15
Expanding the Horizons of Quantum Dot Technology for Biomedical and Environmental Applications

M. Boopalan, M. J. Umapathy, P. Siva Karthik, C. Revathi Ganesh, Rajangam Vinodh, and R. Atchudan

Abstract Quantum dot (QD) technology has emerged as a groundbreaking advancement, transforming multiple domains such as optoelectronics, biomedicine, and sustainable energy. This chapter delves into recent developments and emerging trends in quantum dot research, highlighting their expanding role in cutting-edge technological applications. Emphasis is placed on innovative strategies to improve the performance, optical tunability, and structural stability of quantum dots, thereby enhancing their effectiveness across diverse platforms. Significant focus is also given to the integration of quantum dots into next-generation devices and systems, unlocking their potential in real-world applications. In particular, biomedical and environmental uses are discussed in detail, showcasing the versatility of quantum dots in diagnostics, drug delivery, imaging, sensing, and pollutant detection. These applications underline the importance of designing quantum dots with high quantum yield, precise optical properties, biocompatibility, long-term stability, and environmental durability. Despite their promising capabilities, several challenges remain in optimizing quantum dot systems for practical use. Addressing issues related to toxicity, degradation, and cost-effective synthesis remains a crucial area of ongoing research. Furthermore, strategies for large-scale production and sustainable implementation are also explored. Overall,

M. Boopalan (✉)
Pachaiyappa's College for Men, Kancheepuram, Tamil Nadu, India

M. J. Umapathy
College of Engineering Guindy, Anna University, Chennai, Tamil Nadu, India

P. S. Karthik
Department of Chemistry, University College of Engineering Panruti, Panruti, Tamilnadu, India

C. R. Ganesh
Chellammal Women's College of the Pachaiyappa's Trust, Guindy, Chennai, Tamil Nadu, India

R. Vinodh
Green Hydrogen Lab (GH2Lab), Institute for Hydrogen Research (IHR), Université du Québec à Trois-Rivières (UQTR), Trois-Rivières, Québec, Canada

R. Atchudan
School of Chemical Engineering, Yeungnam University, Gyeongsan-si, Republic of Korea

T. Bhuyan et al. (eds.), *Multifunctional Quantum Dots*, Materials Horizons: From Nature to Nanomaterials, https://doi.org/10.1007/978-981-92-0172-3_15

the chapter provides a comprehensive overview of current trends and future directions in quantum dot technology. By overcoming existing limitations and pushing the boundaries of material science, quantum dots are poised to play a pivotal role in driving technological progress, offering innovative solutions across scientific and industrial landscapes.

Keywords Quantum dots · Optoelectronics · Biomedicine · Energy applications · Quantum dot performance · Stability · and Functionality

1 Introduction

Quantum dots (QDs) have emerged as a revolutionary class of nanomaterials with extraordinary optical and electronic properties. These nanoscale semiconductor particles—typically ranging from 2 to 10 nm in diameter—exhibit size-dependent photoluminescence, high brightness, broad absorption spectra, and exceptional photostability. These unique features make QDs highly valuable in a diverse array of scientific and industrial applications. Of particular importance are their transformative roles in biomedical and environmental domains, where their contributions continue to expand the frontiers of diagnosis, therapy, sensing, and sustainability.

The conceptualization of quantum dots dates back to 1985, when Alexei Ekimov first reported their quantum confinement properties. This pivotal discovery initiated a new era in nanoscience. Progress in the 1990s saw advancements in synthetic techniques and integration of QDs into polymer composites, resulting in enhanced electronic functionalities. By the early 2000s, QDs had permeated biomedical research, primarily in targeted imaging and diagnostics. Their capabilities later extended into optoelectronics, quantum computing, cryptography, energy harvesting, and environmental sensing.

The impact of QDs was formally recognized in 2023 when M.G. Bawendi, L.E. Brus, and A.I. Ekimov were awarded the Nobel Prize in Chemistry for their work on the synthesis and characterization of quantum dots. These foundational studies laid the groundwork for QD deployment across numerous scientific disciplines. Quantum dots possess quantum mechanical properties governed by their size and shape. When excited by photons, they emit light at specific wavelengths determined by quantum confinement effects [1, 2]. This tunable emission is especially advantageous for applications in bioimaging and light-based technologies [3, 4].

In biomedicine, quantum dots have ushered in novel strategies for diagnostics, imaging, and targeted therapies [5]. Their photostability and narrow, tunable emission profiles offer clear advantages over conventional fluorophores [6]. QDs are particularly valuable in multiplexed imaging, where simultaneous detection of multiple biological markers enhances diagnostic accuracy [7, 8].

Quantum dots are widely used in molecular and cellular imaging due to their ability to emit stable and intense light signals. Their broad excitation and narrow emission spectra enable high-resolution, multi-color imaging, crucial for cancer diagnostics and real-time tracking of cellular processes [9]. Functionalization of QDs with antibodies, ligands, or peptides facilitates selective binding to target cells or tissues [10]. This specificity makes them suitable carriers for therapeutic agents, ensuring precise drug delivery with reduced systemic toxicity [6, 11].

Quantum dots also play a critical role in photodynamic therapy (PDT), where they act as energy donors to activate photosensitizers. Upon light activation, these sensitizers produce reactive oxygen species capable of destroying cancer cells [12, 13]. This localized treatment reduces damage to healthy tissues and enhances therapeutic precision.

Quantum dots are proving indispensable in addressing environmental challenges through their application in sensing, remediation, and renewable energy systems [14]. QDs function as highly sensitive fluorescent sensors for detecting pollutants, heavy metals, and toxins in air and water [15]. Their photoluminescent properties shift in the presence of specific analytes, enabling real-time and on-site monitoring [16]. This application holds promises for early detection of contamination and enhanced public health protection.

In energy conversion, QDs have been integrated into quantum dot photovoltaics (QDPVs), which offer advantages such as low-cost production, mechanical flexibility, and higher efficiency via multiple exciton generation (MEG) [17, 18]. These attributes make QDPVs a compelling alternative to traditional silicon solar cells, contributing to clean and sustainable energy initiatives [19]. The photocatalytic and adsorptive capabilities of QDs are being leveraged for the removal of heavy metals and organic pollutants from wastewater [20]. Functionalized QDs can degrade contaminants under light exposure, making them effective agents in environmental remediation [21].

Despite their promise, several challenges limit the broader use of quantum dots in biomedical and environmental fields. One of the primary concerns is **toxicity**—many QDs contain heavy metals like cadmium, which pose risks to biological systems [14, 22]. Research is ongoing to develop **biocompatible alternatives**, such as carbon and graphene quantum dots (Baker and Baker, 2010), offering safer human and environmental profiles. Another key issue is the **stability** of QDs in biological and environmental matrices. Surface modifications and encapsulation techniques are being explored to improve resistance to degradation and leaching [5, 23]. **Scalability** is another concern—producing high-quality, monodisperse QDs on an industrial scale remains technically challenging [24]. With increasing integration into healthcare and environmental systems, QD raises regulatory and ethical questions. Establishing standardized protocols for synthesis, application, disposal, and safety is vital to prevent unintended consequences [25]. Ethical considerations also include informed consent in medical use, potential environmental risks, and equitable access to nanotechnological advancements [26].

Quantum dots represent a transformative nanotechnology with far-reaching implications for biomedicine and environmental science. Their distinct optical properties, functional versatility, and growing adaptability position them at the forefront of innovation in diagnostics, imaging, therapy, energy, and environmental monitoring.

However, realizing their full potential requires addressing concerns around toxicity, stability, large-scale manufacturing, and regulation. As ongoing research refines their design and improves safety profiles, quantum dots are poised to revolutionize personalized medicine and sustainable technologies alike.

This review chapter underscores the importance of continued investigation into the biomedical and environmental applications of QDs. Special emphasis is placed on improving optical performance, enhancing biocompatibility, and meeting real-world challenges of deployment [24, 27]. The future of quantum dot technology is bright, promising breakthroughs that could redefine approaches to health, energy, and environmental stewardship.

2 Properties of Quantum Dots

Quantum dots (QDs) are semiconductor nanocrystals that exhibit unique optical and electronic properties arising from quantum confinement effects. These properties become evident when the size of the nanocrystal approaches or falls below the exciton Bohr radius, leading to discrete energy levels rather than continuous bands [28]. Such behavior has attracted considerable attention across scientific disciplines, particularly in biomedical and environmental applications due to the precision, sensitivity, and multifunctionality they offer [29].

One of the most remarkable features of QDs is their size-tunable fluorescence. The emission wavelength of a QD is directly influenced by its size—smaller dots emit at shorter wavelengths (e.g., blue), whereas larger dots emit at longer wavelengths (e.g., red) [30]. This tunability allows researchers to select specific emission profiles suitable for particular imaging or sensing tasks. For example, in bioimaging, this enables the labeling of different cellular structures with distinct colors in multiplexed imaging platforms, allowing simultaneous tracking of multiple biological targets within a single sample [31].

Another significant advantage of quantum dots lies in their high quantum yield and photostability. Unlike conventional organic dyes, QDs can fluoresce intensely and remain stable under prolonged exposure to excitation light, exhibiting minimal photobleaching [32]. This makes them particularly suitable for long-term imaging studies in living cells, tissues, or organisms, where signal consistency over time is critical. These properties also enhance their usefulness in environmental monitoring, where continuous or time-lapse data acquisition is often required [33].

Quantum dots also demonstrate a broad absorption spectrum along with a narrow, symmetric emission peak. This means that multiple QDs of different sizes can be excited using a single light source, while each emits at its characteristic wavelength [34]. Such multiplexing capabilities are essential in high-throughput diagnostics and biosensing, where detecting multiple analytes simultaneously with high accuracy is desirable. In medical diagnostics, for instance, this property allows for the detection of several disease biomarkers in a single assay, improving diagnostic efficiency and reliability [35].

Surface functionalization further extends the utility of quantum dots. By modifying their surface with specific ligands, polymers, or biomolecules, QDs can be tailored for solubility in aqueous environments, enhanced biocompatibility, and target specificity [36]. In biomedical contexts, this enables their use in targeted drug delivery systems, where QDs are conjugated with antibodies or peptides that bind selectively to diseased cells. This approach not only improves therapeutic efficacy but also minimizes off-target effects and toxicity [37].

The electronic properties of QDs, such as their tunable bandgap, add another layer of functionality. By altering the size and composition of the quantum dots, researchers can engineer materials with desirable electronic responses for applications in photovoltaics and sensing [38]. For example, QDs can be embedded in solar cells to broaden the spectrum of absorbed light, thus enhancing the overall power conversion efficiency [39]. In sensors, QDs' electronic sensitivity enables the detection of trace levels of gases, ions, or biological molecules, which is vital for both clinical diagnostics and environmental pollutant detection [40].

Furthermore, the high surface area-to-volume ratio of QDs enhances their interaction with surrounding molecules, increasing their reactivity. This makes them attractive candidates for catalytic applications and chemical sensing in environmental science [41]. For instance, QDs incorporated into photocatalysts can facilitate the degradation of persistent organic pollutants under light exposure, contributing to water purification efforts [42].

In practical applications, quantum dots have demonstrated transformative potential across multiple domains. In bioimaging, QDs serve as high-resolution fluorescent probes capable of visualizing intricate cellular processes with minimal photodamage [43]. In drug delivery, functionalized QDs allow for the controlled release of therapeutics at specific sites within the body [44]. For diagnostics, their high sensitivity enables the early detection of disease markers, which is crucial for preventive healthcare [45]. From an environmental perspective, QDs are employed in pollutant detection, where their optical sensitivity aids in identifying heavy metals and organic contaminants at very low concentrations [46]. In water purification, QDs integrated into photocatalytic systems break down toxic compounds, contributing to safer water supplies [47]. In the area of solar energy, quantum dots are integrated into next-generation photovoltaic devices that aim to surpass the efficiency limits of traditional silicon-based solar cells [48].

In summary, quantum dots represent a dynamic and multifaceted technology platform with the potential to revolutionize both biomedical and environmental fields. Their customizable optical and electronic properties, combined with surface versatility and nanoscale dimensions, enable a wide range of cutting-edge applications—from precise medical diagnostics to sustainable energy solutions [49]. Continued advancements in synthesis, surface engineering, and biocompatibility will further enhance their practical implementation and pave the way for innovative solutions to some of the most pressing challenges in health and environmental sustainability [50].

3 Types of Quantum Dots

Quantum dots (QDs) are nanoscale semiconductor particles that exhibit unique optical and electronic properties due to quantum mechanics. These properties make them useful in various applications such as displays, solar cells, and medical imaging. Quantum dots are classified based on several criteria, including their composition, structure, and method of synthesis. Below is an in-depth exploration of different types of quantum dots.

3.1 Composition-Based Classification

Quantum Dots (QDs) can be classified based on their **core elemental composition**, which directly influences their **optical, electronic, and toxicological properties**. The primary categories are **II–VI, III–V, and IV–VI semiconductor QDs**, as well as **perovskite** and **carbon-based QDs (CQDs)**. The following section explores these classes in detail, maintaining a focus on their synthesis, advantages, applications, and limitations.

3.1.1 II–VI Semiconductor Quantum Dots

Composed of group II and VI elements such as **CdSe, CdS, and CdTe**, II–VI QDs have been widely investigated for their outstanding **photoluminescence quantum yield, size-dependent emission tunability**, and **well-established synthesis protocols** [51]. Their optical behavior is typically fine-tuned by manipulating the **quantum confinement effect**, allowing emissions across the visible to NIR spectrum.

- **Advantages**: High brightness, narrow emission bandwidths, and well-developed hot-injection and colloidal synthesis techniques.
- **Applications**: Widely used in **bioimaging**, **QLEDs**, **solar cells**, and **photodetectors**.
- **Limitations:** Practical deployment is constrained by the **toxicity** associated with heavy metal content, particularly **cadmium and tellurium** [52]. Surface modifications and **core-shell architectures** (e.g., CdSe/ZnS) are often required to enhance biocompatibility and stability.

3.1.2 III–V Semiconductor Quantum Dots

Quantum dots composed of group III and V elements such as **InP, InAs, and GaAs** are considered **less toxic** alternatives to their II–VI counterparts. These QDs offer **relatively safer profiles**, making them suitable for biomedical research and in vivo applications [53].

- **Advantages**: Narrow emission lines, reduced photobleaching, and moderate to good PLQY with proper passivation.
- **Applications**: Ideal for **biomedical imaging**, **quantum computing**, **LEDs**, and **photovoltaics**.
- **Limitations**: These QDs generally require **more complex synthesis procedures**, including stringent control of precursor reactivity. Additionally, they often exhibit **lower PLQY** than II–VI QDs without proper surface engineering [53].

3.1.3 IV–VI Semiconductor Quantum Dots

This class includes **PbS and PbSe QDs**, known for their **strong absorption in the infrared (IR)** and **short-wave infrared (SWIR)** regions [54]. The **large exciton Bohr radius** allows for strong quantum confinement even at larger particle sizes.

- **Advantages**: Their IR-tunability makes them useful for **telecommunications**, **night vision**, **optical sensing**, and **SWIR imaging**.
- **Limitations**: The presence of **lead** and **limited air stability** significantly hinder their real-world usage, especially in biological contexts [55]. Strategies such as core-shell encapsulation and polymer coatings are being explored to address these concerns.

3.1.4 Perovskite Quantum Dots

Typically based on **cesium lead halides** (e.g., $CsPbBr_3$, $CsPbI_3$), perovskite QDs represent a newer generation of nanomaterials characterized by **exceptional optical properties** and **high PLQY** [56]. They offer **defect tolerance**, **sharp emission**, and **broad spectral tunability**.

- **Advantages**: Synthesized under **mild conditions**, often via **ligand-assisted reprecipitation**, resulting in high reproducibility and scalability.
- **Applications**: Extensively applied in **perovskite LEDs (PeLEDs)**, **solar cells**, **photodetectors**, and **lasers**.
- **Limitations**: Their high performance is offset by their **sensitivity to air, moisture, and heat**, as well as the presence of **lead**, raising long-term **environmental and stability concerns** [57]. Research into **lead-free alternatives** and **encapsulation** is ongoing.

3.1.5 Carbon Quantum Dots (CQDs)

CQDs are **metal-free nanostructures**, composed mainly of **carbon, oxygen, hydrogen**, and sometimes **nitrogen** or **sulfur**. These dots are gaining popularity due to their **excellent water solubility**, **low cytotoxicity**, **biocompatibility**, and **cost-effective green synthesis** methods [58].

- **Advantages**: Environmentally benign, chemically stable, and surface-functionalizable for **targeted bioimaging**, **drug delivery**, and **sensor applications**.
- **Synthesis**: Produced using **hydrothermal, microwave-assisted**, or **electrochemical methods** from organic precursors such as citric acid or glucose.
- **Limitations**: CQDs tend to have **lower PLQY** and **limited emission tunability** compared to metal-based QDs, though this can be enhanced through **surface passivation or heteroatom doping** [59].

3.2 Structure-Based Classification

3.2.1 Core Quantum Dots

Core QDs consist solely of a single semiconductor material. Their properties depend on the core's size and composition, offering strong quantum confinement effects and simple synthesis, though with lower stability and surface defect risks [60].

3.2.2 Core-Shell Quantum Dots

Core-shell structures comprise a semiconductor core surrounded by a shell of a different material. This design improves photostability and quantum yield by passivating surface defects [61], though synthesis is more complex and can be hindered by lattice mismatch between the core and shell [62].

3.2.3 Alloyed Quantum Dots

Formed by combining multiple semiconductors at the atomic level, alloyed QDs offer continuous tunability and improved stability [63]. Despite these benefits, synthesis can be intricate and may result in phase segregation [64].

3.2.4 Doped Quantum Dots

Doping involves introducing foreign atoms into QDs to alter their optical and electronic characteristics. These structures enable property tuning for specific applications, including magnetic and luminescent enhancements [65]. However, maintaining controlled doping levels is technically challenging and may affect quantum yield [66].

3.3 Synthesis and Application-Based Classification of Quantum Dots

Quantum dots (QDs) can be classified based on their synthesis methods and the nature of their applications. These classifications are essential for understanding their structural features, performance characteristics, and end-use suitability.

Among various synthetic strategies, colloidal synthesis is widely adopted due to its solution-based nature, cost-effectiveness, and scalability. It offers precise control over size and morphology [67], although issues such as solvent residues and sensitivity to environmental factors persist [68]. Epitaxial quantum dots, synthesized using high-precision techniques like molecular beam epitaxy (MBE) or metal-organic chemical vapor deposition (MOCVD), provide excellent uniformity and spatial precision, which are particularly valuable for quantum computing and optoelectronic devices [69]. However, these methods are expensive and lack scalability [70]. Electrochemical synthesis offers control over quantum dot size and composition through redox reactions [71]. This method opens avenues for novel material development, but challenges in achieving uniformity and scale-up still remain [72].

Photoluminescent quantum dots, which emit light when excited by photons, are integral to display technologies, lighting, and bioimaging due to their brightness and tunable emission [52]. However, prolonged exposure to light and oxygen can reduce their stability [73]. Electroluminescent QDs emit light in response to electrical stimulation and are pivotal in QD-based LED (QD-LED) development. These dots offer energy efficiency and a broad color range, though their operational stability still needs enhancement [74, 75]. Quantum dots are also applied in photovoltaics to improve solar energy conversion efficiency through broadened absorption spectra and the possibility of multiple exciton generation [76]. However, concerns related to their toxicity and difficulty in large-scale production must be addressed [77]. In the biomedical domain, QDs are used extensively for bioimaging, offering high-resolution, multiplexed visualization due to their photostability and size-tunable fluorescence. Functionalization enhances targeting capability, yet issues like toxicity, complex surface chemistry, and biocompatibility remain significant hurdles [78, 79].

Overall, quantum dots are a diverse and adaptable class of nanomaterials. Their synthesis route and application context determine their optical behavior, stability, and environmental compatibility [80]. With ongoing research, new QD types with superior performance and reduced ecological impact are expected to broaden their utility [68].

4 Design and Development of Quantum Dots

Quantum dots (QDs) are nanoscale semiconductor particles that exhibit distinctive optical and electronic characteristics as a result of quantum mechanical effects [81]. Their ability to confine electrons in three dimensions leads to discrete energy levels, giving them size-dependent optical properties. This section explores the design, synthesis, characterization, and applications of quantum dots, emphasizing their potential in various fields.

4.1 Properties of Quantum Dots

Quantum dots (QDs) exhibit unique optical and electronic properties due to **quantum confinement**, a phenomenon that occurs when the electronic states are restricted within nanoscale dimensions [82, 83]. Typically ranging from 2 to 10 nm in size, QDs show discrete energy levels, which allow for **tunable emission spectra**. Smaller QDs have larger bandgaps and emit shorter wavelengths (blue shift), while larger QDs emit at longer wavelengths (red shift). This size-dependent behavior forms the foundation of their use in **optoelectronics, biomedical imaging, and sensing** [84, 85].

A hallmark feature of QDs is their **efficient photoluminescence**. When excited by light, they absorb and re-emit photons at specific wavelengths, a result of their discrete energy states and customizable surface chemistry [86, 87]. The intensity and wavelength of this emission are influenced by the QD's size and composition, making them highly adaptable for **sensors, imaging probes, and display technologies** [88].

Another key advantage is their **superior photostability**. Unlike traditional organic dyes, quantum dots resist photobleaching due to their robust crystalline structure, maintaining their luminescence over extended periods of illumination. This makes them ideal for **long-duration applications** such as advanced microscopy, fluorescence-based diagnostics, and high-performance display systems [89]. The synthesis method significantly determines the physicochemical characteristics of quantum dots—including their size, morphology, crystallinity, and surface functionalization—which in turn influence their optical performance and application scope.

Among various techniques, **colloidal synthesis** remains the most widely used due to its **simplicity, scalability**, and **precise control over QD dimensions and composition** [90]. This method involves reacting chemical precursors in a solvent under controlled temperature and time, producing high-quality nanocrystals suitable for various applications.

On the other hand, **top-down approaches**, such as lithography and etching, fabricate quantum dots by carving them from bulk materials. These techniques offer high precision and structural uniformity but are limited by **high costs and complex instrumentation**, restricting their routine use [91].

Bottom-up approaches, including molecular beam epitaxy (MBE) and chemical vapor deposition (CVD), build quantum dots through **self-assembly at the atomic or molecular level** [92]. These methods can produce highly uniform and crystalline QDs, making them ideal for **high-end optoelectronic and quantum computing applications**, though they often demand stringent processing conditions and sophisticated equipment.

Together, these synthesis techniques cater to different application demands, balancing trade-offs between precision, scalability, cost, and material quality.

4.2 Characterization of Quantum Dots

Characterizing quantum dots ensures their suitability for targeted applications by examining their structural, compositional, and optical properties. TEM is a crucial technique providing direct visualization of QD size, morphology, and crystallinity [93]. It involves transmitting electrons through an ultrathin sample to generate detailed images but requires meticulous sample preparation. XRD offers information about the crystal structure, phase purity, and strain within QDs by analyzing how X-rays scatter from the material's atomic planes [94]. It is indispensable for confirming QD crystallinity and identifying structural defects. PL spectroscopy evaluates the emission characteristics and quantum efficiency of QDs by measuring the light emitted after excitation [95]. It provides insights into the electronic band structure, defect states, and surface properties of QDs.

Quantum dots (QDs) require precise and comprehensive characterization to analyze their physical, chemical, optical, and electrical properties [96, 97]. A variety of analytical tools are employed to obtain detailed insights into their morphology, structure, composition, and performance characteristics.

Structural characterization techniques provide vital information on the morphology and crystallinity of QDs. Transmission Electron Microscopy (TEM) offers high-resolution images that reveal the size, shape, and internal structure of quantum dots, enabling direct visualization of their morphology [98]. Scanning Electron Microscopy (SEM) complements this by providing surface morphology and aggregation behavior [99]. X-ray Diffraction (XRD) is widely used to determine the crystal structure, phase composition, and lattice parameters, offering insight into the material's degree of crystallinity [100, 101]. Additionally, Atomic Force Microscopy (AFM) is employed to measure surface topography and roughness, enabling three-dimensional imaging at the nanoscale [102].

Chemical characterization focuses on identifying the elemental composition and surface chemistry of QDs. X-ray Photoelectron Spectroscopy (XPS) provides information on surface elements and oxidation states, revealing bonding environments [96, 97]. Energy-Dispersive Spectroscopy (EDS), often paired with SEM, allows spatially resolved analysis of elemental distribution [98]. Inductively Coupled Plasma Mass Spectrometry (ICP-MS) offers extremely sensitive detection and quantification of both trace and bulk elements in QDs [103]. Fourier Transform Infrared Spectroscopy (FTIR), meanwhile, identifies surface functional groups by detecting their characteristic vibrational frequencies [100, 101].

Optical characterization is essential for evaluating the photophysical properties of QDs. Ultraviolet-Visible (UV-Vis) Spectroscopy determines absorption spectra and extinction coefficients, providing insights into bandgap energy and particle size [102]. Photoluminescence (PL) Spectroscopy is employed to assess emission characteristics and quantum yield, revealing information about electronic states and defect levels [96, 97]. Fluorescence Spectroscopy extends this analysis by measuring emission intensity and lifetime, critical for applications in bioimaging and optical sensing [98]. Raman Spectroscopy is also used to study vibrational modes and structural features, offering insights into crystallinity and mechanical stress within the nanocrystals [104].

Electrical characterization is crucial for evaluating QDs in electronic and optoelectronic applications. Current-voltage (I–V) measurements are performed to assess electrical conductivity and carrier transport properties, which are vital for device performance [100, 101]. Electrical Impedance Spectroscopy (EIS) helps analyze the capacitive and resistive behavior of QD-based systems, shedding light on charge transfer mechanisms [102]. Hall Effect Measurements are used to determine charge carrier concentration and mobility, providing critical parameters for semiconductor functionality [96, 97].

Additional characterization techniques provide complementary data for QD evaluation. Dynamic Light Scattering (DLS) is widely employed to measure particle size distribution and colloidal stability in suspension [98]. Zeta Potential analysis evaluates surface charge and electrostatic stability, which can predict aggregation tendencies [103]. Thermogravimetric Analysis (TGA) assesses thermal stability by recording weight changes under heating conditions [100, 101], while Differential Scanning Calorimetry (DSC) investigates phase transitions and thermal properties by monitoring heat flow during material changes [102].

5 Need for Quantum Dots and their Technology

Quantum dots (QDs) are increasingly recognized for their potential to revolutionize biomedical and environmental applications. Their unique optical and electronic properties, such as size-tunable emission, high photoluminescence, and stability, make them ideal for various innovative solutions in these fields. This chapter explores their application in biomedical and environmental sectors, addressing their advantages, challenges, and prospects.

5.1 Advanced Bio-Imaging

Quantum dots (QDs) have transformed the field of bio-imaging due to their unique optical properties, including exceptionally high brightness, superior photostability, and size-tunable emission spectra. Unlike traditional organic fluorophores, QDs exhibit broad absorption and narrow, symmetric emission spectra, allowing simultaneous excitation of multiple QDs with a single light source and enabling multiplex imaging of various biological targets within the same sample. This capacity is particularly useful in complex cellular environments where multiple biomarkers need to be tracked simultaneously with high specificity and minimal spectral overlap. One of the most critical advantages of QDs is their resistance to photobleaching, which allows for long-term real-time imaging of live cells, tissues, or organisms. Their stable fluorescence signals make them especially suitable for extended-duration studies, such as monitoring intracellular trafficking, signal transduction pathways, or tumor progression over time. In pioneering work by Gao et al., QDs were conjugated with tumor-targeting peptides and used to selectively bind to cancer cells, achieving highly specific imaging and early detection of malignant tumors. The functionalized QDs demonstrated enhanced accumulation in tumor tissues via active targeting mechanisms, offering significantly improved contrast over non-targeted imaging agents. Building on these findings, Zhou et al. [105] reported the development of multifunctional QDs capable of both imaging and delivering therapeutic agents, establishing a platform for theranostic applications that integrate diagnosis and treatment. Moreover, QDs have been effectively utilized for deep tissue imaging due to their emission in the near-infrared (NIR) region, which allows for greater tissue penetration and minimal autofluorescence interference. This makes them valuable in in vivo imaging of tumors, vasculature, and organs, providing precise anatomical and functional insights that are not achievable with conventional dyes. Overall, quantum dots represent a significant advancement in biomedical imaging technologies. Their combination of high signal-to-noise ratio, targeting capability, and long-term stability continues to drive innovation in cellular imaging, cancer diagnostics, and integrated therapeutic platforms, paving the way for next-generation personalized medicine.

5.1.1 Drug Delivery Systems

Quantum dots (QDs) are increasingly being engineered as dual-functional platforms that integrate both targeted drug delivery and real-time in vivo imaging, offering a powerful tool for precision medicine. Their nanoscale size, tunable surface chemistry, and exceptional optical properties make them ideal candidates for the development of theranostic systems—nanostructures that simultaneously diagnose and treat diseases.

One of the major advantages of QD-based delivery systems is their ability to achieve controlled and localized drug release. By functionalizing their surfaces with biomolecular ligands such as antibodies, aptamers, or peptides, QDs can selectively bind to disease-specific receptors on the surfaces of target cells (e.g., tumor cells). This active targeting approach ensures that therapeutic agents are delivered specifically to diseased tissues, thereby minimizing systemic toxicity and side effects commonly associated with conventional chemotherapy.

Moreover, QDs serve as optical beacons that allow for real-time tracking of drug biodistribution and accumulation at the target site. Their bright and photostable fluorescence enables continuous imaging over extended periods, providing dynamic insights into the pharmacokinetics and therapeutic responses at a cellular or tissue level.

In a landmark study, Bagalkot et al. demonstrated a multifunctional QD-based platform where quantum dots were conjugated with both a chemotherapeutic drug (doxorubicin) and a targeting ligand (aptamer) that binds specifically to prostate cancer cells. This system enabled the simultaneous imaging of tumor tissues and stimuli-responsive drug release, significantly enhancing treatment precision. More recently, Mehta et al. [106] reported the use of QDs for delivering anticancer agents in a pH-sensitive manner, releasing the drug preferentially in the acidic tumor microenvironment while tracking the therapeutic progress using fluorescence imaging.

These advances highlight the potential of QDs not only to improve therapeutic efficacy through site-specific delivery but also to serve as non-invasive tools for monitoring treatment response. Such dual-functional nanocarriers are paving the way for personalized treatment regimens, reducing the trial-and-error approach of conventional therapies, and offering early feedback on drug efficacy in real-time. As research progresses, QD-based theranostic systems are expected to play a central role in next-generation oncology, infectious disease treatment, and regenerative medicine, combining diagnostic precision with targeted therapeutic action in a single nanoscale platform.

5.1.2 Biosensors

Quantum dot (QD)-based biosensors have emerged as highly effective tools for ultra-sensitive and highly specific detection of biomolecules, offering significant advantages over traditional detection methods. Their unique optical properties—such as size-tunable fluorescence, high quantum yield, and resistance to photobleaching—allow for enhanced signal intensity and long-term stability, making them ideal for various biosensing applications.

One of the most impactful uses of QD-based biosensors is in point-of-care (POC) diagnostics, especially in resource-limited or field settings where rapid, accurate, and low-cost testing is essential. These sensors can be engineered to detect a broad range of biological targets, including proteins, nucleic acids, pathogens, and small molecules, by conjugating QDs with specific recognition elements such as antibodies, aptamers, or oligonucleotides.

Notably, QD biosensors have been successfully applied to the detection of infectious diseases such as HIV, influenza, and hepatitis, enabling early diagnosis and timely treatment decisions. These sensors allow for rapid, real-time detection with high sensitivity, often capable of identifying biomarkers at picomolar or even femtomolar concentrations.

A landmark advancement in this domain was the multiplexed QD-based biosensor developed by Medintz et al., which enabled the simultaneous detection of multiple disease biomarkers within a single assay. This technology uses QDs of distinct emission wavelengths, allowing each one to be associated with a different target molecule. As a result, clinicians can screen for multiple pathogens or disease indicators at once, improving diagnostic efficiency and reducing testing time. More recent enhancements reported by Sun et al. [107] have integrated these sensors into portable diagnostic platforms, expanding their usability in non-clinical environments and improving public health surveillance efforts.

Overall, QD-based biosensors are poised to revolutionize next-generation diagnostics by combining exceptional optical performance with miniaturization and multiplexing capabilities, supporting the development of smart diagnostic devices that are fast, accurate, and accessible at the point of care. Continued research is expected to further improve their biocompatibility, detection range, and integration with digital health technologies, paving the way for widespread adoption in clinical, environmental, and biomedical applications.

5.2 Water Purification

Quantum dots (QDs) have shown remarkable potential in advanced water purification technologies, particularly due to their ability to both detect and degrade environmental contaminants with high sensitivity and precision. Their tunable optical properties, high surface area, and ease of functionalization make them excellent candidates for use in environmental sensing and photocatalysis. Functionalized QDs can be engineered to selectively bind with toxic heavy metals, such as lead (Pb^{2+}) and mercury (Hg^{2+}), enabling the detection of these pollutants at ultra-trace levels in aqueous environments. These QD-based sensors offer real-time detection with rapid response times, surpassing conventional analytical techniques in terms of portability and ease of use. Their fluorescence emission often changes in response to metal ion binding, providing a clear optical signal that facilitates accurate quantification even in complex water samples. In addition to detection, QDs also play a critical role in photocatalytic degradation of organic pollutants. For example, ZnS quantum dots, as reported by Li et al. [82, 83], have demonstrated the ability to efficiently degrade synthetic dyes and pharmaceutical residues under UV light irradiation. These QDs act as photocatalysts by generating reactive oxygen species (ROS) such as hydroxyl radicals (•OH) upon light exposure, which in turn break down toxic organic compounds into less harmful by-products. This capability is especially beneficial for treating wastewater from textile, pharmaceutical, and agrochemical industries, where conventional treatment methods often fall short.

Furthermore, hybrid systems combining QDs with other nanomaterials, such as graphene oxide or TiO_2 nanoparticles, have shown enhanced photocatalytic activity, broader light absorption spectra (extending into visible light), and improved recyclability, making them attractive for sustainable water treatment solutions. As research progresses, there is a growing focus on developing non-toxic, environmentally benign quantum dots, such as carbon and silicon-based QDs, to mitigate any secondary contamination risks. These eco-friendly alternatives maintain high photocatalytic performance while offering greater biocompatibility and environmental safety. In summary, quantum dots are transforming water purification technologies by providing dual-function capabilities—highly sensitive pollutant detection and efficient contaminant degradation—positioning them as vital components in next-generation smart water treatment and environmental remediation systems.

5.2.1 Environmental Monitoring

Quantum dot (QD)-based sensors are emerging as powerful tools for the real-time monitoring of air and soil pollutants, owing to their exceptional optical sensitivity, tunable emission properties, and rapid response times. These nanoscale sensors are particularly effective in detecting low concentrations of hazardous environmental pollutants, making them invaluable in environmental assessment and management.

In the domain of air quality monitoring, QD-incorporated sensors have demonstrated outstanding performance in detecting nitrogen oxides (NO_n) and volatile organic compounds (VOCs)—two major contributors to air pollution and human health risks. In a notable study, Lee et al. developed QD-based chemiresistive and fluorescence sensors capable of detecting trace levels of NO_2 and VOCs with high specificity and sensitivity [84, 85].These sensors exploit the fluorescence quenching or enhancement behavior of QDs in the presence of specific gas molecules, enabling continuous, real-time tracking of pollutant concentrations. Such capabilities are essential for both urban air quality assessment and industrial emissions monitoring. In soil monitoring, quantum dots offer unique advantages in mapping and assessing contamination. For instance, CdSe quantum dots have been employed to visualize and quantify arsenic contamination in agricultural soils [84, 85]. Through fluorescence-based detection mechanisms, these QDs provided spatially resolved maps of arsenic distribution, supporting targeted remediation strategies. This is particularly important in regions where groundwater and crops are at risk due to heavy metal leaching from industrial or mining activities. Moreover, by conjugating QDs with specific ligands or recognition elements (such as DNA aptamers or antibodies), these sensors can be tailored for selective detection of a wide range of analytes, including pesticides, fertilizers, and polycyclic aromatic hydrocarbons (PAHs) in complex environmental matrices.

In addition to standalone applications, QD-based sensors are being integrated into portable and wireless sensing platforms, including smartphone-based detection systems and IoT-enabled environmental monitors. These advances facilitate on-site, user-friendly, and cost-effective monitoring—especially crucial in low-resource or remote areas. As research continues, efforts are being made to replace traditional cadmium- or lead-based QDs with eco-friendly alternatives like carbon, silicon, or indium phosphide (InP) QDs, which offer reduced toxicity and enhanced environmental compatibility. In summary, the integration of QDs into environmental sensors offers high-resolution, real-time surveillance of air and soil pollutants, providing critical data for pollution control, environmental policy-making, and sustainable land use planning.

5.2.2 Renewable Energy

Quantum dots are significantly enhancing the performance and versatility of next-generation energy devices, particularly in solar cells and thermoelectric systems. In the realm of photovoltaics, quantum dot solar cells (QDSCs) are gaining momentum due to their superior optical properties and flexibility in design. The hallmark of QDs lies in their size-dependent tunable bandgap, allowing for tailored absorption across a broader solar spectrum compared to conventional semiconductors. This characteristic makes it possible to capture both visible and near-infrared light, thereby boosting the overall light-harvesting efficiency [86, 87].

One of the most promising advancements in QDSCs is the phenomenon of **multiple exciton generation (MEG)**, a process wherein a single high-energy photon can produce multiple electron-hole pairs (excitons), significantly increasing the photocurrent output. This contrasts with traditional silicon-based photovoltaics, where excess photon energy is mostly lost as heat. Pioneering work by Chen et al. demonstrated that MEG in colloidal QDs such as PbSe and PbS could theoretically raise the power conversion efficiency beyond the Shockley–Queisser limit (~33%) of traditional single-junction cells [86, 87]. Moreover, QDs can be integrated into **tandem solar cells**—a multilayered configuration that stacks QDs with different bandgaps to optimize the use of the full solar spectrum. This not only enhances efficiency but also offers a pathway to lightweight, flexible, and cost-effective solar panels suitable for portable or wearable electronics and building-integrated photovoltaics. In addition to photovoltaics, quantum dots are making impactful contributions in the field of **thermoelectric energy harvesting**, where they help convert waste heat into usable electrical energy. The integration of QDs into thermoelectric materials introduces **quantum confinement effects** and **energy filtering mechanisms** that can enhance the Seebeck coefficient (a measure of thermoelectric efficiency) while suppressing thermal conductivity. This dual effect leads to an overall improvement in the **thermoelectric figure of merit (ZT)**. For instance, embedding QDs into a bulk matrix can create phonon scattering centers, reducing heat transfer while maintaining or improving charge transport [86, 87]. These advancements position quantum dots as pivotal nanomaterials in the push toward sustainable energy solutions, enabling more efficient solar harvesting and waste heat recovery, both critical for addressing the global energy crisis and meeting clean energy targets.

6 Surface Functionalization of Quantum Dots

Quantum dots (QDs) are tiny semiconductor particles with unique optical and electrical properties. Their small size and high surface-to-volume ratio make them highly susceptible to surface modifications. Surface functionalization of QDs is a crucial step in tailoring their properties for various applications, including biomedical imaging, sensing, and optoelectronics [108].

6.1 Importance of Surface Functionalization

Surface functionalization of QDs is essential for several reasons:

1. **Stability and Solubility**: QDs are prone to aggregation and precipitation in aqueous solutions. Surface functionalization helps to improve their stability and solubility [99].
2. **Biocompatibility**: QDs can be toxic due to their heavy metal composition. Surface functionalization with biocompatible ligands can reduce their toxicity [109].
3. **Targeting and Specificity**: Surface functionalization allows for the attachment of targeting molecules, such as antibodies or peptides, to QDs, enabling specific binding to cells or tissues [110].
4. **Optical Properties**: Surface functionalization can influence the optical properties of QDs, such as their fluorescence intensity and wavelength [111].

6.2 Methods of Surface Functionalization

Several methods have been developed for surface functionalization of QDs:

1. **Ligand Exchange**: Replacing the native ligands on QDs with new ligands that provide desired properties [108].
2. **Silanization**: Coating QDs with a silica shell to improve their stability and biocompatibility [104].
3. **Polymer Coating**: Encapsulating QDs in a polymer matrix to enhance their stability and processability [109].
4. **Bioconjugation**: Attaching biomolecules, such as proteins or nucleic acids, to QDs using various conjugation chemistries [100, 101].

6.3 Types of Surface Functional Groups

Various surface functional groups can be attached to QDs, including:

1. **Carboxyl (-COOH)**: Provides negative charge and enables conjugation to amine-containing molecules [111].
2. **Amine (-NH_2)**: Provides positive charge and enables conjugation to carboxyl-containing molecules [108].
3. **Hydroxyl (-OH)**: Provides neutral charge and enables conjugation to molecules containing complementary functional groups [103].
4. **Thiol (-SH)**: Provides strong binding to metal surfaces and enables conjugation to molecules containing complementary functional groups [109].

6.4 Applications of Surface-Functionalized QDs

Surface-functionalized QDs have numerous applications in:

1. **Biomedical Imaging**: Targeted imaging of cells, tissues, and organs using QDs conjugated to biomolecules [100, 101].
2. **Sensing**: Detection of biomolecules, ions, and other analytes using QDs functionalized with recognition elements [111].
3. **Optoelectronics**: Development of QD-based devices, such as LEDs, solar cells, and photodetectors [108].
4. **Therapeutics**: Delivery of therapeutic agents, such as drugs and genes, using QDs functionalized with targeting molecules [104].

6.5 Optimizing Quantum Dot Surface Chemistry for Scalable and Safe Applications

Surface functionalization of quantum dots (QDs) plays a pivotal role in optimizing their physicochemical properties for a broad range of applications, including bioimaging, photovoltaics, and optoelectronics [112]. Despite considerable progress in the development of surface modification strategies, several critical challenges continue to limit their full potential.

One major challenge is scalability. Current functionalization methods often rely on complex, time-consuming processes that are difficult to scale up for industrial production [108]. Developing robust and reproducible techniques that can be applied on a large scale without compromising the structural and optical integrity of QDs remains a significant hurdle. Another concern is the stability of surface-functionalized QDs under physiological or operational conditions. Long-term colloidal stability and resistance to photobleaching or oxidative degradation are crucial for reliable performance [103]. Enhancing the chemical robustness of QD surfaces through improved ligand chemistry and crosslinking strategies is actively being researched. Additionally, toxicity associated with heavy metal-based QDs raises biosafety concerns, especially for biological applications. Functionalization with biocompatible ligands or encapsulation within inert shells such as silica or polymers helps mitigate these effects while maintaining performance [113, 114]. Standardization of surface modification protocols is another area requiring attention. The diversity of materials and functional groups used in QD surface chemistry has led to variability in experimental outcomes and hindered reproducibility [111]. Establishing universal protocols and characterization standards would facilitate consistency across research findings. Overall, innovations in ligand design, hybrid coating strategies, and scalable manufacturing approaches will be instrumental in overcoming current limitations and unlocking the full potential of QDs for scientific and industrial use [112].

7 Limitations

Despite their tremendous potential, quantum dots (QDs) face significant challenges and limitations that must be addressed for broader technological deployment [109].

7.1 *Scalability and Manufacturing*

1. **Scalable Synthesis Methods**: Developing cost-effective and scalable methods for synthesizing high-quality QDs is crucial for commercial viability [100, 101].
2. **Uniformity and Reproducibility**: Achieving uniformity and reproducibility in QD synthesis and processing remains a major hurdle due to sensitivity to reaction conditions [99].
3. **High-volume Production**: Scaling up production to meet increasing demand while maintaining quality and consistency is a complex challenge [96, 97].

7.2 *Stability and Durability*

1. **Photostability**: Improving QD photostability is necessary to maintain optical performance under prolonged illumination [100, 101].
2. **Chemical Stability**: Enhancing chemical stability ensures that QDs resist degradation in harsh or reactive environments [112].
3. **Thermal Stability**: Developing QDs that can withstand high temperatures without degradation is essential for applications in demanding conditions [110].

7.3 *Toxicity and Biocompatibility*

1. **Toxicity Reduction**: Reducing the inherent toxicity of QDs, especially those containing heavy metals like cadmium, is a priority for safe applications [103].
2. **Biocompatibility**: Surface engineering strategies are being developed to enhance QD biocompatibility for biomedical uses [96, 97].
3. **Environmental Impact**: Addressing the environmental risks associated with QD production and disposal is gaining attention in sustainability research [100, 101].

7.4 *Optical and Electrical Properties*

1. **Optical Properties**: Efforts are ongoing to improve QD quantum yield, absorption, and emission spectra to meet diverse application needs [112].

2. **Electrical Properties**: Enhancing QD conductivity, carrier mobility, and carrier lifetime is vital for optoelectronic device integration [100, 101].
3. **Tuning and Control**: Developing precise methods for tuning and controlling QD optical and electrical properties is necessary for custom device fabrication [103].

7.5 Integration and Interfacing

1. **Integration with Other Materials**: Strategies for integrating QDs with polymers, metals, and other semiconductors are under active investigation to create hybrid systems [96, 97].
2. **Interfacing with Biological Systems**: Designing QDs that can efficiently interact with biological environments without toxicity remains a challenge [100, 101].
3. **Scalable and Flexible Device Fabrication**: New methods are being explored to enable the scalable and flexible fabrication of QD-based devices for commercial applications [112].

7.6 Fundamental Understanding and Modeling

1. **Fundamental Understanding**: Elucidating the physics and chemistry of QD behavior is critical for optimizing performance [100, 101].
2. **Theoretical Modeling**: Accurate theoretical models are needed to predict and improve QD properties and behavior [104].
3. **Simulation and Design Tools**: Advanced simulation and computational tools are facilitating the development of QD-based systems [96, 97].

7.7 Regulatory and Standards Framework

1. **Regulatory Frameworks**: Clear regulatory guidelines are essential to ensure the safe and ethical development of QD technologies [100, 101].
2. **Standards and Certification**: Developing standardized certification processes will help ensure product quality and consistency across the QD industry [112].
3. **Public awareness and Education**: Promoting public understanding of both the benefits and risks associated with QDs is necessary for widespread acceptance [100, 101].

8 Applications

Expanding the horizons of quantum dot (QD) technology has opened up innovative applications across multiple fields [115].

8.1 Biomedical Applications

1. **Imaging and Diagnostics**: Quantum dots are employed for advanced biomedical imaging, diagnostics, and biosensing due to their high photostability and tunable emission [116, 117].
2. **Cancer Treatment**: Quantum dots are engineered for targeted cancer therapy and drug delivery by functionalizing their surfaces with ligands specific to tumor cells [118].
3. **Gene Therapy**: Quantum dots assist in gene therapy and gene editing by serving as delivery vehicles for nucleic acids and tracking gene expression [119].
4. **Biosensing**: Quantum dots are critical in biosensing for the ultrasensitive detection of proteins, nucleic acids, and other biomolecules [120, 121].

8.2 Optoelectronic Applications

1. **LEDs and Displays**: Quantum dots improve color purity, brightness, and energy efficiency in light-emitting diodes (LEDs) and display technologies [115].
2. **Solar Cells**: Quantum dots enhance the efficiency of solar cells by broadening light absorption and improving charge separation [116, 117].
3. **Lasers and Optics**: Quantum dots are utilized in developing compact, tunable lasers and other photonic devices [122].
4. **Telecommunications**: Quantum dots are incorporated into telecommunications systems for high-speed data transmission due to their fast response times [119].

8.3 Energy Applications

1. **Energy Storage**: Quantum dots are integrated into next-generation battery technologies to enhance energy density and cycle stability [120, 121].
2. **Fuel Cells**: Quantum dots contribute to improved electrocatalytic activity in fuel cells, promoting better energy conversion efficiency [115].
3. **Thermoelectrics**: Quantum dots enhance thermoelectric materials, allowing for more efficient waste heat recovery [116, 117].
4. **Catalysis**: Quantum dots serve as highly active catalysts in chemical reactions due to their large surface area and unique electronic properties [123].

8.4 Environmental Applications

1. **Water Purification**: Quantum dots are explored for advanced water purification techniques, including photocatalytic degradation of pollutants [119].
2. **Air Purification**: Quantum dots are applied in air purification systems to capture and neutralize harmful gases and particulates [120, 121].
3. **Soil Remediation**: Quantum dots facilitate soil remediation by breaking down organic contaminants through photocatalysis [115].
4. **Climate Change Mitigation**: Quantum dots are studied for their role in carbon capture technologies and enhancing photosynthetic efficiency [116, 117].

8.5 Consumer Products

1. **Consumer Electronics**: Quantum dots are revolutionizing consumer electronics like smartphones, TVs, and cameras by offering superior display performance [118].
2. **Textiles and Fabrics**: Quantum dots are incorporated into smart textiles for wearable health monitoring and dynamic displays [119].
3. **Food and Agriculture**: Quantum dots assist in food safety monitoring, crop imaging, and animal health diagnostics [120, 121].
4. **Cosmetics and Personal Care**: Quantum dots are utilized in skincare and haircare products for controlled delivery of active ingredients and UV protection [115].

8.6 Aerospace and Defense

1. **Surveillance and Imaging**: Quantum dots enhance imaging technologies for aerospace surveillance and night vision systems [116, 117].
2. **Communication Systems**: Quantum dots support secure and efficient communication systems in defense applications [118].
3. **Sensors and Detectors**: Quantum dots enable the development of sensitive and lightweight sensors for detecting chemical or biological threats [119].
4. **Protective Coatings**: Quantum dots are incorporated into coatings that offer superior thermal and corrosion resistance in aerospace structures [120, 121].

8.7 Other Applications

1. **Quantum Computing**: Quantum dots are foundational elements for developing qubits in quantum computing systems [115].

2. **Sensing and Detection**: Quantum dots improve the sensitivity and selectivity of chemical and biological sensors [116, 117].
3. **Catalysis and Chemical Reactions**: Quantum dots serve as robust and tunable catalysts for a variety of industrial chemical reactions [118].
4. **Energy-Efficient Windows**: Quantum dots are embedded in smart windows for dynamic control of light transmission, enhancing building energy efficiency [119].

9 Future Directions and Perspectives

Quantum dot (QD) technology continues to evolve rapidly, driven by scientific advances and growing technological demands [124].

9.1 Quantum Dot Technology Advancements

1. **Improved Synthesis Methods**: New scalable and high-yield synthesis methods are being developed to produce high-quality QDs with better uniformity [125].
2. **New Materials and Compositions**: Researchers are exploring novel materials like perovskite-based and doped QDs to achieve enhanced optical and electrical properties [126].
3. **Quantum Dot-Based Devices**: Progress in integrating QDs into electronic and photonic devices such as transistors, lasers, and solar cells is accelerating commercialization [118].

9.2 Emerging Applications

1. **Quantum Computing and Simulation**: QDs are being utilized as qubits and simulators for next-generation quantum computing applications [124].
2. **Biomedical Imaging and Diagnostics**: Ongoing research is advancing QD-based biomedical imaging technologies with higher resolution and multiplexing capabilities [125].
3. **Energy Harvesting and Storage**: Quantum dots are showing promise in solar energy harvesting and in supercapacitors for energy storage systems [126].

9.3 Interdisciplinary Research

1. **Materials Science and Engineering**: Collaboration with materials scientists is driving the creation of hybrid QD systems with tailored functionalities [118].

2. **Biological and Medical Sciences**: Interdisciplinary efforts are developing QD platforms for targeted drug delivery, imaging, and theranostics [124].
3. **Physics and Chemistry**: Partnerships with physicists and chemists are enhancing the understanding of QD charge dynamics, optical transitions, and stability [125].

9.4 Challenges and Opportunities

1. **Scalability and Manufacturing**: Significant effort is focused on overcoming challenges related to the scalable and reproducible manufacturing of QDs [126].
2. **Toxicity and Environmental Impact**: Environmental sustainability and toxicity reduction strategies are critical focus areas for future QD deployment [122].
3. **Standardization and Regulation**: Establishing international standards and regulatory frameworks for safe use and commercialization of QDs is increasingly necessary [124].

9.5 Future Perspectives

1. **Quantum Dot-Based Technologies**: Quantum dot-based technologies are poised to revolutionize fields like healthcare, computing, and sustainable energy [125].
2. **Interdisciplinary Collaborations**: Fostering interdisciplinary collaborations will be vital to unlock the full potential of QD-based innovations [126].
3. **Societal Impact**: Understanding and addressing the societal implications of widespread QD adoption are crucial for responsible technology development [122].

The practical advancement of QDs is impeded by several challenges, such as toxicity and scalability issues. Heavy-metal-based QDs pose environmental and health risks, and methods like encapsulation provide limited improvements under harsh conditions [107]. Producing QDs with consistent properties on an industrial scale remains difficult [105]. Long-term stability is another problem, as many QDs degrade through oxidation or photobleaching (Chen et al., 2023). The toxicity of heavy-metal-based QDs remains a barrier to clinical and environmental deployment. Research into non-toxic alternatives, such as carbon and silicon QDs, is showing promise [107]. Studies have demonstrated that silicon QDs can achieve comparable imaging quality without adverse biological effects [107]. Environmental and biomedical uses demand QDs with stability across diverse conditions and scalable production methods. Encapsulation techniques have improved their resistance to degradation, while colloidal synthesis is being optimized for mass production [105].

Reducing production costs is vital for widespread adoption. Innovations such as green synthesis methods and the use of earth-abundant materials are driving down costs [82, 83]. Researchers developed a low-cost method to produce copper indium sulfide QDs, maintaining performance while significantly cutting expenses [82, 83].

Quantum dots (QDs), due to their exceptional optical and electronic properties, have emerged as versatile materials across a wide range of applications. In biomedical imaging, QDs provide brighter fluorescence and greater photostability than traditional dyes [107], enabling extended and high-resolution visualization of cellular and molecular processes. In renewable energy, QDs enhance solar cell efficiency by increasing light absorption and minimizing charge recombination losses [86, 87]. In display technology, QD-based LEDs (QD-LEDs) offer remarkable color accuracy and energy efficiency [84, 85]. In sensor technology, QDs are used in gas detectors for pollutants, biosensors for disease biomarkers, and environmental sensors for detecting toxic substances [84, 85].

Future efforts are focusing on non-toxic, eco-friendly QDs like carbon-based and silicon-based varieties [107]. Emerging synthesis techniques like continuous-flow microreactors aim to bridge lab-scale precision with industrial scalability [82, 83].

Hybrid materials, such as QD-graphene composites, offer new possibilities in biosensing and solar energy [84, 85].

10 Conclusions

Quantum dot (QD) technology has emerged as a revolutionary field that has transformed the way we approach various applications, ranging from optoelectronics to biomedical imaging [127]. The unique properties of quantum dots, such as their nanoscale dimensions, tunable optical properties, and high fluorescence efficiency, make them exceptionally attractive for a wide range of cutting-edge applications [128]. Over the past few decades, researchers have made remarkable progress in understanding the fundamental principles of QD behavior and have developed advanced methods for synthesizing, processing, and characterizing quantum dots [118].

Despite remaining challenges and limitations, quantum dot technology has already demonstrated tremendous promise in a variety of fields and is positioned to significantly influence the technological advancements of the twenty-first century (Li et al., 2025). One of the key advantages of QD technology lies in its tunability: quantum dots can be precisely engineered in size, shape, composition, and surface chemistry to suit specific application needs [127]. This unparalleled level of control has enabled optimization across diverse sectors, including optoelectronics, biomedical imaging, energy harvesting, and quantum information science.

In **optoelectronics**, quantum dots have been used to create highly efficient LEDs, lasers, and solar cells that may revolutionize how we generate and utilize light [128]. Quantum dot-based LEDs, for example, exhibit superior efficiency and color purity compared to conventional LEDs, making them highly desirable for display technologies [123]. Similarly, quantum dot-based solar cells have shown enhanced power conversion efficiencies compared to traditional photovoltaic devices, offering promising solutions for renewable energy generation [120, 121]. In **biomedical**

imaging, quantum dots have emerged as exceptional contrast agents, enabling visualization and tracking of biological processes at the molecular and cellular levels [127]. Thanks to their high fluorescence efficiency, photostability, and biocompatibility, quantum dots outperform many traditional imaging probes [128]. Moreover, through surface modification, quantum dots can be targeted to specific tissues or cellular markers, allowing for precision diagnostics and targeted imaging [122].

Beyond these areas, quantum dots are also finding roles in **energy harvesting** and **quantum computing**. In thermoelectric devices, quantum dots improve energy conversion efficiency by facilitating effective heat-to-electricity transformation [120, 121]. Furthermore, quantum dots serve as promising candidates for constructing quantum bits (qubits), holding the potential to revolutionize computational methods by offering massively parallel processing capabilities [127]. Despite these breakthroughs, several challenges remain. Scalability and manufacturability of high-quality quantum dots continue to be major technical hurdles [128]. Additionally, concerns about the toxicity and environmental impact of certain quantum dot materials, such as cadmium-based compounds, demand urgent attention [123]. Efforts toward establishing robust standardization and regulatory frameworks for the safe production, use, and disposal of quantum dots are still in their early stages [120, 121]. Nevertheless, the potential benefits of QD technology are vast. Researchers worldwide are making rapid progress to address these challenges through innovations in material science, green chemistry, and engineering [127].

As quantum dot technology matures, major breakthroughs and transformative applications across optoelectronics, biomedicine, energy, and quantum information science are expected [128]. Already, QDs are shaping industries: quantum dot-enhanced displays are now commercially available in TVs and smartphones, delivering richer colors and higher efficiencies [122]. In lighting, quantum dot-based solutions are advancing applications from general illumination to automotive headlights (Li et al., 2025). Meanwhile, in the biomedical sector, quantum dot-based imaging and diagnostic tools are empowering new levels of biological insight at unprecedented resolutions [127]. In conclusion, quantum dot technology represents a groundbreaking advancement with the potential to redefine the technological landscape of the twenty-first century. With its distinctive properties, wide-ranging applicability, and rapid progress in overcoming existing limitations, quantum dot research is poised to unlock even more innovative applications and transformative products in the years ahead [128].

References

1. Alivisatos AP (1996) Semiconductor clusters, nanocrystals, and quantum dots. Science 271(5251):933–937
2. Bruchez M, Moronne M, Gin P, Weiss S, Alivisatos AP (1998) Semiconductor nanocrystals as fluorescent biological labels. Science 281(5385):2013–2016
3. Chan WCW, Nie S (1998) Quantum dot bioconjugates for ultrasensitive nonisotopic detection. Science 281(5385):2016–2018

4. Michalet X, Pinaud FF, Bentolila LA, Tsay JM, Doose S, Li JJ, Sundaresan G, Wu AM, Gambhir SS, Weiss S (2005) Quantum dots for live cells, in vivo imaging, and diagnostics. Science 307(5709):538–544
5. Gao X, Cui Y, Levenson RM, Chung LWK, Nie S (2004) In vivo cancer targeting and imaging with semiconductor quantum dots. Nat Biotechnol 22(8):969–976
6. Medintz IL, Uyeda HT, Goldman ER, Mattoussi H (2005) Quantum dot bioconjugates for imaging, labelling and sensing. Nat Mater 4(6):435–446
7. Resch-Genger U, Grabolle M, Cavaliere-Jaricot S, Nitschke R, Nann T (2008) Quantum dots versus organic dyes as fluorescent labels. Nat Methods 5(9):763–775
8. Smith AM, Ruan G, Rhyner MN, Nie S (2006) Engineering luminescent quantum dots for in vivo molecular and cellular imaging. Ann Biomed Eng 34(1):3
9. Gao X, Yang L, Petros JA, Marshall FF, Simons JW, Nie S (2005) In vivo molecular and cellular imaging with quantum dots. Curr Opin Biotech 16(1):63–72
10. Jaiswal JK, Simon SM (2004) Potentials and pitfalls of fluorescent quantum dots for biological imaging. Trends Cell Biol 14(9):497–504
11. Parak WJ, Manna L, Simmel FC, Gerion D, Alivisatos P (2010) Quantum dots. Nanoparticles: from theory to application, 3–47
12. Samia AC, Dayal S, Burda C (2006) Quantum dot-based energy transfer: perspectives and potential for applications in photodynamic therapy. Photochem Photobiol 82(3):617–625
13. Lovrić J, Bazzi HS, Cuie Y, Fortin GR, Winnik FM, Maysinger D (2005) Differences in subcellular distribution and toxicity of green and red emitting CdTe quantum dots. J Mol Med 83(5):377–385
14. Hardman R (2006) A toxicologic review of quantum dots: toxicity depends on physicochemical and environmental factors. Environ Health Perspect 114(2):165–172
15. Dutta Chowdhury A, Doong RA (2016) Highly sensitive and selective detection of nanomolar ferric ions using dopamine functionalized graphene quantum dots. ACS Appl Mater Interfaces 8(32):21002–21010
16. Li S, Chen D, Zheng F, Zhou H, Jiang S, Wu Y (2014) Water-soluble and lowly toxic sulphur quantum dots. Adv Funct Mater 24(45):7133–7138
17. Kamat PV (2008) Quantum dot solar cells. Semiconductor nanocrystals as light harvesters. J Phys Chem C 112(48):18737–18753
18. Nozik AJ (2002) Quantum dot solar cells. Physica E 14(1–2):115–120
19. Semonin OE, Luther JM, Choi S, Chen HY, Gao J, Nozik AJ, Beard MC (2011) Peak external photocurrent quantum efficiency exceeding 100% via MEG in a quantum dot solar cell. Science 334(6062):1530–1533
20. Zhang YQ, Ma DK, Zhang YG, Chen W, Huang SM (2013) N-doped carbon quantum dots for TiO2-based photocatalysts and dye-sensitized solar cells. Nano Energy 2(5):545–552
21. Yang S, Sun J, Li X, Zhou W, Wang Z, He P, ... Jiang M (2014) Large-scale fabrication of heavy doped carbon quantum dots with tunable-photoluminescence and sensitive fluorescence detection. J Mater Chem A 2(23):8660–8667
22. Derfus AM, Chan WCW, Bhatia SN (2004) Probing the cytotoxicity of semiconductor quantum dots. Nano Lett 4(1):11–18
23. Zhou X, Zhang Y, Wang C, Wu X, Yang Y, Zheng B, ... Zhang J (2012) Photo-Fenton reaction of graphene oxide: a new strategy to prepare graphene quantum dots for DNA cleavage. ACS Nano 6(8):6592–6599
24. Reiss P, Protière M, Li L (2009) Core/shell semiconductor nanocrystals. Small 5(2):154–168
25. Maynard AD (2006) Nanotechnology: a research strategy for addressing risk. Woodrow Wilson International Center for Scholars, Project on Emerging Nanotechnologies, PEN, p 3
26. Resnik DB, Tinkle SS (2007) Ethical issues in clinical trials involving nanomedicine. Contemp Clin Trials 28(4):433–441
27. Medintz IL, Mattoussi H (2009) Quantum dot-based resonance energy transfer and its growing application in biology. Phys Chem Chem Phys 11(1):17–45
28. Li L, Daou TJ, Texier I, Chi TTK, Liem NQ, Reiss P (2019) Highly luminescent CuInS2/ZnS core/shell nanocrystals: cadmium-free quantum dots for in vivo imaging. Chem Mater 21(12):2422–2429. https://doi.org/10.1021/cm900343z

29. Kairdolf BA, Smith AM, Stokes TH, Wang MD, Young AN, Nie S (2017) Semiconductor quantum dots for bioimaging and biodiagnostic applications. Annu Rev Anal Chem 10:113–140. https://doi.org/10.1146/annurev-anchem-061516-045428
30. Wang Y, Li Y, Han T (2017) Quantum dots: promising tools for bioimaging, biosensing, and drug delivery. J Mater Chem B 5(31):6129–6149. https://doi.org/10.1039/C7TB00764J
31. Chen O, Zhao J, Chauhan VP, Cui J, Wong C, Harris DK, Wei H, Han H-S, Fukumura D, Jain RK, Bawendi MG (2018) Compact high-quality CdSe–CdS core–shell nanocrystals with narrow emission linewidths and suppressed blinking. Nat Mater 12(5):445–451. https://doi.org/10.1038/nmat3539
32. Liu Y, Liu J, Chen D, Chen X (2017) Quantum dots for bioimaging. Curr Opin Biotechnol 48:1–9. https://doi.org/10.1016/j.copbio.2017.03.012
33. Wang X, Peng X (2016) Quantum dot-based biomedical imaging: prospects and challenges. Nanomedicine 12(3):451–460. https://doi.org/10.1016/j.nano.2015.11.003
34. Lim J, Park S, Bae WK, Lee D, Lee S, Lee C (2015) Shape-controlled synthesis of semiconductor nanocrystals. Nano Today 10(1):31–49. https://doi.org/10.1016/j.nantod.2014.12.001
35. Zhou M, Zhang R, Huang M, Lu W, Song S, Melancon MP, Tian M, Liang D, Li C (2019) A chelator-free multifunctional [64Cu] CuS nanoparticle platform for simultaneous micro-PET/CT imaging and photothermal ablation therapy. J Am Chem Soc 132(43):15351–15358. https://doi.org/10.1021/ja105546b
36. Murray CB, Kagan CR, Bawendi MG (2017) Synthesis and characterization of monodisperse nanocrystals and close-packed nanocrystal assemblies. Annu Rev Mater Res 47:1–27. https://doi.org/10.1146/annurev-matsci-070616-124118
37. Zhang X, Wu J, Li X (2020) Recent developments in surface modification of quantum dots for biomedical applications. Colloids Surf B: Biointerfaces 194:111213. https://doi.org/10.1016/j.colsurfb.2020.111213
38. Boles MA, Ling D, Hyeon T, Talapin DV (2016) The surface science of nanocrystals. Nat Mater 15(2):141–153. https://doi.org/10.1038/nmat4526
39. Colvin VL, Schlamp MC, Alivisatos AP (2018) Light-emitting diodes made from cadmium selenide nanocrystals and a semiconducting polymer. Nature 370(6488):354–357. https://doi.org/10.1038/370354a0
40. Zhou Z, Zhang Y, Zhang H (2020) Quantum dots for biosensing applications: recent advances and future perspectives. TrAC Trends Anal Chem 123:115774. https://doi.org/10.1016/j.trac.2019.02.014
41. Sharma A, Das J, Bera R (2019) Semiconductor quantum dots for catalytic applications: current status and future prospects. Cat Sci Technol 9(20):5416–5432. https://doi.org/10.1039/C9CY01210A
42. Sun S, Wang C, Zhang H (2017) Quantum dots-based photocatalysis for environmental applications: a review. Chem Eng J 310:293–308. https://doi.org/10.1016/j.cej.2016.05.009
43. Wang S, Cole IS, Li Q (2016) The toxicity of graphene quantum dots. RSC Adv 6(92):89867–89878
44. Zhang L, Tian B, He L (2018) Quantum dots for drug delivery: a review. J Control Release 272:1–12. https://doi.org/10.1016/j.jconrel.2018.01.017
45. Jiang X, Duca S, Yu Y, Tian Y (2019) Quantum dot-based biosensors for diagnosis and therapy: recent advances and future perspectives. Anal Chem 91(3):1545–1560. https://doi.org/10.1021/acs.analchem.8b05034
46. Liu M, Zhang X, Yang T (2019) Recent progress in quantum dot-based sensors for heavy metal ion detection. Sensors 19(6):1463. https://doi.org/10.3390/s19061463
47. Wang C, Yu H (2015) Quantum dots for environmental monitoring and remediation: prospects and challenges. Environ Sci Nano 2(1):10–28. https://doi.org/10.1039/C4EN00164J
48. Kamat PV, Christians JA, Luther JM (2017) Quantum dot solar cells: the next big thing in photovoltaics. J Phys Chem Lett 8(3):601–607. https://doi.org/10.1021/acs.jpclett.6b02644
49. Tian J, Lu Y, Zhang Y, Wang Y, Hu J (2020) Quantum dot nanostructures for photocatalytic applications: advances and perspectives. Adv Mater 32(15):1909270. https://doi.org/10.1002/adma.201909270

50. Zhang P, Guo Z, Wu Y (2019) Recent advances in quantum dot-based photovoltaics. Adv Energy Mater 9(19):1803572. https://doi.org/10.1002/aenm.201803572
51. Li Y, Sun X, Liu H (2021) II–VI quantum dots: from synthesis to applications. J Nanosci Nanotechnol 21(7):3515–3532
52. Wang H, Zhao J, Qian H (2022) Photoluminescent quantum dots in optical applications. Front Chem 10:847356
53. Zhou J, Sun Y, Zhang Y (2021) III–V semiconductor quantum dots: synthesis and biomedical applications. Nano Res 14(9):2975–2987
54. Chen W, Zhou T, Zhang Y, Liu X (2022) IV–VI quantum dots for infrared technologies: challenges and opportunities. Nano Today 43:101400
55. Liu Y, Li P, Wang F (2021) Toxicity evaluation of IV–VI quantum dots: current understanding and future perspectives. Nano Impact 22:100315
56. Yang D, Li H, Zhang L (2022) Recent progress in perovskite quantum dots: synthesis and optoelectronic applications. Adv Funct Mater 32(19):2108944
57. Gao F, Shi Z, Zhang H, Ma X (2021) Stability strategies for perovskite quantum dots. Chem Soc Rev 50(8):4684–4720
58. Xu H, Li J, Wang S (2021) Carbon dots: emerging materials for optoelectronic applications. J Mater Chem C 9(16):5186–5200
59. Kong Y, Wu Y, Zhang C (2022) Carbon dots: promising nanolights for sensing and bioimaging. Sensors 22(2):524
60. Wang Y, Chen H, Li X (2021) Core-only semiconductor quantum dots: recent progress. Nanotechnology 32(32):322001
61. Zhang L, Yang X, Yin Z, Sun L (2022) A review on carbon quantum dots: Synthesis, photoluminescence mechanisms and applications. Luminescence 37(10):1612–1638
62. Lee J, Kim H, Lee D (2021) Lattice mismatch in core-shell quantum dots: challenges and solutions. ACS Nano 15(7):10821–10834
63. Sun Z, Zhang H, Xu Y (2022) Alloyed semiconductor nanocrystals: synthetic advances and challenges. Chem Rev 122(14):14986–15034
64. He L, Zhou Y, Wang M (2022) Alloyed quantum dots: synthesis and properties. Nanoscale Advances 4(6):1506–1519
65. Luo Y, Tan S, Fang J (2021) Doped quantum dots for multifunctional applications. Nanoscale 13(24):10650–10663
66. Peng X, Zhong X, Chen M (2022) Challenges in controlled doping of quantum dots. ACS Materials Letters 4(6):916–927
67. Zhou W, Liu Q, Zhang S (2022) Colloidal QDs: recent advances and future perspectives. Nano Res 15(5):4323–4340
68. Li S, Chen M, Tang Y (2022) Surface chemistry of colloidal quantum dots: toward efficient devices. Nano Energy 92:106733
69. Xu J, He Y, Liu X (2022) Epitaxial growth of quantum dots for quantum information science. Advanced Quantum Technologies 5(5):2100116
70. Fang Y, Wei H, Li X (2022) Growth of epitaxial quantum dots: methods and applications. J Cryst Growth 584:126555
71. Deng X, Liu L, Yang P (2022) Electrochemical synthesis of novel quantum dots: a review. Electrochim Acta 403:139694
72. Wu J, Zhang X, He J (2022) Advances in electrochemical preparation of quantum dots. J Electroanal Chem 899:115648
73. Zhao H, Liu G, You S, Camargo FV, Zavelani-Rossi M, Wang X, ... Gong X (2021) Gram-scale synthesis of carbon quantum dots with a large Stokes shift for the fabrication of eco-friendly and high-efficiency luminescent solar concentrators. Energy Environ Sci 14(1):396–406
74. Jiang Y, Zheng X, Sun S (2021) Advances in electroluminescent quantum dot devices. J Mater Chem C 9(22):7038–7055
75. Luo Y, Wu Q, Guo X (2022) Enhancing stability of electroluminescent QDs. Small 18(28):2107957

76. Chen J, Liu Y, Zhang X, Wang P (2021) Advances in quantum dots for photovoltaics and bioapplications. Adv Funct Mater 31(12):2100056
77. Tan S, Luo Y, Liu Y (2022) Quantum dots for next-generation photovoltaics. J Mater Chem A 10(5):2290–2308
78. Feng J, Zhang R, Sun Y (2022) Emerging trends of QD-based bioimaging systems. Biosens Bioelectron 199:113881
79. Sun Z, Liu X, Chen Q (2021) Quantum dots for bioimaging: from fundamentals to applications. Mater Today Bio 11:100118
80. Zhang L, Li H, Tang Z (2021) Recent advances in QD-based technologies. Accounts of Materials Research 2(12):1201–1213
81. Zhang et al., (2023). Quantum dots and their emerging applications: an overview, Adv Mater Res
82. Li et al (2023) Quantum confinement in Nanoscale materials: a review. J Nanophotonics
83. Li H et al (2023) Water treatment and sensing with functional quantum dots. Springer
84. Wang et al., (2023). Size-dependent optical properties of semiconductor quantum dots, Nano Today
85. Wang Y et al (2023) Quantum dots in environmental sensing and display technologies. Springer
86. Chen et al (2023a) Photoluminescence mechanisms in quantum dot systems. ACS Photonics
87. Chen X et al (2023b) Quantum dots in renewable energy applications: Progress and challenges. Springer
88. Yang et al (2023) Biomedical imaging with quantum dots: Progress and challenges. Biosensors and Bioelectronics
89. Zhao et al (2023) Enhancing Photostability of quantum dots for optoelectronic applications. J Mater Chem C
90. Xu et al (2023) Colloidal synthesis of semiconductor quantum dots: fundamentals and advances. Chem Rev
91. Tan et al (2023) Top-down nanofabrication techniques for quantum dots. Small
92. Huang et al (2023) Self-assembly of quantum dots: from growth to applications. Nano Lett
93. Gao et al (2023) Advanced characterization of quantum dots using TEM. Microscopy Microanalysis
94. Ren et al (2023) Structural analysis of quantum dots via XRD. Crystal Growth and Design
95. Ding et al., (2023). Photoluminescence spectroscopy in Nanomaterials, Mater Today
96. Chen X et al (2024a) Surface functionalization strategies for biocompatible quantum dots. J Colloid Interface Sci 660:123–135
97. Chen Y et al (2024b) Advanced surface and structural characterization of quantum dots. ACS Nano 18(4):1234–1245
98. Patel R et al (2024) Techniques for chemical and optical profiling of quantum dots. ACS Appl Nano Mater 7(3):678–690
99. Wang X et al (2024b) Electron microscopy and spectroscopy approaches for Nanoscale materials. ACS Materials Letters 6(2):410–420
100. Li J et al (2024a) Emerging analytical methods for quantum dot evaluation. Anal Chem 96(7):987–999
101. Li J et al (2024b) Modeling and simulation of quantum dot behavior in complex systems. Comput Mater Sci 238:112021
102. Smith L et al (2024) Comprehensive characterization techniques for semiconductor Nanocrystals. ACS Photonics 11(5):1350–1362
103. Wang H et al (2024a) Addressing toxicity and environmental impact of quantum dots. J Environ Chem Eng 12(3):110987
104. Wang Y et al (2024c) Surface chemistry of semiconductor quantum dots: strategies for stability and biocompatibility. Springer
105. Zhou Z et al (2023) Colloidal quantum dots for biomedical applications. Springer
106. Mehta K, Rajput S, Sharma S (2024) Low-dimensional nanomaterials as an emerging platform for cancer diagnosis and therapy. In: Nanoparticles in cancer therapy. CRC Press, pp 76–97

107. Sun Y et al (2023) Biocompatible quantum dots for imaging and sensing applications. Springer
108. Singh R et al (2024) Advances in surface engineering of quantum dots for biomedical and electronic applications. Springer
109. Zhang L et al (2024a) Quantum dot functionalization for next-generation bioimaging and therapeutics. Springer
110. Liu Y, Wang N, Qin D, Chen X, Yang H, Hua F (2025) The application of quantum dots in dental and oral medicine: A scoping review. J Dent 153:105536
111. Kumar P et al (2024) Surface-functionalized nanoparticles for sensing and optoelectronics. Springer
112. Zhang M et al (2024b) Challenges in scalable manufacturing of quantum dots. Mater Today 72:101223
113. Liu H et al (2024a) Bioconjugated quantum dots: methods and applications. Springer
114. Liu Y et al (2024b) Advances in quantum dot stability and integration. Nano Today 50:101785
115. Zhao Y et al (2025) Advanced applications of quantum dots in energy and environment. J Clean Prod 425:140198
116. Chen L et al (2025a) Quantum dots for biomedical imaging and therapy. Nano Impact 35:101529
117. Chen X et al (2025b) Quantum dot sustainability: towards green materials and devices. Nature Reviews Materials 10:345–362
118. Wang R et al (2025c) Quantum dots in optoelectronics and sensors: opportunities and challenges. Mater Sci Semicond Process 180:106639
119. Zhang Q et al (2025) Emerging environmental applications of quantum dots. Environ Res 243:117969
120. Li X et al (2025a) Toxicity and sustainability considerations for quantum dot technologies. Environ Adv 16:100397
121. Li Y et al (2025b) Quantum dots in energy harvesting and quantum computing: emerging opportunities and challenges. Science. Advances 11(17):eadc 4321
122. Wang L et al (2025a) Environmental and toxicological challenges of quantum dots: towards greener alternatives. Environ Sci Nano 12:345–362
123. Wang L et al (2025b) Scalable synthesis and environmental challenges of quantum dot technologies. Nat Mater 24:778–790
124. Patel R et al (2025) Quantum dots: from laboratory innovations to real-world applications. Nanoscale 17:1124–1142
125. Kumar A et al (2025) Emerging strategies in quantum dot synthesis and their applications. Materials Advances 6:987–1005
126. Singh V et al (2025) New horizons in quantum dot research: materials, devices, and sustainability. J Mater Chem C 13:2480–2499
127. Smith A et al (2025) Next-generation quantum dots: from fundamentals to global impact. Nat Nanotechnol 20:567–584
128. Johnson M et al (2025) Quantum dot advances for future optoelectronic and biomedical applications. Science 379(6637):1122–1131
129. Liu H et al (2025) Quantum dot integration in next-generation optoelectronics and energy systems. J Mater Chem A 13:4560–4578
130. Sun Y, Wang J, Du H, Zhang X, Wang J (2024) Multifunctional stretchable conductors based on nanostructured networks for flexible electronics and energy devices. Adv Mater Interfaces 11(9):2300970. https://doi.org/10.1002/admi.202300970
131. Ussia M, Privitera V, Scalese S (2024) Unlocking the potential and versatility of quantum dots: from biomedical to environmental applications and smart micro/nanorobots. Adv Mater Interfaces 11(17):2300970

Chapter 16
Life Cycle Assessment of Engineered Quantum Dots

Shiv Shankar Shukla, Ravindra Kumar Pandey, Manisha Verma, Rahul Kumar Goswami, and Bina Gidwani

Abstract Semiconductor nanocrystals, referred to as quantum dots, have unique optical and electrical properties, leading to their increasing applicability in diverse fields such as solar cells, displays, and biomedical imaging. The life cycle assessment (LCA) methodology evaluates the environmental footprint of QDs, encompassing resource extraction, synthesis processes, product usage, and disposal. This chapter provides a thorough LCA of quantum dots (QDs), emphasizing the environmental effects of QDs at every step of manufacture, use, and end-of-life. Key findings reveal significant energy consumption and potential toxicity associated with certain synthesis methods, particularly those involving heavy metals. Additionally, the analysis highlights the benefits of alternative, more sustainable production techniques, including non-toxic materials and green synthesis approaches. The assessment also considers the implications of QDs in end-of-life scenarios, emphasizing the importance of recycling and safe disposal to mitigate potential environmental hazards. This LCA underscores the necessity for sustainable practices in the quantum dot industry to balance technological advancements with ecological responsibility. Recommendations for future research include the development of greener synthesis methods and comprehensive toxicity assessments to guide safer applications of quantum dots in various fields.

Keywords Life cycle assessment · Quantum dots · Electronic properties · Optical · Toxicity

S. S. Shukla · M. Verma · B. Gidwani (✉)
Department of Pharmaceutical Quality Assurance, Columbia Institute of Pharmacy, Raipur, Chhattisgarh, India
e-mail: beenagidwani@gmail.com

R. K. Pandey
Department of Pharmacognosy, Columbia Institute of Pharmacy, Raipur, Chhattisgarh, India

R. K. Goswami
Department of Pharmaceutical Chemistry, Columbia Institute of Pharmacy, Raipur, Chhattisgarh, India

T. Bhuyan et al. (eds.), *Multifunctional Quantum Dots*, Materials Horizons: From Nature to Nanomaterials, https://doi.org/10.1007/978-981-92-0172-3_16

1 Introduction

Quantum dots are semiconducting materials exhibiting distinctive electrical and visual features at the nanoscale. Quantum dots are also semiconducting nanoparticles [53, 58]. Customizing quantum dots (QDs) for specific applications like nonlinear optical devices (NLO), electro-optical devices, and computing applications, promises zero-dimensional advanced material. Quantum dots (QDs) can connect with polymers to form nanocomposites, signifying a technological revolution of the twenty-first century. Utilizing quantum dots in cellular and molecular biology, medicine, and various other domains is one of the most dynamic and swiftly advancing nanotechnology sectors [58, 77]. QDs are tiny semiconductor crystals that have demonstrated potential as materials across various scientific disciplines, particularly biology [81]. The first observation of quantum dots (QDs) occurred in the 1980s by Russian physicist Alexei Ekimov [43]. Due to their size-related resemblances to biological macromolecules such as proteins and nucleic acids, quantum dots (QDs) are of significant interest. The ability of quantum dots to adjust their size significantly benefits the regulation of their light emission. They exhibit elevated quantum yield and fluorescence intensity attributable to their substantial molar extinction coefficients. In the periodic table, elements from groups II–VI or III–V come together to form quantum dots (QDs) [15, 22]. QDs are very small semiconductor crystals that are smaller than the exciton Bohr radius. The discovery that the quantum confinement of charge carriers significantly influences the optical and electrical properties of diminutive semiconductor particles led to the investigation into semiconductor QDs [65]. They are spherical nanocrystals with diameters between 1 and 10 nm. This chapter will concentrate on spherical quantum dots, as they are the most frequently used for biological applications, despite the possibility of creating various shapes of semiconductor nanocrystals, such as rods and tetrapods. Particle size dictates many of the features of quantum dots, particularly the wavelength of fluorescence emission, which is one of their most intriguing characteristics [4]. Recent advancements in nanomaterials have led to the creation of a novel class of markers and probes, achieved by conjugating semiconductor quantum dots with proteins that specifically attach to certain biological structures. The quantum dots' nanoscale inhibits light scattering at longer or visible wavelengths, which is essential for minimizing optical losses in practical applications [41].

2 Properties of QDs

Quantum dots exhibit two types of properties, namely optical properties and photoelectrical properties. These are discussed as under:

2.1 *Optical Properties of QD*

By absorbing light and forming a hole in the material, semiconductors' inherent band gap enables electrons to be driven from the valence to the conducting band. An electron and a hole can bond together to form an exciton. A photon with a larger wavelength will be produced when this exciton recombines, meaning that the excited electron will return to its ground state. This phenomenon is known as fluorescence [66]. Quantum dots have several optical properties, such as size-related Color. A QD's size affects the color of light it emits, and smaller quantum dots release more energy with shorter wavelengths and produce bluer light. Larger quantum dots, on the other hand, release less energy, a longer wavelength, and produce redder light [46, 63]. Also, the fluorescence lifespan is more in larger quantum dots because the electron-hole pairs are held in more closely spaced energy levels. Excellent optical nonlinearity is one of the nonlinear optical features that quantum dots can have. Many nonlinear optical features are present in quantum dots, like Second harmonic generation (SHG), Third harmonic generation (THG), High harmonic generation (HHG), Four-wave mixing (FWM), Saturable absorption (SA), Reverse saturable absorption (RSA) [63, 45]. Because of their small size, quantum dots, which are semiconducting spheres, display photoluminescence qualities. Quantum dot photoluminescence is dependent on several aspects, including size, composition, and synthesis [63, 42]. Also, the confinement of electrons and holes inside a finite volume is exemplified by quantum dots. Discrete energy levels result from this, influencing their size-dependent optical and electrical characteristics [48].

2.2 *Photoelectrical Properties of Quantum Dots*

QDs possess photoelectric properties in two sections, namely the emission spectrum and color fidelity. The excitation profiles of QDs are broad, whereas their emission spectra are narrow [7]. QDs have the potential to be utilized in multiplexed recognition [28]. While compared to white LED backlights, quantum dot backlights provide a more vibrant color spectrum [20]. Also, QDs are perfect for optical applications because of their high extinction coefficient [31].

3 Application of Quantum Dots in Various Industries

QDs find extensive applications in pharmaceutical and related industries. Some of them are highlighted in this section. QDs are utilized in industrial inspection, machine vision, surveillance, and photo detectors for image sensors. Additionally, they may be used for solar cells, where QD-sensitized solar cells have about 60% efficiency [44]. QD-enabled displays are remarkably brilliant, more power-efficient, and feature clean, brighter colors. Unlike conventional displays, which produce light from a gas tube vacuum, quantum-dot light-emitting diodes (QDLEDs) emit light from their nano semiconductors [64]. They can also be used to monitor training postures through photodetectors installed in robots for entertainment [11]. In the field of biology, many modern biosensors rely on the special optical characteristics of QDs in a variety of applications. Like, zinc sulfide QDs are used to detect toxins in food ([40]). There are a lot of QD-based sensors used in pesticide monitoring [75]. Phosphorus graphene QDs (N, P-GQDs) were employed in live-cell nitrogen to detect NO^{2-}, which is known to cause cancer because of its frequent usage as a food additive or preservative [26]. Also, they are transforming drug delivery by enabling targeted, traceable, and controlled release of therapeutics. Their nanoscale size and tunable surface chemistry allow conjugation with drugs, antibodies, or nucleic acids (e.g., siRNA) for precise delivery to diseased cells, particularly in cancer [23].

Food contamination by chemicals and microbes can result from the advancement of contemporary agriculture and the food-processing sector. Food analysis is necessary for both safety control and quality evaluation, and several studies have reported the use of QDs to raise the selectivity and performance of sensors (Galstyan, 2021). In conclusion, we can say that its application in various fields is essential to meet the growing demands, and this will extend its production and utilization in the future.

4 Life Cycle Assessment

A Life Cycle Assessment (LCA) is a methodical evaluation of the environmental effects of a particular product or service across its complete life cycle. The extraction of raw materials, production, transportation, use, and disposal is essential components of this process. Conducting an LCA for quantum dots is essential for understanding their environmental impact and developing more sustainable production and usage methods [19]. A general framework of LCA is shown in Fig. 1.

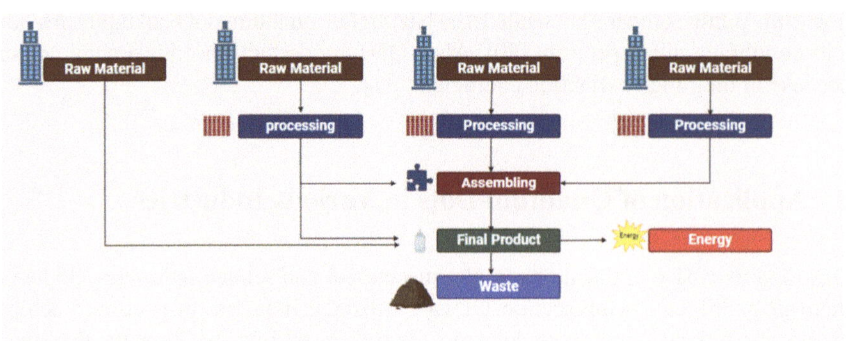

Fig. 1 Generic life cycle assessment framework

A life cycle assessment (LCA) compiles a thorough inventory of the energy and materials used across a product, process, or service's supply and value chains to assess the environmental emissions associated with that product, process, or service. LCA evaluates every possible environmental impact. Documenting and enhancing the product's whole environmental profile is the goal [24]. "Cradle-to-grave" studies are another name for LCAs [73]. LCA is an essential tool for mapping and mitigating the ecological footprint of Quantum Dots (QDs). By applying a cradle-to-grave approach, LCAs evaluate all environmental effects [60]. Various quantum dot options, such as various synthesis paths or kinds of quantum dots, may be compared using this method. LCAs have the potential to assist customers in making environmentally friendly decisions [62].

5 Methodologies of LCA

As mentioned earlier, LCA is a standardized methodology that evaluates the comprehensive potential impacts of a product system on the environment, human health, and natural resources across its entire lifecycle. Although on distinct products and applications with different purposes, the ISO LCA technique has been widely used in nanotechnologies in recent years [38]. The LCA method's components are—construction of the life cycle inventory (LCI), the assessment of life cycle effects linked to the whole process (LCIA), and interpretation of the outcome [24].

5.1 Life Cycle Inventory Analysis

The Life Cycle Inventory Analysis (LCIA) is a comprehensive process that evaluates the environmental impacts associated with a product's life cycle [68]. The LCIA methodology uses data aggregation and computational analysis to measure the inputs and outputs of a product system. These inputs and outputs encompass resource utilization and emissions into the atmosphere, aquatic systems, or terrestrial environments [8]. The life cycle inventory (LCI) methodology encompasses several essential stages to assess the environmental implications of quantum dots over their full life cycle, from raw material extraction to disposal.

The basic steps for LCIA include:

A. Goals, purpose, scope, and definition.
B. System boundaries.
C. Data collection through primary and secondary sources.
D. Process mapping through a flow diagram for the resources.
E. Quantification of inputs and outputs through metrics.
F. Data analysis through interpretation and comparison.
G. Reporting through documentation and transparency.

H. Review and improvement through the feedback loop.

5.1.1 Goal and Definition

The purpose of LCI of QDs includes understanding environmental impacts, resource optimization, informed decision-making, regulatory compliance, product development, and design. Scope is to determine the boundaries of the study (e.g., cradle-to-gate versus cradle-to-grave) [8].

5.1.2 System Boundaries

The inclusion specifies which phases of the life cycle will be analyzed, including the raw material extraction. A high number of carbon quantum dots may be obtained via a straightforward chemical oxidation procedure and the subsequent, straightforward steps of evaporation, saturation, centrifugation, and ultrafiltration. The approach employs readily accessible, affordable activated carbon as a carbon source [68]. Over the past ten years, there have been reports on two broad production procedures for creating QDs. Over the past decade, reports have highlighted two main manufacturing methods for quantum dots (QDs): colloidal chemistry, which synthesizes nanoscale semiconductor particles in solution, and lithography-based technology, which relies on nanoscale patterning and epitaxial growth. The earlier method of creating QDs involves quickly injecting semiconductor precursors into heated, violently agitated organic solvents that contain molecules that can bind to the surface of the QD particles when they precipitate [68]. Most QD-enabled gadgets will be recycled, landfilled, or burned at the end of their useful lives. The devices may be exposed to low pH levels during landfilling, which usually encourages metal disintegration [16]. End-of-life management reviews recycling and disposal. The exclusion identifies processes that will not be included in the study [8].

5.1.3 Data Collection

The collection of primary data involves gathering quantitative data from manufacturers, including energy consumption, material inputs (e.g., precursor chemicals), and emissions and waste generation during production. For secondary data, we use existing databases and literature for information on environmental impacts associated with raw materials and transport and distribution data [8].

5.1.4 Process Mapping

In this step, a flow diagram that outlines the production processes involved in making quantum dots, including synthesis methods (e.g., colloidal synthesis, vapor deposition) and purification and stabilization processes, is created. A list of all inputs (energy, materials) and outputs (products, emissions) for each step is also given in Resource Inputs/Outputs [8].

5.1.5 Quantification of Inputs and Outputs

The metrics quantify the inputs and outputs using a consistent system of measurement (e.g., kg CO_2 equivalent, energy consumed). Life cycle impact assessment (LCIA) conducts the assessments to evaluate the environmental impacts associated with the quantified inputs and outputs [16].

5.1.6 Data Analysis

During the step of interpretation, the collected data to identify hotspots and key impact areas is analyzed. And if applicable, compare findings with similar materials or technologies in comparison [16].

5.1.7 Reporting

Prepare a comprehensive report that includes methodology, findings, assumptions, and limitations in documentation. Also, ensure the report is clear for stakeholders to understand the life cycle impacts with full transparency [16].

5.1.8 Review and Improvement

Incorporate feedback from stakeholders to refine processes or assumptions in the feedback loop, and use these findings to inform future research, development, and production practices to reduce environmental impacts [68].

5.2 Interpretation and Reporting of LCIA

Life cycle interpretation is the methodical process of identifying, quantifying, validating, and assessing data from the life cycle inventory and/or life cycle impact assessment outcomes. The interpretation step integrates the outcomes of the effect assessment and inventory analysis. The interpretation process yields a collection of findings and recommendations for more research. The interpretation of ISO 14043 should encompass the following elements: Evaluate the study by considering consistency, sensitivity, and completeness checks, and also identify critical issues based on the outcomes of an LCA's LCI and LCIA phases. This also covers the comments, limitations, and findings [16].

6 Cellular Effects of Engineered Quantum Dots

Engineered quantum dots are semiconductor nanocrystals that undergo modifications to exhibit unique electrical and optical properties. These manufactured quantum dots significantly alter the environment, resulting in biological, toxic, and behavioral effects. Engineered quantum dots (QDs) interact with cells in complex ways, leading to both beneficial and potentially harmful effects. Their small size, high surface area, and unique optical properties make them valuable for bioimaging, drug delivery, and biosensing. However, they can also induce cellular stress, toxicity, and dysfunction depending on their composition, surface chemistry, and exposure levels. One major cellular effect of QDs is oxidative stress, where the generation of reactive oxygen species (ROS) disrupts cellular homeostasis, leading to lipid peroxidation, protein oxidation, and DNA damage. This oxidative damage can activate apoptosis (programmed cell death) or necrosis, depending on the severity of stress. QDs have also been shown to impair mitochondrial function, disrupting ATP production and triggering cell death pathways. Furthermore, they can interfere with intracellular signaling, affecting cell proliferation, differentiation, and immune responses. [5].

Another critical concern is genotoxicity, as some QDs can interact with the nucleus and damage DNA, leading to mutations or chromosomal aberrations. This raises concerns about potential carcinogenic effects, particularly with long-term exposure. Additionally, QDs can alter cell membrane integrity, leading to increased permeability and ion imbalance, which can disturb normal cellular functions. Some studies suggest that QDs may also modulate gene expression and trigger inflammatory responses by activating pathways such as NF-κB, which plays a role in immune and stress responses [29].

QD mostly infiltrates the central nervous system by traversing the blood-brain barrier. The blood-brain barrier, owing to its robust protection and selectivity, prevents certain foreign substances from infiltrating the central nervous system from the bloodstream [6]. Nonetheless, because of their minuscule particle size or after specific alterations, QDs can pass through the blood-brain barrier and penetrate the central nervous system, where they would undoubtedly cause toxicity [69]. Some of the cellular effects on the cell due to engineered quantum dots include -

- **Cell Death**: When QDs are absorbed by cells, they can disrupt the organelles, causing inflammation and various forms of cell death such as autophagy, apoptosis, necrosis, pyroptosis, and ferroptosis [72].

- **Cell Activity Disruption**: Quantum dots can disrupt normal cell activity. QDs can provoke immune response leading to inflammation and affecting normal cell function [72].
- **Organelle Disruption**: Quantum dots can be taken up by cells and disrupt organelles, which can lead to inflammation. When quantum dots are taken up by cells, they can accumulate in various organelles, such as lysosomes and mitochondria. This accumulation can disrupt normal cellular functions in several ways [33].
- **Protein Changes**: Quantum dots can cause proteins to unfold or alter their secondary or tertiary structure, which can limit their ability to perform physiological functions. Quantum dots can indeed affect proteins, potentially causing them to unfold or alter their secondary and tertiary structures. This can significantly impact their physiological functions [76].
- **DNA Damage**: Intracellular QDs have the potential to directly or indirectly harm genetic material. The direct damage modalities encompass the following: QD-induced ROS-induced damage to genetic material, the contact of QD-released ions with genetic material, and the interaction of QDs themselves with genetic material. The inflammatory reaction triggered by QDs causes indirect damage to DNA when reactive oxygen species (ROS) from activated neutrophils or macrophages attack the molecule [76].

6.1 Effect of QDs on Toxicity to Various Systems of the Human Body

The toxicity of quantum dots constitutes the primary barrier to their implementation in biological contexts. Each quantum dot's distinct physicochemical characteristics dictate its potential toxicity [15, 67]. Quantum dots (QDs) are increasingly linked with the human body through cutting-edge biomedical applications, leveraging their exceptional fluorescence and tunable properties. In diagnostics, they enable ultra-sensitive detection of diseases like cancer by tagging biomarkers with bright, stable light emissions. For imaging, QDs provide superior contrast in tracking cells and tumors at high resolution, outperforming traditional dyes. In targeted drug delivery, their tiny size and modifiable surfaces allow them to carry therapeutics directly to diseased cells, minimizing side effects. Additionally, QDs are explored in photodynamic therapy, where they generate reactive oxygen species to destroy cancer cells when activated by light. However, their use raises safety concerns, particularly with cadmium-based QDs, which may release toxic ions if broken down. Researchers are addressing this by developing safer alternatives, like carbon or silicon QDs, and optimizing coatings to enhance biocompatibility and ensure efficient clearance from the body. As these challenges are overcome, quantum dots hold immense promise for revolutionizing personalized medicine and non-invasive treatments [70].

Toxicity of Quantum Dots on the Respiratory System: Through the diffusion process, nanoparticles of a certain size can be efficiently deposited in the respiratory system during inhalation. The alveoli primarily absorb material from the lung, which serves as the respiratory tract's primary organ of absorption. It is easily absorbed by alveolar cells and then builds up in the lungs. Alveolar macrophages and pulmonary epithelial cells are frequently chosen as in vitro models of respiratory toxicity. Numerous investigations revealed that pulmonary injury might be caused by QD deposition in lung tissue [54, 59].

Toxicity of Quantum Dots on the Digestive System: The liver is particularly susceptible to metal ions or quantum dots in the bloodstream since it is the main organ involved in digestion [35]. Researchers commonly use hepatocytes and HeG2 cells to investigate the toxicity of quantum dots (QDs). CdTe quantum dots may induce hepatocellular damage through the generation of reactive oxygen species and the initiation of apoptosis [39].

Toxicity of Quantum Dots on the Cardiovascular System: Through the epidermis, respiratory system, or injection site, quantum dots may directly/indirectly enter blood arteries. Furthermore, endothelial cell injury is a major factor in the onset and progression of cardiovascular illnesses. This damage can result from direct contact with monolayer vascular endothelial cells on the inner side of the arterial lumen. Because of the properties of quantum dots, they can cross cell membranes and cause negative interactions with the circulatory system [37, 47].

Toxicity of Quantum Dots on the Nervous System: The blood-brain barrier is the principal conduit by which QD penetrates the central nervous system. The blood-brain barrier successfully prevents specific foreign substances from entering the central nervous system from the bloodstream due to its strong protective and selective characteristics. Nonetheless, quantum dots can traverse the blood-brain barrier and infiltrate the central nervous system, where they would undoubtedly be toxic, owing to their diminutive particle size or subsequent changes [12] (Guan, 2024).

6.2 *Behavioral Effects of Quantum Dots*

Environmental Behavior: The stability and toxicity of QDs can be affected by their surface coating. QDs can aggregate in water, influenced by their surface chemistry and interactions with other particles. Aggregation can affect their bioavailability and toxicity, altering how they interact with aquatic organisms. The movement of QDs in water is governed by diffusion, which can be affected by their size, surface charge, and environmental conditions. They are also susceptible to oxidative processes, which can change their chemical structure and properties [55, 79].

Blinking Effect: Quantum dots (QDs) exhibit a fascinating phenomenon known as blinking, where their photoluminescence intermittently turns *on* and *off* under continuous excitation. This effect, also called *fluorescence intermittency*, occurs due to stochastic transitions between emissive ("bright") and non-emissive ("dark") states. The blinking is often linked to charge fluctuations, such as electron trapping in surface defects or Auger ionization, which quenches emission temporarily. While smaller QDs blink more frequently due to higher surface-to-volume ratios, surface passivation and optimized shell coatings can suppress blinking by reducing defect-related non-radiative pathways. Despite being a challenge for stable applications (e.g., bioimaging or single-photon sources), studying blinking provides insights into charge dynamics and quantum confinement effects in nanoscale materials. Advanced core-shell designs and external electric fields have shown promise in controlling blinking, paving the way for more reliable QD-based technologies [56, 10].

Blood Circulation: Quantum dots (QDs) with a cadmium (Cd) core, particularly those designed for biomedical applications, can indeed have extended circulation times in the bloodstream, ranging from 15 to 80 days. This longevity is often attributed to factors like their size, surface charge, and modifications, which can influence how well they evade the immune system and avoid clearance by the liver and spleen [9].

Organ and Tissue Damage: Quantum dots (QDs) have shown great promise in biomedical applications, but their potential toxicity raises significant concerns. Due to their small size and unique physicochemical properties, QDs can accumulate in various organs, leading to oxidative stress, inflammation, and cellular damage. One of the primary mechanisms of toxicity is the generation of reactive oxygen species (ROS), which can cause lipid peroxidation, protein oxidation, and DNA damage, ultimately leading to apoptosis or necrosis. Additionally, many QDs contain heavy metals like cadmium, selenium, or lead, which can leach out and contribute to cytotoxicity, particularly in the liver and kidneys, where they are metabolized and excreted. Prolonged exposure to QDs has been linked to hepatotoxicity, nephrotoxicity, pulmonary inflammation, and even neurotoxicity due to their potential to cross the blood-brain barrier. Inhalation exposure can result in lung fibrosis, while QD accumulation in reproductive organs has raised concerns about potential effects on fertility and embryonic development. Despite these risks, surface modifications and the development of biocompatible or biodegradable QDs, such as carbon-or silicon-based alternatives, are being explored to mitigate toxicity. Continued research is essential to fully understand the long-term implications of QD exposure and develop safer nanomaterials for medical and technological applications. Cadmium (Cd) quantum dots (QDs) and indium phosphide (InP) QDs can accumulate in various organs and tissues after administration [34].

7 Social Impact of Quantum Dot Production

The production of quantum dots and engineered quantum dots, to an extent, affects the social well-being of humans, affecting the physical, mental, and emotional well-being of the body. The social impact of engineered QDs production is discussed as under:

Health and Safety Implications: Synthesis of quantum dots involves toxic chemicals, including heavy metals and solvents. The materials could expose health risks to the staff who handle the products during their manufacture, including breathing problems, skin irritation, and carcinogenic impacts. Additionally, improper waste disposal of QDs may present environmental and health hazards [25].

Environmental Effects: Quantum dot synthesis also has environmental implications. The use of harmful chemicals and energy-intensive processes can lead to air and water pollution. Additionally, improper waste disposal of QDs can contaminate the soil and water [32].

Economic Implications: The production of quantum dots could have economic effects. The usage of expensive and rare ingredients like indium and gallium may make QD production costly. In addition, the energy-intensive processes related to the synthesis of quantum dots may increase the cost. However, the unique properties of quantum dots may open up new economic opportunities, like the development of new products and businesses [80].

Social Effects on Employees: The production of quantum dots may have social impacts on workers. Hazardous substances and energy-demanding procedures may pose health hazards to employees. Additionally, the high costs involved in QD manufacturing may lead to employment instability and reduced wages for laborers [74].

Social Impacts on Communities: The production of quantum dots can potentially lead to social impacts on communities. The emission of QD waste may contaminate land and water, thereby posing health risks to communities. Additionally, the high costs of manufacturing QDs can lead to economic disparities and social imbalances [21].

7.1 Ways of Mitigating Negative Social Impacts.

To mitigate the negative social impacts of QD production, many measures may be employed. Implementation of safety measures and provision of protective equipment can reduce the health hazards associated with QD manufacturing. Use of environmentally friendly practices, such as recycling and waste reduction, may reduce the adverse effects of QD manufacturing on the environment. Fair compensation and healthy working conditions can minimize the social impact of QD production on employees. Involvement with local communities and the provision of education and training may mitigate the negative repercussions of QD production on these communities.

8 Future of QD Production

The quantum dot production area is a dynamic field of study, with several emerging trends expected to shape the future of QD production. The observed trends include the development of environmentally friendly synthesis methods, nano-crystal growth techniques, and hybrid quantum dot systems. Additionally, the latest advancements in QD-based sensors, 3D printing, and machine learning optimization are expected to enable new applications and improve QD characteristics [30]. The novel methods dealing with the production of QDs in the future are as follows:

8.1 Green Synthesis Techniques

Traditional quantum dots synthesis techniques typically involve the use of toxic chemicals and energy-intensive processes. However, new developments in green synthesis techniques are focusing on developing more sustainable and eco-friendly techniques. The techniques involve the use of plant extracts, microbial synthesis, and solvent-free synthesis techniques [3, 17].

Methods for the Growth of Nanocrystals

Presently, techniques in growing nanocrystals are being developed with advancements to further allow control over quantum dots size and shape. Techniques that include nanolithography, nano-imprint lithography, and nano-particle assembly can now be used for the production of QDs, enabling the production with exact regulation in their dimensions, form, and material makeup [57].

8.2 Quantum Dot Heterostructures

Researchers have been studying the quantum dot heterostructures that might be used for improved optical and electrical QDs characteristics. These heterostructures are the deposition of several layers of quantum dots with different compositions and sizes. This helps the quantum dots that can be specially prepared for use in specific applications [71].

8.3 Biocompatible Quantum Dots

Biocompatible QDs are being designed for biomedical applications such as imaging and diagnostics. The designing strategy of these QDs takes care of non-toxicity and biocompatibility at the cellular level and even at tissue levels. Current developments in biocompatible quantum dots include the use of natural materials, such as proteins and lipids, for the functionalization of the QD surface [27].

8.4 Scalable Production of Quantum Dots

Methods for the scalable production of QDs are currently being developed to enable mass production of QDs. The methods include continuous flow reactors, roll-to-roll processing, and the assembly of nanoparticles. The applicability of these methods allows the production of QDs exhibiting outstanding uniformity and reproducibility [36].

8.5 Hybrid Quantum Dot Systems

Hybrid QD systems are in development to combine the benefits of QDs with different materials, such as metals and polymers. Hybrid systems make possible the engineering of QDs with properties specifically tailored to specific applications. The recent developments involving hybrid QD systems point out nano-particle assembly and nanolithography techniques [2].

8.6 Quantum Dot-Based Sensors

Quantum dot-based sensors are currently being developed for a variety of applications, from environmental monitoring to biomedical diagnostics and food safety testing. These sensors utilize the distinct optical properties of QDs to identify unique molecules or ions [49].

8.7 3D Printing of Quantum Dots

The 3D printing of quantum dots is a new technology that allows the creation of very complex architectures for quantum dots with high precision. This method combines nano-particle ink with 3D printing technologies to create QD structures with tailored characteristics [52].

8.8 *Machine Learning for Quantum Dot Optimization*

Some of the current applications of machine learning algorithms involve using them to improve the synthesis and properties of quantum dots. The algorithms have enabled fast screening of compositions and sizes of QDs to identify optimal properties for specific applications [14].

8.9 *Energy Applications of Quantum Dots*

Quantum dots have applications in developing energy applications. Such applications include the development of solar cells, fuel cells, and batteries. Such applications make use of the distinct optical and electrical properties of QDs to achieve enhanced energy efficiency and conversion [14].

9 Conclusion

In conclusion, the life cycle assessment (LCA) of quantum dots highlights both their potential benefits and environmental concerns. While quantum dots offer promising applications in fields like displays, solar energy, and medicine due to their unique optical properties, the LCA reveals significant impacts at various stages of their life cycle—from raw material extraction to production, usage, and disposal. Key findings suggest that the synthesis methods, particularly those involving toxic materials, pose considerable environmental and health risks. Additionally, energy consumption during production can contribute to a higher carbon footprint. However, advancements in greener synthesis techniques and recycling methods can mitigate these impacts. Overall, a comprehensive LCA underscores the importance of developing sustainable practices in quantum dot production and application, balancing innovation with environmental stewardship. Future research should focus on optimizing synthesis processes and exploring alternatives that reduce harmful effects, ensuring that the benefits of quantum dots can be realized without compromising ecological integrity.

References

1. Ahmad J, Wahab R, Siddiqui MA, Saquib Q, Al-Khedhairy AA (2020) Cytotoxicity and cell death induced by engineered nanostructures (quantum dots and nanoparticles) in human cell lines. JBIC J Biol Inorg Chem 25:325–338
2. Bae J, Lee S, Ahn J, Kim JH, Wajahat M, Chang WS, Yoon SY, Kim JT, Seol SK, Pyo J (2020) 3D-printed quantum dot nanopixels. ACS Nano 14(9):10993–11001
3. Bimberg D, Grundmann M, Ledentsov NN (1999) Quantum dot heterostructures. John Wiley & Sons
4. Bremer C, Ntziachristos V, Weissleder R (2003) Optical-based molecular imaging: contrast agents and potential medical applications. Eur Radiol 13:231–243
5. Buckley S, Rivoire K, Vučković J (2012) Engineered quantum dot single-photon sources. Rep Prog Physics 75(12):126503
6. Chen T, Li L, Xu G, Wang X, Wang J, Chen Y, Jiang W, Yang Z, Lin G (2018) Cytotoxicity of InP/ZnS quantum dots with different surface functional groups toward two lung-derived cell lines. Front Pharmacol 13(9):763
7. Chilton A. (2014) The properties and applications of quantum dots. Azo Quantum. AZoQuantum. com
8. Chopra SS, Bi Y, Brown FC, Theis TL, Hristovski KD, Westerhoff P (2019) Interdisciplinary collaborations to address the uncertainty problem in life cycle assessment of nano-enabled products: case of the quantum dot-enabled display. Environ Sci Nano 6(11):3256–3267
9. Clift MJ, Stone V (2012) Quantum dots: an insight and perspective of their biological interaction and how this relates to their relevance for clinical use. Theranostics 2(7):668
10. Efros AL, Nesbitt DJ (2016) Origin and control of blinking in quantum dots. Nat Nanotechnol 11(8):661–671
11. Fan LT, Zhang T (2012) Life cycle assessment (LCA). In: Recent advances in sustainable process design and optimization: (with CD-ROM), pp 65–78
12. Feng X, Chen A, Zhang Y, Wang J, Shao L, Wei L (2015) Central nervous system toxicity of metallic nanoparticles. Int J Nanomedicine 3:4321–4340
13. Ferguson AB, Akahoshi Y, Laing PG, Hodge ES (1962) Trace metal ion concentration in the liver, kidney, spleen, and lung of normal rabbits. JBJS 44(2):317–322
14. García de Arquer FP, Talapin DV, Klimov VI, Arakawa Y, Bayer M, Sargent EH (2021) Semiconductor quantum dots: technological progress and future challenges. Science 373(6555):eaaz8541
15. Gidwani B, Sahu V, Shukla SS, Pandey R, Joshi V, Jain VK, Vyas A (2021) Quantum dots: Prospectives, toxicity, advances and applications. J Drug Deliv Sci Technology 61:102308
16. Giroux MS, Zahra Z, Salawu OA, Burgess RM, Ho KT, Adeleye AS (2022) Assessing the environmental effects related to quantum dot structure, function, synthesis and exposure. Environ Sci Nano 9(3):867–910
17. Gour A, Jain NK (2019) Advances in green synthesis of nanoparticles. Artif Cells Nanomedicine Biotechnol 47(1):844–851
18. Gursel AP, Masanet E, Horvath A, Stadel A (2014) Life-cycle inventory analysis of concrete production: a critical review. Cem Concr Composites 51:38–48
19. Hauschild MZ, Rosenbaum RK, Olsen SI (2018) Life cycle assessment. Springer International Publishing, Cham. https://doi.org/10.1007/978-3-319-56475-3
20. Holban AM, Grumezescu AM (2016) Nanoarchitectonics for smart delivery and drug targeting. William Andrew
21. Hussain I, Singh NB, Singh A, Singh H, Singh SC (2016) Green synthesis of nanoparticles and its potential application. Biotechnol Lett 38:545–560
22. Jamieson T, Bakhshi R, Petrova D, Pocock R, Imani M, Seifalian AM (2007) Biological applications of quantum dots. Biomaterials 28(31):4717–4732
23. Jana P, Dev A (2022) Carbon quantum dots: a promising nanocarrier for bioimaging and drug delivery in cancer. Mater Today Commun 32:104068

24. Jeevitha M, Gupta JK, Sahithi A, Tallam AK, Dantinapalli VLS, Elumalai S, Devadhas J, Muzib YI (2024) Life cycle assessment of quantum dots and its composites. In: Quantum dots based nanocomposites: design, fabrication and emerging applications. Springer, pp 455–463
25. Jouyandeh M, Khadem SS, Habibzadeh S, Esmaeili A, Abida O, Vatanpour V, Rabiee N, Bagherzadeh M, Iravani S, Saeb MR, Varma RS (2021) Quantum dots for photocatalysis: synthesis and environmental applications. Green Chem 23(14):4931–4954
26. Kalyani NT, Dhoble SJ, Domanska MM, Vengadaesvaran B, Nagabhushana H, Arof AK (eds) (2023) Quantum dots: emerging materials for versatile applications. Woodhead Publishing
27. Kantner M, Mittnenzweig M, Koprucki T (2017) Hybrid quantum-classical modeling of quantum dot devices. Phys Rev B 96(20):205301
28. Kesrevani RK, Sharma AK (2016) Nanoarchitectured biomaterials: present status and future prospects in drug delivery. Nanoarchitectonics Smart Deliv Drug Target 35–66
29. Kirsch-Volders M, Vanhauwaert A, Eichenlaub-Ritter U, Decordier I (2003) Indirect mechanisms of genotoxicity. Toxicol Lett 140:63–74
30. Klimov VI (ed) (2017) Nanocrystal quantum dots. CRC press
31. Kumar YA, Koyyada G, Ramachandran T, Kim JH, Hegazy HH, Singh S, Moniruzzaman M (2023) Recent advancement in quantum dot-based materials for energy storage applications: a review. Dalton Trans 52(25):8580–8600
32. Labatt S, White RR (2011) Carbon finance: the financial implications of climate change. John Wiley & Sons
33. Le N, Zhang M, Kim K (2022) Quantum dots and their interaction with biological systems. Int J Mol Sciences 23(18):10763
34. Lee J, Mahendra S, Alvarez PJ (2010) Nanomaterials in the construction industry: a review of their applications and environmental health and safety considerations. ACS Nano 4(7):3580–3590
35. Li J, Chen C, Xia T (2022) Understanding nanomaterial–liver interactions to facilitate the development of safer nanoapplications. Adv Mater 34:2106456
36. Li T, Li Z, Huang T, Tian L (2021) Carbon quantum dot-based sensors for food safety. SensS Actuators A: Physical 1(331):113003
37. Liang Y, Zhang T, Tang M (2022) Toxicity of quantum dots on target organs and immune system. J Appl Toxicol 42:17–40
38. Lima AR, Cristofoli NL, Vieira MC (2023) Life-cycle assessment: a tool to evaluate the environmental impact of nanomaterials used in packaging. In: Postharvest nanotechnology for fresh horticultural produce. CRC Press, pp 347–367
39. Lin X, Chen T (2023) A review of in vivo toxicity of quantum dots in animal models. Int J Nanomedicine 31:8143–8168
40. Ma F, Li CC, Zhang CY (2018) Development of quantum dot-based biosensors: principles and applications. J Mater Chem B 6(39):6173–6190
41. Mansur HS (2010) Quantum dots and nanocomposites. Wiley Interdiscip Rev Nanomedicine Nanobiotechnology 2(2):113–129
42. Mintz KJ, Zhou Y, Leblanc RM (2019) Recent development of carbon quantum dots regarding their optical properties, photoluminescence mechanism, and core structure. Nanoscale 11(11):4634–4652
43. Miroslav P (2017) Quantum dots in the therapy: current trends and perspectives. Mini Rev Med Chem 2017:17. https://doi.org/10.2174/1389557517666170120153342
44. Mohamed WA, Abd El-Gawad H, Mekkey S, Galal H, Handal H, Mousa H, Labib A (2021) Quantum dots synthetization and future prospect applications. Nanotechnol Rev 10(1):1926–1940
45. Mohammadi R, Naderi-Manesh H, Farzin L, Vaezi Z, Ayarri N, Samandari L, Shamsipur M (2022) Fluorescence sensing and imaging with carbon-based quantum dots for early diagnosis of cancer: a review. J Pharm Biomed Analysis 212:114628
46. Moreels I, Lambert K, Smeets D, De Muynck D, Nollet T, Martins JC, Vanhaecke F, Vantomme A, Delerue C, Allan G, Hens Z (2009) Size-dependent optical properties of colloidal PbS quantum dots. ACS Nano 3(10):3023–3030

47. Oberdörster G, Oberdörster E, Oberdörster J (2005) Nanotoxicology: an emerging discipline evolving from studies of ultrafine particles. Environ Health Perspect 113(7):823–839
48. Onyia AI, Ikeri HI, Nwobodo AN (2018) Theoretical study of the quantum confinement effects on quantum dots using particle in a box model. J Ovonic Research 14(1):49–54
49. Peng J, Muhammad R, Wang SL, Zhong HZ (2021) How machine learning accelerates the development of quantum dots? Chin J Chemistry 39(1):181–188
50. Pereira MC, Pereira ML, Sousa JP (1998) Evaluation of nickel toxicity on liver, spleen, and kidney of mice after administration of high-dose metal ion. J Biomed Mater Res Off J Soc Biomater Jpn Soc Biomater Aust Soc Biomaterials 40(1):40–47
51. Qi L, Gao X (2008) Emerging application of quantum dots for drug delivery and therapy. Expert Opin Drug Deliv 5(3):263–267
52. Rasal AS, Yadav S, Yadav A, Kashale AA, Manjunatha ST, Altaee A, Chang JY (2021) Carbon quantum dots for energy applications: a review. ACS Appl Nano Materials 4(7):6515–6541
53. Reddy KT, Madhavi Latha B, Gupta JK, Chaitanya A, Srinivasa Babu P, Manikanta Y, Sahithi A (2024Jun 2) Discovery and history of quantum dots. In: Quantum dots based nanocomposites: design, fabrication and emerging applications. Springer Nature Switzerland, Cham, pp 13–34
54. Roberts JR, Antonini JM, Porter DW, Chapman RS, Scabilloni JF, Young SH, Schwegler-Berry D, Castranova V, Mercer RR (2013) Lung toxicity and biodistribution of cd/se-ZnS quantum dots with different surface functional groups after pulmonary exposure in rats. Part Fibre Toxicol 10:1–7
55. Rocha TL, Mestre NC, Sabóia-Morais SM, Bebianno MJ (2017) Environmental behaviour and ecotoxicity of quantum dots at various trophic levels: a review. Environ Int 98:1–7
56. Rombach-Riegraf V, Oswald P, Bienert R, Petersen J, Domingo MP, Pardo J, Gräber P, Galvez EM (2013) Blinking effect and the use of quantum dots in single molecule spectroscopy. Biochem Biophys Res Commun 430(1):260–264
57. Rosenthal SJ, Chang JC, Kovtun O, McBride JR, Tomlinson ID (2011) Biocompatible quantum dots for biological applications. Chem Biol 18(1):10–24
58. Sahoo SK, Parveen S, Panda JJ (2017) The present and future of nanotechnology in human health care. Nanomedicine Cancer 1:775–806
59. Simeonova PP, Erdely A (2009) Engineered nanoparticle respiratory exposure and potential risks for cardiovascular toxicity: predictive tests and biomarkers. Inhal Toxicol 21(sup1):68–73
60. Simonen K (2014) Life cycle assessment. Routledge
61. Sistani S, Shekarchizadeh H (2022) Applications of quantum dots in the food industry. In: Quantum dots-recent advances, new perspectives and contemporary applications. IntechOpen
62. Som S, Aggarwal SK (2010) Effects of primary breakup modeling on spray and combustion characteristics of compression ignition engines. Combust Flame 157(6):1179–1193
63. Stier O. Electronic and optical properties of quantum dots and wires; 2005
64. Su Z, Shen H, Wang H, Wang J, Li J, Nienhaus GU, Shang L, Wei G (2015) Motif-designed peptide nanofibers decorated with graphene quantum dots for simultaneous targeting and imaging of tumor cells. Adv Funct Materials 25(34):5472–5478
65. Sutherland M, Hawthorn DG, Hill RW, Ronning F, Wakimoto S, Zhang H, Proust C, Boaknin E, Lupien C, Taillefer L, Liang R (2003) Thermal conductivity across the phase diagram of cuprates: low-energy quasiparticles and doping dependence of the superconducting gap. Phys Rev B 67(17):174520
66. Troparevsky MC, Kronik L, Chelikowsky JR (2003) Optical properties of CdSe quantum dots. J Chem Phys 119(4):2284–2287
67. Tsoi KM, Dai QI, Alman BA, Chan WC (2013) Are quantum dots toxic? Exploring the discrepancy between cell culture and animal studies. Acc Chem Res 46(3):662–671
68. Varisco M, Zufferey D, Ruggi A, Zhang Y, Erni R, Mamula O (2017) Synthesis of hydrophilic and hydrophobic carbon quantum dots from waste of wine fermentation. R Soc Open Sci 4(12):170900
69. Walling MA, Novak JA, Shepard JR (2009) Quantum dots for live cell and in vivo imaging. Int J Mol Sci 10(2):441–491

70. Wang Z, Tang M (2021) The cytotoxicity of core-shell or non-shell structure quantum dots and reflection on environmental friendly: a review. Environ Research 194:110593
71. Wang D, Wang Z, Zhan Q, Pu Y, Wang JX, Foster NR, Dai L (2017) Facile and scalable preparation of fluorescent carbon dots for multifunctional applications. Engineering 3(3):402–408
72. Wang X, Wu T (2023) An update on the biological effects of quantum dots: from environmental fate to risk assessment based on multiple biological models. Sci Total Environment 879:163166
73. Wickerts S, Arvidsson R, Sanden BA, Peters G, Hou L, Albinsson B (2021) Prospective life-cycle modeling of quantum dot nanoparticles for use in photon upconversion devices. ACS Sustain Chem Engineering 9(14):5187–5195
74. Winnik FM, Maysinger D (2013) Quantum dot cytotoxicity and ways to reduce it. Acc Chem Res 46(3):672–680
75. Yang S, Sun J, He P, Deng X, Wang Z, Hu C, Ding G, Xie X (2015) Selenium doped graphene quantum dots as an ultrasensitive redox fluorescent switch. Chem Materials 27(6):2004–2011
76. Yao Y, Zhang T, Tang M (2023) The DNA damage potential of quantum dots: toxicity, mechanism, and challenge. Environ Pollution 317:120676
77. Yih TC, Moudgil VK (2007) Nanotechnology comes of age to trigger the third industrial revolution. Nanomed Nanotechnol Biol Med 3(4):245
78. Zhang T, Stilwell JL, Gerion D, Ding L, Elboudwarej O, Cooke PA, Gray JW, Alivisatos AP, Chen FF (2006) Cellular effect of high doses of silica-coated quantum dot profiled with high throughput gene expression analysis and high content cellomics measurements. Nano Lett 6(4):800–808
79. Zhang T, Wang B, Cheng Q, Wang Q, Zhou Q, Li L, ... Tang Z (2024) Polaron engineering promotes NIR-II absorption of carbon quantum dots for bioimaging and cancer therapy. Sci Adv 10(27):eadn7896
80. Zhang Y, Wang Z, Zhou G (2014 Mar) Determinants of employee electricity saving: the role of social benefits, personal benefits and organizational electricity saving climate. J Clean Prod 1(66):280–287
81. Zheng XT, Ananthanarayanan A, Luo KQ, Chen P (2015) Glowing graphene quantum dots and carbon dots: properties, syntheses, and biological applications. Small 11(14):1620–1636
82. Zhou JW, Zou XM, Song SH, Chen GH (2018) Quantum dots applied to methodology on detection of pesticide and veterinary drug residues. J Agric Food Chemistry 66(6):1307–1319
83. Şengül H, Theis TL (2011) An environmental impact assessment of quantum dot photovoltaics (QDPV) from raw material acquisition through use. J Clean Production 19(1):21–31

Chapter 17
Emerging Applications of Graphene Quantum Dots in Supercapacitors

Marcos Jefferson Moreno Baque, Tania C. Valdiviezo-Abarca, Nelly Ma. Rosas Laverde, and Alina Pruna

Abstract One of the most promising energy storage devices is represented by supercapacitors, thanks to their attractive features. On the other hand, the requirements for improved performance of supercapacitors triggered the research of varying nanomaterials. The rise of the graphene family gained increased attention in this respect. From the graphene nanomaterials, the graphene quantum dots exhibit excellent characteristics, including non-toxicity, high surface area, chemical stability, tunable chemistry, and abundance of active sites that are attractive for supercapacitor applications. This chapter aims to provide an overview of the state of the art in graphene quantum dots' synthesis, characterization, and supercapacitor application.

Keywords Graphene · Carbon allotropes · Quantum dots · Synthesis · Properties · Supercapacitors · Conductivity

1 Introduction

Over the last few years, there has been a surge in demand for high-energy storage solutions worldwide, driven by the gradual shift toward clean global power [57]. Supercapacitors (SCs) are one of the most promising energy storage technologies due to their unique features, such as quick-discharge rates, long cycling life, and high-power density [1]. These characteristics make them applicable in a wide range of systems, from portable electronics to electric vehicles and renewable energy implementations [103].

One of the key areas of research is the development of new nanomaterials that could improve supercapacitor performance and help establish them as a mainstream

M. J. Moreno Baque · T. C. Valdiviezo-Abarca · N. Ma. Rosas Laverde
Departamento de Materiales, Escuela Politécnica Nacional, Quito, Ecuador

A. Pruna (✉)
Instituto Universitario de Investigación de Tecnología de los Materiales, Universitat Politècnica de València, Valencia, Spain
e-mail: apruna@itm.upv.es

T. Bhuyan et al. (eds.), *Multifunctional Quantum Dots*, Materials Horizons: From Nature to Nanomaterials, https://doi.org/10.1007/978-981-92-0172-3_17

technology [99]. Among various materials, carbon-based nanomaterials are extensively studied for their excellent electrical conductivity, high surface area, and chemical stability [147]. Graphene, a two-dimensional material with unique properties, has become a major focus of research; besides, zero-dimensional graphene derivatives, specifically graphene quantum dots (GQDs), have gained significant attention in the field of SCs. GQDs combine the beneficial properties of graphene with quantum confinement and edge effects, which can be leveraged to enhance the electrochemical performance of SCs. Their tunable band gap, photoluminescence, and remarkable electron transport properties make them promising candidates for advancing energy storage technologies [36, 67].

For a better understanding, it is necessary to note that all published articles in Scopus dealing with this topic were considered in the development of this chapter. A bibliometric study was carried out and even though the GQDs topic seems relatively new, statistics show that since 1995 worldwide researchers have put interest in Quantum Dots, later, in 2004, when Andre Geim and Konstantin Novoselov isolated for the first time graphene monolayer using graphite and scotch tape, the topic of QD reemerged as a hope for innovation and technological evolution. Over time, QDs have gained prominence, and by around 2010, the term graphene quantum dots (GQDs) was already in use.

Nowadays, the Scopus platform lists approximately 14,259 available documents on this subject, as illustrated in Fig. 1. From 2010 to 2020, the number of GQD-related publications increased steadily, reaching a peak in 2020 with over 1,500 documents. However, from 2021 onward, the number of publications has shown a slight decline. This trend may reflect shifts in research focus or funding, but GQDs remain an active research area. Notably, in 2023, the Nobel Prize in Chemistry was awarded to Moungi G. Bawendi, Louis E. Brus, and Alexei I. Ekimov for their groundbreaking discovery and synthesis of quantum dots, further underscoring the significance of this research. Nonetheless, it is important to recognize the achievements that each country has reached (see Fig. 2).

Thus, the papers published in journals indexed in Scopus are not only classified per year of publication, but also the origin is important. Figure 2 displays the distribution of GQD-related publications by country, with China leading the research efforts, contributing over 5,986 documents. India, the United States, and Iran follow, with 1,983, 1,520, and 1,071 documents, respectively. Other countries, such as South Korea, Germany, and the United Kingdom, also show notable contributions. These statistics highlight China's dominant role in GQD research and the strong international interest in this field.

Applying classification techniques from other scientific areas is another method. The distribution of GQD applications is shown in Fig. 3, with materials science accounting for the largest share at 22.48%, followed by chemistry at 18.97% and physics and astronomy at 15.13%. This implies that most of the research on GQDs is focused on their chemical interactions and material characteristics, which make them essential for creating new materials and technologies. Chemical engineering (9.52%), engineering (13.72%), and other fields are also heavily represented, suggesting that GQDs are being investigated for industrial uses more and more. There is growing

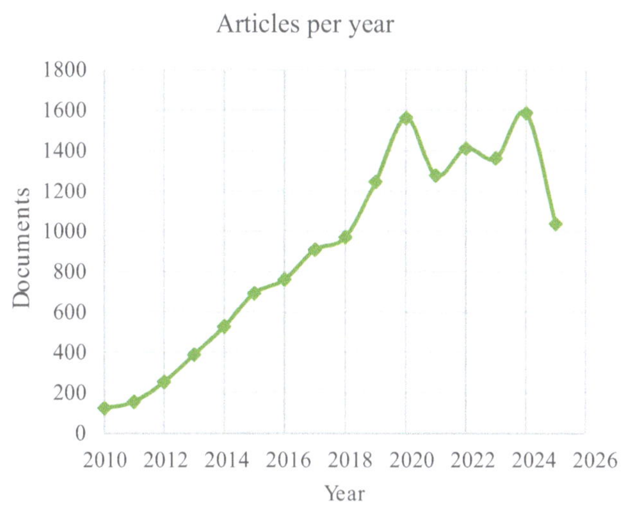

Fig. 1 Yearly GQD- and CQD-related articles as indexed in Scopus in 2024

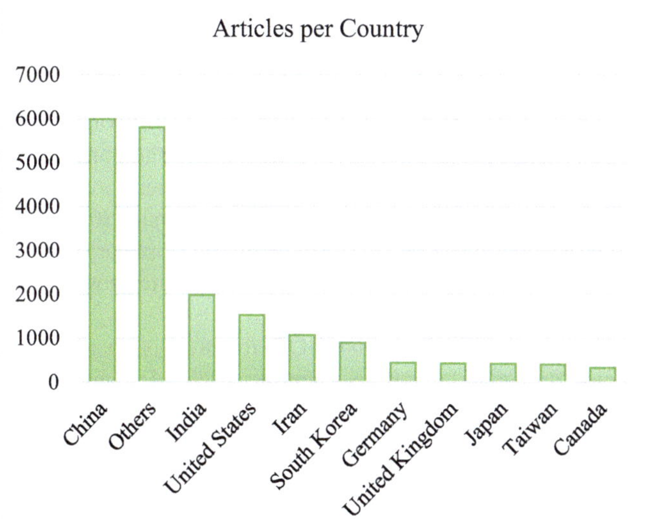

Fig. 2 Geographic distribution of GQDs-related articles as indexed in Scopus in 2024

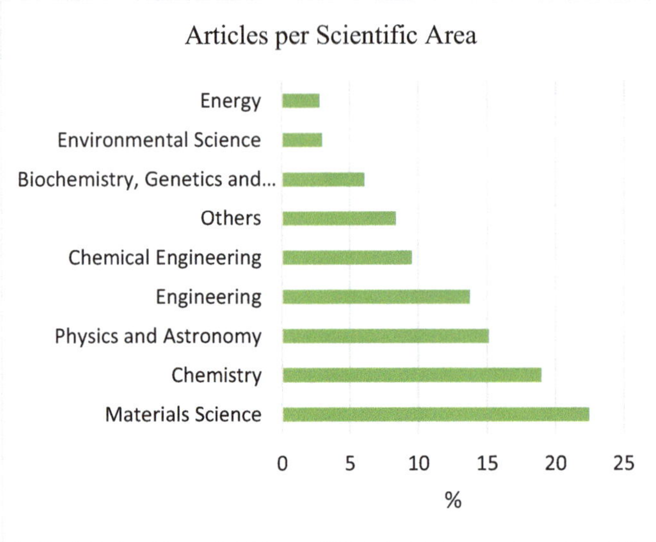

Fig. 3 Percentage of GQD-related articles as indexed in Scopus in 2024

interest in other fields as well, such as environmental science (2.94%) and energy (2.77%), especially in environmental sustainability and energy storage.

Moreover, GQDs' potential for biological applications is highlighted by biochemistry, genetics, and molecular biology (5.1%); pharmacology, toxicology and pharmaceutics (2.4%); and health professions (0.5%), all of which show an increasing interest in applications connected to medicine and health. Furthermore, even in subjects not typically connected with nanomaterials, fields like nursing (0.1%) and business, management, and accounting (0.3%) imply more multidisciplinary approaches which are being pursued. This wide range of studies demonstrates how adaptable GQDs are in a variety of contexts.

Together, the two graphs show that GQD research is an active and quickly developing sector, propelled by various scientific disciplines and focused efforts in strategic areas such as China. The consistent increase in publications until 2020 indicates an increasing interest worldwide, especially in physics, chemistry, and materials science applications. This increasing trend demonstrates the interest in investigating the adaptability of GQD, which may be used to improve energy storage systems or enhance optical qualities. However, the drop after 2020 points to a possible change in the field's research objectives or difficulties.

Several factors, such as the saturation of specific research avenues, challenges in scaling up GQD production for commercial applications, or the emergence of competing materials with similar properties—such as new nanotechnologies or other carbon-based nanomaterials—could be responsible for this decline. Furthermore, international calamities like the COVID-19 pandemic could have momentarily shifted attention and funds from experimental nanomaterial research to more pressing issues like epidemiological and medical research. Notwithstanding this decline, GQD research is still very promising, particularly in multidisciplinary domains where novel applications are currently being investigated, such as biotechnology, environmental science, and energy.

Geographically, China leads the charge in GQD research, far surpassing other countries. India, the United States, and several others also contribute significantly, though with varying degrees of focus. The wide range of disciplines engaging with GQD research, from materials science to energy and environmental applications, underscores the broad potential of these materials across scientific and industrial applications.

Future research may focus on stabilizing and increasing publication rates while exploring untapped applications, particularly in energy and environmental sustainability, which are gaining importance globally. In this chapter, the authors examine the role of GQDs in supercapacitor technology by reviewing their synthesis methods and integration into supercapacitive systems, along with their impact on device performance. In the next phase, the capacitance of GQD-based SCs will be evaluated through various techniques, including cyclic voltammetry (CV) and charge–discharge (CD) tests using different electrolytes. In this evolving exploration, we anticipate the potential of GQDs for energy storage applications and provide recommendations to accelerate future research initiatives in this domain.

2 Carbon Materials

Carbon is one of the most versatile elements in the periodic table, known for its ability to form various allotropes, each with unique properties and applications. Our lives are composed of carbon; even the most microscopic form of life is integrated by carbon. Research has developed multiple incredible forms to supply the growing need for sustainable materials to obtain, e.g., renewable energy. Over the years, carbon-based materials have played a pivotal role in technological advancements, particularly in fields such as electronics, energy storage, and nanotechnology [65, 129].

Carbon exists in several allotropes, including graphite, diamond, carbon nanotubes (CNTs), graphene, and fullerenes. Each one of these forms exhibits distinct structural, electronic, and mechanical properties, which are determined by the type of carbon–carbon bonds and the arrangement of carbon atoms [130]. For instance, graphite is composed of layers of hexagonally arranged carbon atoms with weak van der Waals forces between them, making it an excellent lubricant [89]. In contrast, diamond, with its strong covalent bonding in a tetrahedral lattice, is one of the hardest materials known [41]. Below, we will briefly review the most common allotropes of carbon, their properties, and applications in energy storage technologies.

2.1 Coal

Coal is the most rudimentary form of carbon. Cave paintings dating back to 30,000 B.C. [47] have seen the light of day, thanks to the skill of their creators, and how could they not? If everything is due to the softness of coal to be impregnated on surfaces, as if it were an exfoliant. Coal has an amorphous structure as a carbon allotrope, meaning it lacks the well-ordered arrangement of atoms found in crystalline carbon forms like graphite or diamond. In its chemical structure, coal is composed primarily of carbon atoms arranged in a disordered, irregular fashion [64, 82]. Like graphite hexagonal sheets or diamond tetrahedral lattice, coal's carbon atoms are bonded in a mixture of single, double, and sometimes even aromatic ring structures [64]. This irregular bonding results from the incomplete combustion of organic matter during its production by nature, leaving behind carbon and insignificant amounts of hydrogen, oxygen, and ash, depending on the source material and production process [86].

The high porosity of coal is directly related to its structure, with the irregular arrangement creating numerous micropores and macropores throughout the material. These pores enhance its surface area, making it highly effective for adsorption applications [79]. Coal's lack of long-range order differentiates it chemically from more structured allotropes. Yet, this disordered arrangement gives it flexibility in terms of reactivity, allowing it to interact with a wide range of substances, which is particularly valuable in applications like water filtration, air purification, and chemical processing [108]. In the field of energy storage, activated charcoal is the most commonly used form of coal. It is obtained from organic waste through many synthetic methods. Specifically, activated charcoal is present as electrodes in two of several studies about it.

In India [54] energy-storing supercapacitor devices were fabricated from coconut shell-based activated charcoal. The main achievement of the research described in the article was the synthesis and characterization of activated carbon from coconut shells and its successful use as an electrode material for electrochemical double-layer SCs (EDLC). The activated carbon showed a specific surface area of 1640 m^2/g and a pore size of 3 nm, which contributed to its high performance in terms of energy storage capacity. A specific capacitance of 356.2 F/g was achieved for a single electrode and an energy density of 88.8 Wh/kg, making it competitive compared to other materials. Furthermore, the results showed stability up to 600 charge–discharge cycles, which is crucial for applications in energy storage devices.

Later, Volperts et al. [133] effectively synthesized microporous carbon materials from wood using thermochemical techniques, demonstrating good electrode properties for SCs. Volperts A manufactured supercapacitor electrodes from activated wood charcoal. Sulfuric acid was used as the electrolyte, and the study found that the maximum capacitance of about 330 F/g was reached at an activation temperature of 650 °C with a carbonate-to-activator ratio of 1:2. The results demonstrated how crucial it is to maximize activation temperature and micropore shape to improve supercapacitor performance.

Activated charcoal is still used in materials science today to fabricate supercapacitors; as a result, its large specific surface area and porous structure make it perfect for creating a double electric layer in electrodes. Because of their strong reactivity and advantageous solubility of reaction products, carbon sources generated from lignite present a significant opportunity for the creation of composite carbon fibers. A recent work by Liu et al. [76] described an eco-friendly method for creating coal-based composite carbon fibers using the water-soluble fraction that is produced when lignite is thermally dissolved with methanol in mild circumstances, with the addition of sodium hydroxide. The resulting fraction was electrospun, pre-oxidized, carbonized, acid-washed, and activated after being mixed with a polyvinyl alcohol solution. Interestingly, by effectively removing surface impurities and increasing the specific surface area, acid washing enhanced the exposure of electrochemically active spots. At a current density of 1 A/g, this treatment dramatically increased the fibers' specific capacitance from 73.5 to 197.1 F/g. Furthermore, with an energy density of 8.5 Wh/kg and a power density of 250 W/kg, the activated fibers showed outstanding performance as symmetric supercapacitor electrodes. This process offers a valuable and sustainable way to turn lignite into superior electrode materials for energy storage applications. However, recently, activated charcoal use has focused on improving its performance for future uses.

2.2 Graphite

Chelgani et al. [23] set graphite forms through the contact or regional metamorphism of sedimentary organic matter, where organic material transforms graphite due to rising temperature and/or pressure within the Earth's crust. Graphite is a remarkable allotrope of carbon, characterized by its unique layered structure. In this form, carbon atoms are arranged in hexagonal sheets, where each atom is bonded to three others, forming strong covalent bonds within the planes. These layers are held together by weaker van der Waals forces, allowing them to easily slide over one another, which gives graphite its lubricating properties [30]. Additionally, graphite exhibits excellent electrical conductivity due to the presence of delocalized electrons between the layers. This combination of properties makes graphite invaluable in various applications, ranging from its use in lubricants, batteries, and SCs to electrodes and nuclear reactors [135].

Wu et al. [137] conducted a redox deposition of manganese oxide on graphite electrodes for SCs. The main results showed increased capacitance over time, following a logarithmic law and reaching values of up to 45 mF/cm^2. Furthermore, the coating maintained good operational stability over 1000 potential cycles, suggesting its potential for supercapacitor applications. Nonetheless, it is worth noting that despite the age of this research, graphite continues to be used as an electrode. For instance, [144] explored expanded graphite as an electrode material due to its properties, including layer structure, conductivity, and redox sites that promote Faraday reactions. A solid and flexible supercapacitor was developed from polyaniline-carbon nanotube/expanded graphite (PANI-CNT/EG), which exhibited high performance. This supercapacitor showed a specific capacitance of 826.7 F/g, good flexibility, an energy/power output of 7.1 kW/kg, an energy density of 12.0 W h/kg, and excellent cycle retention, with 77.6% capacitance retention after 3000 cycles. In summary, graphite is a versatile and promising material for energy storage due to its unique structure and enhanced properties.

2.3 *Diamond*

Physically, diamonds are transparent, with a high refractive index that gives them their characteristic sparkle when cut and polished [6]. They are also excellent thermal conductors, efficiently dispersing heat due to the strong atomic bonds that allow phonons (vibrations of the lattice) to travel quickly [18]. However, despite being excellent heat conductors, diamonds are electrical insulators since they lack free-moving electrons within their structure. This combination of thermal conductivity and electrical insulation makes diamonds useful in high-performance semiconductor technology, particularly for heat sinks in electronic devices where heat dissipation is necessary [58].

Beyond industrial and technological uses, diamonds are highly valued for their beauty, particularly when used in jewelry. Their optical brilliance, combined with their durability, has made diamonds highly sought after for engagement rings, necklaces, and other ornamental pieces [43]. Furthermore, synthetic diamonds, produced through high-pressure, high-temperature (HPHT) methods or chemical vapor deposition (CVD), have become vital in various applications, such as in precision optics, electronics, and quantum computing [125]. Diamonds also have potential in cutting-edge technologies. In quantum computing, they serve as hosts for nitrogen-vacancy (NV) centers, which are defects in the diamond lattice that can be manipulated at the quantum level for advanced computing and sensing applications [27]. In medicine, diamonds are currently under exploration for drug delivery systems and bioimaging due to their biocompatibility [131]. Diamond is also used in energy storage devices due to its extreme chemical stability, wide potential window, and oxidation resistance.

Another study [39] used diamond as SC electrodes; specifically, utilizing boron-doped polycrystalline diamond (BDD) which made it a highly attractive material for enhancing energy storage devices. This research achieved a significant milestone by fabricating the first diamond-based supercapacitor prototype. The free-standing diamond paper developed in this study demonstrated a high potential window of up to 2.5 V in aqueous electrolytes, offering a specific capacitance of 0.688 mF/cm^2 per layer. Additionally, the supercapacitor exhibited a remarkable specific power of 1×10^5 W/kg, bridging the gap between traditional capacitors and high-power SCs.

Furthermore, a review by [142] states that diamonds are used in conductive film forms synthesized by introducing boron, nitrogen, or phosphorus as dopants. Particularly, boron-doped diamond (BDD) has high potential as an electrode material for supercapacitors. One of the main reasons is its great stability in harsh environments, such as corrosive solutions and high currents, voltages, and temperatures. Furthermore, when the boron doping level is appropriate, BDD exhibits good metallic conductivity, making it a perfect material for these devices. The author states that diamond-based supercapacitors have demonstrated considerable performance increases when utilized in devices such as electrical double-layer capacitors (EDLCs) and pseudocapacitors.

An important step forward in the creation of high-performance energy storage devices is the incorporation of diamond-based composite materials, such as boron-doped diamond (BDD), into supercapacitor electrodes. Specifically, the creation of "diamond cloth," a two-dimensional, flexible substance made by applying a thin layer of BDD onto carbon fabric using microwave plasma-enhanced chemical vapor deposition, combines the mechanical flexibility of carbon fabric with the electrical conductivity and chemical stability of diamond. A high specific capacitance of 81.94 mF cm^{-2} at a scan rate of 10 mV s^{-1}, outstanding cycling stability with 99.59% capacitance retention after 10,000 cycles, and enhanced mechanical durability with repetitive bending are only a few of the remarkable electrochemical qualities produced by this hybrid structure. Diamond cloth performs better than many conventional carbon-based supercapacitors, with a high energy density of 45.96 μWh cm^{-2} and a power density of 67.93 mW cm^{-2}. These qualities make BDD-based materials attractive options for wearable and flexible energy storage devices of the future [69]. All things considered, diamond-related materials offer a balance of high energy and power densities, stability, and mechanical durability, making them appealing candidates for high-performance supercapacitors. To fully realize their potential in commercial applications and get past the remaining challenges, more research is needed.

2.4 Carbon Nanotubes

Carbon nanotubes (CNTs) are a unique class of nanomaterials composed of rolled-up sheets of single-layer carbon atoms arranged in a hexagonal lattice, known as graphene [100]. Depending on the number of rolled layers, CNTs can be categorized as single-walled carbon nanotubes (SWCNTs) or multi-walled carbon nanotubes (MWCNTs) [107]. CNTs possess extraordinary mechanical, electrical, and thermal properties. Their tensile strength is about 100 times that of steel, while their density is much lower, making them ideal for lightweight and high-strength applications [31, 50]. Additionally, CNTs have remarkable electrical conductivity, which can be metallic or semiconducting depending on their chirality, and they exhibit excellent thermal conductivity, with values as high as 3,000 W/m K, surpassing most metals [53].

The synthesis of carbon nanotubes can be achieved through various methods, including arc discharge, laser ablation, and chemical vapor deposition (CVD), with the latter being the most employed due to its scalability and control over nanotube properties [101]. In the CVD process, a hydrocarbon gas, such as methane, is decomposed at high temperatures over a metal catalyst (e.g., iron, cobalt, or nickel), resulting in the formation of CNTs. Parameters such as temperature, catalyst type, and gas flow rate can significantly influence the quality, diameter, and yield of the nanotubes [46, 118]. CNTs are typically characterized by techniques like transmission electron microscopy (TEM) and Raman spectroscopy [12]. TEM allows for the observation of individual CNTs' structure and wall thickness, while Raman spectroscopy provides insights into the nanotubes' diameter distribution, crystallinity, and electronic properties [20].

Due to their exceptional properties, CNTs have a wide range of applications across multiple fields. In electronics, CNTs are used in the development of transistors and sensors due to their ability to conduct electricity efficiently at the nanoscale [11]. Their high mechanical strength and lightweight nature make them suitable for reinforcing composites in the aerospace and automotive industries [110]. Furthermore, CNTs are explored for use in energy storage devices like batteries and SCs, where their high surface area enhances charge storage capacity [123]. The versatility and multifunctional nature of CNTs make them a key material in the advancement of nanotechnology and materials science.

As a special use, CNTs play a critical role in the development of high-performance supercapacitors due to their distinct material features. In a research study conducted by Binfaris et al. [16], the fundamental role of CNTs was to increase the energy and power density of supercapacitors. The CNT yarn (CNTY) employed in the study has good physical durability, electrical conductivity, and a wide surface area for charge storage, considerably improving the capacitive behavior of supercapacitors. The primary result of this work is the effective manufacturing and characterization of CNTY-based supercapacitors with excellent performance metrics. These include a large specific capacitance (up to 636.75 F/g), energy density (up to 6.7 Wh/kg), and power density (up to 13.20 W/kg). The study shows that CNTY supercapacitors can effectively store and release energy, making them very suited.

Kwon et al. [63] reported on a scalable, binder-free method to overcome the drawbacks of binder-containing electrodes in electrochemical double-layer capacitors (EDLCs) by employing hybrid scaffolds of hierarchical carbon nanoplates (HCN) and functionalized carbon nanotubes (FCNT). The HCNs, which are made from ZIF-8 nanoplates by thermal treatment, have high aspect ratios and hierarchical porosity, which greatly increase surface area and ion transport. This architecture not only removes the insulating effects of polymeric binders but also contributes to improved charge storage and mechanical integrity via the entangled FCNT network. Electrochemical evaluations showed high specific capacitances of up to 206 F/g at 0.8 A/g and excellent rate retention (72%) at 8 A/g, with symmetric cells achieving energy and power densities of 16.2 Wh/kg and 3880 W/kg, respectively. It is combined with FCNT hydrogels, which are made by oxidizing MWCNTs to create a shear-thinning slurry that can be used for scalable electrode fabrication via bar coating. The study emphasizes the potential for industrial-scale use of binder-free supercapacitor electrodes and the complementary interaction between HCN and FCNT components in fostering rapid charge–discharge dynamics.

In the search for cutting-edge materials for high-performance supercapacitors, [56] carbon nanotubes (CNTs) were combined with a brand-new bimetallic vanadium–yttrium metal–organic framework (VY-MOF) to create a hybrid electrode material. By utilizing the remarkable conductivity of carbon nanotubes and the synergistic contributions of redox-active bimetallic centers, the VY-MOF@CNT composite was created through a hydrothermal process aided by probe sonication. Characterizations by structure and electrochemistry showed an impressive energy density of 60 Wh/kg at a power density of 1200 W/kg and a high specific capacitance of 1661 F/g at 2 A/g. With 68% of the contribution attributable to surface-controlled processes, Trasatti and Dunn's analyses validated the hybrid mechanism of charge storage, which combines electric double-layer capacitance with pseudocapacitance. 92% of the capacitance was retained after 10,000 cycles, demonstrating the symmetric supercapacitor device's exceptional long-term stability. These findings highlight the promise of CNT-interwoven bimetallic MOF composites as long-lasting and scalable electrode materials for energy storage devices of the future.

2.5 Fullerene

Fullerenes are a unique class of carbon allotropes consisting of carbon atoms arranged in a spherical, ellipsoidal, or tubular shape, with the most famous structure being buckminsterfullerene (C_{60}). This molecule resembles a soccer ball, composed of 60 carbon atoms linked in a pattern of 12 pentagons and 20 hexagons [61]. Fullerenes exhibit remarkable stability due to their delocalized π-electrons and symmetrical structure, which makes them resistant to high temperatures and chemical reactions. One of the key properties of fullerenes is their ability to accept and donate electrons, making them highly useful in applications involving redox reactions and electron transport. Additionally, fullerenes show potential for use in superconducting materials when doped with alkali metals [49].

The unique electronic and structural properties of fullerenes have led to numerous applications in various fields, including materials science, electronics, and medicine. In organic photovoltaics, fullerenes are used as electron acceptors in solar cells, enhancing the efficiency of energy conversion [21]. In the biomedical field, they are explored as drug delivery vehicles due to their ability to trap and carry molecules within their hollow structure, as well as their antioxidant properties, which can neutralize free radicals. Fullerenes are also being studied for their potential use in nanomedicine for imaging and targeted cancer therapies [81].

To improve the electrochemical performance of supercapacitor electrodes, other works [106] have presented a novel ternary composite material made of polyaniline (PANI), fullerene C_{60}, and cesium indium chloride ($CsInCl_3$) halide perovskite. The goal of their synergistic integration was to bring together the redox-active properties of halide perovskites, the remarkable electron-accepting capacity and structural stability of fullerene, and the high pseudocapacitive behavior of PANI. Electrochemical experiments showed an exceptional specific capacitance of 382 F/g at 1 A/g and excellent capacitance retention of more than 90% after 5000 cycles, while structural and morphological characterizations confirmed the formation of a homogenous hybrid network. Improved performance was attributed to the components' increased charge transport, stable interfacial connection, and efficient ion diffusion pathways. This composite system demonstrates the fascinating role of fullerene-based materials in next-generation supercapacitors, where multifunctional hybrid designs can be tailored for improved energy storage efficiency and long-term durability.

2.6 Graphene

One of the most innovative materials to be found in the twenty-first century is graphene, which is a single layer of carbon atoms organized in a hexagonal lattice. It exhibits extraordinary properties, including high mechanical strength, flexibility, and electrical conductivity. Graphene is often considered the thinnest material ever made, yet it is incredibly strong, with a tensile strength about 200 times that of steel. Its electrical conductivity surpasses that of copper, making it an ideal candidate for electronic applications. In addition, graphene has remarkable thermal conductivity and is nearly transparent, allowing up to 97% of light to pass through its atomic layer [40]. On the other hand, graphene synthesis can be achieved through various methods, such as mechanical exfoliation of graphite, chemical vapor deposition (CVD), and chemical reduction of graphene oxide [94]. Graphene's potential applications are vast, including flexible electronics, high-frequency transistors, and transparent conductive films used in touch screens [5]. In the energy sector, graphene is being explored for use in batteries and SCs due to its large surface area and conductivity, which enhance charge storage and transfer capabilities. Additionally, its biocompatibility makes it promising for biomedical applications such as biosensors, drug delivery, and tissue engineering [19].

According to [8], ternary composites, which comprise conducting polymers, metal oxides, and graphene, show higher specific capacitance than binary composites and single components. Examples of capacitance values that have been published are as high as 1247 F/g, demonstrating the usefulness of these composites in energy storage applications. The study emphasizes how mixing several materials might have a synergistic impact that improves electrochemical characteristics. The development of flexible and stable electrode materials is another important accomplishment. The combination of materials like MnO_2 and Co_3O_4 with graphene and conducting polymers improves charge storage capabilities and lowers ion diffusion resistance, leading to better overall performance. Excellent cycling stability has been demonstrated by the ternary composites, which have been able to retain capacitance even after prolonged cycling (up to 3500 cycles). The development of high-performance supercapacitors, which are necessary to fulfill the rising demand for energy storage in the world, is aided by this research.

Liu et al. [75] suggested the use of three-dimensional graphene papers made from CO_2 laser pyrolysis of polyimide (PI) paper to create high-voltage thin-film supercapacitors (TFSCs) in a scalable and metal-free manner. High-quality laser-induced graphene (LIG) papers with ultralow sheet resistance (4.88 $\Omega\cdot\text{square}^{-1}$), low longitudinal resistance (3.46 Ω), and big crystalline size (96.1 nm) were created by optimizing laser parameters such as scan density, power, defocus, and speed. The LIG paper reduces the need for commercial metal components and separators by acting as the electrode material, current collector, and interconnector all at once. After 15,000 cycles, the resultant TFSCs retained 86.9% capacitance retention, showed an energy density of 10.9 $\mu\text{Wh}\cdot\text{cm}^{-2}$, and had an aerial capacitance of 54.5 $\text{mF}\cdot\text{cm}^{-2}$. In an impressively small 3.8 cm^3 space, up to 160 TFSC units were put together in series to achieve a linear voltage increase from 1.2 to 200 V. These results demonstrate how 3D LIG paper designs can be included in flexible, high-performance, and environmentally friendly energy storage systems.

2.7 *Carbon Quantum Dots*

Carbon quantum dots (CQDs) are small, carbon-based nanoparticles with sizes typically below 10 nm. These zero-dimensional materials have garnered significant interest due to their excellent optical properties, such as strong photoluminescence, tunable emission spectra, and high photostability. CQDs also exhibit good biocompatibility and low toxicity, making them particularly appealing for biological applications. Their emission properties can be tuned by modifying their size or surface functionalities, allowing them to be used in bioimaging, sensing, and optoelectronics [124].

According to Nesakumar [91], CQDs are an intriguing family of nanomaterials distinguished by their adaptability to a wide range of applications, biocompatibility, and distinctive optical features. Since carbonaceous materials make up most of the CQDs' construction, they are more ecologically benign and appropriate for use in biomedical applications than typical semiconductor quantum dots, which are frequently composed of heavy metals and cause environmental issues. In a study made by [126], it states that due to quantum confinement phenomena, these nanostructures usually have diameters of fewer than 10 nm and unique electrical characteristics. By adjusting their size and surface functionalization, CQDs' extraordinary fluorescence may be controlled to produce a range of emission colors that are essential for use in bioimaging, sensing, and light-emitting devices.

CQDs can be categorized according to their structural shapes, surface functions, and synthesis techniques. Top-down and bottom-up strategies are the two main synthetic methodologies. Larger carbon materials, such as graphite or carbon nanotubes, are broken down into smaller CQDs using top-down techniques, whereas bottom-up techniques use chemical processes to create CQDs from smaller precursors. CQDs can display a variety of surface states, which affect their optical and electrical characteristics, depending on the synthesis technique used. For example, the comparatively defect-free structures of CQDs produced by hydrothermal techniques frequently result in enhanced photoluminescence capabilities [33].

Furthermore, CQDs can be divided into three main groups according to their structure: multishell quantum dots, which use multiple layers to further optimize optical and electronic properties; core–shell quantum dots, which have a solid carbon-based core that exhibits strong fluorescence due to quantum confinement effects; and core–shell quantum dots, which have a shell made of another material enclosing the core to enhance photostability and quantum yield while providing protection. These structural differences improve solubility, biocompatibility, charge transport, and photoluminescence, demonstrating the potential applications of CQDs in several areas of materials science and nanotechnology [17].

Regarding synthesis methods, it can be achieved through various paths, including electrochemical synthesis, laser ablation, and hydrothermal treatments. These methods allow for control over particle size, shape, and surface chemistry [116]. Due to their luminescent properties and excellent electron transfer capabilities, CQDs are used in fields like bioimaging and drug delivery, where they serve as fluorescent labels or carriers. In optoelectronics, CQDs are applied in light-emitting diodes (LEDs) and solar cells due to their efficient light absorption and emission. Additionally, CQDs have potential applications in environmental sensing, where they can detect contaminants in water or air [9].

By using a solvothermal method and electrophoretic deposition to assemble carbon quantum dots (CQDs) into a layered structure on interdigital finger electrodes, [24] created a micro-supercapacitor that exhibited an impressive 534.7 μF cm^{-2} specific capacitance at an electrode loading area and a rate capability of up to 1000 V s^{-1}. This creative method not only improved the supercapacitor's power responsiveness and cycle stability, but it also demonstrated the promise of CQDs in energy storage applications. Future developments indicate that even higher performance metrics may result from further refining the CQD assembly procedure and investigating hybrid materials. This opened the door to the creation of next-generation supercapacitors with enhanced energy density and efficiency for a range of uses in portable electronics and renewable energy systems.

Akram and Kayani [2] developed a metal-free, scalable process for creating high-voltage TFSCs using three-dimensional graphene papers made from polyimide (PI) paper that has been pyrolyzed by a CO_2 laser. They created superior laser-induced graphene (LIG) papers with ultralow sheet resistance (4.88 $\Omega \cdot square^{-1}$), low longitudinal resistance (3.46 Ω), and big crystalline size (96.1 nm) by fine-tuning laser parameters such as scan density, power, defocus, and speed. Commercial metal parts and separators were not necessary because the LIG paper served as the electrode material, current collector, and interconnector all at once. The resultant TFSCs showed an energy density of 10.9 $\mu Wh \cdot cm^{-2}$, an aerial capacitance of 54.5 $mF \cdot cm^{-2}$, and an 86.9% capacitance retention after 15,000 cycles. Notably, a linear voltage increase from 1.2 to 200 V was achieved in a small 3.8 cm^3 container by assembling up to 160 TFSC units in series. These results highlight how 3D LIG paper designs can be included into energy storage systems that are versatile, high-performing, and sustainable.

2.8 Graphene Quantum Dots (GQDs)

GQDs are nanoscale fragments of graphene, typically below 20 nm in size, which retain the unique electronic and optical properties of graphene while exhibiting quantum confinement effects due to their small size [149]. The first synthesis of GQDs was reported in 2008 by the team led by Phaedon A. Vouris at the IBM Thomas J. Watson Research Center in Yorktown Heights, United States. In this work, the researchers used a method of cutting graphene nanoribbons through chemical etching and lithography to obtain GQDs. The discovery of GQDs was key, as it allowed for the utilization of the unique quantum confinement properties of graphene at the nanoscale, opening up new applications in fields such as optoelectronics, bioimaging, and supercapacitors [39, 54].

GQDs possess tunable photoluminescence, where the emission wavelength can be controlled by the size and edge structure of the dots [71]. Additionally, GQDs have excellent chemical stability, high surface area, and good dispersibility in various solvents. Their biocompatibility and low toxicity make them attractive for biomedical applications, particularly in bioimaging and drug delivery, where they serve as fluorescent markers or drug carrier [96]; nevertheless, the use of GQD in SC is the main focus of this work therefore, it is crucial to highlight some important details.

Some researchers [77] applied electrodeposition (EPD) to assemble GQDs into a symmetric micro-supercapacitor. X-ray diffraction and Fourier transform infrared spectroscopy analyses confirmed that the electrodeposition process had little effect on the intrinsic crystallinity and carbon–carbon conjugated backbones of the GQDs. Cyclic voltammetry was used to assess the electrochemical performance of the GQDs/GQDs micro-supercapacitor. A characteristic rectangular CV curve was seen when high scan rates of up to 1000 V/s were reached, indicating perfect capacitive behavior and quick charge transfer capabilities. With a high area-specific capacitance of 468.1 $\mu F/cm^2$ and a small RC time constant of 103.6 μs in a 0.5M Na_2SO_4 aqueous solution, the device demonstrated improved rate capability. In a 2M $EMIMBF_4$/AN electrolyte, it achieved a broader voltage window and enhanced energy density. These results imply that GQDs are attractive candidates for high-performance energy storage applications because of their special qualities, which include a large specific surface area and an abundance of active sites. To further improve their performance and investigate their potential in next-generation micro-supercapacitors and other energy-harvesting systems, future research may concentrate on refining the synthesis and integration of GQDs in a variety of energy storage devices.

Thus far, every substance generated from carbon that is used in the construction and operation of supercapacitors has been examined, and several instances of how researchers are using carbon allotrope-based energy storage technology have been provided. In particular, the synthesis of GQD must be emphasized, and techniques for developing them must be learned. Future views must also be considered to maximize and use current resources and incorporate them into the technological advancement of mankind.

3 Synthesis of Quantum Dots

Quantum dots (QDs) are semiconductor nanoparticles with unique optoelectronic properties, including size-dependent emission and high photostability [124]. Shi et al. [74] set that these characteristics make them attractive for applications in fields such as bioimaging, solar energy, and optoelectronics. Nevertheless, the synthesis of quantum dots can be tailored to control their size, shape, surface characteristics, and chemical composition, which directly affect their optical properties [132]. The most common methods of QD synthesis are colloidal synthesis, chemical vapor deposition (CVD), molecular beam epitaxy (MBE), and hydrothermal methods. Each technique offers distinct advantages and limitations depending on the desired application and the specific type of QD being produced [134].

3.1 Colloidal Synthesis

Colloidal synthesis is one of the most widely used methods for preparing quantum dots due to its simplicity and scalability. This method involves the reaction of precursors, such as metal salts and organic ligands, in a solvent at elevated temperatures to nucleate and grow nanoparticles. The process is typically carried out under controlled conditions, where factors such as temperature, precursor concentration, and reaction time influence the size and crystallinity of the QDs. The size of the nanoparticles can be fine-tuned by varying the reaction parameters, enabling precise control over the optical properties of the quantum dots [87]. One of the primary advantages of colloidal synthesis is the ability to produce high-quality QDs with narrow size distributions, which is critical for applications that require precise emission characteristics, such as display technologies and lighting.

Several studies have looked at the colloidal manufacturing of graphene quantum dots (GQDs) for usage in supercapacitors; certain approaches result in improved electrochemical performance. Citric acid (2 g) was utilized as the carbon precursor in a hydrothermal synthesis described by [73]. The carbon precursor was heated to 180 °C for 12 h while under autogenous pressure in an autoclave lined with Teflon. The generated GQDs were claimed to have 350 F/g of specific capacitance in supercapacitors at a current density of 1 A/g. On average, their size was 3.5 nm. The electrochemical tests showed that the GQDs' high surface area (250 m^2/g) and exceptional conductivity allowed for effective charge storage and quick charge/discharge cycles; after 5000 cycles, the retention rate was 92%, suggesting good performance.

In other work, chemical reduction was employed [150] to create GQDs. Specifically, 0.5 g of graphene oxide was reduced for 24 h at 95 °C in an aqueous solution containing 5 mL of hydrazine hydrate. The resultant GQDs showed size-dependent photoluminescence and had an average diameter of 5 nm. In a three-electrode setup used to assess their electrochemical characteristics, they showed 210 F/g of specific capacitance at 0.5 A/g of current density and an 85% retention of capacitance after 2000 cycles. The consistent charge/discharge behavior of GQDs in supercapacitors can be attributed to their conductive nature, with a conductivity of around 15 S/cm.

Another work [71] used a microwave-assisted synthetic method to create GQDs with electro-chemiluminescent characteristics. The precursor was microwave-irradiated in the presence of a solvent, such as water or ethanol, to create the GQDs using glucose as the carbon source. GQDs with an even size distribution and superior luminous characteristics were produced using this quick and energy-efficient process. These GQDs demonstrated improved specific capacitance and energy density in supercapacitors when tested. This was because of their large specific surface area and strong electrical conductivity, which encouraged effective ion transport and charge buildup.

In summary, several precursors like citric acid, glucose, or graphene oxide are used in the colloidal synthesis of GQDs for supercapacitors, and techniques including hydrothermal treatment, microwave-assisted synthesis, and chemical reduction are employed. The size of the produced GQDs is usually between 3 and 5 nm, and depending on the synthesis technique and supercapacitor arrangement, the specific capacitance can range from 150 to 350 F/g. Their exceptional electrical conductivities (between 10 and 800 S/cm) and large surface areas help them perform better in energy storage applications.

3.2 Chemical Vapor Deposition

Chemical vapor deposition (CVD) is another prominent method used for synthesizing high-quality quantum dots. In CVD, gases are injected into a reaction chamber, leading to chemical reactions that create a thin film or nanoparticles on a substrate. This technique is particularly useful to produce quantum dots with well-defined shapes and sizes, as the growth process can be precisely controlled by adjusting the reaction conditions. CVD is commonly used to produce quantum dots of materials such as CdSe, ZnS, and PbS, which have applications in optoelectronics and photovoltaic devices [85].

A special mention is given to the production of GQDs by CVD when [37] produced GQDs on polycrystalline copper foils using a modified CVD technique. During the atmospheric pressure synthesis, argon functioned as the carrier gas and methane as the carbon source. The copper substrate (99.98% purity) was first cleaned by washing in alcohol and hydrochloric acid to remove surface oxidation, which provided a homogenous surface for GQD development.

For 40 min, a steady flow of 200 mL/min of argon and 10 mL/min of hydrogen was used to heat the copper foil to 1000 °C. To avoid hydrogen etching during the GQD development, the hydrogen flow was then stopped, and the argon was kept constant. To initiate the synthesis of GQDs, 2 mL/min of methane was added to the reaction chamber for a short while, 3 s. With haste, the copper foil was taken out of the heating area and placed under argon protection to cool.

The size and dispersion of the GQDs were carefully controlled by adjusting the growth parameters. By changing the flow rates of methane and hydrogen, as well as the growth period, the GQDs were created with diameters ranging from 5 to 15 nm. While consistent deposition of smaller GQDs was assured by lowering the methane flow to 2 mL/min, bigger graphene domains were formed at higher methane flow rates (5–10 mL/min). The growing conditions were tuned to create well-dispersed, uniformly sized, and defect-free GQDs. Images obtained with a transmission electron microscope (TEM) showed that the hexagonal graphene structures were compatible with the lattice spacings of the GQDs, which demonstrated excellent crystallinity. The height of the GQDs was determined by atomic force microscopy (AFM) to be between 1 and 3 nm, indicating that they had a few layers of structure. For the synthesis of GQDs with regulated sizes, this CVD technique offers a scalable, repeatable procedure appropriate for use in optoelectronics and bioimaging.

One of the key advantages of CVD is its ability to produce quantum dots with high crystallinity and uniformity. The process allows for the deposition of quantum dots on various substrates, making it suitable for the integration of QDs into electronic devices such as solar cells and light-emitting diodes (LEDs). However, CVD typically requires high temperatures and specialized equipment, which can limit its scalability compared to colloidal synthesis. Additionally, the use of toxic precursors, such as cadmium or lead, in some CVD processes raises environmental and health concerns, leading to ongoing research into alternative materials for QD production [105].

3.3 Molecular Beam Epitaxy

Molecular beam epitaxy (MBE) is a highly precise method for quantum dot synthesis that involves the deposition of atoms or molecules onto a substrate in an ultra-high vacuum environment. This technique allows for atomic-level control over the growth of quantum dots, making it ideal for producing semiconductor heterostructures and quantum wells. MBE is widely used in the production of III–V semiconductor quantum dots, such as those made from gallium arsenide (GaAs) and indium arsenide (InAs), which are used in advanced optoelectronic devices, including lasers and quantum computing components [68].

The process of MBE begins by evaporating or sublimating the source material. It is then directed as a beam onto a heated substrate. The atoms or molecules condense on the substrate, leading to a layer-by-layer growth of the material. By carefully controlling the flux of the source material and the substrate temperature, quantum dots with precise size and composition can be synthesized. MBE enables the production of quantum dots with sharp interfaces and minimal defects, making it essential for high-performance optoelectronic devices. However, the process is slow and expensive, which limits its use to high-value applications where precision is crucial [122].

Even though there are disadvantages, Tatti et al. [127] combined the fundamentals of supersonic molecular beam methods and MBE to facilitate the accurate synthesis of

GQDs from fullerene precursors. Based on testing results, certain annealing temperatures, kinetic energy ranges, or equipment configurations can be adjusted as needed. To ensure contamination-free conditions, the synthesis is usually carried out in a UHV system, attaining base pressures below 6×10^{-11} mbar.

The MBE system comprises a molecular beam source and a primary analysis chamber. Fullerenes like C_{60} are the carbon precursor for GQDs. A highly collimated beam of this precursor is directed toward the substrate after it has evaporated. To retain the fullerene vapors in a supersonic beam condition, a quartz capillary tube is heated resistively at around 500 °C. Depending on the source temperature and gas pressure, a supersonic molecular beam with kinetic energies of between 15 and 35 eV may be produced using either hydrogen (H_2) or helium (He) as the carrier gas.

Substrates such as high-purity polycrystalline copper (99.98%) or single-crystal copper (Cu(111)) are cleaned using an argon ion sputtering/annealing cycle and polished to a roughness of less than 0.03 μm. The substrate's temperature is tracked by using an infrared pyrometer and a thermocouple. Substrate annealing is done at temperatures as high as 800 °C to get rid of impurities and prepare the surface for epitaxial growth. It uses a low-energy electron diffraction (LEED) pattern to confirm crystallographic alignment and surface cleanliness.

At room temperature, fullerene molecules are oriented normally to the substrate surface throughout the deposition process. To guarantee that a single monolayer (1ML) of fullerene forms on the substrate surface, the deposition rate is measured. The substrate temperature is progressively raised sequentially from ambient temperature to around 645 °C to improve the breakdown of C_{60}. This annealing process makes it easier for fullerene cages to break, which creates graphene sheets.

High-energy collisions between the fullerene molecules and the substrate cause cage opening, which in turn causes the rearrangement of carbon atoms into a graphene lattice, which is necessary for the formation of GQDs from fullerene precursors in MBE. Fullerene cages undergo considerable distortion upon contact at kinetic energies less than 40 eV, but greater energy levels or thermal treatment are necessary to accomplish full cage rupture and graphene production.

In situ analysis methods including ultraviolet photoelectron spectroscopy (UPS), X-ray photoelectron spectroscopy (XPS), and LEED are used to evaluate the structural and chemical characteristics of the deposited layers both during and after deposition. The electrical structure and bonding environment are revealed by XPS and UPS, which also show charge transfer between the substrate and the carbon species. Superstructures are observed on the Cu(111) surface according to LEED analysis, which suggests that the substrate has been rebuilt as a result of robust interactions with fullerene molecules.

At temperatures as high as 795 °C, post-deposition annealing is carried out to guarantee full transformation into GQDs. Through this thermal treatment, fullerene molecules break down into structures resembling graphene, as demonstrated by scanning electron microscopy (SEM), Raman spectroscopy, and XPS. The annealed samples' Raman spectra show the distinctive defective graphene D, G, and 2D bands; the 2D band verifies the existence of single-layer graphene domains.

A possible method for creating high-quality graphene layers with regulated size and defect density is the production of GQDs by MBE. Growth of defective but uniform GQD structures can be accomplished by optimizing the deposition parameters, such as substrate temperature, precursor kinetic energy, and post-deposition annealing conditions. This technology is appropriate for applications where substrate thermal stability is a problem since it provides a lower temperature alternative to traditional chemical vapor deposition (CVD) procedures.

3.4 Hydrothermal and Solvothermal Methods

Hydrothermal and solvothermal methods are widely used for the synthesis of quantum dots, particularly for materials that are difficult to produce using other techniques. These methods involve the reaction of precursors in water or organic solvents under high temperature and pressure in a sealed vessel, often referred to as an autoclave. The high-pressure environment promotes the formation of crystalline nanoparticles with controlled size and morphology. Hydrothermal synthesis is particularly useful to produce oxide-based quantum dots, such as ZnO and TiO_2, which have applications in photocatalysis and sensors [139].

Solvothermal methods, which use organic solvents instead of water, are employed for the synthesis of quantum dots from a wider range of materials, including chalcogenides and other metal compounds. These methods offer advantages in terms of control over particle size and shape, as well as the ability to produce quantum dots with high crystallinity and purity. One of the key benefits of hydrothermal and solvothermal methods is their scalability, making them attractive for large-scale production of QDs for commercial applications [15].

Luo et al. [78] synthesized graphene quantum dots/three-dimensional graphene (GQDs/3DG) composites for supercapacitor-improved electrode materials. Graphene oxide (GO) aqueous dispersion (2 $mg \cdot mL^{-1}$) was the main precursor used to create the 3DG framework. GQDs were produced by chemically oxidizing polyacrylonitrile carbon fibers, which gave precise control over the size and structure of the resulting quantum dots. Precursor preparation is significant because it influences the way GQDs and the graphene matrix interact during hydrothermal treatment.

GO dispersion (14 mL) was added to a 22 mL Teflon-lined autoclave in the one-step hydrothermal synthesis procedure, and then a predefined amount of GQDs was added. To obtain various GQD loadings in the 3DG framework, the GQDs/GO feeding ratios were adjusted between 10, 20, 40, and 80% by weight. After sealing, the mixture was heated to 180 °C for 4 h. Following the hydrothermal process, any remaining GQDs not incorporated into the structure were completely removed from the composite hydrogel by vigorously washing it in distilled water for 5 days. This method made it easier to create a highly porous 3DG structure with a consistent distribution of GQDs. Through their oxygenated functional groups, the GQDs established robust chemical bonds with the GO matrix, which boosted electron transport and improved stability in the composite material.

On the other hand, the two-step hydrothermal technique required pre-synthesizing the 3DG hydrogel before submerging it in an aqueous GQD dispersion. The identical hydrothermal treatment was then applied to this composite for 4 h at 180 °C. Because fewer GQDs were chemically bonded to the framework, this approach was less effective in integrating GQDs into the 3DG network. In this instance, weaker π–π interactions dominated the contact between GQDs and 3DG so than more stable covalent bonding were shown in the one-step technique.

Using a hydrothermal technique, graphene quantum dots (GQDs) were effectively incorporated into a three-dimensional graphene (3DG) framework. Composites with varied GQD loadings were produced by altering the GQDs/GO feeding ratios (10, 20, 40, and 80%). The one-step hydrothermal procedure ensured uniform dispersion of GQDs inside the graphene matrix. SEM and TEM structural characterization verified that the GQDs/3DG-40 composite had an evenly dispersed quantum dot morphology and was more porous and fluffier than pure 3DG. Further, TEM research revealed that modest GQD loading promoted increased surface area and electron transport by contributing to a more disordered crystalline structure.

Through electrochemical testing, the maximum specific capacitance of 242 $F \cdot g^{-1}$ was reached by the GQDs/3DG-40 composite, with a 40% GQD loading. This represents a 22% improvement above pure 3DG. Better electrical conductivity and a bigger surface area were credited for improved performance. The composite also showed outstanding cycling stability, holding onto 93% of its capacitance after 10,000 cycles. The denser structure and lower capacitance resulting from excessive GQD loading (80%) highlight the significance of optimizing GQD content to optimal performance in supercapacitor applications.

In conclusion, a reliable technique for creating sophisticated electrode materials for supercapacitors is the hydrothermal synthesis of GQDs/3DG composites. GQDs can be effectively incorporated into the graphene framework using, for example, the one-step hydrothermal process. This produces composites with improved electrochemical properties, such as longer cycling stability and better capacitance. The scalability of the hydrothermal process and this approach's adaptability make GQDs/3DG composites an attractive option for next-generation energy storage devices.

To improve the electrochemical performance of supercapacitors, [38] nitrogen-doped graphene quantum dots (N-GQDs) have been combined with $Ta_4C_3T_x$ MXene using solvothermal and chemical etching to create a novel electrode material. In order to create $Ta_4C_3T_x$ MXene nanosheets, the Ta_4AlC_3 MAX phase powder was first chemically etched with an HCl/NaF solution to remove aluminum layers. To enhance exfoliation and functionalization, a 48 h wet etching procedure was carried out at 60 °C. Then, using a solvothermal synthesis technique, nitrogen-doped graphene quantum dots (N-GQDs) were applied to the exfoliated MXene nanosheets. This allowed the N-GQDs to be integrated between MXene layers, preventing restacking and enhancing electrochemical performance. By acting as efficient spacers between MXene layers, the N-GQDs helped with ion transport and prevented restacking. With 92% capacitance retention after 10,000 cycles, this composite structure demonstrated exceptional cycling stability and a high specific capacitance of 450 $F \cdot g^{-1}$ at 1 $A \cdot g^{-1}$. The promise of this composite for high-performance energy storage applications was highlighted by the enhanced electrical conductivity and electrochemical stability brought about by the synergistic interaction between N-GQDs and MXene.

3.5 Green Synthesis

In response to growing environmental concerns, green synthesis methods for quantum dots have emerged as a sustainable alternative to traditional synthesis techniques. These methods aim to reduce the use of toxic chemicals and solvents by employing environmentally friendly precursors, such as plant extracts or biomolecules. Green synthesis typically occurs under mild reaction conditions, which reduces energy consumption and minimizes hazardous waste. Quantum dots produced via green synthesis have shown promise in applications such as bioimaging, where low toxicity and biocompatibility are critical [115].

Chen et al. [26] states that GQDs are produced from cellulose using a straightforward but efficient green synthesis method that mainly uses cotton cellulose as the precursor. First, a cellulose and deionized water mixture is made for the experiment. There is no need for strong oxidizing agents, acids, or organic solvents in this environmentally friendly synthesis. After being dissolved in water, the cellulose is exposed to hydrothermal conditions. The reaction usually takes place for 6–8 h at high temperatures of 200–220 °C, which creates an environment conducive to the hydrolysis of cellulose.

The process of hydrolyzation breaks down the cellulose into glucose units during the reaction. The hydrothermal conditions' high temperature and pressure encourage the breaking of glycosidic bonds in the cellulose, which releases glucose molecules. Glucose passes through a cyclic condensation process as the reaction goes on, building aromatic rings and ultimately producing GQDs. The graphitic structure of the glucose molecule-bound carbon atoms forms the GQDs, which usually have a limited size dispersion and a particle size distribution between 1.0 and 5.0 nm.

Following the reaction, a mixture of GQDs, water, and carbide precipitates is present in the resultant solution. The separation of GQDs is performed using centrifugation, which isolates the GQDs from any byproducts. Following their reaction, the GQDs exhibit remarkable hydrophilicity and photoluminescent (PL) characteristics, rendering them very well-suited for biological applications, including bioimaging. The structural and chemical integrity of the GQDs is confirmed by characterization using methods such as transmission electron microscopy (TEM), Fourier transform infrared (FTIR) spectroscopy, and Raman spectroscopy.

The technique produces a GQD yield of about 57.7%, indicating good manufacturing efficiency. This green synthesis method has the advantages of being straightforward, requiring little post-processing, and not involving any hazardous chemicals. In addition, the synthesized GQDs show minimal cytotoxicity, which makes them perfect for use in biological applications. The entire process offers a scalable method for creating high-quality GQDs from renewable resources like cellulose in addition to adhering to sustainability guidelines.

Song et al. [120] used waste eggshell membranes as the carbon precursor in their sustainable synthesis method for creating nitrogen-sulfur co-doped hierarchical porous carbon (NS-HPC) electrodes. The synthesis was carried out in two stages: first, the eggshell membranes were hydrothermally carbonized, and then they were pyrolyzed. The electrochemical performance of the electrodes was then improved by adding vanadium oxide (VxOy) quantum dots to the NS-HPC matrix by hydrothermal treatment. The resultant VxOy/NS-HPC composites showed outstanding cycling stability with 92.3% capacitance retention over 10,000 cycles and a high specific capacitance of 355 $F \cdot g^{-1}$ at 0.5 $A \cdot g^{-1}$. The potential of turning biowaste into high-performance electrode materials for flexible supercapacitor applications is highlighted by this work.

While green synthesis methods are still in the early stages of development, they hold significant potential for reducing the environmental impact of quantum dot production. However, achieving the same level of control over QD size and optical properties as traditional methods remains a challenge. Ongoing research is focused on optimizing green synthesis techniques to produce high-quality quantum dots with comparable performance to those synthesized via conventional methods [45].

3.6 Synthesis from Compounds

A well-documented top-down approach for the synthesis of GQDs has been developed by Shin et al. [117]. It involves the oxidative cleavage of GO as a compound, which can be achieved using various methods, including strong acid treatments. However, a novel and more environmentally friendly method utilizes oxone (potassium monopersulfate) as an oxidant, which avoids the use of harmful acids. In this approach, the GO sheets are subjected to sonication or UV irradiation in the presence of oxone, generating sulfate and hydroxyl radicals that cleave the carbon–carbon bonds in the graphene structure. This process results in the formation of GQDs that exhibit strong photoluminescent properties and are efficiently produced with minimal purification steps.

In this approach, the GO is dispersed in an organic solvent such as dimethylformamide (DMF), and 500 mg of oxone was added to the solution, which is then subjected to either sonication (500 W) or UV irradiation (100 W) for 1 h at room temperature. Oxone's free radicals degrade the GO, resulting in the production of GQDs that were between two and six nanometers in size. With the use of a 200 nm membrane for filtering and subsequent dialysis for additional purification, the resultant GQDs may be readily distinguished from larger graphene pieces. This process offers an environmentally friendly substitute for GQD synthesis by drastically cutting down on reaction time and eliminating the requirement for strong acids or bases.

Comparing oxone-based synthesis of GQDs to conventional acid-based techniques reveals several benefits. The procedure not only reduces the reaction time but also makes purification easier by doing away with the need for neutralizing procedures that are normally necessary when working with strong acids like nitric or sulfuric acid. Additionally, the GQDs generated using this technique exhibit homogeneous particle sizes and high fluorescence, which qualifies them for use in optoelectronics and bioimaging. The crystalline structure of the GQDs had been verified by Raman spectroscopy and high-resolution transmission electron microscopy (HR-TEM). They typically measure 4–5 nm in size.

This oxone-mediated synthesis represents a significant step forward in the green production of GQDs throughout compounds. By eliminating the use of hazardous chemicals and reducing the complexity of the reaction, this method provides an efficient and scalable route for producing high-quality GQDs, paving the way for their broader use in environmentally sensitive applications such as biological labeling and nanomedicine.

To sum up, the use of oxone in the production of graphene quantum dots presents a viable and effective substitute for conventional techniques that depend on hazardous substances. This method is appropriate for a variety of applications since it improves scalability and streamlines the production process. This approach is a useful tool for improving green chemistry since it produces high-quality GQD output under environmentally friendly circumstances. This is especially important in domains that minimize environmental impact, such as bioimaging and nanomedicine.

Compounds involving organic molecules and waste are the basis of numerous synthesis techniques. While mentioned in the section 3.5 an example [44], using citric acid and urea as precursors for the synthesis of GQD by infrared-assisted pyrolysis is worth highlighting. The authors heated a solid mixture of urea and citric acid in an infrared (IR) furnace as a straightforward, economical, and scalable method to create GQDs. With no need for extra chemical reagents or solvents, this technology reduces the reaction time to just 20 min at 200 °C, offering a substantial advantage over conventional techniques. This study examined how different ratios of urea to citric acid as precursors affect the optical and chemical properties of the resulting GQDs. The GQDs with the best quantum yield (22.2%) and best fluorescence characteristics were those that were produced using a 3:1 citric acid to urea ratio. GQDs with a quasi-spherical form and an average size of 5–10 nm were produced by the infrared carbonization (IRC) technique, which also resulted in the inclusion of nitrogen into the graphene lattice. It's interesting to note that the GQDs also included crystalline cyanuric acid at greater urea contents, creating a binary crystalline structure that influenced optical characteristics like red-shift behavior and fluorescence quenching.

Using Fourier transform infrared spectroscopy (FTIR), X-ray photoelectron spectroscopy (XPS), and high-resolution transmission electron microscopy (HR-TEM), the structural and chemical properties of the GQDs were verified. According to these investigations, the GQDs' variable fluorescence emissions were caused by a range of functional groups, including hydroxyl, carboxyl, and nitrogen-based groups (such as pyridinic and graphitic nitrogen). Functional group distributions and defect-related emissive traps on the GQD surfaces were found to be responsible for the optical characteristics of the GQDs, such as absorption and fluorescence.

As a result of adjusting the precursor ratios of citric acid and urea, Gu et al.'s infrared-assisted pyrolysis process offers a novel and effective way to create nitrogen-doped GQDs with adjustable fluorescence features. This process is extremely scalable and ecologically benign, and it improves manufacturing efficiency by lowering reaction durations to a mere 20 min at low temperatures. It also exhibits exquisite control over the optical properties and chemical composition of the GQDs. Depending on the nitrogen concentration and crystalline structure, the integration of cyanuric acid and nitrogen into the GQDs contributes to a multiple chromophoric band gap structure that enables tunable emissions ranging from blue to green fluorescence.

Connected to earlier studies on GQD synthesis via graphene oxide oxidative cleavage, the work emphasizes the adaptability of carbon-based precursors and synthesis techniques in generating GQDs with tunable characteristics for a range of uses, from energy storage to bioimaging. When taken as a whole, these studies show the variety of GQD synthesis methods available, ranging from fast, effective pyrolysis to environmentally friendly green methods utilizing oxidants like oxone. This opens up new avenues for the functionalization and use of GQDs in a variety of scientific and industrial domains.

4 Properties and Applications of Quantum Dots in Supercapacitors

This phenomenon establishes the conditions necessary to modify the energy band gap based on the size and chemical composition of the QDs. These nanocrystals typically range in size from 1 to 12.5 nm [60]. However, recent advances have allowed the identification of QDs with sizes exceeding 31 nm [34] or even sizes exceeding 50 nm [111]. Furthermore, QDs are zero-dimensional nanoparticles (dots) relative to their volume and exhibit specific light absorption and emission behaviors that are dependent on the size of the nanocrystal. These changes in properties can be observed in QDs with particle sizes smaller than <~30 nm [13]. Additionally, QDs are also referred to as artificial atoms because the number of atoms in such confined spaces generates wavelengths similar to those observed in atomic systems [4]. Unlike conventional semiconductors, QDs enable controlled photon emission across different wavelengths, from ultraviolet to infrared, simply by varying their size [109].

QDs exhibit outstanding optical properties due to the previously mentioned quantum confinement effect. Notably, factors such as size, chemical composition, and crystalline structure directly influence their optical properties, including light absorption and emission. Managing these properties is highly valuable in various applications such as bioimaging, high-resolution displays, and solar cells [98].

4.1 Light Absorption

QDs exhibit discrete optical absorption behavior, attributed to quantum confinement. This results in a series of well-defined absorption peaks, rather than a continuous spectrum as seen in bulk materials [35]. Electrons in QDs are confined to a small space, leading to the quantization of energy levels [48]. The absorption peaks of QDs are size-dependent: the smaller the QD ($\leq$10 nm in general), the greater the energy required to excite electrons from the valence band to the conduction band, causing a shift toward shorter wavelengths (ultraviolet or blue) in the absorption spectrum [138].

4.2 Light Emission

The light emission process in QDs occurs when excited electrons in the conduction band return to the valence band, releasing photons in the process. Due to quantum confinement, the emission wavelength is strongly related to the size of the QD [93]. Smaller QDs ($\leq$10 nm in general) emit photons with higher energy, resulting in shorter emission wavelengths. This causes a blue shift in the visible spectrum as the size of the QDs decreases [112]. Graphene and Graphene Oxide QDs combined with graphene particles are primarily used in solar cell applications due to their enhanced optical and photoelectric properties. The synthesis of these hybrid structures is achieved through chemical deposition and mechanical exfoliation methods, enabling the assembly of QDs on graphene surfaces or its derivatives. The QDs employed range in size from a few nanometers to tens of nanometers, allowing efficient absorption of solar radiation across a broad spectral range. This combination with graphene improves charge carrier mobility, which is crucial for optimizing energy conversion efficiency in solar devices [119]. Additionally, another application of QDs with graphene particles is in light-emitting devices (QLEDs) to enhance hole injection efficiency, achieving greater stability and brightness. For example, core–shell CdSe/ZnS quantum dots synthesized using the thermal cycle technique assisted by TBP (tributylphosphine) and coupled with the single precursor method [70] are combined with UV-ozone-treated partially reduced graphene oxide (rGO), enabling a stepwise energy level alignment and a high-quality interface. These quantum dots have a size of 10 nm and exhibit an emission peak in the red region, specifically at 630 nm. The synthesis technique involves solution-based layer deposition, optimizing both conductivity and charge injection balance, which increases the device's brightness and external quantum efficiency (EQE) to up to 17.3% [66].

Quantum dots, specifically carbon dots (C-dots) synthesized by hydrothermal method, combined with zinc oxide nanoparticles (ZnO–NPs), are being explored as materials that enhance photocatalytic activity, in particular for the degradation of pollutants such as methylene blue dye. ZnO nanoparticles were synthesized using sol–gel method, yielding particles within the size range of 1–100 nm. These particles exhibit high stability and possess antimicrobial and antioxidant capabilities. However, their large band gap (3.37 eV) restricts their photocatalytic efficiency. By incorporating 10% C-dots into ZnO nanoparticles, the band gap is reduced to 2.8 eV, which enhances visible light absorption and increases the dye degradation efficiency to 98.4% in 60 min. These nanocomposites demonstrate significant potential for environmental decontamination applications under solar irradiation and can be reused for up to four cycles with minimal loss of efficiency [128].

QDs are semiconductor nanocrystals that exhibit unique electronic properties due to quantum confinement; these properties depend on particle size, chemical composition, and interactions with their surroundings, all these making QDs highly attractive for various technological applications, particularly in electronics, biological, photovoltaics etc. [111]. Below, key electronic properties of QDs are detailed, along with specific examples from recent studies. One of the most defining characteristics of QDs is quantum confinement, which occurs when the dimensions of the material become comparable to the electron's wavelength. In this state, electrons are confined within a small spatial region, leading to discrete energy levels instead of the continuous spectrum seen in bulk materials [143]. This confinement alters the band gap, which can be tuned by adjusting the QD size, the smaller the QD ($\leq$10 nm), the larger the band gap, directly influencing its electronic properties [4]. Another important electronic property is quantum tunneling, where electrons can pass through energy barriers that they could not classically overcome. In QD systems, quantum tunneling plays a crucial role in electronic transport [60].

QDs are semiconductor nanocrystals that, due to quantum size effects, enable band gap tuning and enhance charge carrier mobility in electronic devices such as field-effect transistors (FETs). This behavior is observed across various types of QDs [62]. In semiconductor materials, high electron mobility is essential for the performance of electronic and optoelectronic devices, as it enables more efficient and faster charge transport [121]. The electronic properties of QDs are highly versatile and can be finely tuned by controlling their size, composition, and environment. This has led to advancements in electronics, optoelectronics, and energy-related technologies. Ongoing research continues to focus on enhancing electronic transport and charge injection efficiency in nanoscale devices based on QDs.

4.3 Interaction-Induced Changes in the Electrical Properties of QDs

The combination of QDs with polymers can modify their electrical properties, enhancing conductivity or introducing semiconductor behavior. Polymers typically act as dielectric matrices that electrically insulate the QDs. However, when QDs are uniformly dispersed within the polymer matrix, conductive pathways can be formed throughout the material [32]. Moreover, it is worth noting that combining QDs with polymers can induce electroluminescent properties, which have been leveraged in the development of polymer-based light-emitting diodes (LEDs). In these devices, the integration of QDs enhances the energy conversion efficiency and improves the electrical characteristics of the polymer [59].

When QDs are incorporated into metals, their electrical properties can be substantially enhanced, particularly in terms of conductivity and electron transport capacity. The inclusion of QDs in metals such as gold or silver can reduce the metal's resistivity, which is crucial for applications in high-speed electronics and nanocircuits [83]. The integration of QDs into ceramics is typically aimed at enhancing their optoelectrical properties and photoluminescence. Ceramic materials are inherently insulating, however, the addition of QDs can increase their ability to release energy [42]. When QDs are combined with organic materials, improvements in charge transport and efficiency in organic optoelectronic devices can be observed. Due to their high electron mobility, QDs can facilitate charge carrier transport in thin films of organic materials [140]. These combinations are commonly used in the fabrication of organic solar cells, where QDs serve as charge collection centers, thereby improving photovoltaic conversion efficiency. QDs can also enhance the quantum efficiency of organic materials in devices such as LEDs and organic transistors [114].

Graphene is an exceptional conductive material, and when combined with QDs, the electrical properties of both materials are enhanced. QDs can act as scattering centers or additional electron sources, improving charge carrier mobility in graphene; in applications where graphene is mixed with QDs, enhanced electrical response has been observed, making these materials suitable for high-precision sensors and field-effect transistors [136]. For example, molybdenum disulfide (MoS_2)-QDs exhibit altered electrical properties when hybridized with reduced graphene oxide (RGO) particles functionalized with 3-mercaptopropyl-trimethoxysilane (MPTS). The MoS_2-QDs are synthesized via liquid-phase exfoliation and a solvothermal process, resulting in particles with an average lateral size of 2.6 nm and a thickness of 1.6 nm. The RGO is chemically functionalized with 3-mercaptopropyl-trimethoxysilane (MPTS). This hybridization enables an energy conversion efficiency of up to 20.12%, with an average efficiency of 18.8% [88].

The electrical properties of QDs significantly vary depending on the material they are combined with. In polymers, they enhance conductivity and introduce electroluminescent properties. In metals, they reduce resistivity and improve electron transport. In ceramics, they improve dielectric properties and energy storage capacity. In organic materials, they increase charge transport efficiency and quantum efficiency. Finally, when combined with graphene, QDs boost electrical conductivity and charge carrier mobility, making these hybrid materials promising for advanced electronic devices.

5 Methods for Quantifying Energy in Capacitors

The LCR bridge method is crucial for measuring the electrical parameters of inductance, capacitance, and resistance in electronic components. This method applies an alternating current (AC) signal to the component or electrode under test and analyzes the relationship between voltage and current [104]. By varying the signal frequency, it is possible to fully characterize the component's inductance and capacitance. At higher frequencies, capacitance becomes more prominent due to the decrease in capacitive reactance [55].

The LCR bridge is particularly useful for evaluating electronic components that operate across a wide frequency range, such as in telecommunications circuits and high-precision devices. Through this method, it is possible to study dielectric losses and how they affect system stability under high-frequency conditions. Additionally, it is used to assess the reliability and durability of electrical materials by conducting measurements at different frequency loads [55].

Frequency response analysis (FRA) is a critical technique used for diagnosing structural and electrical faults in power transformers. This method injects a variable frequency signal and measures the system's response, which helps identify deviations in the mechanical or electrical properties of the equipment. Specifically, FRA can detect winding deformations, internal displacements, and partial short circuits [10].

At lower frequencies, FRA responses are dominated by the magnetic interaction between the core and windings, while at higher frequencies, variations in capacitance and leakage inductance become evident, enabling the detection of internal transformer defects. FRA can be conducted in two configurations: open-circuit and short-circuit. In the open-circuit configuration, core magnetization impedance is analyzed, while the short-circuit configuration evaluates leakage inductance, reflecting winding-related problems [80].

International standards such as IEC 60,076-18 and IEEE 1036-2020 define the procedures and best practices for conducting FRA tests, offering detailed guidelines for interpreting results. This type of analysis plays a key role in predictive maintenance programs, as it allows fault detection before they escalate into major issues, optimizing the lifespan of transformers and capacitors [51, 52].

EIS is essential for analyzing batteries, supercapacitors, sensors, and corrosion studies, as it provides insights into charge transfer kinetics, interfacial capacitance, and electrolyte resistance. Moreover, it is critical for characterizing interfaces, detecting material defects or variations, and improving the efficiency and longevity of electrochemical devices [84].

This method maintains a constant current in the system while monitoring potential over time, offering crucial information about reaction kinetics and ionic transport characteristics in the electrode material [25]. Chronopotentiometry enables the determination of key parameters such as ionic diffusivity and specific material capacity, which are essential for high-energy–density storage devices. The transition time, representing the time needed for the potential to significantly change, correlates with the kinetics of redox reactions and the efficiency of energy storage [28].

The process involves repeated cycles of charge and discharge at constant current, which allows for the evaluation of electrode material stability over multiple cycles [95]. This is particularly relevant in the research of lithium-ion batteries and supercapacitors, as it quantifies coulombic efficiency and observes material capacity degradation [148]. This analysis can detect factors such as passive layer formation and electrode structural degradation, which impact device lifespan and performance [95].

Ragone plots are critical tools for evaluating the relationship between energy density and power density in electrochemical devices such as batteries and supercapacitors [22]. These plots efficiently visualize and compare the performance characteristics of different energy storage systems in terms of the amount of energy they can deliver (energy density) and the rate at which they can be delivered (power density) [14].

To apply Ragone plots to batteries and capacitors, charge and discharge measurements are recorded at different current rates. Energy density is determined from the total energy stored in the device, calculated by integrating the area under the voltage curve during discharge [29]. Power density is obtained by measuring the rate at which this energy is delivered, relating the available energy to the discharge time at a specific current [14].

A supercapacitor developed by [141] consisted of an activated carbon (AC) cathode that was purposefully electrochemically treated throughout the supercapacitor activation process, greatly improving its electrochemical performance. To accomplish this, a high voltage (>2.2 V) is applied to the cathode, which causes electrolyte breakdown and the consequent evolution of chlorine gas. The ensuing chemical reactions increase the AC structure's porosity and electrochemically active surface area by facilitating the creation of new pores. After activation, cyclic voltammetry studies show that the electrode capacitance significantly rises from about 114 F g^{-1} to over 417 F g^{-1}, indicating a significant expansion of its ion-accessible surface. By enhancing electrolyte ion penetration and transport, this pore formation raises the device's gravimetric capacitance and energy storage capacity. Additionally, following extended cycling (up to 10,000 cycles), the activated cathodes show extraordinary stability, retaining 100% of their initial capacity, highlighting the effectiveness and longevity of the activation technique in maximizing supercapacitor performance.

6 Applications of Quantum Dots (QDs) in Capacitors

QDs play a crucial role in supercapacitors due to their ability to significantly enhance the electrochemical performance of these energy storage devices. QDs serve as fluorescent electrolytes and electrode materials, providing higher specific capacitance and improving energy storage efficiency. Moreover, the integration of QDs with metals and polymers leads to the formation of electrodes with outstanding electrochemical properties [92].

Graphene quantum dots (GQDs) are widely utilized in supercapacitors due to their remarkable ability to significantly enhance energy storage. GQDs synthesized via a simple hydrothermal method using nitric acid yield nanocrystals with sizes smaller than 5 nm. These GQDs exhibit a high specific capacitance of 296.7 F g^{-1} and an energy density of 41.2 Wh kg^{-1} at a current density of 1 A g^{-1}. Additionally, GQDs demonstrate low internal resistance, short relaxation times, and excellent cycle stability, resulting in high electrochemical activity and conductivity. These characteristics enable efficient charge storage, prevent graphene sheet stacking, and maximize active surface area, making GQDs ideal for energy storage applications [146].

Moreover, various elements can be doped into graphene to improve QDs structure and properties. Nitrogen-doped GQDs (N-GQDs) are used to enhance polyaniline (PANI) in the production of flexible, printed supercapacitors. The integration of N-GQDs into the PANI matrix significantly increases the surface area and energy storage efficiency. As a result, the supercapacitor achieves a high capacitance of 128.01 mF/ cm^2 and retains 76.70% of its capacity after 5000 charge–discharge cycles. Additionally, it exhibits excellent mechanical stability, making it ideal for portable devices requiring flexibility and durability [102].

Similarly, GQDs measuring between 2 and 5 nm, combined with acidic functional groups such as –COOH and –OH, can also be synthesized through the chemical oxidation of graphite oxide. These functionalized GQDs are used as electrolytes in supercapacitors, either in solution or solid-state form. By neutralizing these acidic groups with KOH, both ionic conductivity and ion donation capacity are significantly enhanced. As electrolytes, these neutralized GQDs demonstrate superior performance and capacitance compared to non-neutralized GQDs [145].

Carbon-QDs (CQDs) are used in supercapacitors due to their superior optoelectronic properties derived from quantum confinement effects. CQDs enhance supercapacitor performance by providing higher specific capacitance, energy density, and durability. These improvements result in significant gains in energy storage capacity and device stability, whether CQDs are employed as standalone electrodes or in combination with other materials. The structural and electrical properties of CQDs contribute to the superior performance of supercapacitors. Additionally, CQDs can be synthesized through various methods, including laser ablation, chemical oxidation, chemical exfoliation, arc discharge, ultrasonication, hydrothermal synthesis, microwave synthesis, and pyrolysis [97].

One of the applications of these nanocrystals includes the use of 17 nm carbon CQDs doped combined with cerium-nickel oxide (Ce–NiO) in supercapacitors to enhance electrochemical performance. This nanocomposite exhibits a high specific capacitance of approximately 2487 F g^{-1} via cyclic voltammetry and 2513 F g^{-1} via galvanostatic charge–discharge, attributed to the redox activity of Ce^{3+}/Ce^{4+} ions and the high conductivity of CQDs. Additionally, it retains 97% of its capacitance after 2000 charge–discharge cycles, highlighting its durability and efficiency for energy storage applications [3].

It's also important to note that the raw materials for CQDs can be sourced from organic waste, such as coffee grounds, which are utilized in flexible carbon electrodes and electric double-layer capacitors (EDLCs) with PVA gel electrolytes [7].

7 Conclusions

QDs exhibit exceptional optical and electronic properties, primarily influenced by quantum confinement effects. As QD size decreases, the energy band gap widens, causing a blue shift in absorption and emission peaks. This precise tuning of optical properties through size and morphology control is a significant advantage in applications like bioimaging, optoelectronics, and high-efficiency displays. The interaction of QDs with other materials also plays a crucial role in modifying their properties. When incorporated into polymer matrices, metal nanoparticles, or graphene, QDs exhibit significant changes in their optical and electronic characteristics. Plasmonic effects induced by metallic nanoparticles, for instance, can enhance QD emission intensity by increasing radiative recombination rates. Similarly, incorporating QDs into graphene has demonstrated enhanced charge transport, boosting efficiency in photovoltaic devices and energy conversion technologies. This capacity to modify QD properties through material integration broadens their potential for advanced applications in photonics and sensing.

In electronic terms, QDs exhibit discrete energy levels due to quantum confinement, resulting in unique electronic characteristics that differ from bulk semiconductors. This uniqueness allows fine-tuning of energy bands and charge transport in QDs, making them especially useful in electronic and optoelectronic applications. However, electron mobility tends to decrease as QD size is reduced due to increased boundary scattering, which can impact device performance, requiring high charge mobility. In energy storage systems, QD integration in supercapacitors has shown notable impacts on electrochemical performance. QDs such as graphene QDs have improved specific capacitance, energy density, and cycle stability in supercapacitors. This is due to the high surface-area-to-volume ratio of QDs, which provides a larger active area for energy storage reactions, as well as their ability to enhance electron mobility and conductivity. Doped QDs have also shown increased mechanical and electrochemical stability, making them ideal for portable and high-energy-efficiency devices. In summary, QDs offer a range of adjustable properties through size and composition control, making them essential components in the development of advanced devices for optoelectronics, energy storage, and electronics. Interactions with materials like polymers, metals, and ceramics optimize these properties, opening up new possibilities for applications in emerging technologies. Improvements in energy conversion efficiency and charge mobility highlight the potential of QDs to revolutionize sectors such as energy storage and display technologies.

References

1. Afif A, Rahman SM, Tasfiah Azad A, Zaini J, Islan MA, Azad AK (2019) Advanced materials and technologies for hybrid supercapacitors for energy storage—a review. J Energy Storage 25:100852

2. Akram A, Kayani ZN (2025) Fabrication of carbon quantum dots by microemulsion route: dual functional materials for energy storage and environmental treatment. Mater Res Bull 190:113494
3. Al-Enizi AM, Ubaidullah M, Kumar D (2021) Carbon quantum dots (CQDs)/Ce doped NiO nanocomposite for high performance supercapacitor. Mater Today Commun 27:102340. Elsevier
4. Alivisatos AP (1979) Semiconductor clusters, nanocrystals, and quantum dots. Science (1979) [Internet]. American Association for the Advancement of Science; 1996 Feb 16 [cited 2024 Sep 16]; 271(5251):933–7. https://doi.org/10.1126/science.271.5251.933
5. Allen MJ, Tung VC, Kaner RB (2010) Honeycomb carbon: a review of graphene. Chem Rev 110(1):132–145
6. Angel RJ, Alvaro M, Nestola F (2022) Crystallographic methods for non-destructive characterization of mineral inclusions in diamonds. Rev Mineral Geochem 88(1):257–305
7. Arora G, Sabran NS, Jun HK (2024) flexible electrodes coated with carbon quantum dots derived from spent coffee grounds for electric double-layer capacitors. In: 2024 IEEE international conference on flexible and printable sensors and systems (FLEPS). Institute of Electrical and Electronics Engineers Inc.
8. Azman NHN, Mamat Mat Nazir MS, Ngee LH, Sulaiman Y (2018) Graphene-based ternary composites for supercapacitors. Int J Energy Res 42(6):2104–2116
9. Baker SN, Baker GA (2010) Luminescent carbon nanodots: emergent nanolights. Angew Chem Int Ed 49(38):6726–6744
10. Banaszak S, Szoka W (2020) Transformer frequency response analysis with the grouped indices method in end-to-end and capacitive inter-winding measurement configurations. IEEE Trans Power Deliv. 35(2):571–579. Institute of Electrical and Electronics Engineers Inc.
11. Bati ASR, Yu L, Batmunkh M, Shapter JG (2018) Synthesis, purification, properties and characterization of sorted single-walled carbon nanotubes. Nanoscale 10(47):22087–22139
12. Belin T, Epron F (2005) Characterization methods of carbon nanotubes: a review. Mater Sci Eng, B 119(2):105–118
13. Bera D, Qian L, Tseng TK, Holloway PH (2010) Quantum dots and their multimodal applications: a review. Materials [Internet]. Multidisciplinary Digital Publishing Institute (MDPI); 2010 [cited 2024 Sep 16]; 3(4):2260. https://www.pmc/articles/PMC5445848
14. Beyers I, Bensmann A, Hanke-Rauschenbach R (2023) Ragone plots revisited: a review of methodology and application across energy storage technologies. J Energy Storage 73:109097. Elsevier
15. Bian Z-Y, Gu X-D, Zhang W (2016) Solvothermal synthesis of chalcogenide quantum dots: applications in energy storage. Nano Energy 20:274–283. Elsevier
16. Binfaris AS, Zestos AG, Abot JL (2023a) Development of carbon nanotube yarn supercapacitors and energy storage for integrated structural health monitoring. Energies (Basel) 16(15)
17. Binfaris AS, Zestos AG, Abot JL (2023) Development of carbon nanotube yarn supercapacitors and energy storage for integrated structural health monitoring. Energies (Basel) 16(15):5736
18. Blank V, Popov M, Pivovarov G, Lvova N, Terentev S (1999) Mechanical properties of different types of diamond. Diam Relat Mater 8(8–9):1531–1535
19. Bonaccorso F et al (2015) Graphene, related two-dimensional crystals, and hybrid systems for energy conversion and storage. Science (1979) 347(6217):1246501
20. Branca C, Frusteri F, Magazù V, Mangione A (2004) Characterization of carbon nanotubes by TEM and infrared spectroscopy. J Phys Chem B 108(11):3469–3473
21. Campbell EB, Hirsch A (2000) Fullerenes and their applications in materials science. J Mater Sci 35(1):1–22
22. Cericola D, Ruch PW, Kötz R, Novák P, Wokaun A (2010) Simulation of a supercapacitor/Li-ion battery hybrid for pulsed applications. J Power Sources 195(9):2731–2736
23. Chehreh Chelgani S, Rudolph M, Kratzsch R, Sandmann D, Gutzmer J (2016) A review of graphite beneficiation techniques. Miner Process Extr Metall Rev 37(1):58–68

24. Chen G, Wu S, Hui L, Zhao Y, Ye J, Tan Z et al (2016) Assembling carbon quantum dots to a layered carbon for high-density supercapacitor electrodes. Sci Rep 6(1):19028
25. Chen L, Lv C, Chen J, Bi S (2013) Numerical simulation study on cyclic reciprocal derivative chronopotentiometry of reversible electrode reaction coupled with Langmuir adsorption. Electrochim Acta 93:222–229. Pergamon
26. Chen W, Shen J, Lv G, Li D, Hu Y, Zhou C et al (2019) Green synthesis of graphene quantum dots from cotton cellulose. ChemistrySelect 4(10):2898–2902
27. Childress L, Hanson R (2013) Diamond NV centers for quantum computing and quantum networks. MRS Bull 38(2):134–138
28. Chowdhury NR, Kumar R, Kant R (2017) Theory for the chronopotentiometry on rough and finite fractal electrode: generalized sand equation. J Electroanal Chem 802:64–77. Elsevier
29. Christen T (2020) Ragone plots and discharge efficiency-power relations of electric and thermal energy storage devices. J Energy Storage 27:101084. Elsevier
30. Chung DDL (2002) Review graphite
31. Dai H (2002) Carbon nanotubes: synthesis, integration, and properties. Acc Chem Res 35(12):1035–1044
32. Dananjaya V, Marimuthu S, Yang R (Chunhui), Grace AN, Abeykoon C (2024) Synthesis, properties, applications, 3D printing and machine learning of graphene quantum dots in polymer nanocomposites. Prog Mater Sci 144:101282. Pergamon
33. Dorfs D, Krahne R, Falqui A, Manna L, Giannini C, Zanchet D (2011) Quantum dots: synthesis and characterization. Compr Nanosci Technol 219–270. Elsevier
34. Efros AL, Brus LE (2021) Nanocrystal quantum dots: from discovery to modern development. ACS Nano [Internet]. American Chemical Society; 2021 Apr 27 [cited 2024 Nov 14]; 15(4):6192–210. https://doi.org/10.1021/acsnano.1c01399
35. Ekimov AI, Efros AL, Onushchenko AA (1985) Quantum size effect in semiconductor microcrystals. Solid State Commun [Internet]. 1985 [cited 2024 Sep 17]; 56(11):921–4. https://www.researchgate.net/publication/222931758_Quantum_Size_Effect_in_Semiconductor_Microcrystals
36. El-Kady MF, Shao Y, Kaner RB (2016) Graphene for batteries, supercapacitors and beyond. Nat Rev Mater 1(7):16033
37. Fan L, Zhu M, Lee X, Zhang R, Wang K, Wei J et al (2013) Direct synthesis of graphene quantum dots by chemical vapor deposition. Part Part Syst Charact 30(9):764–769
38. Fatima S, Waqas Hakim M, Zheng X, Sun Y, Li Z, Han N et al (2025) Nitrogen doped graphene quantum dots (NGQD) pillared Ta4C3Tx MXene as high-performance electrochemical supercapacitors. J Power Sources 645:237190
39. Gao F, Nebel CE (2016) Diamond-based supercapacitors: realization and properties. ACS Appl Mater Interfaces 8(42):28244–28254
40. Geim AK, Novoselov KS (2007) The rise of graphene. Nat Mater 6(3):183–191
41. Gerberich WW, Ballarini R, Hintsala ED, Mishra M, Molinari JF, Szlufarska I (2015) Toward demystifying the mohs hardness scale. J Am Ceramic Soc 98(9):2681–2688. Blackwell Publishing Inc.
42. Ghaemi B, Zhao G, Jie G, Xi H, Li X, Wang J et al (2011) A study of formation and photoluminescence properties of ZnO quantum dot doped zinc-alumino-silicate glass ceramic. Opt Mater (Amst) 33(6):827–830. North-Holland
43. Grill A (1999) Electrical and optical properties of diamond-like carbon. Thin Solid Films 355–356:189–193
44. Gu S, Hsieh C-T, Yuan C-Y, Ashraf Gandomi Y, Chang J-K, Fu C-C et al (2020) Fluorescence of functionalized graphene quantum dots prepared from infrared-assisted pyrolysis of citric acid and urea. J Lumin 217:116774
45. Gupta A, Chaudhary V (2020) Optimizing green synthesis of quantum dots: challenges and future perspectives. ACS Sustain Chem Eng 8(5):1791–1803. ACS Publications
46. Gupta N, Gupta SM, Sharma SK (2019) Carbon nanotubes: synthesis, properties and engineering applications. Carbon Lett 29(5):419–447
47. Harris P (1999) On charcoal. Interdisc Sci Rev 24(4):301–306

48. Hbibi M, Mommadi O, Chouef S, Boussetta R, Belamkadem L, Moussaouy A El et al (2022) Finite confinement potentials, core and shell size effects on excitonic and electron-atom properties in cylindrical core/shell/shell quantum dots. Sci Rep 12:1 [Internet]. Nature Publishing Group; 2022 Sep 1 [cited 2024 Oct 22]; 12(1):1–19. https://www.nature.com/articles/s41598-022-19118-3
49. Hirsch A (2005) The chemistry of the fullerenes. Wiley-VCH
50. Ibrahim KS (2013) Carbon nanotubes-properties and applications: a review. Carbon Lett 14(3):131–144
51. IEC. IEC 60076-18:2012 | IEC [Internet]. 2012 [cited 2024 Oct 19]. https://webstore.iec.ch/en/publication/597
52. IEEE. IEEE SA—IEEE 1036-2020 [Internet]. 2020 [cited 2024 Oct 19]. https://standards.ieee.org/ieee/1036/5912
53. Iijima S (1991) Helical microtubules of graphitic carbon. Nature 354(6348):56–58
54. Jain A, Tripathi SK (2014) Fabrication and characterization of energy storing supercapacitor devices using coconut shell based activated charcoal electrode. Mater Sci Eng, B 183:54–60
55. Kanakaraju P, Rao MP (2016) Design and development of portable digital LCR meter by auto balancing bridge method-channel analog multiplexer/demultiplexer, operational amplifier and PCB with minimum components. Int J Innovat Eng Technol (IJIET) 130
56. Khan MN, Shoeb M, Mashkoor F, Kim KJ, Kang MS, Yu J et al (2025) Exploring the integration of CNT interwoven VY-MOF nanostructured for enhanced electrochemical performance supercapacitor device and their charge storage mechanism. J Energy Storage 127. Elsevier Ltd.
57. Khan N, Dilshad S, Khalid R, Kalair AR, Abas N (2019) Review of energy storage and transportation of energy. Energy Storage 1(3)
58. Kidalov SV, Shakhov FM (2009) Thermal conductivity of diamond composites. Materials 2(4):2467–2495
59. Kim D, Lee S, Kim J, Lee H (2023) Understanding the electroluminescence mechanism of CdSe/ZnS quantum-dot light-emitting diodes with a focus on charge carrier behavior in quantum-dot emissive layers. IEEE Electron Dev Lett 44(6):959–962. Institute of Electrical and Electronics Engineers Inc.
60. Klimov Victor I (2003) Semiconductor and metal nanocrystals. Semicond Metal Nanocrystals [Internet]. CRC Press; 2003 Nov 7 [cited 2024 Sep 16]; https://doi.org/10.1201/9780203913260/semiconductor-metal-nanocrystals-victor-klimov
61. Kroto HW, Heath JR, O'Brien SC, Curl RF, Smalley RE (1985) C60: Buckminsterfullerene. Nature 318(6042):162–163
62. Kumar Singhal V, Kumar Verma U, Joshi M, Kumar B (2022) Charge transport in quantum dot sensitized solar cells: a mathematical model. Solar Energy 245:120–126. Pergamon
63. Kwon O, Bae JH, Kim JY, Kim M, Ji Y, Kim J et al (2025) High aspect ratio hierarchical carbon nanoplate/functionalized carbon nanotube scaffolds for scalable binder-free ultrafast-charging supercapacitors. Mater Today Sustain 31. Elsevier Ltd.
64. Laubach SE, Marrett RA, Olson JE, Scott AR (1998) Characteristics and origins of coal cleat: a review. Int J Coal Geol 35(1–4):175–207
65. Lee J, Kim J, Hyeon T (2006) Recent progress in the synthesis of porous carbon materials. Adv Mater 18(16):2073–2094
66. Lei Y, Zhao Y, Zhang Q, Xiong Z, Chen L (2020) Highly efficient and bright red quantum dot light-emitting diodes with balanced charge injection. Org Electron 81:105683. North-Holland
67. Lemine AS, Zagho MM, Altahtamouni TM, Bensalah N (2018) Graphene a promising electrode material for supercapacitors-a review. Int J Energy Res 42(14):4284–4300
68. Leonard D, Krishnamurthy M, Petroff PM (1993) MBE growth of quantum dots for optoelectronic devices. Appl Phys Lett 63(23):3203–3205. AIP Publishing
69. Li C, Dong X, Zhai Z, Huang N, Jiang X, Yang N (2025) High-performance flexible supercapacitors using diamond cloth. J Phys Chem C 129(14):6618–6627
70. Li J, Zheng H, Zheng Z, Rong H, Zeng Z, Zeng H (2022) Synthesis of CdSe and CdSe/ZnS quantum dots with tunable crystal structure and photoluminescent properties. Nanomaterials

12:2969 [Internet]. Multidisciplinary Digital Publishing Institute; 2022 Aug 27 [cited 2024 Nov 15]; 12(17):2969. https://www.mdpi.com/2079-4991/12/17/2969/htm
71. Li L et al (2012) Electrochemical performance of graphene quantum dots as anode for lithium-ion batteries. J Mater Chem 22(14):22802–22806
72. Li LL, Ji J, Fei R, Wang CZ, Lu Q, Zhang JR et al (2012) A facile microwave avenue to electrochemiluminescent two-color graphene quantum dots. Adv Funct Mater 22(14):2971–2979
73. Li X, Rui M, Song J, Shen Z, Zeng H (2015) Carbon and graphene quantum dots for optoelectronic and energy devices: a review. Adv Funct Mater 25(31):4929–4947. Wiley-VCH Verlag
74. Lim SY, Shen W, Gao Z (2015) Carbon quantum dots and their applications. Chem Soc Rev 44(1):362–381
75. Liu H, Gan L, Chen R, Xiao Z, Chen Y, Yang H et al (2025a) 3D graphene paper-based tandem metal-free thin-film supercapacitors with integrated 200 V output. Int J Extreme Manuf 7(5). Institute of Physics
76. Liu H, Hao L, Zhou G, Teng D, Li P, Cao Y et al (2025b) Preparation and electrochemical performance of coal-based composite carbon fibers using water-soluble fraction from lignite by thermal dissolution. Fuel [Internet]. Elsevier; 2025b Aug 15 [cited 2025 May 28]; 394:135085. https://www.sciencedirect.com/science/article/abs/pii/S0016236125008105
77. Liu W, Feng Y, Yan X, Chen J, Xue Q (2013) Superior micro-supercapacitors based on graphene quantum dots. Adv Funct Mater 23(33):4111–4122
78. Luo P, Guan X, Yu Y, Li X, Yan F (2019) Hydrothermal synthesis of graphene quantum dots supported on three-dimensional graphene for supercapacitors. Nanomaterials 9(2):201
79. Mahajan OP, Walker PL Jr (1978) Porosity of coals and coal products. DC (United States), Washington
80. Mahfouz S, Ei-Nemr MK, Abu-Siada A (2022) Review: frequency response analysis for transformer's status diagnoses. In: 2022 23rd international middle east power systems conference, MEPCON 2022. Institute of Electrical and Electronics Engineers Inc.
81. Martin N et al (2012) Applications of fullerenes in drug delivery. Nano Res 5(7):498–510
82. Mathews JP, Krishnamoorthy V, Louw E, Tchapda AHN, Castro-Marcano F, Karri V et al (2014) A review of the correlations of coal properties with elemental composition. Fuel Process Technol 121:104–113
83. May BM, Zikhalala N, Fakayode OJ, Bambo MF, Mkhohlakali A, Maiga DT (2024) Quantum dot-based nanocomposite as metal sensors. Eng Mater [Internet]. Springer, Cham; 2024 [cited 2024 Sep 27]; Part F2942:257–86. https://doi.org/10.1007/978-3-031-54779-9_13
84. Middlemiss LA, Rennie AJR, Sayers R, West AR (2020) Characterisation of batteries by electrochemical impedance spectroscopy. Energy Reports 6:232–241. Elsevier
85. Mohamed M, Xu Y, Zhao X (2012) Synthesis and application of quantum dots in optoelectronics and photovoltaic devices. Adv Funct Mater 22(14):2921–2927. Wiley Online Library
86. Mohr SH, Evans GM (2009) Forecasting coal production until 2100. Fuel 88(11):2059–2067
87. Murray CB, Norris DJ, Bawendi MG (1993) Synthesis and optical properties of quantum dots. J Am Chem Soc 115(19):8706–8715. ACS Publications
88. Najafi L, Taheri B, Martín-García B, Bellani S, Di Girolamo D, Agresti A et al (2018) MoS2 quantum dot/graphene hybrids for advanced interface engineering of a CH3NH3PbI3 Perovskite solar cell with an efficiency of over 20%. ACS Nano [Internet]. American Chemical Society; 2018 Nov 27 [cited 2024 Sep 27]; 12(11):10736–54. https://doi.org/10.1021/acsnano.8b05514
89. Nasir S, Hussein M, Zainal Z, Yusof N (2018) Carbon-based nanomaterials/allotropes: a glimpse of their synthesis, properties and some applications. Materials 11(2):295
90. Negri V, Pacheco-Torres J, Calle D, López-Larrubia P (2020) Carbon nanotubes in biomedicine, pp 177–217
91. Nesakumar N, Srinivasan S, Alwarappan S (2022) Graphene quantum dots: synthesis, properties, and applications to the development of optical and electrochemical sensors for chemical sensing. Springer, Microchimica Acta

92. Nguyen QH, Le TH, Tran QN (2024) Quantum dots as fabricating materials for supercapacitors. J Alloys Compd 985:173947. Elsevier
93. Nirmal M, Brus L (1999) Luminescence photophysics in semiconductor nanocrystals. Acc Chem Res [Internet]. Am Chem Soc 1999 [cited 2024 Oct 20]; 32(5):407–14. https://doi.org/10.1021/ar9700320
94. Novoselov KS et al (2004) Electric field effect in atomically thin carbon films. Science (1979) 306(5696):666–669
95. Palagonia MS, Erinmwingbovo C, Brogioli D, La Mantia F (2019) Comparison between cyclic voltammetry and differential charge plots from galvanostatic cycling. J Electroanal Chem 847:113170. Elsevier
96. Pan D et al (2010) Hydrothermal route for cutting graphene sheets into blue-luminescent graphene quantum dots. Adv Mater 22(6):734–738
97. Permatasari FA, Irham MA, Bisri SZ, Iskandar F (2021) Carbon-based quantum dots for supercapacitors: recent advances and future challenges. Nanomaterials 11:91 [Internet]. Multidisciplinary Digital Publishing Institute; 2021 Jan 3 [cited 2024 Oct 17]; 11(1):91. https://www.mdpi.com/2079-4991/11/1/91/htm
98. Pietryga JM, Park YS, Lim J, Fidler AF, Bae WK, Brovelli S et al (2016) Spectroscopic and device aspects of nanocrystal quantum dots. Chem Rev [Internet]. American Chemical Society; 2016 Sep 28 [cited 2024 Sep 17]; 116(18):10513–622. https://www.researchgate.net/publication/309067152_Spectroscopic_and_Device_Aspects_of_Nanocrystal_Quantum_Dots
99. Poonam, Sharma K, Arora A, Tripathi SK (2019) Review of supercapacitors: materials and devices. J Energy Storage 21:801–825
100. Popov V (2004) Carbon nanotubes: properties and application. Mater Sci Eng R Rep 43(3):61–102
101. Prasek J, Drbohlavova J, Chomoucka J, Hubalek J, Jasek O, Adam V et al (2011) Methods for carbon nanotubes synthesis—review. J Mater Chem 21(40):15872
102. Punnakkal N, Naneena S, C P SL, Pradeep A, T G SB, Suneesh PV (2024) Nitrogen-doped graphene quantum dot embedded polyaniline for the fabrication of high-performance flexible supercapacitor with enhanced cycling stability. J Energy Storage 100:113527. Elsevier
103. Rajagopal S, Pulapparambil Vallikkattil R, Mohamed Ibrahim M, Velev DG (2022) Electrode materials for supercapacitors in hybrid electric vehicles: challenges and current progress. Condens Matter. MDPI
104. Ralston P, Duong TH, Yang N, Berning DW, Hood C, Hefner AR et al (2009) High-voltage capacitance measurement system for SiC power MOSFETs. In: 2009 IEEE energy conversion congress and exposition, ECCE 2009, pp 1472–1479
105. Reiss P, Protiere M, Li L (2011) Environmental concerns in the synthesis of quantum dots: alternative materials and greener methods. Chem Mater 23(3):398–408. ACS Publications
106. Riaz M, Ali SM, Alotaibi R, Ali SD, Ullah J (2025) Synergistic electrochemical performance of PANI and fullerene (C60) based novel composites with halide perovskites (CsInCl<inf>3</inf>) for advanced supercapacitors electrodes. J Inorg Organomet Polym Mater 35(3):1891–1899
107. Rode A, Sharma S, Mishra DK (2018) Carbon Nanotubes: classification, method of preparation and pharmaceutical application. Curr Drug Deliv 15(5):620–629
108. Rodrigues CF, Lemos de Sousa MJ (2002) The measurement of coal porosity with different gases. Int J Coal Geol 48(3–4):245–251
109. Rooj B, Mandal U (2023) A review on characterization of carbon quantum dots. Vietnam J Chem [Internet]. Wiley; 2023 Dec 1 [cited 2024 Oct 22];61(6):693–718. https://doi.org/10.1002/vjch.202300022
110. Ruoff RS, Lorents DC (1995) Mechanical and thermal properties of carbon nanotubes. Carbon N Y 33(7):925–930
111. Sanmartín-Matalobos J, Bermejo-Barrera P, Aboal-Somoza M, Fondo M, García-Deibe AM, Correra-Vázquez J et al (2022) Semiconductor quantum dots as target analytes: properties, surface chemistry and detection. Nanomaterials 12:2501 [Internet]. Multidisciplinary Digital

Publishing Institute; 2022 Jul 21 [cited 2024 Oct 21]; 12(14):2501. https://www.mdpi.com/2079-4991/12/14/2501/htm

112. Schaller RD, Klimov VI (2004) High efficiency carrier multiplication in PbSe nanocrystals: implications for solar energy conversion. Phys Rev Lett [Internet]. Phys Rev Lett; 2004 May 7 [cited 2024 Oct 20]; 92(18). https://pubmed.ncbi.nlm.nih.gov/15169518/
113. Scopus (2025) Graphene quantum. Dots Publications
114. Shabbir H, Wojnicki M (2023) Recent progress of non-cadmium and organic quantum dots for optoelectronic applications with a focus on photodetector devices. Electronics 12:1327 [Internet]. Multidisciplinary Digital Publishing Institute; 2023 Mar 10 [cited 2024 Sep 27]; 12(6):1327. https://www.mdpi.com/2079-9292/12/6/1327/htm
115. Sharma N, Singh S, Singh R (2017) Green synthesis of quantum dots: Biocompatibility and environmental impact. J Mater Chem B R Soc Chem 5(28):5604–5615
116. Shen J et al (2012) Synthesis and modification of luminescent carbon quantum dots and their application in bioimaging and biosensing. Nanotechnology 23(10):105601
117. Shin Y, Park J, Hyun D, Yang J, Lee H (2015) Generation of graphene quantum dots by the oxidative cleavage of graphene oxide using the oxone oxidant. New J Chem 39(4):2425–2428
118. Shoukat R, Khan MI (2021) Carbon nanotubes: a review on properties, synthesis methods and applications in micro and nanotechnology. Microsyst Technol 27(12):4183–4192
119. Sokolov PM, Zvaigzne MA, Krivenkov VA, Litvin AP, Baranov AV., Fedorov AV et al (2019) Graphene−quantum dot hybrid nanostructures with controlled optical and photoelectric properties for solar cell applications. Russian Chem Rev Auton Non-Profit Organ Editorial Board J Uspekhi Khimii 88(4):370–386
120. Song A, Li Y, Zhu S, Zhang G, Tang M (2025) VxOy quantum dot-enhanced nitrogen-sulfur dual-doped hierarchical porous carbon electrodes from waste eggshell membranes for advanced flexible supercapacitors. J Colloid Interface Sci 688:526–539
121. Song HZ, Lan S, Akahane K, Jang KY, Okada Y, Kawabe M (2000) Effect of charge distribution in quantum dots array on two-dimensional electron gas. Jpn J Appl Phys Part 1: Regular Papers Short Notes Rev Papers [Internet]. JJAP; 2000 Oct 1 [cited 2024 Nov 7]; 39(10):5746–50. https://doi.org/10.1143/JJAP.39.5746
122. Stiff-Roberts AD (2009) Advances in molecular beam epitaxy for quantum dot synthesis. J Vacuum Sci Technol B 27(3):410–418. AIP Publishing
123. Sun L, Wang X, Wang Y, Zhang Q (2017) Roles of carbon nanotubes in novel energy storage devices. Carbon N Y 122:462–474
124. Sun YP et al (2006) Quantum-sized carbon dots for bright and colorful photoluminescence. J Am Chem Soc 128(24):7756–7757
125. Sung J (2021) Handbook of industrial diamonds. Jenny Stanford Publishing
126. Tang X, Yang F (2022) Quantum dots: synthesis, characterization, and applications. In: Handbook of energy materials. Springer Nature, Singapore, pp 1–30
127. Tatti R, Aversa L, Verucchi R, Cavaliere E, Garberoglio G, Pugno NM et al (2016) Synthesis of single layer graphene on Cu(111) by C 60 supersonic molecular beam epitaxy. RSC Adv 6(44):37982–37993
128. Tegenaw AB, Yimer AA, Beyene TT (2023) Boosting the photocatalytic activity of ZnO–NPs through the incorporation of C-dot and preparation of nanocomposite materials. Heliyon [Internet]. Elsevier; 2023 Oct 1 [cited 2024 Sep 17]; 9(10):2405–8440. https://www.pmc/articles/PMC10570576/
129. Titirici M-M, White RJ, Brun N, Budarin VL, Su DS, del Monte F et al (2015) Sustainable carbon materials. Chem Soc Rev 44(1):250–290
130. Tiwari SK, Kumar V, Huczko A, Oraon R, De AA, Nayak GC (2016) Magical allotropes of carbon: prospects and applications. Crit Rev Solid State Mater Sci 41(4):257–317
131. Turcheniuk K, Mochalin VN (2017) Biomedical applications of nanodiamond (Review). Nanotechnology 28(25):252001
132. Valizadeh A, Mikaeili H, Samiei M, Farkhani SM, Zarghami N, kouhi M et al (2012) Quantum dots: synthesis, bioapplications, and toxicity. Nanoscale Res Lett 7(1):480

133. Volperts A, Dobele G, Zhurinsh A, Zalane Z, Ozolinsh J, Kleperis J et al (2016) Supercapacitor electrodes from activated wood charcoal. Bulgarian Chem Commun
134. Wang Y, Hu A (2014) Carbon quantum dots: synthesis, properties and applications. J Mater Chem C Mater. 2(34):6921
135. Wissler M (2006) Graphite and carbon powders for electrochemical applications. J Power Sources 156(2):142–150
136. Wu J, Lu Y, Feng S, Wu Z, Lin S, Hao Z et al (2018) The interaction between quantum dots and graphene: the applications in graphene-based solar cells and photodetectors. Adv Funct Mater [Internet]. Wiley; 2018 Dec 1 [cited 2024 Sep 27]; 28(50):1804712. https://doi.org/10.1002/adfm.201804712
137. Wu M, Snook GA, Chen GZ, Fray DJ (2004) Redox deposition of manganese oxide on graphite for supercapacitors. Electrochem commun 6(5):499–504
138. Xue X, Hao Q, Chen M (2024) Very long wave infrared quantum dot photodetector up to 18 μm. Light: Sci Appl 13:1 [Internet]. Nature Publishing Group; 2024 Apr 12 [cited 2024 Oct 22]; 13(1):1–11. https://www.nature.com/articles/s41377-024-01436-y
139. Yang W, Zhao C, Li J (2014) Hydrothermal synthesis of oxide-based quantum dots: applications in photocatalysis and sensors. Mater Today 17(2):58–66. Elsevier
140. Yang Z, Lin S, Liu J, Zheng K, Lu G, Ye B et al (2020) High performance phototransistors with organic/quantum dot composite materials channels. Org Electron 78:105565. North-Holland
141. Yao L, Koripally N, Shin C, Mu A, Chen Z, Wang K et al (2025) Engineering electro-crystallization orientation and surface activation in wide-temperature zinc ion supercapacitors. Nat Commun Nat Res 16(1)
142. Yu S, Yang N, Liu S, Jiang X (2021) Diamond supercapacitors: progress and perspectives. Curr Opin Solid State Mater Sci 25(3):100922
143. Zare S, Edalatpour S (2021) The quantum confinement effect on the spectrum of near-field thermal radiation by quantum dots. J Appl Phys [Internet]. American Institute of Physics Inc.; 2021 Jul 7 [cited 2024 Oct 21] 130(1):15108. https://www.aip/jap/article/130/1/015108/158541/The-quantum-confinement-effect-on-the-spectrum-of
144. Zhang D, Tan C, Zhang W, Pan W, Wang Q, Li L (2022) Expanded graphite-based materials for supercapacitors: a review. Molecules 27(3):716
145. Zhang S, Li Y, Song H, Chen X, Zhou J, Hong S et al (2016) Graphene quantum dots as the electrolyte for solid state supercapacitors. Sci Rep 6:1 [Internet]. Nature Publishing Group; 2016 Jan 14 [cited 2024 Oct 17]; 6(1):1–7. https://www.nature.com/articles/srep19292
146. Zhang S, Sui L, Dong H, He W, Dong L, Yu L (2018) High-performance supercapacitor of graphene quantum dots with uniform sizes. ACS Appl Mater Interfaces [Internet]. American Chemical Society; 2018 Apr 18 [cited 2024 Oct 17] 10(15):12983–12991. https://doi.org/10.1021/acsami.8b00323
147. Zhong M, Zhang M, Li X (2022) Carbon nanomaterials and their composites for supercapacitors. Carbon Energy 4(5):950–985
148. Zhou H, Gao LT, Li Y, Lyu Y, Guo ZS (2024) Electrochemical performance of lithium-ion batteries with two-layer gradient electrode architectures. Electrochim Acta Pergamon 476:143656
149. Zhu S et al (2014) The photoluminescence mechanism in carbon dots (graphene quantum dots, carbon nanodots, and polymer dots): current state and future perspective. Nano Res 7(3):355–381
150. Zhu S, Zhang J, Tang S, Qiao C, Wang L, Wang H et al (2012) Surface chemistry routes to modulate the photoluminescence of graphene quantum dots: from fluorescence mechanism to up-conversion bioimaging applications. Adv Funct Mater 22(22):4732–4740

Chapter 18
Quantum Dots: Transforming Cancer Diagnosis and Treatment Through Advanced Theranostics

Simran Gupta, Avantika Patil, and Govinda Kapusetti

Abstract Quantum dots (QDs) are one of those new tools that are changing our understanding of cancer diagnosis and treatment. These tools share the characteristic of precision with unparalleled versatility. Quantum dots (QDs) are nanoscale, light-emitting particles known for their exceptional brightness, stability, and tunable properties, which enable them to significantly outperform traditional imaging agents. Real-time monitoring of biological activity and drug delivery directly to cancerous sites can reduce side effects dramatically. This chapter focuses on the synthesis and surface modification of quantum dots (QDs) to enhance their suitability for medical applications and to optimize their structural engineering. It demonstrates the involvement of QDs in modern imaging techniques such as Magnetic Resonance Imaging (MRI), Positron Emission Tomography (PET), and Computed Tomography (CT), and its improvement in photothermal and photodynamic therapy beyond imaging. It also revolutionizes drug delivery to give targeted controlled release, therefore, increasing the treatment success rate. Yet, much has yet to be done, particularly concerning toxicity and safe application in a clinical environment. The way for the safe and efficient application of QDs is paved by innovative coatings and AI-based safety analyses. In the near future, quantum dot-based theranostic platforms are expected to transform cancer management by simultaneously enabling early diagnosis, targeted drug delivery, photothermal therapy, and real-time therapeutic monitoring within a single integrated system. These multifunctional nanomaterials may pave the way for highly effective and personalized oncological interventions tailored to the specific needs of individual patients.

Keywords Quantum dots (QDs) · Cancer theranostics · Targeted drug delivery · Photothermal therapy (PTT) · Photodynamic therapy (PDT) · Nanomedicine · Biomedical imaging · Biocompatibility and toxicity

---

S. Gupta · A. Patil · G. Kapusetti (✉)
National Institute of Pharmaceutical Education and Research - Kolkata, Kankurgachi, Kolkata, West Bengal, India
e-mail: govindphysics@gmail.com

T. Bhuyan et al. (eds.), *Multifunctional Quantum Dots*, Materials Horizons: From Nature to Nanomaterials, https://doi.org/10.1007/978-981-92-0172-3_18

1 Introduction

Quantum dots (QDs) are a semiconductor nanocrystal family with optoelectronic properties dominated by quantum confinement effects. The tunability of their fluorescence emission spectra can thus be determined by adjusting their size alone, with an unprecedented degree of tunability that cannot be obtained using traditional imaging probes [1]. With their high quantum yield, photostability, and symmetrically narrow emission profiles as well, QDs represent a revolutionary platform for drug delivery and biomedical imaging [63].

Ever since their discovery in the 1980s, QDs have been an area of interest because they are able to surpass the long-standing limitations of conventional fluorophores like organic dyes and fluorescent proteins, for instance, quick photobleaching, low brightness, and limited penetration in tissue. Their wide absorption spectra and tight, size-tunable emission peaks allow for simultaneous excitation and multiplexed detection of numerous biological targets—a convenient advantage in sophisticated diagnostic and research environments [41].

In biomedicine and biomedical imaging, QDs surpass conventional agents by enabling real-time, high-resolution visualization of dynamic cellular and molecular processes. The straightforward conjugation of QDs with a wide range of biological ligands, such as antibodies, peptides, and nucleic acids, also renders them specific and functionally diverse for site-specific drug delivery and targeted imaging [63]. These surface functionalization strategies not only make the selective targeting of pathologic tissue possible but also minimize the off-target effects, paving the way for safer and more effective therapeutic strategies [60].

In tumor imaging and cancer therapy, QDs have markedly transformed current paradigms. Their intrinsic luminescence and deep tissue penetration have rendered them indispensable for tumor border delineation, metastatic tumor growth monitoring, and intraoperative real-time navigation with unparalleled precision [56, 91]. In addition, they recently achieved utility in gene therapy—e.g., delivery of nucleic acids targeting, and theranostic platforms integrating diagnosis and drug release control demonstrate their wide application in personalized medicine [10, 29].

Together, the coherent photophysics of QDs and surface engineering and bioconjugation technology position them as perhaps the most exciting nanotechnological technologies for targeted therapeutics and diagnostics of the future. Further development of their promise is certain to redefine the order of magnitude of sensitivity, specificity, and multifunctionality of contemporary biomedicine [14, 38].

2 Synthesis and Surface Engineering of Quantum Dots for Biocompatibility

The degree of biocompatibility for a range of biomedical applications can be achieved by the preparation and surface engineering of quantum dots. Some of the common chemical methods for the synthesis of QDs are colloidal, solvothermal, and hydrothermal synthesis (Fig. 1), as reported by Wagner et al. Colloidal synthesis is suited for uniform nanocrystals with great precision, while solvothermal and hydrothermal methods exploit hot and high-pressure conditions to result in QDs with unique structural properties. Integrated QDs with biological systems will then require surface engineering through functionalization and chemical modification [5]. Functionalization is the strategy of tailoring the surface chemistry to enhance the QDs' ability to dissolve better in water and their stability within the biological fluid and interaction ability with other biomolecules. The replacement of hydrophobic surface ligands with hydrophilic or bio-functional ligands allows QDs to be operated in an aqueous environment [8]. Another is surface capping with PEG, silica, or lipid layers to minimize toxicity and improve stability for targeted functionality. The core–shell structure with nontoxic materials combined with bio-functional coatings provides complete safety in the implementation of QDs-based applications for biomedical imaging, diagnostics, and targeted drug delivery, therefore, QDs find a critical tool in the scenario of modern medicines [71].

Major steps for functionality and biomedical applications with biocompatibility are the synthesis of QD and surface engineering. Many colloidal, solvothermal, and hydrothermal methods have been adapted for the synthesis of high-quality QDs applicable to many applications [1]. The most common method among those colloidal syntheses involves a reaction of precursors in solution that produce QDs with tightly controlled size and uniformity. Solvothermal and hydrothermal are synthesis methods. The synthesis of QDs is performed in a solvent medium or water at high temperature and pressure [71]. Furthermore, the QDs could also show extra advantages in terms of biocompatibility through surface engineering techniques such as polymer coatings, biocompatible in nature, silicon dioxide, and other functional groups. Such modifications increase the stability of QDs in biological media, decrease toxicity, and allow conjugation to biomolecules like antibodies or peptides for target applications [37]. Altogether, these approaches ensure that QDs fulfill all the stringent demands posed by biomedical applications, imaging, to drug delivery, preserved with excellent optical and electronic properties.

Synthesis and surface engineering of quantum dots constitute a pivotal area of focus for enhancing their biocompatibility and functional performance in biomedical applications. The colloidal, solvothermal, and hydrothermal techniques provide significant control over the nanoscale size, shape, and composition of QDs with tailored optical and electronic properties for a given application [71]. Nevertheless, interactions at the level of intrinsic surface chemistry continue to play an important role in determining their interaction with biological systems. Advancement in the strategy through surface engineering includes functionalization-coating of QDs

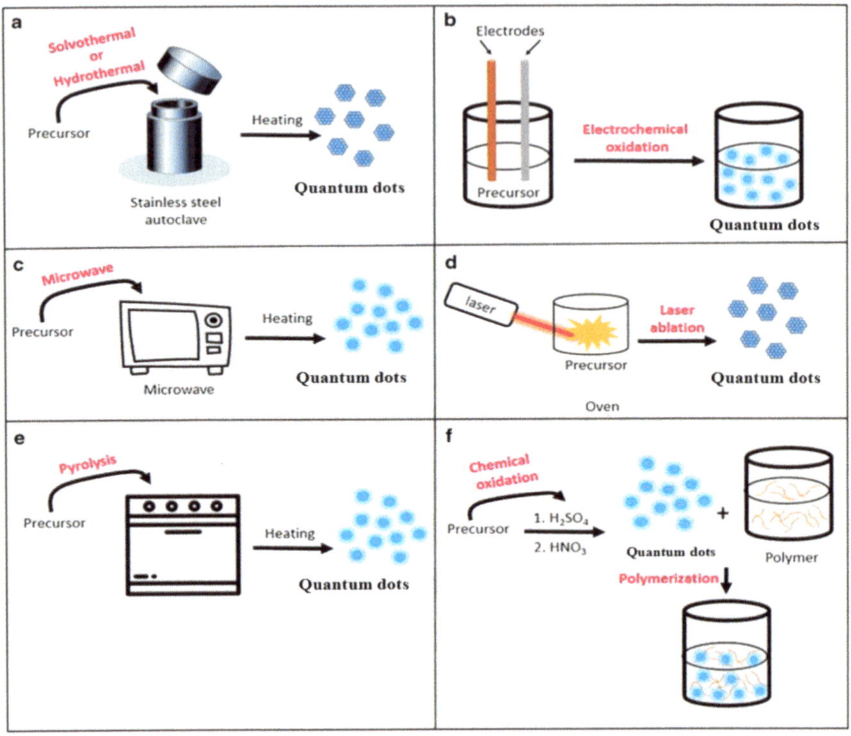

Fig. 1 Synthesis steps for quantum dots. **a** Solvothermal/Hydrothermal: precursors are heated under high pressure inside a sealed chamber-also called an autoclave. **b** Electrochemical oxidation: precursors are oxidized using electrodes. **c** Microwave-Assisted: rapid heating of precursors by microwave radiation. **d** Laser Ablation: QDs are produced through the laser irradiation of precursors. **e** Pyrolysis: precursors are decomposed by heat in an oven. **f** Chemical oxidation and polymerization: chemical oxidation of precursors, where QDs are embedded in polymers. These approaches allow for controlling the size and properties of QD for most applications

with biocompatible materials, including silica, PEG, or lipid-based layers [94]. These have enhanced stability, reduced toxicity, and active sites for conjugation of targeting ligands, antibodies, or nucleic acids that enhance specificity and efficacy for both diagnostics and therapeutics. Their value lies in the modification of the surface of QDs to generate targeted applications, such as targeted drug delivery or multimodal imaging, for personalized medicine with advanced capabilities for clinical safety and efficacy [60].

Synthesis and surface engineering of quantum dots are elementary steps for improving their biocompatibility and further developing their biomedical applications. The most important techniques for modulating the QD's surface properties, hence maintaining stability and functionality within a biological environment, are ligand exchange and surface capping techniques [60]. Ligand exchange involves replacing the native surface ligands with other new ones, which might offer water solubility, less toxicity, or even targeted binding properties. Such an exchange process can easily help the QDs transfer from hydrophobic solvents into aqueous solutions and then find potential use in biology. Surface capping strategies include the addition of a protective layer of silica, PEG, or zwitterionic coatings, which improve biocompatibility and prevent aggregation. Such capping materials can also present reactive groups for conjugating biomolecules to the QDs, enhancing targeting specificity and therapeutic potential. Ligand exchange and surface capping play an important role together in adjusting QDs so that they might be used appropriately for biomedical applications, such as imaging and drug delivery [51].

Quantum dots (QDs) have been an important candidate for safe and effective biomedical applications, which inherently require enhancement in biocompatibility. At times, the inherent properties of QDs, like their size and surface chemistry, pose problems related to toxicity and also provoke an immune response; in this context, surface engineering techniques are helpful by enhancing compatibility with the biological system [10]. QDs coated with biocompatible materials like PEG, silica, or lipids show higher biocompatibility. The outcomes are decreased toxicity but increased stability under physiological conditions. Functionalizing QDs with biomolecules, such as antibodies, peptides, or nucleic acids, adds specificity and limits the interaction of QDs with healthy tissues [55]. Primarily, the core–shell structures and the non-toxic material used during synthesis reduce adverse effects. Advanced biological compatibility has placed QDs among many applications used for imaging purposes, targeted drug delivery, and theranostics where a new set of personalized as well as more secure medical practices may be envisioned [87].

3 Quantum Dots in Advanced Cancer Imaging and Diagnostics

Quantum dots (QDs) are a revolutionary tool for advanced imaging and diagnostics of cancer due to their unique properties of optics and electronics. Quantum dots are brilliant fluorescent probes, with excellent photostability as probes for molecular imaging, offering precise cellular and molecular target visualization [43]. QDs allow multiplex imaging, that is, imaging of multiple biomarkers simultaneously. Other than their fluorescence, QDs are coupled into hybrid imaging modalities—including MRI, PET, CT, and optical imaging—and enhance the sensitivity and specificity. In in vivo applications, QDs are important in the imaging of tumors: preclinical and clinical diagnostics already utilize them, but have a higher prospect in later clinical diagnostics [91]. More importantly, QDs enable real-time imaging of tumor margins, providing critical information during surgery by defining the boundary of the tumor with high accuracy. These developments indicate that QDs hold the promise to transform cancer diagnostics and eventually patients' lives [27]. Figure 2 illustrates the major biomedical applications of QDs, including imaging, therapy, and molecular sensing.

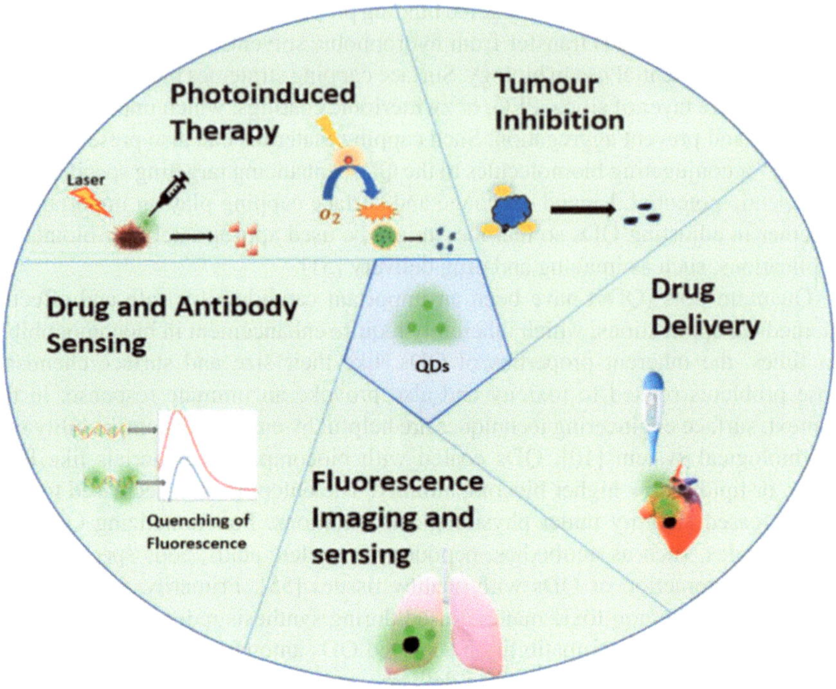

Fig. 2 Multifunctional applications of quantum dots (QDs) in biomedicine and therapy depicting the multifunctional biomedical applications of QDs, including photo-induced therapy, tumor inhibition, fluorescence imaging, and drug/antibody sensing. The circular layout highlights their combined diagnostic and therapeutic potential

QDs transform molecular imaging into highly efficient fluorescent probes with unique optical properties. These were engineered to have nanoscale sizes to produce high photostability and to be engineered so that a wide range of emission spectra could be accessed, thus qualifying them for the precise visualization of molecular targets. Unlike traditional fluorophores, QDs are characterized by very little photobleaching and blinking, thus sustaining signal intensity over long imaging periods. They are extremely useful in the real-time monitoring of dynamic cellular processes. Many QDs can be excited with the same light source because of their broad absorption spectra and narrow emission peaks. Through this method, multiplexed imaging of different molecular targets is achievable. Antibody conjugation to QD-based probes increases the selective targeting of tumor-specific markers, hence improving the sensitivity and specificity of the diagnosis for cancer by the increased detection capability [12]. They provide exceptional resolution capabilities for deep tissue penetration and thus present critical novel tools in earlier diagnoses of cancer or the monitoring of the responses toward therapies. Fluorescent QD probes then form a position of strategic strength, moving advanced methods of molecular imaging forward toward improved diagnosis protocols as well as better drug therapies against this very deadly cancerous disease.

This has essentially replaced multiplex microscopy in the diagnosis of cancer by introducing QDs that could enable the simultaneous detection of multiple biomarkers. It is possible to excite more than one QD using a single light source as several QDs emit at different colors due to their broad absorption spectra and narrow, tunable emission peaks. This capability makes QDs indispensable in the study of tumor heterogeneity and the complex interactions of various biomarkers in the context of cancer progression and metastasis. After conjugating specific ligands, antibodies, or peptides, QDs enable them to specifically bind to any given molecular targets, such as surface proteins, nucleic acids, or enzymes, for an overall representation of the microenvironment of a tumor [48]. This also improves diagnosis accuracy and aids in personalized medicine through the detection of biomarkers that predict a therapeutic response. In addition, photostability coupled with a high signal-to-noise ratio offers reliable detection in challenging biological samples. This will open the pathway to more profound insights into the biology of cancer and more effective early diagnosis and treatment strategies [90].

QDs have surpassed the traditional fluorescence imaging technique by incorporating advanced imaging modalities like MRI, PET, CT, and optical imaging into the system, which greatly extends their range of applications in cancer diagnostics. Their wide flexibility in surface chemistry also allows them to be functionalized with contrast agents, radionuclides, or targeting ligands and hence makes multimodal imaging possible where all the benefits acquired from various imaging technologies can be utilized. For instance, QDs can be conjugated with MRI contrast agents such as gadolinium or iron oxide, and they retain their fluorescent properties but gain high-resolution anatomical details [64]. The conjugation of PET isotopes further increases the sensitivity for the early detection of molecular changes at cancer stages. Functionalized QDs with high-atomic-number materials, such as gold, enhance contrast in CT imaging for precise localization of tumors [82]. Simultaneously, their intrinsic fluorescence properties still shine very brightly in the use of optical imaging toward real-time visualization of cellular and molecular processes. This multi-dimensional approach, thereby not only increases the accuracy in diagnosis but also characterizes in detail tumors between anatomical and functional imaging, this bridges gaps that bring higher efficiency into the treatment of cancers [24].

QDs are among the most effective agents currently in use for in vivo tumor imaging, with very high applications in both preclinical research and clinical diagnostics. Their superior brightness, photostability, and emission spectra tunability make QDs an optimal agent for direct visualization of tumors and deep tissues [56]. In preclinical studies, they were widely applied in the mapping of growth, metastasis, and angiogenesis of tumors, this gave key information on biology and therapeutic efficiency in cancer treatments. They have high specificity in binding with cancer biomarkers once conjugated with various tumor-targeting ligands including antibodies, peptides, or aptamers; thus, ensuring correct imaging [30]. QDs are envisioned to capture real-time live images of the tumor and the margin during surgery with the goal that the surgeon be able to fully excise tumors. Besides this, their integration into multimodal imaging devices allows the simultaneous acquisition of structural, functional, and molecular data, hence, findings are about to pave the way for personalized treatments. Though issues related to biocompatibility and toxicity are under active investigation, the progress in surface modification and encapsulation strategies on QDs is bringing QDs closer to routine clinical use in cancer treatment [20].

QDs revolutionize the tools for in vivo real-time imaging and visualization of tumor margins during cancer surgery to give precision inaccuracy that could not be given by any tool. These also have excellent fluorescence properties, such as brightness and photostability, and are thus able to define the borders of the tumors even in adverse situations at the surgical sites [79]. The QDs conjugated with tumor-specific targeting agents via antibodies or peptides selectively bind to cancer cells and by this, the surgeon can differentiate precisely between malignant and benign tissues. Real-time visualization minimizes the chances of incomplete removal of the tumor, which is critical in reducing recurrence rates and improving patient outcomes [21]. The emission spectra of QDs can be tuned to allow for multiplexed imaging in the simultaneous detection of several markers that may further delineate the margin, but as an improvement on QD biocompatibility and safety continues to advance, these particles will increasingly be integrated into surgical workflows, helping in increasing precision in oncologic surgery and revolutionize intraoperative cancer management [26].

4 Targeted Delivery Mechanisms Using Quantum Dots

Quantum dots are revolutionizing the targeted drug delivery system based on their unique properties and innovative engineering. These structures can be formed with extreme precision and can be established to have high surface area-to-volume ratios and custom functionalization with the loading of drugs could be done to target the molecules [84]. The conjugation of targeting ligands such as antibodies, peptides, or aptamers on QDs yields impressive specificity, which targets selectively the cancer biomarkers and ensures direct delivery to the tumor site of therapeutic agents. Further accuracy is brought by the design of stimuli-responsive QDs that will consider the release of a drug in response to specific triggers including pH-altering, temperature-altering, or enzymatic alterations, that will ensure effective, controlled, and localized drug delivery while reducing the side effects [19]. The major biological barriers include the vascular endothelium and tumor microenvironment, and the QDs are optimized through surface coatings and size adjustments for biocompatibility and tissue penetration enhancement. Case studies have proven that QD-based delivery systems should have the efficacy to target cancers, which include breast, lung, and pancreatic tumors, with a better therapeutic effect without higher toxicity. In turn, such developments will project great potential for QDs in changing the treatment of cancer with improved efficient targeted drug-delivery systems [68].

The design and engineering of QDs for targeted drug delivery is based on the optimization of their physical, chemical, and functional properties to ensure precision in therapeutic applications. QDs are engineered at nanoscale precision, which provides them with a high surface-area-to-volume ratio, thus carrying substantial drug payloads. These exhibit biocompatibility by coating PEG, which increases stability with reduced immunogenicity in living systems. Then, QDs are designed in such a manner that they help in the selective targeting of cells: they incorporate functional groups to conjugate with specific ligands like antibodies, peptides, or aptamers. Thereby, with the help of their tunable size and charge, they evade off-target effects and maximize blood circulation, penetration of tissues and cells. These design features ensure not only the effective delivery of therapeutic agents but also control their release at the target site. The convergence of these engineering strategies holds great promise for the advancement of precision medicine, particularly in the targeted treatment of cancer and other complex diseases [2].

Most strategies employed in obtaining specificity in targeted drug delivery are the conjugation of targeting ligands such as antibodies, peptides, and aptamers onto the quantum dots. This involves the modification of the functional groups that appear on the QD surface using chemical groups that can accommodate the covalent or non-covalent attachment of the ligands. Antibodies are highly specific because they selectively bind to specific antigens expressed on the surface of cancer cells to specifically target tumors. Peptides are short and flexible, making them capable of targeting a particular receptor with better tissue penetration and reduced immunogenicity. Aptamers, as synthetic single-stranded nucleic acids, would bring stability, cost-effectiveness, and ease of synthesis benefits that would be perfect in targeting a broad range of biomarkers [11]. The conjugation of the ligands does not change their bioactivity, whereas the optical and structural properties of QDs were retained. The QD-based systems therefore displayed selective targeting and binding to diseased cells to ensure enhanced therapeutic efficacy with minimal off-target effects.

Controlled release of drugs, through the help of environmental cues such as pH, temperature, and enzymatic activity, to deliver therapeutic products in a well-targeted way is one of the latest strategies via the use of stimuli-responsive quantum dots (QDs). These functional QDs are synthesized with coatings or linkers to respond to unique stimuli within the tumor microenvironment [65]. For example, pH-responsive quantum dots (QDs) are in the process of being designed as nanocarriers for multi-drug cancer treatment that may leverage the mildly acidic tumoral microenvironment toward site-specific drug delivery and the avoidance of normal tissues. QDs are designed to be dormant under physiological pH (~7.4) but active in mildly acidic extracellular tumor pH between (~6.5 and 6.8) with conformational change or bond breakage triggering therapeutic release [17]. AlSawaftah et al. [6] also recently prepared pH-responsive carbon quantum dot-magnetic nanoparticle conjugates as a dual photodynamic and hyperthermia therapy agent, intriguingly, the agent was pH-sensitive for use toward site-specific drug delivery to the acidic tumor microenvironment and reduction of normal tissue toxicity. The dual-response agent also enabled enhanced imaging and therapy efficacy with potential for dual therapy and diagnosis. They are proof of state-of-the-art nanostructure genius engineering with the ability to initiate a cancer nanomedicine revolution with stimulus-responsive, target therapeutic therapy. Equally important are temperature-responsive QDs as they are designed to release their payloads upon coming into contact with hyperthermia, which in practice often occurs under conditions of treatments against cancer diseases [25]. Enzyme-responsive QDs rely on the overexpression of certain enzymes like matrix metalloproteinases during cancer that serve to activate drugs in the disease target site. Such mechanisms not only ensure enhanced efficacy because of direct and controlled release at the action site but also eliminate the unwanted effects caused by the systemic route. Combining the stimulus-responsive properties with inherent optical and targeting capabilities of QDs does offer much promise for developing personalized, efficient cancer treatments [39].

The biggest barrier to be overcome for effective targeted delivery by quantum dots is biological barriers, although new strategies are being developed to enhance their efficacy. The vascular endothelium, the immune system's defenses, and the tumor microenvironment present a barrier through which QDs have to breach to reach the target [54]. This can be achieved by engineering QDs with a biocompatible coating, for example, PEG, which would help in avoiding immune recognition and prolong their circulation time. Size and surface charge are optimized for passive targeting using the enhanced permeability and retention effect and, therefore, their tendency to accumulate in tumor tissue due to leaky vasculature. Ligands like antibodies, peptides, or aptamers conjugated to QDs will allow the QDs for active targeting such that selective bindings of the molecules can occur only with the available receptors of the cancer cells [87]. Further development may increase tissue penetration and cellular uptake without any extracellular matrices serving as a high barrier. This is achieved toward further designing QD-based delivery systems capable of navigating far more complex biological environments with a much-improved level of precision and efficacy in delivering therapeutic agents at the site of action with decreased side effects [32]. Extensive research has been carried out on QDs to be applied for the purpose of targeted drug delivery in cancer, especially breast, lung, and pancreatic cancer. Quantum dots conjugated to specific ligands have been used to target cancer cells effectively in the case of breast cancer. For instance, photothermal and photodynamic therapy by graphene quantum dots (GQDs) have shown excellent theranostic performances in treating breast cancer [18]. QDs have also been applied in lung cancer studies as drug delivery vehicles for chemotherapeutic agents. Targeted drug delivery has been achieved using QDs conjugated with doxorubicin (DOX) in human lung cancer cells, and it holds great promise for the enhancement of therapeutic efficacy through QD-based systems [18]. Application principles developed through breast and lung cancer studies have an underpinning basis in research involving targeting mechanisms of QD-based methods on pancreatic cancer. Functionalization aspects of QDs render them more flexible than many of the other available techniques, with specific optical properties of QDs rendering them excellent candidates for the formulation of targeted delivery across the huge category of diseases referred to as cancer [77].

5 Therapeutic Applications of Quantum Dots in Cancer Treatment

A new minimally invasive treatment for cancer called photothermal therapy (PTT) uses agents that absorb light to transform optical energy into thermal energy for the thermal annihilation of malignant cells [85]. The proposed method has several advantages. It can be operated remotely, offers increased specificity, has limited drug resistance, and has fewer side effects [85]. PTT is essential because it may be able to minimize some of the drawbacks associated with more traditional therapeutic approaches. This technique has been promising for the treatment of a wide variety of cancers, including certain breast cancers and Triple-Negative Breast Cancer (TNBC), the treatments of which often require a more specific approach [22]. Moreover, PTT can be combined with other forms of therapy to obtain synergistic beneficial therapeutic effects in both primary and secondary tumors [75, 85]. Interestingly, recent studies have demonstrated that the combination of PTT with chemotherapy can induce strong antitumor immune responses against widespread tumors, which might be a solution to the problem of treating cancer that has spread throughout the body. This photothermal agent converts NIR radiation to thermal energy when it is used. The increased temperature in this localized region further causes apoptosis among the cancerous cells in this region [88]. When excited, the jump from the ground state to the excited state releases the absorbed energy in radiative (light) or non-radiative (heat). Heat generation is the basic mechanism underlying the photothermal action of these materials. PTT is a promising treatment modality for several reasons. 1. Targeted precision: PTT enables focused light application, which minimizes damage to adjacent healthy tissue 2. Low invasiveness: This method is less invasive than traditional surgical methods 3. Shorter duration of treatment: PTT is completed relatively sooner, so the comfort of the patient improves, and the length of stay in the hospital is decreased. 4. Synergistic potential: PTT may synergistically act with other treatments, including chemotherapy or immunotherapy, thus potentially increasing the final effectiveness of treatment [45]. Their small size promotes higher penetrations and uptakes in tumors, while the high surface-area-to-volume ratio and distinctive optoelectronic properties may facilitate efficient light-to-heat conversion [49]. Many nanoscale photothermal agents have specific advantages. For example, gold nanoshells and nanorods are uniquely effective at converting near-infrared radiation, whereas very small spherical gold nanoparticles demonstrate superior photoconversion capabilities within the visible light spectrum [57].

Two-dimensional nanomaterials are very promising in the context of combinatorial therapies that include photothermal treatment and other modalities, such as chemotherapy, photodynamic therapy, and immunotherapy [49]. Nanoscale photothermal agents, therefore, critically play an important role in advancing selectivity, efficacy, and safety in PTT. These agents are important in cancer treatment strategies due to their ability to selectively target malignant cells, transform light energy into thermal energy, and combine it with another treatment method. These include high absorption of near-infrared (NIR) light and high photothermal conversion efficiency, which are essential for successful PTT. For instance, polylysine-modified carbon dots exhibited a photothermal conversion efficiency of 54.9%. This will lead to the targeted delivery of multiple therapeutic agents, enabling synergistic multimodal cancer therapy using quantum dots. The $MoSe_2$ nanodots have approximately 46.5% efficacy. High biocompatibility and minimal toxicity are necessary to overcome these side effects. In vitro studies have demonstrated the low cytotoxicity of $MoSe_2$ nanodots [89]. Carbon nanoparticles (CNPs) extracted from activated carbon have higher photothermal conversion efficiencies than carbon dots while still showing acceptable biocompatibility. Quantum dots are nanoscale semiconductor crystals whose optical and electronic properties depend on the quantum confinement effects. These quantum dots have very bright and photoluminescent fluorescence with a high quantum yield and wavelength-tunable emission.

Quantum dots show great promise in the area of oncological treatment as therapeutic agents. They demonstrate a lot of advantages in performing photothermal therapy over other possible therapeutic methods due to their intrinsic physicochemical properties, attributed to their nanoscale nature and large surfaces combined with quantum confinement effect features. These effects greatly enhance their efficiency in applications pertaining to PTT (Fig. 3).

Photodynamic therapy (PDT) is a noninvasive therapy in which photosensitizer agents, light, and oxygen initiate the management of various forms of cancer urging the execution of this procedure, the photosensitizers are excited by light, which promotes the formation of ROS, mainly singlet oxygen that can kill cancer cells. In PDT, Type 1 and Type 2 reactions are different interactions of a special molecule, a

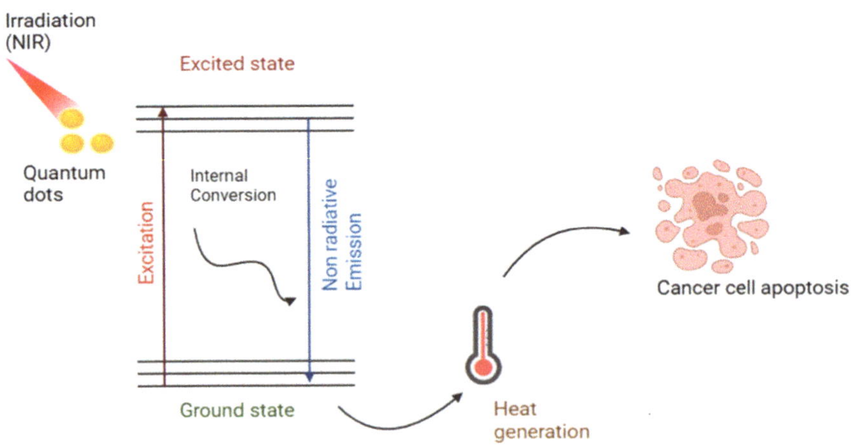

Fig. 3 General mechanism of photothermal action of quantum dots. The heat in Photothermal therapy occurs due to the photothermal effect of nanoparticles. When these nanoparticles absorb specific wavelengths of light, they get excited to higher energy states. As they relax to the ground state, they release the absorbed energy in the form of heat, which is responsible for the photothermal effect

light-activated molecule, with its surrounding tissues in dealing with diseases such as cancers. Type 1 reactions involve the generation of free radicals when the molecule absorbs light, as it transfers some of the light energy to nearby molecules, damaging all cellular structures present, such as the cell membrane and DNA, leading to eventual cell death. The Type 2 reaction occurs through the excited photosensitizer and its interaction with oxygen in the tissue to produce singlet oxygen. This is an extremely reactive form of oxygen that is capable of attacking and destroying the very important components of cells, resulting in cell death. These two reactions in PDT are potent because they preferentially target and destroy cancerous or abnormal cells when exposed to light. These have been identified as one of the most outstanding optically active nanomaterials for molecular imaging and PDT due to their high brightness, stability, and potential for biomedical applications. QDs are a new kind of photosensitizer and nanoplatform for delivering photosensitizers, thereby bypassing the several limitations that face the traditional PDT agents, like low singlet oxygen quantum yields, photobleaching, and poor biocompatibility. The integration of QDs into PDT has opened new avenues for cancer diagnosis and treatment, which may lead to more effective and targeted therapies. Quantum dots (QDs) play important roles in PDT as highly effective photosensitizers that lead to the generation of ROS upon irradiation. Most of the QDs, including graphitic carbon nitride (g-C_3N_4), ZnO, and CdSe/ZnS, have been proven to be efficient in applications associated with PDT. For instance, g-C_3N_4 QDs emit ROS under UV irradiation, and thus their efficacy as photosensitizers for PDT depends on it [13]. These QDs can be conjugated with upconversion nanoparticles to overcome the limited tissue penetration of UV light, enabling NIR irradiation to induce ROS production. Similarly, ZnO nanoparticles have outstanding applications in PDT since they are capable of generating ROS upon irradiation with light [86]. The efficiency of the production of ROS and the excitation wavelength of ZnO nanoparticles can be modified by chemical modification, including organic photosensitizers, dopants, or chemotherapeutic drugs. Although most QDs are very efficient at producing ROS, the mechanism of action may be different for different QDs, and the quantum yields may be very low. For example, PEGylated CdSe/ZnS QDs primarily produce superoxide anion radicals with a quantum efficiency of 0.35% under visible-light excitation. Electron donors, such as NADH Graphene QDs have also been proven to be efficient in the generation of singlet oxygen and ROS, which makes them very promising candidates for PDT [76]. To put it in simple words, QDs are multifunctional photosensitizers that produce all kinds of ROS under any given conditions of illumination. Their superior ROS production capability with tunability and functionalization puts them on an advantageous scale toward making PDT a successful cancer treatment. Still, more research must be done to perfect them, especially regarding the limitations involved, which include penetration through tissues and minimal side effects. QDs have several advantages over traditional photosensitizers and are hence of great interest for various biomedical applications. Some of their unique optical properties are size-tunable emission spectra, broad excitation ranges, and narrow emission peaks allowing for precision control over fluorescence characteristics [34].

One of the key advantages of QDs is their multifunctional property. Ligands and various types of biomolecules can be easily used on their surfaces, thus conferring targeted delivery, improved solubility, and reduced toxicity. Quantum dots, or QDs, are a versatile multifunctional nanomaterial that might revolutionize chemotherapy and immunotherapy efficacy through synergistic combination therapy approaches. QDs can also be used in hybrid platforms by combining with other nanomaterials such as noble metal nanoparticles or upconversion nanoparticles that enhance functionalities for cancer theranostics. QDs can also be used in hybrid platforms by combining with other nanomaterials such as noble metal nanoparticles or upconversion nanoparticles that enhance functionalities for cancer theranostics Optical properties of QDs makes them ideal for noninvasive imaging and long-term observation of vaccine dynamics, giving crucial information on the spatial and temporal distribution of the vaccine. For instance, QDs conjugated with OVA increased the levels of antibodies, cytokines, and cytotoxic T lymphocytes compared to OVA alone and improved OVA-specific immune responses [74]. It showed that QDs can be used for the delivery of siRNA to down-regulate the expression of the EGFRvIII gene in brain cancer cells, which would further suppress the PI3K-Akt signaling pathway to induce apoptosis in cancer cells [36]. These examples demonstrate versatility of QDs as a foundation for immunoimaging and efficient antigen delivery, thus improving cancer immunotherapy and vaccine development.

Synergic action can be achieved by designing and fabricating QDs to contain chemo therapeutic and immune-modulatory drugs. In one recent study, Ag_2Se QDs, encapsulated with DOX and chemokine SDF-1α, were injected into tumor areas, recruiting natural killer cells labeled with Ag_2S QDs through the application of SDF-1α-based chemotaxis [29]. This approach shows how QDs can be used in the coordinated delivery of chemotherapy and immune cells to improve antitumor effects. In another study, special tiny particles called PEGylated Ag_2S QDs were attached to a cetuximab antibody and filled with 5-fluorouracil (5FU) to deliver medicine directly to lung cancer cells. This system improves how the drug is delivered and also helps to overcome drug resistance by stopping autophagy, which is a process in cells that can make them resistant to dying [23]. This is vital because the QD-based systems integrate imaging capabilities with treatment functions. For instance, zinc oxide quantum dots were applied in bioimaging and in transporting doxorubicin inside cells, where the pH level matters [46]. These examples show how QDs can be used in real-time tracking of drug delivery and efficacy of the treatment to provide care tailored to their needs. In summary, QD-based nanoplatforms represent a powerful tool combining chemotherapy and immunotherapy with the possibility of precise delivery, improved efficacy of the combined treatment, and real-time monitoring. Taking into account the prospects of the union of several therapeutic functions with imaging in one system, QDs offer promise for some important gains in outcomes in patients with cancer (Fig. 4).

Quantum dots (QDs) have been identified as flexible and potential nanocarriers for target drug delivery and precision medicine. QDs have been engineered with various modifications on the surface according to the application, allowing simultaneous detection of more than one target. As research continues to advance, quantum dots

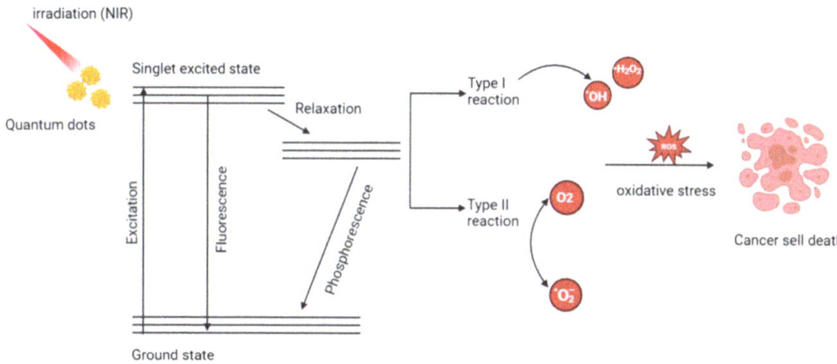

Fig. 4 General mechanism of photodynamic action of quantum dots, where heat is not generated but a photosensitizer molecule absorbs light and gets excited to higher energy states, and the energy released is transferred to the oxygen molecule nearby, which is responsible for the generation of reactive oxygen species

(QDs) will play an increasingly critical role in personalized nanomedicine and potentially change the nature of treatment for diseases that exhibit pathologies targeted at specific organelles, thereby improving general therapeutic efficacy. Quantum dots have emerged as a useful nanocarrier that could deliver pharmaceuticals in a targeted fashion and improve therapeutic efficacy with minimal off-target side effects.

The unique properties of QDs will enable targeted delivery, real-time imaging, and drug release on exposure to certain stimuli. Numerous studies have been fruitful in the design of an efficient QD-based drug delivery system. For example, carbon nanomedicine for the delivery into the tumor site of HER2-positive breast cancer cells was successful after pH-triggering. This nanosystem effectively acted as a delivery vehicle for docetaxel in the interstitial fluid microenvironment within the tumor site at pH 6.5 with a cumulative release percent of 96.13% during the time course of 72 h, thereby increasing the concentration of the drug at the tumor location and enhancing therapeutic efficacy. In a similar vein, a multifunctional nanocarrier system that integrates quantum dots (QDs), mesoporous silica nanoparticles, and gold nanoparticles exhibited selective cytotoxic effects against colorectal cancer cells when linked with Epithelial Cell Adhesion Molecule (EpCAM) aptamers [83]. Additionally, another investigation crafted a multifunctional anticancer nanomedicine utilizing biotin-PEG-PCDA polymeric nanocarriers that were loaded with paclitaxel, magnetic nanoparticles, and QDs, demonstrating substantial efficacy against multidrug-resistant breast cancer cells [78].

The potential of detecting several signals simultaneously makes QDs of importance for multiplexed imaging and identification of molecular biomarkers with potential use for cancer diagnosis, treatment, and prognosis. QDs can be engineered as multifunctional nanoplatforms to integrate imaging and therapeutic functions. For example, the Ag_2S quantum dots conjugated with targeting peptides and chemotherapy drugs enable selective tumor imaging and monitoring of drug delivery [15]. Similarly, Zn-doped Cu_2S QDs combine near-infrared fluorescence imaging with chemodynamic therapy in one single ultrasmall nanoparticle [44]. In light of their real-time visualization ability of medication distribution and therapeutic reactions, these methods have great promise for individualized cancer medicine.

They can provide real-time tracking of biological processes and therapeutic agents with a high resolution. These are important tools for elucidating disease mechanisms and optimizing treatment. Overcoming toxicity and improving biocompatibility are among the most challenging tasks for QD-based platforms to be used clinically. Quantum dots (QDs) have emerged as an extremely versatile nano platform that can deliver both diagnostic and therapeutic functions as a powerful tool for personalized cancer theranostics. Luminescent nanocrystals, such as QDs, support a multifunctional platform that integrates simultaneous imaging, drug delivery, and therapy while providing real-time monitoring of treatment efficacy. QDs are excellent candidates for targeted administration of medicinal compounds as well as molecular diagnostics, thanks to their optical characteristics, surface chemistry, and sensitivity. Hybrid systems composed of combinations of other nanostructures, such as dendrimers or liposomes, and QDs significantly enhance their theranostic potential. Fe_3O_4–ZnO core–shell magnetic QDs, for instance, are presently being investigated for deep-tumor targeting with external magnetic fields and cancer markers, which could revolutionize cancer theranostics. In summary, QD-based theranostics may prove to be a new approach to achieving personalized medicine because it can combine imaging, drug delivery, and therapy into a nano platform. The recently newly designed near-infrared fluorescent QDs for bio-imaging with enhanced sensitivity also include the use of endogenous neutrophils as natural drug carriers However, challenges like toxicity, biocompatibility, and crossing the biological barriers have to be overcome to achieve clinical translation [80]. As research advances, QD-based theranostics are on the way to revolutionize cancer treatment protocols and patient survival.

6 Quantum Dots in Theranostic Nanomedicine

Quantum dots, or QDs, are increasingly becoming very important as effective tools in theranostics. Theranostics includes a set of diagnostic functions within a single nano platform with therapeutic ones. QDs can be used as fluorescence contrast agent for imaging and delivery agents for therapy, simultaneous diagnosis and targeted treatment. There have been various types of research devised to overcome these problems, such as encapsulating QDs in liposomal hybrids to form QD-liposome complexes and QLCs with much-improved biocompatibility and stability [69]. Another approach to improve the biostability of Ag_2S QDs using long single-stranded DNA as the template is combined with an ability to resist bleaching, thereby enhancing the potential development of a theranostic platform that targets the PD-L1 high-expressing inflamed tumors. QDs appear to be promising for developing theranostics for diseases, such as cancer. Such nanoengineered particles possess the capability to integrate several functionalities, such as imaging agents, drug delivery systems, and targeting moieties, which present new advanced tools for personalized medicine [62]. QDs appear to be promising for developing theranostics for diseases, such as cancer. Such nano-engineered particles possess the capability to integrate several functionalities, such as imaging agents, drug delivery systems, and targeting moieties, which present new advanced tools for personalized medicine. These properties, combined with their flexible surface chemistry, make QDs uniquely suitable for a wide range of in vitro and in vivo applications including cellular labeling, tumor diagnostic imaging, and targeted therapeutic interventions.

However, to make QDs a part of daily practice in a clinical setup, there is a need for further developments in appropriate health and safety regulations and resolution of the issues of toxicity. Quantum dots have emerged as versatile nanoplatforms for the integration of both diagnostic and therapeutic attributes into a single nanoscale architecture designed for application in cancer theranostics. The blending of QDs with other nanoparticles along with functionalized molecules has enabled the development of multifunctional nanocomposites for concurrent imaging and targeted delivery. These QD-based nanocomposites offer several benefits for theranostics: they can, for example, provide multimodal imaging capabilities, thereby combining fluorescence imaging from QDs with other modalities such as MRI from iron oxide nanoparticles [58]. The fluorescence of quantum dots can also be engineered to report drug-release events in cellular environments. Additionally, nanocomposites can be functionalized with targeting ligands for active tumor targeting and loaded with anticancer drugs for therapy.

This provides unique optical properties with high brightness and long-time stability, hence applicable for all those applications requiring adjustable emission spectra. Quantum dots enable the concurrent visualization of tumors, treatment, and pharmacokinetic evaluation, and therefore offer a safe and effective approach to cancer therapy. For example, fluorescently illuminated nanomicelles incorporating quantum dots can concentrate at tumor locations and facilitate real-time observation of drug delivery mechanisms. Interestingly, QDs can be used to create multifunctional nanoplatforms that can be used for diagnosis, monitoring, and treatment [83]. With fluorescence imaging, not only can drug-loaded theranostic systems selectively deliver drugs at target sites but also offer information in real-time on the effect of treatment, while these many advantages notwithstanding, QDs, for instance, are still found to have certain related challenges with their toxicity and inability to biodegrade.

Inorganic QDs, including heavy metal cadmium, provide better fluorescence and optical stability but have big issues with toxicity in biological applications [70]. For the issue of toxicity, different types of biocompatible coatings have been made. For example, a nanocomposite emulsion of a POSS-PCU polymer is capable of encapsulating QDs with their photophysical properties still being maintained while also reducing the cytotoxicity. Organic quantum dots (QDs) and organic–inorganic hybrid perovskite quantum dots (OIHP-QDs) exhibit improved biocompatibility and less toxicity. More specifically, the OIHP-QDs exhibit excellent photoelectric properties, color-tunability, and a high photoluminescence quantum yield [28]. However, these materials typically suffer from the problems of instability and complexity associated with their synthesis. Recent advances, such as the gram-scale synthesis of Mn-doped all-inorganic perovskite quantum dots, suggest promise for overcoming some of these issues. In addition, 2D inorganic material-based QDs are promising because of their excellent chemical stability, favorable aqueous dispersion, and biocompatibility. Theranostic quantum dots have been highly influencing personalized medicine because of their unique functions, especially for both diagnostics and therapy. In turn, nanomaterials have provided opportunities for accurate imaging, as well as targeted delivery, and have helped tailor drug administrations based on certain patient-related conditions. CuInS/ZnS quantum dots, or ZCIS QDs, have proven to be promising as "all-in-one" theranostic nanomedicines. Imaging and therapeutic functionalities were integrated into the QDs as part of a well-defined nanostructure, thereby enabling MSOT imaging and synergistic PTT and PDT. The size-dependent distribution profiles of ZCIS QDs can be visualized at tumor sites in vivo, smaller nanomedicines retain longer at tumors, show higher uptake at tumors, and penetrate deeper into tumors [52]. Thus, theranostic QDs represent a huge step forward in personalized medicine: diagnostics could be combined with targeted therapeutic interventions. This opens up great possibilities for optimizing outcomes in cancer therapy based on real-time imaging with tailored treatments by their genetic profiles and tumor characteristics.

However, further research is needed on data analysis, interpretation, and protocol standardization to address these challenges so that these nanomedicines can be fully implemented in clinical settings. Nano constructs with QDs have emerged as important tools for targeted cancer therapy, demonstrating their potential for multimodal applications including imaging, drug delivery, and sensing. These systems merge QD-based distinctive optical features with ligands selective to targeting cancer, thus providing a highly versatile set of platforms for therapy against cancer. The QD-aptamer-doxorubicin conjugate is one of the excellent examples that illustrates the utility of QD-based nanosystems in targeted cancer therapy. This innovative system employs an A10 RNA aptamer to selectively target prostate-specific membrane antigens that are markers of prostate cancer cells. Insertion of doxorubicin into the aptamer framework forms a self-quenching mechanism through bi-directional fluorescence resonance energy transfer (Bi-FRET), allowing for concurrent delivery of the drug together with imaging and detection of drug release within the targeted cancer cell [9]. Another interesting application of QD-based nanosystems is the development of RGD-QD-MoS_2 nanosheets for targeted fluorescence imaging and photothermal therapy of cervical cancer. This will allow imaging via the fluorescence properties of QDs and the photothermal conversion ability of MoS_2 nanosheets, as the system will be delivered to the site of action via the RGD peptide, targeting cells that overexpress integrin receptors [92]. In summary, QD-based nanosystems represent a highly versatile platform for targeted cancer therapy, integrating imaging, drug delivery, and sensing functionalities. These systems have an extremely high level of specificity toward cancer cells and could eventually translate into improved therapeutic efficacy with fewer side effects. However, more intensive studies should be directed at toxicity issues and performance optimization to facilitate clinical translation.

7 Safety, Toxicity, and Regulatory Challenges of Quantum Dots in Cancer Theranostics

Quantum dots are newly developed multifunctional nanoparticles with superior physicochemical and biological properties and have been applied in cancer theranostics. These semiconductor nanocrystals possess a high brightness, long stability, and adjustable emission spectra, thus suitable for practical application in diagnosis and treatment approaches. Although QDs have many advantages major issues, primarily toxicity, greatly limit their practical application in the clinic. The inherent cytotoxicity of some types of QDs, especially those containing cadmium, selenium, or tellurium, has led to the search for less toxic alternatives, such as type I–III–VI2 QDs ($AgInS_2$, $CuInS_2$, and ZnS–$AgInS_2$) [59]. In addition, the bigger size of QDs than small molecules could hinder their accumulation and penetration in tumors, and novel strategies like charge reversal nanoparticles are needed to improve their performance. QDs are highly promising for cancer theranostics where diagnosis, targeting, and treatment can be simultaneously carried out, but their clinical acceptance is still pending on safety concerns and optimization of their in vivo performance. Advanced research may bring biocompatible and multifunctional QDs to offer new opportunities for more efficient and personalized cancer management, thereby changing oncology. Quantum dots (QDs) exhibit cellular toxicity through multiple mechanisms, including oxidative stress and the generation of reactive oxygen species, which play central roles. QDs can cause the intracellular production of ROS, leading to oxidative damage of cellular components [31]. This ROS generation could be because of QDs and cell interaction resulting in dysfunction. Furthermore, QDs reduce cellular GSH levels that contribute to further oxidative stress, and the mechanisms of toxicity of QDs can be cell type-specific, that is, between normal cells and cancerous cells.

For instance, QDs cause chronic endoplasmic reticulum stress and autophagosome accumulation, leading to apoptosis in normal liver cells. On the other hand, protective autophagy in liver cancer cells may be induced, which would protect against QD-induced apoptosis. Thus, such a differential response underlines the complexity of the mechanisms of QD toxicity. Accumulation of QDs in the different organs leads to their major systemic toxicity. Especially, the release of heavy-metal ions is related to the most prevalent systemic toxicity mechanism. QDs are very readily transported to the other organs; the liver represents the most accumulated organ. These processes can lead to morphological changes in the hepatic lobules, increased oxidative stress, and liberation of cadmium ions from CdS QDs, with a significant effect on their toxicities [50]. Of critical importance is the fact that QD-induced in vitro and in vivo toxicity has been demonstrated to be more pronounced than that induced by cadmium ions at equivalent or even higher doses. This would mean that QD toxicity does not depend on the toxicity of the released metal ions, but is also due to the nanoscale particles. Nanotoxicology of QDs involves cellular as well as systemic mechanisms. Key contributors to QD toxicity include accumulation of the dose in organs, ROS generation, the release of heavy metal ions, and oxidative stress.

The properties of QDs, such as size, concentration, surface charge, and coating, may play a determining role in their toxicity profile. Further research is needed to fully elucidate the molecular mechanisms underlying QD-induced toxicity and develop strategies for mitigating these effects in biomedical applications. The in vivo clearance and metabolism of QDs are very complex, showing the involvement of both renal and hepatic pathways. Size, surface chemistry, and charge significantly influence the biodistribution and clearance of QDs. Renal clearance is usually determined for smaller QDs with hydrodynamic diameters of less than 5.5 nm. However, ultrasmall QDs are not necessarily efficiently removed by the kidneys in all cases. For example, negatively charged 3.7 nm QDs accumulated within kidney mesangial cells for up to 30 days, while only traces appeared in the urine. In contrast, cationic QDs of similar size (~5.67 nm) are readily excreted in urine, highlighting the significant impact of surface charge on renal clearance [47]. The major excretion pathway for QD, which is of secondary importance, is hepatic clearance. Several studies have shown that a major portion of QDs is deposited in the liver. Its uptake depends on surface modification. For example, QDs conjugated with BSA showed 99% uptake after 90 min, whereas QDs coated with MUA and cross-linked with lysine showed 40% uptake. Interestingly, the relationship between QD size and liver accumulation has yielded contradictory results. While studies are showing that smaller QDs (<10 nm) can be stealthy to reticuloendothelial system organs, others have reported a significant accumulation of AuNPs even at a size of 1.4 nm in the liver. Such long retention times may be associated with toxicity and interference with other diagnostic tests. Thus, solutions for these problems should be developed, including enhancement of renal clearance and minimization of non-specific accumulation in the organs [93]. In summary, the clearance and in vivo metabolism of QDs involve a complex interplay between the renal and hepatic pathways, which varies with size, surface chemistry, and charge. Long-term exposure to CdSe–ZnS QDs with different surface coatings significantly affected the lifespan, induced a marked increase in ROS levels, and caused severe genotoxic effects and increased apoptosis rates in the hemocytes of Drosophila melanogaster. In summary, the chronic effects of QD on tissues involved inflammation, altered immune responses, and genotoxicity. The data presented above are of considerable value for the safety of QD applications in biomedicine and other fields. The fact that the lowest achievable doses may already provoke chronic tissue and organ damage raises a need for a complete assessment of the long-term toxicity of QDs and thus to make cautionary recommendations about the proper usage of QDs in biological media [81].

Several strategies have been devised to decrease the toxicity of QDs and nanoparticles for biomedical applications. PEGylation is the most popular method for minimizing QD toxicity, prolonging blood circulation time, and preventing non-specific uptake. For instance, thiolated PEGs have recently been employed to coat Ag_2S near-infrared QDs, thereby yielding highly cytocompatible nanoparticles that can be used for bioimaging applications as well as drug delivery purposes [7]. Similar neutralization of the toxicity effects of QD has been demonstrated with the help of polymer encapsulation using POSS-PCU materials, which improve the photostability of QD without significantly affecting their optical properties. Remarkably, although PEG is widely utilized, the growing seroprevalence of PEG-specific antibodies in humans necessitates alternative means of coating. Bioinspired nanoparticle shielding-based methods are among those tested for possible superiority over PEG and other synthetic polymer coatings. Apart from the coating, biodegradable nanoparticles and stimuli-responsive drug delivery systems have also emerged with the potential to mitigate toxicity. For instance, biocompatible polyphenol-coated intracellular glutathione-sensitive MSNs degradable at a pH level as low as acid pH can be designed to trigger the controlled delivery of anticancer drugs. Similarly, numerous controlled gatekeepers of MSNs have been developed to reach "zero premature release" drug delivery, which even further reduces the possible toxicity problems. In summary, a multidimensional approach toward the minimization of nanoparticle toxicity is realized by combining biocompatible coatings, biodegradable materials, and targeted delivery systems.

Interestingly, some papers note the challenges of specific areas; for example, genomics implementation entails issues in patient and clinician education, the integration of technology into electronic health records, and regulatory and reimbursement considerations. The translation of methods such as fMRI and DTI into clinical use is problematic in neuroimaging, despite the potential benefits of neurosurgery, diagnosis, and rehabilitation. In the final analysis, the specific issues alluded to in the question were not directly addressed, although the papers were unanimous in pointing out that clinical translation is a rather complex process impeded by several barriers, ranging from individual-level challenges such as lack of knowledge and skills, organizational factors involving poor access to evidence in research, and lack of equipment. Multidisciplinary approaches with effective collaboration from different disciplines and effective participation of stakeholders are needed to overcome these challenges [3]. QD-based cancer theranostics has been promising in preclinical as well as early clinical studies and confers targeted drug delivery as well as imaging capabilities. QD-based nanocarriers have proven to be particularly cytotoxic toward cancer cells while enhancing the antitumor effects in in vivo animal models. For instance [2], synthesized a multimodal nanoparticle using magnetic GZCIS/ZnS QDs coated with an epirubicin-loaded mesoporous silica. The authors reported that such a system manifested selective cytotoxicity in vitro against HT-29 colorectal cancer cells by enhancing the in vivo antitumor effects but with fewer side effects. Similarly, Reports upon a folate-modified theranostic liposome (FL/QD-TK) containing near-infrared fluorescent CdSeTe/ZnS core/shell QDs that showed highly specific tumor imaging and strong inhibition of folate receptor-overexpressed Bel-7402 mouse xenografts without systemic toxicity [73]. However, it is to be noted that although these studies provide promising outcomes concerning tumor targeting and localized treatments, there is a long way before widespread clinical application is achieved, with obstacles to toxicity, clearance by the immune system, and targeted delivery to the tumor sites [72]. The methodologies that have been done as scholarly research include surface modification, biodegradable nanoparticles, and targeted methods to enhance biocompatibility and accumulation within the tumor. Thus, in the concluding part of this chapter, QD-based cancer theranostics is promising for the treatment of a tumor, but more work needs to be done to eliminate issues related to safety and further enhance clinical efficacy. Future: Given the development pace of nanoformulations and targeted delivery systems, QD-based theranostics can be expected to make a lot of contributions toward cancer treatment. However, further clinical trials need to be carried out to study their safety profile and therapeutic results.

Quantum dots (QDs) have shown promising potential in different biomedical applications. However, several challenges await its clinical translation. The main barriers to its widespread use in the clinic are scalability, reproducibility, and manufacturing costs. Scalability is one of the biggest concerns in the production of QD. While batch reactors are widely applied for QD synthesis, this reactor type suffers from scaling limitations. To overcome the limitations of batch reactors, continuous synthetic techniques have been recently developed in microfluidic devices, high-gravity reactors, as well as spray-based techniques. These are supposed to offer improved reproducibility and enhanced large-scale manufacturing [67]. The lack of control measures and standard production methods further adds to the existing problem of reproducibility that seriously limits the huge potential applications of QDs in a clinical environment. Fabrication of QDs is expensive and one of the barriers to their clinical translation. This is because the synthesis of QDs is supposed to be complicated and demands special equipment and materials [53].

Indeed, future trends in quantum dot research will focus on the development of nontoxic materials, advanced coatings, and AI-driven toxicity prediction for safety enhancement to ensure regulatory compliance. There is an increasing trend in research on the application of nontoxic materials to QD synthesis, for example, the study of InP quantum dots [61]. The primary motivator is the potential adverse effects of QD materials on human health and the environment, leading to their substitution with safer counterparts. Research on developing advanced coatings and surface modifications is required to make dramatic changes to the properties and behavior of QDs to reduce their toxicity and increase their functionality. Interestingly, it can be seen that the desire for nontoxic materials is paradoxically in conflict with the present scenario of toxicity prediction. Though researchers are developing safer QDs, predicting the toxicity of the QDs remains challenging, as nano-bio interactions are complex and standard data is scarce [35]. This method calls for enhanced and standardized measurements in the examination of nanotoxicity. To overcome these challenges, AI and machine learning have emerged as powerful tools for QD properties and toxicity prediction. These approaches will facilitate the design of QDs with the desired properties and avoid toxic side effects. The integration of biological data with AI will enable scientists to gain better insights into the nano-bio interactions and direction toward safer QDs [4]. As this area of research keeps on evolving, standardized methodologies and databases will prove to be extremely important in allowing the reliability and comparability of AI-based toxicity predictions, thus being able to comply with regulations and support the safe progression of QD technologies.

8 Challenges and Future Directions in Quantum Dot Theranostics

The main problem is toxicity, which remains the main challenge for clinical translation. Quantum dots generally contain heavy metals such as cadmium, known for their ability to be toxic to biological systems. This toxicity limitation limits their use in vivo and makes long-term experimentation difficult. In addition, the biocompatibility and potential long-term effects of quantum dots in the human body are not well understood, which increases safety concerns regarding clinical applications. Interestingly, although QDs exhibit unique optical properties and tunability, their application in PDT is problematic. Photo physics for the formation of singlet oxygen in energy transfer systems based on QDs is highly complex. Yet, in use cases properly researched for PDT as photosensitizers by QDs are lacking in the biological system. This controversy of promising properties related to optoelectronic in PDT-related applications calls for additional research. In conclusion, though QDs possess tremendous promise for applications related to cancer, surmounting the challenges posed by toxicity and enhancing the efficiency of PDT among therapeutic uses is important.

Future research on QDs could be directed to develop nontoxic QD formulations, which enhance their biocompatibility to optimize performance across various applications used in cancer management for successful transfer from the lab to clinics. CRISPR-Cas9 technology is the most powerful tool in cancer gene therapy, and recent advances have used quantum dots for better delivery and tracking. The cationic glucosamine/PEI-derived graphene quantum dots, PEI-GQDs, have been applied successfully to deliver CRISPR ribonucleic protein complexes targeting the over-expressed TP53 mutation present in approximately 50% of cancers. This approach has shown promising results, such as a 40% rate of apoptotic cell death in PC3 prostate cancer cells, without affecting non-cancerous cells. Quantum dots can not only enhance the delivery of CRISPR-Cas9 components but can also be used to guide image gene therapy. Since GQDs have intrinsic fluorescence, the translocation of PEI-GQD/RNP complexes into the cytoplasm [42]. Such roles of quantum dots as delivery mechanisms and imaging agents mark a very significant step forward in the arena of cancer-targeted gene therapy. This incorporation of quantum dots with CRISPR-Cas9 technology introduces a new development thought to be more promising for advancing gene therapy in cancer treatments. This method allows for overcoming physiological barriers that otherwise make this type of methodology only more precise and effective for cancer therapies. Further studies in this area may also bring forth further breakthroughs and advancements in the quantum-dot-based CRISPR delivery systems for cancer therapy. Quantum dots (QDs) have emerged as significant tools in immuno-oncology and vaccine formulations, owing to their unique optical and electronic properties.

QDs have been regarded for a long as an agent with enormous promise in cancer biomarker's biomolecular profiling, in vivo tumor imaging, and targeted drug delivery in oncology Interestingly, though, QDs offer immense promise towards the application for the diagnostics of cancer and cancer therapy, in this context less attention has been paid towards developing a vaccine against this deadly cancer. However, the fact that QDs can be used in immunolabeling, cell tracking, and in situ hybridization has shown promise for this kind of research in vaccines. A convenient aqueous route to antibody-derivatized ZnS immuno-quantum dots offers valuable access to this potential application for vaccines. Even though the direct application of QDs in vaccine design is not a subject of the given context, their potential in biological labeling and imaging shows an opportunity for more research in the field. The improvement of the diagnostic and therapeutic functions of future generations of QDs while simultaneously improving their safety and biocompatibility stands at the forefront. The development of multifunctional QD-based nanoplatforms that integrate the functionalities of imaging, drug delivery, and therapeutics into one single system offers promising avenues for QD research. For instance, scientists are designing hybrid systems using a combination of QDs and other nanomaterials, such as graphene oxide, by leveraging synergies. The complex GO-PEI-GQDs exhibited a strong photothermal response coupled with increased cytotoxic activities toward cancer cells; hence, there is the promise of such a hybrid material in theranostics [40].

Greater interest has also focused on bioinspired quantum dots (BQDs), especially with enhanced aqueous solubility, virtually negligible toxicity, and selective targeting at a cancer site. Conclusion: As a future step in QD-based cancer theranostics, multifunctional, biocompatible, biodegradable, and nontoxic materials could be used as the basis. These next-generation QDs will address the combination of sophisticated imaging capability with targeted drug delivery and therapeutic functions, areas that could be revolutionary in personalizing cancer treatments. QDs could also be integrated with new technologies, including optical coherence tomography (OCT), to offer new diagnostic and therapeutic approaches [16]. Platforms consisting of quantum dots in combination with other nanomaterials have emerged at the cutting edge of nanotheranostic applications in biomedicine. These platforms merge the specific optical properties of QDs with the functional properties of other nanomaterials to design reconfigurable platforms for detection, imaging, and drug delivery. This is the underlying reason why quantum dots, such as silicon quantum dots, are considered very valuable and known for their stability, optical properties, and even surface modification. These have been applied to produce nanocomposites of high performance, such as graphene, carbon nanoparticles, and noble metal nanoparticle-based nanocomposites that present excellent sensing properties [33].

In summary, quantum dots have been integrated with other nanomaterials into the nanotheranostic platform, accelerating the pace of nanomedicine advances. Such hybrid systems are inherently multifunctional because they marry unique properties from quantum dots with the best advantages of other nanomaterials. With the advancement of research, it is expected that more will be developed in areas such as targeted drug delivery, multimodal imaging modalities, and personalized medicine, which will lead to advanced diagnostic and therapeutic methods.

9 Conclusion

Quantum dots have been the ultimate game-changer in the treatment of cancer as they provide diagnosis and treatment under one roof by their outstanding properties. The ability to emit bright, tunable fluorescence and resistance against photobleaching make QDs of great value in the identification of tumors, the monitoring of the progression of diseases, and exceptional guidance during surgery. On the therapeutic side, QDs have unlocked new avenues and opened a new possibility wherein drugs or even genetic material could be delivered in a very specific manner to the cancer cells themselves while having minimal side effects. Improvements in their design and engineering to make them safer and more applicable for medical uses have allowed the possibility of delivering personalized care tailored to an individual patient's needs. Examples include stimulus-responsive mechanisms and the integration of multimodal imaging systems, among many others, making them versatile. While much promise lies in them, many challenges such as safety, lowering toxicity, and translation of lab research to the clinic remain to be overcome. Research is on the development of safer materials and their design, which will be the first step toward making QDs a standard tool in cancer care. Quantum dots are sure to revolutionize the diagnosis and treatment of cancer as we continue pushing the boundaries of technology and giving patients a future with more precise and effective therapies.

Acknowledgements The authors are grateful to the National Institute of Pharmaceutical Education and Research, Kolkata, Kolkata 700054, and the Department of Science and Technology for supporting the students through fellowships.

Competing Interests None declared.

Ethical Approval Not required.

Conflicts of Interest The authors declare that they have no known competing financial interests or personal relationships that could have appeared to influence the work reported in this paper.

References

1. Abdellatif AAH, Younis MA, Alsharidah M, Al Rugaie O, Tawfeek HM (2022) Biomedical applications of quantum dots: overview, challenges, and clinical potential. Int J Nanomed 17:1951–1970. https://doi.org/10.2147/IJN.S357980
2. Abrishami A, Bahrami AR, Nekooei S, Sh. Saljooghi A, Matin MM (2024a) Hybridized quantum dot, silica, and gold nanoparticles for targeted chemo-radiotherapy in colorectal cancer theranostics. Commun Biol 7(1):1–19. https://doi.org/10.1038/s42003-024-06043-6
3. Abu-Odah H, Said NB, Nair SC, Allsop MJ, Currow DC, Salah MS, Hamad BA, Elessi K, Alkhatib A, ElMokhallalati Y, Bayuo J, AlKhaldi M (2022) Identifying barriers and facilitators of translating research evidence into clinical practice: a systematic review of reviews. Health Soc Care Community 30(6):e3265–e3276. https://doi.org/10.1111/hsc.13898
4. Ahmad F, Mahmood A, Muhmood T (2021) Machine learning-integrated omics for the risk and safety assessment of nanomaterials. Biomater Sci 9(5):1598–1608. https://doi.org/10.1039/D0BM01672A

5. Al-Muhanna MKA, Alghamdi AA, Alrfaei B, Afzal M, Al-Subaiee R, Haddadi R (2024) Surface-engineered quantum dot nanocomposites for neurodegenerative disorder remediation and avenue for neuroimaging. Nanotechnol Rev 13(1). https://doi.org/10.1515/ntrev-2024-0128
6. AlSawaftah NM, Awad NS, Pitt WG, Husseini GA (2022) PH-Responsive nanocarriers in cancer therapy. Polymers 14(5):936. https://doi.org/10.3390/polym14050936
7. Asik D, Yagci MB, Duman FD, Acar HY (2016) One-step emission tunable synthesis of PEG-coated Ag2S NIR quantum dots and the development of receptor-targeted drug delivery vehicles thereof. J Mater Chem B 4(11):1941–1950. https://doi.org/10.1039/C5TB02599K
8. Azam N, Najabat Ali M, Javaid Khan T (2021) Carbon quantum dots for biomedical applications: review and analysis. Front Mater 8. https://doi.org/10.3389/fmats.2021.700403
9. Bagalkot V, Zhang L, Levy-Nissenbaum E, Jon S, Kantoff PW, Langer R, Farokhzad OC (2007) Quantum dot−aptamer conjugates for synchronous cancer imaging, therapy, and sensing of drug delivery based on Bi-fluorescence resonance energy transfer. Nano Lett 7(10):3065–3070. https://doi.org/10.1021/nl071546n
10. Bajwa N, Mehra NK, Jain K, Jain NK (2016) Pharmaceutical and biomedical applications of quantum dots. Artif Cells Nanomed Biotechnol 44(3):758–768. https://doi.org/10.3109/21691401.2015.1052468
11. Bojarska J, Wolf WM (2024) Short peptides as powerful arsenal for smart fighting cancer. Cancers 16(19):19. https://doi.org/10.3390/cancers16193254
12. Bruns OT, Bischof TS, Harris DK, Franke D, Shi Y, Riedemann L, Bartelt A, Jaworski FB, Carr JA, Rowlands CJ, Wilson MWB, Chen O, Wei H, Hwang GW, Montana DM, Coropceanu I, Achorn OB, Kloepper J, Heeren J et al (2017) Next-generation in vivo optical imaging with short-wave infrared quantum dots. Nat Biomed Eng 1(4):1–11. https://doi.org/10.1038/s41551-017-0056
13. Chan M-H, Chen C-W, Lee I-J, Chan Y-C, Tu D, Hsiao M, Chen C-H, Chen X, Liu R-S (2016) Near-infrared light-mediated photodynamic therapy nanoplatform by the electrostatic assembly of upconversion nanoparticles with graphitic carbon nitride quantum dots. Inorg Chem 55(20):10267–10277. https://doi.org/10.1021/acs.inorgchem.6b01522
14. Chehelgerdi M, Chehelgerdi M, Allela OQB, Pecho RDC, Jayasankar N, Rao DP, Thamaraikani T, Vasanthan M, Viktor P, Lakshmaiya N, Saadh MJ, Amajd A, Abo-Zaid MA, Castillo-Acobo RY, Ismail AH, Amin AH, Akhavan-Sigari R (2023) Progressing nanotechnology to improve targeted cancer treatment: overcoming hurdles in its clinical implementation. Mol Cancer 22(1):169. https://doi.org/10.1186/s12943-023-01865-0
15. Chen H, Li B, Zhang M, Sun K, Wang Y, Peng K, Ao M, Guo Y, Gu Y (2014) Characterization of tumor-targeting Ag_2S quantum dots for cancer imaging and therapy in vivo. Nanoscale 6(21):12580–12590. https://doi.org/10.1039/C4NR03613A
16. Chen H, Liu Z, Wei B, Huang J, You X, Zhang J, Yuan Z, Tang Z, Guo Z, Wu J (2021) Redox responsive nanoparticle encapsulating black phosphorus quantum dots for cancer theranostics. Bioact Mater 6(3):655–665. https://doi.org/10.1016/j.bioactmat.2020.08.034
17. Choppadandi M, Soumya K, Ghosh S, Balu A, Shingote T, Babu SS, Prasanna VS, Arumugam S, Velyutham R, Yallapu MM, Kapusetti G (2024) Dual functional magnetic nanoparticles conjugated with carbon quantum dots for hyperthermia and photodynamic therapy for cancer. Nanotheranostics 8(4):442–457. https://doi.org/10.7150/ntno.91871
18. Dar MS, Tabish TA, Thorat ND, Swati G, Sahu NK (2023) Photothermal therapy using graphene quantum dots. APL Bioeng 7(3):031502. https://doi.org/10.1063/5.0160324
19. Davodabadi F, Mirinejad S, Fathi-Karkan S, Majidpour M, Ajalli N, Sheervalilou R, Sargazi S, Rozmus D, Rahdar A, Diez-Pascual AM (2023) Aptamer-functionalized quantum dots as theranostic nanotools against cancer and bacterial infections: a comprehensive overview of recent trends. Biotechnol Prog 39(5):e3366. https://doi.org/10.1002/btpr.3366
20. Devi S, Kumar M, Tiwari A, Tiwari V, Kaushik D, Verma R, Bhatt S, Sahoo BM, Bhattacharya T, Alshehri S, Ghoneim MM, Babalghith AO, Batiha GE-S (2022) Quantum dots: an emerging approach for cancer therapy. Front Mater 8. https://doi.org/10.3389/fmats.2021.798440

21. Díaz-García D, Díaz-Sánchez M, Álvarez-Conde J, Gómez-Ruiz S (2024) Emergence of quantum dots as innovative tools for early diagnosis and advanced treatment of breast cancer. ChemMedChem 19(16):e202400172. https://doi.org/10.1002/cmdc.202400172
22. Dong L, Chen Z, Ye L, Yu Y, Zhang J, Liu H, Zang J (2019) Gram-scale synthesis of all-inorganic perovskite quantum dots with high Mn substitution ratio and enhanced dual-color emission. Nano Res 12(7):1733–1738. https://doi.org/10.1007/s12274-019-2430-8
23. Duman FD, Akkoc Y, Demirci G, Bavili N, Kiraz A, Gozuacik D, Acar HY (2019) Bypassing pro-survival and resistance mechanisms of autophagy in EGFR-positive lung cancer cells by targeted delivery of 5FU using theranostic Ag2S quantum dots. J Mater Chem B 7(46):7363–7376. https://doi.org/10.1039/C9TB01602C
24. Ejeromedoghene O, Omoniyi AO, Akor E, Alowakennu M, Samson KA, Abesa S, Zhang Z (2024) Progress in stimuli-responsive hydrogel composites for digital technologies. Appl Mater Today 37:102088. https://doi.org/10.1016/j.apmt.2024.102088
25. Elsherbeny A, Bayraktutan H, Oz UC, Moloney C, Ashworth JC, Grabowska AM, Alexander C (2024) Responsive nanomaterial delivery systems for pancreatic cancer management. Adv Therapeutics 7(3):2300330. https://doi.org/10.1002/adtp.202300330
26. Fang M, Peng C, Pang D-W, Li Y (2012) Quantum dots for cancer research: current status, remaining issues, and future perspectives. Cancer Biol Med 9(3):151–163. https://doi.org/10.7497/j.issn.2095-3941.2012.03.001
27. Gao X, Dave SR (2007) Quantum dots for cancer molecular imaging. In: Chan WCW (ed) Bio-applications of nanoparticles. Springer, pp 57–73. https://doi.org/10.1007/978-0-387-76713-0_5
28. Han Y, Chang X, Cheng X, Lin Y, Cui B-B (2023) Recent progress of organic-inorganic hybrid perovskite quantum dots: preparation, optical regulation, and application in light-emitting diodes. Laser Photonics Rev 17:2300383. https://doi.org/10.1002/lpor.202300383
29. Hao X, Li C, Zhang Y, Wang H, Chen G, Wang M, Wang Q (2018) Programmable chemotherapy and immunotherapy against breast cancer guided by multiplexed fluorescence imaging in the second near-infrared window. Adv Mater 30(51):1804437. https://doi.org/10.1002/adma.201804437
30. Henke E, Nandigama R, Ergün S (2020) Extracellular matrix in the tumor microenvironment and its impact on cancer therapy. Front Mol Biosci 6. https://doi.org/10.3389/fmolb.2019.00160
31. Horie M, Tabei Y (2021) Role of oxidative stress in nanoparticles toxicity. Free Radical Res 55(4):331–342. https://doi.org/10.1080/10715762.2020.1859108
32. Hu Y, Wang Y, Ye D (2020) Semiconductor quantum dots for cell imaging. In: Wu F-G (ed) Fluorescent materials for cell imaging. Springer, pp 17–48. https://doi.org/10.1007/978-981-15-5062-1_2
33. Ja'farawy MSA, Thirumalai D, Lee J, Jung HS, Chang S-C, Yoon J-H, Kim D-H (2023) Graphene quantum dot nanocomposites: electroanalytical and optical sensor technology perspective. J Anal Sci Technol 14(1):29. https://doi.org/10.1186/s40543-023-00393-2
34. Ji X, Peng F, Zhong Y, Su Y, He Y (2014) Fluorescent quantum dots: synthesis, biomedical optical imaging, and biosafety assessment. Colloids Surf B 124:132–139. https://doi.org/10.1016/j.colsurfb.2014.08.036
35. Ji Z, Guo W, Wood EL, Liu J, Sakkiah S, Xu X, Patterson TA, Hong H (2022) Machine learning models for predicting cytotoxicity of nanomaterials. Chem Res Toxicol 35(2):125–139. https://doi.org/10.1021/acs.chemrestox.1c00310
36. Jung J, Solanki A, Memoli KA, Kamei K, Kim H, Drahl MA, Williams LJ, Tseng H-R, Lee K-B (2010) Selective inhibition of human brain tumor cell proliferation via multifunctional quantum dot-based siRNA delivery. Angew Chemie (International Ed. in English) 49(1):103–107. https://doi.org/10.1002/anie.200905126
37. Karakoti AS, Shukla R, Shanker R, Singh S (2015) Surface functionalization of quantum dots for biological applications. Adv Coll Interface Sci 215:28–45. https://doi.org/10.1016/j.cis.2014.11.004
38. Kashyap BK, Singh VV, Solanki MK, Kumar A, Ruokolainen J, Kesari KK (2023) Smart nanomaterials in cancer theranostics: challenges and opportunities. ACS Omega 8(16):14290–14320. https://doi.org/10.1021/acsomega.2c07840

39. Kaushik N, Borkar SB, Nandanwar SK, Panda PK, Choi EH, Kaushik NK (2022) Nanocarrier cancer therapeutics with functional stimuli-responsive mechanisms. J Nanobiotechnol 20(1):152. https://doi.org/10.1186/s12951-022-01364-2
40. Kumawat MK, Thakur M, Bahadur R, Kaku T, RS P, Suchitta A, Srivastava R (2019) Preparation of graphene oxide-graphene quantum dots hybrid and its application in cancer theranostics. Mater Sci Eng: C 103:109774. https://doi.org/10.1016/j.msec.2019.109774
41. Le N, Kim K (2023) Current advances in the biomedical applications of quantum dots: promises and challenges. Int J Mol Sci 24(16):16. https://doi.org/10.3390/ijms241612682
42. Lee B, Gries K, Valimukhametova AR, McKinney RL, Gonzalez-Rodriguez R, Topkiran UC, Coffer J, Akkaraju GR, Naumov AV (2023) In vitro prostate cancer treatment via CRISPR-Cas9 gene editing facilitated by polyethyleneimine-derived graphene quantum dots. Adv Func Mater 33(48):2305506. https://doi.org/10.1002/adfm.202305506
43. Li M, Huang Y, Shen C, Wang Y, Lin Y, Wang Z, Chen N, Luo Y (2025) Application of quantum dots in cancer diagnosis and treatment: Advances and perspectives. Nano Res 18(1). https://doi.org/10.26599/NR.2025.94907163
44. Li S-L, Jiang P, Hua S, Jiang F-L, Liu Y (2021) Near-infrared Zn-doped Cu_2S quantum dots: an ultrasmall theranostic agent for tumor cell imaging and chemodynamic therapy. Nanoscale 13(6):3673–3685. https://doi.org/10.1039/D0NR07537J
45. Li Y, Dang G, Rizwan Younis M, Cao Y, Wang K, Sun X, Zhang W, Zou X, Shen H, An R, Dong L, Dong J (2023) Peptide functionalized actively targeted MoS_2 nanospheres for fluorescence imaging-guided controllable pH-responsive drug delivery and collaborative chemo/photodynamic therapy. J Colloid Interface Sci 639:302–313. https://doi.org/10.1016/j.jcis.2023.02.027
46. Li Z, Fu S, Li H, Chen B, Xie D, Fu D, Feng Y, Gao C, Liu S, Wilson DA, Tu Y, Peng F (2023) Light-driven micromotor swarm induced in-situ polymerization and synergistic photothermal therapy. Chem Eng J 468:143393. https://doi.org/10.1016/j.cej.2023.143393
47. Liang X, Wang H, Zhu Y, Zhang R, Cogger VC, Liu X, Xu ZP, Grice JE, Roberts MS (2016) Short- and long-term tracking of anionic ultrasmall nanoparticles in kidney. ACS Nano 10(1):387–395. https://doi.org/10.1021/acsnano.5b05066
48. Liang Z, Khawar MB, Liang J, Sun H (2021) Bio-conjugated quantum dots for cancer research: detection and imaging. Front Oncol 11. https://doi.org/10.3389/fonc.2021.749970
49. Liu S, Pan X, Liu H (2020) Two-dimensional nanomaterials for photothermal therapy. Angew Chem 132(15):5943–5953. https://doi.org/10.1002/ange.201911477
50. Liu W, Zhang S, Wang L, Qu C, Zhang C, Hong L, Yuan L, Huang Z, Wang Z, Liu S, Jiang G (2011) CdSe quantum dot (QD)-induced morphological and functional impairments to liver in mice. PLoS ONE 6(9):e24406. https://doi.org/10.1371/journal.pone.0024406
51. Luchini A, Vitiello G (2019) Understanding the nano-bio interfaces: lipid-coatings for inorganic nanoparticles as promising strategy for biomedical applications. Front Chem 7. https://doi.org/10.3389/fchem.2019.00343
52. Lv G, Guo W, Zhang W, Zhang T, Li S, Chen S, Eltahan AS, Wang D, Wang Y, Zhang J, Wang PC, Chang J, Liang X-J (2016) Near-infrared emission CuInS/ZnS quantum dots: all-in-one theranostic nanomedicines with intrinsic fluorescence/photoacoustic imaging for tumor phototherapy. ACS Nano 10(10):9637–9645. https://doi.org/10.1021/acsnano.6b05419
53. Ma P, Jia X, He Y, Tao J, Wang Q, Wei C-I (2024) Recent progress of quantum dots for food safety assessment: a review. Trends Food Sci Technol 143:104310. https://doi.org/10.1016/j.tifs.2023.104310
54. Mandal AK, Mandal A (2024) Quantum dots as theranostic nano delivery system in combating various diseases. J Drug Deliv Ther 14(9):9. https://doi.org/10.22270/jddt.v14i9.6766
55. Mazumder S, Dey R, Mitra MK, Mukherjee S, Das GC (2009) Review: biofunctionalized quantum dots in biology and medicine. J Nanomater 2009(1):815734. https://doi.org/10.1155/2009/815734
56. McHugh KJ, Jing L, Behrens AM, Jayawardena S, Tang W, Gao M, Langer R, Jaklenec A (2018) Biocompatible semiconductor quantum dots as cancer imaging agents. Adv Mater 30(18):1706356. https://doi.org/10.1002/adma.201706356

57. Mendes R, Pedrosa P, Lima JC, Fernandes AR, Baptista PV (2017) Photothermal enhancement of chemotherapy in breast cancer by visible irradiation of gold nanoparticles. Sci Rep 7(1):10872. https://doi.org/10.1038/s41598-017-11491-8
58. Mitra RN, Doshi M, Zhang X, Tyus JC, Bengtsson N, Fletcher S, Page BDG, Turkson J, Gesquiere AJ, Gunning PT, Walter GA, Santra S (2012) An activatable multimodal/multifunctional nanoprobe for direct imaging of intracellular drug delivery. Biomaterials 33(5):1500–1508. https://doi.org/10.1016/j.biomaterials.2011.10.068
59. Mohammadi J, Hheidari A, Sardari S, Nouri M, Ebrahimi S, Rahdar A, Pishbin E (2024) A comprehensive review of lessons learned from quantum dots in cancer therapy. Biomed Mater 19(5):052004. https://doi.org/10.1088/1748-605X/ad68af
60. Mohkam M, Sadraeian M, Lauto A, Gholami A, Nabavizadeh SH, Esmaeilzadeh H, Alyasin S (2023) Exploring the potential and safety of quantum dots in allergy diagnostics. Microsyst Nanoeng 9(1):1–23. https://doi.org/10.1038/s41378-023-00608-x
61. Nguyen HA, Dou FY, Park N, Wu S, Sarsito H, Diakubama B, Larson H, Nishiwaki E, Homer M, Cash M, Cossairt BM (2022) Predicting indium phosphide quantum dot properties from synthetic procedures using machine learning. Chem Mater 34(14):6296–6311. https://doi.org/10.1021/acs.chemmater.2c00640
62. Panda S, Hajra S, Kaushik A, Rubahn HG, Mishra YK, Kim HJ (2022) Smart nanomaterials as the foundation of a combination approach for efficient cancer theranostics. Mater Today Chem 26:101182. https://doi.org/10.1016/j.mtchem.2022.101182
63. Panja A, Patra P (2023) A review on quantum dots (QDs) and their biomedical applications. 4open 6:1. https://doi.org/10.1051/fopen/2022020
64. Pellico J, Gawne PJ, de Rosales RTM (2021) Radiolabelling of nanomaterials for medical imaging and therapy. Chem Soc Rev. https://doi.org/10.1039/d0cs00384k
65. Petrovic S, Bita B, Barbinta-Patrascu M-E (2024) Nanoformulations in pharmaceutical and biomedical applications: green perspectives. Int J Mol Sci 25(11):11. https://doi.org/10.3390/ijms25115842
66. Probst CE, Zrazhevskiy P, Bagalkot V, Gao X (2013) Quantum dots as a platform for nanoparticle drug delivery vehicle design. Adv Drug Deliv Rev 65(5):703–718. https://doi.org/10.1016/j.addr.2012.09.036
67. Pu Y, Cai F, Wang D, Wang J-X, Chen J-F (2018) Colloidal synthesis of semiconductor quantum dots toward large-scale production: a review. Ind Eng Chem Res 57(6):1790–1802. https://doi.org/10.1021/acs.iecr.7b04836
68. Qi L, Pan T, Ou L, Ye Z, Yu C, Bao B, Wu Z, Cao D, Dai L (2021) Biocompatible nucleus-targeted graphene quantum dots for selective killing of cancer cells via DNA damage. Commun Biol 4(1):1–12. https://doi.org/10.1038/s42003-021-01713-1
69. Qu W, Zuo W, Li N, Hou Y, Song Z, Gou G, Yang J (2017) Design of multifunctional liposome-quantum dot hybrid nanocarriers and their biomedical application. J Drug Target 25(8):661–672. https://doi.org/10.1080/1061186X.2017.1323334
70. Robidillo CJT, Veinot JGC (2020) Functional bio-inorganic hybrids from silicon quantum dots and biological molecules. ACS Appl Mater Interfaces 12(47):52251–52270. https://doi.org/10.1021/acsami.0c14199
71. Rosenthal SJ, Chang JC, Kovtun O, McBride JR, Tomlinson ID (2011) Biocompatible quantum dots for biological applications. Chem Biol 18(1):10–24. https://doi.org/10.1016/j.chembiol.2010.11.013
72. Shahcheraghi SH, Shahcheraghi SH, Lotfi M, Lotfi M, Khaleghinejad SH, Tambuwala ZM, Mishra V, Mishra Y, Serrano-Aroca Á, A Aljabali AA, El-Tanani M, Naikoo GA, Chava SR, Charbe NB, Bharti S, Jaganathan SK, Goyal R, Negi P, Tambuwala MM, Folorunso O (2023) Photonic nanoparticles: emerging theranostics in cancer treatment. Ther Deliv 14(4):311–329. https://doi.org/10.4155/tde-2023-0011
73. Shao D, Li J, Pan Y, Zhang X, Zheng X, Wang Z, Zhang M, Zhang H, Chen L (2015) Noninvasive theranostic imaging of HSV-TK/GCV suicide gene therapy in liver cancer by folate-targeted quantum dot-based liposomes. Biomater Sci 3(6):833–841. https://doi.org/10.1039/C5BM00077G

74. Sun J, Liu F, Yu W, Fu D, Jiang Q, Mo F, Wang X, Shi T, Wang F, Pang D-W, Liu X (2021) Visualization of vaccine dynamics with quantum dots for immunotherapy. Angew Chem 133(45):24477–24485. https://doi.org/10.1002/ange.202111093
75. Sun J, Zhao H, Xu W, Jiang G-Q (2022) Recent advances in photothermal therapy-based multifunctional nanoplatforms for breast cancer. Front Chem 10. https://doi.org/10.3389/fchem.2022.1024177
76. Tabish TA, Scotton CJ, J Ferguson DC, Lin L, der Veen AV, Lowry S, Ali M, Jabeen F, Ali M, Winyard PG, Zhang S (2018) Biocompatibility and toxicity of graphene quantum dots for potential application in photodynamic therapy. Nanomedicine 13(15):1923–1937. https://doi.org/10.2217/nnm-2018-0018
77. Tade RS, Patil PO (2020) Theranostic prospects of graphene quantum dots in breast cancer. ACS Biomater Sci Eng 6(11):5987–6008. https://doi.org/10.1021/acsbiomaterials.0c01045
78. Wang J, Wang F, Li F, Zhang W, Shen Y, Zhou D, Guo S (2016) A multifunctional poly(curcumin) nanomedicine for dual-modal targeted delivery, intracellular responsive release, dual-drug treatment and imaging of multidrug resistant cancer cells. J Mater Chem B 4(17):2954–2962. https://doi.org/10.1039/C5TB02450A
79. Wang L-W, Peng C-W, Chen C, Li Y (2015) Quantum dots-based tissue and in vivo imaging in breast cancer researches: current status and future perspectives. Breast Cancer Res Treat 151(1):7–17. https://doi.org/10.1007/s10549-015-3363-x
80. Wang Q, Chao Y (2018) Multifunctional quantum dots and liposome complexes in drug delivery. J Biomed Res 32(2):91–106. https://doi.org/10.7555/JBR.31.20160146
81. Wang Q, Zhou Y, Song B, Zhong Y, Wu S, Cui R, Cong H, Su Y, Zhang H, He Y (2016) Linking Subcellular disturbance to physiological behavior and toxicity induced by quantum dots in Caenorhabditis elegans. Small 12(23):3143–3154. https://doi.org/10.1002/smll.201600766
82. Wu R, Tian G, Zhang S, Zhang P, Lei X (2024) A comprehensive review: versatile imaging probe based on chemical materials for biomedical applications. Appl Biochem Biotechnol. https://doi.org/10.1007/s12010-024-05043-w
83. Xu D, Guo D, Zhang J, Tan X, Deng Z, Hou X, Wang S (2024) Innovative tumor interstitial fluid-triggered carbon dot-docetaxel nanoassemblies for targeted drug delivery and imaging of HER2-positive breast cancer. Int J Pharm 657:124145. https://doi.org/10.1016/j.ijpharm.2024.124145
84. Yadav S, Jadaun SS, Suttee A (2023) Possible role of quantum dots as nanocarriers for targeted drug delivery. AIP Conf Proc 2800(1):020187. https://doi.org/10.1063/5.0162913
85. Yang N, Cao C, Li H, Hong Y, Cai Y, Song X, Wang W, Mou X, Dong X (2021) Polymer-based therapeutic nanoagents for photothermal-enhanced combination cancer therapy. Small Struct 2(11):2100110. https://doi.org/10.1002/sstr.202100110
86. Yi C, Yu Z, Ren Q, Liu X, Wang Y, Sun X, Yin S, Pan J, Huang X (2020) Nanoscale ZnO-based photosensitizers for photodynamic therapy. Photodiagn Photodyn Ther 30:101694. https://doi.org/10.1016/j.pdpdt.2020.101694
87. Yong K-T, Wang Y, Roy I, Rui H, Swihart MT, Law W-C, Kwak SK, Ye L, Liu J, Mahajan SD, Reynolds JL (2012) Preparation of quantum dot/drug nanoparticle formulations for traceable targeted delivery and therapy. Theranostics 2(7):681–694. https://doi.org/10.7150/thno.3692
88. Yu S, Xia G, Yang N, Yuan L, Li J, Wang Q, Li D, Ding L, Fan Z, Li J (2024) Noble metal nanoparticle-based photothermal therapy: development and application in effective cancer therapy. Int J Mol Sci 25(11):5632. https://doi.org/10.3390/ijms25115632
89. Yuwen L, Zhou J, Zhang Y, Zhang Q, Shan J, Luo Z, Weng L, Teng Z, Wang L (2016) Aqueous phase preparation of ultrasmall $MoSe_2$ nanodots for efficient photothermal therapy of cancer cells. Nanoscale 8(5):2720–2726. https://doi.org/10.1039/C5NR08166A
90. Zdobnova TA, Lebedenko EN, Deyev SM (2011) Quantum dots for molecular diagnostics of tumors. Acta Naturae 3(1):29–47
91. Zhang T, Guo S, Li F, Lan X, Jia Y, Zhang J, Huang Y, Liang X-J (2022) Imaging-guided/improved diseases management for immune-strategies and beyond. Adv Drug Deliv Rev 188:114446. https://doi.org/10.1016/j.addr.2022.114446

92. Zhang Y, Xiu W, Sun Y, Zhu D, Zhang Q, Yuwen L, Weng L, Teng Z, Wang L (2017) RGD-QD-MoS_2 nanosheets for targeted fluorescent imaging and photothermal therapy of cancer. Nanoscale 9(41):15835–15845. https://doi.org/10.1039/C7NR05278B
93. Zhou C, Long M, Qin Y, Sun X, Zheng J (2011) Luminescent gold nanoparticles with efficient renal clearance. Angew Chem (International Ed. in English) 50(14):3168–3172. https://doi.org/10.1002/anie.201007321
94. Zhou J, Yang Y, Zhang C (2015) Toward biocompatible semiconductor quantum dots: from biosynthesis and bioconjugation to biomedical application. Chem Rev 115(21):11669–11717. https://doi.org/10.1021/acs.chemrev.5b00049

Chapter 19
Microfluidic Approaches for the Controlled Synthesis and Multifunctional Applications of Quantum Dots

Saurabh Dubey, Srijita De, Rahul Deka, Jiwajyoti Mahanta, Tanusree Ghoshal, Prathu Raja Parmar, Nayanjyoti Kakati, Sanbed Das, Amit Kumar Singh, Omkar Suresh Deshmukh, and Dipankar Bandyopadhyay

Abstract Quantum dots (QDs) are nanomaterials with different size-dependent optoelectronic characteristics. These characteristics make them one of the essential materials for applications in health care, the environment, and energy. The present chapter talks in detail about the microfluidic platforms as an approach for the controlled synthesis of QDs, as it allows for systematic regulation of reaction parameters such as flow rates, mixing, and temperature, which ensures better control over the size of the QDs, their crystallinity, and their surface chemistry. The chapter starts by explaining the basic properties of QDs, microfluidic systems, and controlling parameters and illustrates how the micro-scale method improves nucleation and growth from the reagents to give QDs with preferred optical, electric, magnetic, and structural properties. Various QDs include semiconductors, carbon, metal, metal alloy, core–shell, core double shell, and perovskite can also be surface-modified and tuned based on the application requirement. The ease of adjusting QD properties like quantum confinement and photoluminescence by microfluidic production

---

S. Dubey · S. De · D. Bandyopadhyay (✉)
Centre for Nanotechnology, Indian Institute of Technology Guwahati, Assam, India
e-mail: dipban@iitg.ac.in

R. Deka · D. Bandyopadhyay
Jyoti and Bhupat Mehta School of Health Sciences and Technology, Indian Institute of Technology Guwahati, Assam, India

J. Mahanta · T. Ghoshal · P. R. Parmar · N. Kakati · O. S. Deshmukh · D. Bandyopadhyay
Department of Chemical Engineering, Indian Institute of Technology Guwahati, Assam, India

S. Das
Department of Chemical Engineering, Indian Institute of Technology Bombay, Mumbai, Maharashtra, India

A. K. Singh
School of Science (Biotechnology), Woxsen University, Telangana, India

T. Bhuyan et al. (eds.), *Multifunctional Quantum Dots*, Materials Horizons: From Nature to Nanomaterials, https://doi.org/10.1007/978-981-92-0172-3_19

of QDs helps in many applications. Additionally, the chapter addresses the various applications of microfluidic-synthesized QDs, especially biomedical, environment, and energy. In environmental research, QDs are key in contaminant detection and catalytic processes for water purification. In the energy sector, QDs are important in technologies such as QD solar cells (QDSCs) and QD LEDs (QLEDs), where their tunable properties increase device performance. Furthermore, the chapter discusses future approaches, including issues in scaling microfluidic synthesis for industrial applications, toxicity concerns, and the utilization of AI to optimize QD synthesis.

Keywords Microfluidics synthesis · Synthesis · QDs · Biomedical · Energy · Environment

1 Introduction to Quantum Dots

1.1 Overview of Quantum Dots (QD)

With the advancements in nanotechnology, in the age of rapid developments, nanoparticles (NPs) have been important in many applications for decades, whether it is medical, such as disease diagnosis and treatment, environment, such as detection of pathogens or energy, solar panels, energy storage, batteries, and fuel cells [1, 15, 94]. Nanomaterials have played a central role in the fabrication of new biosensors and in improving existing biosensing techniques for improved sensor characteristics such as specificity and sensitivity in diagnostics. QDs have progressed beyond laboratory research, affecting the commercial sector and transforming innovations into real-world applications. QDs are tiny particle sizes ranging from 5 to 20 nm, noted for their excellent photophysical properties due to their quantized energy states. These photophysical properties include the ability to emanate wavelengths of light determined by the size of QDs, along with high brightness and stability [72]. Significant advancements have been made in both theoretical and translational research with QDs in cellular, preclinical, and clinical settings [9]. These nanomaterials are employed in health diagnostics for fluorescent labeling and imaging [39, 70, 95, 112]. QDs exhibit a long fluorescence lifetime and superior fluorescence stability. Therefore, due to their excellent photoluminescence and quantum efficiency, QDs are used in biological applications [21].

1.2 Unique Optical and Electronic Properties of QDs

QDs display unusual optical and electrical capabilities due to quantum confinement phenomena, giving unique photonic and electronic behavior, which are highly dependent on the size and shape of the QDs. The quantum confinement effect develops

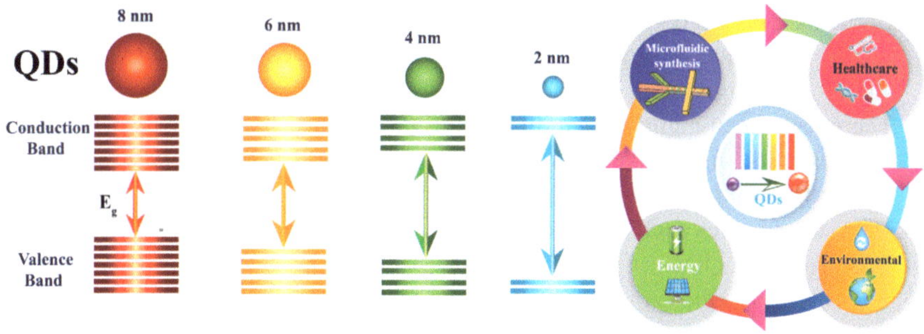

Fig. 1 Figure shows optoelectronic and size-based properties of microfluidic-synthesized QDs and their utilization in biology, environment, and energy applications

because the size of the particles is comparable to or less than the Bohr radius of electrons in the material, altering the behavior of electrons [31, 32]. These qualities include size-dependent emission, where smaller QDs emit light at shorter wavelengths (blue) and bigger QDs emit at longer wavelengths [99]. QDs also demonstrate high brightness and photostability, sustaining their fluorescence over time without appreciable photobleaching. They exhibit narrow and symmetrical emission peaks, allowing for fine color management with minimum overlap between various QD emissions, which is particularly important in biomedical and electronics [127, 155]. QDs have a broad absorption spectrum so that they can be activated by various wavelengths. Where smaller QDs emit light at shorter wavelengths (blue) and larger QDs emit at longer wavelengths (red), this makes them acceptable for many industrial applications, such as health care, environment, electronics, and energy [73]. Additionally, in biomedical applications, QDs' notable Stokes shift decreases excitation light interference, thus improving the signal-to-noise ratio for medical imaging [12, 140], diagnostics, optoelectronics, and quantum computing [31, 34, 99]. Lastly, QDs have high quantum yields and molar absorptivity, making them especially useful in light-emitting applications. The unique optical and electronic properties of QDs are shown in Fig. 1, illustrating color change concerning changing size and the corresponding change in the band gap.

1.3 Synthesis Challenges in Conventional Methods

The synthesis of QDs has undergone several developments in the last decade or so. Some of these challenges include:

a. The conventional methods face obstacles that affect their efficacy and application. One important challenge with the synthesis method is economics (e.g., cost) and complexity. The Cd-based chalcogenide QDs are one such example because of their complex core–shell arrangement [126].

b. Carbonaceous aggregation while carbonization in carbon quantum dots (CQDs), can be addressed through electrochemical synthesis or restricted pyrolysis but still requires significant attention.
c. There have been numerous studies investigating the CdX (X = S, Se, Te) group of QDs. Still, its possible toxicity and danger from heavy metal constituents hinder its employment in biological applications. Research and progress in developing alternatives such as cadmium-free QDs, perovskite QDs, carbon QDs, and biocompatible QDs.
d. The solution to these problems lies in the advancement of biocompatible QDs.
e. Careful synthesis and post-treatment changes are required for tuning surface properties of QDs, which are critical for their specific functioning.
f. Post-treatment procedures such as centrifugation and gel electrophoresis may be used for uniformity and size control.
g. However, one of the tough challenges with the conventional method of QD production includes restricted control over variable reaction parameters between batches and scalability concerns. On the other hand, microfluidics provides improved control over reaction, better control over QD photophysical properties, and minimizes the constraints associated with conventional approaches [5, 73].

2 Microfluidic Platforms for Quantum Dot Synthesis

2.1 Principles of Microfluidics: Fundamentals and Mechanisms

In nanotechnology, microfluidics channels are devices with very small diameters or cross-section capillaries where various unit operations or unit processes can be carried out simultaneously. These devices are recognized for their tiny size, low reagent consumption, and ease of incorporating multiple processes on a single chip. The overall transport phenomenon (e.g., control over fluid dynamics, heat transfer, mass transfer, and reaction) can be achieved efficiently in these devices. In these channels, physical, chemical, and biological interactions can be manipulated [139]. Microfluidic synthesis utilizes a bottom-up fabrication technique. This strategy relies on carefully regulating the synthesis conditions, where precursor components mix to form nanoparticles. The main issues with microfluidic synthesis are,

a. Keeping the balance between solvent and anti-solvent interactions (hydrophobicity).
b. The formation of ion pairs through charge interactions.
c. Easy hydrophobic chelation to happen through coordinate covalent bonds.

In recent advancements, microfluidics has been exploited as a synthesis platform for QDs [105]. The coordination between microfluidics and QD synthesis provides significant benefits, including improved control over particle size, composition, and surface features, hence enabling the creation of high-quality QDs for varied applications [21, 70, 71].

Here is a brief timeline to inform readers of the progress in the microfluidic synthesis of QDs. The first report on microfluidic CdSe QDs synthesis in continuous phase came in 2002 [76]. Subsequently, other chip-type microreactors were used for QD synthesis in 2004 [48]. Microfluidic synthesis of QDs of III–V group was achieved in 2009 [11, 82], followed by other discoveries such as Cu–In–Zn–S/ZnS core–shell QDs in 2013 [60]. This is the same year when dual stage microreactors have been developed to synthesize QDs with high PL. Subsequently, the microfluidic synthesis of perovskite QDs in 2016 and microfluidic chip reactor to produce various types of QDs in 2018 joined the queue of advancement. Furthermore, millisecond CdS QDs' nucleation and growth studies in microfluidics with in-situ spectroscopy in 2019 [102], and very recently on demand microfluidic synthesis of QDs in digital droplet reactors [98]. The microfluidic-based synthesis mechanisms have been extensively employed for the synthesis of CdS, CdSe [77], ZnO [87], ZnSe [147], InP [79], Ag [41], Au [129], Co [113, 114], TiO_2 [33], SiO_2 [48], Fe_xO_y [33], $LaPO_4$ [2, 3], $BaSO_4$ and many core–shell QDs (e.g., Fe_2O_3/SiO_2 [2, 3], CdSe/ZnS [50], and CdSe/ZnSe/ZnS [59]). There are some Core/Shell1/Shell2 type syntheses of QDs, such as InAs/CdSe/ZnSe [77]. The intermediate CdSe buffer layer minimizes the strain between the InAs core and the ZnSe outer shell, and this helps in achieving higher fluorescence performance [7].

2.2 *Microreactor Design for QD Synthesis*

Microfluidic reactors have emerged as a viable method for nanocrystal production since their inception in 2002 [91]. The use of microfluidics is an efficient way of preparing QDs [135]. A considerable body of literature exists where surface modification and functionalization have increased the use of QDs in specific therapeutic, drug delivery, and diagnostic systems [89]. Among many conventional synthesis methods for making QDs of semiconductors, metals, metal alloys, perovskite, carbon, graphene, and surface modifications, microfluidics proved to be the best [122, 139]. As discussed in the last section, it provides better control over various QD properties by controlling parameters [21]. The base materials for making microfluidic channels/reactors can include Polydimethylsiloxane (PDMS), glass, silicon, thermoplastics, PMMA, PTFE, and PC.

Polydimethylsiloxane (PDMS) is used because of its flexibility, biocompatibility, and transparency [28]. Silicon and glass are preferred because of their good mechanical strength, chemical resistance, thermal stability, transparency, and easy handling [8]. Mostly, thermoplastics are economical compared to glass or silicon.

Furthermore, they can be easily fabricated. Most previous works showed that surface modification and functionalization have applications of QDs in healthcare, drug delivery, and diagnostic imaging. Mostly, these modifications can be quickly

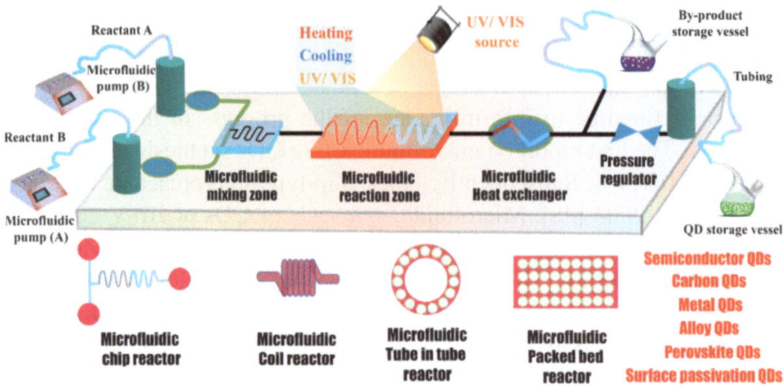

Fig. 2 The illustration shows the various microfluidic synthesis approaches, types of reactors, and sub-integrations for various QD synthesis

done using microfluidic reactors. Consequently, microfluidics has become a central step in the production of QDs. We have already discussed many facets of microfluidic reactors and why they are so important in real-world applications. Typically, microfluidic microreactor designs integrate elements like micromixers, temperature controllers, and pressure controllers to maintain consistent synthesis conditions. We will see in the upcoming section how, by fine-tuning flow rates and other reaction variables, microfluidics may efficiently manufacture good QDs while minimizing waste [126]. Figure 2 shows the various microfluidic synthesis approaches, types of reactors, and sub-integrations for various QD syntheses.

Continuous microfluidic reaction systems can be categorized into two types, which are illustrated as follows:

Capillaries Type: Simple capillaries of inner diameter 100–1000 μm, with multiple inlets for reactants infused through multiple microfluidic pumps. As discussed above, the materials for the construction of capillaries are PDMS, glass, PTFE, silicone, PMMA, and PC due to their chemical inertness and thermal stability. Mixing the reactants can be achieved through various active methods (e.g., acoustic, electric, or magnetic stirrers), and passive (e.g., tributary junctions or serpentine channels). Thermal management is done by heating oil baths and allowing the microfluidic coils containing the reaction mixture to pass through.

Solid host or "chip" to house the channels: This is also known as host or chip-type channels. These microreactors have dimensions similar to the capillary type (e.g., inner diameter 100–1000 μm). These microfluidic reaction systems can be made using materials like glass, silicon, PDMS, PMMA, PVDF, and PC.

2.3 *Controlling Reaction Parameters: Flow Rates, Temperature, and Mixing*

In microreactors for QD synthesis, reaction factors such as flow rates, temperature, and input mixing play a part in determining the quality of QDs. The design of microfluidic microreactors plays a key role in improving these factors [80]. The volumetric flow rate of the reactants through the microfluidic channels is a key measure, as it affects the residence time of the reactants in the reaction region, affecting the nucleation and growth of the QDs. Reactant infusion speed allows for the tight control of the residence time of reactants within the microreactor. The volumetric flow rate determines the residence time for contacting the reaction mixture, so it controls the size and dispersity of the QDs. A higher flow rate means less contact time, and vice versa [53]. Thermal management can also be done through modulating the flow rate along with the reaction time.

The main reason for the dependence of the temperature on the QDs' properties is that it directly alters the reaction mechanism, changes the rate constant, changes the reaction equilibrium, changes the equilibrium constant, and, in some cases, the catalytic activity can also be altered. That finally alters the kinetics and thermodynamics of the process and determines the yield and selectivity of synthesized QDs. Microfluidic systems involve localized heating elements, hot plates, and external laser power to control temperature conditions. The yield and selectivity depend on the nucleation and growth of the synthesized QDs. Usually, higher temperatures stimulate faster nucleation and growth, big size QDs, while lower temperatures slow down the processes of stimulation, resulting in smaller QDs [18, 26, 66, 69, 103]. Apart from flow throughput and temperature, the contacting pattern of the reactants also plays an important role in deciding the performance of the synthesis method. Proper mixing of the reaction mixture is important for cluster nucleation and growth [121, 147]. The microfluidic systems equipped with various mixing strategies, such as passive mixers (e.g., the serpentine channels) and active mixers (e.g., acoustic, magnetic stirrers, electric) [29], provide consistent mixing and help in the tuning synthesis process. Additionally, the suitable mixing within microfluidic configurations helps in the proper spread of reactants, which improves the quality and output of the synthesized QDs. Mixing confirms that the whole reaction mixer is evenly distributed, leading to consistent quality QDs [106]. These reaction parameters are optimized to achieve the desired physicochemical properties and unique photonic and electronic properties of the QDs [58, 101, 136].

2.4 *Comparative Analysis: Batch Versus Microfluidic Synthesis of QDs*

The microfluidic reactors have advantages over typical tank reactors, including shorter reaction times and easier separation and purification operations. In all cases, microfluidics has been advantageous in comparison to bulk synthesis. Its ability is to fine-tune the physical, chemical, optical, electronic, and magnetic properties of the synthesized QDs [27]. The traditional method includes Batch synthesis, which involves mixing reactants in a large vessel. Batch method suffers from the main difficulties that are variability in particle size and quality due to less control over reaction conditions [51, 87]. Non-uniform mixing and thermal management will directly affect the consistent properties of the QDs produced. In this problem, microfluidic synthesis provides reasonable control over reaction parameters like flow rates, temperature, and mixing [115]. This control ensures the synthesis with homogeneous size and properties, improving repeatability and consistency. Additionally, microfluidic systems are scalable without compromising quality [6]. The continuous nature of microfluidic synthesis also ensures live monitoring and surface modifications, significantly providing great efficiency and quality of the synthesis process [51, 148]. The uniform and consistent ligand coverage during the synthesis minimizes batch-to-batch variability, resulting in good-quality QDs with controlled surface functionalities [87].

Feature	Microfluidic synthesis	Batch synthesis (traditional)
Reaction time	Shorter	Longer
Separation and purification	Easier	More challenging
Control over properties	High (fine-tuning of physical, chemical, optical, electronic, and magnetic properties)	Low (less control over reaction conditions)
Particle size and quality	Uniform, consistent	Variable, inconsistent
Mixing efficiency	Uniform	Non-uniform
Thermal management	Better	Poor, leading to inconsistencies
Reaction control	Precise (flow rates, temperature, mixing)	Limited control
Repeatability and consistency	High	Low
Scalability	Scalable without quality loss	May compromise quality
Live monitoring	Possible	Not possible
Surface modification	Enhanced (better ligand coverage)	Less controlled
Batch-to-batch variability	Minimal	High
Overall efficiency	High	Lower

2.5 *Tailoring Quantum Dot Properties Using Microfluidics*

In microfluidic QD synthesis is achieved by controlling flow rates, reaction times, and reactant concentrations within microfluidic channels. The surface chemistry of

QDs, which affects their stability and interactions, can also be modified [21]. As we know, surface chemistry modification is an interplay of surface energy and the wettability of the various analytes towards each other. Using these interfacial factors, the synthesis of surface-functionalized QDs can be achieved [70, 112]. This makes them more soluble and biocompatible for medicinal purposes and electronic devices. Fine-tuning of bandgaps and photoluminescence features enables excellent control of properties. Additionally, the method ensures better crystallinity by minimizing structural defects and irregularities [71].

3 Microfluidic Synthesis of Different Types of Quantum Dots

3.1 Semiconductor Quantum Dots

In this section, we will discuss semiconducting QDs. They can be classified by their composition, for example, the Group II-VI QDs (e.g., CdSe, CdS, ZnS) [77], Group III-V QDs (e.g., InAs, InP) [79], Group IV QDs (e.g., Si, Ge) [23, 138], and Perovskite QDs (e.g., $CsPbBr_3$, $MAPbI_3$). These QDs absorb light and re-emit it in a distinct color, with this emitted color being dictated by their size and composition, as the detailed mechanism of which is discussed in the previous Sect. 2.1 [99]. The microfluidic synthesis has been instrumental, providing exact control over reaction settings to optimize the size and optical properties of the dots [51]. As previously discussed, the microfluidic platform helps control reaction conditions (thermal control, mixing, and residence time). Among the above-discussed semiconducting QDs. Cadmium-based QDs, including CdSe and CdS, are among the most thoroughly researched [77]. Their popularity arises from their highly adjustable optical characteristics, rendering them suitable for various applications, including displays and imaging [77]. Silicon QDs, generally produced with silane precursors, also specialize in microfluidics [74]. Microfluidic systems under controlled conditions provide exact size regulation and consistency. These characteristics are important for Si QDs in bioimaging and optoelectronics applications [23]. QDs made up of Indium Arsenide (InAs) and Germanium (Ge) face challenges in microfluidic synthesis. Their complex precursor needs and opposing reaction kinetics render them less frequently examined with this method [13]. With the development in microfluidics, many such limitations may become more controllable, providing unique opportunities in synthesizing such complex QDs [22].

3.2 Carbon Dots and Graphene Quantum Dots

Carbon quantum dots (CQDs) are mostly adopted for their biocompatibility, photoluminescence, low cost, and reduced toxicity. Conventional methods for making CQD are microwave, ultrasonic, plasma, magnetic, and hyperthermia [87]. In the last few sections, we have already discussed the importance of microfluidic synthesis, so now we will discuss how this technology has been used for carbon dots in recent decades. In a previous study, Xiang-Yun Du and the team devised methods for producing CQDs on a large scale and conjugating them with polymers [27]. Research groups have widely explored graphene-based nanomaterials, including graphene oxide (GO), reduced graphene oxide (rGO), and graphene-based quantum dots (GQDs/GOQDs) for healthcare applications. The unique properties of the GQDs/GOQDs are tunable fluorescence, their quenching, reactivity with oxygen-containing functional groups, π–π orbital interactions at interfaces, small size (<100 nm), high surface-to-volume ratio, affordability, nontoxicity, biocompatibility, and environmental friendliness [144]. With the decades of advantages now microfluidic methods are now advantageous for synthesizing both CQDs and GQDs/GOQDs [78, 94, 124, 130]. Furthermore, the conjugation of CQDs and GQDs/GOQDs with polymers creates a composite material with improved properties [27].

3.3 Metal and Alloy Quantum Dots

The unique properties that a metal or alloy-based QDs exhibit are the plasmonic properties that can be utilized for the surface-enhanced Raman scattering. This happens as a result of molecular level light matter interaction. Plasmonic properties arise due to the strong collective oscillation of surface conduction electrons in metallic nanostructures when excited by light [14, 143]. In surface plasmon resonance (SPR), electrons oscillate coherently at a metal–dielectric interface, creating evanescent waves sensitive to changes in the local environment. This provides SPR and localized surface plasmon resonance (LSPR) properties that, in turn, change the absorbance and Raman scattering [14]. LSPR occurs in nanoparticles (e.g., gold, silver) where confined electrons generate intense, localized electromagnetic fields near the surface, enhancing light absorption and scattering. These resonances depend on the metal's dielectric function, nanoparticle geometry, and surrounding medium [14, 143]. The MQDs and AQDs, specially made of gold, silver, platinum, copper, palladium, aluminum, or their combination, show excellent surface-enhanced Raman scattering properties [118]. The enhanced near-field effects from LSPR amplify Raman scattering (SERS) by increasing interaction cross-sections, making metallic quantum dots (MQDs) and alloy quantum dots (AQDs) effective sensors [118].

As discussed earlier, the microfluidic synthesis method can be adapted to produce metal and alloy quantum dots (AQDs). MQDs, like those made from Ag, Au, Cu, and Pt, exhibit strong plasmonic effects, making them useful for applications in diagnostics and imaging. Recently, many applications involving these metal nanoparticles revolve around diagnostics and environmental applications. The shift from the colloidal synthesis to microfluidic synthesis will serve as an alternate method for the synthesis, increasing tunability and reaction control. Alloy QDs, which combine different metals, can be engineered to have tunable band gaps and enhanced stability, broadening their use in optoelectronics and catalysis. Much previous literature shows the synthesis and application of MQDs of a single metal, like Au [129], Ag [41], or Co [114]. However, it is not only the plasmonic effect that makes them unique, but also their photothermal effects that are utilized for various healthcare applications.

Now, we will discuss AQDs made of a combination of metals. A series of previous studies indicated AQDs such as Cu-doped Zn–In–Se [17], CdSeTe [30], and Co-doping of Ag into Mn:ZnSe [42]. An easy one-spot synthesis method was proposed by Wei and the team to manufacture Ag–Cu–Ga–Se/ZnSe (ACGSe/ZnSe) QDs. These AQDs showcase an enhanced photoluminescence quantum yield (PLQY) of 71.9% with a spectral emission ranging from 510 to 620 nm. This experiment reveals the dependence of QD qualities on the composition of metals. The researchers have concluded that the defect levels decrease with increasing Ag concentration, which slows down the ZnSe shelling process, owing to increased lattice distortion. Furthermore, these ACGSe/ZnSe QDs are employed in light-emitting diodes for various lighting applications [137].

3.4 *Perovskite Quantum Dots*

Another class of QDs exhibiting excellent optoelectronic properties is called Perovskite quantum dots (PQDs). These PQDs render some unique qualities, including good defect tolerance, adjustable emission wavelengths, long charge carrier lifetime, and easy charge mobility. These superior qualities of PQDs outperform the chalcogenide QDs in numerous applications [40], such as light-emitting diodes (LEDs), solar cells, and photodetectors.

The chemistry of PQDs is also fascinating.

- A is a monovalent cation,
- B is a divalent metal cation, and
- X is a halide anion (Cl, Br, or I).

As PQDs can withstand imperfections, they have exceptional light-emission efficiency and high open-circuit potential [40]. As discussed earlier, the microfluidic approach is superior to conventional approaches. Therefore, microfluidic techniques provide precise control over the reaction environment, enabling continuous and scalable production of PQDs with uniform size and composition for their vast applications [19]. Abdel-Latif and colleagues demonstrate a method for tuning the bandgap of PQDs at room temperature using a single solvent. Modular flow chemistry involves halide exchange reaction and colloidal synthesis, where the method analyzes the effect of ligand composition and halide salt transfer on anion exchange reaction kinetics and tuning of the bandgap. Cesium lead tribromide ($CsPbBr_3$) solution and in situ optical spectroscopy have been used for the analysis [1].

Ning et al. synthesized blue-emitting perovskite $CsPbBr_3$ QDs utilizing a microfluidic chip technique. Microfluidics allowed for the precise control of various controlling parameters, including the ratio of Cs to Pb. Consequently, blue-emitting $CsPbBr_3$ QDs with a photoluminescence quantum yield (PLQY) of 41.9% were synthesized. These QDs benefited from enhanced homogeneity, scalability, and accurate emission spectrum tuning from blue to green (463–522 nm) [81]. Another literature work by Cheng et al. [20] synthesized PQDs using the microfluidic electrospinning approach where PQDs are embedded in polymer core–shell nanofiber films. This method made red, green, and blue emissive nanofiber films with high photoluminescence quantum yield and stability. Fabricated color-tunable fluorescent formamidinium lead halogen ($FAPbX_3$, X = Cl, Br, I) PQDs embedded in polymer core–shell nanofiber films can be used for a variety of applications, including displays and optoelectronics [20].

3.5 *Surface Chemistry, Ligand Exchange, Surface Passivation, and Functionalization in Microfluidic Synthesis*

The QDs have a big problem with stable fluorescence and how they conduct electricity because of the oxidation that might harm them. The surface of the QDs is crucial as it dictates all the QD's properties. As we know, QDs can be modified by various chemical and biological entities. To prevent such oxidative damage, they frequently undergo surface chemistry modifications via ligand exchange, surface passivation, and functionalization. These types of customizations by attachment of molecules (ligands) are known as functionalization [89]. Researchers have utilized functionalized QDs with specific affinity ligands to detect biomolecules or imaging targets as fluorescent labels. Du and the team mixed CQDs with polymers by combining CDs with polymers. The team builds composites beyond the sum of their components [27]. These hybrids boost mechanical strength and enhance photoluminescence, making them perfect for LEDs or display applications. The stability of QDs is critical for long-term usage in biomedicine or energy storage [83]. Imagine flexible, brighter, and practically indestructible LED screens, or think about printing stunning images that will not fade or lose their vibrancy with age [27, 89].

4 Recent Advancements in Microfluidic Quantum Dot Synthesis

4.1 Integration of AI and Machine Learning for Process Optimization

The breakthroughs that have taken place in the previous few decades are currently increasing further with the aid of the recent integration of AI and Machine Learning (ML). A recent work by Chen and colleagues merged AI/ML with microfluidics to manufacture PQDs. They found that this combination helps fine-tune QD synthesis factors to regulate QD characteristics precisely. The process enables scalability and opens a way to utilize new material synthesis technologies [19]. These integrations and automatic control help monitor and enhance the synthesized QD characteristics. The merger of microfluidic technology with AI-driven development marks a considerable advance in nanomaterial manufacturing. AI algorithms examine massive datasets from microfluidic experiments. They determine the ideal circumstances for quantum dot (QD) synthesis and anticipate results reliably. Machine learning models respond to fresh data, enhancing the synthesis process. This connection enhances the efficiency and accuracy of QD manufacturing.

4.2 Sustainable and Green Synthesis Approaches

Recent advancements in microfluidic QD synthesis have increasingly focused on sustainable and green approaches. Researchers are developing methods that minimize toxic solvents and heavy metals, opting for environmentally friendly materials and processes. For instance, water-based synthesis and biodegradable polymers are becoming more prevalent, reducing the environmental impact of QD production because of the toxicity issues associated with chemically synthesized QDs. Green methods for synthesizing biocompatible QDs are preferred. Using living organisms (prokaryotes to eukaryotes) for QD synthesis has already been explored to synthesize biocompatible QDs so they can be comfortably used in biological and biomedical applications. The less harsh synthesis condition of biosynthesis methods paves the way for better biostability and other properties of QDs. Additional encapsulation or cap exchange steps are unnecessary in this green synthesis route. So, various steps like water solubilization and biofunctionalization of chemically synthesized QDs are carried out [159].

There are three ways to make biocompatible QDs,

a. Synthesis of QDs in biological systems
b. Synthesis of QDs in biomimetic systems
c. Biofunctionalization of chemically synthesized QDs

On the other hand, carbon quantum dots (CQDs) and graphene quantum dots (GQDs) are another class of biocompatible QDs. Ultimately nonvulnerable green methods can quickly produce these. These methods require one eco-friendly material as a carbon source. Previous research utilization of tomato, cotton linter, lemon and onion juice, pomelo peels, finger millet ragi, broccoli, and sweet pepper have been used as carbon sources [24].

4.3 Advanced Characterization Techniques in Microfluidic QD Synthesis

The development of characterization tools has strengthened recent advances in microfluidic QD production. Accurate and consistent manufacturing is ensured by using these approaches, which allow for the real-time monitoring and study of QD characteristics. To get real-time information on the size, structure, and composition of QDs, researchers may use in-situ spectroscopy to watch their optical characteristics as they develop. Furthermore, the structural and morphological properties of QDs at the nanoscale may be thoroughly understood with the use of modern imaging techniques like atomic force microscopy (AFM) and transmission electron microscopy (TEM) [126, 155].

5 Applications of Microfluidic-Synthesized Quantum Dots

Microfluidic-synthesized QDs offer various applications owing to their carefully regulated characteristics and high-quality manufacture. In bioimaging, these QDs provide higher brightness and stability, making them suitable for monitoring cellular activities and detecting illnesses. The versatility and precision of microfluidic synthesis make these QDs ideal for complex technological applications, boosting innovation across several fields. Figure 2 illustrates different microfluidic synthesis methods, types of reactors, and sub-integrated systems used in the synthesis of various quantum dots (QDs). These methods are used in applications such as biology, the environment, and energy.

5.1 Biological Applications

5.1.1 Bioimaging and Fluorescence Labeling

The QDs synthesized from the microfluidic pathways are changing the disciplines of bioimaging and fluorescent labeling due to their exceptional optical properties and biocompatibility. These QDs have outstanding brightness, photostability, and customizable emission wavelengths, making them appropriate for comprehensive and long-term imaging of biological processes at the cellular and molecular levels. In bioimaging, QDs may be used to observe the movement and interaction of cells, monitor gene expression, and examine the distribution of biomolecules with exceptional precision.

The changes in the levels of cellular pH may be a detrimental cause for many different ailments in the human system. A study wherein a solution of long-sustained fluorescence-based QDs was generated employing $AgInS_2$ (AIS)/ZnS, which was able to discern the changes in the cellular pH levels. The living cells are collected from different sources and cultured in varied pH settings. After that, AIS/ZnS-QDs are integrated to image living cells in fluorescence lifetime imaging microscopy (FLIM). Normal cells and cancer cells are investigated simultaneously in fluorescence lifetime imaging incorporating AIS/ZnS-QDs, and a comparison between them indicates a difference of approximately 100 ± 7 ns. Further, it is found that the fluorescence lifetime difference is 115 ± 9 ns compared to normal cells when actual cervical cell samples are taken from 20 patients. The real sample line denotes a fluorescence lifetime difference of 64 ± 9 ns when captured in AIS/ZnS-QDs-based FLIM [119].

Another study focused on the CD59 cancer marker, which is important for the tumor cells to bypass immune cell surveillance, instigate poor prognosis, and become resistant to cancer-based therapies. Their detection was facilitated by preparing polyethyleneimine (PEI) functionalized graphene quantum dots from weathered lemon leaf extract. The graphene-based quantum dots facilitated the detection of CD59 in the sample range of 0.01–40.0 ng mL^{-1} in the detection limit at 5.3 pg mL^{-1}. The cellular bioimaging using graphene-based quantum dots was tested in the A549 cell line using different lasers. Wavelength ranges of 540 nm (red), 405 nm (blue), and 488 nm (green) showed well-favored accumulation of the QDs in the nucleus and cytoplasm, showing the new potential for using QDs in bioimaging [123]. Similarly, in another study, quantum dots synthesized from carbon skeleton, i.e., carbon dots, have been designed to target CD71, a vital biomarker to target cancer cells precisely due to their alleged role in iron uptake. The quantum yield was around 8% when the carbon dots were tested against the HL60 cell line expressing the CD71 marker [117].

5.1.2 Multiplexed Biological Assays

Microfluidic synthesis of QDs has revolutionized the field of multiplexed biological assays by enabling the production of high-quality QDs with exceptional optical properties. Wang et al. in 2021 created Graphene Oxide Quantum Dot-based Microfluidic Immunofluorescence Assay (GIMIFA), a universal biosensing technology merging solid-phase GOQDs with a microfluidic immunofluorescence assay chip for ultra-sensitive, fast, and high-throughput detection of numerous serum biomarkers [135]. By integrating the nanostructure and fluorescence quenching capabilities of GO QDs with the efficiency of microfluidics, GIMIFA optimizes analyte mass transfer, improves the signal-to-noise ratio, and enables multiplex detection with minimal time and reagent use. Clinical validation with samples from breast cancer patients and healthy individuals proved its accuracy, establishing GIMIFA as a versatile tool for disease detection and biological research. The GIMIFA achieves strong and uniform immobilization of antibody/antigen barcode arrays through simple π–π stacking with GOQDs assembled on the substrate. It can simultaneously detect multiple biomarkers from at least 60 clinical samples in just 40 min, offering a wide linear quantitative range (5 pg–0.5 mg or 0.5–5000 U/mL), a low detection limit (1 pg/mL or 0.01 U/mL), exceptional repeatability with a relative standard deviation under 5% across 10 tests, and minimal consumption of detection reagents and test samples (2 mL) [135].

Importantly, the point-of-care testing (POCT) devices are important for faster illness diagnostics and reaching a broader population restricted from superior medical facilities. A paper-based analytical device was developed to detect lung cancer biomarkers, namely carcinoembryonic antigen (CEA) and neuron-specific enolase (NSE), in a multiplexed fashion. $CsPbI_3$ perovskite quantum dots were conjugated with streptavidin and specific antibodies for multiplexed detection via the sandwich immunoassay principle through phase engineering, as shown in Fig. 3. The detection limits for CEA and NSE were recorded to be 0.095 ng/mL and 30.0 ng/mL, respectively, in a single trial and 0.12 ng/mL and 32 ng/mL simultaneously [100].

Mental disorders such as Attention-deficit/hyperactivity disorder (ADHD) can compromise childhood, along with adulthood, in some people. ADHD is such a disorder wherein the target is multifaceted [92]. The multiplexing and droplet manipulation can be assisted by microfluidic assays. One such study examines multiple biomarkers that have to be targeted simultaneously to confirm the disorder, for which the QD-based fluorescent immunoassay-based platform was developed. The multiplexed detection of ferritin, interleukin-6, tumor necrosis factor-alpha, and brain-derived neurotrophic factor was possible using different spectral regions of fluorescence of QDs. The detection limit for the assay platform came to be in the range of 0.021–0.068 pg/mL. The QDs-based assay platform showed good sensitivity in terms of the biomarkers, and the reproducibility rate for the platform was more than 90%, giving insights into its use in the clinical setup [107]. However, there is less literature on microfluidic synthesis with in situ biological application. Usually, already synthesized QDs are employed for the analysis. A more generic approach would be to combine both synthesis and applications on a single chip, which remains an open area of research for the future.

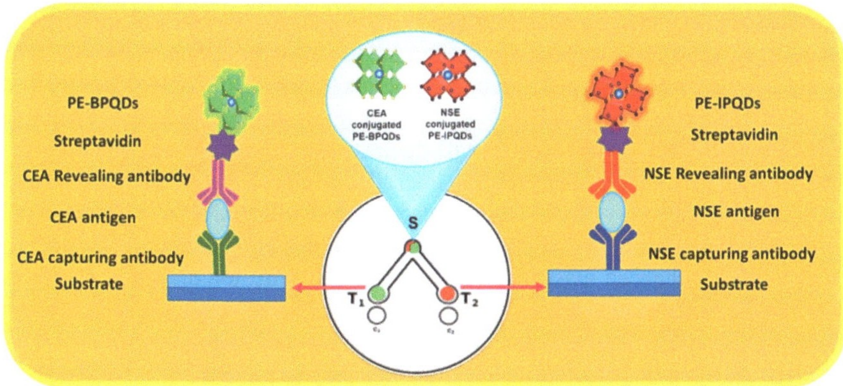

Fig. 3 The illustration shows the multiplexed detection of lung cancer biomarkers carcinoembryonic antigen (CEA) and neuron-specific enolase (NSE). Reprinted from Sanjayan et al. [100], Copyright (2023), with permission from Elsevier

Magnetic QDs barcode is a unique arrangement of magnetic nanoparticles (MNPs) combined with the fluorescence capability of QDs for enhanced sensitivity and multiplexed detection of various analytes/biomarkers. A study used Fe_3O_4/TiO_2 core/shell MNPs and CdSe/ZnS QDs to design and develop such barcodes. The QDs-MNPs barcode system showed a high sensitivity for the simultaneous detection of four cancer-related markers namely cancer antigen 199 (CA199), cancer antigen 125 (CA125), alpha-fetoprotein (AFP), and neuron-specific enolase (NSE) with detection limits of around 0.72 KU/L, 0.89 KU/L, 0.05 ng/mL, and 0.15 ng/mL, respectively [154]. Similarly, the multiplexed screening of respiratory-related viruses, namely, SARS-CoV-2, influenza A virus and human adenovirus, was facilitated by developing a lateral flow immunoassay (LFA) biosensing platform. The LFA was designed by continuously layering highly photostable CdSe@ZnS–COOH QDs onto a graphene oxide substrate, providing an enhanced reaction surface. They have been reported to be 8 pg/mL, 488 copies/mL, and 471 copies/mL for SARS-CoV-2, influenza A virus, and human adenovirus, respectively. The procedure gave results within 20 min, with specificity, reliability, and repeatability [133].

5.1.3 Drug Delivery and Therapeutic Monitoring

The microfluidic production of QDs has opened new vistas in drug delivery and therapeutic monitoring. These QDs, created using precise microfluidic processes, have exceptional qualities such as excellent stability, customizable size, and surface functionality, making them great candidates for targeted drug delivery systems. Researchers have created microfluidic devices for multiplex detection of antibody-antigen interactions, utilizing microsphere light scattering for identification and QD emission for coding [62, 90, 111]. Microfluidic systems with bead-based microarrays have also been explored for gene mutation analysis and other biomedical applications. Zhang et al. [156] have created a microfluidic device integrating a microbead array for sensitive detection and genotyping of hepatitis B virus (HBV) DNA directly from the serum extract as shown in Fig. 4. This device integrates rapid microbead binding, efficient microfluidics, and sensitive QD fluorescence detection to create a universal platform for precise DNA analysis. The microfluidic chip achieved a detection sensitivity of 0.03 nM for synthetic DNA targets labeled with TAMRA, equivalent to 1 $\times$ 10^{-16} moles in a 10 μL sample, beating off-chip hybridization, which detected only 0.1 nM. Enhanced mass transport and reduced diffusion distances within the microfluidic microbead array significantly improved sensitivity. QDs as fluorescent labels significantly enhanced assay sensitivity, achieving a detection limit of 4 pM for synthetic DNA targets. This enabled the device to detect the HBV virus at concentrations as low as 1×10^3 copies/mL and successfully identify genotypes in serum samples by targeting in vitro transcribed RNA labeled with QDs [156].

Kim et al. [52] have developed an innovative way to capture and image HIV using a microfluidic device that requires only 10 μL of whole blood from HIV-infected patients. This clever system captures HIV particles directly on-chip and allows rapid imaging in 10 min. The device and surface mechanism are shown in Fig. 5. To ensure the presence of HIV particles, they employed two types of fluorescent nanoparticles for dual staining,

a. **Qdot525 streptavidin**
   - Conjugated to a biotinylated anti-gp120 antibody,
   - Targeting the HIV envelope glycoprotein.
b. **Qdot655 streptavidin**
   - Conjugated to biotinylated Concanavalin A (ConA) lectin,
   - Binds specifically to mannose residues on the virus surface.

The microfluidic fabricated using polymethylmethacrylate (PMMA) on glass created a chip for capturing and imaging HIV particles from unprocessed patient blood. This dual-labeling strategy gives clear, countable fluorescence microscopy images of viruses [52].

In another work, researchers used a Y-shaped microfluidic channel where the DMSO/water interface was utilized as a precursor to produce a dipeptide hydrogel. These hydrogels are further embedded in premixed porphyrins and water-soluble QDs. The confined microfluidic channel gelation and entrapment happen, leading to well-controlled shape and distribution. This method was better than the test tube-based method, with better self-assembly and fewer structural flaws [63]. Localized delivery of the drugs to the affected tumor tissues is known to be the most effective way of treating cancer. Chitosan-nitrogen embedded graphene QDs have recently been utilized as a nanocarrier because of their pH responsiveness. The nanocarriers loaded with doxorubicin were utilized and tested with the A549 lung cancer cell line. It has been noted that within 12 h, the drug added internalized into the nucleus because of gradual cellular uptake [57].

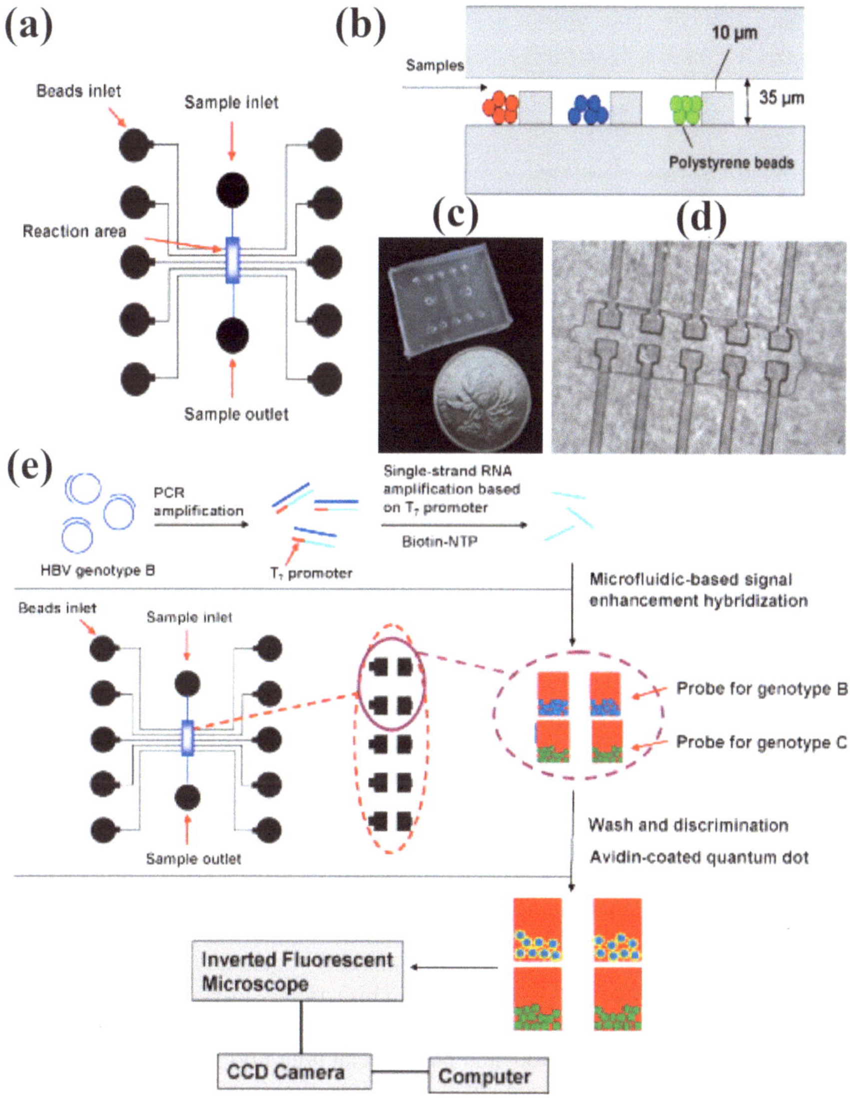

Fig. 4 The figure shows a microfluidic device integrating a microbead array for sensitive detection and genotyping of hepatitis B virus (HBV) DNA directly from the serum extract. **a** Channel schematic, **b** a side view of the channel, **c** a real fabricated channel, **d** shows an optical microscope image of the array in a microfluidic device, and **e** the mechanism of sensing and genotyping of hepatitis B virus (HBV) DNA directly from the serum extract. Reprinted from Zhang et al. [156], Copyright (2010), with permission from Elsevier

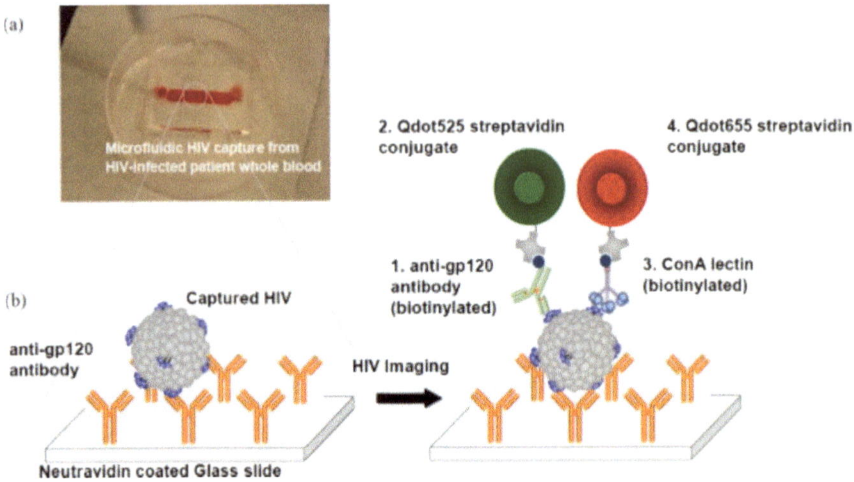

Fig. 5 The figure shows an integrated microfluidic device for HIV capture and imaging from whole blood. **a** HIV patient blood flowing through the microfluidic device (**b**), a side view of the microfluidic device showing capturing and imaging method via dual strain imaging technique. Reprinted from Kim et al. [52], Copyright (2009), with permission from Elsevier

In a similar study, N-doped-carbon QDs were decorated with folic acid and doxorubicin on their surfaces. In-vitro testing was conducted in MCF7, PC12, and 4T1 cell lines. In-vivo testing was carried out in BALB/c breast tumor-induced mice models. Both in-vitro and in-vivo testing showed carbon QDs conjugated with both folic acid and doxorubicin had enhanced anticancer activity and cytotoxicity analysis confirmed the same in comparison to the control group administered with doxorubicin [49].

Cancer therapy and the advent of nanotechnology have brought mankind close to targeted delivery at the site of tumor outgrowth. An experimental study involved the development of a drug delivery platform composed of quantum dot β-cyclodextrin folic acid loaded with C-2028 unsymmetrical bisacridine. In-vitro testing was done in cancer cell lines, namely, human prostate cancer Du-145 and LNCaP, human non-small-cell lung cancer H460, and human fetal lung fibroblasts MRC-5 and the standard cell line, human prostate PNT1A cell line. The C-2028 loaded QDs-based drug delivery platform showed enhanced cytotoxicity compared to the control drug state [86].

5.2 Environmental Applications

The microfluidic production of QDs offers significant environmental benefits. These precisely engineered QDs possess unique optical properties and a high surface area, making them ideal for environmental monitoring. They can detect specific water

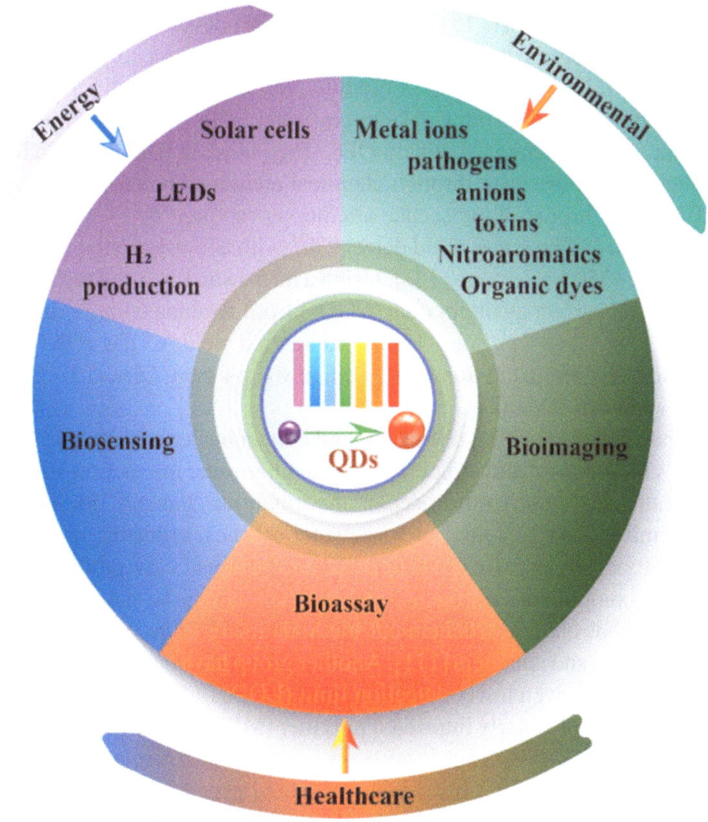

Fig. 6 The figure shows various applications of microfluidic-synthesized QDs. Specific targets in different fields such as healthcare, environment, and energy

contaminants, providing real-time data on pollution levels. Furthermore, their ability to generate reactive oxygen species under light exposure can be harnessed to degrade harmful pollutants, offering a sustainable solution for pollution detection and remediation. Figure 6 shows the specific application of the microfluidic-synthesized QDs in healthcare, environment, and energy applications.

The detection of heavy metals [16, 35, 88, 131, 141, 151], anions [93], pharmaceuticals [25], pesticides [65, 75], organic dyes [149], toxic gases [64], nitroaromatics [146], and pathogens [84] have been thoroughly studied in literature[36].

5.2.1 Sensors for Environmental Pollutant Detection

Environmental pollution detection is being improved by using microfluidic synthesis of QDs. QDs have been used to detect various environmental pollutants in fluorescent nanosensors. Specifically, metal ions (such as Pb^{2+}, Cu^{2+}, Cd^{2+}, Ni^{2+}, Hg^{2+}, and

Ag^{+}) [16, 35, 88, 131, 141, 151]. Wang et al. [131] developed a microfluidic-based fluorescent electronic eye (E-eye). The device was equipped with a microfluidic chip, an ultraviolet light excitation module, an optical lens, and a smartphone for photographing and analyzing Cd^{2+} ions in water. This eye uses tetrasodium iminodisuccinate (IDS)-etched CdTe/CdS QDs. They utilized IDS to the CdTe/CdS QDs to dimmer the fluorescence through chemical etching. The etched QDs will detect the Cd^{2+} added through fluorescence alterations captured by E-eye for quantitative detection of Cd^{2+}. The E eye could detect Cd^{2+} with a low detection limit of 0.26 μg/L within the 1–250 μg/L range [131]. Using a microwave-assisted technique, another work proposed carbon quantum dots (CQDs) from Arabica coffee grinds utilized to detect Fe^{3+}, Pb^{2+}, and Cr^{3+} ions. This rapid method yields CQDs with blue emission (455 nm) on 365 nm excitation. The average size of the CQDs is 10.12 nm, with a quantum yield of 6.01% [78].

The endocrine-disrupting chemicals and exogenous chemicals that interact with our body's metabolism and affect receptor and hormone interaction adversely invite several health issues like cardiovascular abnormalities, diabetes, obesity, and cancer. These chemicals can be found in wastewater sources, effluents, micro–nano plastics [128], food preservatives, plasticizers, and personal care products. EDCs such as phenolic compounds, estrogens, and bisphenols can be detected through QDs using fluorescent and electrochemical methods using molecularly imprinted polymers, antibodies, and aptamers [111]. Another group have developed an electrochemical sensor with an even lower detection limit (LOD) of 0.015 nmol/L for bisphenol A (BPA) [157]. Molecularly imprinted polymer (MIP), CdTe QDs, and multi-walled carbon nanotubes (MWCNTs) have been used to make nanosensors for sensitive and specific detection of 2,2-bis(4-hydroxyphenyl)propane (BPA). Here, MIPs have been constructed around combining CdTe QDs and MWCNTs for the nanosensor fabrication [157]. In another similar work, Shi et al. [110] demonstrated an ECL sensor that blends CdSe QDs with MIPs, resulting in one of the lowest LODs for 2,2-bis(4-hydroxyphenyl)propane (bisphenol A—BPA) at 3.4×10^{-10} mol/L. The sensor fabrication consists of magnetic nanoparticles (MNPs) with CdSe quantum dots (QDs) both encapsulated within a molecularly imprinted polymer (MIP) shell, resulting in a CdSe-MNPs-MIP composite for the detection of bisphenol A (BPA).

Pesticides are chemical substances that prevent or remove pests such as insects, weeds, fungi, and other hazardous organisms in food and agricultural items [145]. While they play a significant function in enhancing agricultural yield, their excessive utilization brings serious threats to human health and ecosystems [85]. Pesticides may contaminate food, water sources, and soil, producing environmental and health hazards. As a result, there is a developing need for trustworthy technology to detect and measure these chemicals promptly and effectively in the environment [85, 125, 145]. Liu et al. [65] created a very sensitive sensor for detecting pretilachlor by constructing a double-layer structure integrating SiO_2@CdSe/ZnS QDs and nitrogen-doped graphene QDs (GQDs) covered with molecularly imprinted polymers (MIPs). This structure has two different emission bands. In the absence of pretilachlor, the sensor emits blue light from the MIP@GQDs, which quenches as the herbicide concentration rises, while the red emission from SiO_2@QDs stays constant.

The approach obtained an outstanding detection limit of 0.42 mg/L, ranking among the lowest LODs recorded for QD-based pesticide sensors [65]. Muhammad et al. [75] devised a straightforward spectrofluorimetric approach for CYP's quantitative and qualitative measurement. The detection procedure comprised mixing thioglycolic acid (TGA)-stabilized ZnS:Mn QDs with CYP solutions at pH 6, where a fluorescence quenching effect was detected. The sensor obtained a detection limit (LOD) of 0.132 μg/mL with a linear detection range of 0.5–12 μg/mL for CYP. Additionally, the scientists determined the limit of quantification (LOQ) at 0.401 μg/mL, a significant metric for demonstrating the dependability of QDs as accurate sensors for both qualitative and quantitative applications [75].

Pathogens are often found in soils, food, and water sources, presenting substantial dangers to human health. These infectious agents—such as bacteria, viruses, parasites, fungi, and protozoa—can lead to illnesses like hepatitis, malaria, dysentery, cholera, and typhoid fever, causing severe disturbances in the human body. Pandit et al. [84] created a fluorescent sensor for detecting *E. coli* in water samples, employing a combination of magnetic nanoparticles (SPIONs) and CdTe-QD-aptamer conjugates. In this system, superparamagnetic iron oxide nanoparticles (SPIONs) were coupled with *E. coli* aptamer I (SPIONs-AptI), whereas CdTe QDs were partnered with aptamer II (QDs-AptII). The detection procedure involves adding SPIONs-AptI to *E. coli*-infected water, generating a bacteria-SPION complex, which was then magnetically separated. Next, QDs-AptII were introduced to connect with the bacteria confined by SPIONs, allowing fluorescence-based detection. This approach acquired a sensitivity capable of detecting as low as 100 CFU of *E. coli* in water. Additionally, the scientists constructed a biosensor prototype employing a UV LED as the excitation source. The fluorescence intensity of the QDs was transformed into a numerical number, which was presented on a monitor, enabling a user-friendly platform for practical applications [84]. One frequent nitro-aromatic molecule in soil and water is 4-nitrophenol (4-NP, or p-nitrophenol) [104]. Known for its high toxicity, carcinogenicity, and environmental persistence, 4-NP has been categorized as a "Priority Pollutant" by the United States Environmental Protection Agency (USEPA) [104, 120]. To detect 4-nitrophenol (4-NP), Xu et al. [146] developed a basic fluorescent sensor based on CeO_2 QDs stabilized with polyvinyl alcohol. This sensor displayed a linear detection range of 0.005–75 μmol/L and obtained a limit of detection (LOD) of 1.50 nmol/L in a tris–HCl buffer at pH 6.0 [146].

Microplastic contamination in human samples is a significant environmental and health research problem. They are another type of contaminants that need to be recognized effectively. The QDs have been exploited lately for various applications [4, 108, 124]. In the literature study, QD-assisted fluorescence was extensively proven for detecting microplastics in environmental and biological matrices, comparing its performance to Raman spectroscopy. QD-based techniques, especially those employing carbon and copper QDs, demonstrate exceptional sensitivity, achieving low detection limits of 0.2 μm in biological samples and 0.5 μm in environmental samples. Notably, microplastics inside the size range of 0.2–0.5 μm are effectively detected applying QD-assisted fluorescence, as indicated by Adebiyi and Ajana [4]. This property considerably aids Raman spectroscopy, which typically struggles with small particles. For more substantial microplastics (1.0–5.0 μm), even fewer sensitive approaches like Raman spectroscopy remain beneficial, as demonstrated by Lv et al. [67]. QD-assisted fluorescence stands out in its ability to recognize nano- and micro-scale plastic particles, a substantial benefit over conventional techniques, which frequently miss them due to their higher detection thresholds and inferior sensitivity [4, 108, 124].

5.2.2 Water Purification and Photocatalysis

As seen in the previous section, the QDs can be engineered to detect specific contaminants in air, water, and soil, giving real-time data on pollution levels. QDs can generate reactive oxygen species under light exposure. This property can degrade harmful pollutants, offering a sustainable solution for environmental cleanup. QDs have potential in ecological applications, notably in water purification and photocatalysis. These QDs have outstanding optical characteristics and a large surface area, making them particularly efficient in degrading contaminants and cleaning water. The synthesis of some composite QDs such as $MoS_2@Fe_2O_3$ using microfluidic hydrodynamic synthesis pathway will pave the way forward. Some studies in the literature dig into the use of $MoS_2@Fe_2O_3$ core–shell micromotors for microplastic degradation.

Quantum dots (QDs) synthesized via microfluidic platforms, such as carbon dots (CDs), perovskite QDs, and $Cu@Cu_2O$ nanocomposites, have demonstrated significant potential in photocatalytic water purification. Additionally, $Cu@Cu_2O$ nanocomposites synthesized in microfluidic reactors demonstrate rapid degradation of methylene blue via synergistic effects between the metallic Cu core and semiconducting Cu_2O shell, achieving high efficiency under UV irradiation [61]. The photoelectrocatalytic (PEC) microreactor employs a planar design with a nanoporous monoclinic $BiVO_4$ (BVO) layer on an ITO glass substrate for visible-light-driven photocatalysis (<2.7 eV bandgap). Under a negative bias exceeding the water electrolysis potential, oxygen generated at the anode via electrolysis migrates to the BVO surface, resolving oxygen deficiency during electron-driven oxidation. Performance tests degrading methylene blue revealed superior efficiency under negative bias compared to positive bias, with no significant degradation over 200 cycles. This design enables precise control of oxidation pathways through bias polarity adjustment while ensuring durability for scalable applications [134].

Composite quantum dots (QDs) like $MoS_2@Fe_2O_3$ and Sb_2S_3, synthesized via microfluidic platforms, show promise for micro/nanoplastic degradation [44]. While $MoS_2@Fe_2O_3$ core–shell micromotors were initially fabricated hydrothermally, microfluidic synthesis could optimize their structure for solar-driven photophoretic motion (up to 6 mm/s), generating reactive oxygen species (ROS) and localized heat to decompose polystyrene microplastics within 4 h. The Fe_2O_3 shell enables magnetic recovery of both micromotors and degraded plastics. Similarly, Sb_2S_3-based microrobots degrade polyesters like PHB and PLA under UV light without additional fuel, emphasizing sustainability [10]. Microfluidic synthesis of such QDs remains underexplored but offers scalable, tunable solutions for eco-friendly plastic remediation.

5.3 Energy Applications

The controlled synthesis environment of microfluidic devices ensures the production of uniform and reproducible QDs, which is crucial for consistent performance in energy applications. In solar cells, QDs can be engineered to absorb a broader spectrum of sunlight, enhancing the efficiency of photovoltaic devices. Light-emitting diodes (LEDs) can also harness their unique luminescent properties to produce brighter, more energy-efficient lighting solutions. This integration of microfluidic technology with QD synthesis not only improves the efficiency of energy devices but also paves the way for innovative approaches to sustainable energy solutions.

5.3.1 Quantum Dot Solar Cells

These nanoscale semiconductors enable fine adjustment of their electrical and optical features, giving better light absorption, efficient charge carrier dynamics, and the potential to boost photovoltaic efficiency dramatically.

The structural design of Quantum Dot Solar Cells (QDSCs) is broadly categorized into two primary architectures,

a. Schottky Junction QDSCs

In Schottky Junction designs, quantum dots are deposited directly onto a metal electrode, generating a Schottky barrier that promotes charge separation and collection [132].

b. Heterojunction QDSCs

Conversely, Heterojunction QDSCs feature a multilayer structure where quantum dots are carefully inserted between p-n junctions, boosting photoconversion efficiency by facilitating efficient charge separation and decreasing recombination losses [150].

Currently, research is focused on broadening the absorption spectrum of solar cells from the ultraviolet (UV) to the near-infrared (NIR) region, allowing their functioning even under low-light or dark scenarios. Ren et al. [96] have created a TiO_2/PbS/$MAPbI_3$/NiO nanowire array solar cell capable of absorbing solar energy across an extended spectral range, from 200 nm (ultraviolet) to 2500 nm (near-infrared). This breakthrough design generated a fantastic solar conversion efficiency of 16.5% [96]. Hu and his co-workers constructed a flexible solar cell based on colloidal QDs, adding PCBM (phenyl-C61-butyric acid methyl ester) into $CsPbI_3$ QDs to optimize device performance. QDs promote an increase in interface charge transfer. It decreased interfacial recombination losses, considerably boosting the power conversion efficiency (PCE), obtaining an outstanding value of 15.1% [43].

5.3.2 Quantum Dot LEDs (QLEDs)

Microfluidic Quantum Dot-based Organic Light-Emitting Diodes (QD-LEDs) represent a technological leap forward in solid-state lighting and display technologies [152]. These devices employ quantum dots with customized emission characteristics for accurate color reproduction and controllable brightness. They also introduce a unique viewpoint to display technology, delivering practical advantages such as broad viewing angles, bright color accuracy, stability, and adaptability [38, 55]. Kawamura and team demonstrated a PDMS-based OLED incorporating red carboxyl-functionalized CdSeS/ZnS core–shell quantum dots in a water solution, achieving the highest light color purity [47]. However, microfluidic technology-based quantum dots often exhibit larger particle sizes than other synthesis methods [54]. Zhu et al. [160] manufacture a microfluidic quantum dot color conversion layer with a minimum pixel size of 20 μm. This layer's red and green arrays exhibit exceptional light uniformity, attaining 98.6% and 98.7%, respectively, emphasizing the promise of microfluidic quantum dot technology for accurate and efficient color conversion in advanced display systems.

5.3.3 Microfluidic QDs in Hydrogen Production

Along with various applications in different fields, using QDs in hydrogen (H_2) production is another aspect explored in recent years. Hydrogen (H_2) is recognized as a key element in a sustainable energy future, offering high energy storage efficiency and minimal environmental impact [68]. In photoelectrochemical (PEC) water splitting to produce hydrogen, quantum dots (QDs) are applied to the surface of the working electrode to act as photoelectrodes, improving the absorption of light and the transfer of charges.

Recently, water-splitting processes have been facilitated by employing QDs with photoanode as they feature high stability and adjustable bandgap materials [46]. Kumar et al. achieved substantial progress by deploying indium nitride (InN) QDs to adorn an $In_{0.25}Ga_{0.75}N$ on silicon (Si) photoanode. This design resulted in a fivefold improvement in photocurrent density compared to a bare InGaN-based Si photoanode. Under a potential of 0.4 V, the applied-bias photon-to-current efficiency (ABPE) achieved 4.1%. Additionally, with an applied potential of 0.2 V under 100 mW cm^{-2} white light irradiation, the H_2 and O_2 production rates were reported as 75 μmol h^{-1} cm^{-2} and 33 μmol h^{-1} cm^{-2}, respectively. These findings underline the potential of QD-decorated nitride alloys in developing PEC systems for efficient hydrogen production [56].

Zhang and colleagues constructed a two-dimensional atomic-level heterostructure by mixing MoS_2 quantum dots (QDs) at sulfur vacancies on the zinc (Zn) plane of monolayered $ZnIn_2S_4$ for photoelectrochemical (PEC) water splitting, showing efficient hydrogen (H_2) production. Enhanced charge density around Zn atoms by including MoS_2 QDs provides a highly efficient interface with lowered edge contact resistance. This novel photoelectrode obtained an impressive H_2 production rate of 6.884 mmol g^{-1} h^{-1}, with an apparent quantum efficiency (AQE) of 63.87% at 420 nm [153].

6 Future Prospects and Challenges

Microfluidic QDs have a wide variety of applications in different fields, as discussed in the QDs application section of the chapter. However, they also have various challenges and future research is needed to address them.

Despite decades of research, industrial-scale manufacturing of quantum dots remains a significant challenge, primarily due to the stringent demands for uniformity, high yield, and cost-effectiveness. This transition from laboratory-scale to industrial-scale production necessitates overcoming numerous obstacles. Scaling up QD synthesis to industrial levels presents significant challenges, including variability and inefficiencies. Important research areas also include ensuring microfluidic systems are economical and easily integrable to industrial processes [116].

Viscus drag at the channel walls is one of the common issues of microfluidic reactors, which leads to a spread of fluid velocities in the channel, resulting in polydispersity and poorly defined particle properties [91]. Using toxic, hazardous elements such as Cd, Pb, and As in QDs synthesis presents a critical concern. Alternatively, Cd is used as a heavy metal for the QDs synthesis. It has significant environmental [142] and health problems [37].

A new future approach is needed to develop eco-friendly and non-toxic QD alternatives, such as carbon-based or silicon-based QDs, which can give the same benefits without the hazards. Additionally, one of the tough challenges for all researchers in this domain is evaluating the long-term environmental impact of QDs. Toxic spots in the manufacturing and disposal cycles of QD products may be identified by a lifecycle assessment, which producers can use to create more sustainable processes. More ethical and sustainable usage of QDs in various applications can be achieved if these problems are addressed. Navigating the legal and safety difficulties connected with quantum dots (QDs) is vital for their greater adoption and commercialization [97].

Industry-academia collaboration is necessary to boost the broad range applicability of the QDs. Defining safety standards by performing safety studies to setting clear rule. In-depth safety evaluations and long-term research are needed to understand QDs' implications on human health and the environment [142]. Also, discovering nontoxic alternatives might help create awareness among research and industrial communities [37]. Responsible statutory bodies must create collaboration to promote a wide range of applicability. For the wide range of QDs, tracking its toxicity is key. Synthesis of biocompatible QDs by biosynthesis and CQDs/GQDs are some of the currently existing solutions, further future research in this line is highly needed [109].

The PQDs stability is another class of difficulty that inhibits its wide application. Moreover, advancement in surface passivation and functionalization could pave the way forward solution [158]. PQDs also have significant limitations in lead toxicity in healthcare applications. Though fewer toxic alternatives such as $CsPbX_3$ [45] have been given as solutions, further future research is still needed for its wide applications. The environmental toxicity is associated with utilized toxic and hazardous solvents, including octadecene (ODE), dimethyl formamide (DMF), and dimethyl sulfoxide (DMSO) [126]. The QDs offer good sensitivity and selectivity for early illness detection. However, their low throughput and sample processing capabilities inhibit their application for simultaneous multianalyte detection.

On one side, the next-generation solar cells [23, 40] (Du et al. 2016) and LEDs [74] with improved efficacy and reduced energy usage are utilizing QDs. On the other hand, wide use of these QDs invited other toxicity-related environmental issues [37, 73]. To overcome these difficulties, research is now progressing toward finding sustainable synthesis methods and eco-friendly alternatives. As the technology matures, the collaboration between scientists, industry stakeholders, and regulatory bodies will be crucial in overcoming these hurdles and unlocking the full potential of QDs in diverse applications. Progress in automation, artificial intelligence, and machine learning may be crucial for enhancing these processes, guaranteeing uniform quality and scalability. Two general issues should be considered when synthesizing nanoparticles in microfluidic reactors.

The choice of the solvent that remains liquid at ambient as well as at high temperature (>200 °C) is important for easy flow in the microfluidic channel. Unlike the batch method microfluidic-based methods cannot use waxy long chain ligands. Mostly solvents like octadecene, trioctylphosphineoxide, and squalane, are used in microfluidic systems. Fouling problem occurs inside the microfluidic reactor because of precipitation of the nanocrystals in narrow channels. Therefor proper reagent selection and solvent choice could mitigate this issue [80].

7 Summary

This chapter presents a comprehensive overview of different kinds of quantum dots, the microfluidic platforms of their synthesis, and their myriad applications. Depending on their size, shape, and constituent material, quantum dots exhibit robust photophysical properties, such as stable fluorescence, resistance to photobleaching, narrow emission and broad absorption spectra, all of which render them eminently suitable for various electronic and optical applications. Microfluidic QD synthesis platforms offer distinct advantages over bulk synthesis protocols. In bulk synthesis, little local control can be exerted over reaction conditions, temperature, and mixture homogeneity, among others; all of these factors govern the size and shape QDs, and may result in their non-uniformity.

In contrast, microfluidic platforms allow for significantly greater control over synthesis parameters, ensuring uniformity of the physicochemical properties of the QDs, which translates to improved performance. Furthermore, microfluidic platforms can be scaled up to increase throughput, while preserving QD properties. Indeed, in recent decades, microfluidics has only seen increasing use in the synthesis of several different kinds of QDs, such as semiconductor, metal alloy, perovskite, and carbon QDs. QDs so fabricated have been directed towards a wide array of biological, optical, electronic, environmental, energy, and diagnostic applications. However, the toxicity of several popular constituent heavy metal elements in QDs remains a major challenge, which has inspired scientists to seek less toxic, and even biocompatible alternatives to fabricate QDs. Research is now being devoted to formulating sustainable and environment-friendly QD synthesis protocols, replacing toxic elements and harsh solvents with carbonaceous precursors from vegetables. Additionally, Artificial Intelligence and Machine Learning tools are being explored to optimize the use of microfluidic platforms in QD synthesis. This may provide invaluable insights into the feasible translation of microfluidic QD synthesis to an industrial scale, which is yet to be realized, due to regulatory barriers, economic constraints, and environmental concerns. Despite such challenges, microfluidic synthesis of QDs has emerged as the most promising avenue of their production.

References

1. Abdellatif AhAH, Tawfeek HM, Younis MA, Alsharidah M, Al Rugaie O (2022) Biomedical applications of quantum dots: overview, challenges, and clinical potential. Int J Nanomed 17:1951. https://doi.org/10.2147/IJN.S357980
2. Abou-Hassan A, Bazzi R, Cabuil V (2009) Multistep continuous-flow microsynthesis of magnetic and fluorescent γ-Fe_2O_3@SiO_2 core/shell nanoparticles. Angew Chem Int Ed 48(39):7180–7183. https://doi.org/10.1002/ANIE.200902181
3. Abou-Hassan A, Sandre O, Neveu S, Cabuil V (2009) Synthesis of goethite by separation of the nucleation and growth processes of ferrihydrite nanoparticles using microfluidics. Angew Chem 121(13). https://doi.org/10.1002/ange.200805933
4. Adebiyi VA, Ajana MO (2024) Quantum dot fluorescence for high-sensitivity microplastic detection in biological and environmental samples: a comprehensive review. https://doi.org/10.2139/SSRN.5025210
5. Agarwal K, Rai H, Mondal S (2023) Quantum dots: an overview of synthesis, properties, and applications. Mater Res Express 10(6). https://doi.org/10.1088/2053-1591/ACDA17
6. Agha A, Waheed W, Stiharu I, Nerguizian V, Destgeer G, Abu-Nada E, Alazzam A (2023) A review on microfluidic-assisted nanoparticle synthesis, and their applications using multiscale simulation methods. Discover Nano 18(1):1–54. https://doi.org/10.1186/S11671-023-03792-X
7. Aharoni A, Mokari T, Popov I, Banin U (2006) Synthesis of InAs/CdSe/ZnSe core/shell1/shell2 structures with bright and stable near-infrared fluorescence. J Am Chem Soc 128(1):257–264. https://doi.org/10.1021/JA056326V
8. Aralekallu S, Boddula R, Singh V (2023) Development of glass-based microfluidic devices: a review on its fabrication and biologic applications. Mater Des 225. https://doi.org/10.1016/j.matdes.2022.111517
9. Astruc D, Boisselier E, Ornelas C (2010) Dendrimers designed for functions: from physical, photophysical, and supramolecular properties to applications in sensing, catalysis, molecular

electronics, photonics, and nanomedicine. Chem Rev 110(4):1857–1959. https://doi.org/10.1021/CR900327D
10. de la Asunción-Nadal V, Solano E, Jurado-Sánchez B, Escarpa A (2024) Photophoretic MoS_2-Fe_2O_3 piranha micromotors for collective dynamic microplastics removal. ACS Appl Mater Interfaces 16(36):47396–47405. https://doi.org/10.1021/ACSAMI.4C06672/ASSET/IMAGES/LARGE/AM4C06672_0005.JPEG
11. Baek J, Shen Y, Lignos I, Bawendi MG, Jensen KF (2018) Multistage microfluidic platform for the continuous synthesis of III-V core/shell quantum dots. Angew Chem (Int Ed Engl) 57(34):10915–10918. https://doi.org/10.1002/ANIE.201805264
12. Bagga A, Chattopadhyay PK, Ghosh S (2006) Origin of Stokes shift in InAs and CdSe quantum dots: exchange splitting of excitonic states. Phys Rev B - Condens Matter Mater Phys 74(3):035341. https://doi.org/10.1103/PHYSREVB.74.035341
13. Bahmani Jalali H, De Trizio L, Manna L, Di Stasio F (2022) Indium arsenide quantum dots: an alternative to lead-based infrared emitting nanomaterials. Chem Soc Rev 51(24). https://doi.org/10.1039/d2cs00490a
14. Bai J, Wang L, Zhang T, Hou T, Zhang M, Xu B, Li D, Jin X, Li Q, Wang Y, Zhang X, Song Y (2023) Enabling ultranarrow blue emission linewidths in colloidal alloy quantum dots by decreasing the exciton fine structure splitting and exciton-phonon coupling. Nano Res 16(1):1576–1585. https://doi.org/10.1007/S12274-022-4784-6/METRICS
15. Bandara TMWJ, Aththanayake AAAP, Gunasekara LBE, Wijesundara WMCY (2023) Applications of quantum dots in energy conversion and storage devices. In: Quantum dots: emerging materials for versatile applications, pp 383–419. https://doi.org/10.1016/B978-0-323-85278-4.00016-7
16. Boonmee C, Noipa T, Tuntulani T, Ngeontae W (2016) Cysteamine capped CdS quantum dots as a fluorescence sensor for the determination of copper ion exploiting fluorescence enhancement and long-wave spectral shifts. Spectrochim Acta Part A Mol Biomol Spectrosc 169:161–168. https://doi.org/10.1016/J.SAA.2016.05.007
17. Cao S, Ji W, Zhao J, Yang W, Li C, Zheng J (2016) Color-tunable photoluminescence of Cu-doped Zn–In–Se quantum dots and their electroluminescence properties. J Mater Chem C 4(3):581–588. https://doi.org/10.1039/C5TC04019A
18. Chan EM, Alivisatos AP, Mathies RA (2005) High-temperature microfluidic synthesis of CdSe nanocrystals in nanoliter droplets. J Am Chem Soc 127(40):13854–13861. https://doi.org/10.1021/JA051381P
19. Chen G, Zhu X, Xing C, Wang Y, Xu X, Bao J, Huang J, Zhao Y, Wang X, Zhou X, Du X, Wang X (2023) Machine learning-assisted microfluidic synthesis of perovskite quantum dots. Adv Photonics Res 4(1):2200230. https://doi.org/10.1002/ADPR.202200230
20. Cheng R, Liang ZB, Shen H, Guo J, Wang CF, Chen S (2023) In-situ synthesis of stable perovskite quantum dots in core-shell nanofibers via microfluidic electrospinning. Chin Chem Lett 34(3):107384. https://doi.org/10.1016/J.CCLET.2022.03.107
21. Cheng Y, Ling SD, Geng Y, Wang Y, Xu J (2021) Microfluidic synthesis of quantum dots and their applications in bio-sensing and bio-imaging. Nanoscale Adv 3(8):2180–2195. https://doi.org/10.1039/D0NA00933D
22. Cheng R, Ma K, Ye HG, Ling L, Wu G, Wang CF, Chen S (2020) Magnetothermal microfluidic-directed synthesis of quantum dots. J Mater Chem C 8(19). https://doi.org/10.1039/d0tc00305k
23. Conibeer G (2012) Si and other group IV quantum dot based materials for tandem solar cells. Energy Procedia 15. https://doi.org/10.1016/j.egypro.2012.02.023
24. Dhariwal J, Rao GK, Vaya D (2023) Recent advancements towards the green synthesis of carbon quantum dots as an innovative and eco-friendly solution for metal ion sensing and monitoring. RSC Sustain 2(1). https://doi.org/10.1039/d3su00375b
25. Ding R, Chen Y, Wang Q, Wu Z, Zhang X, Li B, Lin L (2022) Recent advances in quantum dots-based biosensors for antibiotics detection. J Pharm Anal 12(3). https://doi.org/10.1016/j.jpha.2021.08.002

26. Dos-Reis-Delgado AA, Carmona-Dominguez A, Sosa-Avalos G, Jimenez-Saaib IH, Villegas-Cantu KE, Gallo-Villanueva RC, Perez-Gonzalez VH (2023) Recent advances and challenges in temperature monitoring and control in microfluidic devices. Electrophoresis 44(1–2). https://doi.org/10.1002/elps.202200162
27. Du XY, Wang CF, Wu G, Chen S (2021) The rapid and large-scale production of carbon quantum dots and their integration with polymers. Angew Chem - Int Ed 60(16). https://doi.org/10.1002/anie.202004109
28. Duffy DC, McDonald JC, Schueller OJA, Whitesides GM (1998) Rapid prototyping of microfluidic systems in poly(dimethylsiloxane). Anal Chem 70(23):4974–4984. https://doi.org/10.1021/AC980656Z
29. Dutta S, Ghosh A, Pattader PSG, Bandyopadhyay D (2019) Electric field mediated von Kármán vortices in stratified microflows: transition from linear instabilities to coherent mixing. J Fluid Mech 865. https://doi.org/10.1017/jfm.2018.1023
30. Ebrahim F, Al-Hartomy O, Wageh S (2023) Cadmium-based quantum dots alloyed structures: synthesis, properties, and applications. Materials 16(17). https://doi.org/10.3390/ma16175877
31. Efros AL (2019) Quantum dots realize their potential. Nature 575(7784):604–605. https://doi.org/10.1038/d41586-019-03607-z
32. El-Sayed MA (2004) Small is different: shape-, size-, and composition-dependent properties of some colloidal semiconductor nanocrystals. Acc Chem Res 37(5):326–333. https://doi.org/10.1021/AR020204F
33. Eun TH, Kim SH, Jeong WJ, Jeon SJ, Kim SH, Yang SM (2009) Single-step fabrication of monodisperse TiO_2 hollow spheres with embedded nanoparticles in microfluidic devices. Chem Mater 21(2):201–203. https://doi.org/10.1021/CM8017133/SUPPL_FILE/CM8017133_SI_001.PDF
34. Feng D, Zhang G, Li Y (2024) Semiconductor quantum dots: synthesis, properties and applications. Nanomaterials 14(22):1825. https://doi.org/10.3390/NANO14221825
35. Freeman R, Finder T, Willner I (2009) Multiplexed analysis of Hg^{2+} and Ag^{+} ions by nucleic acid functionalized CdSe/ZnS quantum dots and their use for logic gate operations. Angew Chem (Int Ed Engl) 48(42):7818–7821. https://doi.org/10.1002/ANIE.200902395
36. Freire MS, Silva HJB, Albuquerque GM, Monte JP, Lima MTA, Silva JJ, Pereira GAL, Pereira G (2024) Advances on chalcogenide quantum dots-based sensors for environmental pollutants monitoring. Sci Total Environ 931:172848. https://doi.org/10.1016/J.SCITOTENV.2024.172848
37. Gidwani B, Sahu V, Shukla SS, Pandey R, Joshi V, Jain VK, Vyas A (2021) Quantum dots: prospectives, toxicity, advances and applications. J Drug Deliv Sci Technol 61. https://doi.org/10.1016/j.jddst.2020.102308
38. Gustafsson G, Cao Y, Treacy GM, Klavetter F, Colaneri N, Heeger AJ (1992) Flexible light-emitting diodes made from soluble conducting polymers. Nature 357(6378):477–479. https://doi.org/10.1038/357477a0
39. Han M, Gao X, Su JZ, Nie S (2001) Quantum-dot-tagged microbeads for multiplexed optical coding of biomolecules. Nat Biotechnol 19(7):631–635. https://doi.org/10.1038/90228
40. Hao M, Ding S, Gaznaghi S, Cheng H, Wang L (2024) Perovskite quantum dot solar cells: current status and future outlook. ACS Energy Lett 9(1):308–322. https://doi.org/10.1021/ACSENERGYLETT.3C01983
41. He S, Kohira T, Uehara M, Kitamura T, Nakamura H, Miyazaki M, Maeda H (2005) Effects of interior wall on continuous fabrication of silver nanoparticles in microcapillary reactor. Chem Lett 34(6). https://doi.org/10.1246/cl.2005.748
42. Hu Z, Xu S, Xu X, Wang Z, Wang Z, Wang C, Cui Y (2015) Co-doping of Ag into Mn:ZnSe quantum dots: giving optical filtering effect with improved monochromaticity. Sci Rep 5(1):1–9. https://doi.org/10.1038/srep14817
43. Hu L, Zhao Q, Huang S, Zheng J, Guan X, Patterson R, Kim J, Shi L, Lin CH, Lei Q, Chu D, Tao W, Cheong S, Tilley RD, Ho-Baillie AWY, Luther JM, Yuan J, Wu T (2021) Flexible and efficient perovskite quantum dot solar cells via hybrid interfacial architecture. Nat Commun 12(1):1–9. https://doi.org/10.1038/s41467-020-20749-1

44. Jancik-Prochazkova A, Jašek V, Figalla S, Pumera M, Jancik-Prochazkova A, Pumera M, Jašek V, Figalla S (2023) Photocatalytic microplastics "On-The-fly" degradation via motile quantum materials-based microrobots. Adv Opt Mater 11(22):2300782. https://doi.org/10.1002/ADOM.202300782
45. Jellicoe TC, Richter JM, Glass HFJ, Tabachnyk M, Brady R, Dutton SE, Rao A, Friend RH, Credgington D, Greenham NC, Böhm ML (2016) Synthesis and optical properties of lead-free cesium tin halide perovskite nanocrystals. J Am Chem Soc 138(9):2941–2944. https://doi.org/10.1021/JACS.5B13470
46. Jin L, Zhao H, Wang ZM, Rosei F, Jin L, Rosei F, Wang ZM, Zhao H (2021) Quantum dots-based photoelectrochemical hydrogen evolution from water splitting. Adv Energy Mater 11(12):2003233. https://doi.org/10.1002/AENM.202003233
47. Kawamura M, Kuwae H, Kamibayashi T, Oshima J, Kasahara T, Shoji S, Mizuno J (2020) Liquid/solution-based microfluidic quantum dots light-emitting diodes for high-colour-purity light emission. Sci Rep 10(1):1–9. https://doi.org/10.1038/s41598-020-70838-w
48. Khan SA, Günther A, Schmidt MA, Jensen KF (2004) Microfluidic synthesis of colloidal silica. Langmuir 20(20). https://doi.org/10.1021/la0499012
49. Khoshnood A, Farhadian N, Abnous K, Matin MM, Ziaee N, Yaghoobi E (2023) N doped-carbon quantum dots with ultra-high quantum yield photoluminescent property conjugated with folic acid for targeted drug delivery and bioimaging applications. J Photochem Photobiol A: Chem 444. https://doi.org/10.1016/j.jphotochem.2023.114972
50. Kikkeri R, Laurino P, Odedra A, Seeberger PH (2010) Synthesis of carbohydrate-functionalized quantum dots in microreactors. Angew Chem - Int Ed 49(11). https://doi.org/10.1002/anie.200905053
51. Kılınçlı B, Çınar AD, Çetin B, Kibar G (2024) Microfluidic versus batch synthesis of fluorescent poly(GMA-co-EGDMA) micro/nanoparticles for biomedical applications. Emerg Mater 1–10. https://doi.org/10.1007/S42247-024-00840-9/FIGURES/6
52. Kim YG, Moon S, Kuritzkes DR, Demirci U (2009) Quantum dot-based HIV capture and imaging in a microfluidic channel. Biosens Bioelectron 25(1):253–258. https://doi.org/10.1016/J.BIOS.2009.06.023
53. Kim EB, Tomczak KM, Chandrasiri HB, Pálmai M, Ghaznavi A, Gritsenko D, Xu J, Snee PT (2023) Nucleation control of quantum dot synthesis in a microfluidic continuous flow reactor. Front Nanotechnol 4. https://doi.org/10.3389/fnano.2022.1096267
54. Kong X, Fan X, Wang Y, Luo Y, Chen Y, Wu T, Chen Z, Lin Y, Wang S (2024) Recent advances of photolithography patterning of quantum dots for micro-display applications. Nano Mater Sci. https://doi.org/10.1016/J.NANOMS.2024.03.005
55. Kubendhiran S, Bao Z, Dave K, Liu RS (2019) Microfluidic synthesis of semiconducting colloidal quantum dots and their applications. ACS Appl Nano Mater 2(4):1773–1790. https://doi.org/10.1021/ACSANM.9B00456
56. Kumar P, Devi P, Jain R, Shivaprasad SM, Sinha RK, Zhou G, Nötzel R (2019) Quantum dot activated indium gallium nitride on silicon as photoanode for solar hydrogen generation. Commun Chem 2(1):1–7. https://doi.org/10.1038/s42004-018-0105-0
57. Kurniawan D, Mathew J, Rahardja MR, Pham HP, Wong PC, Rao NV, Ostrikov K, Chiang WH (2023) Plasma-enabled graphene quantum dot hydrogels as smart anticancer drug nanocarriers. Small 19(20). https://doi.org/10.1002/smll.202206813
58. Lee CY, Chang CL, Wang YN, Fu LM (2011) Microfluidic mixing: a review. Int J Mol Sci 12(5). https://doi.org/10.3390/ijms12053263
59. Lee CG, Uehara M, Nakamura H, Maeda H (2008) High temperature preparation of core and core/shell composite nanocrystals in a multiphase microreactor. J Chem Eng Jpn 41(7). https://doi.org/10.1252/jcej.07we116
60. Li S, Chen Y, Huang L, Pan D (2013) Simple continuous-flow synthesis of Cu–In–Zn–S/ZnS and Ag–In–Zn–S/ZnS core/shell quantum dots. Nanotechnology 24(39):395705. https://doi.org/10.1088/0957-4484/24/39/395705
61. Li S, Lin Q, Liu X, Yang L, Ding J, Dong F, Li Y, Irfan M, Zhang P (2018) Fast photocatalytic degradation of dyes using low-power laser-fabricated Cu_2O–Cu nanocomposites. RSC Adv 8(36):20277–20286. https://doi.org/10.1039/C8RA03117G

62. Li H, Liu X, Zhu F, Ma D, Miao C, Su H, Deng J, Ye H, Dong H, Bai X, Luo Y, Lin B, Liu T, Lu Y (2022) Spatial barcoding-enabled highly multiplexed immunoassay with digital microfluidics. Biosens Bioelectron 215. https://doi.org/10.1016/j.bios.2022.114557
63. Li Y, Männel MJ, Hauck N, Patel HP, Auernhammer GK, Chae S, Fery A, Li J, Thiele J (2021) Embedment of quantum dots and biomolecules in a dipeptide hydrogel formed in situ using microfluidics. Angew Chem - Int Ed 60(12). https://doi.org/10.1002/anie.202015340
64. Lin Q, Zhang F, Zhao N, Zhao L, Wang Z, Yang P, Lu D, Dong T, Jiang Z (2022) A flexible and wearable nylon fiber sensor modified by reduced graphene oxide and ZnO quantum dots for wide-range NO_2 gas detection at room temperature. Materials 15(11). https://doi.org/10.3390/ma15113772
65. Liu C, Liao J, Zheng Y, Chen Y, Liu H, Shi X (2022) Random forest algorithm-enhanced dual-emission molecularly imprinted fluorescence sensing method for rapid detection of pretilachlor in fish and water samples. J Hazard Mater 439. https://doi.org/10.1016/j.jhazmat.2022.129591
66. Liu Z, Sun D, Jiang B, Shen L, Zhou P, Gao C, Jin Z, Liu X, Yang L, Tan S (2023) Continuous gradient temperature control of microfluidic chip based on thermoelectric cooler. Appl Therm Eng 234. https://doi.org/10.1016/j.applthermaleng.2023.121277
67. Lv X, Dong Q, Zuo Z, Liu Y, Huang X, Wu WM (2019) Microplastics in a municipal wastewater treatment plant: fate, dynamic distribution, removal efficiencies, and control strategies. J Clean Prod 225:579–586. https://doi.org/10.1016/J.JCLEPRO.2019.03.321
68. Mahanta J, Gupta S, Parmar PR, Saha DR, Parashar CK, De M, Mandal TK, Bandyopadhyay D (2024) Nanoplasmonic microfluidic electrolyzer for membraneless hydrogen production. Energy Fuels 38(16):15701–15713. https://doi.org/10.1021/ACS.ENERGYFUELS.4C02131/SUPPL_FILE/EF4C02131_SI_004.MP4
69. Martínez-Cisneros CS, de Pedro SG, Puyol M, García-García J, Alonso-Chamarro J (2012) Design, fabrication and characterization of microreactors for high temperature syntheses. Chem Eng J 211–212. https://doi.org/10.1016/j.cej.2012.09.101
70. Medintz IL, Uyeda HT, Goldman ER, Mattoussi H (2005) Quantum dot bioconjugates for imaging, labelling and sensing. Nat Mater 4(6):435–446. https://doi.org/10.1038/nmat1390
71. Michalet X, Pinaud FF, Bentolila LA, Tsay JM, Doose S, Li JJ, Sundaresan G, Wu AM, Gambhir SS, Weiss S (2005) Quantum dots for live cells, in vivo imaging, and diagnostics. Science (New York, NY) 307(5709):538–544. https://doi.org/10.1126/SCIENCE.1104274
72. Michalet X, Pinaud FF, Bentolila LA, Tsay JM, Doose S, Li JJ, Sundaresan G, Wu AM, Gambhir SS, Weiss S (2005) Quantum dots for live cells, in vivo imaging, and diagnostics. Science 307(5709):538–544. https://doi.org/10.1126/SCIENCE.1104274/SUPPL_FILE/MICHALET.SOM.PDF
73. Mohamed WAA, Abd El-Gawad H, Mekkey S, Galal H, Handal H, Mousa H, Labib A (2021) Quantum dots synthetization and future prospect applications. Nanotechnol Rev 10(1):1926–1940. https://doi.org/10.1515/NTREV-2021-0118
74. Morozova S, Alikina M, Vinogradov A, Pagliaro M (2020) Silicon quantum dots: synthesis, encapsulation, and application in light-emitting diodes. Front Chem 8. https://doi.org/10.3389/fchem.2020.00191
75. Muhammad M, Khan S, Rahim G, Alharbi W, Alharbi KH (2022) Highly selective and sensitive spectrofluorimetric method for determination of cypermethrin in different environmental samples. Environ Monit Assess 194(12). https://doi.org/10.1007/s10661-022-10640-5
76. Nakamura H, Yamaguchi Y, Miyazaki M, Maeda H, Uehara M, Mulvaney P (2002) Preparation of CdSe nanocrystals in a micro-flow-reactor. Chem Commun 23. https://doi.org/10.1039/b208992k
77. Naughton MS, Kumar V, Bonita Y, Deshpande K, Kenis PJA (2015) High temperature continuous flow synthesis of CdSe/CdS/ZnS, CdS/ZnS, and CdSeS/ZnS nanocrystals. Nanoscale 7(38):15895–15903. https://doi.org/10.1039/C5NR04510J
78. Nazar M, Hasan M, Wirjosentono B, Gani BA, Nada CE (2024) Microwave synthesis of carbon quantum dots from Arabica coffee ground for fluorescence detection of Fe^{3+}, Pb^{2+}, and Cr^{3+}. ACS Omega 9(18):20571–20581. https://doi.org/10.1021/ACSOMEGA.4C02254

79. Nightingale AM, De Mello JC (2009) Controlled synthesis of III–V quantum dots in microfluidic reactors. ChemPhysChem 10(15):2612–2614. https://doi.org/10.1002/CPHC.200900462
80. Nightingale AM, De Mello JC (2010) Microscale synthesis of quantum dots. J Mater Chem 20(39):8454–8463. https://doi.org/10.1039/C0JM01221A
81. Ning Y, Guan S, Cheng C, Zhang B, Qin B, Huang B (2024) Microfluidic synthesis of monodispersed sharp emitting perovskite $CsPbBr_3$ quantum dots via multidimensional parameterization. J Mater Chem C. https://doi.org/10.1039/D4TC04104F
82. Pan J, El-Ballouli AO, Rollny L, Voznyy O, Burlakov VM, Goriely A, Sargent EH, Bakr OM (2013) Automated synthesis of photovoltaic-quality colloidal quantum dots using separate nucleation and growth stages. ACS Nano 7(11):10158–10166. https://doi.org/10.1021/NN404397D/SUPPL_FILE/NN404397D_SI_001.PDF
83. Pan J, Quan LN, Zhao Y, Peng W, Murali B, Sarmah SP, Yuan M, Sinatra L, Alyami NM, Liu J, Yassitepe E, Yang Z, Voznyy O, Comin R, Hedhili MN, Mohammed OF, Lu ZH, Kim DH, Sargent EH, Bakr OM (2016) Highly efficient perovskite-quantum-dot light-emitting diodes by surface engineering. Adv Mater 28(39):8718–8725. https://doi.org/10.1002/adma.201600784
84. Pandit C, Alajangi HK, Singh J, Khajuria A, Sharma A, Hassan MS, Parida M, Semwal AD, Gopalan N, Sharma RK, Suttee A, Soni U, Singh B, Sapra S, Barnwal RP, Singh G, Kaur IP (2022) Development of magnetic nanoparticle assisted aptamer-quantum dot based biosensor for the detection of Escherichia coli in water samples. Sci Total Environ 831. https://doi.org/10.1016/j.scitotenv.2022.154857
85. Pathak VM, Verma VK, Rawat BS, Kaur B, Babu N, Sharma A, Dewali S, Yadav M, Kumari R, Singh S, Mohapatra A, Pandey V, Rana N, Cunill JM (2022) Current status of pesticide effects on environment, human health and it's eco-friendly management as bioremediation: a comprehensive review. Front Microbiol 13:962619. https://doi.org/10.3389/FMICB.2022.962619/PDF
86. Pilch J, Potęga A, Kowalik P, Kowalczyk A, Bujak P, Kasprzak A, Paluszkiewicz E, Nowicka AM (2024) In vitro biological evaluation of a novel folic acid-targeted receptor quantum dot-β-cyclodextrin carrier for C-2028 unsymmetrical bisacridine in the treatment of human lung and prostate cancers. Pharmacol Rep 76(4):823–837. https://doi.org/10.1007/S43440-024-00606-4
87. Popa ML, Preda MD, Neacşu IA, Grumezescu AM, Ginghină O (2023) Traditional versus microfluidic synthesis of ZnO nanoparticles. Int J Mol Sci 24(3):1875. https://doi.org/10.3390/IJMS24031875
88. Pourghobadi Z, Mirahmadpour P, Zare H (2018) Fluorescent biosensor for the selective determination of dopamine by TGA-capped CdTe quantum dots in human plasma samples. Opt Mater 84:757–762. https://doi.org/10.1016/J.OPTMAT.2018.08.003
89. Pramanik S, Bhandari S (2022) Surface-modified quantum dots for advanced sensing applications. In: Sensing and biosensing with optically active nanomaterials, pp 243–282. https://doi.org/10.1016/B978-0-323-90244-1.00010-0
90. Prasad A, Tran T, Gartia MR (2019) Multiplexed paper microfluidics for titration and detection of ingredients in beverages. Sensors (Switzerland) 19(6). https://doi.org/10.3390/s19061286
91. Pu Y, Cai F, Wang D, Wang JX, Chen JF (2018) Colloidal synthesis of semiconductor quantum dots toward large-scale production: a review. Ind Eng Chem Res 57(6). https://doi.org/10.1021/acs.iecr.7b04836
92. Purper-Ouakil D, Ramoz N, Lepagnol-Bestel AM, Gorwood P, Simonneau M (2011) Neurobiology of attention deficit/hyperactivity disorder. Pediatr Res 69(5 PART 2). https://doi.org/10.1203/PDR.0B013E318212B40F
93. Rahimi F, Anbia M (2022) Determination of cyanide based on a dual-emission ratiometric nanoprobe using silver sulfide quantum dots and silicon nanoparticles. Microchim Acta 189(3):1–10. https://doi.org/10.1007/S00604-022-05209-5
94. Rasal AS, Yadav S, Yadav A, Kashale AA, Manjunatha ST, Altaee A, Chang JY (2021) Carbon quantum dots for energy applications: a review. ACS Appl Nano Mater 4(7):6515–6541. https://doi.org/10.1021/ACSANM.1C01372

95. Reiss P, Protière M, Li L (2009) Core/shell semiconductor nanocrystals. Small 5(2):154–168. https://doi.org/10.1002/SMLL.200800841
96. Ren Y, He Y, Li S, Wang Y, He L, Ye P, Zhou L, Gao R, Chen G, Cai W, Fu C (2024) A new type of core-shell nanowire array structured quantum dot-composite perovskite solar cell with near full-spectrum absorption. Physica E 160:115937. https://doi.org/10.1016/J.PHYSE.2024.115937
97. Reshma VG, Mohanan PV (2019) Quantum dots: applications and safety consequences. J Lumin 205. https://doi.org/10.1016/j.jlumin.2018.09.015
98. Richard C, McGee R, Goenka A, Mukherjee P, Bhargava R (2020) On-demand milifluidic synthesis of quantum dots in digital droplet reactors. Ind Eng Chem Res 59(9):3730–3735. https://doi.org/10.1021/ACS.IECR.9B04230
99. Sanderson K, Castelvecchi D (2023) Tiny "quantum dot" particles win chemistry Nobel. Nature 622(7982):227–228. https://doi.org/10.1038/D41586-023-03048-9
100. Sanjayan CG, Ravikumar CH, Balakrishna RG (2023) Perovskite QD based paper microfluidic device for simultaneous detection of lung cancer biomarkers—carcinoembryonic antigen and neuron specific enolase. Chem Eng J 464. https://doi.org/10.1016/j.cej.2023.142581
101. Saravanakumar SM, Cicek PV (2023) Microfluidic mixing: a physics-oriented review. Micromachines 14(10). https://doi.org/10.3390/mi14101827
102. Seibt S, Mulvaney P, Förster S (2019) Millisecond CdS nanocrystal nucleation and growth studied by microfluidics with in situ spectroscopy. Colloids Surf, A 562:263–269. https://doi.org/10.1016/J.COLSURFA.2018.10.043
103. Selva B, Marchalot J, Jullien MC (2009) An optimized resistor pattern for temperature gradient control in microfluidics. J Micromechanics Microengineering 19(6). https://doi.org/10.1088/0960-1317/19/6/065002
104. Serrà A, Artal R, Pozo M, Garcia-Amorós J, Gómez E (2020) Simple environmentally-friendly reduction of 4-nitrophenol. Catalysts 10(4):458. https://doi.org/10.3390/CATAL10040458
105. Shang L, Cheng Y, Zhao Y (2017) Emerging droplet microfluidics. Chem Rev 117(12). https://doi.org/10.1021/acs.chemrev.6b00848
106. Shanko ES, van de Burgt Y, Anderson PD, den Toonder JMJ (2019) Microfluidic magnetic mixing at low Reynolds numbers and in stagnant fluids. Micromachines 10(11). https://doi.org/10.3390/mi10110731
107. Shao X, Dong Z, Zhang S, Qiao Y, Zhang H, Guo H (2024) Quantum dots-based multiplexed immunosensors for accurate diagnosis of attention deficit hyperactivity disorder in childhood. J Pharm Biomed Anal 243. https://doi.org/10.1016/j.jpba.2024.116096
108. Shauloff N, Morag A, Yaniv K, Singh S, Malishev R, Paz-Tal O, Rokach L, Jelinek R (2021) Sniffing bacteria with a carbon-dot artificial nose. Nano-Micro Lett 13(1):1–15. https://doi.org/10.1007/S40820-021-00610-W
109. Shen J, Zhu Y, Yang X, Li C (2012) Graphene quantum dots: emergent nanolights for bioimaging, sensors, catalysis and photovoltaic devices. Chem Commun 48(31):3686–3699. https://doi.org/10.1039/C2CC00110A
110. Shi J, Zhang X, Zhang Q, Yang P (2022) Ultrasensitive and highly selective detection of bisphenol a using core-shell magnetic molecularly imprinted quantum dots electrochemiluminescent probe. Bull Environ Contam Toxicol 108(2). https://doi.org/10.1007/s00128-021-03351-z
111. Shi Y, Zhao J, Cao G (2023) Advances in multiplexed microfluidics for infectious disease detection. Infect Microbes Dis 5(2). https://doi.org/10.1097/IM9.0000000000000120
112. Smyder JA, Krauss TD (2011) Coming attractions for semiconductor quantum dots. Mater Today 14(9):382–387. https://doi.org/10.1016/S1369-7021(11)70182-1
113. Song Y, Henrys LL, Yang W (2009) Stable amorphous cobalt nanoparticles formed by an in situ rapidly cooling microfluidic process. Langmuir 25(17). https://doi.org/10.1021/la9009866
114. Song Y, Modrow H, Henry LL, Saw CK, Doomes EE, Palshin V, Hormes J, Kumar CSSR (2006) Microfluidic synthesis of cobalt nanoparticles. Chem Mater 18(12). https://doi.org/10.1021/cm052811d

115. Song Z, Shafiq M, Tian R, Uchida S, Chen H, Ma M (2024) Microfluidic production of inorganic nanoparticles, pp 133–162. https://doi.org/10.1007/978-3-031-60717-2_6
116. Stavis SM, Edel JB, Samiee KT, Craighead HG (2005) Single molecule studies of quantum dot conjugates in a submicrometer fluidic channel. Lab Chip 5(3). https://doi.org/10.1039/b416161k
117. Strickland S, Jorns M, Fourroux L, Heyd L, Pappas D (2024) Cancer cell targeting via selective transferrin receptor labeling using protein-derived carbon dots. ACS Omega 9(2). https://doi.org/10.1021/acsomega.3c07744
118. Stroyuk O, Raievska O, Solonenko D, Kupfer C, Osvet A, Batentschuk M, Brabec CJ, Zahn DRT (2021) Spontaneous alloying of ultrasmall non-stoichiometric Ag–In–S and Cu–In–S quantum dots in aqueous colloidal solutions. RSC Adv 11(34):21145–21152. https://doi.org/10.1039/D1RA03179A
119. Su W, Yang D, Wang Y, Kong Y, Zhang W, Wang J, Fei Y, Guo R, Ma J, Mi L (2022) $AgInS_2$/ZnS quantum dots for noninvasive cervical cancer screening with intracellular pH sensing using fluorescence lifetime imaging microscopy. Nano Res 15(6). https://doi.org/10.1007/s12274-022-4104-1
120. Sudha M, Renu G, Sangeeta G (2021) Mineralization and degradation of 4-nitrophenol using homogeneous Fenton oxidation process. Environ Eng Res 26(3). https://doi.org/10.4491/eer.2019.145
121. Suh YK, Kang S (2010) A review on mixing in microfluidics. Micromachines 1(3). https://doi.org/10.3390/mi1030082
122. Sui J, Yan J, Liu D, Wang K, Luo G (2020) Continuous synthesis of nanocrystals via flow chemistry technology. Small 16(15):1902828. https://doi.org/10.1002/SMLL.201902828
123. Tade RS, Kalkal A, Patil PO (2023) Functionalized graphene quantum dots (GQDs) based label-free optical fluorescence sensor for CD59 antigen detection and cellular bioimaging. J Fluoresc. https://doi.org/10.1007/s10895-023-03501-y
124. Tammina SK, Khan A, Rhim JW (2023) Advances and prospects of carbon dots for microplastic analysis. Chemosphere 313. https://doi.org/10.1016/j.chemosphere.2022.137433
125. Tang FHM, Lenzen M, McBratney A, Maggi F (2021) Risk of pesticide pollution at the global scale. Nat Geosci 14(4):206–210. https://doi.org/10.1038/s41561-021-00712-5
126. Tang X, Yang F (2022) Quantum dots: synthesis, characterization, and applications. In: Handbook of energy materials, pp 1–30. https://doi.org/10.1007/978-981-16-4480-1_27-1
127. Kalsoom Ue, Yi R, Qu J, Liu L (2021) Nonlinear optical properties of CdSe and CdTe core-shell quantum dots and their applications. Front Phys 9:612070. https://doi.org/10.3389/FPHY.2021.612070/BIBTEX
128. Ullah S, Ahmad S, Guo X, Ullah S, Ullah S, Nabi G, Wanghe K (2023) A review of the endocrine disrupting effects of micro and nano plastic and their associated chemicals in mammals. Front Endocrinol 13:1084236. https://doi.org/10.3389/FENDO.2022.1084236
129. Wagner J, Köhler JM (2005) Continuous synthesis of gold nanoparticles in a microreactor. Nano Lett 5(4). https://doi.org/10.1021/nl050097t
130. Wang Y, Hu A (2014) Carbon quantum dots: synthesis, properties and applications. J Mater Chem C 2(34):6921–6939. https://doi.org/10.1039/C4TC00988F
131. Wang X, Kong L, Gan Y, Liang T, Zhou S, Sun J, Wan H, Wang P (2020) Microfluidic-based fluorescent electronic eye with CdTe/CdS core-shell quantum dots for trace detection of cadmium ions. Anal Chim Acta 1131:126–135. https://doi.org/10.1016/J.ACA.2020.06.072
132. Wang A, Wen Y, Zhu H, Liu Z, Wang H, Yao W, Fan Y, Xie G, Chen X, Yan K, Duan Q, Jiang Z, Xu M, Wei Y, Lin H (2024) Oxygen vacancy mediated photocatalysis of Ti_3C_2 MXene quantum dots/$W_{18}O_{49}$ hybrid membrane for peroxymonosulfate-enhanced oxidation degradation. Chem Eng J 496:154264. https://doi.org/10.1016/J.CEJ.2024.154264
133. Wang W, Yang X, Rong Z, Tu Z, Zhang X, Gu B, Wang C, Wang S (2023) Introduction of graphene oxide-supported multilayer-quantum dots nanofilm into multiplex lateral flow immunoassay: a rapid and ultrasensitive point-of-care testing technique for multiple respiratory viruses. Nano Res 16(2). https://doi.org/10.1007/s12274-022-5043-6

134. Wang N, Zhang X, Chen B, Song W, Chan NY, Chan HLW (2012) Microfluidic photoelectrocatalytic reactors for water purification with an integrated visible-light source. Lab Chip 12(20):3983–3990. https://doi.org/10.1039/C2LC40428A
135. Wang C, Zhang Y, Tang W, Wang C, Han Y, Qiang L, Gao J, Liu H, Han L (2021) Ultrasensitive, high-throughput and multiple cancer biomarkers simultaneous detection in serum based on graphene oxide quantum dots integrated microfluidic biosensing platform. Anal Chim Acta 1178:338791. https://doi.org/10.1016/J.ACA.2021.338791
136. Ward K, Fan ZH (2015) Mixing in microfluidic devices and enhancement methods. J Micromechanics Microengineering 25(9). https://doi.org/10.1088/0960-1317/25/9/094001
137. Wei J, Hu Z, Zhou W, Qiu Y, Dai H, Chen Y, Cui Z, Liu S, He H, Zhang W, Xie F, Guo R (2021) Emission tuning of highly efficient quaternary Ag-Cu-Ga-Se/ZnSe quantum dots for white light-emitting diodes. J Colloid Interface Sci 602:307–315. https://doi.org/10.1016/J.JCIS.2021.05.110
138. Werwa E, Seraphin AA, Chiu LA, Zhou C, Kolenbrander KD (1994) Synthesis and processing of silicon nanocrystallites using a pulsed laser ablation supersonic expansion method. Appl Phys Lett 64(14):1821–1823. https://doi.org/10.1063/1.111766
139. Whitesides GM (2006) The origins and the future of microfluidics. Nature 442(7101):368–373. https://doi.org/10.1038/NATURE05058
140. Wu Y, Huang J, Zang J, Zhou J, Cheng C, Hu Z, Shan D, Yang W, Sychugov I, Sun L, Xu B (2024) Enlarging the Stokes shift of $CuInS_2$ quantum dots using thiol–ene polymers for efficient large-area luminescent solar concentrators. Energy Environ Sci 17(17):6338–6349. https://doi.org/10.1039/D4EE02603A
141. Wu H, Liang J, Han H (2008) A novel method for the determination of Pb^{2+} based on the quenching of the fluorescence of CdTe quantum dots. Microchim Acta 161(1–2):81–86. https://doi.org/10.1007/S00604-007-0801-4
142. Wu T, Tang M (2014) Toxicity of quantum dots on respiratory system. Inhal Toxicol 26(2). https://doi.org/10.3109/08958378.2013.871762
143. Xie H, Chen E, Ye Y, Xu S, Guo T (2020) Highly stabilized gradient alloy quantum dots and silica hybrid nanospheres by core double shells for photoluminescence devices. J Phys Chem Lett 11(4):1428–1434. https://doi.org/10.1021/ACS.JPCLETT.9B03578
144. Xie R, Wang Z, Zhou W, Liu Y, Fan L, Li Y, Li X (2016) Graphene quantum dots as smart probes for biosensing. Anal Methods 8(20):4001–4016. https://doi.org/10.1039/C6AY00289G
145. Xu L, Abd El-Aty AM, Eun JB, Shim JH, Zhao J, Lei X, Gao S, She Y, Jin F, Wang J, Jin M, Hammock BD (2022) Recent advances in rapid detection techniques for pesticide residue: a review. J Agric Food Chem 70(41). https://doi.org/10.1021/acs.jafc.2c05284
146. Xu H, Wang H, Lu Y, Zeng Y, Yang Y, Zhang Z, Wang H, Wang X, Li L (2021) CeO_2 quantum dots for highly selective and ultrasensitive fluorescence detection of 4-nitrophenol via the fluorescence resonance energy transfer mechanism. Spectrochim Acta Part A, Mol Biomol Spectrosc 262. https://doi.org/10.1016/j.saa.2021.120115
147. Yang C, Feng G, Dai S, Wang S, Li G, Zhang H, Zhou S (2017) Femtosecond pulsed laser ablation in microfluidics for synthesis of photoluminescent ZnSe quantum dots. Appl Surf Sci 414:205–211. https://doi.org/10.1016/J.APSUSC.2017.03.082
148. Yao F, Zhu P, Chen J, Li S, Sun B, Li Y, Zou M, Qi X, Liang P, Chen Q (2023) Synthesis of nanoparticles via microfluidic devices and integrated applications. Microchim Acta 190(7):1–21. https://doi.org/10.1007/S00604-023-05838-4
149. Yu H, Zhuang Z, Li D, Guo Y, Li Y, Zhong H, Xiong H, Liu Z, Guo Z (2020) Photo-induced synthesis of molybdenum oxide quantum dots for surface-enhanced Raman scattering and photothermal therapy. J Mater Chem B 8(5). https://doi.org/10.1039/c9tb02102g
150. Zandi P, Phani A, Kim S (2024) High-performance P-N junction heterostructure with carbon quantum dot and nitrogen self-doped graphitic carbon nitride for visible light photodetection. Adv Opt Mater 12(29):2401317. https://doi.org/10.1002/ADOM.202401317
151. Zare H, Ghalkhani M, Akhavan O, Taghavinia N, Marandi M (2017) Highly sensitive selective sensing of nickel ions using repeatable fluorescence quenching-emerging of the CdTe quantum dots. Mater Res Bull 95:532–538. https://doi.org/10.1016/J.MATERRESBULL.2017.08.015

152. Zhang W, Li B, Chang C, Chen F, Zhang Q, Lin Q, Wang L, Yan J, Wang F, Chong Y, Du Z, Fan F, Shen H (2024) Stable and efficient pure blue quantum-dot LEDs enabled by inserting an anti-oxidation layer. Nat Commun 15(1):1–7. https://doi.org/10.1038/s41467-024-44894-z
153. Zhang S, Liu X, Liu C, Luo S, Wang L, Cai T, Zeng Y, Yuan J, Dong W, Pei Y, Liu Y (2018) MoS_2 quantum dot growth induced by S vacancies in a $ZnIn_2S_4$ monolayer: atomic-level heterostructure for photocatalytic hydrogen production. ACS Nano 12(1):751–758. https://doi.org/10.1021/ACSNANO.7B07974
154. Zhang B, Tang WS, Ding SN (2023) Magnetic quantum dots barcodes using Fe_3O_4/TiO_2 with weak spectral absorption in the visible region for high-sensitivity multiplex detection of tumor markers. Biosens Bioelectron 227. https://doi.org/10.1016/j.bios.2023.115153
155. Zhang S, Xin Cheng JY, Chua JJ, Olivo M (2022) Characterization of quantum dots with hyperspectral fluorescence microscopy for multiplexed optical imaging of biomolecules. BioRxiv. https://doi.org/10.1101/2022.03.17.484752
156. Zhang H, Xu T, Li CW, Yang M (2010) A microfluidic device with microbead array for sensitive virus detection and genotyping using quantum dots as fluorescence labels. Biosens Bioelectron 25(11):2402–2407. https://doi.org/10.1016/J.BIOS.2010.02.032
157. Zhang Y, Zhang W, Zhang L, Song G, Wang N, Xu W, Huang W (2021) A molecularly imprinted electrochemical BPA sensor based on multi-walled carbon nanotubes modified by CdTe quantum dots for the detection of bisphenol A. Microchem J 170. https://doi.org/10.1016/j.microc.2021.106737
158. Zhou Q, Bai Z, Lu W, Wang Y, Zou B, Zhong H, Zhou QC, Bai ZL, Zhong HZ, Lu WG, Wang YT, Zou BS (2016) In situ fabrication of halide perovskite nanocrystal-embedded polymer composite films with enhanced photoluminescence for display backlights. Adv Mater 28(41):9163–9168. https://doi.org/10.1002/ADMA.201602651
159. Zhou J, Yang Y, Zhang CY (2015) Toward biocompatible semiconductor quantum dots: from biosynthesis and bioconjugation to biomedical application. Chem Rev 115(21). https://doi.org/10.1021/acs.chemrev.5b00049
160. Zhu L, Tao J, Li P, Sun W, Li J, Fan K, Lv J, Qin Y, Zheng K, Zhao B, Zhao Y, Chen Y, Tang Y, Wang W, Liang J (2023) Microfluidic static droplet generated quantum dot arrays as color conversion layers for full-color micro-LED displays. Nanoscale Adv 5(10):2743–2747. https://doi.org/10.1039/d2na00765g

Chapter 20
Carbon Quantum Dots: Structure, Properties, and Applications

Manjusha Chourasia, Tarab Fatima, and Manika Khanuja

Abstract Carbon quantum dots (CQDs) are a novel class of carbon-based nanomaterials with remarkable optical, electrical, and chemical properties. Their exceptional biocompatibility, controllable fluorescence, and nanoscale size have generated a lot of interest in a wide range of applications, such as drug delivery, biosensing, biomedical imaging, and photocatalysis. The basic features of CQDs are examined in this chapter, starting with their structural features and the synthesis techniques. The remarkable photoluminescence of CQDs is a fundamental characteristic that results from surface states, functional groups, and quantum confinement effects. This characteristic is tunable, allowing for emission throughout a broad spectrum by changes in doping, size, or surface chemistry. In addition to contributing to their optical properties, functional groups improve their solubility and facilitate facile conjugation with biomolecules, which increases their suitability for sensing and medicinal applications. Surface states offer control over their optical and chemical characteristics by influencing their fluorescence intensity and quantum yield. An extensive review of CQD structure, synthesis techniques, and changes is given in this chapter. The ability of top-down and bottom-up methods to fine-tune CQD properties is highlighted through a detailed discussion of techniques like laser ablation, hydrothermal synthesis, and microwave irradiation. CQDs are known as a transformative material in photocatalysis, biosensing, drug delivery, and biomedical imaging, providing high-performance and sustainable solutions across a variety of applications because of their exceptional photoluminescence, flexible surface chemistry, and versatile functionalization.

Keywords Quantum dots · Optical properties · Photoluminescence · Surface chemistry · Functionalization

M. Chourasia
Ministry of Electronics and Information Technology, New Delhi, India

T. Fatima · M. Khanuja (✉)
Center for Nanoscience and Nanotechnology, Jamia Millia Islamia, New Delhi, India
e-mail: manikakhanuja@gmail.com

T. Bhuyan et al. (eds.), *Multifunctional Quantum Dots*, Materials Horizons: From Nature to Nanomaterials, https://doi.org/10.1007/978-981-92-0172-3_20

1 Introduction

Carbon-derived nanomaterials, including graphene, carbon nanotubes (CNTs), nanodiamonds, fullerenes, and carbon quantum dots (CQDs), have gained extensive attention owing to their extraordinary structural features and unique optical properties [13, 66]. Carbon dots, or Cdots, are quasi-spherical, zero-dimensional nanoparticles, typically less than 10 nm. They were first identified in 2004 during the purification process of single-walled carbon nanotubes [35]. The word "carbon quantum dots" was first used by Sun and his team. Cdots are structurally made up of carbon atoms with sp^2/sp^3 hybridization. The size of quantum dots (QDs) significantly impacts their characteristics, such as their emission wavelength; shorter emission wavelengths are associated with smaller QD sizes [3, 14, 18]. There are several kinds of QDs, including carbon-based forms like Cdots, semiconductor forms like selenium and cadmium sulfides, and metallic forms like gold and silver [36, 64, 37]. Among them, CQDs are remarkable for their high resistance to photobleaching, chemical stability, strong luminescence, ease of chemical modification and functionalization, exceptional biocompatibility, and outstanding solubility in water [5, 57]. The crystal lattice parameter of CQDs is approximately 0.34 nm, which is similar to the (002) interlayer spacing of graphite [102]. In contrast to other carbon allotropes, CQDs have special chemical and photophysical characteristics, such as strong photoluminescence (PL) [86], broad-spectrum light absorption [30], excellent photostability [58], high solubility in water [2], upconversion capabilities [97], photosensitization [94], chemical stability, ease of functionalization [11], and low toxicity because of their non-metallic composition. Additionally, CQDs can be synthesized using a wide variety of cost-effective precursors and methods. These characteristics have made it possible for them to be widely used in a variety of applications, including drug delivery systems [79], biomedical imaging [56], sensors [17], photovoltaic devices, and photocatalysis [28, 66] as shown in Fig. 1. CQDs can be functionalized with various surface groups, predominantly oxygen-containing groups such as hydroxyl and carboxyl, which can be attributed to their excellent water solubility [99]. These functional groups make the structure of CQDs dissolve readily and form stable colloids in water or polar organic solvents. This property provides an advantage over graphene quantum dots (GQDs), which generally exhibit low solubility in conventional solvents. Additionally, the availability of surface functional groups can enhance the fluorescence properties of CQDs [98].

In this chapter, we reviewed the deep insights into structural characteristics, unique properties, and various applications of carbon quantum dots (CQDs). The discussion will encompass their synthesis methods, morphological features, optical behaviors, and electronic properties, emphasizing their potential in various domains. Additionally, we will explore their role in emerging technologies, particularly in sensing, bioimaging, and environmental remediation.

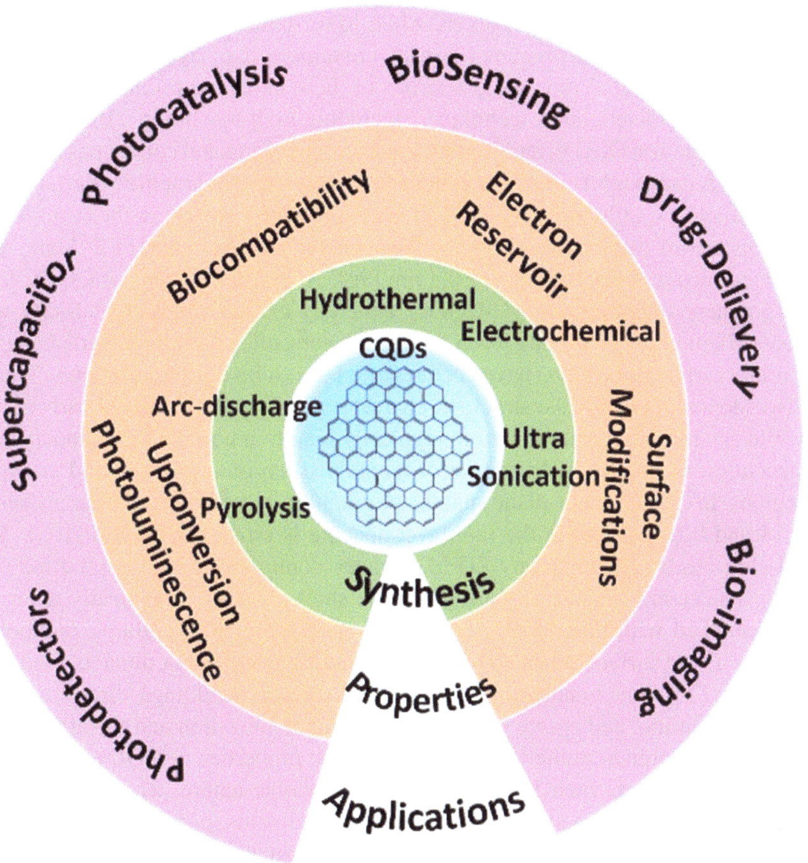

Fig. 1 General structure, properties, and applications of carbon quantum dots (CQDs)

2 Structure of Carbon Quantum Dots

Carbon quantum dots (CQDs) exhibit a core–shell structure that forms through a nucleation process involving the gradual growth of a core and the development of a shell, which contains various functional groups as shown in Fig. 2 [32, 90]. Studies have shown that the core, also referred to as intrinsic states, can either be graphitic crystalline (sp^2) or amorphous (a mixture of sp^2/sp^3), depending on the extent of sp^2 carbon present. Remarkably, graphitic crystalline (sp^2) cores are most commonly reported in the literature [19, 22]. These cores are typically very small, measuring about 2–3 nm in size, and possess a lattice spacing of approximately 0.2 nm [41]. The characteristics of the core are influenced by the synthesis method, precursor materials, and specific synthetic parameters such as reaction temperature, duration, and pH [55]. In general, synthesis temperatures exceeding 300 °C facilitate the formation of graphitic crystalline (sp^2) cores, while lower temperatures often lead to amorphous cores unless the precursor already contains sp^2/sp^3 hybridized carbon [100]. Insights into the structure and properties of these cores are primarily derived from advanced characterization techniques, including high-resolution TEM (HR-TEM), X-ray diffraction (XRD), and Raman spectroscopy. Tang and colleagues highlighted that CQDs can exhibit either amorphous (mixed sp^2/sp^3) or graphitic crystalline (sp^2) core structures, influenced by the degree of sp^2 carbon present in the core [85]. The dimensions of these cores remain small, generally in the range of 2–3 nm, with a lattice spacing of approximately 0.2 nm [96]. A specialized type of CQDs, known as graphene quantum dots (G-CQDs), generally exhibits high crystallinity with a layered structure comprising fewer than four layers, often just a single graphene layer. Certain carbon dots (CDs) have been reported to exhibit hollow structures. They can also adopt a diamond-like structure, characterized by sp^3-hybridized carbon atoms, while sp^2-hybridized carbon atoms typically lead to amorphous or nanocrystalline structures. Generally, CDs consist of a combination of crystalline and amorphous regions. In CQDs, the in-plane spacing of graphitic carbon typically ranges between 0.18 and 0.24 nm, while the interlayer spacing is around 0.334 nm [105]. Transmission electron microscopy (TEM) images commonly reveal crystalline CQDs with diameters of about 5–6 nm [76]. The shell of CQDs is typically amorphous and enriched with functional groups, such as oxygen, amino groups, and polymer chains. These surface states arise from the starting materials or dopants used during synthesis [75]. Functional groups on the CQDs surface, including hydroxyl, carboxyl, carbonyl, amino, and epoxy groups, enable functionalization and the attachment of organic or inorganic moieties [74]. The surface properties of CQDs play a crucial role in determining their hydrophilic or hydrophobic nature, which, in turn, influences their thermodynamic stability in various solvents, particularly in water. The stability of CQDs' aqueous solutions is directly impacted by the surface chemistry of the shell [18].

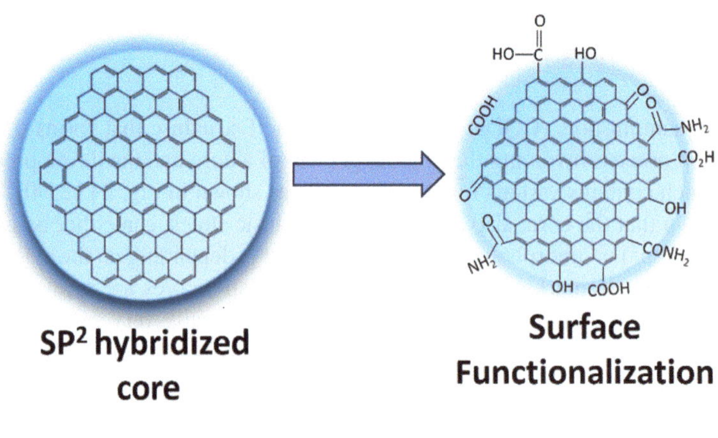

Fig. 2 Structure of carbon quantum dots (CQDs)

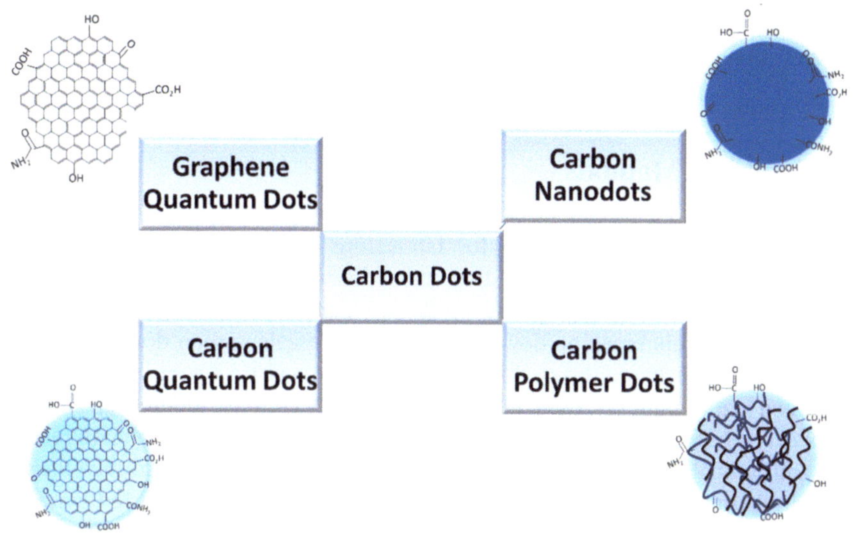

Fig. 3 Classification of carbon dots: graphene quantum dots (GQDs), carbon quantum dots (CQDs), carbon nanodots (CNDs), and carbonized polymer dots (CPDs)

CDs can be broadly divided into graphene quantum dots (GQDs), carbon quantum dots (CQDs), carbon nanodots (CNDs), and carbonized polymer dots (CPDs) based on factors such as size, shape, synthesis mechanisms, the structure of the core, functional groups present on the surface, and other similar properties as shown in Fig. 3 [49]. Carbon dots (CDs) can be classified into several categories based on their structure and properties: GQDs are small graphene fragments typically composed of one or a few graphene layers (Wadhwa et al. 2020). They exhibit distinct graphene lattice patterns and contain chemical groups located at the edges or within interlayer defects.

Carbon Quantum Dots (CQDs): CQDs are typically spherical and characterized by clear crystalline lattice structures. Their surfaces are functionalized with various chemical groups. Carbon Nanodots (CNDs): CNDs have a high degree of carbonization but lack distinct crystalline lattice structures. They may exhibit polymer-like features, with chemical groups distributed on their surface.

Carbonized Polymer Dots (CPDs): CPDs possess a hybrid structure that combines polymer and carbon components [77, 95]. They feature a carbon core surrounded by abundant surface functional groups or polymer chains. The type of CD is determined by its core structure, while its surface chemistry enables subsequent functionalization based on its intended application. The elemental composition (C, N, and O) and physicochemical properties of CDs can vary significantly depending on the synthesis approach, defects, heteroatom incorporation, and surface functional groups [78]. The optical properties, including absorption and emission behavior, are strongly influenced by surface chemistry. Fine-tuning the surface chemistry of CDs can enhance their quantum yield (QY) and solubility [75]. Overall, these nanomaterials, despite their differences in synthesis, functionalization, and surface modifications, are collectively categorized as CDs.

3 Synthesis Methods

Several methods were developed for fabricating carbon quantum dots (CQDs), which are widely classified into two main divisions: top-down and bottom-up approaches, as shown in Fig. 4. The top-down method involves larger carbon-based precursor materials breaking down to produce nanoscale carbon dots. Techniques including arc discharge, laser ablation, electrochemical treatment, and ultrasonic treatment are commonly used in this approach. In contrast, bottom-up methods involve synthesizing CQDs from smaller molecular precursors through polymerization and carbonization processes. This method includes techniques like solvothermal or hydrothermal synthesis, combustion, pyrolysis, and microwave-assisted synthesis. The fluorescence properties of CQDs are influenced by three primary factors: quantum confinement states, defect states, and surface states. By selecting specific precursors and synthesis techniques, these properties can be precisely controlled. Additionally, CQDs can be functionalized during synthesis with single atoms such as sulfur, nitrogen, or with amine groups.

3.1 Top–Down Approach

(a) **Laser Ablation**

Laser ablation (LA) has emerged as an innovative and efficient method for synthesizing nanomaterials, particularly carbon quantum dots (CQDs), due to its rapid processing and straightforward operation. The process generates high local pressures and temperatures, which enable the direct fragmentation of the target material into micro- and nanoparticles without leaving residues [72, 87]. This ensures minimal risk of contamination and eliminates the introduction of impurities. For example, Cui et al. developed CQDs using a custom-built dual-beam pulsed laser ablation (PLA) system with carbon cloth as a cost-effective precursor. By splitting a single laser beam into two, the ablation time was significantly reduced, and the CQD production yield was enhanced. The resulting CQDs exhibited a quantum yield of 35.4%, high uniformity, and stable, excitation-independent photoluminescence (PL), attributed to electron transitions from the π^* state to surface states [15]. Similarly, Thongpool and team synthesized carbon particles through laser ablation of a graphite target in ethanol, investigating the influence of the ethanol medium on the particles' physical, chemical, and optical characteristics [87]. In another report, fluorescence-emitting carbon dots were synthesized using Nd: YAG laser with harmonic generation, enabling the ablation of the carbon target as shown in Fig. 5.

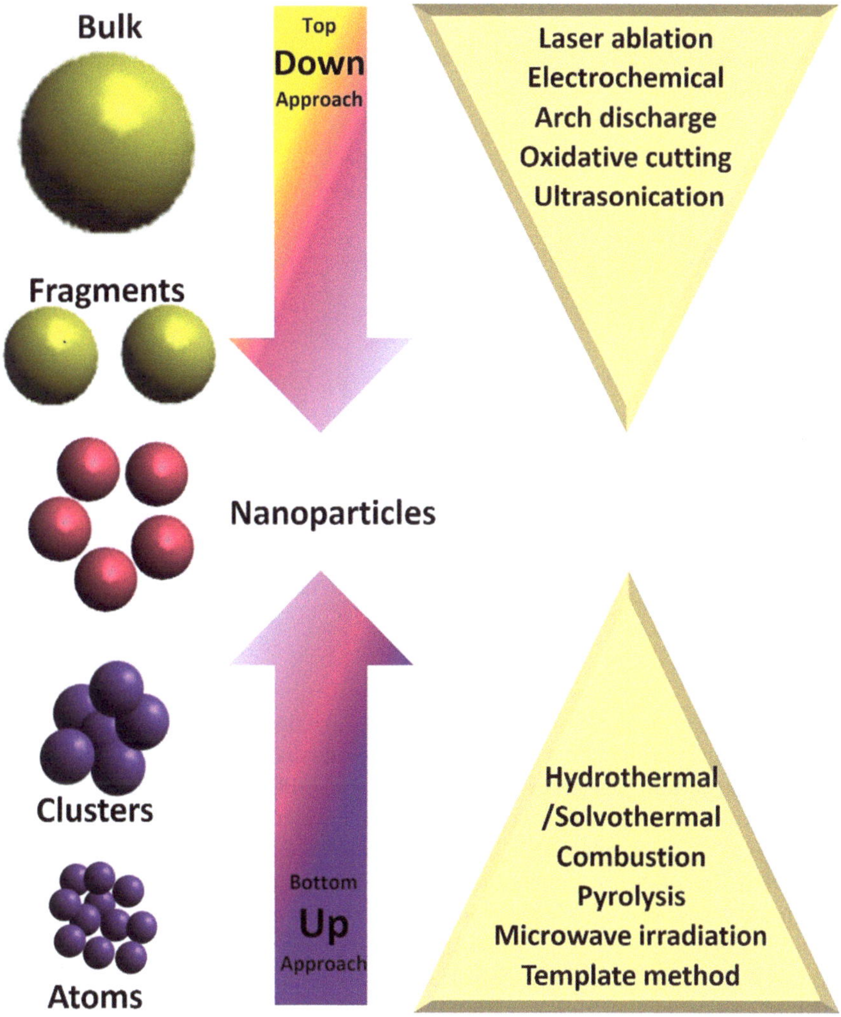

Fig. 4 Synthesis methods of carbon quantum dots

(b) **Electrochemical**

The electrochemical synthesis of carbon dots (CDs) has proven to be an effective top-down approach for producing high-quality CDs, offering a wide range of potential applications [106]. In 2007, Zhou and team pioneered this method by employing electrochemical exfoliation of multi-walled carbon nanotubes (MWCNTs). Their study revealed that cyclic voltammetry (CV) of MWCNTs in degassed acetonitrile (ACN) with tetrabutylammonium perchlorate (TBAP) as the supporting electrolyte caused

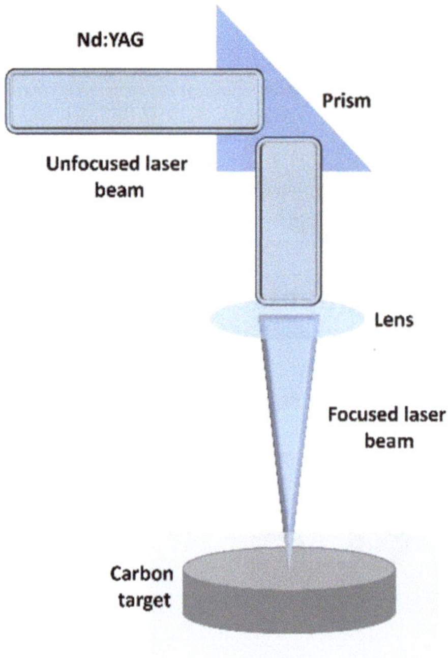

Fig. 5 A schematic representation for the synthesis of the CDs using laser ablation

the solution to turn dark brown, emitting strong blue fluorescence under UV irradiation. Following purification through evaporation and dialysis, the resulting CDs were monodispersed with an average size of 2.8 ± 0.5 nm [106]. This study emphasized the importance of selecting an appropriate supporting electrolyte based on the intended applications of the synthesized CDs. Alkaline conditions were frequently cited as optimal for the synthesis, with key goals being the production of CDs with sizes under 10 nm and desirable fluorescence properties. In another approach, Wang et al. utilized an electrochemical process to synthesize photoluminescent carbon nanodots (Cdots) from glycine in an alkaline medium. The formation mechanism involved electrooxidation, electropolymerization, carbonization, and passivation [92] (Fig. 6). These Cdots exhibited excitation-wavelength-dependent and pH-sensitive photoluminescence (PL), with remarkable stability in high-salt solutions (up to 500 mM NaCl). Furthermore, Yu-Sheng Lee and colleagues fabricated CDs through the electrochemical treatment of diethylenetriaminepentaacetic acid and ethylenediamine as precursors. These CDs displayed blue fluorescence under UV light, with a maximum emission at 407 nm for an excitation wavelength of 325 nm, and a quantum yield of 2.5%. The CDs also showed excellent water solubility, resistance to photobleaching, and good stability, which could be fine-tuned by adjusting the electrochemical parameters [45].

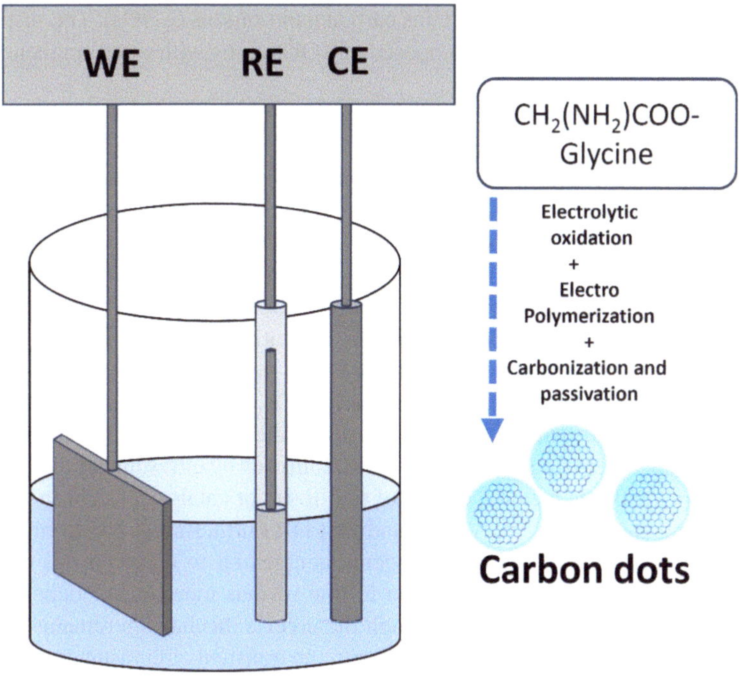

Fig. 6 Electrochemical method for the synthesis of carbon dots

(c) Arc Discharge Method

The arc discharge method is a widely recognized top-down technique for synthesizing CQDs. This method involves generating plasma through an electric arc, which breaks down gas and facilitates the evaporation of carbon atoms from opposing carbon electrodes. While CQDs produced using this technique exhibit favorable fluorescence properties, their non-uniform size and low yield limit their suitability for large-scale production. Interestingly, CQDs were accidentally discovered in 2004 during the purification of single-walled carbon nanotubes (SWNTs) using the arc discharge method. Nitric acid, acting as an oxidizing agent, was employed to oxidize the arc ash, introducing functional groups on the surface of the CQDs. These functional groups enhanced the aqueous solubility of the CQDs, which demonstrated a quantum yield of 1.66% at an excitation wavelength of 366 nm [96]. Su et al. synthesized CQDs exhibiting upconversion fluorescence by utilizing carbon by-products generated during arc synthesis. Similarly, Hamid Delavari and colleagues prepared CQDs using the arc discharge technique in water [8, 83]. Further advancements were reported by F. J. Chao-Mujica et al., who synthesized CQDs through the submerged arc discharge in water (SADW) method [4]. The CQDs produced had a size distribution ranging from 1 to 5 nm. During the SADW process, the products segregate into three distinct phases: floating material, suspended material, and precipitate. The precipitate is enriched with multi-walled carbon nanotubes (MWCNTs), while the floating material primarily contains carbon nano-onions (CNOs). The suspended phase in water is predominantly composed of CQDs along with some graphene oxide (GO) sheets.

(d) **Ultrasonication**

The sonochemical synthesis is an environmentally friendly method characterized by its simplicity, mild operational conditions, and utilization of green energy sources. It also offers the ability to produce carbon dots with controlled physical and chemical properties. The standard procedure for synthesizing carbon dots via sonochemical methods generally involves synthesizing carbon precursors in an appropriate solvent at a defined concentration and then sonicating the mixture for a predetermined amount of time. The energy from ultrasonic radiation facilitates condensation and carbonization, leading to the formation of CDs. Li et al. demonstrated a sonochemical approach to synthesizing CDs by employing glucose as a precursor material in an alkali or acid-assisted ultrasonic process, followed by ultrasonic irradiation [48]. This one-pot synthesis method, devoid of stabilizers or catalysts, has been extended to include metals such as Au, Ag, Sn, and Pb [44]. Furthermore, identifying appropriate carbon precursors and doping agents is essential to the synthesis of doped carbon dots. The formation of CDs can follow various mechanisms depending on the synthesis technique employed, though the precise mechanism remains undetermined. The sonochemical technique uses extreme transient circumstances generated by ultrasound, creating unique hotspots with temperatures exceeding 5000 K, and pressures above 1000 atmospheres. The transient and extreme conditions generated in acoustic cavitation not only enable the synthesis of targeted materials but also allow such processes to occur at room temperature in a benchtop setup [40]. For example, the sonochemical carbonization and degradation of polyethylene glycol (PEG) as a precursor involves a complex series of chemical reactions that remain challenging to pinpoint [42]. The characteristic features of Sn@CDs, which serve as primary carbon particles and secondary metal coatings, are significantly influenced by reaction parameters such as sonication time and the polymer-to-metal ratio (Fig. 7) [43]. By optimizing these parameters, the performance of the synthesized materials can be enhanced.

3.2 Bottom–Up Approach

(a) **Hydrothermal Method**

Hydrothermal synthesis is a widely employed method for fabricating CQDs, with their average size being controlled by varying precursor concentrations, reaction temperatures, and processing durations [21, 60, 103]. A notable method for controlling CQD size involves adjusting the volume of a sucrose solution in the hydrothermal reactor while maintaining other experimental conditions constantly. In cases where homogeneous nucleation dominates the synthesis process, the CQDs

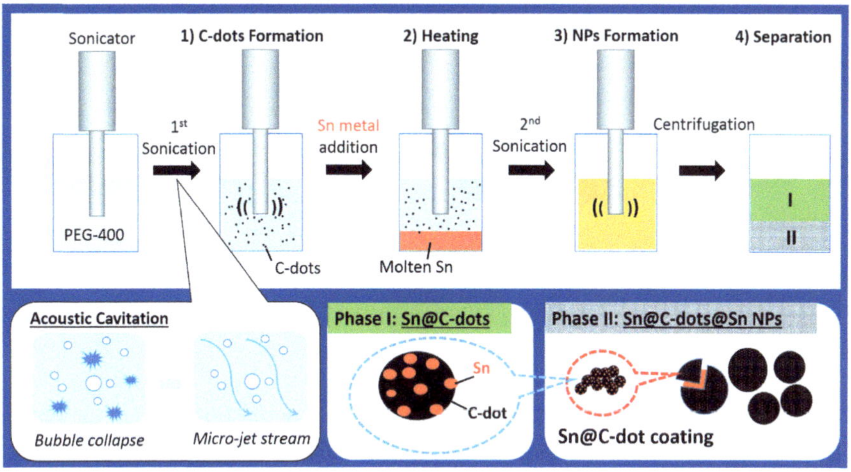

Fig. 7 Schematic representation of the sonochemical synthesis of Cdots, Sn@Cdots, and Sn@Cdots@Sn NPs [43]. Permission from Royal Society of Chemistry

produced under varying volumes would typically exhibit similar size distributions. However, studies have shown that the size of CQDs synthesized from sucrose under hydrothermal conditions is significantly influenced by decomposition and polymerization processes. During decomposition, sucrose in an aqueous medium undergoes hydrolysis to form glucose and fructose, which are subsequently broken down into smaller organic molecules such as furfurals and weak acids. These intermediates undergo polymerization to form larger molecular structures, eventually resulting in CQDs within the hydrothermal reactor. Homogeneous nucleation, which is well-documented for metals, ceramics, and organic compounds, typically involves higher activation-free energy compared to heterogeneous nucleation. Given that the surface of the hydrothermal reactor may act as a catalyst, it is hypothesized that heterogeneous nucleation plays a critical role in CQD formation under these conditions [6]. Classical nucleation theory provides a framework for analyzing the free energy barriers associated with the formation of critical nuclei during heterogeneous nucleation. The presence of foreign surfaces, such as reactor walls or impurities, can significantly reduce these energy barriers, making nucleation on the reactor walls more favorable than in the liquid precursor. Furthermore, the number of nuclei formed heterogeneously is proportional to the surface area of the reactor. To explore the role of heterogeneous nucleation in CQD formation, experiments can be conducted using varying filling volumes of sucrose solution while keeping the reactor geometry, precursor concentration, reaction temperature, and time constant. For example, Tarab et al. successfully synthesized CQDs from lemon juice, achieving particle sizes between 3 and 5 nm using hydrothermal processing at 220 °C for 24 h (Fig. 8) [20]. Similarly, Pradeep et al. reported CQDs synthesized from *Azadirachta indica* (neem) with high fluorescent quantum yields (QYs) of up to 27.2% [84].

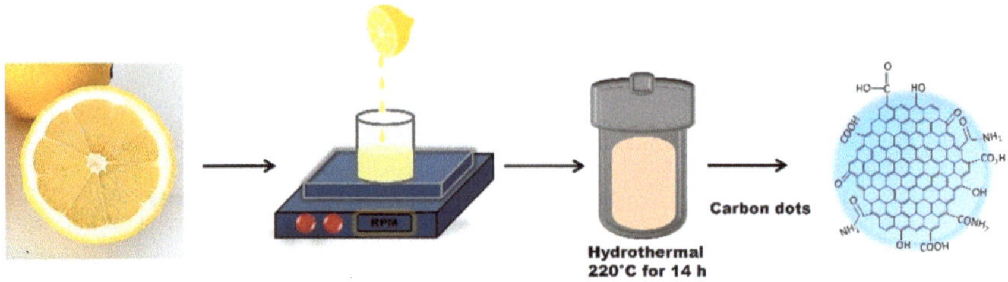

Fig. 8 Hydrothermal synthesis of carbon quantum dots (CQDs) using lemon juice

(b) **Combustion**

In 2007, Liu and colleagues pioneered the use of the combustion method for synthesizing CQDs [52]. This technique employs oxidative acid treatments to aggregate smaller carbon precursors into CQDs, improving their water solubility and enabling control over their fluorescence characteristics. Later, Ming-Cong Rong and colleagues used a combustion synthesis method to produce fluorescent B, N-carbon dots using aminophenylboronic acid (APBA) as the starting material (Fig. 9) [73]. Green fluorescence was seen in the resultant B, N-carbon dots, which had a 1.6% quantum yield (QY). These carbon dots demonstrated excellent resistance to photobleaching and high stability in varying ionic strengths. The proposed synthesis method showed promise for practical applications, particularly in detecting Cu^{2+} ions in natural water samples.

(c) **Pyrolysis**

The pyrolysis method involves the thermal decomposition of a precursor at elevated temperatures (typically above 430 °C) and under pressure in an oxygen-free environment [96]. In this process, the carbon precursor breaks down into nanoscale

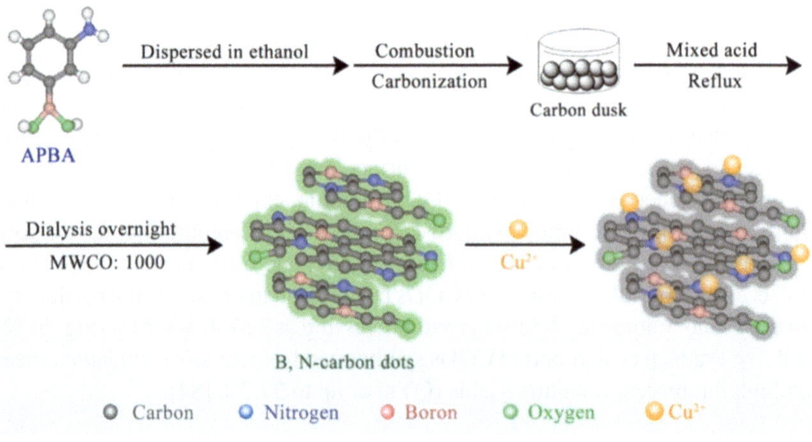

Fig. 9 Synthesis of carbon quantum dots using the combustion method [73]. Permission from Elsevier

colloidal particles, often with the assistance of alkalis or strong acids as catalysts. This technique is recognized for its practicality, repeatability, simplicity, and ability to produce carbon dots (CDs) with high quantum yields (QYs). However, separating smaller precursors from raw materials remains a significant challenge.

In 2009, Liu et al. introduced a novel pyrolysis-based approach for CQD synthesis, utilizing resol as the carbon source and surfactant-modified silica spheres as templates [53]. The CQDs obtained through this method exhibited blue fluorescence, with sizes ranging from 1.5 to 2.5 nm, and achieved a QY of 14.7%. These CQDs demonstrated excellent stability across a broad pH range (pH 5–9). Additionally, Zeinab Bagheri [7] and colleagues demonstrated that direct pyrolysis of citric acid (CA) is an effective bottom-up strategy for synthesizing pristine CDs with uniform size distribution. While systematic studies on optimizing the carbonization degree are limited, they investigated pyrolysis temperatures between 160 and 220 °C and varied heating durations based on the thermal decomposition pathway of CA. The results revealed that CDs prepared at 160 °C for 50 min achieved a remarkable QY of 29%, nearly double the highest reported QY for CA-based pyrolysis. The CDs synthesized through this approach are well-suited for applications such as turn-off sensing.

(d) **Microwave Irradiation**

Microwave synthesis is a fast and economical technique for synthesizing CQDs using microwave heating. This method is simple and convenient, requiring minimal synthesis time while yielding CQDs with enhanced quantum yields. Zhu et al. were among the first to use microwave heating (500 W) to synthesize fluorescent CQDs by heating poly(ethylene glycol) (PEG-200) and saccharides for 2–10 min [107]. This approach is recognized as fast, eco-friendly, and energy-efficient for CQD synthesis. Microwave-assisted techniques for producing carbon-based nanomaterials generally operate at lower growth temperatures and yield products with high surface areas, excellent adsorption capacities, and high purity. These features make microwave synthesis an energy-efficient and practical solution for tackling challenges in industrial and municipal water treatment applications.

In another study, Muhammad Nazar et al. [61] utilized microwave heating for 5–10 min to synthesize CQDs derived from arabica coffee ground-activated carbon. These CQDs exhibited vibrant blue fluorescence at 455 nm under 365 nm excitation. Dynamic light scattering (DLS) analysis showed that the average size of the CQDs was approximately 10.12 nm, with a quantum yield of 6.01%. These CQDs were successfully applied as fluorescent sensors for detecting Fe^{3+} ions, demonstrating their potential in sensing applications.

4 Properties of CQDs

4.1 Optical Properties

(a) **Absorbance**

The optical properties of carbon quantum dots (CQDs) can be explored through UV–Vis absorption and photoluminescence analyses. CQDs exhibit strong absorption in the ultraviolet region, with an absorption tail extending into the visible spectrum. This behavior is attributed to π–π* transitions associated with aromatic C=C bonds and n–π* transitions linked to C=O bonds or other functional groups on their surface [29]. Notably, CQDs can act as photosensitizers when integrated with wide band gap semiconductors like TiO_2, broadening their photoresponse from the UV to visible range. The UV absorption characteristics of CQDs are influenced by several factors, including surface functionalities, synthesis techniques, precursor materials, chemical environments (such as pH and solvents), and the incorporation of heteroatoms like nitrogen or sulfur into their structure. Absorption bands observed beyond 350 nm, typically in the 350–600 nm range, are linked to electronic transitions of surface functional groups. Surface passivation significantly enhances the optical activity of CQDs, enabling them to emit fluorescence across a broad spectrum, from visible to near-infrared wavelengths (Fig. 10). Furthermore, surface passivation can improve the quantum yield (QY) of CQDs, reaching values as high as 55–60% [34, 38, 46]. For example, functionalization with 4,7,10-trioxa-1,13-tridecanediamine (TTDDA) has been shown to shift absorption toward longer wavelengths (350–550 nm), further enhancing their optical attributes [65].

A group of researchers [82] described the preparation of multicolor zinc-doped carbon dots (CDs). TEM analysis demonstrated that the variation in emission colors was not affected by the size of the CDs, as the particles with different emissive properties exhibited comparable sizes. Findings from FTIR and XPS studies indicated

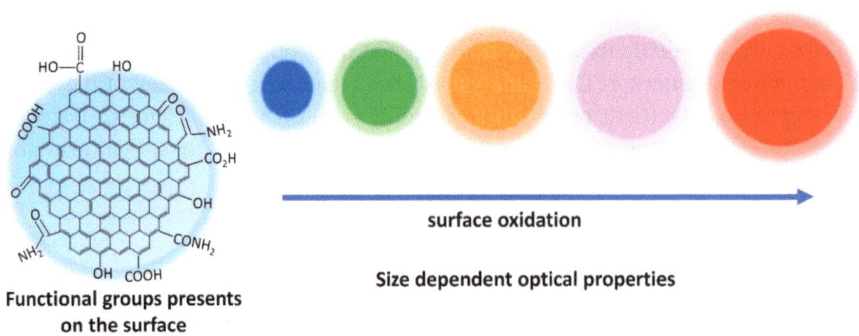

Fig. 10 A schematic illustration depicting the various functional groups that can form on the surface of carbon nanodots (QDs) and their distinctive photoluminescence behavior, influenced by surface oxidation levels

that zinc doping reduces the percentage of graphitic carbon by enhancing oxidation levels, thereby decreasing the abundance of other functional groups. CDs emitting red fluorescence were found to have fewer oxygen-related functional groups. In addition, nitrogen-based functional groups, such as amine groups, nitro, and amide groups, are also the most prevalent groups present on both the core and the surface of the CQDs. Nitrogen doping is particularly useful in modifying the characteristics of CDs because of the same atomic size and electrical arrangement of carbon and nitrogen [23, 63].

(b) **Photoluminescence**

An important optical characteristic of CQDs is their photoluminescence (PL), which occurs when light is emitted following the photon's absorption, starting through photoexcitation. The emission varies with increasing excitation wavelengths [71]. Various mechanisms have been proposed to explain their PL behavior, including size-dependent optical effects (quantum confinement), defect states, trap states, surface energy states, and recombination of surface-localized electron–hole pairs [12, 47, 66]. Regardless of the underlying mechanism, the PL emission wavelength depends on the excitation wavelength. Recent studies have explored the synthesizing mechanisms of CQDs through the hydrothermal method using citric acid (CA) and lemon juice as precursors at different temperatures. At 300 and 400 °C, CQDs emitting PL predominantly from their carbogenic cores were obtained, whereas lower temperatures favored the formation of molecular fluorophores [39]. This suggests that the PL behavior of CQDs is highly sensitive to the synthesis conditions. This adaptability makes CQDs highly versatile for applications in biosensing, photocatalysis, bio-medical imaging, drug delivery, and supercapacitors.

The PL of CQDs provides crucial insights into their optical properties, which are influenced by factors including band gap energy, particle size, pH, chemical environment, synthesis conditions, and surface functionalization. As the size decreases results in discrete energy levels which also affect the wavelength and fluorescence-emitting property of CQDs (Fig. 11). PL arises from the recombination of an excited electron with a hole as it transitions back to the ground state, emitting energy within a specific spectral range from ultraviolet to near-infrared. Generally, PL emission occurs at a longer wavelength than the excitation wavelength (Stokes PL emission or down-conversion) [18]. However, CQDs can also exhibit upconversion photoluminescence (UCPL), where shorter wavelengths are emitted than the excitation wavelength. This phenomenon is attributed to multiphoton absorption processes or anti-Stokes PL emission, where photons of shorter wavelengths are absorbed simultaneously, resulting in unique optical behavior [33, 67, 91].

4.2 Electrical Properties

Quantum dots, also called "artificial atoms" due to their charge carriers, specifically e^- and h^+, are confined to discrete energy levels parallel to the quantized energy

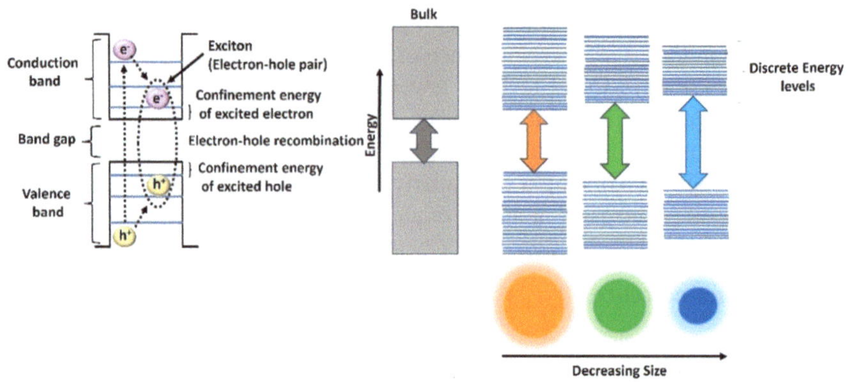

Fig. 11 Size-dependent properties of CQDs

states of electrons in individual atoms. This phenomenon arises due to the spatial confinement of carriers within the dimensions of the nanostructure. In QDs, holes remain in the valence band while electrons are stimulated from the valence band to the conduction band when enough energy is applied. In the valence band, this hole acts as a mobile positive charge. Electrons and holes rapidly release energy when excited, stabilizing at levels close to the top of the valence band and the bottom of the conduction band, respectively. During the recombination process, where electrons return to the valence band to fill the holes, energy is released, as depicted in Fig. 12 [18, 54, 67].

Typically, a single photon can excite only one electron across the band gap, forming a single exciton, with any surplus energy dissipated as heat. However, QDs can produce multiple excitons simultaneously through a process known as multiple exciton generation, enhancing the energy conversion efficiency of these nanocrystals [62].

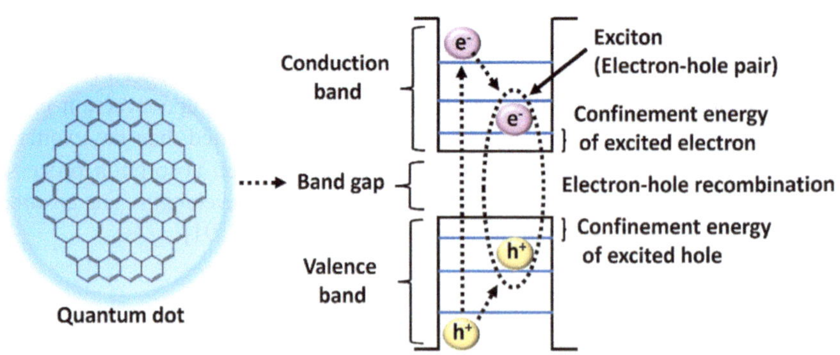

Fig. 12 Schematic representation of the electron–hole recombination and their energy levels

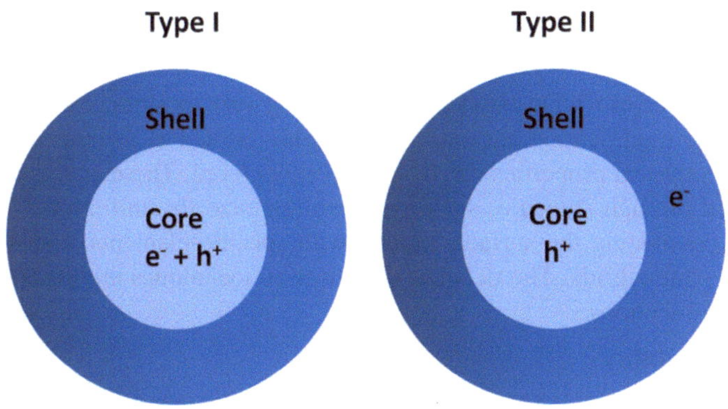

Fig. 13 Types of CQDs, **a** type-I and **b** type-II

The confinement behavior in QD core/shell structures is significantly influenced by the intrinsic properties of the core and shell materials. When the band gap of the shell material exceeds that of the core material, the electron–hole pair remains confined within the core, as shown in Fig. 13. Such QDs, categorized as type-I QDs, exhibit superior PL efficiency compared to type-II QDs. In contrast, type-II QDs have a shell with a smaller band gap than the core, resulting in electrons and holes being localized within the shell and the core, respectively, as illustrated in Fig. 13b. The spatial separation of charge carriers in type-II QDs requires crossing the core–shell interface for radiative recombination, producing fluorescence with broad and tunable wavelengths [9].

It is important to highlight the substantial interaction between charge carriers in QDs and their surrounding environment. These interactions are heavily influenced by the nature of the surrounding materials, thereby significantly impacting the overall electrical properties of the QDs. Due to these complexities, developing a comprehensive many-electron theory for QDs remains a formidable challenge. Jeison Manuel Arroyave et al. experimented to measure electrical resistance using two-point contact configurations for ITO/sample systems. The recorded resistance values for the nanoparticles were approximately 1000 Ω for ITO/Cdots, 850 Ω for ITO/AgNPs, and 600 Ω for ITO/Ag@Cdots. This research emphasizes comparing these electrical resistances, highlighting the significant reduction in resistance observed for ITO/Ag@Cdots. This decrease indicates enhanced electrical conductivity compared to ITO/Cdots and ITO/AgNPs, with conductivity improving by nearly 50% relative to ITO/Cdots and around 30% relative to ITO/AgNPs [54].

5 Modifications of Carbon Quantum Dots

The performance of CQDs can be enhanced using various strategies. Their ease of synthesis and modification has made them highly attractive, as their optical, electronic, and physical properties can be effectively tailored. These enhancements can be achieved through strategies such as controlling size for uniformity, functionalizing and passivating the surface, doping with specific elements, and developing innovative nanohybrids. The detailed modification techniques are discussed in the following subsections.

(a) **Size Tunability**

Size tunability is one of the most significant factors affecting the properties of carbon quantum dots (CQDs), as their quantum confinement effects are size dependent. Smaller CQDs, typically below 5 nm, exhibit stronger quantum confinement, resulting in blue-shifted photoluminescence and higher bandgap energy. In contrast, larger CQDs show red-shifted emission due to reduced confinement effects. This tunability is achieved through precise control of synthesis parameters such as precursor concentration, temperature, reaction time, and pH. For example, hydrothermal and solvothermal methods allow fine control of CQD size by adjusting reaction temperature and time. A study demonstrated that CQDs synthesized using citric acid and ethylenediamine at different temperatures (180–220 °C) exhibited varying sizes and corresponding emission wavelengths ranging from blue to green fluorescence.

The size also affects applications like bioimaging, where smaller CQDs offer higher tissue penetration and less scattering. Similarly, in optoelectronics, size-controlled CQDs enhance charge carrier dynamics, making them suitable for devices like light-emitting diodes (LEDs). Uniform size distribution is often achieved using post-synthesis methods like dialysis or size-selective precipitation to eliminate larger particles. Additionally, the correlation between size and surface states influences the fluorescence quantum yield, which is critical for applications in sensing and catalysis [59, 101].

(b) **Surface Functionalization**

Surface functionalization plays a pivotal role in tailoring the chemical and physical properties of CQDs. Functional groups like –COOH, –OH, and $-NH_2$ can be introduced through chemical treatment or by selecting appropriate precursors during synthesis. These groups improve water solubility, biocompatibility, and stability under various environmental conditions. For example, CQDs functionalized with polyethylene glycol (PEG) exhibit enhanced stability and are widely used in bioimaging and drug delivery applications. Functionalization also enables the incorporation of targeting ligands, enhancing selectivity for biomedical imaging or biosensing. CQDs modified with folic acid can target cancer cells with high folate receptor expression. Furthermore, functionalized CQDs improve their interaction with other nanomaterials, expanding their applications in hybrid systems. For example, CQDs functionalized with thiol groups can form strong bonds with gold nanoparticles, creating nanocomposites for enhanced catalytic and sensing applications [75, 98, 99].

(c) **Elemental Doping**

Elemental doping introduces heteroatoms like nitrogen, sulfur, or phosphorus into the carbon lattice of CQDs, significantly enhancing their properties. Nitrogen doping, one of the most common methods, increases electron density and improves the photoluminescence quantum yield. CQDs doped with nitrogen using urea or ammonia precursors show brighter fluorescence and better stability. Sulfur doping, achieved using thiourea or sulfuric acid, enhances optical absorption and catalytic activity, making CQDs suitable for applications like photocatalysis and energy storage [50].

Phosphorus doping has been reported to improve the electrocatalytic performance of CQDs, particularly in oxygen reduction reactions. A notable study synthesized phosphorus-doped CQDs using phosphoric acid, demonstrating enhanced electrocatalytic activity for fuel cells. Elemental doping also enables excitation-dependent emission behavior, which is crucial for multicolor imaging and tunable light-emitting applications [27, 66].

(d) **Passivation**

Passivation reduces the defects on the CQD surface, enhancing their optical and electronic properties. This process often involves coating the surface with small molecules, polymers, or inorganic shells. Passivated CQDs exhibit higher quantum yields and better resistance to photobleaching and environmental degradation. For instance, CQDs capped with polyethyleneimine (PEI) showed enhanced fluorescence intensity and stability, making them suitable for bioimaging. Passivation is also crucial in reducing surface trap states that cause non-radiative recombination. CQDs coated with silica shells have demonstrated improved optical properties and thermal stability, expanding their use in high-temperature catalytic processes. Additionally, passivated CQDs are more effective in optoelectronic devices, such as solar cells, due to their enhanced charge transfer efficiency [16, 50].

(e) **Nanocomposite Formation**

Synthesizing nanocomposites by integrating CQDs with other materials enhances their properties and broadens their applications. CQDs combined with metal nanoparticles, such as gold or silver, exhibit improved catalytic and sensing performance. For example, a CQD-gold nanocomposite was used for the efficient detection of heavy metal ions due to enhanced electron transfer and stability. CQDs can also form hybrids with semiconductors like TiO_2 or ZnO, improving their photocatalytic efficiency under visible light. A study reported that CQD-TiO_2 nanocomposites degraded organic pollutants more effectively than pure TiO_2 due to enhanced charge separation and visible light absorption. Similarly, CQDs incorporated into 2D materials like MoS_2 have shown improved conductivity and electrocatalytic performance, making them valuable for energy storage and conversion technologies.

6 Applications

(a) **Photocatalysis**

Carbon quantum dots (CQDs) are emerging as effective photocatalysts due to their unique optical and electronic properties, such as broad light absorption, high photostability, and excellent electron transfer capability. CQDs act as photosensitizers or co-catalysts, enhancing the photocatalytic efficiency of other materials by improving charge separation and reducing recombination. Their ability to upconvert light from near-infrared (NIR) to visible wavelengths further expands their application in photocatalysis under solar illumination.

For example, CQDs were incorporated into MIL-53(Fe) and gold nanoparticles for the improved degradation efficiency of chromium (VI) and dyes. The developed photocatalyst exhibited superior degradation efficiency of nearly 100% in 20 min, and the mechanism was given in Fig. 14a [26]. The enhanced performance was attributed to better electron–hole pair separation facilitated by CQDs. Another study demonstrated that CQD-modified WS_2-PANI composites exhibited superior photocatalytic degradation of estradiol due to the synergistic effects of CQDs on WS_2 and polyaniline. Estradiol was reported to be a harmful endocrine-disrupting hormone. The high degradation efficiency was attributed to the low rate of recombination and high visible light capturing activity [20]. Bingqing Wang et al. reported the carbon quantum dot decorated ternary nanocomposite CQDs/MoS_2/$ZnIn_2S_4$ exhibited enhanced photocatalytic behavior for hydrogen evolution than binary MoS_2/$ZnIn_2S_4$ (Fig. 14b). The 3.0 weight% MoS_2/CQDs/$ZnIn_2S_4$ nanocomposite demonstrated remarkable performance with Na_2S/Na_2SO_3 as a sacrificial agent, achieving 750 μmol of hydrogen evolution over 5 h. At a wavelength of 420 nm, the apparent quantum efficiency (AQE) was improved to 25.6%, which is among the highest values recorded for $ZnIn_2S_4$-based photocatalytic systems. These applications highlight the potential of CQDs in addressing environmental and energy challenges, such as water purification and renewable energy generation [93].

(b) **Biomedical Imaging**

CQDs have gained prominence in biomedical imaging because of their unique photoluminescence characteristics, good biocompatibility, and non-toxic nature. Unlike traditional quantum dots containing toxic heavy metals, CQDs are safer alternatives for in vivo as well as in vitro imaging applications. Their tunable fluorescence allows multicolor imaging, enabling detailed visualization of biological structures.

CQDs have recently garnered significant attention as a pleasing alternative to hazardous synthetic organic dyes for various bio-applications, including live cell imaging and diagnostic techniques. In a study, fluorescent CQDs were prepared by a one-pot microwave-assisted green method, utilizing almond resin as the carbon source. Honey was incorporated as an additional reagent to facilitate surface functionalization. The resulting CQDs had a size of less than 10 nm and exhibited a fluorescence with blue color, when excited at 350 nm, with a quantum yield of 61%. These fluorescent CQDs demonstrated rapid cellular uptake without compromising

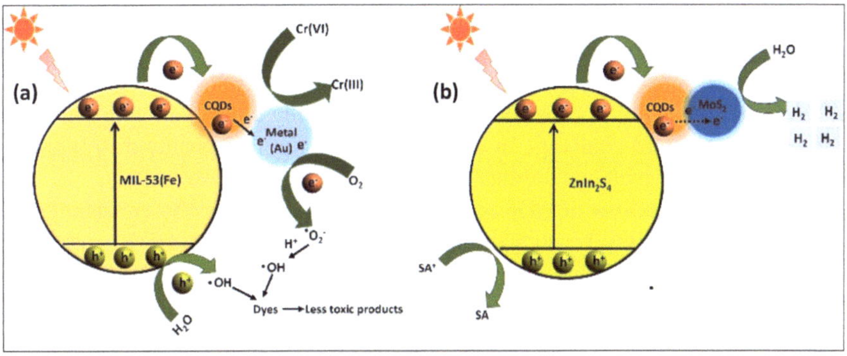

Fig. 14 CQDs-based photocatalytic application. **a** Photocatalytic degradation. **b** Photocatalytic hydrogen evolution

cell stability. Their imaging capability allowed multicolor visualization of cellular structures, including detailed architectures of the nucleus. This performance can be assigned to their unique surface properties, including amphiphilicity and a high density of positive charges, which enhance their interaction with cellular components [70].

Another study represented the synthesis of nitrogen-rich mesoporous carbon nanospheres with uniform sizes ranging from 65 to 70 nm. These nanospheres were prepared using pyrrole and aniline as precursors, with Triton-X serving as the structure-directing agent. The resulting mesoporous hollow carbon spheres (NCQD-HCS) exhibited a fluorescence quantum yield of up to 14.6% when excited at 340 nm. Interestingly, samples that were calcined at temperatures higher than 800 °C showed notable absorption in the 700–1000 nm wavelength range, with a light-to-heat conversion efficiency of up to 52%. Research using human oral cancer cells (FaDu) in vitro showed that the cells successfully ingested NCQD-HCS. When exposed to a 980 nm near-infrared laser, these nanospheres induced a pronounced thermal ablation effect, highlighting their potential for photothermal therapy [56]. CQDs are also ideal for real-time imaging due to their photostability and rapid cellular uptake, offering advantages over conventional dyes in medical diagnostics.

(c) **Drug Delivery**

CQDs are widely explored as nanocarriers for the delivery of various targeted drugs owing to their small size, high surface area, and functionalizable surface. These features allow CQDs to efficiently load, transport, and release therapeutic agents at targeted sites. Surface functionalization with biocompatible polymers or targeting ligands further improves their stability, selectivity, and interaction with biological systems.

Nanocarriers are designed to deliver therapeutic agents directly to the target site, such as tumors, thereby increasing the concentration of the active pharmaceutical ingredient (API) at the diseased location while minimizing exposure to healthy tissues. This targeted delivery approach not only reduces the required drug dosage but also minimizes adverse effects on surrounding healthy cells. Quantum dots (QDs) have emerged as innovative nanocarriers in healthcare, offering a large surface area for loading chemotherapeutic drugs. However, the commonly used drugs (DOX, MTX, 5-FU) often lack specificity and can harm healthy cells, leading to significant side effects. A major challenge in chemotherapy is the nonspecific distribution of anticancer drugs, which can diminish their efficacy and increase toxicity.

Surface modifications of QDs can address these challenges by enhancing their stability, water solubility, biocompatibility, and target specificity. CQDs, in particular, are used to improve the dissolving ability of non-soluble drugs, enhance biocompatibility, and facilitate cellular uptake. Fluorescent QDs offer additional benefits, including multifunctional imaging capabilities, biocompatibility, and controlled drug release. Drug molecules can be incorporated into CQDs through various techniques, including adsorption, dissolution, and non-covalent interactions such as electrostatic forces and hydrogen bonding.

For example, researchers have demonstrated the non-covalent coupling of DOX with CQDs, resulting in a CQD-DOX composite that exhibited improved tumor stability and enhanced drug delivery efficiency in vivo. Yifang et al. synthesized CQDs conjugated with DOX via electrostatic interactions (Fig. 15), resulting in a system with pH-responsive drug release. Compared to free DOX, the CQD-DOX complex demonstrated superior toxicity toward adenoid cystic carcinoma (ACC-2) cells, highlighting its potential for more effective cancer treatment [1, 79, 80, 81].

(d) **Biosensors**

CQDs are highly effective in biosensor applications because of their tunable fluorescence, high sensitivity, and stability. They can detect a wide variety of biomolecules, such as glucose, DNA, proteins, and metal ions, through fluorescence quenching or enhancement mechanisms. The functional groups on CQDs provide sites for binding with analytes, making them versatile sensing platforms [69].

For example, Huang et al. synthesized CQDs and modified them with chitosan on the glassy carbon electrode (CQDs-CS/GCE) for the dopamine detection as shown in Fig. 16. CQDs conjugated with aptamers have been utilized to detect thrombin with

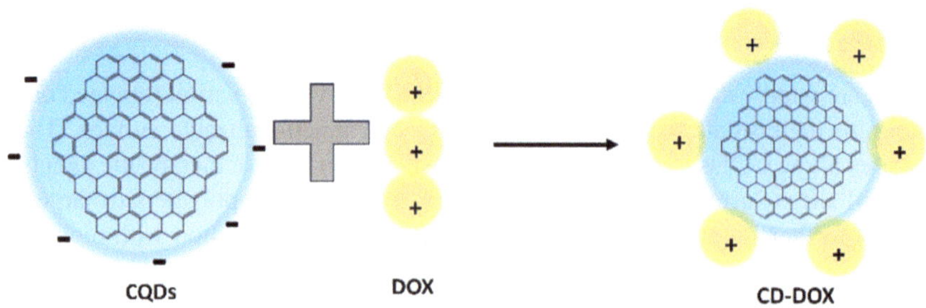

Fig. 15 Carbon quantum dots modified DOX drug delivery platform by electrostatic interaction

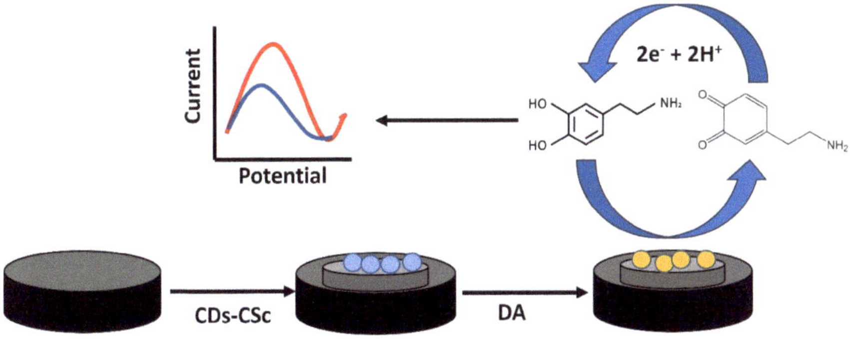

Fig. 16 Carbon dots-modified biosensor for the detection of dopamine (DA)

high specificity and sensitivity [31]. In another study, CQD-based sensors successfully detected heavy metal ions like lead (Pb^{2+}) and mercury (Hg^{2+}), crucial for environmental monitoring. These biosensors offer rapid, cost-effective, and reliable detection methods for clinical, environmental, and industrial applications [104].

Various redox-active enzymes and nanomaterials have been extensively explored in electrochemical biosensors due to their ability to facilitate direct electron transfer [25, 88]. CQDs and GQDs have occurred as promising candidates for integrating with redox-active biological molecules, attributed to their nanoscale dimensions and excellent electrical properties. Immobilized enzymes, commonly used in biosensors, play a crucial role as the biological recognition element. The high surface ratios and superior adsorption properties of CQDs offer significant potential for direct enzyme immobilization in the design of enzymatic biosensors [10, 24]. Aptamers are employed for specific analyte detection, acting as recognition elements with their ability to bind selectively. Similarly, peptide strands can serve this purpose, differing mainly in their chemical makeup of amino acids instead of nucleic acids. Chemical library screenings are frequently utilized to develop aptamers and peptide segments, although these molecules also naturally occur, such as RNA riboswitches found in bacteria. Aptamers possess several advantages, including high stability, specificity, compact structure, and compatibility with diverse chemical compositions. For instance, Liu et al. developed an electrochemiluminescent (ECL) biosensor using nitrogen-doped CQDs functionalized with DNA, achieving an enhanced signal for detecting miRNA-21. Furthermore, the properties of CQDs can be significantly improved by combining them with other nanomaterials. Pourmadadi et al. reported the creation of a CQDs-gold nanoparticle-based electrochemical aptasensor for the selective detection of prostate-specific antigen (PSA). The combination of CQDs and Au nanoparticles can enhance the electrical conductivity of CQDs, as evidenced by an increased current peak in cyclic voltammetry (CV) for the composite compared to CQDs alone [51, 68, 89].

7 Conclusion

Carbon quantum dots presented a novel and promising material in the realm of nanotechnology. They are ideal candidates for novel applications in photocatalysis, biosensing, and the biomedical domains due to their unique properties, which include exceptional photoluminescence, outstanding biocompatibility, and environmental friendliness. As this chapter illustrates, the incorporation of CQDs into modern technology offers great potential to revolutionize areas such as biosensors, pharmaceutical diagnosis and therapeutics, and environmental remediation. Furthermore, the ability to modify their surface through functionalization, doping, and passivation enables the tailoring of their properties for specific applications, making them adaptable to a wide array of technological domains. However, these impressive features are accompanied by certain unresolved challenges. The precise control over size distribution, surface states, and luminescence quantum yield still requires optimization. Furthermore, to support industrial translation, large-scale and economical production techniques must be improved. The long-term biocompatibility and stability of CQDs in physiological environments also demand thorough investigation to ensure safe clinical applications. Further studies aimed at enhancing CQDs' scalability, affordability, and functional diversity will solidify their position as a key component of next-generation nanomaterials.

References

1. Abbaspourrad A, Datta SS, Weitz DA (2013) Controlling release from pH-responsive microcapsules. Langmuir 29(41):12697–12702
2. Aghamali A, Khosravi M, Hamishehkar H, Modirshahla N, Behnajady MA (2018) Synthesis and characterization of high efficient photoluminescent sunlight driven photocatalyst of N-carbon quantum dots. J Lumin 201:265–274
3. Akbar K, Moretti E, Vomiero A (2021) Carbon dots for photocatalytic degradation of aqueous pollutants: recent advancements. Adv Opt Mater 9. https://doi.org/10.1002/adom.202100532
4. Al FJC-M et al (2021) Carbon quantum dots by submerged arc discharge in water: Synthesis, characterization, and mechanism of formation carbon quantum dots by submerged arc discharge in water: synthesis, characterization, and mechanism of formation. J Appl Phys 129:163301. https://doi.org/10.1063/5.0040322
5. Alia N, Omar S, Fen YW et al (2022) A review on carbon dots: synthesis, characterization and its application in optical sensor for environmental monitoring. Nanomaterials 12:2365
6. Ashraf W, Sharma D, Khan A, Khanuja M (2025) Green carbon dots as adsorbents for removal of toxic chemicals. In green carbon dots: sustainable analytical approaches. Am Chem Soc 141–160
7. Bagheri Z, Ehtesabi H, Rahmandoust M, Ahadian MM (2017) New insight into the concept of carbonization degree in synthesis of carbon dots to achieve facile smartphone based sensing platform. Sci Rep 1–11. https://doi.org/10.1038/s41598-017-11572-8
8. Biazar N, Poursalehi R, Delavari H (2018) Optical and structural properties of carbon dots/ TiO_2 nanostructures prepared via DC arc discharge in liquid. IP Conf Proc 1920:020033
9. Boxberg FTJ (2004) The handbook of nanotechnology. nanometer structures: theory, modeling, and simulation

10. Brahim S et al (2003) Chemical and biological sensors based on electrochemical detection using conducting electroactive polymers. Microchim Acta 143:123–137
11. Cailotto S, Mazzaro R, Enrichi F, Vomiero A, Selva M, Cattaruzza E, Cristofori D, Amadio E, Perosa A (2018) Design of carbon dots for metal-free photoredox catalysis. ACS Appl Mater Interfaces 10:40560–40567
12. Cao LI, Meziani MJ, Sahu S, Sun YP (2013) Photoluminescence properties of graphene versus other carbon nanomaterials. Acc Chem Res 46(1):171–180
13. Clancy AJ, Bayazit MK, Hodge SA et al (2018) Charged carbon nanomaterials: redox chemistries of fullerenes. Carbon Nanotubes Graph. https://doi.org/10.1021/acs.chemrev.8b00128
14. Cong S, Zhao Z (2018) Carbon quantum dots: a component of efficient visible light photocatalysts. Visible-Light Photocatal Carbon-Based Mater. https://doi.org/10.5772/intechopen.70801
15. Cui L, Ren X, Wang J, Sun M (2020) Materials today nano synthesis of homogeneous carbon quantum dots by ultrafast dual-beam pulsed laser ablation for bioimaging. Mater Today Nano 12:100091. https://doi.org/10.1016/j.mtnano.2020.100091
16. Dimos K (2016) Carbon quantum dots: surface passivation and functionalization. Curr Organ Chem 20(6):682–695
17. Elugoke SE, Adekunle AS, Fayemi OE et al (2020) Carbon-based quantum dots for electrochemical detection of monoamine neurotransmitters. Review 1–41
18. Elugoke SE, Uwaya GE, Quadri TW, Ebenso EE (2024) Carbon quantum dots: basics, properties, and fundamentals. In: Carbon dots: recent developments and future perspectives, Part 1—Carbon quantum dots: basics, properties, and fundamentals. https://doi.org/10.1021/bk-2024-1465.ch001
19. Fang Y, Guo S, Li D, Zhu C, Ren W, Dong SWE (2012) Easy synthesis and imaging applications of crosslinked green fluorescent hollow carbon nanoparticles. ACS Nano 6:400–409
20. Fatima T, Husain S, Khanuja M (2024a) Nano materials science novel ternary Z scheme carbon quantum dots (CQDs) decorated WS2/PANI ((CQDs@WS2/PANI): 0D: 2D: 1D) nanocomposite for the photocatalytic degradation and electrochemical detection of pharmaceutical drugs. Nano Mater Sci. https://doi.org/10.1016/j.nanoms.2024.04.005
21. Fatima T, Husain S, Khanuja M (2024b) Optimization of WS2 modified polyaniline for superior photocatalytic degradation and electrochemical detection of pharmaceutical drug. Flat Chem 44:100624
22. Georgakilas V, Perman JA, Tucek JZR (2015) Broad family of carbon nanoallotropes: classification, chemistry, and applications of fullerenes, carbon dots, nanotubes, graphene, nanodiamonds, and combined superstructures. Chem Rev 115(11):4744–4822
23. Gong X, Zhang Q, Gao Y, Shuang S, Choi MM, Dong C (2016) Phosphorus and nitrogen dual-doped hollow carbon dot as a nanocarrier for doxorubicin delivery and biological imaging. ACS Appl Mater Interfaces 8:11288–11297
24. Guiseppi-Elie A (1994) Method of measuring an analyte by measuring electrical resistance of a polymer film reacting with the analyte. Google Patents
25. Guiseppi-Elie A, Lei C, Baughman RH (2002) Direct electron transfer of glucose oxidase on carbon nanotubes. Nanotechnology 13:559
26. He Z, Liang R, Zhou C et al (2021) Carbon quantum dots (CQDs)/noble metal co-decorated MIL-53 (Fe) as difunctional photocatalysts for the simultaneous removal of Cr(VI) and dyes. Sep Purif Technol 255:117725. https://doi.org/10.1016/j.seppur.2020.117725
27. Heng ZW, Chong WC, Pang YL, Koo CH (2021) An overview of the recent advances of carbon quantum dots/metal oxides in the application of heterogeneous photocatalysis in photodegradation of pollutants towards visible-light and solar energy exploitation. J Environ Chem Eng 9
28. Hu X, Guo RT, Lin ZD et al (2022) Construction of carbon dot-modified g-C_3N_4/$BiOIO_3$ Z-scheme heterojunction for boosting photocatalytic CO_2 reduction under full spectrum light. ACS Sustain Chem Eng 10:11143–11153. https://doi.org/10.1021/acssuschemeng.2c02042

29. Hu C, Yu C, Li M, Wang X, Yang J, Zhao Z, Eychmüller A, Sun Y-P, Qiu J (2014) Chemically tailoring coal to fluorescent carbon dots with tuned size and their capacity for Cu(II) detection. Small 10:4926–4933
30. Hu X, Zhao H, Liang Y, Chen F, Li J, Chen R (2021) Broad-spectrum response NCQDs/ $Bi_2O_2CO_3$ heterojunction nanosheets for ciprofloxacin photodegradation: unraveling the unique roles of NCQDs upon different light irradiation. Chemosphere 264:128434
31. Huang Q, Hu S, Zhang H et al (2013) Biosensor for the sensitive and selective determination of dopamine. Analyst 5417–5423. https://doi.org/10.1039/c3an00510k
32. Hui S (2023) Carbon dots (CDs): basics, recent potential biomedical applications, challenges, and future perspectives. Springer, Netherlands
33. Huo X, Shen H, Liu R, Shao J (2021) Solvent effects on fluorescence properties of carbon dots: implications for multicolor imaging. ACS Omega 6:26499–26508
34. Jhonsi MA (2018) Carbon quantum dots for bioimaging. In: State of the art in nano-bioimaging. IntechOpen, London, UK, pp 35–55
35. Jiang M, Niu N, Chen Z et al (2018) Review of natural productderived carbon dots: from natural products to functional materials. Chemsuschem 11:11–24. https://doi.org/10.1002/cssc.201701847
36. Joshi PN, Kundu S, Sanghi SK, Sarkar D (2016) Graphene quantum dots—From emergence to nanotheranostic applications. Smart Drug Deliv Syst 7:159–195
37. Kamathe V, Kumar A, Khanuja M, Nagar R (2024) SnO2 quantum dots dispersed TGCN nanosheets for improved ethanol sensing under ambient conditions. Nano Express 5(4):045013
38. Konstantinos D (2016) Carbon quantum dots: Surface passivation and functionalization. Curr Org Chem 6:682–695
39. Krysmann MJ, Kelarakis A, Dallas P, Giannelis EP (2012) Formation mechanism of carbogenic nanoparticles with dual photoluminescence emission. J Am Chem Soc 134(2):747–750
40. Kumar R, Bhooshan V (2020) Sonochemical synthesis of carbon dots, mechanism, effect of parameters, and catalytic, energy, biomedical and tissue engineering applications. Ultrason Sonochem 64:105009. https://doi.org/10.1016/j.ultsonch.2020.105009
41. Kumar P, Dua S, Kaur R et al (2022) RSC advances REVIEW and their application in photovoltaics. RSC Adv 12:4714–4759. https://doi.org/10.1039/D1RA08452F
42. Kumar VB, Porat Z, Gedanken A (2016) Facile one-step sonochemical synthesis of ultrafine and stable fluorescent C-dots. Ultrason Sonochem 28:367–375
43. Kumar VB, Tang J, Lee KJ, Pol VG, Gedanken A (2016) In situ sonochemical synthesis of luminescent Sn@ C-dots and a hybrid Sn@ C-dots@ Sn anode for lithium-ion batteries. RSC Adv 6(70):66256–66265
44. Kumar VB, Kumar R, Gedanken A, Shefi O (2019) Fluorescent metal-doped carbon dots for neuronal manipulations. Ultrason Sonochem 52:205–213
45. Lee YS, Hu CC, Chiu TC (2022) Electrochemical synthesis of fluorescent carbon dots for the selective detection of chlortetracycline. J Environ Chem Eng 10(3):107413
46. Li L, Dong T (2018) Photoluminescence tuning in carbon dots: surface passivation or/and functionalization, heteroatom doping. J Mater Chem C 30:7944–7970
47. Li H, He X, Kang Z, Huang H, Liu Y, Liu J et al (2010) Water-soluble fluorescent carbon quantum dots and photocatalyst design. Angew Chem 26(122):4532–4536
48. Li H, He X, Liu Y, Huang H, Lian S, Lee S-T, Kang Z (2011) One-step ultrasonic synthesis of water-soluble carbon nanoparticles with excellent photoluminescent properties,. Carbon N Y 49:605–609
49. Li G, Xu J, Xu K (2023) Physiological functions of carbon dots and their applications in agriculture: a review. Nanomaterials 13:2684
50. Liang S, Wang M, Gao W, Zhao X (2022) Effects of elemental doping, acid treatment, and passivation on the fluorescence intensity and emission behavior of yellow fluorescence carbon dots. Opt Mater 128:112471
51. Liu Q et al (2017) A novel electrochemiluminescence biosensor for the detection of microRNAs based on a DNA functionalized nitrogen doped carbon quantum dots as signal enhancers. Biosens Bioelectron 92:273–279

52. Liu H, Ye T, Mao C (2007) Fluorescent carbon nanoparticles derived from candle soot. Angew Chem 119:6593–6595
53. Liu R, Wu D, Liu S, Koynov K, Knoll W, Li Q (2009) An aqueous route to multicolor photoluminescent carbon dots using silica spheres as carriers. Angew Chem Int Ed 48:4598–4601
54. Manuel J, Eugenia M, Prías-barragán JJ (2020) Nano-structures & nano-objects comparison between electrical and optical properties of carbon dots, silver nanoparticles and hybrid carbon/silver nanoparticles: experimental evidence of the interaction of carbon dots on silver nanoparticles. Nano-Struct Nano-Objects 24:100596. https://doi.org/10.1016/j.nanoso.2020.100596
55. Martindale BC, Hutton GA, Caputo CA, Prantl S, Godin R, Durrant JR, Reisner E (2017) Enhancing light absorption and charge transfer efficiency in carbon dots through graphitization and core nitrogen doping. Angew Chem Int Ed 56:1–6
56. Mohapatra S (2019) N-doped carbon quantum dot (NCQD)-deposited carbon capsules for synergistic fluorescence imaging and photothermal therapy of oral cancer. Langmuir 35:15320–15329. https://doi.org/10.1021/acs.langmuir.9b03001
57. Molaei MJ (2019) Carbon quantum dots and their biomedical and therapeutic applications: a review. RSC Adv 9:6460–6481. https://doi.org/10.1039/c8ra08088g
58. Mondal S, Yucknovsky A, Akulov K, Ghorai N, Schwartz T, Ghosh HN, Amdursky N (2019) Efficient photosensitizing capabilities and ultrafast carrier dynamics of doped carbon dots. J Am Chem Soc 141:15413
59. Nammahachak N, Aup-ngoen KK (2022) Hydrothermal synthesis of carbon quantum dots with size tunability via heterogeneous nucleation. RSC Adv 12:31729–31733. https://doi.org/10.1039/d2ra05989d
60. Nammahachak N, Aup-Ngoen KK, Asanithi P, Horpratum M, Chuangchote S, Ratanaphan S, Surareungchai W (2022) Hydrothermal synthesis of carbon quantum dots with size tunability via heterogeneous nucleation. RSC Adv 12(49):31729–31733
61. Nazar M, Hasan M, Wirjosentono B et al (2024) Microwave synthesis of carbon quantum dots from arabica coffee ground for fluorescence detection of Fe^{3+}, Pb^{2+}, and Cr^{3+}. ACS Omega. https://doi.org/10.1021/acsomega.4c02254
62. Nozik AJ, Beard MC, Luther JM, Law M, Ellingson RJ, Johnson JC (2010) Semiconductor quantum dots and quantum dot arrays and applications of multiple exciton generation to third-generation photovoltaic solar cells. Chem Rev 110(11):6873–6890
63. Ogi T, Aishima K, Permatasari FA, Iskandar F, Tanabe E, Okuyama K (2016) Kinetics of nitrogen-doped carbon dot formation via hydrothermal synthesis. New J Chem 40:5555–5561
64. Ozyurt D, Al KM, Hocking RK, Fox B (2023) Properties, synthesis, and applications of carbon dots: a review. Carbon Trends 12:100276. https://doi.org/10.1016/j.cartre.2023.100276
65. Peng H, Travas-Sejdic J (2009) Simple aqueous solution route to luminescent carbogenic dots from carbohydrates. Chem Mater 21:5563–5565
66. Phang SJ, Tan LL (2019) Recent advances in carbon quantum dot (CQD)-based two dimensional materials for photocatalytic applications. Catal Sci Technol 9:5882–5905. https://doi.org/10.1039/c9cy01452g
67. Pisanic Ii TR, Zhang Y, Wang TH (2014) Quantum dots in diagnostics and detection: principles and paradigms. Analyst 139:2968–2981. https://doi.org/10.1039/c4an00294f
68. Pourmadadi M et al (2022) An electrochemical aptasensor for detection of prostatespecific antigen-based on carbon quantum dots-gold nanoparticles. In: Biotechnology and applied biochemistry environmental applications of nanomaterials. Elsevier, pp 527–542
69. Pourmadadi M, Rahmani E, Rajabzadeh-khosroshahi M et al (2023) Properties and application of carbon quantum dots (CQDs) in biosensors for disease detection: a comprehensive review. J Drug Deliv Sci Technol 80
70. Praseetha PK, Litany RIJ, Alharbi HM et al (2024) Green synthesis of highly fluorescent carbon quantum dots from almond resin for advanced theranostics in biomedical applications. Sci Rep 1–12

71. Rahil M, Ansari RM, Ahmad S, Islam SS (2013) Nanostructured Ruddlesden–Popper-Layered lead bromide perovskites with stable and selected wavelength for photodetection applications. ACS Appl Nano Mater 6:5187–5199. https://doi.org/10.1021/acsanm.2c05092
72. Reyes D, Camacho M, Camacho M et al (2016) Laser ablated carbon nanodots for light emission. Nanoscale Res Lett 11:424. https://doi.org/10.1186/s11671-016-1638-8
73. Rong M, Zhang K, Wang Y, Chen X (2016) The synthesis of B, N-carbon dots by a combustion method and the application of fluorescence detection for Cu^{2+}. Chin Chem Lett. https://doi.org/10.1016/j.cclet.2016.12.009
74. Semeniuk M, Yi Z, Poursorkhabi V, Tjong J, Jaffer S, Lu Z-H, Sain M (2019) Future perspectives and review on organic carbon dots in electronic applications. ACS Nano 6:6224–6255
75. Shabbir H, Csap E (2023) Carbon quantum dots: the role of surface functional groups and proposed mechanisms for metal ion sensing. Inorganics 11:262
76. Shi R, Dai X, Li W, Lu F, Liu YQH et al (2017) Hydroxyl-group-dominated graphite dots reshape laser desorption/ionization mass spectrometry for small biomolecular analysis and imaging. ACS Nano 11:9500–9513
77. Singh P, Kumar S, Kumar P, Kataria N, Bhankar V, Kumar K, Kumar R, Hsieh CT, Khoo KS (2023) Assessment of biomass-derived carbon dots as highly sensitive and selective templates for the sensing of hazardous ions. Nanoscale 15:16241–16267
78. Singh P, Rani N, Kumar S, Kumar P, Mohan B, Bhankar V, Kataria N, Kumar R, Kumar K (2023) Assessing the biomass-based carbon dots and their composites for photocatalytic treatment of wastewater. J Clean Prod 413:137474
79. Soumya K, More N, Choppadandi M et al (2023) Biomedical technology a comprehensive review on carbon quantum dots as an effective photosensitizer and drug delivery system for cancer treatment. Biomed Technol 4:11–20. https://doi.org/10.1016/j.bmt.2023.01.005
80. Soumya K, More N, Choppadandi M, Aishwarya DA, Singh G, Kapusetti G (2023) A comprehensive review on carbon quantum dots as an effective photosensitizer and drug delivery system for cancer treatment. Biomed Technol 4:11–20
81. Su W et al (2020) Red-emissive carbon quantum dots for nuclear drug delivery in cancer stem cells. J Phys Chem Lett 11(4):1357–1363
82. Su R, Guan Q, Cai W, Yang W, Xu Q, Guo Y, Zhang L, Fei L, Xu M (2019) Multi-color carbon dots for white light-emitting diodes. RSC Adv 9:9700–9708
83. Su Y, Xie M, Lu X, Wei H, Geng H, Yang Z, Zhang Y (2013) Facile synthesis and photoelectric properties of carbon dots with upconversion fluorescence using arc-synthesized carbon by-products. RSC Adv 4:4839–4842
84. Yadav PK, Singh VK, Chandra S, Bano D, Kumar V, Talat M, Hasan SH(2018) Green synthesis of fluorescent carbon quantum dots from Azadirachta Indica leaves and their peroxidase-mimetic activity for the detection of H_2O_2 and ascorbic acid in common fresh fruits. ACS Biomater Sci Eng 5:623–632. https://doi.org/10.1021/acsbiomaterials.8b01528
85. Tang L, Ji R, Cao X, Lin J, Jiang H, Li X, Teng KS, Luk CM, Zeng S, Hao J et al (2012) Deep ultraviolet photoluminescence of water-soluble self-passivated graphene quantum dots. ACS Nano 6:5102–5110
86. Tao S, Zhu S, Feng T et al (2018) The polymeric characteristics and photoluminescence mechanism in polymer carbon dots: a review
87. Thongpool V, Asanithi P, Limsuwan P (2012) Synthesis of carbon particles using laser ablation in ethanol. Procedia Eng 32:1054–1060. https://doi.org/10.1016/j.proeng.2012.02.054
88. Tyagi M, Singh S, Hasan MR, Fatima T, Khanuja M, Narang J (2024) Lab in tube: comparing different morphological dependent gold nanomaterials towards naked eye and optical sensing of dopamine using aptamer. Int J Environ Anal Chem 104(19):8477–8490
89. Wang Y et al (2015) Horseradish peroxidase immobilization on carbon nanodots/CoFe layered double hydroxides: direct electrochemistry and hydrogen peroxide sensing. Biosens Bioelectron 64:57–62
90. Wang Q, Huang X, Long Y, Wang X, Zhang HZR, Al E (2013) Hollow luminescent carbon dots for drug delivery. Carbon 59:192–199

91. Wang Y, Hu A (2014) Carbon quantum dots: synthesis, properties and applications. J Mater Chem C 2(34):6921–6939
92. Wang C, Wu W, Periasamy AP, Chang H (2014) Electrochemical synthesis of photoluminescent carbon nanodots from glycine for highly sensitive detection of hemoglobin. Green Chem 16(5):2509–2514. https://doi.org/10.1039/c3gc42325e
93. Wang B, Deng Z, Fu X, Li Z (2018) MoS2/CQDs obtained by photoreduction for assembly of a ternary MoS2/CQDs/ZnIn2S4 nanocomposite for efficient photocatalytic hydrogen evolution under visible light. J Mater Chem A 6(40):19735–19742
94. Wang R, Lu KQ, Zhang F, Tang ZR, Xu YJ (2018) 3D carbon quantum dots/graphene aerogel as a metal-free catalyst for enhanced photosensitization efficiency. Appl Catal B: Environ 233:11–18
95. Xia C, Zhu S, Feng T, Yang M, Yang B (2019) Evolution and synthesis of carbon dots: from carbon dots to carbonized polymer dots. Adv Sci 6:1901316
96. Yadav PK, Chandra S, Kumar V et al (2023) Carbon quantum dots: synthesis, structure, properties, and catalytic applications for organic synthesis. Catalysts 13:422
97. Yadegari A, Khezri J, Esfandiari S, Mahdavi H, Karkhane AA, Rahighi R, Heidarimoghadam R, Tayebi L, Hashemi E, Farmany A (2019) Bottom-up synthesis of nitrogen and oxygen co-decorated carbon quantum dots with enhanced DNA plasmid expression. Colloids Surf B: Biointerfaces 1(84):110543
98. Yan F, Sun Z, Zhang H et al (2019) The fluorescence mechanism of carbon dots, and methods for tuning their emission color: a review. Microchim Acta 186:583
99. Yang H, Bai L, Geng Z et al (2023) Carbon quantum dots: preparation, optical properties, and biomedical applications. Mater Today Adv 18:100376. https://doi.org/10.1016/j.mtadv.2023.100376
100. Yu T, Wang H, Guo C, Zhai Y, Yang J, Yuan J (2018) A rapid microwave synthesis of green-emissive carbon dots with solid-state fluorescence and pH-sensitive properties. R Soc Open Sci 5:1–15
101. Yuan F, Yuan T, Sui L et al (2018) Engineering triangular carbon quantum dots with unprecedented narrow bandwidth emission for multicolored LEDs. Nat Commun 1–11. https://doi.org/10.1038/s41467-018-04635-5
102. Yuan F et al (2016) Shining carbon dots: synthesis and biomedical and optoelectronic applications. Nano Today 11(5):565–586
103. Zehra F, Fatima T, Khanuja M (2025) Colorimetric sensing of mercury with machine learning based on nanozyme activity of chitosan functionalized MXene/ZIF-L nanocomposite. Adv Mater Technol 10(21):e00519
104. Zhao W, Huang C, Guo X, Zhu Y, Li Y, Duan Y, Gao J (2024) A fluorescence biosensor based on carbon quantum dots prepared from pomegranate peel and T-Hg^{2+}-T mismatch for Hg^{2+} detection. J Fluoresc 1–9
105. Zheng XT, Ananthanarayanan A, Luo KQCP (2015) Glowing graphene quantum dots and carbon dots: properties, syntheses, and biological applications. Small 11(14):1620–1636
106. Zhou J, Booker C, Li R et al (2007) An electrochemical avenue to blue luminescent nanocrystals from multiwalled carbon nanotubes (MWCNTs). J Am Chem Soc 8:744–745
107. Zhu H, Wang X, Li Y, Wang Z, Yang F, Yang X (2009) Microwave synthesis of fluorescent carbon nanoparticles with electrochemiluminescence properties. Chem Commun 34:5118–5120

MIX
Papier aus verantwortungsvollen Quellen
Paper from responsible sources
FSC® C105338

If you have any concerns about our products,
you can contact us on
ProductSafety@springernature.com

In case Publisher is established outside the EU,
the EU authorized representative is:
Springer Nature Customer Service Center GmbH
Europaplatz 3, 69115 Heidelberg, Germany

Printed by Libri Plureos GmbH
in Hamburg, Germany